כתבי האקדמיה הלאומית הישראלית למדעים

PUBLICATIONS OF THE ISRAEL ACADEMY

OF SCIENCES AND HUMANITIES

SECTION OF SCIENCES

FLORA PALAESTINA

THE BRYOPHYTE FLORA
OF ISRAEL AND ADJACENT REGIONS

edited by

C. CLARA HEYN and ILANA HERRNSTADT

FLORA PALAESTINA

THE BRYOPHYTE FLORA
OF ISRAEL AND ADJACENT REGIONS

Scientific Editor

NIR L. GIL-AD

Executive Editor

ZOFIA LASMAN

Professor C. Clara Heyn
passed away on 27 December 1998 following a serious illness.

Clara's co-authors and editors, have completed this Flora
while trying to fulfil her wishes and meet her high standards.
We greatly regret that she was not able to see the work,
which was so dear to her, in print.

Landscape with bryophytes, Bar'am Forest, Upper Galilee

FLORA PALAESTINA

THE BRYOPHYTE FLORA OF ISRAEL AND ADJACENT REGIONS

C. Clara Heyn and Ilana Herrnstadt, editors

BRYOPSIDA (MOSSES)

by Ilana Herrnstadt and C. Clara Heyn

ANTHOCEROTOPSIDA (HORNWORTS) and MARCHANTIOPSIDA (LIVERWORTS)

by Helene Bischler and Suzanne Jovet-Ast

DRAWINGS: Michal Boaz-Yuval and Esther Huber

COLOUR PLATES: David Darom

JERUSALEM 2004

THE ISRAEL ACADEMY OF SCIENCES AND HUMANITIES

Albert Einstein Square, P.O. Box 4040, Jerusalem 91040, Israel
tami@academy.ac.il

Addresses of the authors:

Ilana Herrnstadt
Department of Evolution, Systematics and Ecology
Berman Building, Edmond J. Safra Campus
The Hebrew University of Jerusalem,
Giv'at Ram, Jerusalem 91904, Israel
ilanahs@vms.huji.ac.il

Helene Bischler and Suzanne Jovet-Ast
Laboratoire de Cryptogamie
Muséum National d'Histoire Naturelle
12, Rue Buffon, F-75005 Paris, France
h.bischler@wanadoo.fr

ISBN 965-208-004-4
ISBN 965-208-152-3

Printed in Israel
at Keterpress Enterprises, Jerusalem

This book is dedicated to the memory of

FELIX BILEWSKY (1902–1979)

pioneer of the bryological research in Israel

After arriving in this country in 1933 from Breslau, Germany, and while working as a pharmacist, Bilewsky devoted considerable time to the collection and investigation of mosses. His herbarium and "Moss-Flora of Israel" (Bilewsky 1965) served as the basis for the present publication.

CONTENTS

LIST OF TABLES

LIST OF PLATES OF SEM MICROGRAPHS

Plate XI spore surfaces
Pohlia wahlenbergii; *Bryum cellulare*; *Bryum subapiculatum*; *Bryum caespiticium*; *Bartramia stricta*; *Zygodon viridissimus*

Plate XII spore surfaces
Orthotrichum striatum; *Orthotrichum affine*; *Orthotrichum cupulatum*; *Orthotrichum diaphanum*; *Leucodon sciuroides*

Plate XIII spore surfaces
Leptodon smithii; *Neckera complanata*; *Fabronia ciliaris*; *Amblystegium serpens*; *Leptodictyum riparium*; *Homalothecium sericeum*

Plate XIV spore surfaces
Homalothecium aureum; *Scorpiurium deflexifolium*; *Scorpiurium sendtneri*; *Scleropodium touretii*; *Eurhynchium riparioides*; *Rhynchostegiella tenella*

Plate XV spore surfaces and leaf margins
Phaeoceros bulbiculosus; *Phaeoceros laevis*; *Riella cossoniana*; *Fossombronia caespitiformis*; *Sphaerocarpos michelii*; *Riella helicophylla*; *Southbya nigrella*; *Southbya tophacea*; *Pellia endiviifolia*; *Corsinia coriandrina*; *Targionia hypophylla*; *Lunularia cruciata*

Plate XVI spore surfaces
Plagiochasma rupestre; *Mannia androgyna*; *Reboulia hemisphaerica*; *Athalamia spathysii*; *Marchantia polymorpha*; *Oxymitra incrassata*; *Ricciocarpos natans*; *Riccia crustata*; *Riccia crystallina*; *Riccia frostii*; *Riccia gougetiana*; *Riccia sorocarpa*

LIST OF COLOUR PLATES

PREFACE

This new addition to the Flora Palaestina series represents an up-to-date account of the bryophytes of Israel and some adjacent regions.

The Bryophyte Flora comprises two parts: I. Bryopsida (mosses) by Ilana Herrnstadt and C. Clara Heyn of The Hebrew University of Jerusalem, and II. Anthocerotopsida (hornworts) and Marchantiopsida (liverworts) by Helene Bischler and Suzanne Jovet-Ast of the Muséum National d'Histoire Naturelle, Paris.

The geographical area treated in this Flora comprises the region extending from the Mediterranean east to the Jordan River, and in addition includes the Golan Heights and Mount Hermon (see the topographical map). Thus, it is not identical with the area treated in the four previous volumes of Flora Palaestina (Zohary 1966, 1972; Feinbrun-Dothan 1978, 1986).

The major limiting factors for bryophyte studies in Israel are: (i) the brief wet season of the predominantly Mediterranean and desert ecosystems, and (ii) the small plant size of most bryophytes growing in those ecosystems. Moreover, and very frequently, plants of several taxa are syntopic and are often intertwined in their growth. Such intermingling of several species places effective collecting into the realm of the specialist.

ECOLOGICAL AND HABITAT FACTORS INFLUENCING BRYOPHYTE DISTRIBUTION

The extreme diversity of topography, climate, soil, and floral history is responsible for the wide range of vegetation types and floras typical of Southwest Asia. This holds for higher plants as well as for bryophytes. Although the area of Israel is relatively small, it reflects well the species richness typical of Southwest Asia as a whole.

Topography. The area treated may be divided into four topographical or altitudinal belts (see the topographical map) with a north-to-south elongation (Zohary 1962, 1973, 1982): (i) the narrow, near sea-level Coastal Plain, which extends into the Negev desert; (ii) a broader mountain region in the centre, ranging up to 1200 m; (iii) the Jordan Valley (part of the Syrian-African Rift Valley) in the east, ranging in elevation from ca. -400 m in the Dead Sea area to +200 m in the Dan Valley; and (iv) the high altitude Golan Heights plateau in the northeast, bordering in the north with Mount Hermon (up to 2000 m). This altitudinal division is meaningful when the distribution patterns of the bryophytes in the region are analysed.

Climate. The climate in the greater part of Israel is generally of the Mediterranean type, characterised by a relatively short, mild, rainy, and cool winter, and a prolonged dry and hot summer. The average annual rainfall can reach up to 1000 mm in the more humid north, decreasing down to 15 (25) mm in the arid south. In general, temperatures increase from north to south and from west to east, though in a less regular pattern (Zohary 1962).

Substrates. Bryophytes can establish only on relatively stable substrates because the plants are barely fixed to the soil surface. In Israel, they inhabit mainly soil surfaces and solid surfaces, e.g., rocks, stones, and walls, where shallow pockets of dust and soil accumulate. They are found to a lesser extent on the bark of living or dead trees. Sciophilous species are quite rare because the areas of forests and maquis are relatively small, and hydrophilous species are also rare because aquatic habitats are scarce.

Edaphic factors. Most of the soils in the area are calcareous with a high pH, usually 7–7.5. Consequently, acidophilous species are rare. In general, bryophytes are sensitive to high salt concentrations. In localities with saline soils or water, bryophytes cannot survive (Jovet-Ast *et al*. 1976).

Bryophytes are edaphically specialised. In open habitats of the Dead Sea area, Judean and Negev deserts, they contribute to the formation of soil crusts. Those crusts stabilise the soils from erosion by enriching them with accumulations of organic matter and by retaining humidity for short periods. Their role in soil preservation might not be negligible (Danin 1996).

Moisture factors. Bryophytes are capable of completing their life cycle in a period as short as a few weeks, provided that aquatic fertilisation is made possible and enough moisture is available for growth and reproduction. They can persist and survive as gametophytes and spores when moisture conditions are insufficient for fertilisation or growth, and often survive in unfavourable microhabitats due to metabolism that allows them to survive periodical desiccation. Their ability to regenerate from parts or even cells of the gametophyte, as well as through sexual and asexual reproductive propagation, enables their adaptation to harsh environments (Richardson 1981).

Within the geographical region covered by this Flora, there is a great diversity of habitat conditions and consequently diverse moisture regimes. Bryophytes are able to colonise habitats that are unsuitable for many other plants, e.g., hard surfaces, soil pockets in rock crevices, temporary water ponds, shallow basins accumulating water for short periods of time, or enclaves on runoff slopes. These ephemerally moist microhabitats are available to bryophytes because they occupy only a shallow soil layer and absorb water or atmospheric humidity over their entire surface.

Air humidity and dew play a more important role in bryophytes than in higher plants. The poikilohydric nature of bryophytes enables water absorption and utilisation of dew or atmospheric humidity by the plants. Significant dew deposits occur along the Coastal Plain, increasing towards the south and decreasing towards the east. In the dunes of the Western Negev, I. H. observed growth renewal and development of reproductive organs when plants of *Bryum dunense* were exposed only to dew and fog prior to the first rains (unpubl. obs., 1992).

PHYTOGEOGRAPHY

Southwest Asia is a meeting point of the Paleotropic and Paleoarctic kingdoms. Within this area, four phytogeographical regions were recognised by Zohary (1962, 1963, 1966), and modified by Danin & Plitmann (1987): Mediterranean, Irano-Turanian, Saharo-Arabian, and Nubo-Sindian (= Sudanian). These regions were delineated by the distribution patterns of phanerogams, including a high percentage of endemics, not taking into account the bryophytes.

Frey & Kürschner (1983) suggested changes in the above phytogeographical regions on the basis of distribution patterns of mosses. They proposed the division of the Irano-Turanian region into three provinces (West, Medio-Asiatic, and Centro-Asiatic) as well as the creation of a superordinate floral subkingdom: the Circum-Tethyan. Their phytogeographical approach is problematic because it is based only on the distribution patterns of mosses. According to Bischler & Jovet-Ast (1986), these phytogeographical subdivisions are meaningless in reference to the hepatics, as most species in Southwest Asia have a distribution range extending outside this geographical area.

ENDEMICS

Southwest Asia has a "large percentage of endemics" among the mosses; they generally belong to the Pottiaceae, Grimmiaceae, and Funariaceae (Frey & Kürschner 1983). Conversely, the number of endemics among the hepatics is negligible (only two). Southwest Asia appears more like a meeting point of hepatic floras of different origins than a centre of speciation (Bischler & Jovet-Ast 1986). Endemic mosses in the local flora are discussed in the Introduction to Part I. There are no endemics among the hornworts and liverworts in the local flora.

HISTORY OF BRYOLOGICAL RESEARCH IN SOUTHWEST ASIA AND ISRAEL

The initial accounts of bryophytes from Southwest Asia were short lists of plants found incidentally during collecting excursions for diverse plants and animals. The earliest report of mosses, collected by Ehrenberg between 1820 and 1826, was published by Lorentz (1868). It was followed by Barbey & Barbey (1882), Hart (1891), Geheeb (1903–1904), Schiffner (1908, 1909), Bornmüller, (1914, 1931), Reimers (1927) and Bizot (1942, 1955). The bryological literature pertaining to Southwest Asia was compiled by Frey & Kürschner (1981, 1983), Frey (1986), El-Oqlah *et al.* (1988a), and Greene & Harrington (1989), and summarised by Kürschner (1997). Great parts of Southwest Asia still remain undercollected, and there is a paucity of Floras. Agnew and Vondráček (1975) published "A Moss Flora of Iraq", and Kürschner (2000) the "Bryophyte Flora of the Arabian Peninsula and Socotra".

The history of bryological research in Israel is rather short (Herrnstadt & Heyn 1987), especially when compared with the extensive research on other plant groups. The relative neglect of bryophyte collections may be a consequence of the long dry season during which most of the mosses, hornworts, and liverworts are inconspicuous. The earliest report on local bryophytes was by Rabinovitz-Sereni (1931). More intensive collecting began in the mid-nineteen-forties. Bryological research was initiated only in the mid-nineteen-fifties (for details, see the Introductions to Parts I and II).

HISTORY OF THE BRYOPHYTE FLORA

Discussions with Peter H. Raven during his visit to Israel in 1975 stimulated the initiation of the project. Methodical studies of the mosses of Israel were begun by Herrnstadt and Heyn in 1978. Previous studies by Bilewsky and Nachmony (Bilewsky 1959, 1965, 1970, 1974, 1977; Bilewsky & Nachmony 1955; Nachmony 1961) and collections of the Bilewsky Herbarium, which were donated to the Herbarium of The Hebrew University of Jerusalem (HUJ) by Bilewsky (partly via the Department of Botany of Tel Aviv University), served as the basis for this research.

Thorough studies of the local hornworts and liverworts have been initiated and carried out by Jovet-Ast and Bischler (Jovet-Ast *et al.* 1965; Jovet-Ast & Bischler 1966; Bischler & Jovet-Ast 1975). In the winter of 1982, they joined Herrnstadt and Heyn in preparing the Flora by making additional collections and conducting further studies. Part II of the Flora is based on their own collections and on the revision of all other available material from Israel (for details, see the Introduction to Part II).

The manuscript of the Flora was submitted to The Israel Academy of Sciences and Humanities for publication in March 1998. A few addition were made during proof-reading.

GUIDE TO THE FLORA

This guide covers items common to both Part I and Part II. Remarks pertaining to each part are provided in guides that follow their respective introductions.

Uniformity. Although the two parts of the Flora have different authors and different histories, we have tried to treat the mosses, hornworts, and liverworts as similarly as possible. In Part I, treating 220 taxa of mosses, a synopsis describing the orders, and a key to families, are provided. In Part II, treating 39 taxa of hornworts and liverworts, a key to classes, orders, and families is provided.

The circumscription of families is based on worldwide variability. In Part I, the concepts of some orders and families are still under debate, and their treatment in the literature differs greatly. Therefore, we decided to compile a synopsis of orders, and construct a key representing mainly the variation among the local families. Out of the twenty local families, eight are represented by a single genus. Therefore, keying out to family is straightforward. In the larger families the range of variation in vegetative characters is wider, and sporophytes may be lacking or immature. Thus, keying out is more problematic, and some families are keyed out more than once.

Bibliographic references to taxa. In Part I, references to publications, with a special emphasis on Bilewsky (1965) — the most comprehensive summary of the local flora prior to this publication — are provided in square brackets after the reference to the taxon. In Part II, references to former publications of the authors of Part II are provided similarly.

Abbreviations. Abbreviations of the names of authors of plant names follow Brummitt & Powell (1992). In a few cases when an author was not listed there, the abbreviation has been chosen from Crosby *et al.* (1992) or Sayre *et al.* (1964). Abbreviations of books cited in the synonymies follow Stafleu & Cowan (1976–1988). Abbreviations of periodicals cited in the synonymies follow Crosby *et al.* (1992) and Lawrence *et al.* (1968) in Part I, and Lawrence *et al.* (1968) in Part II. Abbreviations of herbaria follow Holmgren *et al.* (1990).

Synonymy. The synonymy, which follows the binomial and the name(s) of the publishing author(s), is generally limited to names given to taxa from the circum-Mediterranean Region, Europe, and Southwest Asia. The entire nomenclatural history of the taxon is not provided.

General Distribution. In the General Distribution data, Southwest Asia and the Mediterranean Region are listed first (to emphasise the affinities of the local taxa) and are followed, more or less, in the traditional pattern of north to south and west to east. Among the regions listed is Macaronesia, which includes the Azores, Madeira, and the Canary Islands (*sensu* Duell, 1984, 1985, 1992).

Literature. Lists of selected references for additional consultation are provided, when applicable, after generic or species descriptions.

Distribution maps. In the distribution maps, we used a grid system of 10×10 km as shown in the topographical map. Each taxon is marked with circles representing one or more occurrences of the taxon within the square. In the text, the terms "common", "occasional", and "rare" refer to the total number of grids in which the taxon occurs. Occasionally, when a taxon is particularly frequent within a grid, we cite it as being "locally abundant". When the occurrence of a taxon is uncertain, it is indicated by a question mark to the right of its dot on the map as well as in the text ("Habitat and Local Distribution"). We are aware of the subjective nature of this terminology.

Drawings. The drawings of the mosses have been made by two botanical artists (see the Introduction to Part I). The magnifications are provided in the legends. The drawings of the hornworts and liverworts have been made by the authors; the scales are designated by bars.

The drawings of Part I (Figures 1–207) are based on specimens deposited in HUJ. The drawings of Part II (Figures 208–246) are based on specimens deposited in PC, except when otherwise indicated in the legend.

SEM plates. Scanning electron micrographs of spore surfaces represent selected samples only. Correspondingly, we have tried to provide a brief description of the spores in the text following the description based on light-microscope observations. Since the terminology of bryophyte spore texture follows palynological terms, and is applied inconsistently in the bryological literature, we have frequently supplemented descriptive terms not used for pollen.

Colour plates. Colour photographs of selected species, which represent the orders treated in the Flora, are provided in a special section.

Keys. Artificial dichotomous keys are provided within families and genera.

Transliteration. The transliteration of the original Hebrew and Arabic geographic names to English is taken from the Israel (North and South) 1:250,000 maps published by The Survey of Israel in 1976.

Districts. The names and the order of the districts presented in the topographical map follow (Zohary 1962) and those of the map at the end of each of the text volumes of Flora Palaestina (Zohary 1966, 1972; Feinbrun-Dothan, 1978, 1986), with some modifications.

Citations of collections. Collections are cited in the text by using the collector's name and followed by collection number when available.

Hebrew common names. Hebrew common names have been published only by Bilewsky (1963) for 15 mosses.

LIST OF ABBREVIATIONS

aff.	*affinis* (allied to)
auct.	*auctorum* (of authors; a name misapplied)
Div.	Division [in References]
emend.	*emendavit* (emended)
env.	environs
ex	according to
Herb.	Herbarium
HUJ	Herbarium of The Hebrew University of Jerusalem
nom. cons.	*nomen conservandum* (conserved name)
nom. inval.	*nomen invalidum* (name not validly published)
nom. nud.	*nomen nudum* (name published without a description or illustration)
pl.	plural
Pl.	Plate
ser./sér.	series/série
Sect.	Section [in References]
sine dato	without a date
sine loco	without a locality
sing.	singular
vs.	versus
±	more or less, somewhat
μm	micrometre; colloquially micron or mu; 1/1000 mm

ACKNOWLEDGEMENTS

We wish to express our gratitude to the many colleagues and institutions who extended their support and assistance during the different stages of this project.

We are particularly indebted to Nir L. Gil-ad for his superior scientific editing. We greatly appreciate his critical, accurate, and meticulous work. We are very grateful to Francis Dov Por for his optimistic approach and valuable advice. He, together with Yossi Segal of The Israel Academy of Sciences and Humanities, helped in securing financial support during the final stages of the project. Daniel Zohary, with his great interest and enthusiasm, provided helpful and erudite comments as well as financial assistance. Among the staff of The Israel Academy of Sciences and Humanities, particular thanks are due to Zofia Lasman for her abundant assistance, excellent advice, and careful attention to the editing and design of the book, Gideon Stern for thoughtful advice and help with the production of the book, and Peter Grossmann for his graphical assistance and skilful production of the maps.

The superb illustrations as well as the colour plates attest to the artistic and photographic skills of Ester Huber, Michal Boaz-Yuval and David Darom.

Thanks to Robert E. Magill and the Missouri Botanical Garden for granting permission to use parts of Glossarium Polyglottum Bryologiae (Magill 1990).

We wish to tender our deep thanks to our families, who supported us with great patience and encouragement. One of us (I. H.) wishes to express her gratitude to Jerry H. Haas for his valuable assistance, and for devoting his time and skills to the completion of this work.

PART I

Our mutual collaboration, which began as a professor-student relationship, flourished into a symbiotic, stimulating professional interaction in which each of us complemented the other; we are grateful to have had this opportunity.

The successful completion of the manuscript is also due to the help, assistance, and encouragement we were fortunate to receive during those years from many institutions and colleagues. Those contacts were especially valuable because, not only were we the only bryologists in the country, we also unfortunately, had very limited ties with colleagues in neighbouring countries.

Peter H. Raven, Director of the Missouri Botanical Garden, stimulated us to begin the Flora project. He secured financial support, which enabled Ilana Herrnstadt to spend a year (1977) working and studying in Marshall Crosby's laboratory at the Garden. This experience provided her with an introduction to all aspects of collecting and studying mosses. Special thanks are due to Marshall R. Crosby, whose didactic skills and extensive knowledge of mosses were critical to the successful initiation of the project, and whose help has extended throughout these many years.

Ilana Herrnstadt stayed for various lengths of time at the herbaria of Berlin (B), Buffalo (BUF), Corvallis (OSC), Durham (DUKE), Gainesville (FLAS), Geneva (G), London (BM), Paris (PC), and Vancouver (UBC), and is grateful for the award of part-time supporting grants from the Centre National de la Recherche Scientifique, the Deutscher Akademischer Austauschdienst, and the British Council.

We thank The Israel Academy of Sciences and Humanities for supporting this project with an initial research grant and with various grants over the years to cover the research and selected expenses (e.g., drawings, use of SEM, and manuscript editing).

We received a grant from the Israel-America Binational Science Foundation (No. 1607) in 1978 and, at different periods, grants from the Authority for Research and Development of The Hebrew University of Jerusalem. The help of those organisations is hereby acknowledged.

During our studies we received much guidance and help (e.g., with specimen identifications or verifications, or the provision of laboratory workspace) from: Lewis E. Anderson, Alan C. Crundwell, Allan J. Fife, Patricia Geissler, Dana G. Griffin III, Alan J. Harrington, Jean-Pierre Hebrard, Zennoske Iwatsuki, Aaron Liston, Ryszard Ochyra, and Ronald A. Pursell. Thanks are due to the following colleagues for critically reviewing problematic sections of the manuscript: Maria A. Bruggeman-Nannenga (*Fissidens*), Diana G. Horton (*Encalypta*), Wolfgang Frey and Harald Kürschner (mosses of Southwest Asia), Richard H. Zander (Pottiaceae), and Thomas Blockeel (double-checking the illustrations). We are indebted to Wilfred B. Schofield, who greatly contributed his extensive knowledge and experience on mosses, and provided encouragement and guidance, especially with the key to families.

During the first five years, Rivka Ben-Sasson worked with us on this project with great enthusiasm and ability, and we are much indebted to her. We thank Pua Markus and Haim Kutiel, who were students at that time, for collecting mosses for the project throughout Israel during 1978.

We are indebted to our Israeli colleagues Avinoam Danin, Ofer Cohen, Didi Kaplan, and Hanan Dimentman for collecting valuable material, Jacob Lorch and David Heller for double-checking the list of references, Uzi Plitmann for administrative assistance and help in acquiring financial support, and Eitan Tchernov for his recognition of the importance of the project.

PART II

We are indebted to C. Clara Heyn, Ilana Herrnstadt, Avinoam Danin, and Baruch Baum for help during our collection trips. Thanks to Didi Kaplan and Hanan Dimentman for samples of *Ricciocarpos natans,* and to David Apirion, Michal Boaz-Yuval, Aaron Liston, Michael Loria, and Avi Shmida for providing additional specimens. Special thanks are due to Yaacov Lipkin for providing valuable data on *Riella.*

We thank the curators of HUJ, BM, and UC for the loan of specimens.

Our fieldwork was supported by the C.N.R.S. (Centre National de la Recherche Scientifique) and by the France-Israel Scientific Exchange Programs. Their support is hereby acknowledged.

We thank Michèle Dumont, Muséum National d'Histoire Naturelle, Paris for the photographic prints of the SEM micrographs.

We are grateful to Michal Boaz-Yuval for inking and shading some of our sketches of the drawings.

Finally, we are indebted to the editors of A.D.A.C. Cryptogamie, Bryologie–Lichénologie for the permission to use some of the drawings of S. Jovet-Ast published in 1986.

Grimmia pulvinata (p. 296)

Bryum caespiticium (p. 390)

Gigaspermum mouretii (p. 307)

Tortula inermis (p. 188)

Acaulon longifolium (p. 264)

Grimmia mesopotamica (p. 290)

Part I

BRYOPSIDA (MOSSES)

by

Ilana Herrnstadt and C. Clara Heyn

INTRODUCTION

This part of the Flora summarises twenty-two years of fruitful and rewarding research on the mosses native to Israel. Fruitful, because our region has attracted very few moss specialists and there was ample scope for augmenting the existing data and interpretations. Rewarding, because this small country has been found to have an unexpectedly wide diversity of species of mosses.

THE HISTORY OF COLLECTION AND STUDIES OF MOSSES IN ISRAEL

Pioneering efforts to establish a representative collection of mosses was made by D. Rabinovitz-Sereni (1931) and by the French botanist M. Bizot (1945). During the early nineteen-forties, extensive collections were made by T. Kushnir and D. Zohary, and during the following two decades by F. Bilewsky and S. Nachmony (Bilewsky & Nachmony 1955, Bilewsky 1959, Nachmony 1961).

The first compilation and integration of the collected material in a form of a moss flora was made by F. Bilewsky (1965), to whom this Flora is dedicated. In subsequent publications (1970, 1974, 1977) he added more data.

Our contributions to the local moss Flora were published as new records (Herrnstadt *et al.* 1982) and checklists (Herrnstadt *et al.* 1991, Herrnstadt 1992). In addition, Brullo *et al.* (1991) published a list of 21 mosses collected in some desert localities in Israel, apparently only a short time after the publication of our 1991 checklist. That list included one species, *Brachymenium exile*, which has not been found by us, and which was recorded by them with doubts.

THE COMPOSITION OF THE MOSS FLORA OF ISRAEL

The moss flora comprises 220 taxa: 210 species and 10 intraspecific taxa. Of these, 84 have been recorded from Israel for the first time. Four species (*Pterygoneurum crossidioides* W. Frey, Herrnst. & Kurschner, *Phascum galilaeum* Herrnst. & Heyn, *Acaulon longifolium* Herrnst. & Heyn, and *Pottia gemmifera* Herrnst. & Heyn) and three varieties (*Phascum cuspidatum* var. *arcuatum* Herrnst. & Heyn, *P. cuspidatum* var. *marginatum* Herrnst. & Heyn, and *Barbula ehrenbergii* var. *gemmipara* Herrnst. & Heyn) were recently described (Frey *et al.* 1990, Herrnstadt *et al.* 1991, Herrnstadt & Heyn 1999). One new name (*Barbula imbricata* Herrnst. & Heyn), replacing a non-valid name, is added.

The 210 species recognised belong to 64 genera and 20 families (Table 1). Fourteen families are represented by only one or two genera (see the Conspectus of Part I). The most commonly represented family is Pottiaceae (24 genera; 86 species). The richest genera are *Bryum* with 19 species, and *Tortula* with 17 species.

Table 1

A summary of the orders, families, genera, and species of mosses treated in the Flora

Order	Number of families	Number of genera	Number of species	Number of species as percent of total
Fissidentales	1	1	10	5
Dicranales	2	4	5	2
Pottiales	2	25	91	43
Grimmiales	1	3	10	5
Encalyptales	1	1	2	1
Funariales	3	6	16	8
Bryales	3	6	27	13
Orthotrichales	1	2	7	3
Leucodontales	3	6	6	3
Hypnales	3	10	36	17
Total number	20	64	210	

ECOLOGY

The moss flora of Israel includes an abundance of species thriving in open, exposed habitats. The mosses occur on a range of soil types, including those common in steppe or desert environments. In addition to colonising soil surfaces, mosses occur on rocky and stony surfaces, on the bark of living and dead trees, on brick and concrete structures, and in aquatic habitats.

The plants growing in the arid or semi-arid habitats of the Judean and Negev deserts are characteristically small, and frequently grow solitarily or gregariously. In the less xeric habitats of the Mediterranean Region of Israel, they are frequently larger, denser, and commonly form tufts or cushions.

The adaptation of plants to harsh habitats with considerable temperature fluctuations, severe long drought periods, and a season with unreliable precipitation, deserves more attention. Mosses found on exposed surfaces often have morphological and anatomical adaptations that seem useful for survival in those inhospitable niches (Frey & Kürschner 1991b, Herrnstadt & Heyn 1989, Longton 1988, Vitt 1984). The local flora comprises numerous species that have leaves with hyaline hair points, costal lamellae, filaments, revolute or involute leaves, or small and often papillose upper leaf cells. In leaves of plants of the Pottiaceae sampled from dry habitats, the cell walls on the abaxial (dorsal) surface are often thicker than those of the adaxial (ventral) surface (e.g., in *Weissia breutelii*, and to a lesser extent in *Aschisma carniolicum* and *Astomum crispum*).

Additional adaptive specialisations for survival in exposed sites of xeric habitats are: (a) cell walls with black pigmentation (Frey & Kürschner 1991b), which absorb UV radiation and protect the interior of the cells from damage (in *Trichostomopsis aaronis* and some species of *Barbula*); (b) reduction of sporophyte characters (Vitt 1981), e.g., more or less sessile capsules with or without short peristome (*Pterygoneurum subsessile*, *Grimmia mesopotamica*); (c) bud-shaped form of the plants (*Tortula atrovirens*, *Crossidium crassinervium* var. *laevipilum*, *Trichostomopsis aaronis*); (d) subterranean fleshy rhizomes (*Gigaspermum mouretii*).

DISTRIBUTION

About 50% of the local mosses are Mediterranean elements, which also extend north and west as far as the Atlantic belt. Approximately 25% of the species are relatively mesic and show a temperate pattern of distribution. Furthermore, approximately 5% of the 220 taxa are boreal elements and a similar percent are continental western and central Asian plants. A few species of paleotropic origin (Duell 1985) have penetrated north along the Syrian-African Rift Valley (e.g., *Bryum cellulare*).

Species endemic to Southwest Asia (*Gymnostomum mosis*, *Tortula rigescens*, *T. pseudohandelii*, and *Cinclidotus pachyloma*) are also present. One species, *Tortula echinata*, has been recorded only from the eastern Mediterranean Region and does not extend into the eastern parts of Southwest Asia. The recently described species and varieties appearing in this Flora may prove to be endemic.

We have studied most, but not all, of the collections available. Special effort has been made to study and collect from areas that had been undercollected by Bilewsky and others. We are aware that the distribution of the species might be wider than represented here.

DELIMITATION OF ORDERS AND FAMILIES

The delimitation of the higher categories (orders and families) of mosses still poses many problems and their descriptions in literature are often inconsistent. It is expected that their circumscriptions would undergo major revisions in the near future due to new evidence accumulating on: (a) the morphology and development of the peristome in peristomate mosses (Edwards 1979, 1984; Shaw, Anderson & Mishler 1987, 1989; Shaw *et al.* 1989); (b) the morphological characters of the gametophyte that have not been used before (e.g., columella shape and length, patterns of stem epidermis at leaf insertion) (Buck & Goffinet 2000); and (c) molecular characters. These data are expected to shed new light on the evolutionary relationships between large taxonomic groups by enabling the generation of new phylogenies. In the future, classification would be based on those phylogenies (Vitt *et al.* 1998).

GUIDE TO PART I

The Checklist. Most of the species described in this Flora are those given in the Checklist (Herrnstadt *et al.* 1991). When the treatment in the Flora disagrees with that of the Checklist, we cite the Checklist in the synonymy of the species. There are instances where the local distribution of the species deviates from the data in the Checklist because of additional material or re-evaluation of existing specimens.

Descriptions. Family descriptions account for characters of local material as well as for variability throughout the world. Generic descriptions are based on local material as well as on worldwide variability. Species descriptions are based on local material. If a genus is monotypic, the species description usually serves as the generic description. Many descriptions include spores. Unless specifically indicated, descriptions of spores preceding the semicolon are based on light microscope observations whereas those following the semicolon are based on SEM.

Classification. The classification system for orders, families, and genera presented in The Moss Flora of Mexico (Sharp *et al.* 1994) is followed. That classification is a modified version of the system of Brotherus (1924, 1925). The species within the genera are usually arranged in alphabetical order, with the exception of strongly affiliated species (e.g., within groups or complexes).

For the total number of species within genera, we cite recent publications, mainly floras. Comparisons with authentic material (specimens cited by the authors or type specimens), when available, were done for a small number of species.

Keys. The keys to families and genera account for local as well as for worldwide variability. The keys to species are exclusively for local material.

General distribution. To place the local mosses in a wider geographical context, we provide global distribution data. These data are mainly adopted from Duell (1984, 1985, 1992) and Frey & Kürschner (1991c), and stress the distribution around the Mediterranean basin, Sinai, and Southwest Asia (except Israel). When available, special regional or monographic treatments, e.g., Zander (1993) for the Pottiaceae, and Delgadillo (1975) for the *Aloina* and *Crossidium*, were used.

Drawings. The drawings of each taxon are usually based on a single plant. The citation of the herbarium sample on the basis of which the plant was drawn is provided in the legend accompanying the drawing. Great efforts have been made to use the same magnification for each part of the plant, at least within a single genus. In some figures, this has caused overlaps of drawings. Usually, a 1× magnification of the plant drawn is provided. When the plant is minute, an oval outline is provided. The drawings were initially done by Esther Huber and, after her death in 1992, continued by Michal Boaz-Yuval, who changed and adjusted some of the original drawings and added many new ones.

CONSPECTUS

The orders, families (subfamilies and tribes when applicable), and genera of Bryidae included in the Flora are listed below. The number of species in each genus follows in parentheses.

Division BRYOPHYTA

Class BRYOPSIDA

Subclass BRYIDAE

Order FISSIDENTALES
FISSIDENTACEAE: *Fissidens* (10)

Order DICRANALES
DITRICHACEAE: *Ditrichum* (2), *Pleuridium* (1), *Cheilothela* (1)
DICRANACEAE: *Dicranella* (1)

Order POTTIALES
POTTIACEAE:
Subfamily I: TRICHOSTOMOIDEAE:
Weissia (7), *Astomum* (1), *Aschisma* (1), *Trichostomum* (2), *Oxystegus* (1), *Tortella* (6), *Pleurochaete* (1), *Timmiella* (1)
Subfamily II: POTTIOIDEAE:
Tribe *PLEUROWEISIEAE*: *Eucladium* (1), *Gymnostomum* (4), *Gyroweisia* (2), *Hymenostylium* (1), *Anoectangium* (1)
Tribe *BARBULEAE*: *Barbula* (13), *Trichostomopsis* (1), *Leptobarbula* (1)
Tribe *POTTIEAE*: *Tortula* (17), *Desmatodon* (1), *Aloina* (3), *Crossidium* (3), *Pterygoneurum* (3), *Pottia* (8), *Phascum* (4), *Acaulon* (3)
CINCLIDOTACEAE: *Cinclidotus* (5)

Order GRIMMIALES
GRIMMIACEAE: *Schistidium* (1), *Grimmia* (8), *Racomitrium* (1)

Order ENCALYPTALES
ENCALYPTACEAE: *Encalypta* (2)

Order FUNARIALES
GIGASPERMACEAE: *Gigaspermum* (1)
FUNARIACEAE: *Physcomitrium* (1), *Pyramidula* (1), *Funaria* (4), *Entosthodon* (6)
EPHEMERACEAE: *Ephemerum* (3)

Order BRYALES
BRYACEAE: *Pohlia* (2), *Leptobryum* (1), *Bryum* (19)
MNIACEAE: *Plagiomnium* (1)
BARTRAMIACEAE: *Bartramia* (1), *Philonotis* (3)

Order ORTHOTRICHALES

ORTHOTRICHACEAE: *Zygodon* (1), *Orthotrichum* (6)

Order LEUCODONTALES

LEUCODONTACEAE: *Leucodon* (1), *Antitrichia* (1), *Pterogonium* (1)

NECKERACEAE: *Leptodon* (1), *Neckera* (1)

THAMNOBRYACEAE: *Thamnobryum* (1)

Order HYPNALES

FABRONIACEAE: *Fabronia* (2)

AMBLYSTEGIACEAE: *Cratoneuron* (1), *Amblystegium* (3), *Leptodictyum* (1)

BRACHYTHECIACEAE: *Homalothecium* (4), *Scorpiurium* (3), *Brachythecium* (4), *Scleropodium* (3), *Rhynchostegium* (3), *Eurhynchium* (7), *Rhynchostegiella* (5)

BRIEF DESCRIPTIONS OF THE LOCAL ORDERS

Fissidentales

Plants acrocarpous or pleurocarpous. *Stems* simple, sometimes branched, apical cell bilateral. *Leaves* alternate, distichous, in a single plane, each leaf with conduplicate base half-clasping the stem and the base of the leaf above; costa usually present; cells often ± isodiametric. *Capsules* usually exserted, straight and symmetrical, or curved and asymmetrical; peristome single, teeth 16, usually forked in the upper part (to below the middle). *Calyptrae* cucullate or mitrate.

A natural order comprising only one family: Fissidentaceae (p. 25).

Dicranales

Plants acrocarpous. *Stems* with apical cell having three cutting faces. *Leaves* usually narrow, lanceolate, often subulate, in numerous rows; cells long or short, but not isodiametric, often smooth (mammillose or papillose); costa single. *Capsules* mostly exserted and stegocarpous, erect or inclined; peristome usually present, rarely absent, single, teeth 16, usually perforated or cleft, sometimes nearly down to base. *Calyptrae* generally cucullate, smooth.

The order commonly comprises four families, but the delimitation of those families is problematic; two are represented in the local flora: Ditrichaceae (p. 46) and Dicranaceae (p. 55).

Pottiales

Plants generally acrocarpous, sometimes cladocarpous. *Stems* erect, simple, or branched. *Leaves* in numerous rows; costa single; cells isodiametric above, usually papillose, often differentiated below. *Capsules* stegocarpous, rarely cleistocarpous, usually terminal on elongate setae; peristome generally present, teeth 16, papillose, often divided to filiform divisions sometimes down to a basal membrane. *Calyptrae* cucullate, smooth.

The order commonly comprises four families; two are represented in the local flora: Pottiaceae (p. 59) and Cinclidotaceae (p. 269).

Grimmiales

Plants acrocarpous, sometimes cladocarpous, perennial, mainly rupestral, forming usually dark, often hoary, tufts or cushions. *Stems* simple or more often branched. *Leaves* usually lanceolate and strongly hygroscopic, arranged in many rows, often ending in a hair point (piliferous); costa single, cells usually small and smooth, sometimes with thickened sinuose or nodulose walls. *Sporophytes* terminal or lateral. *Setae* straight or sometimes cygneous; capsules immersed or exserted, straight or curved, symmetrical or bulging on one side of base, smooth or striate, peristome usually present, single, of 16 often perforated or variously divided teeth. *Calyptrae* mitrate or cucullate.

An order comprising two families; one is represented in the local flora: Grimmiaceae (p. 280).

Encalyptales

Plants acrocarpous, growing in tufts. *Stems* erect, branched. *Leaves* arranged in many rows; costa single, strongly prominent at back; upper cells usually isodiametric, densely papillose with C-shaped papillae and bulging incrassate cell walls (cells smooth in *Bryobrittonia*); basal cells smooth, with incrassate yellowish transverse walls. *Capsules* terminal, exserted, usually symmetrical, smooth, striate or furrowed; peristome when present long and double with 16 teeth opposite the segments, or short and single. *Calyptrae* large extending below capsule, long-mitrate and long-rostrate, entire, or lobed or fringed at base.

An order comprising one family only: Encalyptaceae (p. 302).

Funariales

Plants acrocarpous, often annual or biennial. *Leaves* usually few, crowded in upper part of stem, usually broad, often obovate; costa usually single (lacking in Gigaspermaceae), thin, slender; cells ± large, lax, thin walled, generally smooth, upper cells oblong-hexagonal, rhomboidal or rectangular. *Capsules* stegocarpous, sometimes cleistocarpous, usually immersed or exserted, often with a distinct neck, peristome absent, single or double; when double the 16 exostome teeth opposite the 16 endostome segments. *Calyptrae* often large (covering ± half length of capsule).

An order comprising five families; three are represented in the local flora: Gigaspermaceae (p. 307), Funariaceae (p. 310), and Ephemeraceae (p. 342).

Bryales

Plants acrocarpous, generally perennial. *Stems* usually with central strand. *Leaves* in numerous rows, often larger and more crowded in stem apex; costa single, usually well developed; cells quadrate to oblong, linear, hexagonal to rhomboidal, and large, usually smooth or mammillose. *Setae* elongate. *Capsules* stegocarpous, exserted, often inclined to pendulous, usually symmetrical, neck generally well developed, peristome mostly double, well developed, exostome with 16 teeth, endostome with 16 segments alternating with the exostome, often arising from a basal membrane; cilia often present. *Calyptrae* cucullate, smooth, naked.

An order comprising eight families (Scott & Stone 1976); three are represented in the local flora: Bryaceae (p. 348), Mniaceae (p. 398), and Bartramiaceae (p. 401).

Orthotrichales

Plants acrocarpous, sometimes apparently pleurocarpous due to pattern of branching; sciophilous or rupestral, forming tufts or cushions. *Stems* erect, ascending, or creeping, simple or branched. *Leaves* lingulate or variously lanceolate, densely arranged in many rows (crowded); costa usually single, strong; upper leaf cells short (mostly round-hexagonal), incrassate, smooth or more often papillose (when papillose mostly unipapillose), basal cells longer ± smooth. *Capsules* stegocarpous, immersed to exserted, symmetrical, often ribbed when dry and empty, stomata superficial or immersed, peristome double, sometimes single or absent; when double, exostome of 16 teeth ± striate or papillose, endostome segments eight or 16, thin, alternating with exostome. *Calyptrae* often mitrate, plicate, and hairy, frequently covering most of the capsule.

The order commonly comprises six families; one is represented in the local flora: Orthotrichaceae (p. 410).

Leucodontales

Plants usually pleurocarpous, primary stems creeping and stoloniform. *Paraphyllia* sometimes present. *Leaves* arranged in many rows; costa single, double, or absent. *Capsules* stegocarpous, immersed, or exserted, often erect and symmetrical; peristome double or single, exostome of 16, mostly papillose teeth, endostome variously developed or absent. *Calyptrae* cucullate or mitrate, more or less hairy, smooth.

A heterogeneous order comprising 15 families; three are represented in the local flora: Leucodontaceae (p. 427), Neckeraceae (p. 434), and Thamnobryaceae (p. 440).

Hypnales

Plants pleurocarpous. *Stems* mostly creeping or creeping to ascending, rarely erect, primary and secondary stems usually identical in shape; costa single, rarely double or absent; leaf cells usually smooth, elongate rhomboidal-hexagonal to linear-flexuose, alar cells often differentiated, setae elongate. *Capsules* stegocarpous, exserted, often inclined to horizontal, asymmetrical; peristome hypnoid, usually double, well developed (except a non-hypnoid peristome in Fabroniaceae). *Calyptrae* cucullate, smooth, usually naked.

An order comprising 18 families; three are represented in the local flora: Fabroniaceae (p. 443), Amblystegiaceae (p. 448), and Brachytheciaceae (p. 460).

KEY TO THE FAMILIES

1 Plants acrocarpous, usually growing in tufts; stems usually erect and sparsely branched; sporophytes at stem apex, or stems prostrate and sporophytes terminal on lateral branches (plants cladocarpous).

2 Leaves distichous in a single plane, each leaf with conduplicate base, sheathing the stem and base of the leaf above **Fissidentaceae**

2 Leaves not distichous, arranged in three or more rows, without conduplicate base of leaf, not sheathing the stem and base of the leaf above.

3 Upper leaf cells subquadrate to linear, or some rhomboidal to hexagonal.

4 Plants slowly absorbing water when dry; stems forked, fastigiately branched, or branched in whorls; upper leaf cells prorate or papillose near one or both ends, or at cell centre; capsules generally sulcate when dry **Bartramiaceae**

4 Plants rapidly absorbing water when dry; stems simple or forked; upper leaf cells generally smooth or with numerous papillae or mammillae; capsules not sulcate when dry.

5 Plants ephemeral; capsules cleistocarpous, immersed to emergent, ± globose, operculum and peristome not differentiated **Ditrichaceae** (*Pleuridium*)

5 Plants not ephemeral; capsules stegocarpous, exerted, usually ± cylindrical, sometimes short-pyriform, operculum and peristome differentiated.

6 Leaves often narrow, tapering to tip, sometimes sheathing at base, leaf cells generally short; peristome usually present, always single.

7 Peristome teeth divided into two filiform divisions and lacking vertical striations **Ditrichaceae** (excluding *Pleuridium*)

7 Peristome teeth divided into flat divisions with pitted vertical striations below **Dicranaceae**

6 Leaves usually broad, often tapering from middle to tip and to base, not sheathing at base, leaf cells often elongate; peristome, when present, often double.

8 Capsules always exerted, often inclined to pendulous, peristome with 16 exostome teeth alternating with 16 endostome segments; costa well developed; leaf cells often ± incrassate.

9 Leaves large, oblong to narrow-elliptical, margins usually clearly bordered by two–five rows of long narrow cells, distinctly toothed; perigonia with clavate paraphyses; stomata usually immersed **Mniaceae**

9 Leaves smaller, ovate-lanceolate to elliptical, margins without a border or with a border of one–two rows of cells, entire or sometimes denticulate or serrate near apex; perigonia with filiform paraphyses; stomata superficial.

10 Leaf cells narrow, elongate-linear, shorter towards apex, margins at least somewhat serrate toward apex; capsule short-pyriform **Bryaceae** (excluding *Bryum*)

10 Leaf cells ± rhomboidal-hexagonal, margins usually not serrate toward apex, capsule ± cylindrical **Bryaceae** (*Bryum*)

8 Capsules immersed or exerted, erect or inclined, peristome absent or present, when double with 16 exostome teeth opposite 16 endostome segments; costa lacking, weak to obsolete, or ± well developed; leaf cells large, pale, thin walled.

11 Plants minute with persistent protonema; sporophytes immersed, capsules cleistocarpous **Ephemeraceae**

11 Plants usually with rapidly disappearing protonema; sporophytes usually exserted*; capsules stegocarpous.

12 Plants often biennial, sometimes perennial, with subterranean fleshy rhizomes; leaves generally with reduced costa; capsules without a neck, without peristome; paraphyses filiform; spores moderate to very large, up to 140 μm in diameter **Gigaspermaceae**

12 Plants annual or biennial, without subterranean rhizomes; leaves commonly with well defined costa; capsules usually with well developed neck, with peristome; paraphyses clavate; spores small, usually up 40 μm in diameter

13 Capsules inclined, asymmetrical, peristome well developed, double **Funariaceae** (*Funaria*)

13 Capsules usually erect, symmetrical, peristome less developed, single, rudimentary or none **Funariaceae** (*Entosthodon, Physcomitrium*)

3 Upper leaf cells ± isodiametric.

14 Basal cells of leaves often clearly differentiated from upper cells, hyaline, thin walled or only transverse walls incrassate.

15 Capsules long exserted; calyptrae large, mitrate, long-rostrate, extending below capsules **Encalyptaceae**

15 Capsules exserted to immersed; calyptrae cucullate, never extending below capsules **Pottiaceae** (Pottioideae)

14 Basal cells of leaves not or hardly differentiated from upper cells, opaque, often ± incrassate.

16 Plants usually robust, often aquatic or periodically submerged, not xerophytic; leaves with strongly thickened margins of several cell layers **Cinclidotaceae**

16 Plants small to moderately large, terrestrial, often xerophytic; leaves without thickened margins of several cell layers.

17 Plants growing mainly on soils; costa differentiated into one or two stereid bands, guide cells, epidermal cells, and sometimes hydroids **Pottiaceae** (Trichostomoideae)

17 Plants growing mainly on rocks and bark of trees; costa often with

* Except in *Gigaspermum*.

± uniform cells, only occasionally differentiated into guide cells, and without hydroids.

18 Plants growing mainly on rocks (rupestral); leaves often piliferous, cells with sinuose or straight walls; calyptrae usually smooth, small, not covering the capsule **Grimmiaceae**

18 Plants growing on bark of trees (corticolous) or on rocks; leaves usually not piliferous, cells generally without sinuose walls; calyptrae frequently hairy, large, covering most of the capsule **Orthotrichaceae**

1 Plants pleurocarpous, growing in mats; stems often prostrate, creeping, and freely branched; sporophytes lateral.

19 Primary and secondary stems usually identical in shape; leaf cells usually long and narrow, alar cells often differentiated; calyptrae mostly cucullate.

20 Plants slender; stems without central strand; leaf margins serrate or with short to long cilia-like teeth; capsules erect, symmetric, ovoid to short-pyriform, peristome non-hypnoid **Fabroniaceae**

20 Plants slender to robust; stems often with central strand; leaf margins entire, serrulate or serrate; capsules inclined to horizontal, asymmetric, ovoid, ellipsoid to cylindrical, peristome hypnoid.

21 Leaves ± concave, plicate, costa well developed, cells elongate-rhomboidal to linear-flexuose, alar cells not or little differentiated; operculum conical, often rostrate **Brachytheciaceae**

21 Leaves usually not concave and plicate, costa often short to elongate, cells generally shorter, oblong-hexagonal not flexuose, alar cells often ± distinctly differentiated; operculum short conical, usually mammillate or apiculate **Amblystegiaceae**

19 Primary and secondary stems usually not identical in shape; leaf cells often short, alar cells generally not differentiated; calyptrae cucullate or mitrate.

22 Leaves usually complanate, apex truncate to rounded or acute; paraphyllia often present; annulus usually not differentiated **Neckeraceae**

22 Leaves not complanate, apex acute or acuminate or acute to obtuse; paraphyllia absent; annulus usually differentiated.

23 Plants growing on rocks, trunks, or barks of trees; stems without or with only rudimentary central strand; costa single, double, or absent, when single with ± supplementary costa near base, when present usually not exceeding ⅔ length of leaf; peristome without cilia **Leucodontaceae**

23 Plants growing on moist rocks, boulders, or tree roots, often on banks of streams; stems with central strand; costa single, ending below apex (subpercurrent); peristome with cilia, usually well developed **Thamnobryaceae**

TAXONOMIC TREATMENT

Bryopsida
Bryidae

Fissidentales

FISSIDENTACEAE

The family is usually considered as comprising only the genus *Fissidens*, though some authors segregate a few small genera. The family Fissidentaceae is clearly distinguishable from other moss families by the distichous leaf arrangement and the unique leaf structure with a typical conduplicate base. The peristome, if present, resembles the peristome of Ditrichaceae and Dicranaceae, i.e., in the low-divided 16 peristome teeth.

For characters of the family, see the description of *Fissidens* below.

Fissidens Hedw.

Plants small (ca. 1 mm) to large (up to 80 mm), growing on various substrates in widely varying habitats. *Stems* branched or not, with central strand or without (in *Fissidens fontanus*). *Leaves* distichous, alternate, complanate, strongly sheathing the stem, differentiated into two vaginant laminae (true laminae) sheathing the stem, a dorsal lamina (opposite the vaginant laminae), and an apical lamina (sometimes called ventral lamina) above the vaginant laminae; margins without border or with unistratose or rarely multistratose border (limbidium) on vaginant laminae or also on other parts of the leaf, completely or partly developed; costa usually ending below apex to shortly excurrent, rarely absent, with two stereid bands; cells usually unistratose, in mid- and upper leaf irregularly hexagonal to round-hexagonal, smooth, papillose or mammillose; basal cells ± enlarged, usually subquadrate to irregularly rectangular; perichaetial and perigonial leaves often differentiated. *Monoicous* or dioicous. *Sporophytes* terminal or lateral, usually one per perichaetium. *Setae* usually exserted, erect or flexuose, sometimes geniculate at base; capsules erect or inclined, symmetric to ± curved, often constricted below mouth; annulus absent; peristome single, rarely rudimentary, teeth 16, usually divided down to below middle, red to reddish-brown; operculum conical to long-rostrate. *Spores* smooth to finely papillose. *Calyptrae* cucullate to mitrate, generally smooth.

A genus with worldwide distribution for which a varying large number of species have been accepted (e.g., 900 — Pursell *in* Sharp *et al.* 1994; 1000 — Bruggeman-Nannenga & Nyholm *in* Nyholm 1986).

Literature. — Potier de la Varde (1956); Smith (1970); Pursell (1976; *in* Sharp *et al.* 1994); Bruggeman-Nannenga (1978, 1982); Corley *et al.* (1981).

The genus *Fissidens* is represented in the local flora by 10 species. Seven species (nos. 3–7) are part of a complex, often treated as the *Fissidens bryoides* complex. The species concepts in this very difficult group differ widely. The whole group is sometimes treated as the single species: *F. bryoides* (e.g., Pursell 1976, Crum & Anderson 1981). Conversely, it is considered by others as a complex comprising several, but not always the same, species (e.g., Corley 1980 — as *F. viridulus* complex; Bruggeman-Nannenga 1978, 1982 — as *F. bryoides* complex). The detailed treatment of the *F. bryoides* complex in Bruggeman-Nannenga (1978, 1982) is mainly, though not in all cases, followed here. Bruggeman-Nannenga (1982) names a group of species within the *F. bryoides* complex "*F. crassipes* subcomplex". Accordingly, in order to facilitate the treatment of the local *Fissidens* species, the remainder of the *F. bryoides* complex is considered here as "*F. bryoides* subcomplex". Five species are included in the *F. bryoides* subcomplex: *F. bryoides*, *F. bambergeri*, *F. bilewskyi*, *F. incurvus*, and *F. viridulus*, and two species (three taxa) are included in the *F. crassipes* subcomplex: *F. crassipes*, and *F. pusillus*.

Taking into account the difficulties of delimitation of species due to intergrading expressions of characters (Pursell *in* Sharp *et al.* 1994), the treatments of individual species in the group seem at present the only feasible option for retaining the extant local information. Thus, those treatments have to be considered as temporary. A better understanding of the complex would depend on future genetic and environmental studies, which may explain the nature of the observed variation of characters among the taxa.

1 Leaves not bordered.
  2 Plants terricolous; stems less than 50 mm long.
    3 Plants large, 10–30(–50) mm; leaves with three–four marginal rows of pellucid, ± paler cells **1. F. adianthoides**
    3 Plants small, 2–3(–4) mm; leaves with marginal rows of cells not pellucid and paler than adjacent laminal cells **2. F. arnoldii**
  2 Plants aquatic; stems over 50 mm long **10. F. fontanus**
1 Leaves bordered [*F. bryoides* complex]
  4 Plants usually saxicolous, hydrophytic or mesophytic; mid-leaf cells of dorsal lamina up to 15 μm long; antheridia terminal on main stem [*F. crassipes* subcomplex]
    5 Plants usually up to 20 mm high, hydrophytic; leaf borders uni- or pluristratose, usually at least partly two–three stratose, often intramarginal.
      6 Plants vivid to dark green; stems usually up to 15 mm long, not encrusted with lime, with up to 18 pairs of leaves; leaves densely spaced **8a. F. crassipes** subsp. **crassipes**
      6 Plants pale green; stems usually longer, encrusted with lime, with up to 35 pairs of leaves; leaves laxly spaced **8b. F. crassipes** subsp. **warnstorfii**
    5 Plants small, 1–3(–5) mm high, mesophytic; leaf borders usually unistratose, not intramarginal **9. F. pusillus**
  4 Plants terricolous, not hydrophytic; mid-leaf cells of dorsal lamina up to 12 μm long; antheridia not terminal on main stem [*F. bryoides* subcomplex]
    7 Capsules inclined to cernuous, asymmetrical.
      8 Mid-leaf cells of dorsal lamina small, 4–6 μm wide **7. F. bilewskyi**
      8 Mid-leaf cells of dorsal lamina large, 7–11 μm wide **6. F. incurvus**
    7 Capsules erect and symmetrical.

9 Dorsal lamina of upper leaves not reaching stem, with cells in mid-leaf 5–8 µm wide **4. F. bambergeri**

9 Dorsal lamina of upper leaves reaching stem, with cells in mid-leaf 8–12 µm wide.

10 Plants up to 8 mm high; border usually complete; antheridia in minute, bud-like axillary branches or naked in leaf axils; leaf apex ± abruptly narrowing into a stout point **3. F. bryoides**

10 Plants up to 6 mm high; border usually not reaching apex; antheridia terminal on dwarf male plants or at plant base, sometimes also in buds or naked in leaf axils; leaf apex gradually narrowing into acute point **5. F. viridulus**

1. Fissidens adianthoides Hedw., Sp. Musc. Frond. 157 (1801).

Figure 1.

Plants large, 10–30(–50) mm high, yellowish to brownish-green. *Stems* erect, simple, or sparsely branched, tomentose below. *Leaves* densely spaced, up to 4 mm long, broadly oblong to lanceolate, acute, undulate in upper part when dry; margins crenate below, irregularly sharply serrate above, not bordered; costa percurrent or subpercurrent; cells ± incrassate, unistratose, hexagonal with rounded corners, opaque; three–four rows of cells at margin pellucid, ± paler; mid-leaf cells 11–17 μm wide; vaginant laminae at least ⅔ the length of leaf, dorsal lamina slightly undulate, round at base, reaching stem; perichaetia and perigonia axillary. *Dioicous* or autoicous. *Sporophytes* single or several from each perichaetium, reddish. *Setae* laterally inserted at mid-stem; capsules erect, straight, ellipsoid; operculum obliquely long-rostrate. *Spores* up to 27 μm in diameter, papillose.

Habitat and Local Distribution. — Only a single collection by Kushnir from Israel (no. 1033 in Herb. Bilewsky; without locality and date) was available.

General Distribution. — Recorded from several localities around the Mediterranean, Europe (throughout, but not widespread), Macaronesia, North Africa, northeastern, eastern and central Asia, Australia, New Zealand, North America, and Antarctica.

Figure 1. *Fissidens adianthoides*: **a** habit (×1); **b** stem (×10); **c** part of stem with sporophyte detached (×10); **d** peristome (×50); **e** leaf (×50); **f** cells at base of vaginant lamina; **g** mid-leaf cells of ventral lamina; **h** leaf apex (all leaf cells ×500); **i** perichaetial leaf (×50); **k** cross-section of costa and portion of upper part of lamina (×500). (*Kushnir sine dato et loco*)

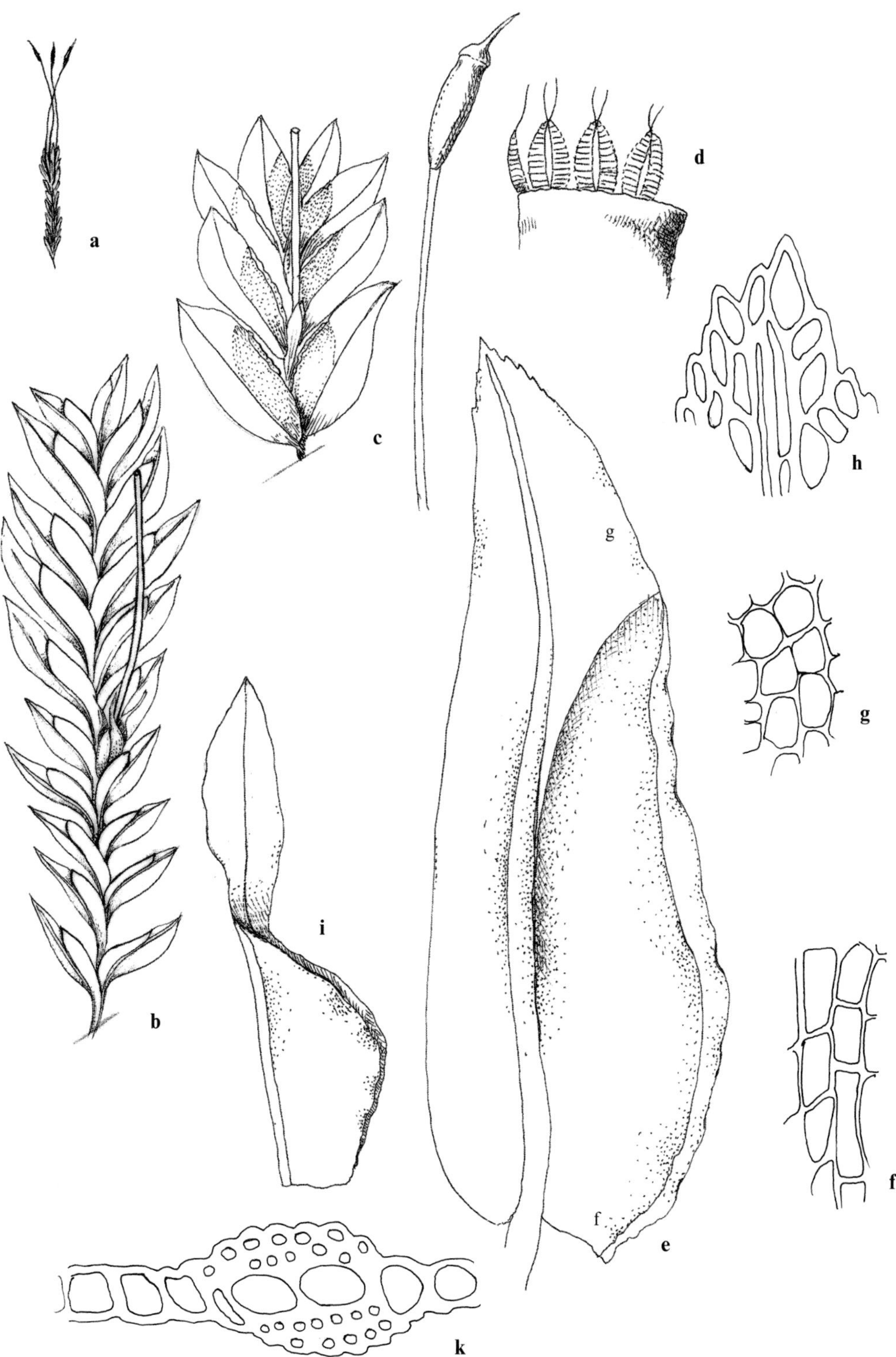
a
b
c
d
e
f
g
h
i
k
g
f

2. Fissidens arnoldii R. Ruthe, Hedwigia 9 : 178 (1870).
F. obtusifolius auct., non Wilson 1845.

Distribution map 1, Figure 2.

Plants very small, up to 2–3(–4) mm high, gregarious, vivid green. *Stems* often branched, branches with up to eight pairs of leaves, upper leaves larger than lower. *Leaves* up to 0.7 mm long, ovate-oblong, obtuse; margins entire, not bordered; costa ± thickened, yellowish, ending below apex; cells thin-walled to ± incrassate, irregularly quadrate-hexagonal, mid-leaf cells of dorsal lamina 8–11 µm wide, not changing in colour towards margins; dorsal lamina not or just reaching stem, vaginant laminae two-thirds the length of leaf. *Sporophytes* not seen in local plants.

Habitat and Local Distribution. — Plants growing on shaded, damp soil in rock crevices or at foot of cliffs. Rare to occasional, locally abundant; occurs in a restricted area of low altitudes: Judean Desert and Dead Sea area.

General Distribution. — Recorded from Southwest Asia (Jordan, Iraq, Kuwait, Saudi Arabia, Yemen, and Oman), no Mediterranean records found, scattered records from Europe (from western through central Europe to Russia in the east, sometimes as *Fissidens obtusifolius*).

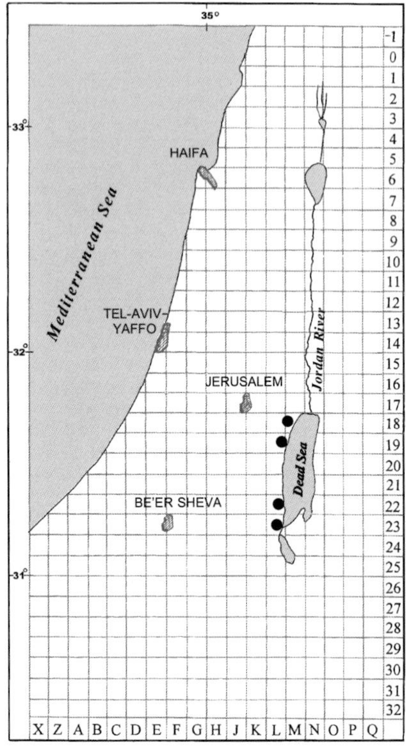

Map 1: *Fissidens arnoldii*

According to Bruggeman-Nannenga (1987), *Fissidens antineae* P. de la Varde (*in* Thér. Hist. Nat. Bull. Soc. Afr. Nord 22 : 158. 1931) from Sahara, Algeria should be considered conspecific with *F. arnoldii*.

Fissidens bryoides complex

Leaves — at least vaginant laminae — bordered.

Records of all species of the *Fissidens bryoides* complex should be taken with some doubts because of the great variation in species concepts among authors.

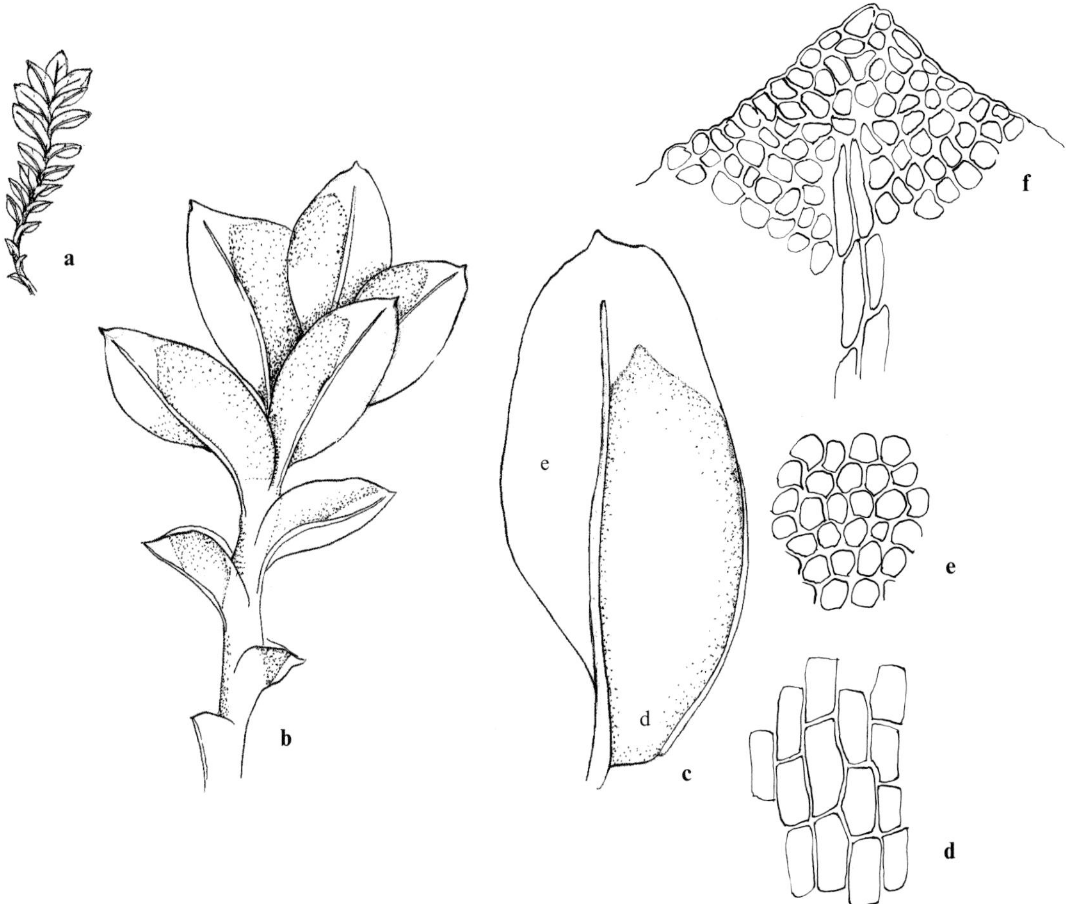

Figure 2. *Fissidens arnoldii*: **a** habit (×10); **b** habit (×50); **c** leaf (×125); **d** cells at base of vaginant lamina; **e** mid-leaf cells of dorsal lamina; **f** leaf apex (all leaf cells × 500). (*Crosby & Herrnstadt 78-46-4*)

A. Fissidens bryoides subcomplex: Species 3–7

Leaves lanceolate to lanceolate-lingulate; costa often percurrent; vaginant laminae ± equal in length to apical lamina; cells irregularly rounded to hexagonal; mid-leaf cells of dorsal lamina up to 12 µm long; border cells unistratose, elongated, with tapering ends, highly variable in shape, sometimes restricted to vaginant laminae. *Autoicous*, polyoicous, or dioicous; antheridia not terminal on main stem. *Capsules* erect and symmetrical to inclined-horizontal, curved and asymmetrical.

Habitat and Local Distribution. — Plants usually terricolous, growing in shaded and moist habitats, in patches or tufts, never hydrophytic.

3. Fissidens bryoides Hedw., Sp. Musc. Frond. 153 (1801). [Bilewsky (1965): 350]

Distribution map 2, Figure 3.

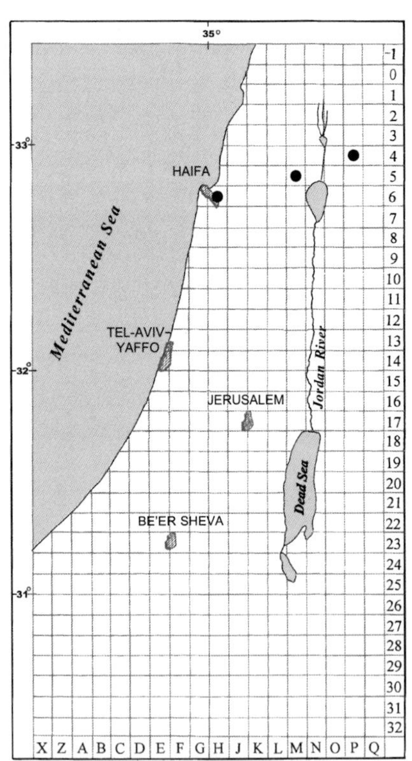

Map 2: *Fissidens bryoides*

Plants up to 8 mm high, vivid green. *Stems* often unbranched, with five–thirteen pairs of leaves. *Leaves*: upper leaves up to 1.5–2 mm long, lanceolate-lingulate, abruptly narrowing into a short, stout, and acute point; margins usually with complete border on all laminae and ending in leaf apex; costa percurrent to short-excurrent; dorsal lamina reaching stem; mid-leaf cells of dorsal lamina 8–12 µm wide. *Autoicous*; antheridia in minute buds or naked in leaf axils. *Setae* up to 9 mm long, turning reddish when mature; capsules usually erect and symmetrical, 0.75–1 mm long without operculum; operculum conical-rostrate, up to half the length of capsule. *Spores* 10–17 µm in diameter, finely papillose to nearly smooth.

Habitat and Local Distribution. — Plants growing in tufts or patches on moist soil at base of limestone rocks, and in basalt rock crevices near a waterfall. Rare: Lower Galilee, Mount Carmel, and Golan Heights.

General Distribution. — Recorded from Southwest Asia (Turkey, Lebanon, Jordan, Iraq, Iran, and Saudi Arabia); around the Mediterranean, Europe (throughout), Macaronesia, tropical and North Africa, western, northeastern, eastern and central Asia, New Zealand, Oceania, and northern, central and southern South America.

Figure 3. *Fissidens bryoides*: **a** habit (×10); **b** deoperculate capsule (×25); **c** leaf (×50); **d** cells at mid-dorsal lamina; **e** leaf apex (all leaf cells ×500); **f** cross-section through lower part of leaf (×500); **g** portion of peristome teeth (×300). (*Sereni*, 11 Apr. 1927)

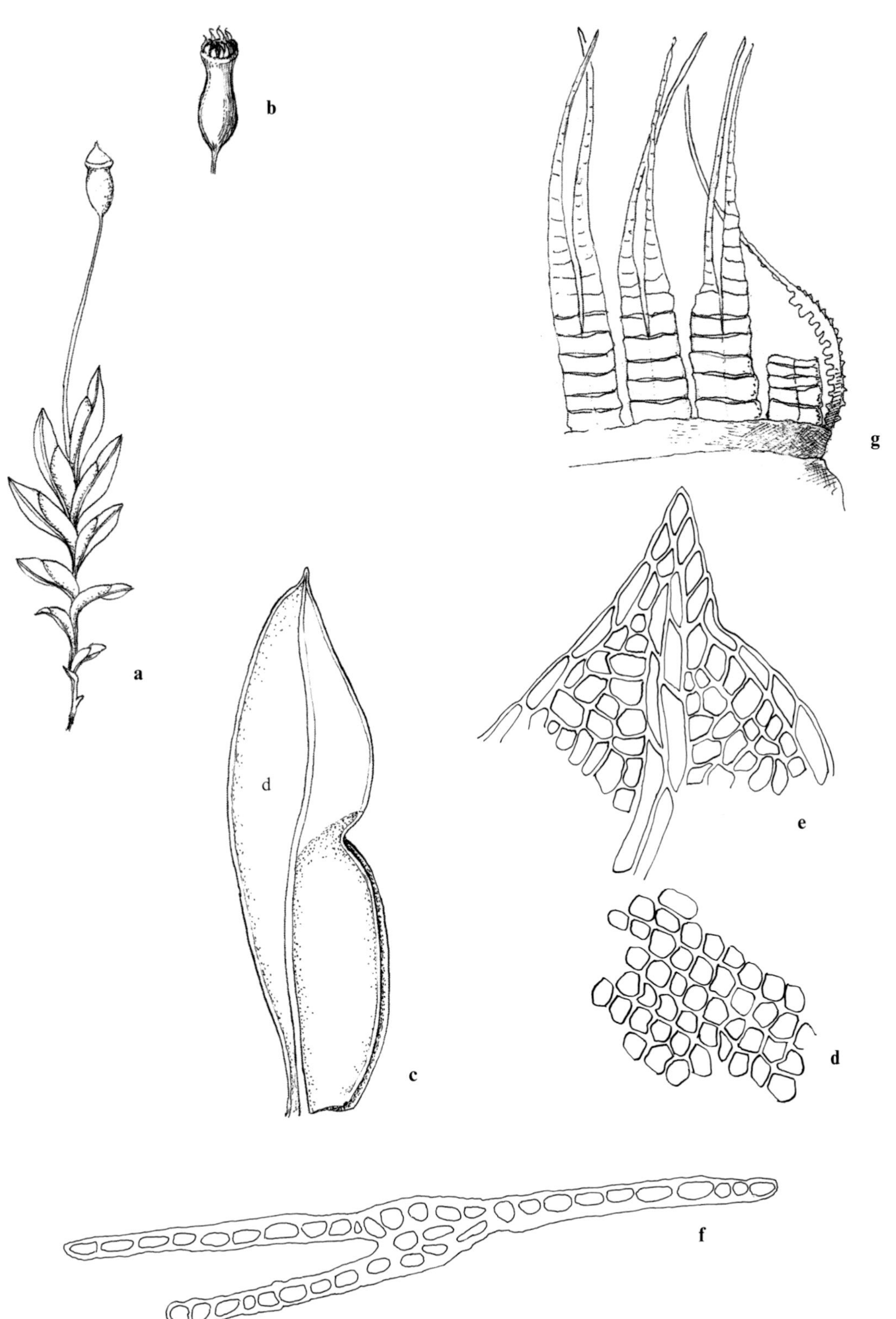
b
g
a
d
e
d
c
f

4. Fissidens bambergeri Schimp. *in* Milde, Bot. Zeit. 22 : 12 (1864).
[Bilewsky (1965) : 351]

Distribution map 3, Figure 4 : a, Plate I : a–c.

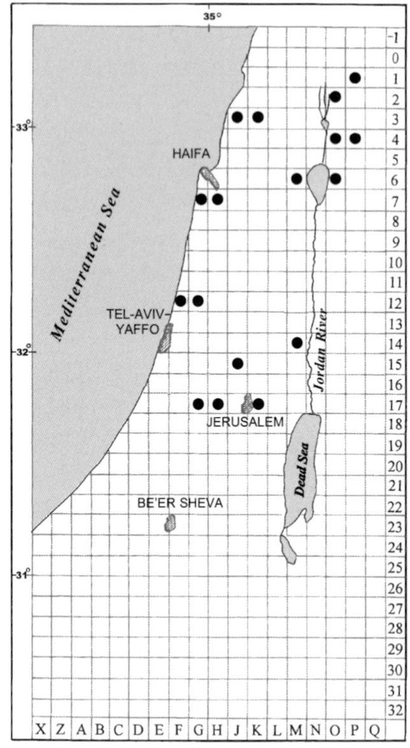

Map 3: *Fissidens bambergeri*

Plants resembling *Fissidens bryoides* but smaller, up to 4(–5) mm high, light green to green. *Leaves*: margins with thin border only on vaginant laminae of upper leaves, dorsal lamina ending far above leaf insertion, or markedly narrowing towards insertion; cells small, in mid-leaf of dorsal lamina 5–8(–9) µm wide. *Synoicous* or autoicous; antheridia terminal, on dwarf shoots at base of stem. *Spores* 10–15 µm in diameter; granulate, granules with rough surface, varying in size (seen with SEM).

Habitat and Local Distribution. — Plants growing in patches or tufts on soil, or on soil on sandstone, limestone, and basalt rocks, in shade of rocks and trees, most often on north-facing slopes. Occasional: Coast of Carmel, Sharon Plain, Upper Galilee, Lower Galilee, Mount Carmel, Samaria, Shefela, Judean Mountains, Judean Desert, and Golan Heights.

General Distribution. — Recorded from Southwest Asia and around the Mediterranean (see the comment about the concept of *Fissidens bryoides* complex above, pp. 25–26).

According to Potier de la Varde (1956), some local specimens should be considered as belonging to *Fissidens bambergeri* var. *aegyptiacus* Renauld & Cardot.

According to Bruggeman-Nannenga (1978), the type specimen of *Fissidens bambergeri* is a "poorly developed and abnormal form of *F. minutulus*" (therefore, *F. bambergeri* was included in that publication in *F. minutulus* with *F. bilewskyi*). *Fissidens bambergeri* and *F. minutulus* are considered as synonyms of *F. limbatus* by Corley *et al.* (1981).

5. Fissidens viridulus (Sw.) Wahlenb., Fl. Lapp. 334 (1812).
Dicranum viridulum Sw., Monthl. Rev. 34 : 538 (1801); *F. impar* Mitt., J. Linn. Soc. Bot. 21 : 554 (1885). — Bilewsky (1965) : 350.

Distribution map 4 (p. 36), Figure 4 : b–h.

Plants closely resembling *Fissidens bryoides*, but smaller, up to 6 mm long, vivid green. *Leaves* lanceolate, gradually narrowing into acute point; border ± weaker, often ending several cells below apex; costa usually not reaching apex; cells in mid-leaf of dorsal lamina 8–12 µm. *Autoicous*, polyoicous (sometimes dioicous, as a secondary condition); antheridia terminal on dwarf male plants or on branches at base of plant, some-

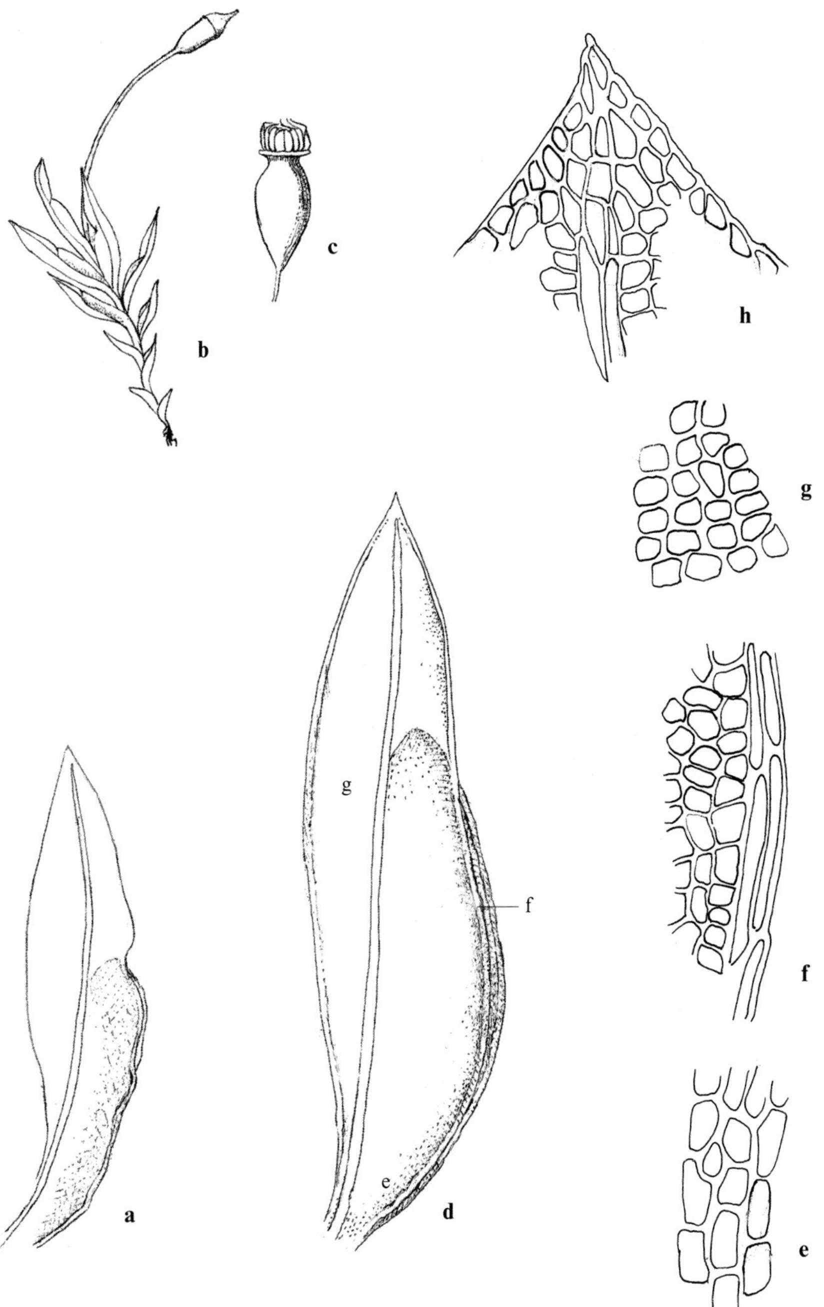

Figure 4. *Fissidens bambergeri*: **a** leaf (×50).
Fissidens viridulus: **b** habit (×10); **c** deoperculate capsule (×25); **d** leaf (×50); **e** cells at base of vaginant lamina; **f** cells at border of vaginant lamina; **g** cells at mid-dorsal lamina; **h** leaf apex (all leaf cells ×500).
(**a**: *Kushnir*, Jan. 1943; **b**–**h**: *Herrnstadt* & *Crosby 78-31-7*)

times also axillary in buds or naked. *Capsules* erect or slightly inclined, symmetrical or asymmetrical (mostly erect and symmetrical). *Spores* 11–20 µm in diameter, nearly smooth.

Habitat and Local Distribution. — Plants growing on calcareous soils rich in clay. The most common local species of the genus: Coast of Carmel, Sharon Plain, Upper Galilee, Lower Galilee, Mount Carmel, Esdraelon Plain, Samaria, Shefela, Judean Mountains, Judean Desert, Central Negev, Bet She'an Valley, Mount Hermon, and Golan Heights.

General Distribution. — Recorded from many parts of Southwest Asia, around the Mediterranean, Europe (throughout), Macaronesia, tropical and North Africa, northeastern and central Asia, and Central America (see comment about the concept of *Fissidens bryoides* complex above, pp. 25–26).

Fissidens *viridulus* is a most variable species considered by some authors as part of *F. bryoides* (e.g., Anderson *et al.* 1990).

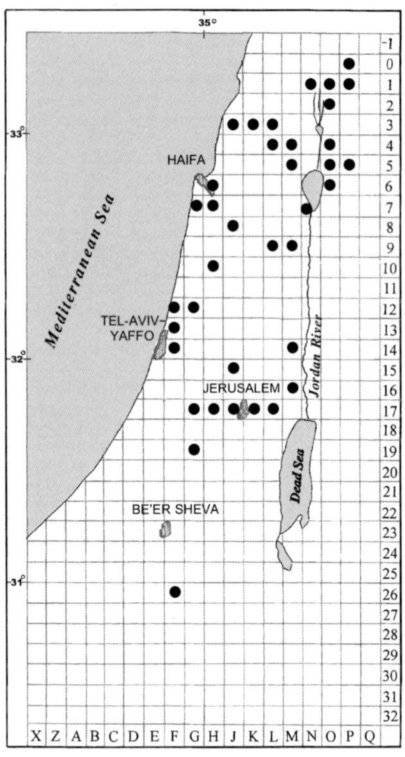

Map 4: *Fissidens viridulus*

6. Fissidens incurvus Starke ex Röhl., Deutschl. Fl. (Edition 2), Kryptog. Gew. 2, 3 : 76 (1813). [Bilewsky (1965) : 351]

Distribution map 5, Figure 5 : a–f.

Plants up to 6(–10) mm high, vivid green. *Stems* decumbent or erect, usually not branched, with 5–10 pairs of leaves. *Leaves*: upper leaves 1–1.75 mm long, lanceolate, abruptly narrowing into short, acute point (as in *Fissidens bryoides*); margins of all laminae bordered, borders complete, usually confluent with costa at apex, dorsal lamina usually reaching stem; costa percurrent or short-excurrent; mid-leaf cells of ventral lamina 7–11 µm wide. *Autoicous*; antheridia terminating basal dwarf shoots, never axillary. *Setae* flexuose, 4–7 mm long, reddish; capsules asymmetrical, curved, inclined to horizontal, 0.75–1 mm long without operculum, oblong-ellipsoid; operculum rostrate, 0.25–0.5 mm long. *Spores* 12–16 µm in diameter, nearly smooth.

Habitat and Local Distribution. — Plants

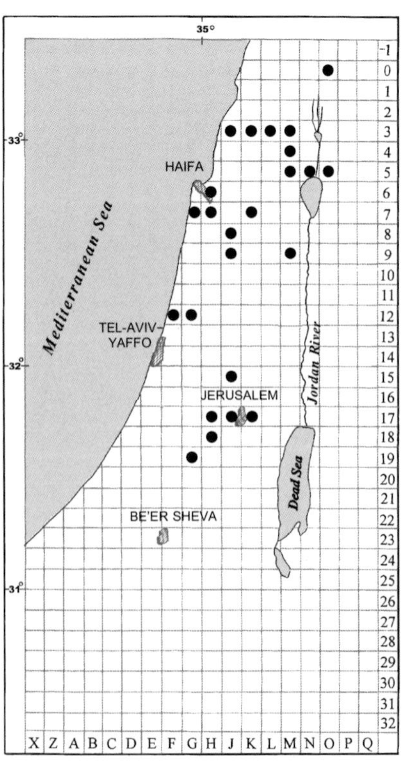

Map 5: *Fissidens incurvus*

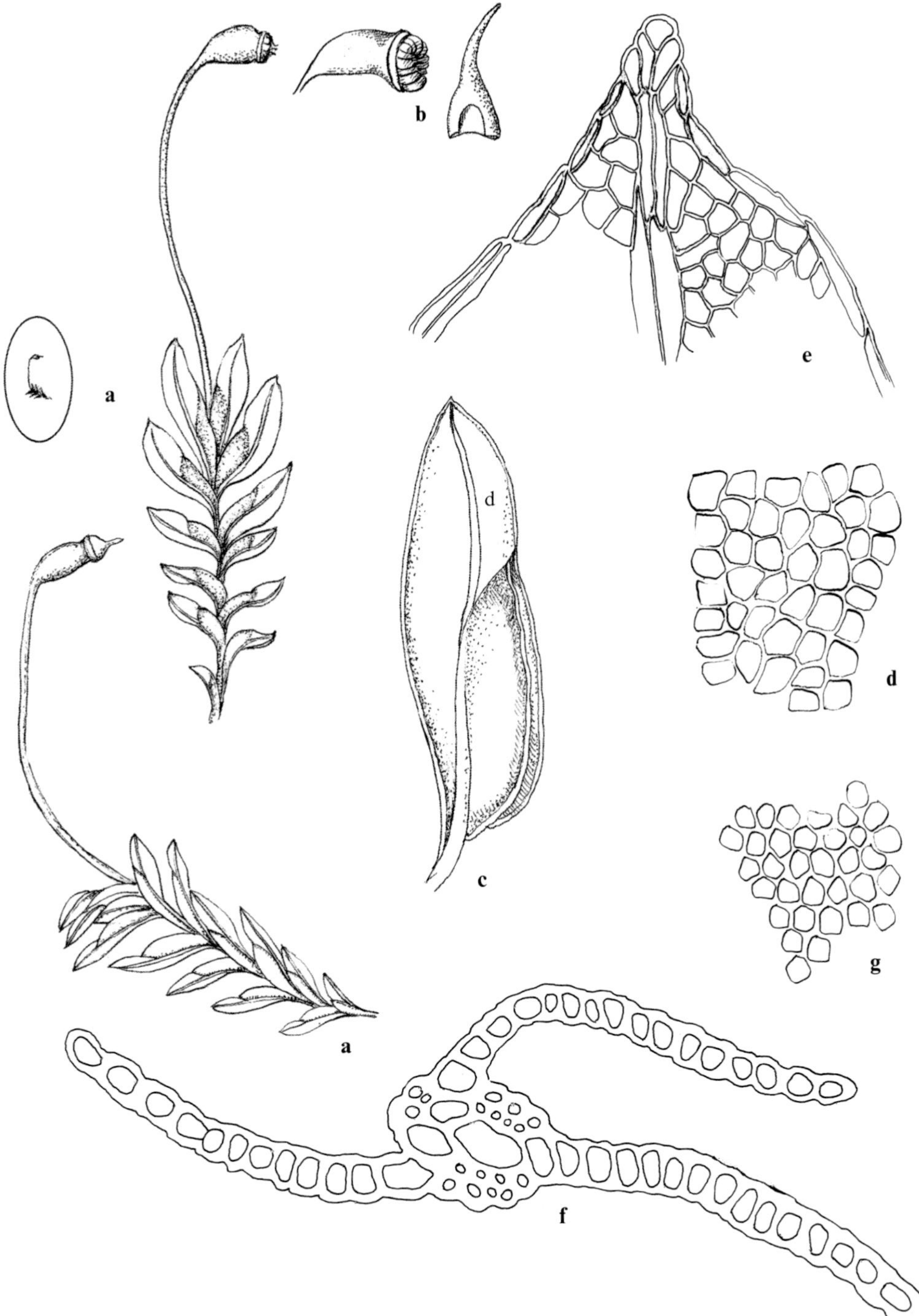

Figure 5. *Fissidens incurvus*: **a** habit of decumbent and erect plants (×10); **b** deoperculate capsule and calyptra (×25); **c** leaf (×50); **d** mid-leaf cells of ventral lamina; **e** leaf apex (all leaf cells ×500); **f** cross-section through lower part of leaf (×500).

Fissidens bilewskyi: **g** mid-leaf cells of ventral lamina (×500).

(**a-f**: *Heyn & Herrnstadt 82-228-5*; **g**: *Friedberg*, 15 Feb. 1952, No. 7 in Herb. Bilewsky)

growing on terra rossa and rendzina, in rock crevices and under rocks of basalt and limestone, in caves, on banks of wadis, near running water, in shade of rocks and trees, and on north- and west-facing slopes. Common: Coast of Carmel, Sharon Plain, Upper Galilee, Lower Galilee, Mount Carmel, Samaria, Shefela, Judean Mountains, Dan Valley, Bet She'an Valley, and Golan Heights.

General Distribution. — Recorded from Southwest Asia, around the Mediterranean, Europe (throughout), Macaronesia, tropical and North Africa, northeast, central, and Southeast Asia, Australia, and North America (see comment about the concept of *Fissidens bryoides* complex above, pp. 25–26).

7. Fissidens bilewskyi P. de la Varde, Rev. Bryol. Lich. 25 : 122 (1956).
[Bilewsky (1965) : 352]

Distribution map 6, Figure 5 : g (p. 37).

Plants resemble *Fissidens incurvus*, except for the smaller leaves and smaller leaf cells; green, turning yellowish when dry. *Leaves* up to 1.25 mm long; mid-leaf cells of ventral lamina 4–6 μm wide; border mainly on vaginant laminae. *Capsules* minute, 0.5–0.75 mm long.

Habitat and Local Distribution. Plants growing on heavy soil. Recorded only from the type locality: Sharon Plain (Ramat-Gan, Napoleon Hill).

Fissidens bilewskyi may represent only a poorly developed form of *F. incurvus* but is retained as a separate species until additional data are obtained.

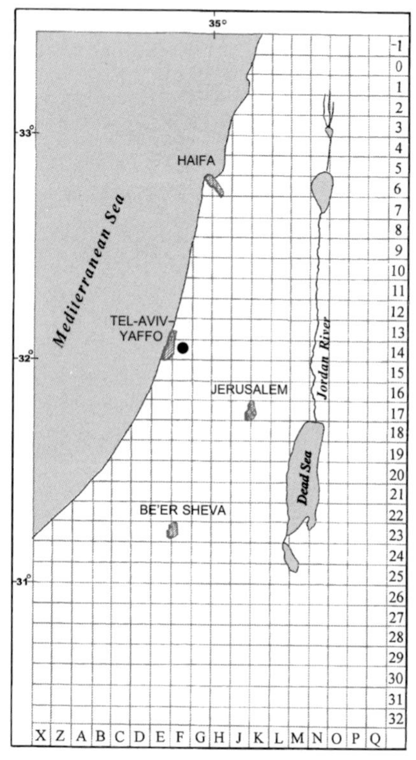

Map 6: *Fissidens bilewskyi*

B. Fissidens crassipes subcomplex

(*sensu* Bruggeman-Nannenga 1982): Species 8–9

Plants usually saxicolous, hydrophytic, or mesophytic, growing in pure mats. *Leaves* lanceolate; costa ending few cells below apex; vaginant laminae ± equal in length to apical lamina or ± longer; cells ± hexagonal; mid-leaf cells of dorsal lamina up to 15 μm long; border cells uni- or pluristratose, usually at least partly bi- to tristratose, often intramarginal. *Dioicous* or often synoicous; antheridia and archegonia terminal. *Capsules* symmetrical, erect or slightly inclined.

8. Fissidens crassipes Wilson ex B.S.G., Bryol. Eur. 1 : 197 (1849).

Plants decumbent to erect. *Leaves* border often two–three-stratose, yellowish, vaginant lamina with a single row of marginal chlorophyllous cells; dorsal lamina (in local plants) reaching stem; mid-leaf cells of vaginant laminae 8–14 μm wide, ± thin-walled, irregularly hexagonal, translucent. *Dioicous*, autoicous, or rarely synoicous. *Setae* reddish; capsules up to 1.25 mm long including operculum, ellipsoid, green; operculum conical-rostrate, up to half the length of capsule; peristome red. *Spores* 15–25(–30) μm in diameter.

Habitat. — Plants growing on calcareous and basaltic rocks near or in water, often below watersheds or under dripping water, rarely on shaded wet soil.

8a. Fissidens crassipes Wilson ex B.S.G. subsp. **crassipes**

F. mildeanus Schimp. *in* Milde, Bot. Zeit. 20 : 459 (1862).

Distribution map 7, Figure 6 : a–e (p. 41).

Plants usually up to 15 mm high, vivid green to dark green. *Stems* with up to 18 pairs of leaves, densely spaced, at least above. *Leaves* up to 2.5 mm long, usually elliptical, rarely ovate, acute; border of all laminae up to 32 μm wide; border of dorsal lamina not reaching apex. *Archegonia* 400–600 μm long. *Peristome* teeth usually over 50 μm wide at base.

Local Distribution. — Occasional: Upper Galilee, Judean Mountains, Hula Plain, and Golan Heights.

General Distribution. — Recorded from Southwest Asia (Turkey, Lebanon, Jordan, Iraq, Iran, and Saudi Arabia), scattered around the Mediterranean, Europe (throughout), tropical and North Africa, eastern and central Asia, and Australia. According to Bruggeman-Nannenga (1982), many of the records are based on misidentifications, and subsp. *crassipes* is mainly a widespread European taxon.

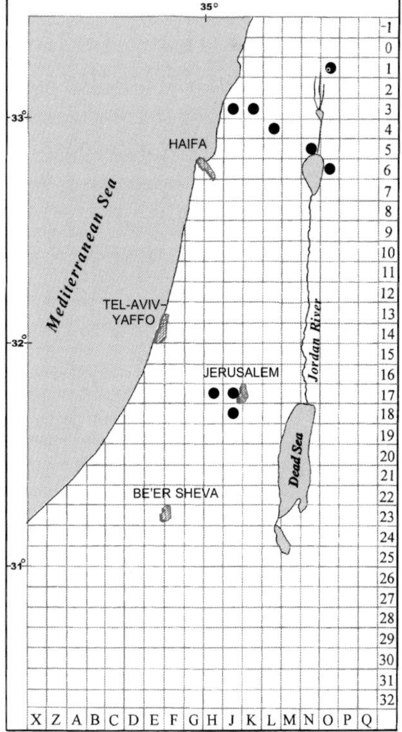

Map 7: *Fissidens crassipes* subsp. *crassipes*

8b. Fissidens crassipes subsp. **warnstorfii** (M. Fleisch.) Brugg.-Nann., Proc. Kon. Ned. Akad. Wetensch., ser. C, Biol. 85 : 74 (1982).

Fissidens warnstorfii M. Fleisch., Bot. Centralbl. 65 : 298 (1896); *F. crassipes* Wilson var. *philibertii* Besch., Cat. Mouss. Algérie 7 (1882); *F. bambergeri* var. *aegyptiacus* Renauld & Cardot, Bull. Soc. Roy. Bot. Belg. 41 : 47 (1905); *F. mnevidis* J. J. Amann, Rev. Bryol. 49 : 51 (1922). — Bilewsky (1965) : 351.

Distribution map 8, Figure 6 : f–k.

Plants longer than subsp. *crassipes*, up to 20 mm high, pale green. *Stems* with up to 35 pairs of leaves, laxly spaced. *Leaves* up to 2 mm long, ± narrower than in typical subspecies, less frequently elliptical, widely acute to obtuse; border distinct mainly on vaginant laminae, up to 32 µm wide. *Archegonia* usually shorter than in typical subspecies. *Peristome* teeth usually less than 50 µm wide at base.

Habitat and Local Distribution. — Plants often encrusted with lime, rarely producing sporophytes. Common, locally more widespread than *Fissidens crassipes* subsp. *crassipes*: Sharon Plain, Upper Galilee, Mount Carmel, Samaria, Judean Mountains, Judean Desert, Dan Valley, and Golan Heights.

General Distribution. — Recorded from Southwest Asia (Turkey, Lebanon, Iraq, and Yemen), around the Mediterranean, Europe (more records from the south), Macaronesia, and Africa (records from throughout). According to Bruggeman-Nannenga (1982) the distribution of this taxon is mainly Mediterranean (see comment on the concept of *Fissidens crassipes* subcomplex above, p. 26).

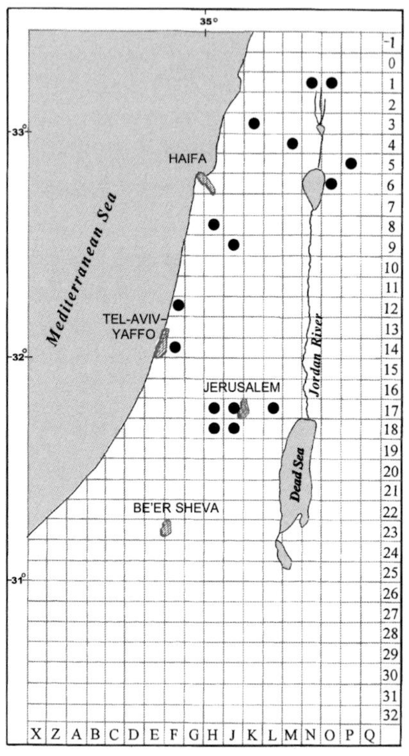

Map 8: *Fissidens crassipes* subsp. *warnstorfii*

Figure 6. *Fissidens crassipes* subsp. *crassipes*: **a** habit, moist (×10); **b** operculate and deoperculate dry capsules (×10); **c** leaf (×50); **d** leaf apex (×500); **e** cross-section through lower part of leaf (×500). *Fissidens crassipes* subsp. *warnstorfii*: **f** habit sterile, moist (×10); **g** leaf (×50); **h** cells of vaginant lamina; **i** cells at border of vaginant lamina; **k** leaf apex (all leaf cells ×500).
(**a**–**e**: *Herrnstadt*, 18 Mar. 1988; **f**–**k**: *Herrnstadt*, 2 Feb. 1984)

b

d

a

c

e

h

i

k

h

f

g

i

9. Fissidens pusillus (Wilson) Milde, Bryol. Siles. 82 (1869).

F. viridulus var. *pusillus* Wilson, Bryol. Brit. 303 (1855); *F. minutulus* auct. Eur., non Sull. 1846; *F. pusillus* var. *madidus* Spruce, J. Bot. 18 : 361 (1880). — Bilewsky (1965) : 352.

Distribution map 9, Figure 7.

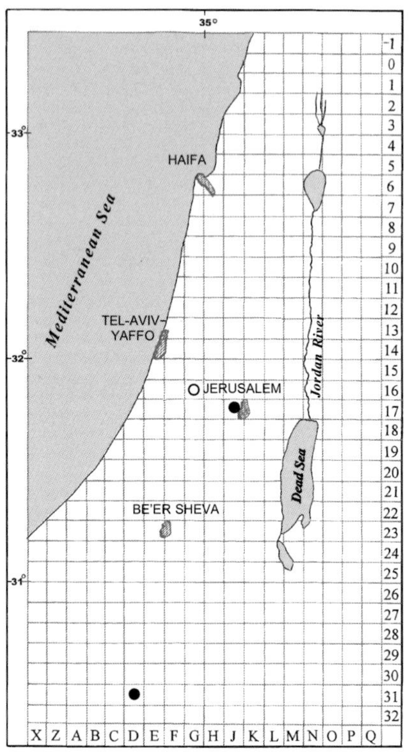

Map 9: ● *Fissidens pusillus*
○ *F. fontanus*

Plants very small, 1–3(–5) mm high (resembling in habit miniature plants of *Fissidens crassipes*), green. *Stems* simple or branched, with three–five pairs of leaves, lower leaves very small. *Leaves* 0.7–1.7 mm long, oblong-lanceolate, acute; margins with border usually unistratose, sometimes bistratose near insertion, in dorsal lamina vanishing above mid-leaf, up to 19 μm wide in mid-leaf of vaginant laminae, dorsal lamina often incomplete in lower leaves, dorsal lamina of upper leaves usually reaching stem; all cells ± thin-walled, irregularly hexagonal to subquadrate, cells in mid-leaf of vaginant laminae 6–10 μm wide. *Archegonia* (250–)270–350 μm long. *Setae* 2–3(–4) mm; capsules 0.75–1 mm long including operculum, ovate-oblong; operculum ± oblique-rostrate, half the length of capsule; peristome teeth 30–41 μm wide at base.

Habitat and Local Distribution. — Plants mesophytic, gregarious, or solitary among other mosses (as *Barbula* species), on limestone rocks in shaded moist habitats. Rare: Judean Mountains (Jerusalem) and Central Negev (Mount Ramon).

General Distribution. — Recorded from Southwest Asia (Turkey, Lebanon, and Iran), around the Mediterranean, Europe (throughout), Macaronesia, North Africa, and North and Central America (as *Fissidens bryoides*, "*pusillus* expression" — Pursell *in* Sharp *et al.* 1994).

Fissidens pusillus is often recorded from wet habitats, growing on acidic rocks. The less vigorous appearance of the local plants may be perhaps explained by the comparatively dryer habitats and the more basic substrate on which they grow in Israel.

The taxon is named "*Fissidens pusillus* Wils. ex Milde, var. *madicus* Spruce" in Bilewsky (1965, p. 352), and is also recorded there from Esdraelon Plain (Emeq Yizre'el). However, no corresponding specimen was found in Herb. Bilewsky.

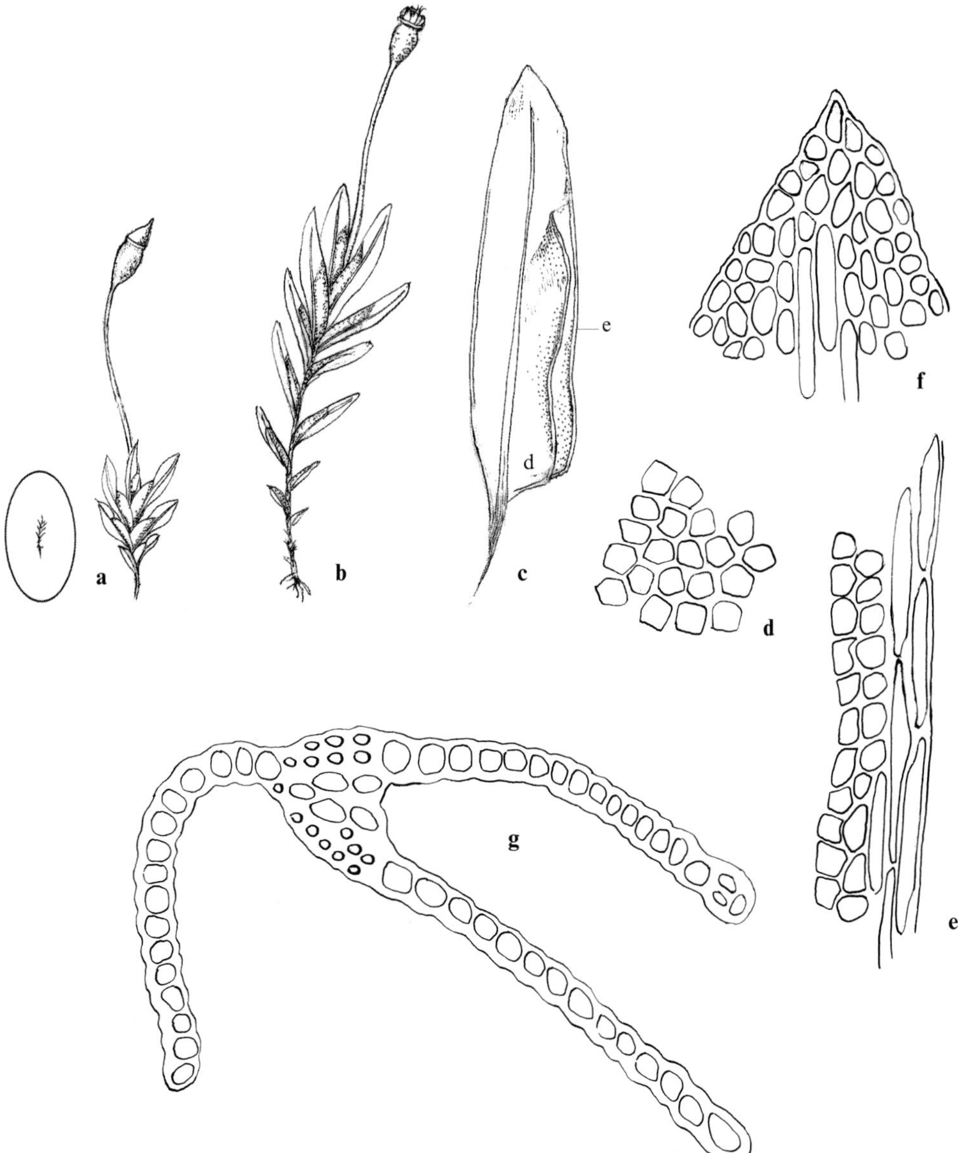

Figure 7. *Fissidens pusillus*: **a** habit (×10); **b** habit with deoperculate capsule (×10); **c** leaf (×50); **d** basal cells of vaginant lamina; **e** cells at border of vaginant lamina; **f** leaf apex (all leaf cells ×500); **g** cross-section through lower part of leaf (×500). (*Bilewsky*, Jan. 1954, No. 6 in Herb. Bilewsky)

10. Fissidens fontanus (Bach. Pyl.) Steud., Nomencl. Bot. 2 : 166 (1824).
Skitophyllum fontanum Bach. Pyl., J. Bot. (Paris), ser. 2, 4 : 158 (1815); *F. julianus* (DC.) Schimp., Flora (Jena) 21 : 271 (1838); *Octodiceras fontanum* (Bach. Pyl.) Lindb., Öfvers. Förh. Konfl. Svenska Vetensk.-Akad. 20 : 405 (1863).

Distribution map 9 (p. 42), Figure 8.

Aquatic plants, large, over 50 mm long, green to brownish-green, feather-like in appearance. *Stems* slender, irregularly branched. *Leaves* spreading when dry and moist, distant, up to 4(–5) mm long, linear to linear-lanceolate, acute to subobtuse; margins entire, not bordered; costa ending shortly below apex; dorsal lamina narrowing towards base of leaf, not reaching stem; vaginant lamina less than half the length of leaf; all cells thin-walled to incrassate, irregularly hexagonal to subquadrate, in mid-leaf 11–17 µm wide, much smaller towards margins, smooth. *Gametangia* on short axillary branches. *Autoicous. Sporophytes* not seen in a single local population; recorded as having short setae, erect and symmetric capsules, and conical-mitrate calyptrae.

Habitat and Local Distribution. — Plants submerged in standing water. A single population seen (Shefela, 'En Yered, *Minis*, 23 Mar. 1996).

General Distribution. — Recorded as scattered around the Mediterranean, Europe (throughout), Macaronesia (few records), North Africa, Madagascar, northern and central Caribbean area, and southern South America.

Fissidens fontanus is often considered as part of subgenus *Octodiceras* (Brid.) Broth. (e.g., Brotherus 1924 — as *F. julianus*; Magill 1981; Pursell 1987 — as *F. fontanus*) or separated from *Fissidens* and treated as *Octodiceras fontanum* (e.g., Corley *et al.* 1981; Bruggeman-Nannenga & Nyholm *in* Nyholm 1986; Duell 1992).

Figure 8. *Fissidens fontanus*: **a** part of habit (×1); **b** stem bearing gametangia (×10); **c** perichaetium terminal on short axillary branch (×50); **d** leaf (×50); **e** leaf apex (×500). (*Minis*, 23 Mar. 1996)

Dicranales

DITRICHACEAE

Plants acrocarpous, very small, small, or medium-sized, usually growing in loose tufts. *Stems* erect, simple, or sparsely branched; central strand present or absent. *Leaves* usually lanceolate, tapering to a long-acuminate or subulate tip, sometimes sheathing at base; costa wide, strong, subpercurrent to excurrent, often with two stereid bands; cells smooth, subquadrate, rectangular to linear, at basal angles without differentiated alar cells. *Capsules* usually long-exserted and stegocarpous, sometimes immersed to emergent and cleistocarpous; erect to inclined, ± cylindrical, often ± asymmetrical; peristome single, teeth 16, usually perforated or cleft nearly down to base into filiform divisions. *Calyptrae* cucullate, smooth.

The Ditrichaceae are represented in the local flora by three genera and four species. The distinction between the Ditrichaceae and the Dicranaceae is discussed in the treatment of Dicranaceae (p. 55).

Literature. — Matsui & Iwatsuki (1990).

1 Plants very small, up to 7 mm high, ephemeral; capsules immersed, subglobose to wide-ellipsoid, cleistocarpous **2. Pleuridium**

1 Plants small, up to 10 mm high, not ephemeral; capsules exserted, ovoid-oblong to cylindrical, stegocarpous.

  2 Leaves unistratose or with bistratose margins above, upper cells usually subquadrate to linear, smooth; capsules smooth when dry **1. Ditrichum**

  2 Leaves bistratose above, except at margins, upper cells short-rectangular with one prominent mammilla at each end of cell; capsules irregularly striate when dry **3. Cheilothela**

1. *Ditrichum* Hampe

Plants small. *Stems* with central strand. *Leaves* ovate-lanceolate or lanceolate with a wider base, narrowing into a long, often curved, acuminate or subulate apex; lamina unistratose, margins often bistratose above; costa subpercurrent to excurrent, with or without ventral stereid band; upper cells subquadrate to linear, smooth; basal cells larger, oblong to linear; perichaetial leaves ± longer than other leaves. *Dioicous* (often recorded as paroicous in *D. subulata*). *Setae* recorded as elongate, straight, or flexuose; capsules recorded as exserted, suberect or inclined, cylindrical or ellipsoid, often curved, smooth when dry; annulus differentiated; peristome teeth cleft into papillose, filiform terete divisions; operculum conical to conical-rostrate.

A terricolous cosmopolitan genus comprising ca. 90 species. Two species occur in the local flora of which only a few plants have been found.

1 Leaves gradually tapering into a long, acuminate apex or subula; margins plane throughout **1. D. heteromallum**

1 Leaves abruptly narrowing into a long, acuminate apex or subula; margins plane below, erect or inflexed above **2. D. subulatum**

1. Ditrichum heteromallum (Hedw.) E. Britton, N. Amer. Fl. 15:64 (1913).
Weissia heteromalla Hedw., Sp. Musc. Frond. 71 (1801); *Ditrichum homomallum* (Hedw.) Hampe, Flora 50:182 (1867).

Distribution map 10, Figure 9:a–e (p. 49).

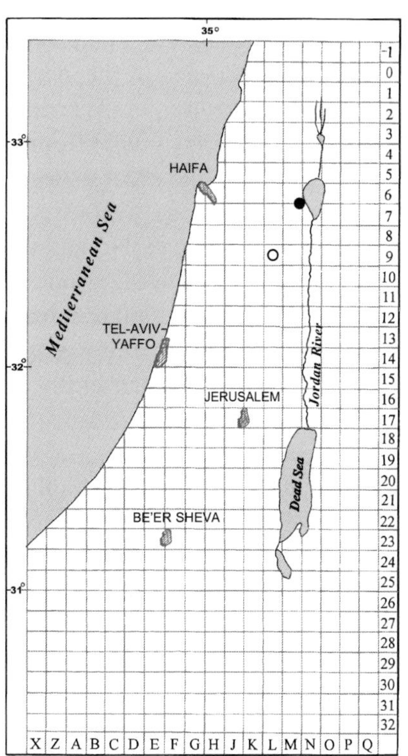

Map 10: ● *Ditrichum heteromallum*
○ *D. subulatum*

Plants up to 8 mm high, green to yellowish-green, brownish at base. *Stems* simple or sometimes branched. *Leaves* laxly spaced near stem base, more crowded above, longer, often subsecund, erect to erectopatent, ± flexuose when dry, hardly changing when moist, up to 1 mm long, with an ovate-lanceolate base, gradually tapering into a long acuminate apex or subula; apex narrow, sometimes channelled; margins plane, entire, slightly denticulate near apex; costa excurrent, wide, composing nearly the whole subula; cells ± incrassate, smooth; lowest basal cells linear to narrow-rectangular; upper cells shorter, rectangular, linear towards costa, mid-leaf cells 5.5–9 µm wide. *Sporophytes* not seen in local plants.

Habitat and Local Distribution. — Plants growing in small tufts and patches on basaltic soil (together with species of *Pottia* and *Fissidens*, and *Phaeoceros laevis*). Very rare, locally abundant: collected in a single locality — Lower Galilee, Tiberias.

General Distribution. — Recorded from Southwest Asia only from Turkey, scattered around the Mediterranean, Europe (throughout), North Africa, Asia (throughout, except the southeast), and North and Central America.

2. Ditrichum subulatum Hampe, Flora 50 : 182 (1867). [Bilewsky (1965) : 353]

Distribution map 10 (p. 47), Figure 9 : f–h.

Ditrichum subulatum differs from *D. heteromallum* in the shape of the leaves. *Leaves* abruptly narrowing into a long-acuminate apex; margins plane below, erect or inflexed above. Often recorded in the literature as paroicous.

Local Distribution. — The only available records of *Ditrichum subulatum* from Israel are of Bilewsky (1965) from Esdraelon Plain (Emeq Yizre'el) and Upper Jordan Valley, 'En Gev. The latter collection is a misidentified sample of *Tortella inflexa.* A small sample from the Esdraelon Plain, Hefziba was annotated by Reimers in 1954 as *Ditrichum subulatum.* No additional plants of *D. subulatum* have been collected in Israel, though it is safe to assume that these small plants grow in additional localities, perhaps hidden among other mosses.

General Distribution. — Recorded from Southwest Asia only from Turkey, scattered around the Mediterranean, Europe (few records), Macaronesia, North Africa, and North America.

Some doubts concerning the validity of the name *Ditrichum subulatum* were expressed (Corley *et al.* 1981).

2. *Pleuridium* Rabenh.

Plants small, ephemeral. *Stems* simple or sometimes forked, with central strand. *Leaves* with an ovate base narrowing into a subula; perichaetial leaves much longer than other leaves. *Dioicous. Setae* very short; capsules immersed to emergent, subglobose to wide-elliptical, cleistocarpous, peristome and operculum not differentiated. *Calyptrae* cucullate.

A genus comprising ca. 30 species of gregarious, terricolous ephemerals, growing in disturbed and weedy habitats, occurring mainly in the Northern Hemisphere. One species occurs in the local flora.

Literature. — Snider & Margadant (1973).

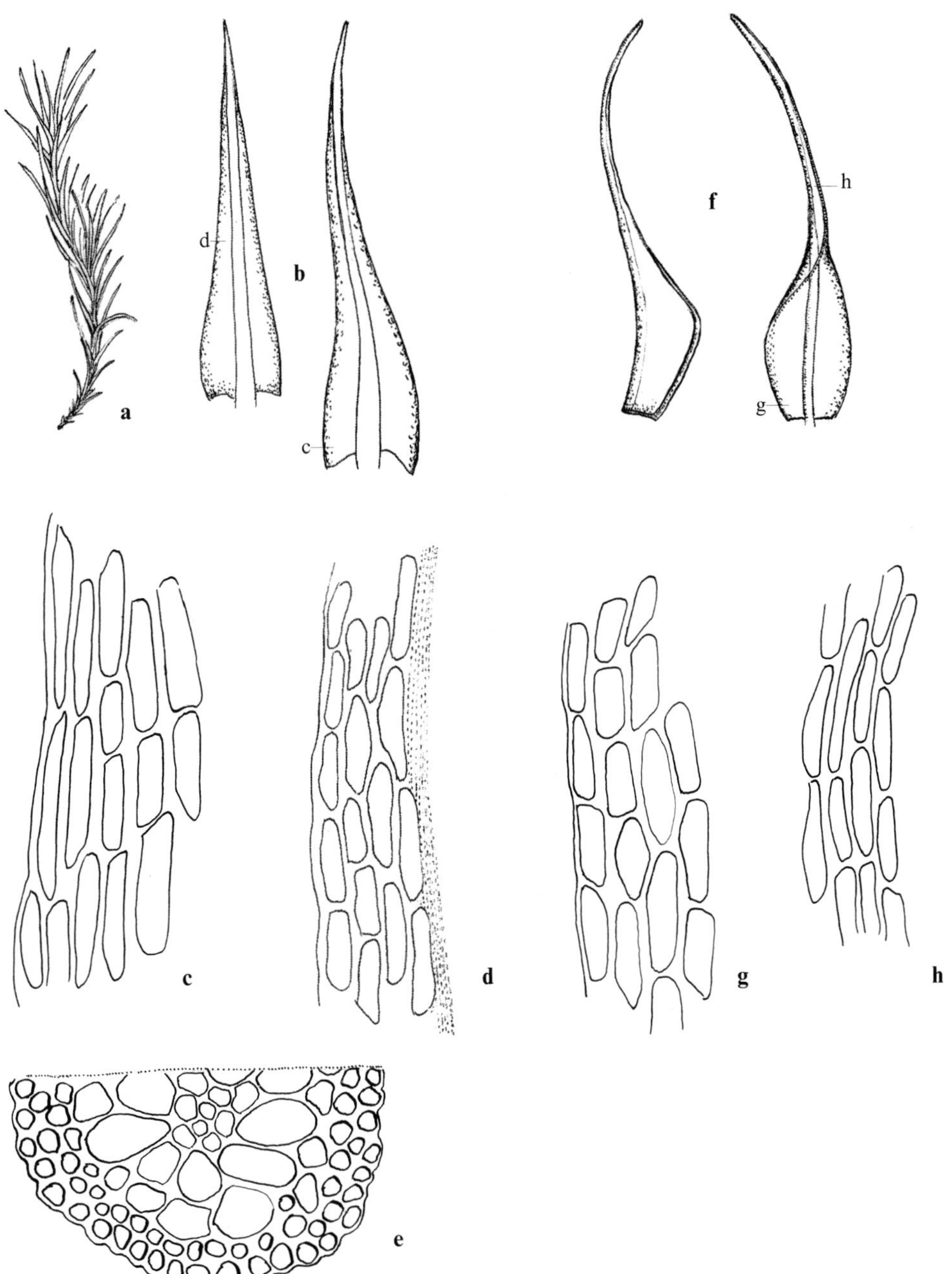

Figure 9. *Ditrichum heteromallum*: **a** habit, moist (×10); **b** leaves (×50); **c** basal cells; **d** mid-leaf cells (all leaf cells ×500); **e** part of cross-section of stem (×500).
Ditrichum subulatum: **f** leaves (×50); **g** basal cells; **h** mid-leaf cells (all leaf cells ×500).
(**a**–**d**: *Herrnstadt*, 10 Mar. 1990; **e**–**h**: *Kushnir*, 11 Sep. 1943)

Pleuridium subulatum (Hedw.) Rabenh., Deutschl. Krypt.-Fl. 2:79 (1848). [Bilewsky (1965):354]

Phascum subulatum Hedw., Sp. Musc. Frond. 19 (1801); *Pleuridium acuminatum* Lindb., Öfvers. Förh. Kongl. Svenska Vetensk.-Akad. 20:406 (1863). — Bilewsky (1965):354; "*Pleuridium alternifolium* (Dicks. ex Hedw.) Rabenh."— Bilewsky & Nachmony (1955) *sensu* Dixon, non (Hedw.) Brid. — Snider & Margadant (1973).

Distribution map 11, Figure 10, Plate I:d.

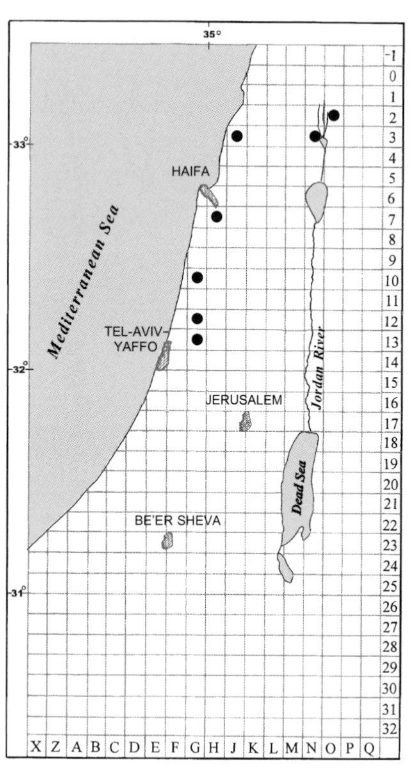

Map 11: *Pleuridium subulatum*

Plants up to 7 mm high, usually bearing sporophytes, yellowish-green. *Stems* turning brownish-red at base. *Leaves* in lower part laxly spaced, erect to spreading when dry, hardly changing when moist, 1–1.5 mm long, ovate-lanceolate, acuminate; margins entire or slightly denticulate at leaf shoulders and upwards, erect or incurved above but not channelled; costa percurrent or excurrent, composing most of the subula; cells often bistratose above, ± incrassate, smooth; basal cells short to long-rectangular; mid-leaf and upper cells elongate, rectangular to linear; mid-leaf cells ca. 15 μm wide; perichaetial leaves ± erect, more densely spaced, up to three times as long as other leaves, with an ovate-oblong base tapering abruptly ("*subulatum* type", Figure 10:f) or gradually ("*acuminatum* type", Figure 10:c) into a subulate apex, channelled above. *Autoicous* or paroicous; antheridia naked and usually few, sometimes in small buds in leaf axils. *Setae* 0.4–0.8 mm long; capsules ca. 1 mm long, ovoid to wide ellipsoid, orange-brown, ending in a small, slightly oblique apiculus. *Spores* ca. 30 μm in diameter, papillose; surface granulate with baculae, baculae up to 2 μm long, sometimes fused, irregularly lobed at apex (seen with SEM).

Habitat and Local Distribution. — Plants growing in tufts on sand, sandy loam, and calcareous soils, in exposed or partly shaded habitats, on road banks and forest clearings. Occasional, sometimes locally abundant: Coastal Galilee, Sharon Plain, Mount Carmel, Hula Plain and Golan Heights.

General Distribution. — Recorded as *Pleuridium subulatum* and *P. acuminatum* from Southwest Asia (Cyprus, Turkey, and Lebanon), around the Mediterranean, Europe (throughout), Macaronesia, North Africa, eastern and central Asia, New Zealand, Oceania, and North and Central America.

In populations of *Pleuridium subulatum* growing on sand and sandy loam in the Sharon Plain, the perichaetial leaves are abruptly narrowing and have a somewhat shorter sheathing base. Thus, they agree with the description of *P. subulatum sensu stricto*. In all other populations, the perichaetial leaves are gradually narrowing and have a

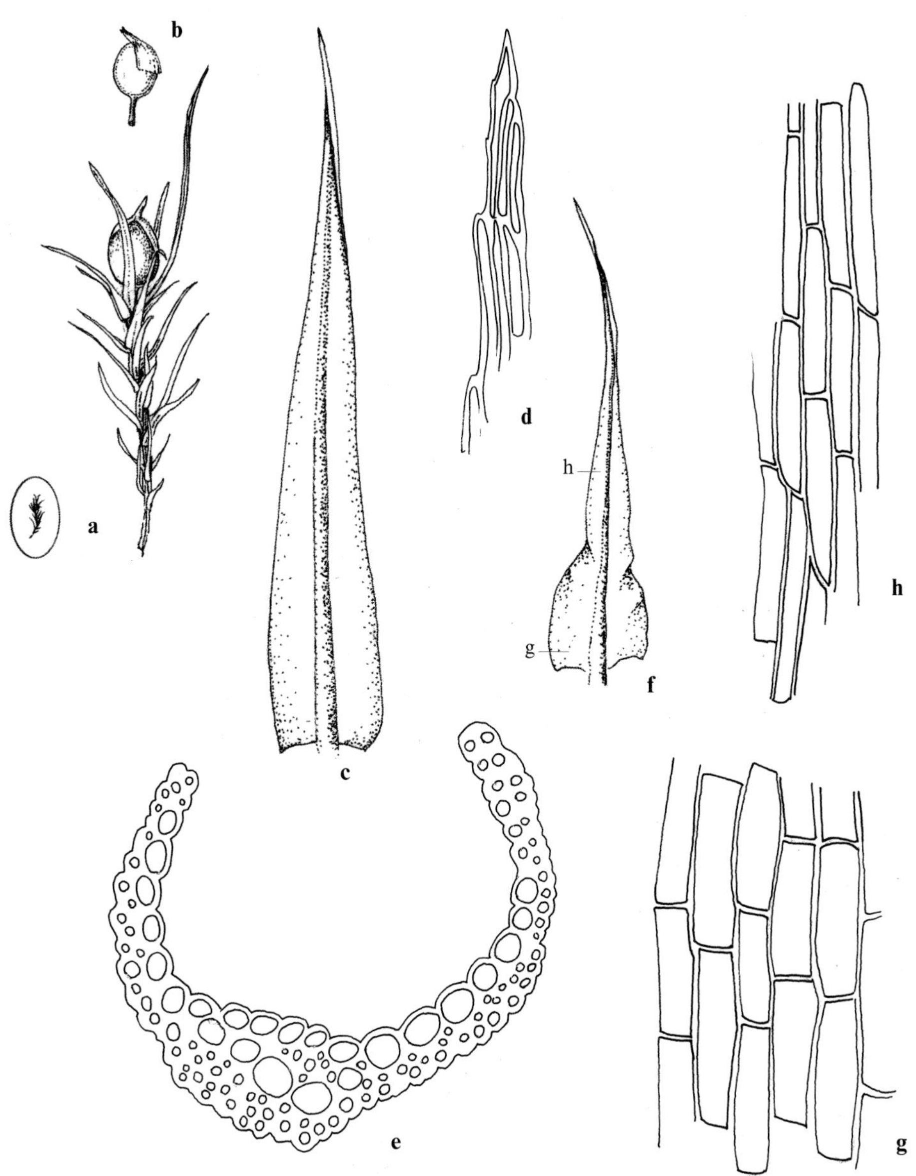

Figure 10. *Pleuridium subulatum*: **a** habit, moist (×10); **b** sporophyte (×10); **c** perichaetial leaf (×50); **d** apical cells; **e** cross-section of leaf (×500); **f** perichaetial leaf (×50); **g** basal cells; **h** mid-leaf cells (all leaf cells ×500).
(**a**–**e**: "*acuminatum* type", *Kushnir*, 7 Feb. 1943; **f**–**h**: "*subulatum* type", *Kushnir*, 17 Feb. 1943)

longer sheathing base, as assumed typical of *P. acuminatum*. However, the character usually associated with *P. acuminatum* — the naked antheridia — was found in the majority of populations with either type of leaves; antheridia forming buds in the leaf axils, associated with *P. subulatum sensu stricto*, were found only in a single population on Mount Carmel.

There does not seem to be a clear correlation between the leaf shape and the arrangement of antheridia in local plants. Therefore, we have preferred not to subdivide the local populations between both species as previously done (Herrnstadt *et al.* 1991). The treatment of Crum & Anderson (1981), considering *Pleuridium subulatum* and *P. acuminatum* as conspecific, was adopted, and *P. subulatum* is used as the valid name for the combined species.

3. *Cheilothela* (Lindb.) ex Broth.

Plants small. *Stems* without differentiated central strand (in local plants). *Leaves* rigid, non-hygroscopic, with a wide-ovate base, long-acuminate; lamina at least partly bistratose; costa wide, not clearly defined, especially in upper part; perichaetial leaves larger than other leaves, the inner with a sheathing base half of its length, abruptly narrowing into a subula. *Dioicous*. *Capsules* exserted, ± inclined, ovoid-oblong to cylindrical, slightly curved, irregularly striate when dry; peristome teeth bifid into papillose filiform divisions; operculum conical-rostrate.

According to Buck (1981) *Cheilothela* should be considered as a monotypic genus. One species occurs in the local flora.

Cheilothela chloropus (Brid.) Lindb., Utkast Eur. Bladmoss. 34 (1878). [Bilewsky (1970)]
Dicranum chloropus Brid., Mant. Musc. 1 : 70 (1819); *Ceratodon chloropus* (Brid.) Brid., Bryol. Univ. 1 : 486 (1826).

Distribution map 12, Figure 11.

Plants up to 10 mm high, yellowish-green. *Stems* simple or branched. *Leaves* laxly spaced, appressed when dry, nearly erect, sometimes slightly contorted when dry, erect-opatent to spreading when moist, 1–2 mm long, triangular in cross-section, with a wide-ovate base gradually tapering to a long, acuminate to subulate apex; margins plane, sometimes partly incurved above, crenulate-papillose in upper half; costa wide, excurrent; cells quadrate to short-rectangular, obscure, with a prominent mammilla, up to 6 µm high at each end of cell, except at extreme base; basal cells unistratose, upper cells bistratose, except at margins; mid-leaf cells ca. 7 µm wide. *Sporophytes* not

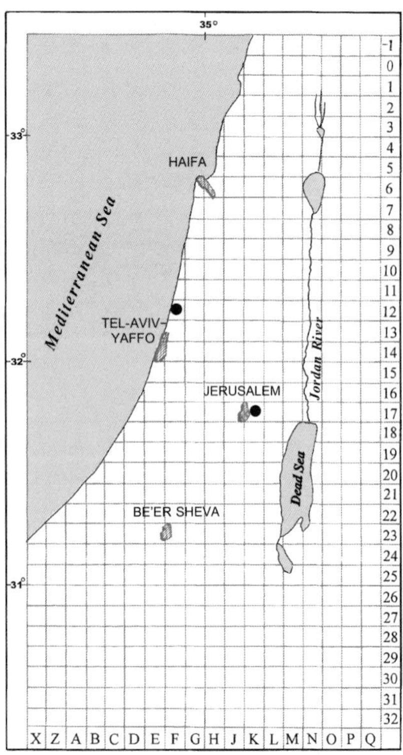

Map 12: *Cheilothela chloropus*

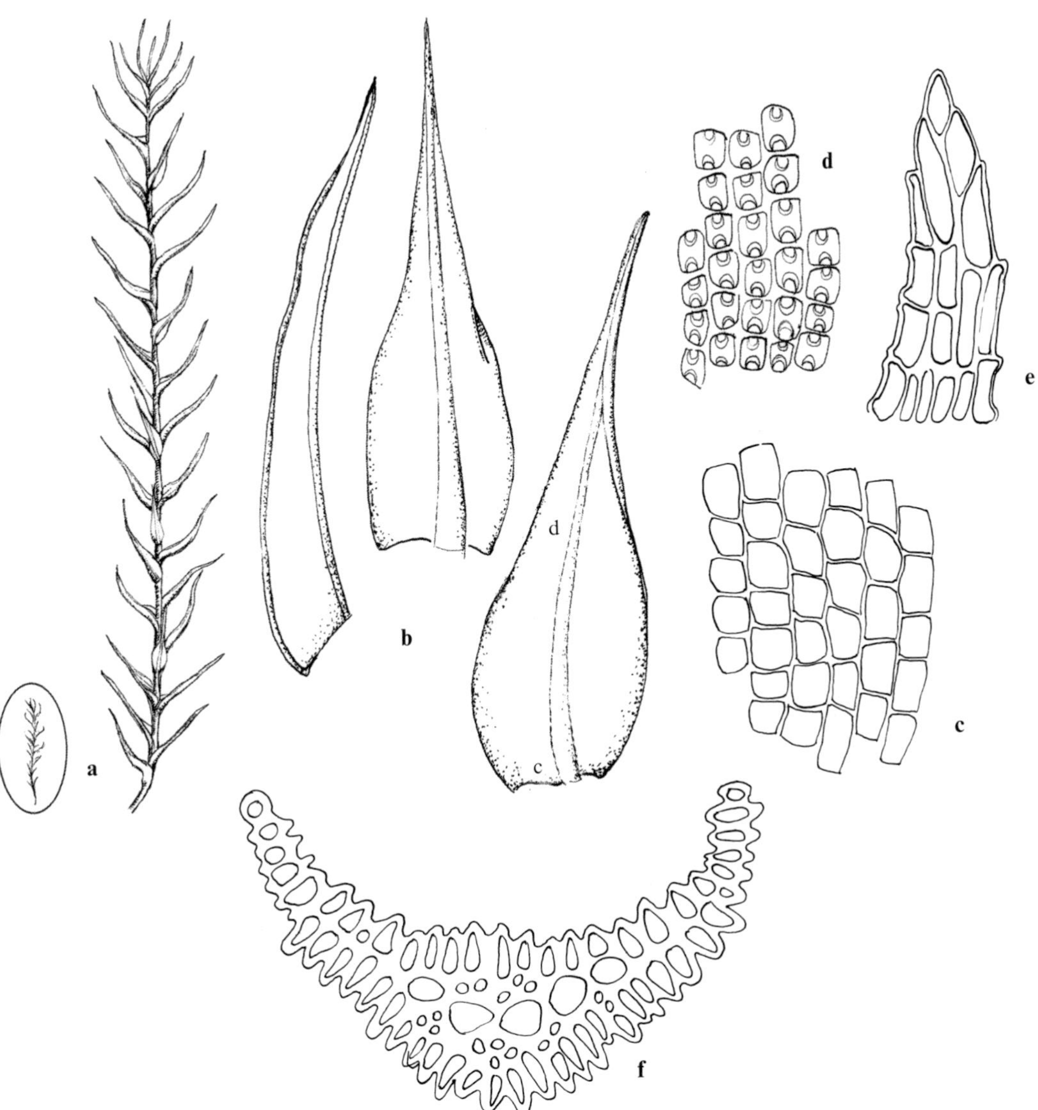

Figure 11. *Cheilothela chloropus*: **a** habit, moist (×10); **b** leaves (×50); **c** basal cells; **d** mid-leaf cells; **e** apical cells (all leaf cells ×500); **f** cross-section of leaf (×500). (*Kushnir*, 15 Feb. 1943)

seen in local plants; only sterile plants were recorded from Iraq and the United Kingdom (Agnew & Vondráček 1975; Smith 1978).

Habitat and Local Distribution. — Plants growing in tufts on shaded calcareous soil (in one population mixed with *Funaria muhlenbergii* R. Hedw. ex Lam. & DC.), sandy loam and rendzina. Rare: found so far only in two localities — Sharon Plain and Judean Desert.

General Distribution. — Recorded from Southwest Asia (Cyprus, Turkey, Iraq, and Iran), scattered around the Mediterranean, Europe (mainly southern Europe), Macaronesia, and North Africa.

Gametophytes of *Cheilothela chloropus* can be distinguished by the markedly nonhygroscopic rigid leaves and the prominent mammillae of leaf cells. This species is perhaps more widespread locally than recorded, as plants may have been overlooked in additional localities because of their small size and sterility.

DICRANACEAE

Plants acrocarpous, minute to large. *Stems* erect, simple or forked, often tomentose below, central strand present or absent. *Leaves* appressed to spreading, squarrose or secund, mostly narrow, long-lanceolate, often tapering from a wide base to a subulate tip; costa usually strong and wide below, sometimes weak, ending below apex to excurrent; ventral stereid band present or absent; upper cells subquadrate to linear, usually smooth, rarely papillose or mammillose, sometimes differing from other leaf cells; basal cells large, rectangular or elongate; alar cells at basal angles often clearly differentiated (not differentiated or rarely slightly differentiated in *Dicranella*). *Capsules* usually long-exserted and stegocarpous, or rarely ± immersed and cleistocarpous, erect or inclined, symmetrical or asymmetrical, ellipsoid to short cylindrical; peristome single, absent or present, usually well-developed, teeth 16, flat, deeply divided, bi- or trifid, with pitted vertical striations below, papillose above. *Calyptra* cucullate, smooth, usually with entire base.

Plants of the Ditrichaceae closely resemble those of Dicranaceae and are sometimes considered as part of that family (e.g., Bilewsky 1965; Corley *et al.* 1981). They differ, however, from the Dicranaceae in not having differentiated alar cells; their terete — not flat — peristome teeth are cleft into filiform divisions or perforated nearly down to base, lacking the vertical pitted striations of the Dicranaceae.

The Dicranaceae are represented in the local flora by the genus *Dicranella*.

Dicranella (Müll. Hal.) Schimp.

Plants small, yellowish-green. *Stems* with central strand. *Leaves* ± squarrose, with a wide, ± sheathing base, abruptly narrowed or gradually tapering to a subulate apex; margins unistratose or bistratose; costa ending below apex to excurrent; cells smooth, upper subquadrate to linear, incrassate; basal cells larger, rectangular and incrassate; alar cells usually not differentiated, sometimes few, ± differentiated cells present; perichaetial leaves usually not different from stem leaves, sometimes more sheathing. *Dioicous* or autoicous. *Setae* elongate; capsules erect or inclined, subglobose to cylindrical, often curved and asymmetric, smooth or furrowed when dry, annulus differentiated or not; operculum conical, long-rostrate.

A cosmopolitan genus for which various numbers of species (60–90) have been recorded. Plants are mainly terrestrial. One, perhaps two, species occur in the local flora (see the discussion of *Dicranella howei* below).

The identification of local plants of *Dicranella* poses many problems. Previous records of *D. heteromalla* (Hedw.) Schimp. from Israel (Herrnstadt *et al.* 1991) were mainly based on the yellow-coloured seta, which is generally considered as characteristic of this species. However, further studies showed that the colour of the seta cannot serve as a differential character in local populations. Plants with yellow setae did not have the very long-subulate leaves of *D. heteromalla* but resembled in all other characters plants of *D. howei* Renauld & Cardot. The inclusion of such plants in the latter species is further supported by the fact that they grow on calcareous soils, whereas *D. heteromalla* occurs on acidic substrates.

Dicranella howei Renauld & Cardot, Rev. Bryol. 20 : 30 (1893).

Distribution map 13, Figure 12, Plate I : e.

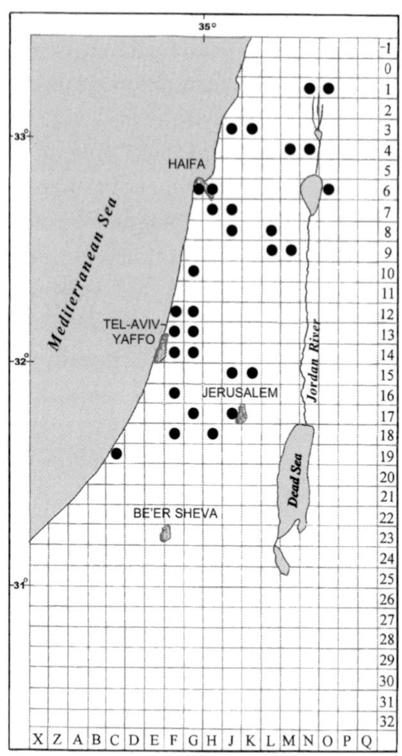

Map 13: *Dicranella howei*

Plants small, up to 10(–15) mm high, light green to yellowish-green. *Stems* simple or sometimes forked. *Leaves* rigid, erect to patent, sometimes slightly secund when dry; lower leaves up to 1 mm long, upper leaves up to 1.5(–2) mm long, lanceolate-linear, gradually tapering into a long acuminate apex; margins plane or narrowly recurved partly or throughout, entire or sometimes irregularly denticulate at tip; costa wide at base, comprising $\frac{1}{4}$–$\frac{1}{3}$ of width at leaf base (rarely only $\frac{1}{5}$), in some leaves not well defined above, more often percurrent to excurrent; basal cells rectangular, alar cells few, quadrate at base of leaf angles; upper cells narrow-rectangular to linear, in mid-leaf 3–7 µm wide; bistratose in patches or throughout; perichaetial leaves hardly differentiated. *Dioicous*. *Setae* red, sometimes yellow, when mature; capsules usually inclined, slightly curved and asymmetrical, ca. 1.5 mm long including operculum, short-elliptical to ovoid, smooth, contracted below mouth; operculum broadly rostrate from a wide conical base; peristome teeth cleft in the upper one third. *Spores* 14–19 µm in diameter, nearly smooth; surface granulate, groups of granules forming verrucae (seen with SEM).

Habitat and Local Distribution. — Plants growing in tufts on calcareous soils and marl, often in moist, open habitats near running water. They are widespread in such habitats throughout the Mediterranean and the Irano-Turanian districts of the local flora. Very common: Sharon Plain, Philistean Plain, Upper Galilee, Lower Galilee, Mount Carmel, Esdraelon Plain, Mount Gilboa, Samaria, Shefela, Judean Mountains, Dan Valley, Hula Plain, Bet She'an Valley, and Golan Heights.

General Distribution. — Recorded from Southwest Asia (Turkey and Iran), around the Mediterranean, southern Europe, Macaronesia, North Africa, and western North America.

Dicranella varia (Hedw.) Schimp. was recorded from Israel by Bilewsky (1965, p. 355). *Dicranella howei* and *D. varia* are treated as separate species in Crundwell & Nyholm (1977) with occasional intermediates. *Dicranella howei* is considered by Crundwell & Nyholm as a Mediterranean species, which grows also in the local flora, replacing the more northern *D. varia*. In the majority of the local collections, the identifications of plants agree in general with Crundwell & Nyholm's concept of *D. howei*. However, some deviations do occur: the width of the costa is usually narrower than stated by Crundwell & Nyholm as characteristic of *D. howei* (ca. one-third of lamina width) and somewhat wider than one-fifth of lamina, which is characteristic of *D. varia*. In

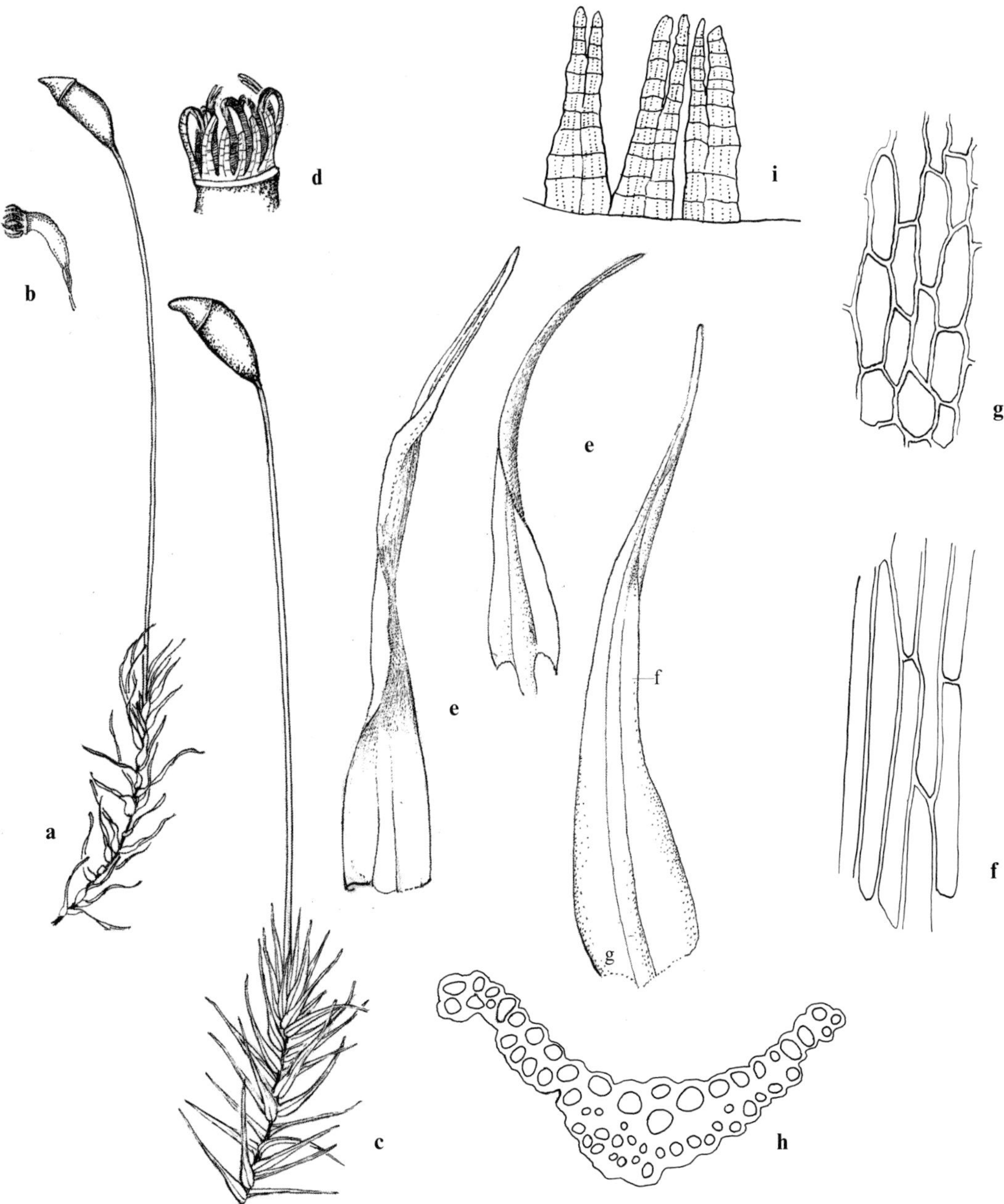

Figure 12. *Dicranella howei*: **a** habit, dry (×10); **b** deoperculate capsule (×10); **c** habit, moist (×10); **d** peristome (×25); **e** leaves (×50); **f** mid-leaf cells; **g** basal cells (all leaf cells ×500); **h** cross-section of leaf (×500); **i** peristome teeth (×50). (*Nachmony*, 19 Mar. 1955)

addition, leaves may be bistratose throughout or may vary in the number and size of bistratose patches, exceptionally bistratose only at margins (assumed by Crum & Anderson 1981 as characteristic of *D. varia*). Therefore, though the name *Dicranella howei* is accepted here for the local collections, the possibility that among them there are at least some plants that should be referred to as *D. varia* cannot be excluded.

Dicranella varia var. "*tenuifolis* B.S.G." (= var. *tenuifolia* (B.S.G.) Schimp.), recorded by Bilewsky (1965, p. 355) as also growing in several districts of Israel, was considered by Crundwell & Nyholm (1977) as synonymous with *D. howei*. The single sample in the Bilewsky herbarium named *Anisothecium varium* var. *tenuifolis* (no. 13, Sharon Plain: "Sheva-Takhanot, on sandy bank of the Yarkon River") was found to agree with *D. howei*. The samples in the Bilewsky herbarium, named *D. varia*, *Anisothecium varium*, or *A. rubrum*, also agree with the present concept of *D. howei*. *Anisothecium* is usually reduced to a synonym of *Dicranella* in recent literature, though sometimes the two genera are separated and delimited by several characters (cf. *Anisothecium varium* (Hedw.) Schimp. *in* Allen 1994).

Literature. — Crundwell & Nyholm (1977), Allen (1994).

Pottiales

POTTIACEAE

Plants acrocarpous, very rarely cladocarpous, usually small, green above, brownish below, growing usually in tufts, sometimes gregarious, or turf-forming. *Stems* generally erect, simple, or branched, central strand usually present; axillary hairs hyaline throughout or with one–three brownish basal cells. *Leaves* lanceolate, lingulate, or spathulate, usually contorted when dry; costa single, well developed, with one or two stereid bands, sometimes with hydroids; basal cells often differentiated, mostly smooth, hyaline; upper cells small, subquadrate or hexagonal, usually papillose. *Dioicous* or monoicous. *Setae* of most capsules elongate; capsules usually exserted, rarely immersed or emergent, usually stegocarpous, erect and symmetric, ovoid to cylindrical, rarely cleistocarpous and cylindrical; annulus often differentiated; peristome absent, rudimentary to well developed, single, consisting of 16 perforated or cleft teeth or 32 filiform divisions; operculum conical to rostrate. *Spores* with variously sculptured exine. *Calyptra* nearly always cucullate, smooth.

The majority of species grow on frequently desiccated calcareous substrates, mainly soils. The widespread occurrence of such habitats in Israel may explain the high representation of the Pottiaceae (24 genera with 86 species), and hence its status as the largest local family.

The Pottiaceae is a large, widely distributed, heterogeneous family in which delimitation of genera may often be rather difficult. For that reason, various characters (e.g., SEM of spores — Saito & Hirohama (1974); axillary hairs — Saito (1975); KOH laminal colour reaction — Zander 1993), in addition to the usually employed morphological and anatomical characters, have been proposed for generic delimitation.

The heterogeneity of the family has led to various subdivisions into subfamilies and tribes. The genera of Pottiaceae that occur in the local flora are most often divided in the literature (e.g., Smith 1978, Walther 1983) among three subfamilies: Pottioideae, Trichostomoideae, and Cinclidotoideae (the latter is considered here as a separate family). Chen (1941) recognised six subfamilies (including Cinclidotoideae), five represented in the local flora. Crum & Anderson (1981), who recognised four subfamilies (without Cinclidotoideae), accommodated the genera that occur in the local flora in the Pottioideae, Trichostomoideae, and Pleuroweisieae. There are several cases of genera treated by different authors as belonging to different subfamilies or tribes — compare Corley *et al.* (1981), Smith (1978), Crum & Anderson (1981), and Sharp *et al.* (1994).

The most thorough treatment of the Pottiaceae is the recently published monograph of the family by Zander (1993). In that detailed study, the division of the family into subfamilies and tribes often differs from previous treatments. The concepts of several genera have also been changed; the most notable among those genera is *Tortula.*

Of the seven subfamilies accepted by Zander (1993), the 24 genera of Pottiaceae that occur in our local flora belong to the following four subfamilies: (a) Timmielloideae (a new monotypic subfamily): *Timmiella*; (b) Trichostomoideae: *Eucladium*, *Pleurochaete*, *Tortella*, and *Trichostomum* (including *Oxystegus*); (c) Merceyoideae (a subfamily with four tribes, of which two occur in our flora): Leptodontieae: *Hymenostylium*, and Barbuleae: *Anoectangium*, *Barbula*, *Gymnostomum*, *Gyroweisia*, and *Trichostomopsis* as part of *Didymodon*; (d) Pottioideae (a subfamily with two tribes): Hyophileae: *Aloina*, *Crossidium*, *Pterygoneurum*, and *Weissia* (including *Astomum*), and Pottieae: *Acaulon*, *Aschisma*, *Leptobarbula*, and *Tortula* (including *Desmatodon*, *Phascum* and *Pottia*).

Zander *et al.* (*in* Sharp *et al.* 1994) — though published at a later date than Zander's 1993 monograph and apparently based on an earlier manuscript — subdivide the family exclusively into tribes. The very radical change in the species concept of *Tortula* (Zander 1993) is also not yet accepted in the Flora of Mexico.

Because of the many conflicting subdivisions of the Pottiaceae, we have preferred here to adopt a rather conservative treatment of the family (as done by many others, e.g., Corley *et al.* 1981, Nyholm 1989, Sharp *et al.* 1994). The local genera of the Pottiaceae are included in two subfamilies: Trichostomoideae and Pottioideae. The sixteen genera of the Pottioideae are distributed among three tribes: Pleuroweisieae, Barbuleae, and Pottieae. The genera are affiliated as follows (in alphabetical order within each taxonomic tribe):

Subfamily I. Trichostomoideae (eight genera): *Aschisma*, *Astomum*, *Oxystegus*, *Pleurochaete*, *Timmiella*, *Tortella*, *Trichostomum*, and *Weissia*;

Subfamily II. Pottioideae:

Tribe Pleuroweisieae (five genera): *Anoectangium*, *Eucladium*, *Gymnostomum*, *Gyroweisia*, and *Hymenostylium*;

Tribe Barbuleae (three genera): *Barbula*, *Leptobarbula*, and *Trichostomopsis*;

Tribe Pottieae (eight genera): *Acaulon*, *Aloina*, *Crossidium*, *Desmatodon*, *Phascum*, *Pottia*, *Pterygoneurum*, and *Tortula*.

Unlike Corley *et al.* (1981), we separate *Cinclidotus* from the Pottiaceae and place it in the monogeneric family Cinclidotaceae following several authors (e.g., Hilpert 1933, Zander 1993).

Literature. — Hilpert (1933), Chen (1941), Saito (1975), Zander (1993), Zander *et al. in* Sharp *et al.* (1994).

Key to the Local Subfamilies, Tribes, and Genera of the Pottiaceae (adapted from Zander *et al. in* Sharp *et al.* 1994)

1 Leaves lanceolate; margins plane to incurved; costa shortly excurrent, usually with two stereid bands and no hydroids — Subfamily I. TRICHOSTOMOIDEAE
- 2 Leaves bistratose, cells without papillae — **8. Timmiella**
- 2 Leaves unistratose, cells papillose.
  - 3 Hyaline basal cells extending upwards along margins, abruptly differentiated to various degrees from chlorophyllous cells.
    - 4 Plants large, up to 60 mm high; leaf margins denticulate, at least in upper part; hyaline border not V-shaped; perichaetia lateral — **7. Pleurochaete**
    - 4 Plants small, up to 20 mm high; leaf margins usually entire; hyaline border usually V-shaped; perichaetia terminal — **6. Tortella**
  - 3 Hyaline basal cells not or scarcely extending upwards, not abruptly differentiated from chlorophyllous cells.
    - 5 Plants usually monoicous; leaf margins usually strongly incurved, rarely plane.

6 Capsules exserted, stegocarpous **1. Weissia**

6 Capsules immersed, cleistocarpous.

7 Leaves incurved when dry, without sheathing base; exothecial cells circling the capsule in horizontal rows **3. Aschisma**

7 Leaves crisped and contorted when dry, with sheathing base; exothecial cells not arranged as above **2. Astomum**

5 Plants usually dioicous; leaf margins plane or slightly incurved.

8 Leaf margins notched to denticulate; peristome without a basal membrane; teeth entire **5. Oxystegus**

8 Leaf margins entire to papillose-crenulate; peristome usually with a basal membrane; teeth deeply cleft **4. Trichostomum**

1 Leaves lanceolate to lingulate or spathulate; margins plane or recurved; costa ending below apex, with one or two stereid bands, sometimes with hydroids Subfamily II. POTTIOIDEAE

9 Leaves usually spathulate to lingulate; costa usually with one stereid band; upper lamina cells large, often exceeding 12 µm Tribe POTTIEAE

10 Leaves bearing photosynthetic outgrowths on ventral surface of costa.

11 Costal outgrowths filamentous.

12 Plants fleshy; leaves with sheathing base; upper lamina infolded, nearly covering costal filaments **19. Aloina**

12 Plants not fleshy; leaves without sheathing base; upper lamina not infolded **20. Crossidium**

11 Costal outgrowths lamellate or rarely both lamellate and filamentous **21. Pterygoneurum**

10 Leaves not bearing photosynthetic outgrowths as above.

13 Plants minute to small 0.5–3 (7) mm high with immersed to slightly exserted cleistocarpous capsules.

14 Plants bud-like; upper leaf cells usually smooth; calyptrae conical **24. Acaulon**

14 Plants not bud-like; upper leaf cells often papillose; calyptrae cucullate or mitriform **23. Phascum**

13 Plants small to large (1) 1.5–30 (40) mm high with exserted stegocarpous capsules.

15 Peristome present consisting of 32 filaments, spirally twisted, or rarely straight* **17. Tortula**

15 Peristome absent, rudimentary, or consisting of 16 straight teeth (entire, regularly, or irregularly divided).

16 Peristome absent, rudimentary, or consisting of 16 entire or irregularly divided teeth **22. Pottia**

16 Peristome present consisting of 16 nearly regularly divided teeth **18. Desmatodon**

9 Leaves lanceolate to lingulate; costa usually with two stereid bands; upper lamina cells smaller, at most up to 12 µm.

---

* In *Tortula atrovirens*, for example.

17 Leaf margins recurved (rarely plane), perichaetia terminal, peristome present Tribe BARBULEAE
18 Costa with a single stereid band; margins bistratose in upper part; upper lamina with bistratose longitudinal patches **15. Trichostomopsis**
18 Costa usually with two stereid bands (except *Barbula rigidula* and sometimes *B. imbricata* from dry habitats); margins and lamina unistratose.
19 Plants very small, up to 3 mm; margins plane; perichaetial leaves convolute-sheathing **16. Leptobarbula**
19 Plants larger, usually 7–20(–50) mm; margins usually recurved; perichaetial leaves usually little differentiated (except *Barbula convoluta* and *B. revoluta*) **14. Barbula**
17 Leaf margins plane to weakly recurved; perichaetia terminal or lateral; peristome usually absent* Tribe PLEUROWEISIEAE
20 Perichaetia lateral **13. Anoectangium**
20 Perichaetia terminal.
21 Annulus comprising two–three rows of large cells, often revoluble **11. Gyroweisia**
21 Annulus without large cells, not revoluble.
22 Peristome present; leaf margin denticulate at upper part of base **9. Eucladium**
22 Peristome absent; leaf margins entire throughout.
23 Stem with weak central strand; upper leaf cells uniform in shape, arranged in rows that meet at right angles; operculum separated from columella in mature capsules **10. Gymnostomum**
23 Stem without central strand; upper leaf cells not uniform in shape, not arranged as above; operculum often still attached to columella in mature capsules **12. Hymenostylium**

SUBFAMILY I. TRICHOSTOMOIDEAE

1. *Weissia* Hedw.

Plants small, growing in tufts. *Stems* simple or branched with a central strand. *Leaves* often crisped when dry, erectopatent to spreading when moist, channelled, lingulate to narrow linear-lanceolate, usually with a wide base, acute or acuminate; margins entire, usually incurved above base, rarely plane; costa excurrent in a short mucro, often strong, with two stereid bands; basal cells rectangular, smooth, hyaline, differentiated across leaf; upper cells small, irregularly quadrate to hexagonal, opaque, densely papillose with bifid papillae. *Monoicous*. *Capsules* exserted (in all local species), erect, ovoid, usually ellipsoid or cylindrical, stegocarpous, exceptionally cleistocarpous; operculum conical to long-rostrate; peristome short, rudimentary or absent, hymenium sometimes present. *Spores* papillose. *Calyptrae* cucullate.

* See the discussion of *Anoectangium* on the affiliation of *Anoectangium handelii*.

A large world-wide genus of mainly terrestrial plants; the number of recorded species in the genus varies according to different generic concepts. In this treatment, *Hymenostomum* is included in *Weissia*, though it differs from *Weissia sensu stricto* in the hymenium at the mouth of the gymnostomous capsule. *Astomum*, with immersed cleistocarpous capsules, is treated as a separate genus (see the discussion of *Astomum*). The genus thus circumscribed comprises 113 species (Zander 1993). Seven species occur in the local flora. Recently a new species, characterised by short, broad-ovoid leaves, was described as *W. ovatifolia* (Kürschner 1995).

Natural hybrids between *Weissia controversa* and *Astomum muehlenbergianum* were recorded, though their spores did not reach maturity (Anderson & Lemmon 1972). There are also scattered records of hybrids between pairs of species of *Weissia* and between species of *Weissia* and *Tortella* (see Smith 1978). The delimitation of *Weissia* from *Trichostomum* is described in the discussion of *Trichostomum*.

The main differences between species in *Weissia* are in sporophytic characters. The identification of sterile plants may pose considerable difficulties.

Literature. — Anderson & Lemmon (1972), Saito (1975), Hill (1981).

1 Peristome absent; mouth of capsule at least partly covered by a membrane (hymenium).
- 2 Leaves squarrose when moist; margins usually plane, sometimes narrowly incurved above in upper leaves **6. W. squarrosa**
- 2 Leaves patent to spreading when moist; margins distinctly incurved above in upper leaves.
  - 3 Leaf cells with smooth, thickened dorsal walls and thinner and bulging ventral walls **2. W. breutelii**
  - 3 Leaf cells with similar dorsal and ventral walls, papillose.
    - 4 Plants up to 15 mm high; costa 60–100(–120) µm wide at base; capsules not constricted at mouth **3. W. condensa**
    - 4 Plants usually 2–6(–10) mm high; costa up to 55 µm wide at base; capsules constricted at mouth **1. W. brachycarpa**

1 Peristome present, developed to various degrees, sometimes rudimentary; mouth of capsule without a membrane.
- 5 Leaf margins almost plane, undulate **5. W. rutilans**
- 5 Leaf margins strongly incurved above.
  - 6 Costa 60–80(–90) µm wide below; mucro 40–55 µm long **7. W. triumphans**
  - 6 Costa 30–60 µm wide below; mucro shorter, less than 40 µm long **4. W. controversa**

1. Weissia brachycarpa (Nees & Hornsch.) Jur., Laubm.-Fl. Oesterr.-Ung. 9 (1882).
Hymenostomum brachycarpum Nees & Hornsch., Bryol. Germ. 1:196 (1823); *W. microstoma* (Hedw.) Müll. Hal., Syn. Musc. Frond. 1:660 (1849); *W. hedwigii* H. A. Crum, Bryologist 74:169 (1971).

Distribution map 14, Figure 13.

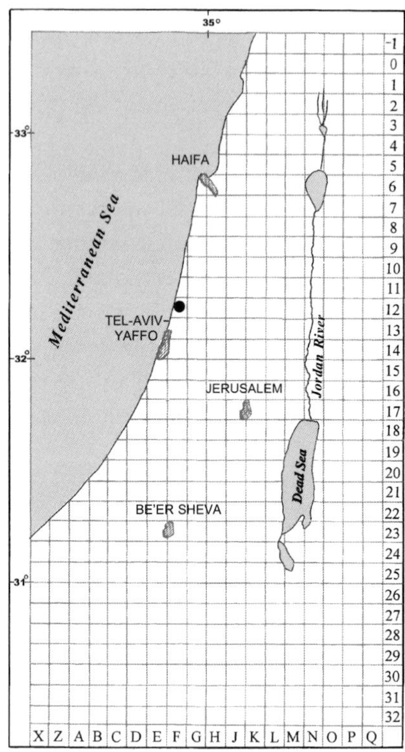

Map 14: *Weissia brachycarpa*

Plants 2–6(–10) mm high, green, turning yellowish-green. *Stems* erect. *Leaves* up to 2 mm long, strongly crisped when dry, patent to spreading when moist, concave, narrowly lanceolate, acute; margins plane below, incurved above; costa up to 55 µm wide near base with rectangular cells on ventral surface; mid- and upper leaf cells round-quadrate, densely papillose; mid-leaf cells 8–10 µm wide. *Autoicous*. *Capsules* ca. 1.5 mm long with operculum, ovate ellipsoid, brown, constricted at mouth; peristome absent, mouth covered by imperfect hymenium. *Spores* ca. 20 µm in diameter.

Habitat and Local Distribution. — Plants growing in tufts on exposed sandy loam. Several herbarium collections collected in the Sharon Plain in the same unspecified locality, "The Hill", have been seen.

General Distribution. — Recorded from Southwest Asia (Cyprus, Turkey, Syria, and Lebanon), around the Mediterranean, Europe (throughout), Macaronesia (in part), North Africa, northeastern, eastern, and central Asia, and North America.

Literature. — Crum & Anderson (1981), Hill (1981).

Following Hill (1981), the name *Weissia brachycarpa* is accepted here for this taxon instead of the invalid *W. microstoma*. Crum & Anderson (1981) cited a correspondence with E. F. Warburg who considered *W. brachycarpa* an "uncommon variety". They chose to continue the usage of *W. hedwigii* H. A. Crum "as a temporary convenience". The plants from the single locality seen by us agree with the description of the typical "*microstoma*" plants in the incurved margins of the leaves.

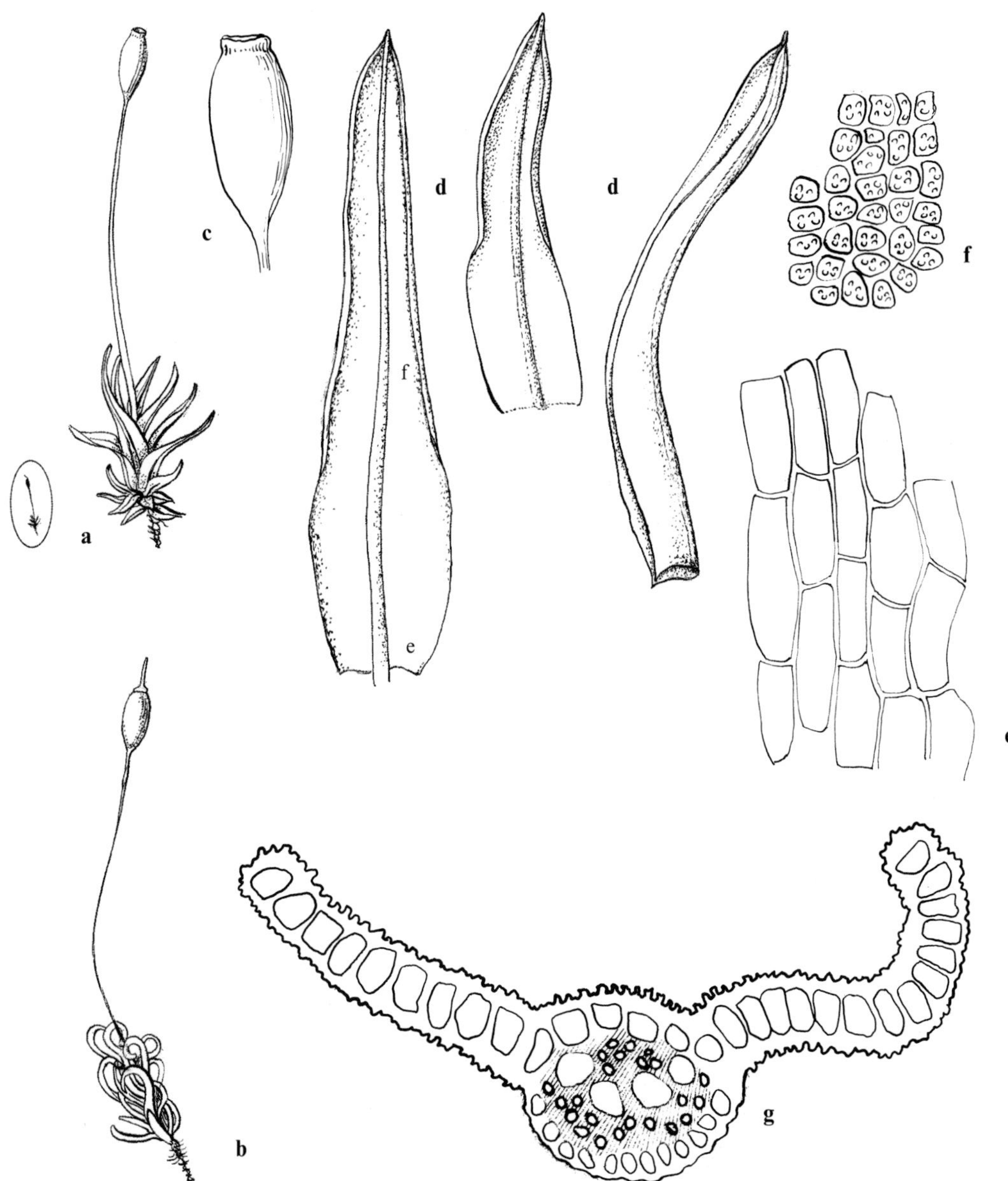

Figure 13. *Weissia brachycarpa*: **a** habit, moist (×10); **b** habit, dry (×10); **c** deoperculate capsule (×25); **d** leaves (×50); **e** basal cells; **f** mid-leaf cells (all leaf cells ×500); **g** cross-section of leaf (×500). (*Kushnir*, 12 Feb. 1943)

2. Weissia breutelii Müll. Hal., Syn. Musc. Frond. 1 : 664 (1849).
Hymenostomum breutelii (Müll. Hal.) Kindb., Enum. Bryin. Exot. 62 (1888). — Herrnstadt *et al.* 1982.

Distribution map 15, Figure 14, Plate I : h.

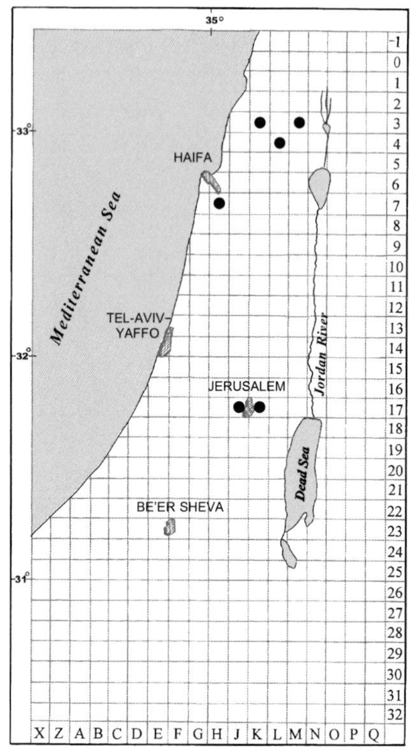

Map 15: *Weissia breutelii*

Plants up to 5(–10) mm high. *Stems* erect. *Leaves* 1.5–2 mm long, crisped when dry, patent to spreading from erect base when moist, concave, narrowly lanceolate, apex acute to acuminate, subcucullate; margins plane below, incurved above; costa up to 55 μm wide below; cross-section without dorsal epidermis and with round to subquadrate cells on ventral surface of costa in upper part of leaf; mid- and upper leaf cells irregularly round to subquadrate with thick, smooth dorsal walls and thinner, bulging ventral walls; mid-leaf cells 6–8 μm wide. *Setae* yellow; capsules ± erect, 1.5–2 mm long including operculum, cylindrical to ellipsoid, brownish; peristome absent; hymenium present. *Spores* 16–20 μm in diameter (24 μm in the isotype cited below); surface with minute baculum-like projections, projections covered by large gemmae, gemmae often fused (seen with SEM).

Habitat and Local Distribution. — Plants small, green to yellowish-green, growing in loose to dense tufts on soil, under rocks, and in rock crevices, in at least partly shaded habitats. Occasional: Upper Galilee, Mount Carmel, and Judean Mountains.

General Distribution. — Recorded so far only from Central America, the Caribbean, and northeastern and northwestern South America.

Weissia breutelii is characterised by the anatomy of the leaves: thick dorsal walls and bulging ventral walls. These characters can be seen in local plants, which were identified as *W. breutelii* by R. Zander. A comparison between local plants and the isotype deposited in the British Museum (West Indies: St. Thomas, *Breutel*) has shown their similarity. Nevertheless, the geographical disjunction of *W. breutelii* seems rather puzzling. It might be an example of a Circum-Tethyan distribution pattern following Frey & Kürschner (1983). On the other hand, the possibility that the thickened cell walls of the leaves are correlated with xeric habitats and may have occurred more than once in *Weissia*, cannot be entirely excluded. This is further supported by the similarity in the texture of the spore surface between *W. breutelii* and *W. controversa*.

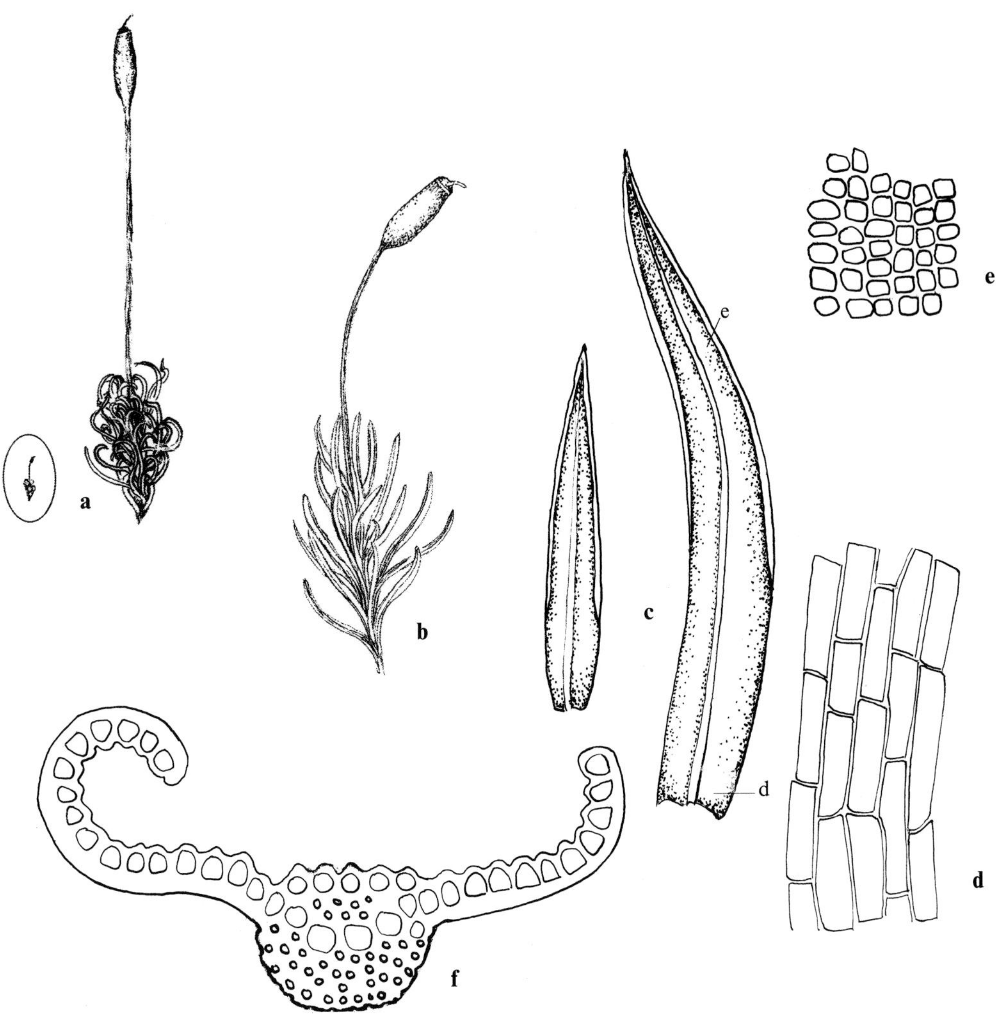

Figure 14. *Weissia breutelii*: **a** habit, dry (×10); **b** habit, moist (×10); **c** leaves (×50); **d** basal cells; **e** upper cells (all leaf cells ×500); **f** cross-section of leaf (×500). (*Kushnir*, 10 Jan. 1943)

3. Weissia condensa (Voit) Lindb., Öfvers. Förh. Kongl. Svenska Vetensk.-Akad. 21 : 230 (1863).

Gymnostomum condensum Voit *in* Sturm, Deutschl. Fl. 2 : Ic. (1811); *Hymenostomum tortile* (Schwägr.) B.S.G., Bryol. Eur. 1 : 56 (1846); *W. tortilis* (Schwägr.) Müll. Hal., Syn. Musc. Frond. 1 : 661 (1849). — Bilewsky (1965) : 358.

Distribution map 16, Figure 15, Plate I : g.

Plants up to 15 mm high, green in upper part, dark green to brown below. *Stems* erect. *Leaves* up to 1.5 mm long, upper leaves up to 2.5 mm long, crisped when dry, patent to spreading when moist, carinate, oblong to narrowly lanceolate, acute; margins plane below, strongly incurved above; costa strong, 60–100 (–120) μm wide below, with quadrate cells on upper ventral surface; basal cells ± incrassate; mid- and upper leaf cells irregularly quadrate, densely papillose, evenly thickened, occasionally bulging ventrally; mid-leaf cells 6–9 μm wide. *Autoicous. Setae* yellowish; capsules up to 2 mm long including operculum, ovoid, yellow to yellowish-brown, ± darker at mouth; peristome absent; mouth often partly covered by a very thin imperfect hymenium. *Spores* 16–19 μm in diameter; densely covered by baculum-like projections, projections often fused into striae (seen with SEM).

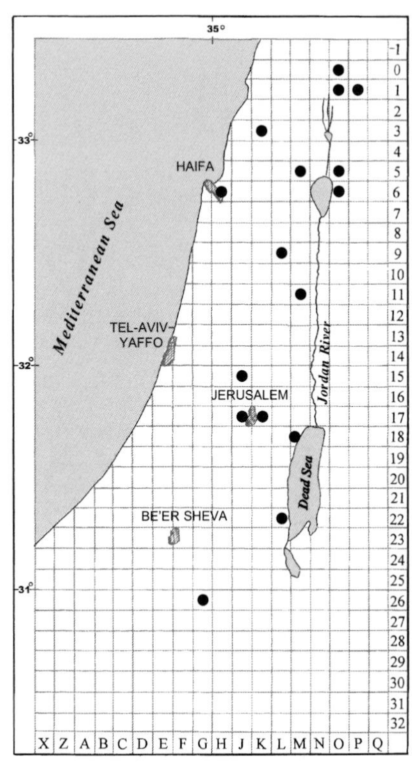

Map 16: *Weissia condensa*

Habitat and Local Distribution. — Plants growing in dense tufts, readily detached from the substrate, on exposed soil, in rock crevices, and among stones. Common: Upper Galilee, Lower Galilee, Mount Carmel, Mount Gilboa, Samaria, Judean Mountains, Judean Desert, Central Negev, Dead Sea area, Mount Hermon, and Golan Heights.

General Distribution. — Recorded from Southwest Asia (Cyprus, Turkey, Syria, Lebanon, Jordan, Iraq, Iran, Afghanistan, Saudi Arabia, and Oman), around the Mediterranean, Europe (the south in particular), Macaronesia, North Africa, northeastern and central Asia, New Zealand, and North and Central America.

Plants of *Weissia condensa* are the largest among the local *Weissia* species. They can be also distinguished by the strong costa and capsules without peristome.

According to Corley *et al.* (1981), *Weissia condensa* and *W. controversa* "are connected by a confusing series of forms". Among these, they name *W. fallax*, which is often treated as *W. controversa* var. *crispata* (e.g., Smith 1978) or as a valid species (e.g., Zander 1993).

Figure 15. *Weissia condensa*: **a** habit, dry (×10); **b** habit, moist (×10); **c** capsule (×10); **d** leaves (×50); **e** basal cells; **f** upper cells (all leaf cells ×500); **g** cross-section of costa and portion of lamina (×500). (*Herrnstadt & Crosby 79–99–16*)

4. Weissia controversa Hedw., Spec. Musc. Frond. 67 (1801).
W. crispata (Nees & Hornsch.) Müll. Hal., Syn. Musc. Frond. 1:662 (1849). — Bilewsky (1965):358; *W. viridula* Hedw. ex Brid., Mant. Musc. 38 (1819). — Bilewsky (1965):359.

Distribution map 17, Figure 16, Plate I:f.

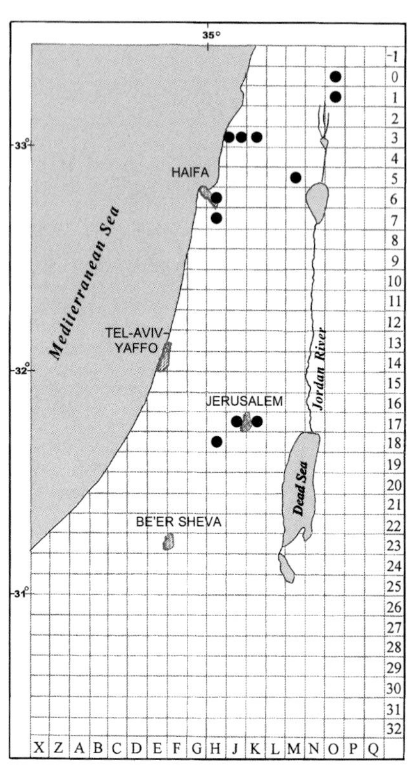

Map 17: *Weissia controversa*

Plants up to 5(–10) mm high, yellowish-green to green. *Leaves* up to 2 mm long, strongly crisped when dry, patent when moist, distant and shorter towards base, narrowly lanceolate, acute, shortly mucronate; margins plane below, strongly incurved above; costa 30–60 µm wide below, with quadrate cells on upper ventral surface of costa; mid- and upper leaf cells quadrate-hexagonal, densely papillose, mid-leaf cells 6–9 µm wide. *Autoicous. Setae* yellowish; capsules up to 1.5 mm long including operculum, ovoid to narrow-ellipsoid, yellowish-brown; peristome teeth very short or rudimentary. *Spores* 15–20 µm in diameter; surface with minute baculum-like projections, irregularly covered by large gemmae, gemmae varying in size (seen with SEM).

Habitat and Local Distribution. — Plants growing on soil on calcareous rocks and in rock crevices in shaded habitats. Occasional: Coastal Galilee, Upper Galilee, Lower Galilee, Mount Carmel, Judean Mountains, Judean Desert, and Golan Heights.

General Distribution. — Recorded as a nearly cosmopolitan species of temperate regions with many records from Southwest Asia (Cyprus, Turkey, Syria, Lebanon, Iraq, Iran, Afghanistan, and Saudi Arabia), and from around the Mediterranean.

Comments on the *Weissia controversa* / *W. condensa* complex are given in the discussion of *W. condensa*.

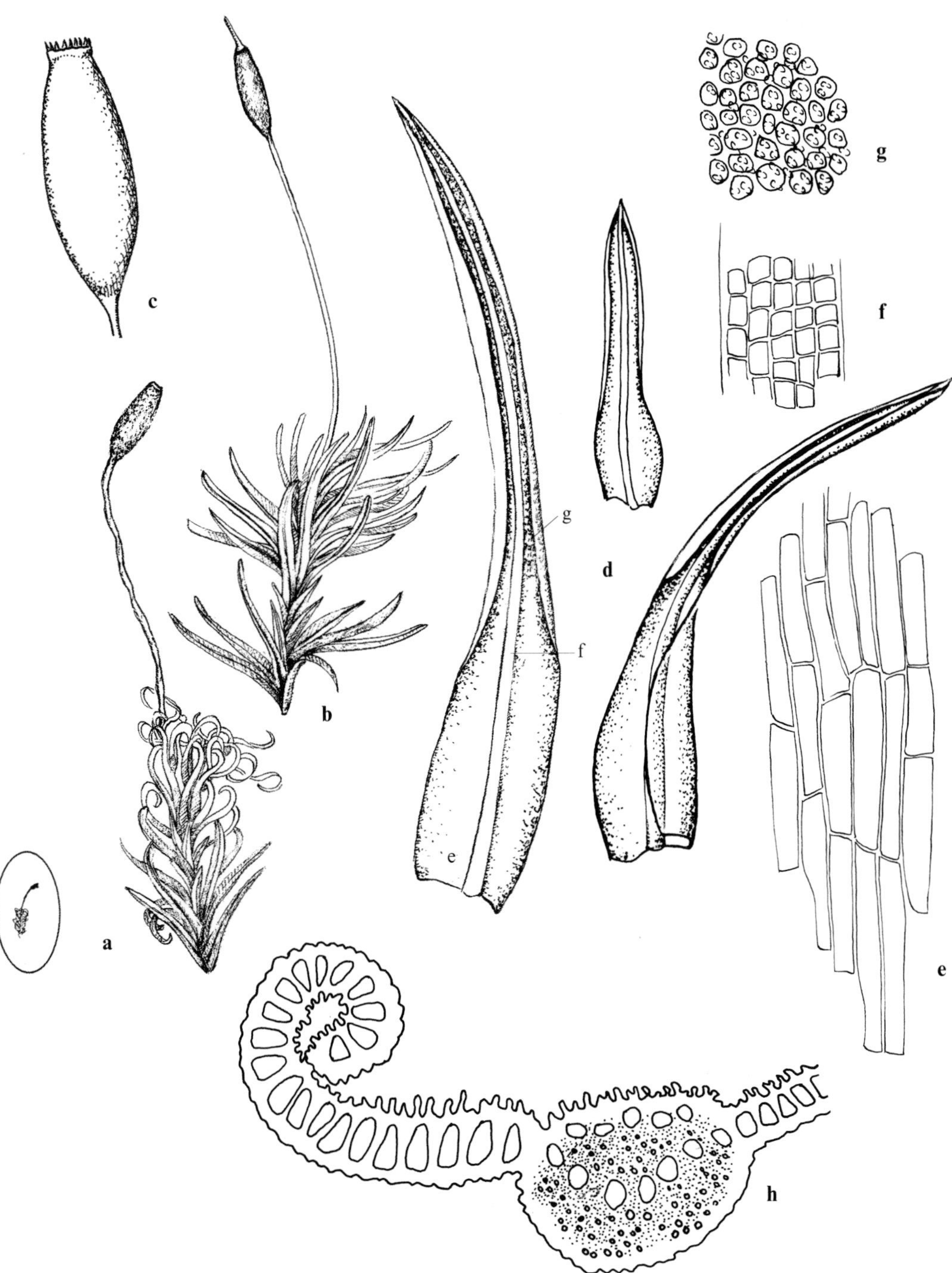

Figure 16. *Weissia controversa*: **a** habit, dry (×10); **b** habit, moist (×10); **c** deoperculate capsule (×25); **d** leaves (×50); **e** basal cells; **f** cells on ventral surface of costa; **g** mid-leaf cells (all leaf cells ×500); **h** cross-section of costa and portion of lamina (×500). (**a**, **b**, **d**–**g**: *Herrnstadt* & *Crosby 78–51–2*; **c**, **h**: *Kushnir*, 10 Oct. 1943)

5. Weissia rutilans (Hedw.) Lindb., Ögfvers. Förh. Kongl. Svenska Vetensk.-Akad. 20 : 417 (1863).

Distribution map 18, Figure 17, Plate I : i.

Plants green above, brown below, up to 15 mm high. *Leaves* up to 2 mm long, upper leaves up to 3 mm long, crisped when dry, patent to spreading when moist, oblong-lanceolate to narrowly lanceolate, acute or acuminate; margins in the majority of leaves almost plane, undulate, rarely slightly incurved in upper leaves; costa strong, up to 120 μm wide below with quadrate cells on upper ventral surface; mid- and upper leaf cells incrassate, round-quadrate, densely papillose; mid-leaf cells 5–8 μm wide. *Setae* yellowish; capsules ca. 1.5 mm long including operculum, ellipsoid, brown; peristome rudimentary, fugacious. *Spores* 20–25 μm in diameter; verrucate, verrucae with rough surface (seen with SEM).

Habitat and Local Distribution. — Plants growing in dense tufts on soil and among rocks in partly shaded habitats. Rare: found growing only in several localities on Mount Carmel.

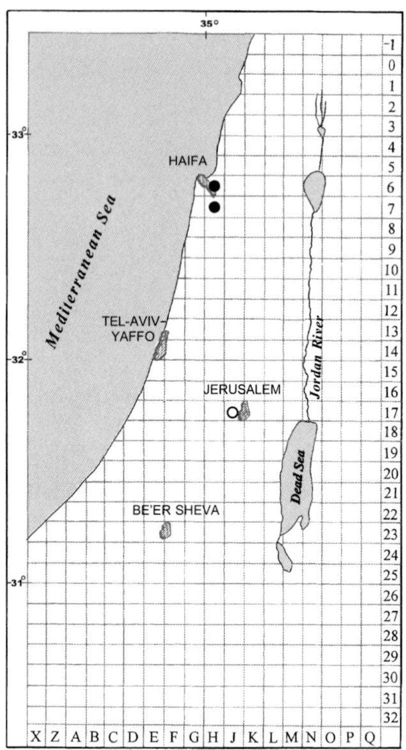

Map 18: ● *Weissia rutilans*
○ *Weissia squarrosa*

General Distribution. — Not previously recorded from Southwest Asia, a few records from around the Mediterranean; recorded from Europe, tropical and North Africa, northeastern and eastern Asia, Australia, and North America.

Weissia rutilans has not been recorded previously from Israel. It can be distinguished from all other local species of the genus by the large spores and the usually plane margins of the leaves. The latter character is often considered to be associated with *Trichostomum*.

Figure 17. *Weissia rutilans*: **a** habit, moist (×10); **b** habit, dry (×10); **c** deoperculate capsule and operculum (×25); **d** leaves (×50); **e** basal cells; **f** mid-leaf cells, (all leaf cells ×500); **g** portion of peristome (×125). (**a–f**: *Herrnstadt*, 15 Mar. 1987; **g**: *Kushnir*, 21 Mar. 1943)

6. Weissia squarrosa (Nees & Hornsch.) Müll. Hal., Syn. Musc. Frond. 1 : 663 (1849).
Hymenostomum squarrosum Nees & Hornsch., Bryol. Germ. 1 : 193 (1823).

Distribution map 18 (p. 72), Figure 18.

Plants dark green, turning olive-green when dry, resembling *Weissia brachycarpa*, except for the strongly spreading and recurved or squarrose leaves when moist; margins plane in lower leaves, narrowly incurved in upper half in upper leaves; spores are released from capsules by rupturing of the thin-walled exothecium.

Habitat and Local Distribution. — Plants growing in tufts on calcareous soil. Seen only at a single locality: Judean Mountains, Jerusalem.

General Distribution. — Available records are mainly from northern and central Europe, few from the Mediterranean Region. No previous records from Southwest Asia.

7. Weissia triumphans (De Not. *in* Schimp.) M. O. Hill, J. Bryol. 11 : 600 (1981).
Trichostomum triumphans De Not. *in* Schimp., Syn. 690 (1860). — Bilewsky (1965) : 363;
T. pallidisetum H. Müll., Verh. Naturhist. Vereines Rheinl. Westphalens 22 : 292 (1865);
T. triumphans var. *pallidisetum* (H. Müll.) Husn., Musc. Gall. 89 (1885). — Bilewsky (1965) : 363.

Distribution map 19, Figure 19 (p. 77).

Plants up to 8(–10) mm high, yellowish-green to green. *Leaves* — Upper leaves up to 2.5 mm long, crisped when dry, patent to spreading when moist, linear to linear-lanceolate, acute; lower leaves ± smaller, lanceolate; margins plane below, incurved above; costa excurrent in a mucro up to 55 μm long with quadrate cells on the ventral surface of costa; mid- and upper leaf cells subquadrate to hexagonal, densely papillose; mid-leaf cells 5–8 μm wide.

Habitat and Local Distribution. — Plants growing in dense tufts, on calcareous soils. Rare to occasional: Sharon Plain, Upper Galilee, Mount Carmel, and Judean Mountains.

General Distribution. — Recorded from Southwest Asia (Iran and Jordan), around the Mediterranean, Europe, Macaronesia (in part), and North Africa.

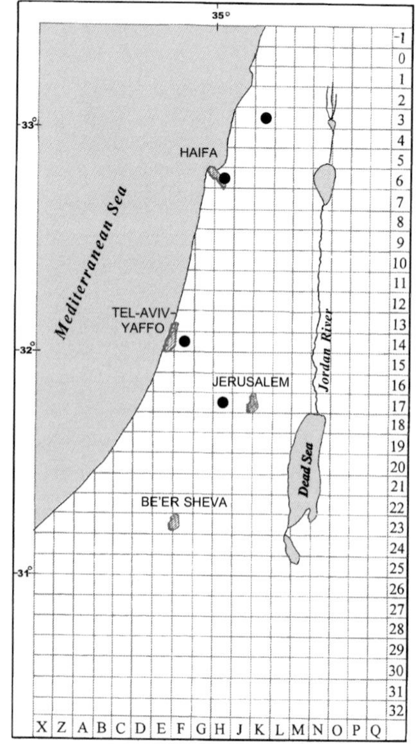

Map 19: *Weissia triumphans*

Bilewsky (1965) recorded sporophytes with reddish (in the typical variety) or yellow (in var. *pallidisetum*) setae, and often stunted peristome teeth. Sporophytes have not been seen by us.

The collections from Israel (*Bilewsky 32, 386, 691*; *Weitz 1141*) and from Jordan (*El-Oqlah & Lahham 415*, *El-Oqlah 417*), identified as *Trichostomum triumphans* by Bilewsky (1965) and El-Oqlah *et al.* (1988a), do not include any sporophytes. Because

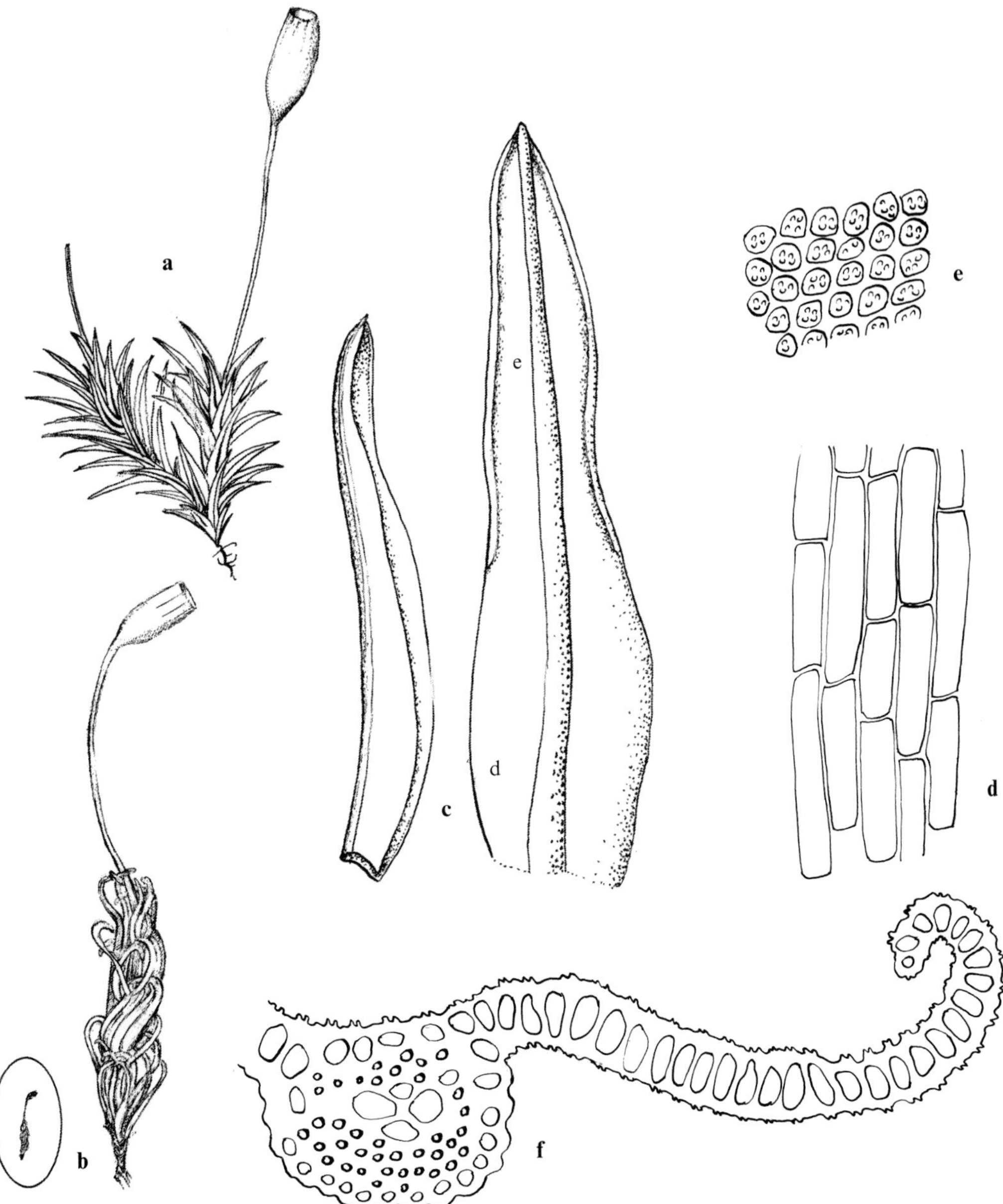

Figure 18. *Weissia squarrosa*: **a** habit, moist (×10); **b** habit, dry (×10); **c** leaves (×50); **d** basal cells; **e** upper cells (all leaf cells ×500); **f** cross-section of costa and portion of lamina (×500). (*D. Zohary*, 9 Jan. 1942)

of the overall similarity of the gametophytes of *W. triumphans* and *Trichostomum brachydontium*, some doubt remains about the local distribution of these two species.

Trichostomum triumphans was transferred to *Weissia* by Hill (1981) following the generic concept of Saito (1975). Thus, the small and monoicous *T. triumphans* is placed in *Weissia* rather than with the larger and dioicous species of *Trichostomum*. *Weissia triumphans* differs from *W. controversa* mainly in the longer leaves with longer mucro and wider costa.

2. *Astomum* Hampe

Plants small, resembling *Weissia*, but with immersed cleistocarpous capsules. A genus comprising 11 species according to Zander (1993), though considered there as part of *Weissia*.

Astomum is often considered as part of *Weissia* (see the treatments by Crundwell & Nyholm 1972, Stoneburner 1985, Zander 1993 — and the literature cited there), usually as subgenus *Astomum* (Hampe) Kindb. (e.g., Corley *et al.* 1981). We prefer here to treat *Astomum* as a separate genus because of the cleistocarpous capsules. In local plants, no intermediate forms between *Astomum* and any *Weissia* species have been observed. As pointed out in the discussion of *Weissia* above, hybrids between species of the two genera have been reported from other regions. Hybridisation between genera of *Trichostomoideae* does not seem to carry enough weight for the inclusion of *Astomum* in *Weissia*. It seems (cf. Smith & Newton 1968, Stoneburner 1985) that the majority of the genera of the *Trichostomoideae* form a natural cytological association and consequently hybrids between them are not infrequent.

Literature. — Smith & Newton (1968), Anderson & Lemmon (1972), Crundwell & Nyholm (1972), Stoneburner (1985), Herrnstadt & Heyn (1989).

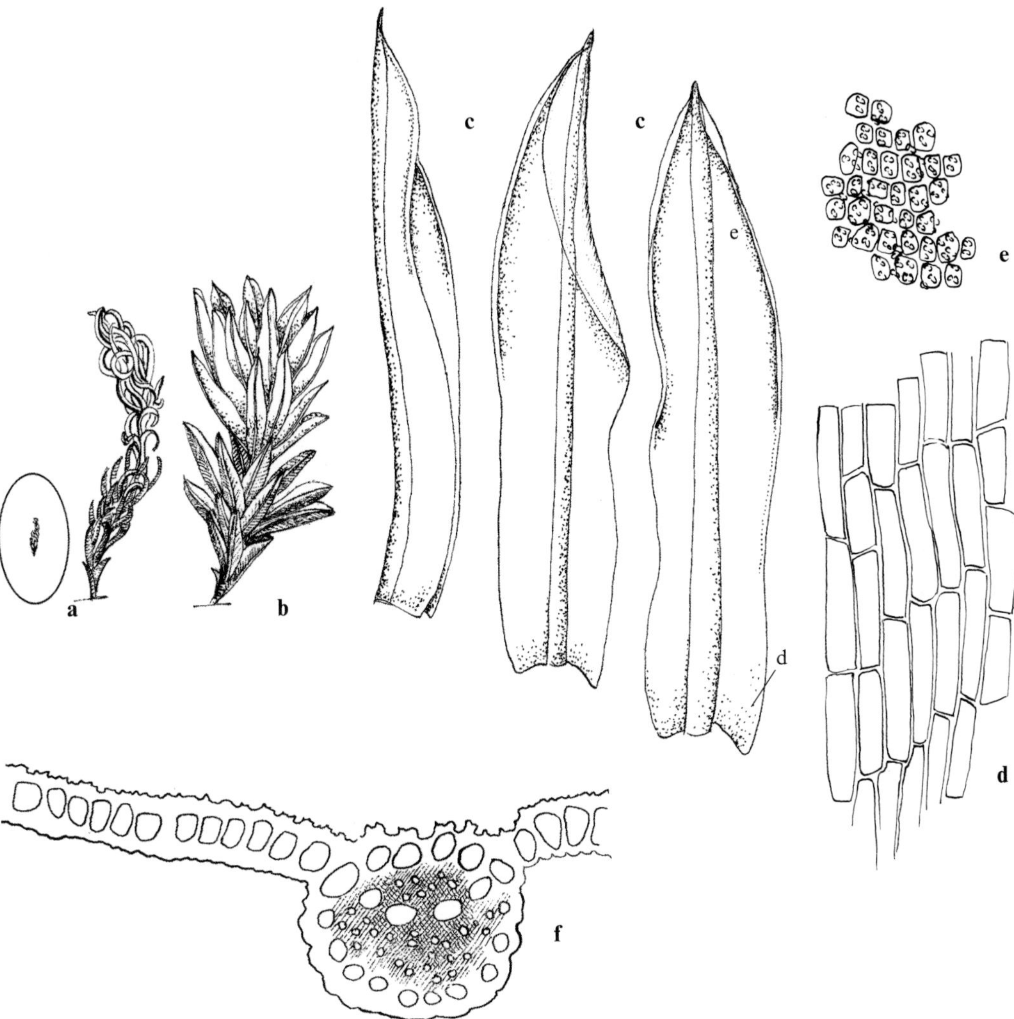

Figure 19. *Weissia triumphans*: **a** habit, dry (×10); **b** habit, moist (×10); **c** leaves (×50); **d** basal cells; **e** upper cells (all leaf cells ×500); **f** cross-section of costa and portion of lamina (×500). (*Bilewsky*, Oct. 1958, No. 691 in Herb. Bilewsky)

Astomum crispum (Hedw.) Hampe, Flora 20 : 285 (1837). [Bilewsky 1965 : 357]
Phascum crispum Hedw., Sp. Musc. Frond. 21 (1801); *Weissia crispa* (Hedw.) Mitt., Ann. Mag. Nat. Hist., ser. 2, 8 : 316 (1851); *Weissia longifolia* Mitt., Ann. Mag. Nat. Hist., ser. 2, 8 : 317 (1851).

Distribution map 20, Figure 20, Plate I : k, l.

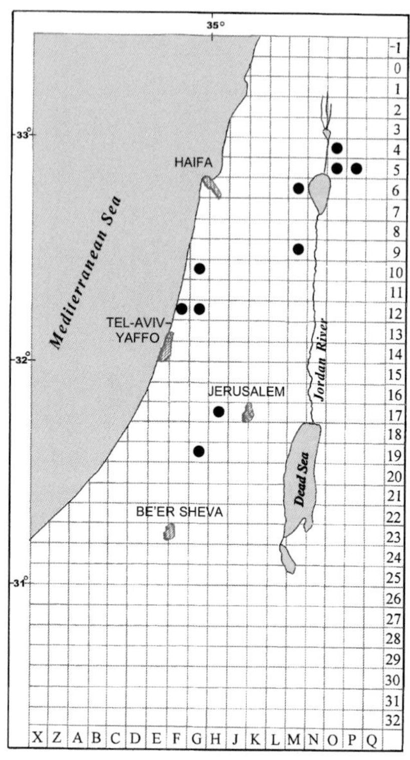

Map 20: *Astomum crispum*

Plants small, up to 3(–6) mm high, green to yellowish-green. *Stems* usually simple with a central strand. *Leaves* strongly crisped and contorted when dry, erectopatent when moist, up to 3(–4) mm long, greatly varying in length above wider leaf base, narrowly lanceolate to linear-lanceolate, acute or acuminate; perichaetial leaves with broad-sheathing hyaline base, ⅓–½ length of leaf, enclosing the capsule; margins plain or narrowly incurved, slightly crenulate, sometimes inconspicuously toothed above base; costa 50–70 µm wide at base, excurrent into a mucro, up to 100 µm long; basal cells rectangular; mid- and upper leaf cells irregularly quadrate, mid-leaf cells 8–11 µm wide. *Monoicous*. *Setae* very short, about ⅓ length of capsules; capsules immersed, 0.5–1.5 mm long, subglobose, short-rostrate, usually indehiscent. *Spores* yellowish-brown, sometimes visible through thin exothecium, 25–30 µm in diameter, densely papillose; baculate, some bacula with irregularly lobed granulate apex (seen with SEM). *Calyptrae* short, covering only the rostrum of the capsules.

Habitat and Local Distribution. — Plants growing in loose or dense tufts on partly exposed sandy loam, terra rossa and basalt soils. Occasional: Sharon Plain, Lower Galilee, Shefela, Judean Mountains, Bet She'an Valley, and Golan Heights.

General Distribution. — Recorded (often as *Weissia longifolia*) from Southwest Asia from Turkey only, around the Mediterranean, Europe (throughout), Macaronesia (in part), North Africa, and northeastern, eastern, and central Asia.

Astomum crispum is often found growing together with *Acaulon muticum*, *A. longifolium*, *Aschisma carniolicum* (see also the discussion of that species), and various *Pottia* and *Bryum* species.

In one collection from the Golan Heights (Nahal Meshushim, *Herrnstadt*, 4 Jul. 1987) plants differ in having persistent protonemata, narrower leaves, which are only slightly crispate when dry, and less conspicuously differentiated perichaetial leaves; they may belong to a different taxon.

The spore texture of this species (observed with SEM) distinctly differs from that found in our studies in *Weissia* and other allied genera. It seems noteworthy that the lobed bacula resemble those of some other cleistocarpous, terrestrial Pottiaceae (e.g., *Acaulon muticum*, *Aschisma carniolicum*, *Phascum cuspidatum* var. *arcuatum*, *P. galilaeum* — see Herrnstadt & Heyn 1989 and Plates below I : m, V : e, f, and VI : c, d).

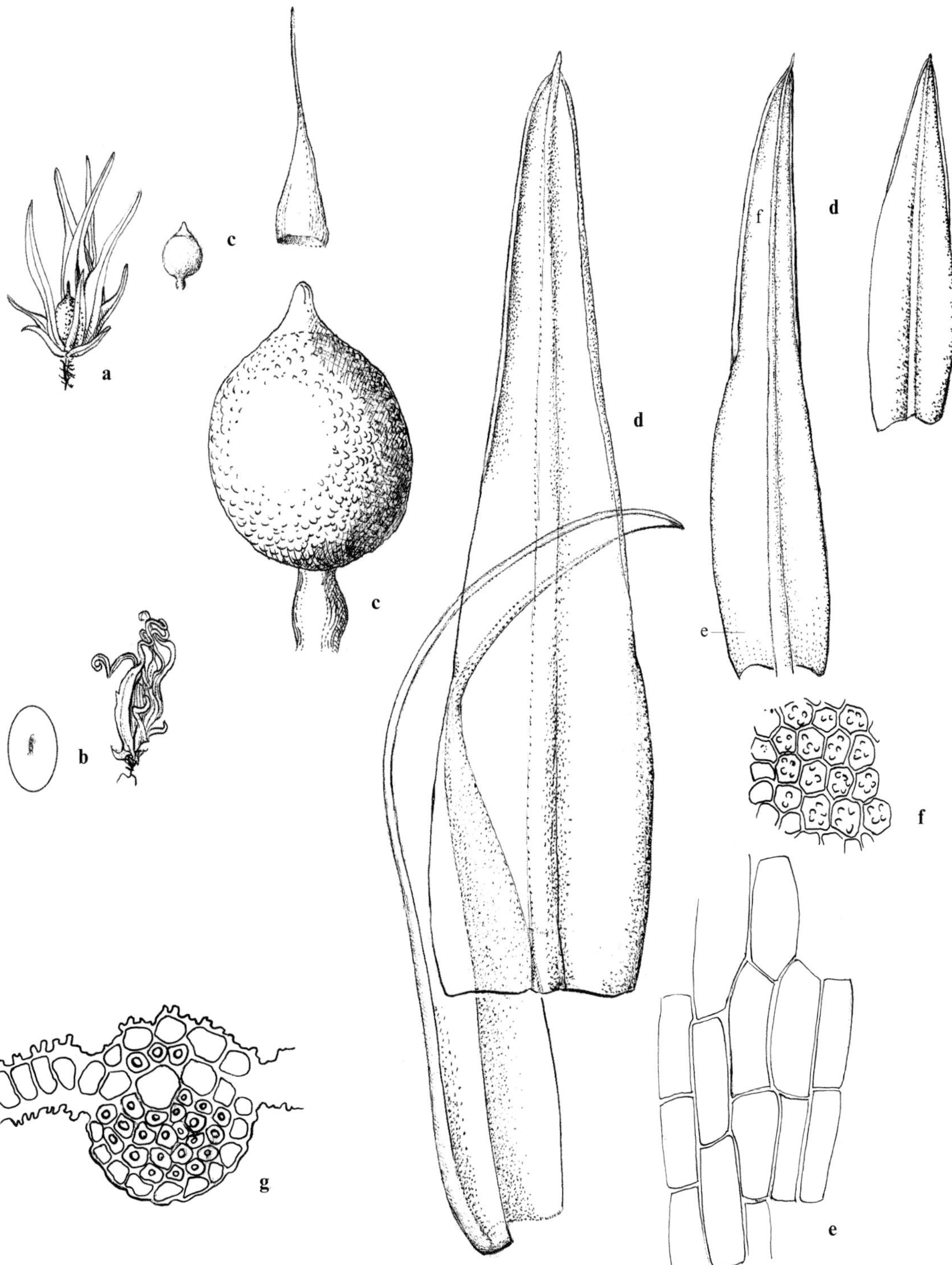

Figure 20. *Astomum crispum*: **a** habit, moist (×10); **b** habit, dry (×10); **c** sporophyte with calyptra (×50); **d** leaves (×50); **e** basal cells; **f** upper cells (all leaf cells ×500); **g** cross-section of costa and portion of lamina (×500). (*Herrnstadt & Crosby 78-18-7* — sample mixed with *Aschisma carniolicum*)

3. *Aschisma* Lindb.

A small genus comprising only two species (*sensu* Zander 1993). For generic characters see the description of *Aschisma carniolicum.*

Aschisma carniolicum (F. Weber & D. Mohr) Lindb., Utkast Eur. Bladmoss. 28 (1878).
Phascum carniolicum F. Weber & D. Mohr, Bot. Taschenb. 69 : 450 (1807).

Distribution map 21, Figure 21, Plate I : m.

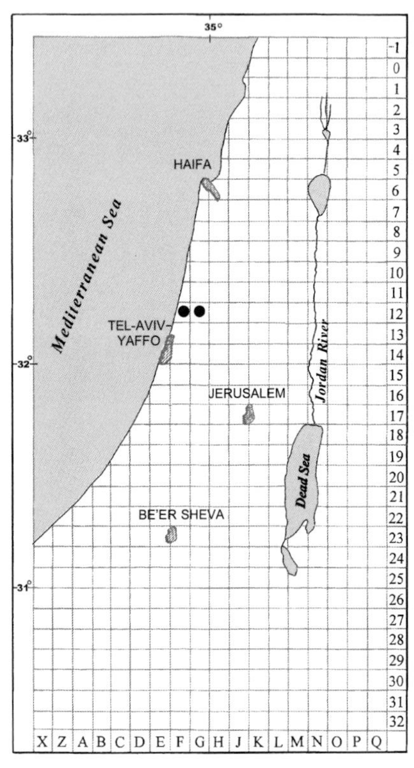

Map 21: *Aschisma carniolicum*

Plants very small, up to 1.2 mm high, gregarious, green, turning yellowish-brown, revealing the brown capsules when dry. *Stem* simple, with weak central strand. *Leaves* 0.5–1 mm long, incurved when dry, erect-spreading when moist, slightly keeled, lower leaves ovate, upper lanceolate, abruptly ending in acute apex; margins plane or weakly incurved, entire, or crenulate above; costa very strong, ending in apex to excurrent in a short mucro, usually with two stereid bands; basal cells rectangular, smooth, rising along margins in a weak V, upper cells irregularly quadrate or round-hexagonal, ± incrassate, papillose with C-shaped, bifid papillae, in mid-leaf 8–11 µm wide. *Monoicous. Setae* straight, shorter than capsules; capsules cleistocarpous, ca. 0.4 mm long, with rectangular exothecial cells circling the capsule in horizontal rows, globose, with short, blunt apiculus, brown; operculum not differentiated. *Spores* 24–27 µm in diameter, spherical, spinulose-papillose; baculate, bacula slightly branched at apex, often several joined (seen with SEM). *Calyptrae* smooth, conical-cucullate, very small, covering only the upper part of capsules.

Habitat and Local Distribution. — Plants found, so far, growing only in two localities on shaded sandy loam in the Sharon Plain. Plants of this species may have been overlooked elsewhere because of their small size.

General Distribution. — Recorded from Southwest Asia (Turkey), scattered around the Mediterranean, Europe, and North Africa.

Aschisma carniolicum was found growing together with *Astomum crispum*, but can be distinguished from *A. crispum* by the very small, not strongly twisted leaves lacking a sheathing base, and by the capsules that have the special arrangement of the exothecial cells. It differs from *Acaulon*, which it resembles, by the denser areolation of the leaf cells and the greater density of the papillae.

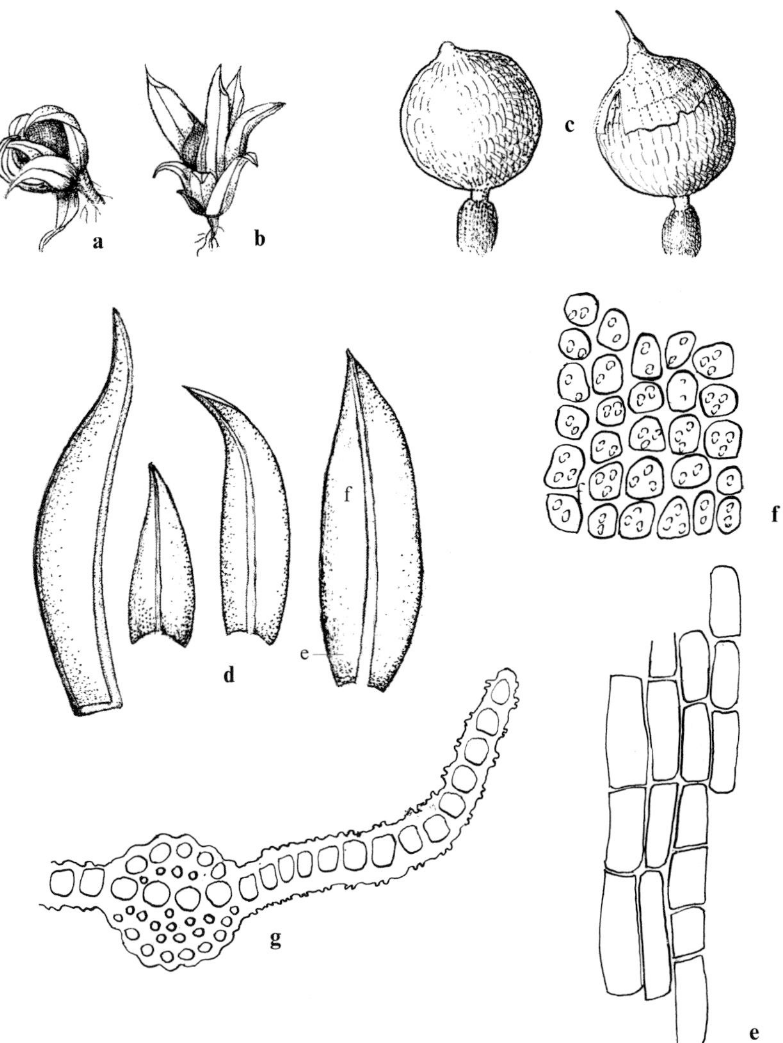

Figure 21. *Aschisma carniolicum*: **a** habit, dry (×20); **b** habit, moist (×20); **c** sporophytes with and without calyptra (×50); **d** leaves (×50); **e** basal cells; **f** mid-leaf cells (all leaf cells ×500); **g** cross-section of costa and portion of lamina (×500). (*Herrnstadt* & *Crosby 78-18-7* — sample mixed with *Astomum crispum*)

4. *Trichostomum* Bruch

Plants small, usually larger than in *Weissia*. *Stems* simple or branched, with a central strand. *Leaves* incurved and usually crisped when dry, erectopatent to spreading when moist, lanceolate to lingulate, acute or obtuse, sometimes cucullate; margins entire to papillose-crenulate, plane or narrowly incurved; costa percurrent to excurrent, with two stereid bands, basal cells irregularly rectangular, hyaline or yellowish, differentiated across leaf or rising weakly along margins; upper cells round quadrate to subhexagonal, densely papillose with usually bifid papillae, opaque. *Dioicous*. *Capsules* exserted on an elongated seta, erect, ellipsoid or cylindrical; operculum slightly oblique, rostrate; peristome usually with a short basal membrane, teeth deeply cleft, straight. *Calyptra* cucullate.

A large genus comprising about 100 species, variously delimited from closely related genera (e.g., *Weissia*, *Oxystegus*), and widely distributed.

Trichostomum and *Weissia* are usually considered as very closely related (Chen 1941). Their delimitation, as accepted here, assigns to *Weissia* slender, usually monoicous, plants with short or reduced peristome and leaves with often more strongly incurved margins. However, some of the generic characters may be partly overlapping in species of both genera. Consequently, the delimitation between them seems sometimes rather awkward (see also the discussion of *Weissia triumphans*). The status of *Oxystegus*, which is also part of this complex of genera, is discussed under that genus (p. 86).

Trichostomum resembles the genus *Tortella*, but basal cells, when ascending along margins, are never abruptly differentiated from adjacent cells.

1 Leaf apex not cucullate, margins usually plane in upper part **1. T. brachydontium**
1 Leaf apex cucullate, margins incurved in upper part **2. T. crispulum**

1. Trichostomum brachydontium Bruch *in* F. A. Müll., Flora 12:393 (1829). [Bilewsky (1974)]
T. mutabile Bruch *in* De Not., Syll. Pl. Nov. 192 (1838). — Bilewsky (1965):363.

Distribution map 22 (p. 84), Figure 22.

Plants 0.5–2 cm high, yellowish-green. *Leaves* up to 2 mm long, lingulate to linear-lanceolate; apex acute to obtuse; margins usually plane, slightly undulate, papillose-crenulate; costa strong, up to 80 µm wide below, excurrent in a stout mucro; cells incrassate, basal cells rectangular, hyaline; mid-leaf cells round-quadrate, 5.5–8 µm wide, densely papillose. *Sporophytes* not seen in local populations.

Habitat and Local Distribution. — Plants growing in dense tufts on soil, stones or in rock crevices. Rare to occasional: Lower Galilee, Mount Carmel, and Judean Mountains.

General Distribution. — Recorded from Southwest Asia (Cyprus, Turkey, Lebanon, Iran, Saudi Arabia, and Yemen), around the Mediterranean, Europe (widespread throughout), Macaronesia, Africa (throughout), Asia (throughout), Texas, Caribbean central, and South America; a widespread species.

For the similarity between sterile materals of *Trichostomum brachydontium* and *Weissia triumphans*, see the discussion of *W. triumphans*.

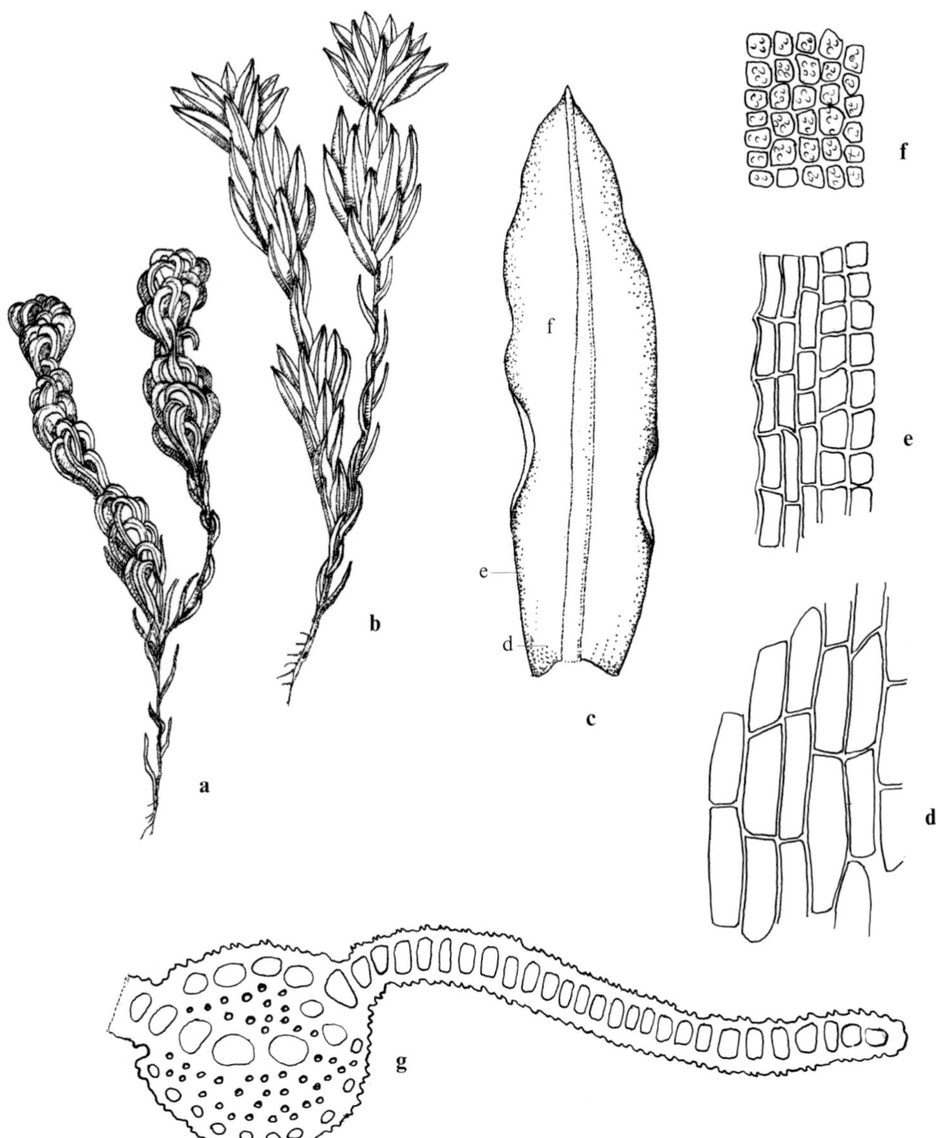

Figure 22. *Trichostomum brachydontium*: **a** habit, dry (×10); **b** habit, moist (×10); **c** leaf (×50); **d** basal cells; **e** marginal cells below mid-leaf; **f** upper cells (all leaf cells ×500); **g** cross-section of costa and portion of lamina (×500). (*Bilewsky*, 20 Feb. 1954, No. 30 in Herb. Bilewsky)

2. Trichostomum crispulum Bruch, *in* F. A. Müll., Flora 12 : 395 (1829).
[Bilewsky (1965) : 363]

T. viridulum Bruch *in* F. A. Müll., Flora 12 : 401 (1829); *T. brevifolium* Sendtn. ex Müll. Hal., Syn Musc. Frond. 1 : 572 (1849).

Distribution map 23, Figure 23, Plate II : a.

Plants up to 2 cm long, yellowish-green to green. *Leaves* 2.5–3.5 mm long, concave, narrowly linear-lanceolate to lanceolate; apex acute, cucullate; margins incurved in upper part, papillose-crenulate above; costa strong, 60–70 µm wide below, ending in apex or excurrent in a ± hyaline mucro; cells incrassate, basal cells rectangular, hyaline; mid-leaf cells round-quadrate to subhexagonal, 5–7 µm wide, densely papillose. *Sporophytes* seen only in one local population. *Setae* yellowish-orange, ca. 6 mm long; capsules 1.5 mm long including operculum, cylindrical to ellipsoid; peristome teeth short, entire, straight. *Spores* 16–17 µm in diameter; surface baculate, some baculae covered by large gemmae (seen with SEM).

Habitat and Local Distribution. — Plants growing in dense tufts on calcareous rocks, in rock crevices, and on soil on north-facing slopes. Occasional: Dan Valley, Lower Galilee, and Judean Mountains.

General Distribution. — Recorded from Southwest Asia (Cyprus, Turkey, Syria, Lebanon, Jordan, Iran, Saudi Arabia, and Yemen), around the Mediterranean, Europe, Macaronesia, North and central Africa, northeastern, eastern, central, and Southeast Asia, Central and South America; a widespread species of the Northern Hemisphere.

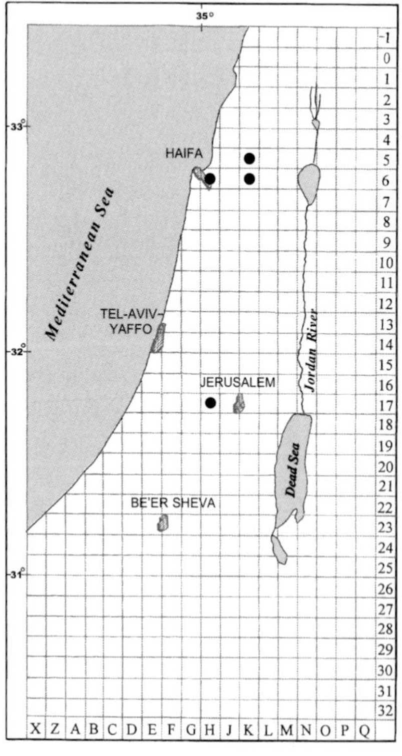

Map 22: *Trichostomum brachydontium*

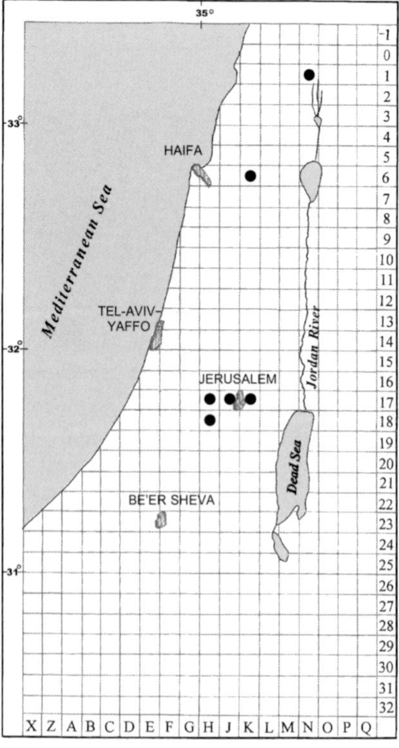

Map 23: *Trichostomum crispulum*

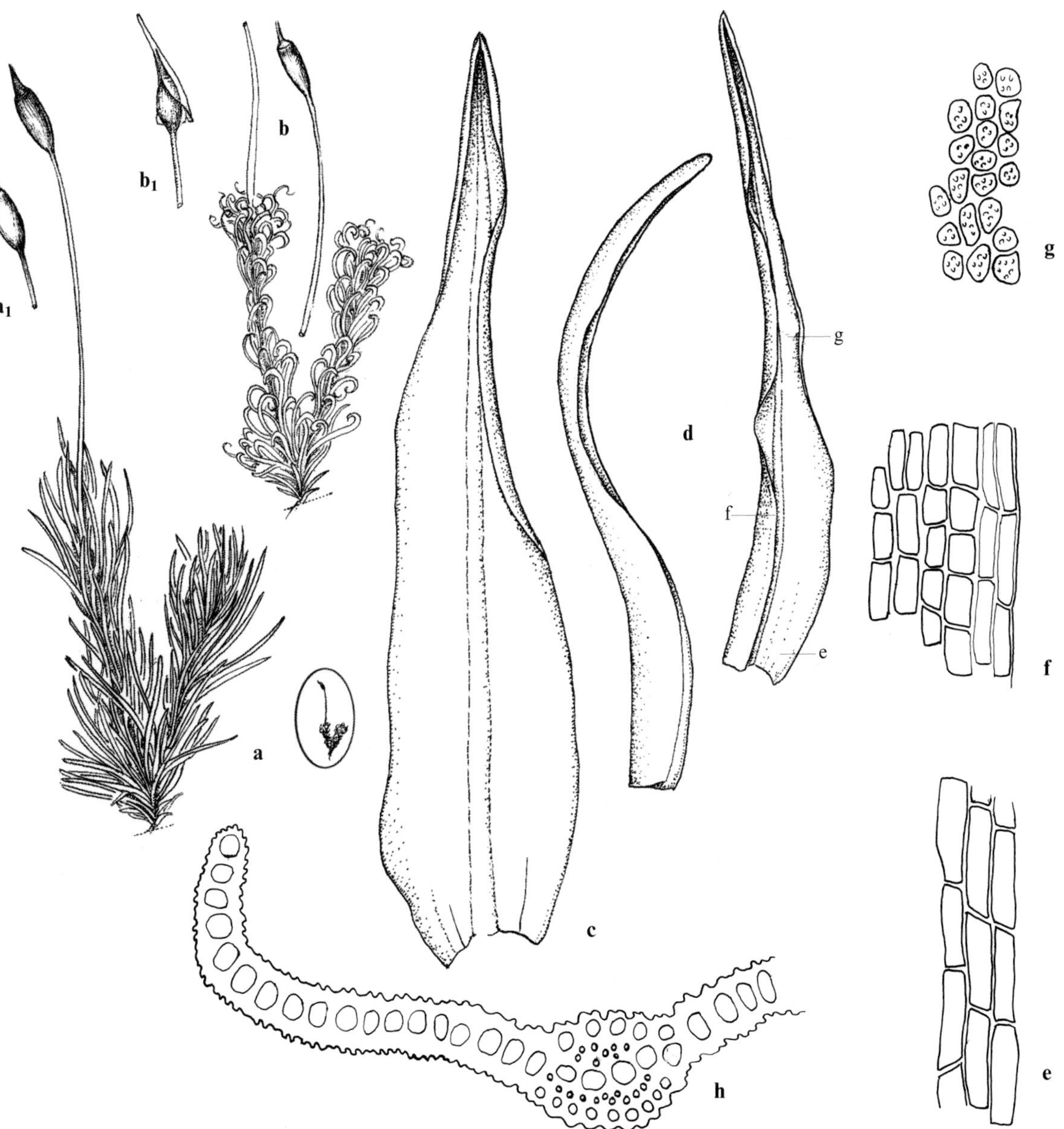

Figure 23. *Trichostomum crispulum*: **a** habit, moist (×10); $\mathbf{a_1}$ deoperculate capsule; **b** habit, dry (×10); $\mathbf{b_1}$ capsule with calyptra; **c**, **d** leaves (×50); **e** basal cells; **f** marginal cells below mid-leaf; **g** upper cells (all leaf cells ×500); **h** cross-section of costa and portion of lamina (×500). (**a**, **b**, **d**–**g**: *Kushnir*, 8 Apr. 1945; **c**, **h**: *Herrnstadt*, 14 May 1976)

5. *Oxystegus* (Limpr.) Hilp.

This small genus comprises, according to different treatments, up to ten species, which are closely related to *Weissia* and *Trichostomum*. Various species are treated in the literature as belonging to any of the three genera. The only species in the local flora considered here as belonging to *Oxystegus* — *O. tenuirostris* — is often included in *Trichostomum* (e.g., Dixon 1924, Nyholm 1956, Crum & Anderson 1981, Zander 1993). The delimitation of *Oxystegus* from *Trichostomum* does not seem to be clear-cut, and is often based on overlapping diagnostic characters. Brotherus (1924) distinguished between them by characters of the peristome: absence of the basal membrane and of cleft teeth in *Oxystegus*. Other authors (Smith 1978, Frahm 1995) use leaf characters, e.g., sinuose, notched to denticulate margins. Magill (1981) proposed the presence of a central strand as characteristic for *Oxystegus* but this character has not been found to be constant in local populations.

Oxystegus is treated here, though with some doubts, as a separate genus. The generic characters are included in the description of *O. tenuirostris*.

Oxystegus tenuirostris (Hook. & Taylor) A. J. E. Sm., J. Bryol. 9:393 (1977).
Weissia tenuirostris Hook. & Taylor, Muscol. Brit., Edition 2:83 (1827); *Trichostomum cylindricum* (Bruch ex Brid.) Müll. Hal., Syn. Musc. Frond. 1:586 (1849). — Bilewsky & Nachmony (1955); *O. cylindricus* (Bruch ex Brid.) Hilp. Beih. Bot. Centralbl. 50:620 (1933); *O. tenuirostris* (Bruch ex Brid.) Hilp. var. *gemmiparus* (Schimp.) R. H. Zander, Lindbergia 4:285 (1978).

Distribution map 24, Figure 24, Plate II: b.

Plants up to 1.5(–2) cm long, yellowish-green to green. *Rhizoids* reddish, papillose, sometimes with gemmae, comprising several protuberant cells. *Stems* without central strand. *Leaves* 3–6 mm long, often fragile, curled to crisped when dry, patent to spreading when moist from a plane, ± wider sheathing base, linear-lanceolate, acute; margins crenulate-papillose, plane, undulate at base, notched to denticulate above; costa strong, up to 300 μm wide below, whitish, shortly excurrent, with two stereid bands; basal cells rectangular, hyaline, differentiated across leaf, sometimes rising along margins, ± resembling those of *Tortella*; upper cells round-quadrate with bifid papillae; mid-leaf cells 7–9 μm wide. *Dioicous*. *Setae* yellowish, flexuose, up to 25 mm long; capsules brownish, up to 5 mm long including operculum, narrow-cylindrical, finely sulcate when empty; operculum ± obliquely long-rostrate; peristome without a basal membrane, teeth reddish, short, entire, straight. *Spores* weakly

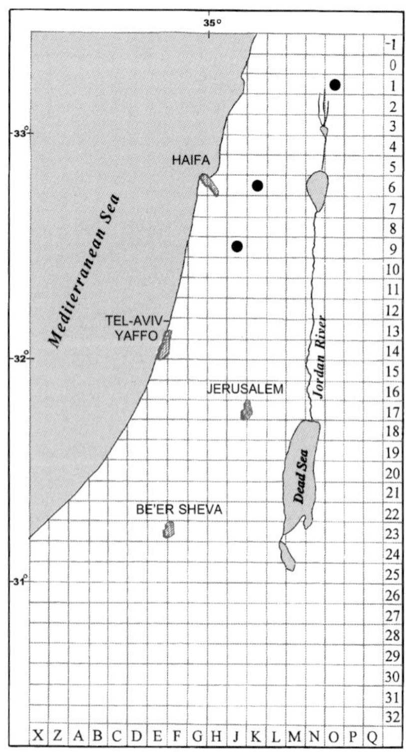

Map 24: *Oxystegus tenuirostris*

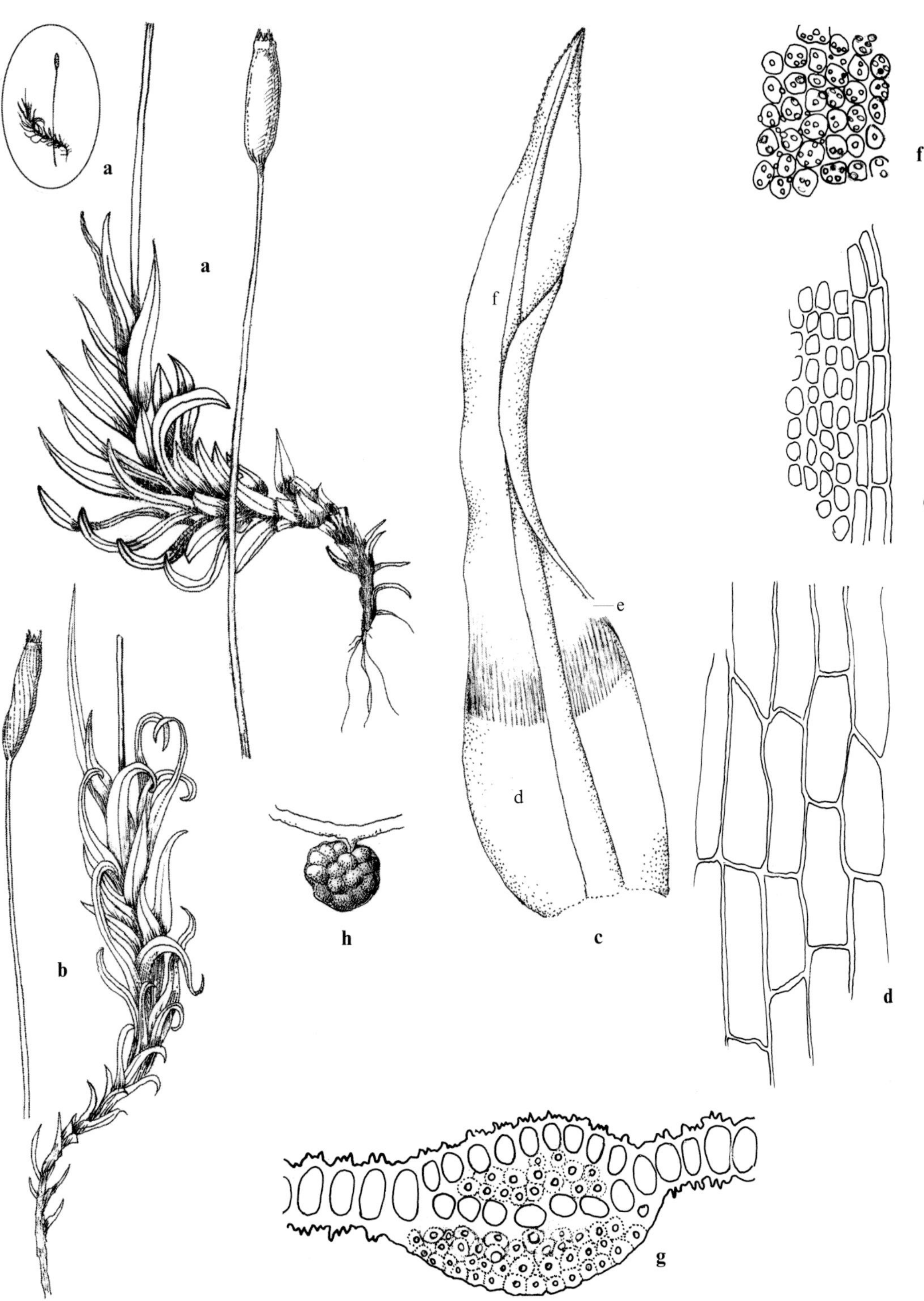

Figure 24. *Oxystegus tenuirostris*: **a** habit, moist (×10); **b** habit, dry (×10); **c** leaf (×50); **d** basal cells; **e** marginal cells below mid-leaf; **f** upper cells (all leaf cells ×500); **g** cross-section of costa and portion of lamina (×500); **h** rhizoidal gemma (×250). (*Bilewsky*, Apr. 1953, No. 230 in Herb. Bilewsky)

papillose, 11–13 μm in diameter; gemmate, processes often joined into small, irregular groups (seen with SEM).

Habitat and Local Distribution. — Plants growing in tufts on soil in limestone rock crevices, and in one locality near a waterfall. Rare: Lower Galilee, Samaria, and Golan Heights.

General Distribution. — Recorded from Southwest Asia (Turkey, Lebanon, Iran, and Yemen), around the Mediterranean, Europe (throughout), Macaronesia (in part), Africa (throughout), Asia (throughout), and northern (widespread), Central, and South America.

Literature. — Smith (1977).

When *Oxystegus tenuirostris* is subdivided into varieties (see Zander *in* Sharp *et al.* 1994), the local plants should perhaps be referred to as var. *gemmiparus* as they often bear rhizoidal gemmae and have narrower upper leaf cells than usually assumed for the typical *O. tenuirostris*.

6. *Tortella* (Lindb.) Limpr.

Plants from 2 mm to about 20 mm high (in local collections), green to yellowish green, brownish below, growing in loose or dense tufts. *Stems* erect, simple, or branched, often tomentose, sometimes with central strand (central strand usually present in all local species). *Leaves* curled and crisped when dry, erect to spreading when moist, often fragile and torn when old, lanceolate to linear-lanceolate with a wider hyaline base, obtuse, acute to long-acuminate, usually mucronate; margins plane or sometimes incurved, usually entire, more rarely crenulate-papillose or weakly denticulate in part of margin; costa strong, sometimes glossy, percurrent or excurrent (in all local species), cross section semicircular or ovate, always with ventral epidermis, sometimes without dorsal epidermis, with two stereid bands, one–two rows of guide cells; basal cells rectangular, often bulging, usually smooth, hyaline, ascending upwards along the margins, usually ± abruptly differentiated across the leaf in the shape of a V; mid- and upper leaf cells round-quadrate to hexagonal, distinctly smaller than basal cells with bifid or simple papillae; perichaetial leaves little differentiated. *Dioicous* usually (among local species only *T. humilis* autoicous). *Setae* straight to slightly curved, 10–30 mm long (in local plants); capsules straight, symmetrical, sometimes slightly asymmetrical, erect, cylindrical, elliptical to ovoid-cylindrical; operculum long-conical to rostrate; peristome of 32 filiform, articulated divisions, straight or one–three times twisted, readily broken; basal membrane absent or low. *Calyptrae* cucullate, smooth. *Spores* papillose or smooth.

A genus comprising 53 species (Zander 1993) with worldwide distribution. Six species occur in the local flora.

The delimitation between species in the literature is partly due to the posture of leaves and the extent and nature of their twisting in dry plants. The different habit descriptions of some of the species by different authors (cf. Mönkemeyer 1927, Smith 1978, and Zander *in* Sharp *et al.* 1994) may be related to the relative age of the plants and local habitat conditions. As observed by us in the local flora, the time and pattern of desiccation may greatly influence the habit of the ensuing dry plants. See the discussion of *Timmiella barbuloides* (below, p. 106) on the resemblance in general habit between species of *Tortella* and *Timmiella*.

Literature. — Crundwell & Nyholm (1962).

1 Plants small, 2–6 mm high; leaves usually with inflexed to strongly incurved margins in upper part and distinctly inflexed apex **4. T. inflexa**

1 Plants larger, usually more than 6 mm, up to 20 mm high; leaves usually with plane to weakly incurved margins in upper part, apex not or sometimes very slightly inflexed.

2 Leaves long-acuminate with tips spirally curled when dry, flexuose when moist; margins undulate **6. T. tortuosa**

2 Leaves obtuse, acute or short-acuminate with tips only ± curled when dry, not flexuose when moist; margins only rarely slightly undulate.

3 Leaves subcucullate; mid-leaf cells up to 11 μm wide.

4 Plants up to 20 mm high; terminal shoots readily detached; cells on ventral side of costa elongate, differing from adjacent cells; plants growing on thin soil layer on exposed limestone rocks; plants without sporophytes* **3. T. inclinata**

4 Plants smaller, up to 15 mm high; terminal shoots not readily detached; cells on ventral side of costa not differing from adjacent cells; plants growing on sandy soils; sporophytes frequent **1. T. flavovirens**

3 Leaves not subcucullate; mid-leaf cells small, up to 8 μm wide.

5 Leaves readily torn with terminal part often missing; transition from hyaline base to chlorophyllous part not distinctly abrupt; recorded as dioicous; no sporophytes seen in local plants **5. T. nitida**

5 Leaves not readily torn as above; transition from hyaline base to chlorophyllous part distinctly abrupt; autoicous; sporophytes seen in local plants **2. T. humilis**

* A single population has been found — Karme Yosef.

1. Tortella flavovirens (Bruch) Broth., Nat. Pflanzenfam. 1 : 397 (1902). [Bilewsky (1965) : 364]

Trichostomum flavovirens Bruch *in* F. A. Müll. Flora 12 : 404 (1829).

Distribution map 25, Figure 25.

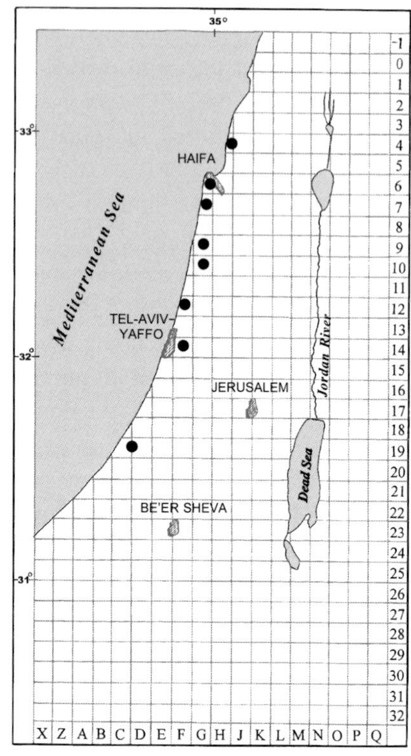

Map 25: *Tortella flavovirens*

Plants up to 10(–15) mm high, yellowish brown-green; annual growth clearly discernible. *Stem* with weak central strand, more clearly discernible in older parts. *Leaves* crisped, contorted and incurved with tips weakly curled when dry, spreading when moist, 2–3 mm long, narrow lanceolate from a wider base, concave, acute to sub-obtuse, subcucullate; leaves sometimes torn, in particular along margins; margins plane below, incurved in upper part, often inflexed at apex; costa excurrent in a short mucro with cells on ventral surface similar to adjacent cells (sometimes few elongated cells towards leaf apex); transition from large hyaline base (¼–⅓ length of leaf) to upper chlorophyllous part abrupt; upper cells round-quadrate, densely papillose, mid-leaf cells 8–11 µm wide. *Sporophytes* frequently borne. *Setae* yellowish to reddish brown; capsules up to 3.5 mm long including operculum, ellipsoid to cylindrical, reddish at mouth, with operculum up to ⅓ length of capsule; peristome teeth nearly straight to once twisted, fugacious; annulus absent. *Spores* 13–14 µm in diameter, coarsely papillose; irregularly verrucose (verrucae larger than in *Tortella inflexa*; seen with SEM).

Habitat and Local Distribution. — Plants growing in somewhat loose tufts on sand and sandy loam along the sea coast. Occasional, locally abundant: Coastal Galilee, Coast of Carmel, Sharon Plain, and Philistean Plain.

General Distribution. — Recorded from Southwest Asia (Cyprus, Turkey, and Lebanon), around the Mediterranean, Europe (throughout), Macaronesia, and North Africa.

Tortella flavovirens is characterised by the yellowish brown-green colour of the tufts and the subcucullate leaves with a large hyaline base abruptly delimited from the chlorophyllous part. It is the only local *Tortella* species of the sandy seacoast.

Several populations were found growing in different parts of Mount Carmel on calcareous soil (rendzina), not on the typical sandy soil. The plants resemble *Tortella flavovirens* in habit and the frequent production of sporophytes. However, their colour is more olive-green, and they tend more often to have extensively torn leaves with shorter and less widened hyaline base (from which the transition to the chlorophyllous part seems less abrupt). At present, it seems preferable not to include these plants in any taxon until more detailed studies are available.

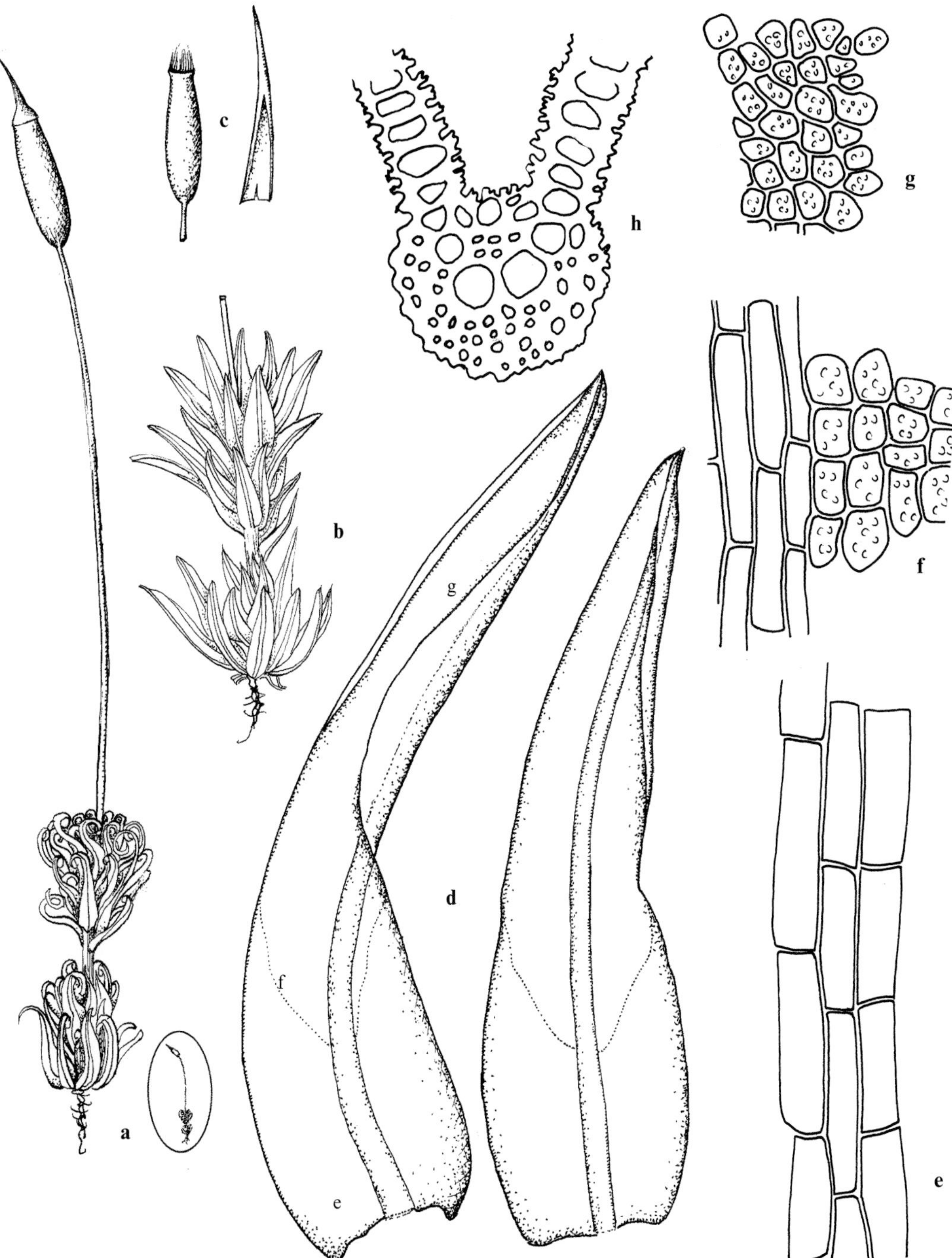

Figure 25. *Tortella flavovirens*: **a** habit, dry (×10); **b** habit, moist (×10); **c** deoperculate capsule and calyptra (×10); **d** leaves (×50); **e** basal cells; **f** cells at the transition from hyaline base to chlorophyllous part; **g** upper cells (all leaf cells ×500); **h** cross-section of costa and portion of lamina (×500). (*Danin*, 7 Dec. 1978)

2. Tortella humilis (Hedw.) Jenn., Man. Mosses W. Pennsylvania 96 (1913).
Barbula humilis Hedw., Sp. Musc. Frond. 116 (1801); *Tortella caespitosa* (Schwägr.) Limpr., *Laubm. Deutschl.* 1 : 600 (1888). — Bilewsky (1965) : 365.

Distribution map 26, Figure 26.

Plants, 7–10 mm high, yellowish-green to olive, sometimes with a persistent protonema. *Leaves* crisped, contorted, and incurved, with tips weakly curled when dry, spreading when moist, 2–3.5(4.5) mm, lanceolate to oblong-lanceolate, concave, acute; margins plane, sometimes slightly undulate (in one collection — Mount Meron, *Herrnstadt & Crosby 78-48-8* — margins denticulate at border of ascending hyaline base of leaf; Figure 26h); costa excurrent in a short mucro with elongated cells on ventral surface at upper ¼ of leaf, a single row usually at mid-leaf; transition from large hyaline base (¼–⅓ length of leaf) to upper chlorophyllous part abrupt; upper cells round-quadrate, densely papillose, mid-leaf cells small, 5.5–7 μm wide. *Autoicous*, antheridia in readily detached axillary perigonial buds. *Setae* yellowish to brown; capsules up to 2 mm long without operculum, cylindrical (in local collections, only very young sporophytes or capsules from previous seasons without operculum and with broken peristome could be observed).

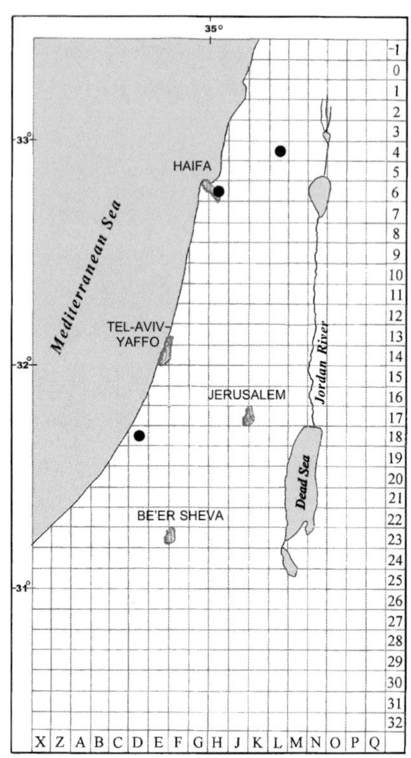

Map 26: *Tortella humilis*

Habitat and Local Distribution. — Plants growing in tufts on humus and soil layers on rocks and walls. Rare: Philistean Plain, Upper Galilee, and Mount Carmel.

General Distribution. — Recorded from Southwest Asia (Cyprus, Turkey, Lebanon, Iran, Saudi Arabia, and Yemen), around the Mediterranean, Europe (mainly the south), Africa (throughout), eastern Asia, Oceania, and America (throughout).

Plants of *Tortella humilis* are characterised by the lanceolate, not cucullate, acute leaves with small mid-leaf cells and elongated cells on upper part of ventral surface of costa. It is the only autoicous local *Tortella* species.

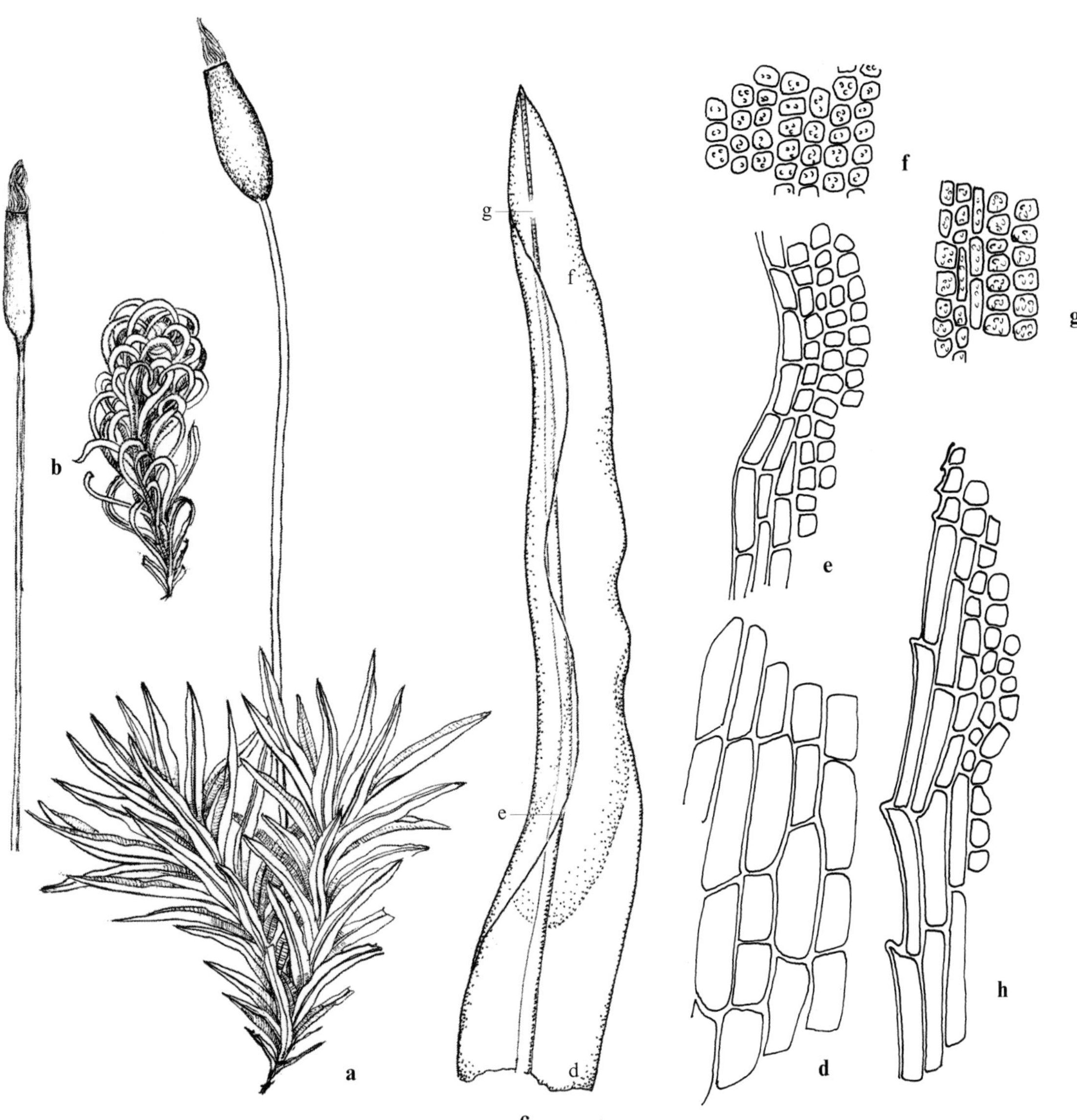

Figure 26. *Tortella humilis*: **a** habit, moist (×10); **b** habit, dry (×10); **c** leaf (×50); **d** basal cells; **e** plane, slightly undulate marginal cells below mid-leaf; **f** upper cells; **g** cells on ventral surface of upper part of costa; **h** denticulate marginal cells below mid-leaf (all leaf cells ×500).
(**a**–**g**: *Bilewsky*, Feb. 1955, No. 26 in Herb. Bilewsky; **h**: *Herrnstadt* & *Crosby 78-48-8*)

3. Tortella inclinata (Hedw. f.) Limpr., Laubm. Deutschl. 1 : 602 (1888).
Tortula inclinata Hedw. f. *in* F. Weber & D. Mohr, Beitr. Naturk. 123 (1806).

Distribution map 27, Figure 27.

Plants rather robust, up to 20 mm high, green to olive; readily detached, annual growth 5–10 mm. *Leaves* twisted and incurved with weakly curled tips when dry, patent to spreading when moist, 2–3.5 mm long, oblong-lanceolate, concave, acute to nearly obtuse, often subcucullate; margins plane or slightly undulate; costa strong, ± glossy at back, excurrent in a short mucro, with elongated cells on ventral surface differing in shape from adjacent cells; transition from hyaline base (ca. ¼ length of leaf) to upper chlorophyllous part abrupt; upper cells irregularly hexagonal to round-quadrate, densely papillose, in mid-leaf 8–11 µm wide. *Sporophytes* not found in the only collection of this species (*Herrnstadt*, 3 Feb. 1990).

Habitat and Local Distribution. — Plants growing on thin soil layer on exposed limestone rocks. Shefela (Karme Yosef).

General Distribution. — Recorded from Southwest Asia (Turkey, Iran, and Saudi Arabia), around the Mediterranean, Europe (throughout), Macaronesia, North Africa, northeastern and central Asia, Australia, and northern and northwestern South America.

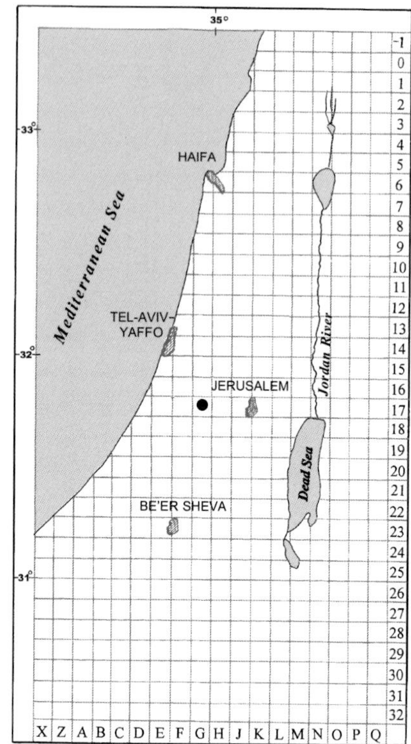

Map 27: *Tortella inclinata*

Figure 27. *Tortella inclinata*: **a** habit, moist (×10); **b** habit, dry (×10); **c** detached shoot apex (×10); **d** leaves (×50); **e** basal cells; **f** cells at the transiton from hyaline base to chlorophylous part; **g** cells on ventral surface of costa; **h** mid-leaf cells (all leaf cells ×500). (*Herrnstadt*, 3 Feb. 1990)

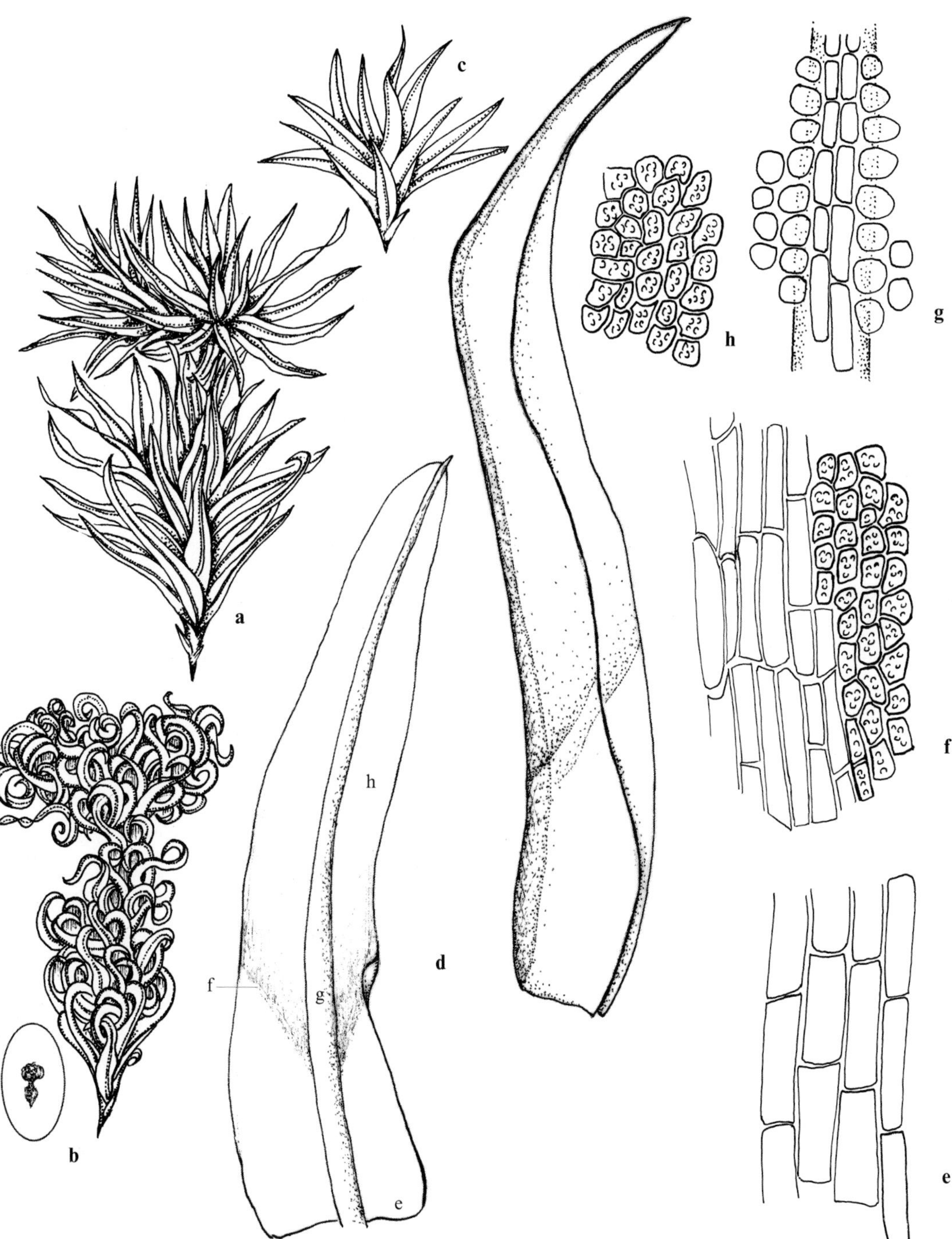
a
b
c
d
e
f
g
h
f
g
h
e

4. Tortella inflexa (Bruch) Broth., Nat. Pflanzenfam. 1:397 (1902).
[Bilewsky (1965):364; Rabinovitz-Sereni (1931)]
Trichostomum inflexum Bruch *in* F. A. Müll. Flora 12:402 (1829).

Distribution map 28, Figure 28.

Plants small, 2–6 mm high, yellowish-green. *Leaves* curled when dry, erect to patent when moist, 1–2(–3) mm long, basal leaves lanceolate, upper leaves linear-lanceolate, with a ± wider base, usually acute, ending in hyaline mucro, subcucullate with inflexed apex; margins plane below, inflexed to strongly incurved above; costa percurrent (upper part not easily observed), cells on ventral surface similar to adjacent cells; basal cells incrassate; transition from hyaline base to upper chlorophyllous part gradual; upper cells quadrate, densely papillose, obscure, in mid-leaf 6–9 μm wide. *Setae* straight to slightly curved, yellowish brown; capsules up to 2.25 mm long including operculum, ovoid to ellipsoid; operculum ca. ⅓ length of capsule, long-rostrate; annulus absent; peristome teeth about once twisted. *Spores* 8–9 μm in diameter, finely papillose; densely verrucose; verrucae small, all equal in height (seen with SEM).

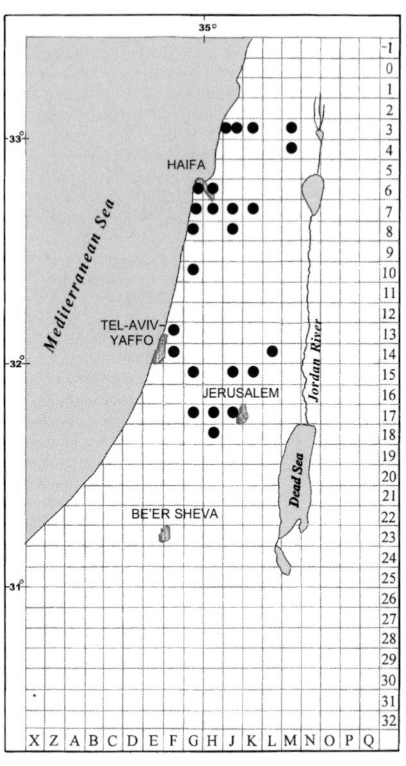

Map 28: *Tortella inflexa*

Habitat and Local Distribution. — Plants growing in dense patches on thin soil layers on limestone rocks or chalk or in rock crevices, exposed or shaded, often facing north. Common: Coastal Galilee, Coast of Carmel, Sharon Plain, Upper Galilee, Lower Galilee, Mount Carmel, Esdraelon Plain, Samaria, Shefela, and Judean Mountains.

General Distribution. — Recorded from few localities: Southwest Asia (Cyprus and Lebanon), seems to occur sporadically in some parts of the Mediterranean Region, Europe (mainly in the south), North Africa, and eastern Asia.

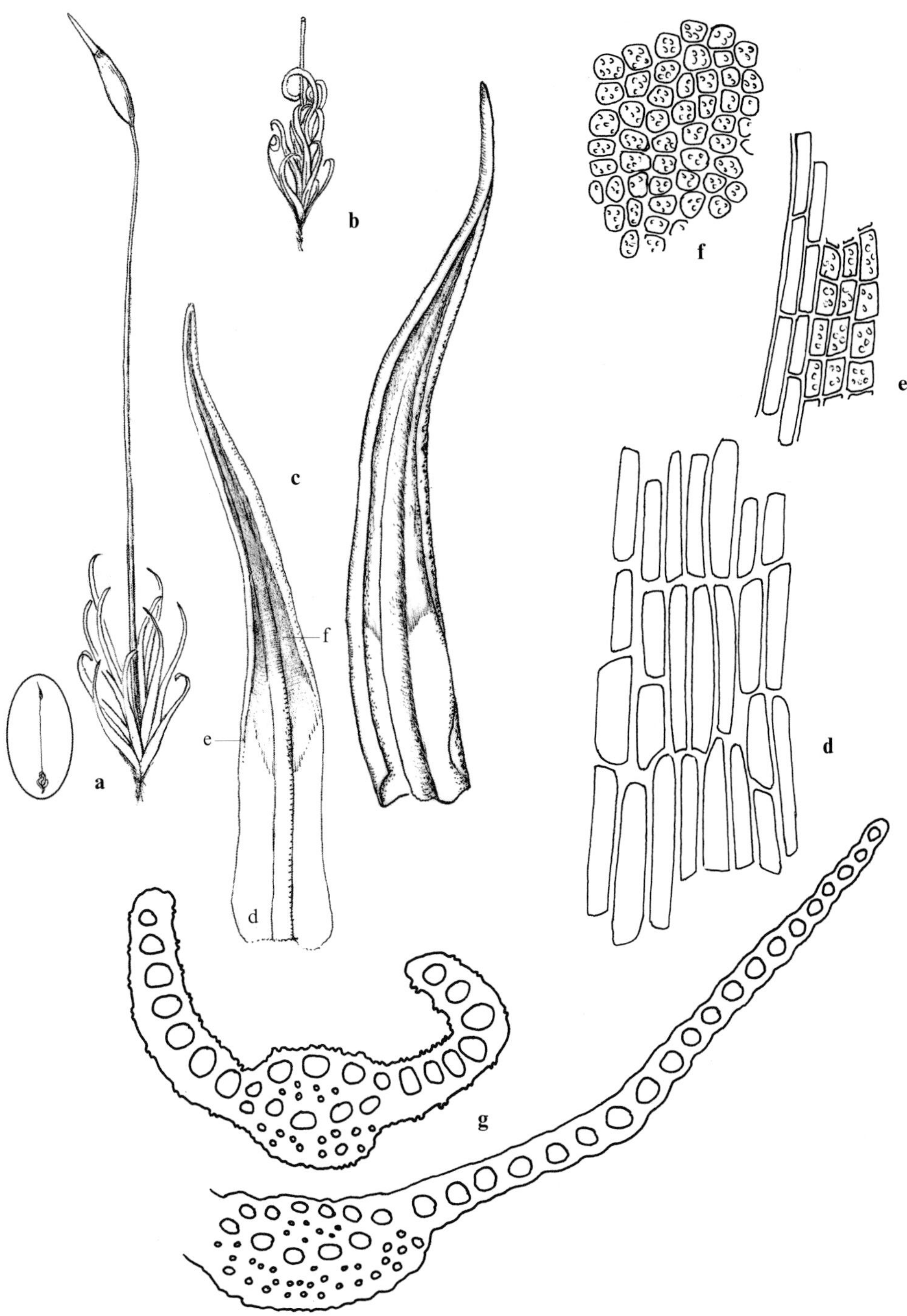

Figure 28. *Tortella inflexa*: **a** habit, moist (×10); **b** habit, dry (×10); **c** leaves (×50); **d** basal cells; **e** marginal cells below mid-leaf; **f** upper cells (all leaf cells ×500); **g** cross-sections of leaves (×500). (*D. Zohary*, 16 Mar. 1951)

5. Tortella nitida (Lindb.) Broth., Nat. Pflanzenfam. 1 : 397 (1902).
[Bilewsky (1965) : 365]
Tortula nitida Lindb., Öfvers. Förh. Kongl. Svenska Vetensk.-Akad. 21 : 252 (1864).

Distribution map 29, Figure 29.

Plants up to 15 mm long, green to olive. *Leaves* tightly incurved and twisted with weakly curled tips when dry, erectopatent to spreading when moist, 2–4(–5) mm long, linear-lanceolate, acute to ± obtuse, readily torn, in particular uppermost part of leaf often missing; margins plane and ± undulate, ± incurved above; costa strong, yellowish to ± brown, glossy at back, excurrent in short mucro; cells on ventral surface similar to adjacent cells; basal cells thin-walled; transition from short hyaline base to upper chlorophyllous part ± gradual, not very distinctly abrupt; upper cells round-quadrate, ± incrassate, densely papillose, in mid-leaf 6.5–8 μm wide. *Sporophytes* not found in local plants.

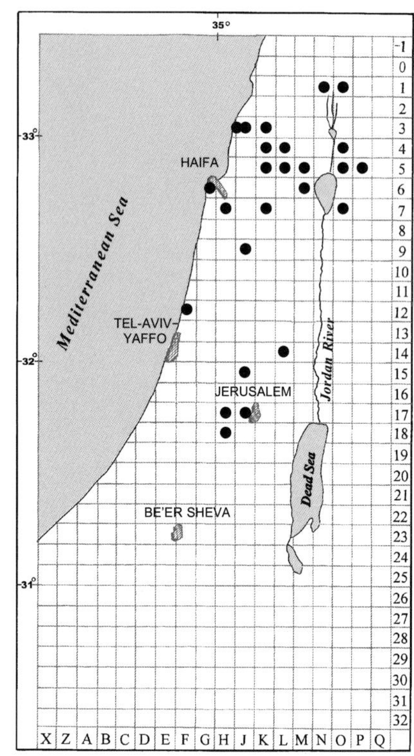

Map 29: *Tortella nitida*

Habitat and Local Distribution. — Plants growing in dense round tufts on soil on calcareous and basaltic rocks or in rock crevices, often facing north, in shade of trees. Common: Coastal Galilee, Sharon Plain, Upper Galilee, Lower Galilee, Mount Carmel, Samaria, Judean Mountains, Dan Valley and Golan Heights.

General Distribution. — Recorded from Southwest Asia (Cyprus, Turkey, Lebanon, and Jordan), around the Mediterranean, Europe (throughout), Macaronesia, and North Africa.

Tortella nitida can be confused with *T. flavovirens*, in particular when the latter has partly torn leaves. However, *T. flavovirens* has subcucullate leaf apices, larger mid-leaf cells, and the transition from the wider and larger hyaline base to the upper part is more distinctly abrupt. In addition, the torn parts of the leaves do not usually include the apex. Plants of *T. flavovirens* seem to be confined to sandy soils.

Dry plants of *Tortella nitida* have a superficial resemblance to *Timmiella* in the glossy costa and the incurved twisted leaves. However, they may be easily distinguished by the less robust stature and the often torn leaves, i.e., the apices (see also the discussion of *Timmiella*).

Figure 29. *Tortella nitida*: **a** habit, dry (×10); **b** habit, moist (×10); **c** leaves (×50); **d** basal cells; **e** marginal cells at the transition from hyaline base to chlorophyllous part; **f** upper cells (all leaf cells ×500); **g** cross-section of costa and portion of lamina (×500). (*Herrnstadt*, 28 Feb. 1982)

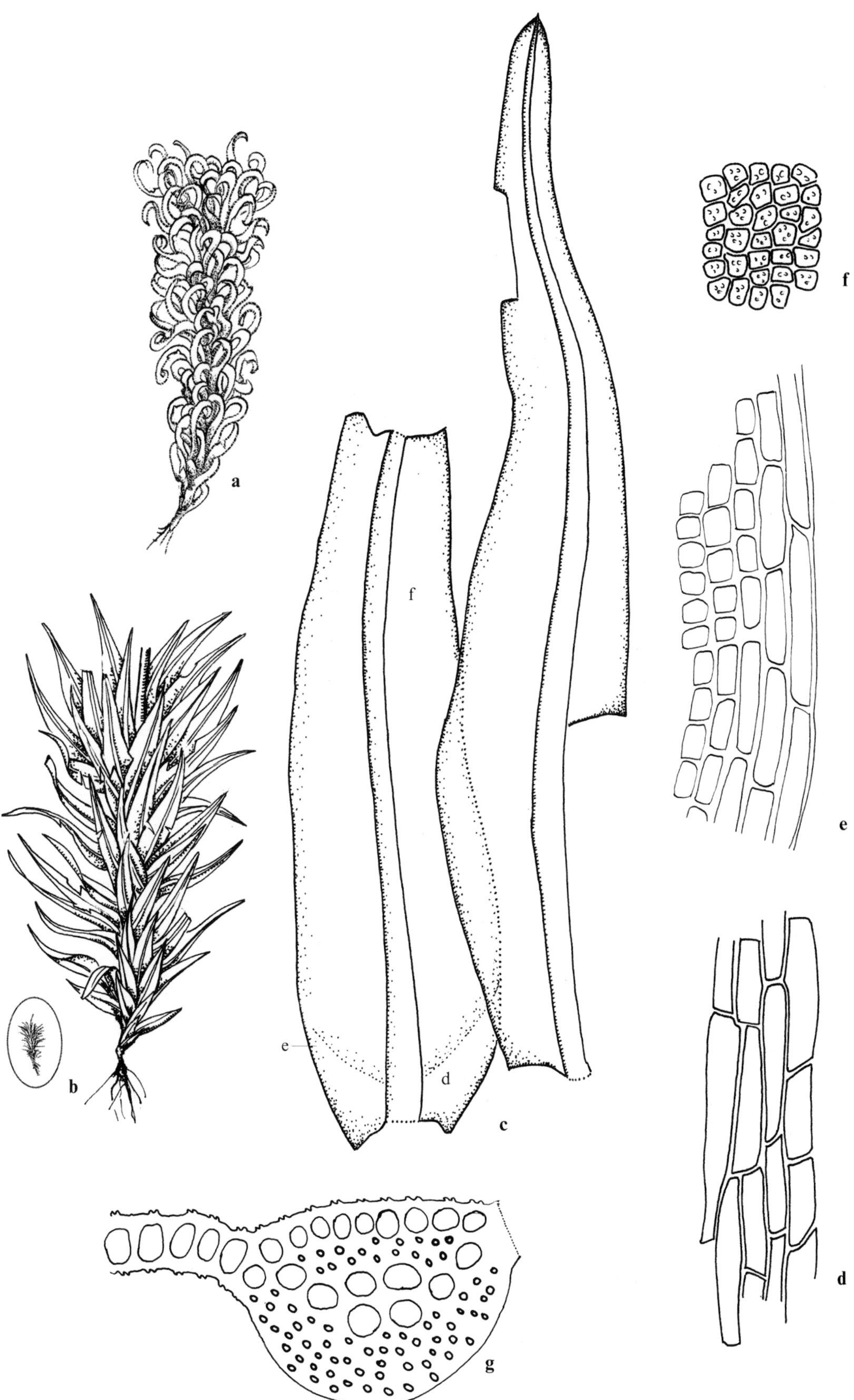
a
b
c
d
e
f
g
f
e
d

6. Tortella tortuosa (Hedw.) Limpr., Laubm. Deutschl. 1:604 (1888). [Nachmony (1961)]

Tortula tortuosa Hedw., Sp. Musc. Frond. 124 (1801).

Distribution map 30, Figure 30.

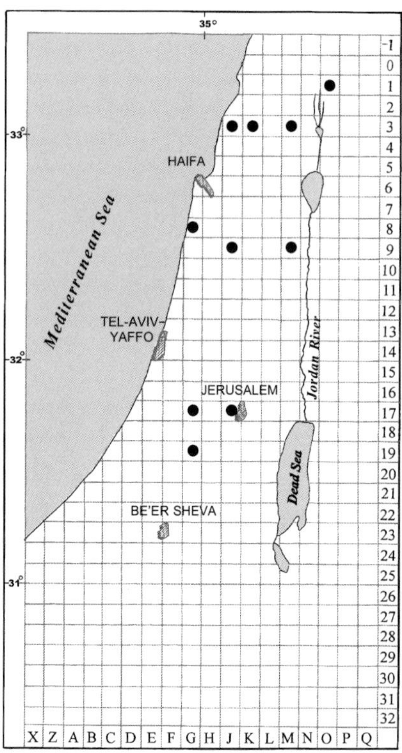

Map 30: *Tortella tortuosa*

Plants (8–)10–20 mm high, green to yellowish-green. *Stems* sometimes branched, with or without a central strand. *Leaves* strongly contorted and curled with spirally curled tips when dry, patent to spreading and flexuose when moist, ± laxly spaced, up to 5(–6) mm long, linear-lanceolate, gradually long-acuminate from a wider base, concave, some leaves torn along margins; margins plane, undulate; costa often excurrent in a long mucro; cells on ventral surface, at least in mid-leaf, similar to adjacent cells; hyaline base varying in size from ¼–⅓ length of leaf, transition to upper chlorophyllous part abrupt to various degrees; cells in mid-leaf 7–11 μm wide. *Sporophytes* not found in local populations.

Habitat and Local Distribution. — Plants growing in tufts on thin soil layers on calcareous rocks, usually in partly shaded, mainly montane localities. Occasional: Upper Galilee, Mount Carmel, Mount Gilboa, Samaria, Shefela, Judean Mountains, and Mount Hermon.

General Distribution. — Recorded from Southwest Asia (Cyprus, Turkey, Iraq, and Iran), around the Mediterranean, Europe (widespread throughout), Macaronesia, North Africa, Asia (except the southeast), and North, Central, and South America; a widespread species in the Northern Hemisphere.

Tortella tortuosa is distinguished from other local species of the genus by the long-acuminate leaves, which are strongly contorted and curled when dry and flexuose when moist. In the local flora, there is a wide range of inter-populational variation in the size of plants, leaves, relative size of the hyaline leaf part, and the degree of abruptness of the transition between the hyaline and chlorophyllous part of the leaves.

Figure 30. *Tortella tortuosa*: **a** habit, moist (×10); **b** habit, dry (×10); **c** leaves (×50); **d** basal cells; **e** marginal cells below mid-leaf; **f** mid-leaf cells (all leaf cells ×500); **g** cross-section of costa and portions of lamina (×500). (*Nachmony*, 7 Jan. 1954)

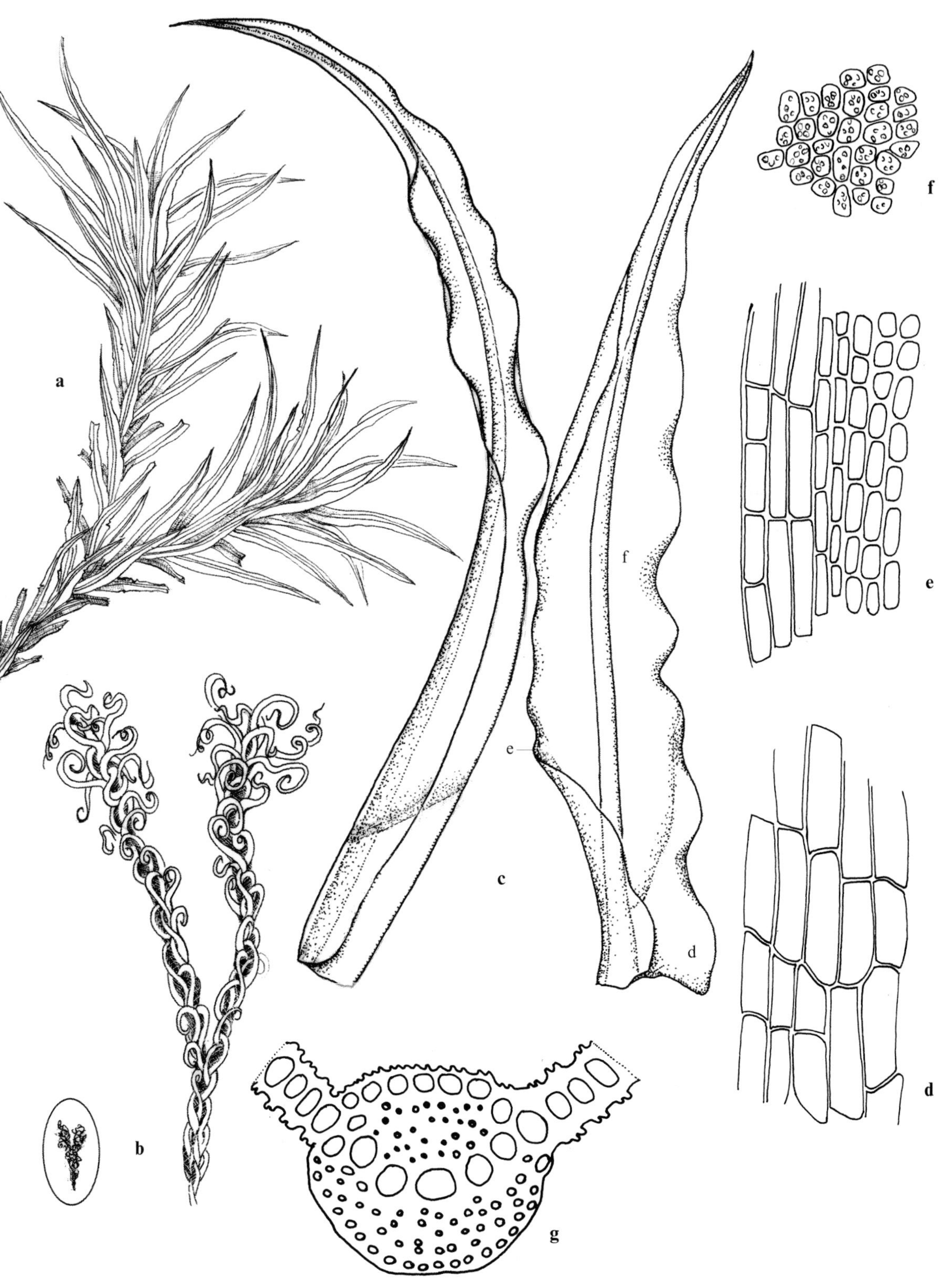
a
b
c
d
e
f
g
e
f
d
f
e
d

7. *Pleurochaete* Lindb.

Plants fairly robust. *Stems* with weak central strand of few thin-walled cells. *Leaves* crisped and twisted when dry, squarrose-recurved from a sheathing base when moist; margins plane, sometimes recurved to revolute along leaf base, denticulate, at least in upper part; costa percurrent to short-excurrent with two stereid bands; hyaline border of four–six rows of thin-walled, smooth, rectangular cells extends from leaf base to beyond shoulders; upper cells papillose. *Dioicous*, gametangia lateral. *Capsules* oblong-cylindrical; annulus persistent; peristome with 32 filiform, papillose, ± twisted teeth; operculum long-conical. *Calyptrae* cucullate.

A small genus comprising four species (Zander 1993). *Pleurochaete squarrosa* — the only local species — is the most widespread species of the genus.

Sterile plants of *Pleurochaete* may superficially resemble *Tortella*, but are more robust and can be distinguished by the leaves with denticulate margins and thin-walled marginal cells.

Pleurochaete squarrosa (Brid.) Lindb., Öfvers. Förh. Kongl. Svenska Vetensk.-Akad. 21 : 253 (1864).

Barbula squarrosa Brid., Bryol. Univ. 1 : 833 (1827).

Distribution map 31, Figure 31.

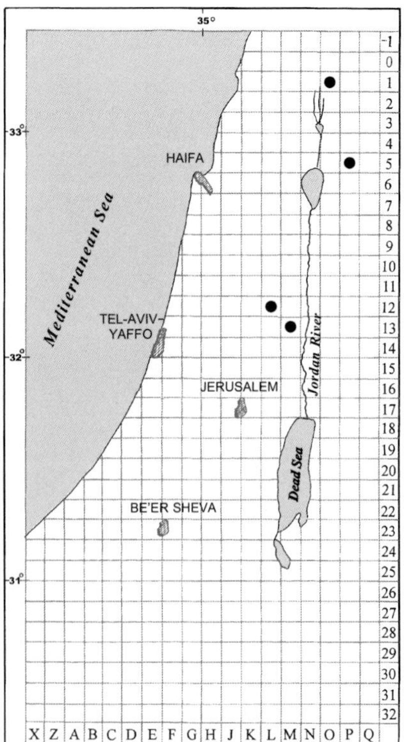

Map 31: *Pleurochaete squarrosa*

Plants large, up to 6 cm high, yellowish-green. *Stems* irregularly branched above. *Leaves* channelled, 3–4 mm long, lanceolate or oblong-lanceolate with an ovate base, acuminate with acute apex; margins plane to slightly undulate, entire at base, crenulate to irregularly denticulate above; costa percurrent to shortly excurrent; basal cells rectangular, ± incrassate; mid-leaf cells subquadrate, 8–10 µm wide with several C-shaped, bifid papillae per cell. *Sporophytes* not found in local plants so far.

Habitat and Local Distribution. — Plants growing mostly in lax tufts on shaded calcareous and basalt soil, on banks of wadis and streams or near waterfalls. Rare to occasional: Samaria and Golan Heights.

General Distribution. — Recorded from Southwest Asia (Cyprus, Turkey, Syria, Iraq, Iran, Saudi Arabia, and Yemen), around the Mediterranean, Europe (mainly the south), Macaronesia, North and central Africa, north-eastern, eastern, and central Asia, North and Central America; also recorded by Zander (*in* Sharp *et al.* 1994) from the Caribbean and northwestern South America.

Figure 31. *Pleurochaete squarrosa*: **a** habit, dry (×10); **b** stem, moist (×10); **c** leaves (×50); **d** marginal cells below mid-leaf; **e** marginal upper cells (all leaf cells ×500); **f** cross-section of costa and portion of lamina (×250). (*Herrnstadt* & *Crosby 79-77-2*)

Records of distribution vary according to the species concept of the authors. The species concept in its wider sense includes *Pleurochaete luteola* (Besch.) Thér. from South America. *Pleurochaete malacophylla* (Müll. Hal.) Broth. described from the Red Sea region (apparently from Saudi Arabia) is considered by Frey (pers. comm.) as comprising large plants of *P. squarrosa*.

8. *Timmiella* (De Not.) Limpr.

Plants robust, comose, green to dark green above, brownish below, growing in tufts or cushions. *Stems* erect, simple, or branched, tomentose with a strong central strand. *Leaves* incurved and contorted, crisped when dry, spreading when moist, long-lanceolate, elliptical to ligulate, acute; margins plane or weakly incurved, irregularly dentate, especially in upper part, bistratose, costa very wide, especially below middle, ending below apex or percurrent; in cross section flattened or elliptical, with bi(–tri)-stratose ventral and unistratose dorsal epidermis and two stereid bands; guide cells in one row with hydroids on one or both sides; all lamina cells without papillae; basal cells unistratose, thin-walled, rectangular, hyaline, clearly delimited from upper part of leaf by a nearly straight line; upper cells bistratose except at margins, thick-walled, quadrate to rounded-hexagonal, bulging ventrally, nearly flat dorsally, without papillae; towards costa one layer of cells not directly situated over the other; perichaetial leaves hardly differentiated. *Dioicous* or monoicous. *Setae* long, straight, or slightly bent; capsules straight, symmetrical, sometimes slightly asymmetrical, erect, long-elliptical to long-cylindrical; operculum long-conical to rostrate; annulus weakly or distinctly differentiated, revoluble or not; peristome of 32 linear-lanceolate to filiform teeth, straight or slightly twisted, basal membrane absent or low. *Calyptrae* cucullate, smooth. *Spores* weakly papillose.

A genus comprising 13 species (*sensu* Zander 1993). One (or two) species occur(s) in the local flora. *Timmiella* is considered by Zander (1993) as the single genus of a separate subfamily of the Pottiaceae — *Timmielloideae* — with the unique character of bistratose laminal cells not situated directly over one another near the costa.

Timmiella barbuloides (Brid.) Mönk., Laubm. Eur. 273 (1927).
[Bilewsky & Nachmony (1955)]

Trichostomum barbuloides Brid., Muscol. Recent., Suppl. 1 : 233 (1806); *Timmiella barbula* Limpr., Laubm. Deutschl. 1 : 594 (1888). — Bilewsky (1965) : 364.

Distribution map 32 (p. 106), Figure 32, Plate II : c.

Plants up to 20 mm high, bright green to brownish-green, brown below. *Stems* weakly tomentose below; rhizoids dark red-brown, smooth. *Leaves* strongly crisped when dry, up to 5 mm long, long-lanceolate to ligulate, with a ± narrower sheathing base, slightly undulate in lower part; margins weakly incurved; costa wide, in lower part up to ⅓ width of lamina, shiny and yellowish in dry plants, stereids sometimes between rows of ventral epidermis, hydroids on both sides of guide cells; upper cells in mid-leaf 9–13 µm wide. *Monoicous* (apparently paroicous in local plants). *Setae* nearly straight, yellowish red; capsules ± symmetrical, 5–6 mm long with operculum; operculum over ⅓ of capsule; peristome teeth filiform, papillose, nearly straight; no revoluble annulus, mouth of theca with small reddish cells. *Spores* 13–16 µm in

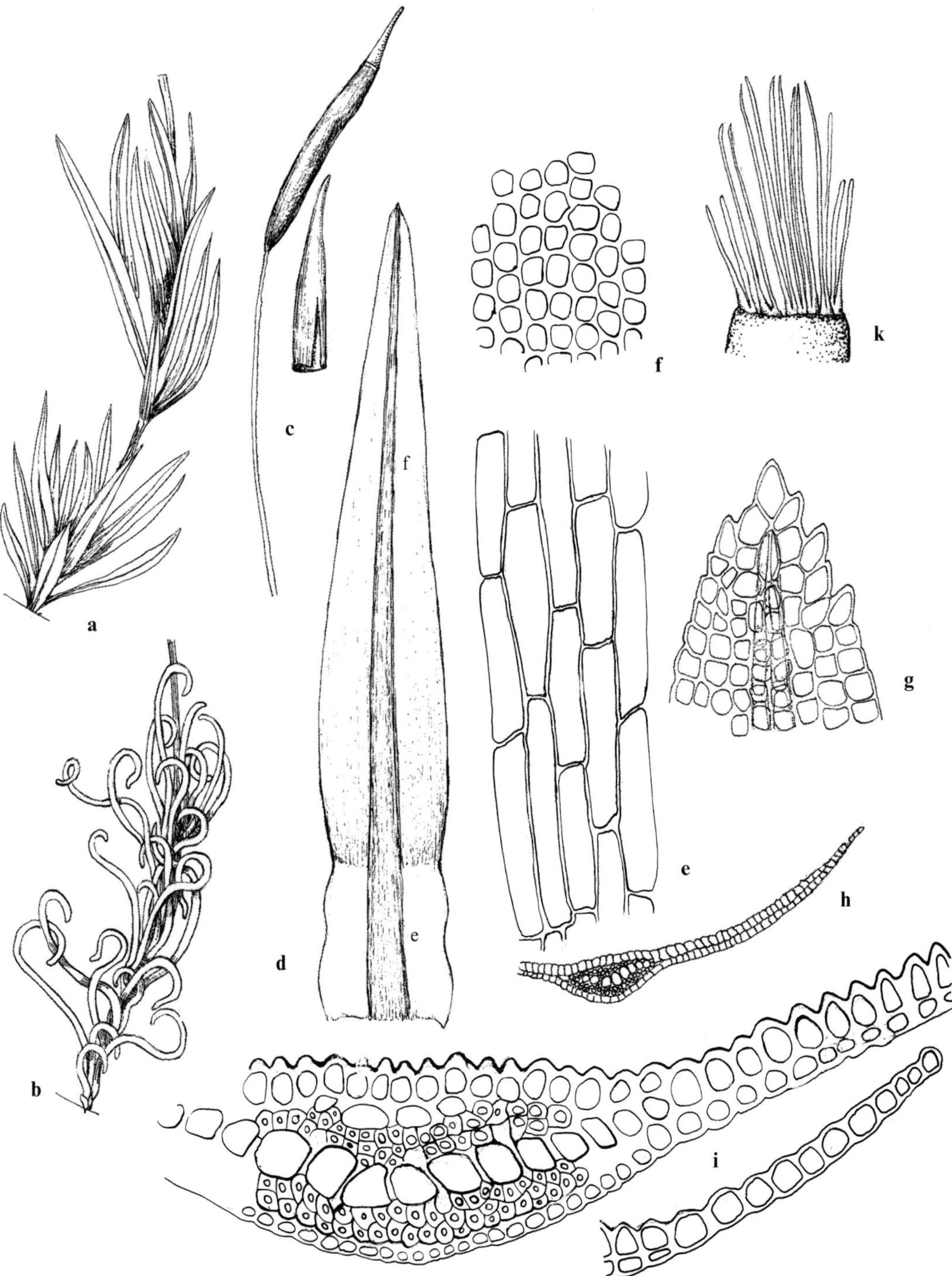

Figure 32. *Timmiella barbuloides*: **a** habit, moist (×10); **b** habit, dry (×10); **c** sporophyte and calyptra (×10); **d** leaf (×50); **e** basal cells; **f** upper cells; **g** apical cells (all leaf cells ×500); **h** cross-section of costa and portion of lamina (×200); **i** cross-section of costa and portion of lamina (×500); **k** peristome (×50). (*Markus & Kutiel 77-582-1*)

diameter, weakly papillose; irregularly gemmate, processes with rough surface (seen with SEM).

Habitat and Local Distribution. — Plants growing in loose tufts on calcareous soils, shaded rocks and in rock crevices. Very common: Coastal Galilee, Sharon Plain, Upper Galilee, Lower Galilee, Mount Carmel, Esdraelon Plain, Mount Gilboa, Samaria, Shefela, Judean Mountains, Judean Desert, Central Negev, Hula Plain, Bet She'an Valley, and Golan Heights.

General Distribution. — Recorded from throughout Southwest Asia (Cyprus, Lebanon, Jordan, Turkey, Iraq, Iran, Afghanistan, Saudi Arabia, Oman, and the Arab Emirates), around the Mediterranean, Europe (mainly the south), Macaronesia, tropical and North Africa, northeastern and eastern Asia, and South America.

Literature. — Castaldo-Cobianchi *et al.* (1982).

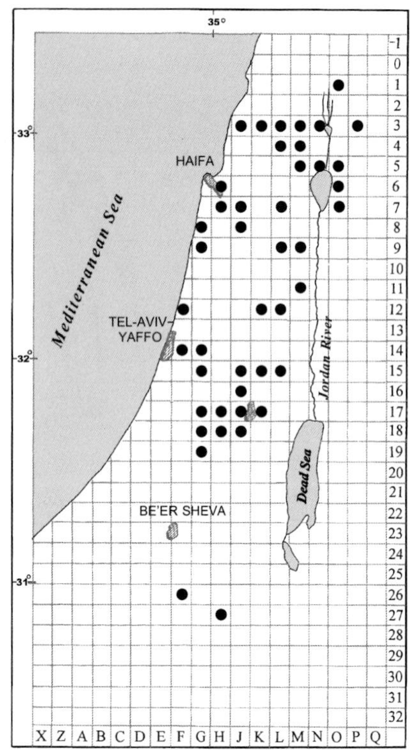

Map 32: *Timmiella barbuloides*

When plants are dry, the leaves are strongly incurved and only the wide, shiny, yellowish costa can be seen from the outside. Perhaps the strong curvature of the lamina and the costa, which is the only surface area in contact with the exterior, may be the reason for the slow uptake of water.

Previous records of *Timmiella anomala* (B.S.G.) Limpr. from Israel (Herrnstadt *et al.* 1991) are based on the publication of two collections of Rabinovitz-Sereni from Jerusalem (6 Jun. 1927) and Mount Carmel (14 May 1927), later verified by M. Bizot (in Herb. Bilewsky). *Timmiella anomala* was not included in Bilewsky (1965) and was only mentioned by him in a list of plants published later (Bilewsky 1974). The determination of the species of *Timmiella* is mainly based on variations in the sexual condition and sporophytic characters. However, it was not possible to verify the identity of the plants in the collections mentioned above because gametangia or sporophytes were not available. *Timmiella anomala* is described as differing from *T. barbuloides* in having a revoluble annulus and ± twisted peristome teeth. Brotherus (1893) considered *T. anomala* as autoicous, not paroicous as *T. barbuloides*. It seems likely that *T. anomala* does indeed grow in Israel. It has been repeatedly recorded from Southwest Asia (Turkey, Lebanon, Iraq, and Afghanistan) (Frey & Kürschner 1991c).

Plants of some local *Tortella* species may resemble *Timmiella barbuloides* in general habit. However, *Timmiella* plants are larger, more robust, and have leaves with the typical, much wider glossy costa. In addition, the leaves do not have papillae, are bistratose, and the border between the hyaline base and the upper chlorophyllous part of the lamina forms a straight — not V-shaped — line.

SUBFAMILY II. POTTIOIDEAE

TRIBE PLEUROWEISIEAE

9. *Eucladium* Bruch & Schimp

A monotypic genus comprising a single widespread species (*sensu* Zander 1993).

Eucladium verticillatum (Brid.) B.S.G., Bryol. Eur. 1 : 9, Tab. 40 (1846).
[Bilewsky (1965) : 361; Rabinovitz-Sereni (1931)]
Weissia verticillata Brid., J. Bot. (Göttingen) 1800 : 283 (1801).

Distribution map 33, Figure 33 (p. 109), Plate II : d.

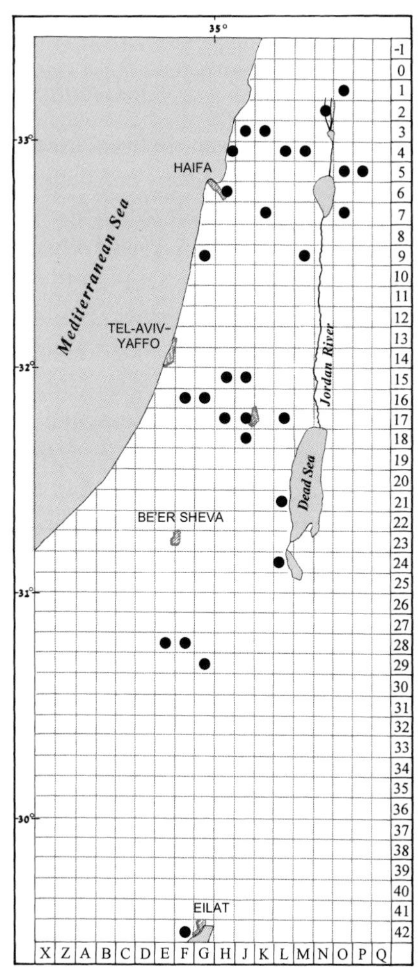

Map 33: *Eucladium verticillatum*

Plants small to medium, 5–20 mm high, bright green to bluish-green. *Stems* irregularly branched, without a central strand. *Leaves* appressed, erect with incurved tips when dry, erect to spreading, sometimes recurved when moist, 1–2.5 mm long, concave, linear to linear-lanceolate, acute; margins plane, irregularly toothed at upper part of base, crenulate above; costa stout, 40–60 µm below, ending below apex or percurrent, with cells on ventral surface slightly longer than adjacent lamina cells; basal cells narrow-rectangular, hyaline, smooth, thin-walled, nearly five times longer than upper cells, distinctly differentiated across leaf; upper cells subquadrate, incrassate, in mid-leaf 8–13 µm wide, papillose with two–five simple, low papillae per cell. *Dioicous*. *Setae* 3–5 mm long, yellow to reddish; capsules 1–1.25 mm long, ellipsoid-ovoid, brown; peristome teeth 16, entire to cleft. *Spores* 11–14 µm in diameter, granulate; gemmate, processes solitary, in small groups, or joined in bead-like structures (seen with SEM). *Calyptrae* cucullate, smooth.

Habitat and Local Distribution. — Plants growing in large, dense tufts, often encrusted with lime, forming tufa; mainly growing on wet, shaded calcareous rocks and soils, on loess, marl, travertine, often on dripping cliffs and in seepage near waterfalls, springs, or cave entrances. Very common, scattered all over the country in wet habitats: Acco Plain, Sharon Plain, Philistean Plain, Upper Galilee, Lower Galilee, Mount Carmel, Shefela, Judean Mountains, Judean Desert, Central Negev, Southern Negev, Dan Valley, Bet She'an Valley, Dead Sea area, and Golan Heights.

General Distribution. — A widespread species, recorded from Southwest Asia (throughout), around the Mediterranean, Europe (throughout), Macaronesia, North, tropical, and South Africa, northeastern, eastern, and central Asia, and North and Central America.

Rhizoidal gemmae have been described in *Eucladium verticillatum* plants growing in caves (see von der Dunk & von der Dunk 1973). Frequently, plants of *E. verticillatum* may grow in very wet habitats together with *Barbula tophacea* (Brid.) Mitt. and *B. ehrenbergii* (Lorentz) M. Fleisch. They can be easily recognised by the narrow leaves, which are irregularly toothed at upper part of base.

10. *Gymnostomum* Nees & Hornsch.

Plants bright green to dull green, glaucous or yellowish, brown below, growing in ± dense tufts. *Stems* erect, branched, usually with a weak central strand. *Leaves* appressed to ± incurved, sometimes contorted when dry, erectopatent or slightly recurved, occasionally keeled when moist, linear-lanceolate to ligulate or ovate-lanceolate, rounded, obtuse to broadly acute; margins plane, entire to crenulate-papillose; costa strong, ending below apex, sometimes with some elongated cells on ventral surface, with one or two stereid bands; cells at leaf apex often in rows that meet at right angles; basal cells rectangular, smooth; mid-leaf and upper cells usually small, irregularly quadrate to subrectangular, sometimes rounded-hexagonal, usually papillose, with simple or bifid papillae. Gemmae occasionally on branching stalks in leaf axils; perichaetial leaves ± larger than other leaves. *Dioicous. Setae* long, straight; capsules erect, symmetrical, ovoid to ellipsoid-cylindrical; operculum obliquely long-rostrate; annulus (in all local species) of two–three(–four) rows of more or less regularly arranged, rectangular, laterally elongated cells; peristome absent. *Calyptrae* cucullate, smooth. *Spores* nearly smooth.

A widely distributed genus, variously delimited, and consequently considered as comprising a different number of species. According to Zander (1993) it comprises 24 species. Four species of *Gymnostomum* occur in the local flora. All belong to a group in which the delimitation of taxa poses many problems and consequently treatments often vary (cf. Townsend 1977, Whitehouse & Crundwell 1991).

Gymnostomum is perceived here as separate from *Hymenostylium* (see the discussion below, p. 121). *Gyroweisia*, which also seems to be a related genus, differs in the well-developed, often revoluble annulus of two–three rows of large cells and in the occurrence of short, rudimentary peristome teeth in some species, among them *Gyroweisia reflexa* in the local flora (cf. Pierrot 1976 for a comparison of the mouth of the capsule of *Gymnostomum calcareum* and *Gyroweisia tenuis*).

Literature. — Whitehouse & Crundwell (1991).

1 Plants 5–7 mm high; leaves contorted when dry, up to 1.5(–2) mm long, apex often acute; cells in mid-leaf 8–13 μm wide; costa usually with both dorsal and ventral stereid bands **1. G. aeruginosum**

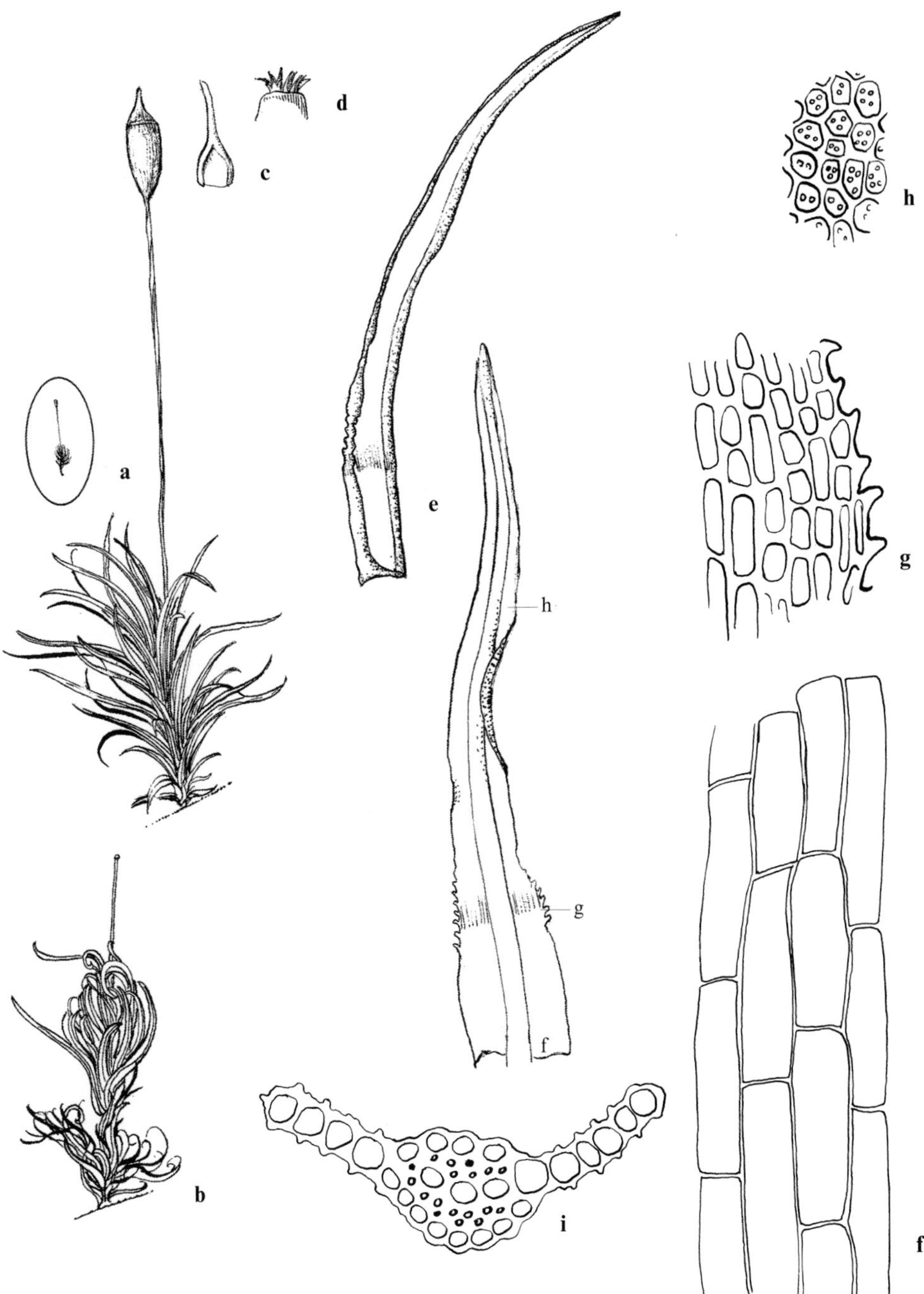

Figure 33. *Eucladium verticillatum*: **a** habit, moist (×10); **b** habit, dry (×10); **c** calyptra (×10); **d** peristome (×20); **e** leaves (×50); **f** basal cells; **g** cells at upper part of leaf base including marginal cells; **h** upper cells (all leaf cells ×500); **i** cross-section of leaf (×500). (*Crosby & Herrnstadt 78-6-6*)

1 Plants 1–4 mm high; leaves slightly contorted when dry, up to 1 mm long (including perichaetial leaves), apex usually round to obtuse; cells in mid-leaf 5–9 µm wide; costa usually with only a dorsal stereid band.
  2 Leaves ligulate; without axillary gemmae **2. G. calcareum**
  2 Leaves ovate to ovate-lanceolate; with or without axillary gemmae.
    3 Plants frequently bearing gemmae; leaves with unistratose margins throughout **4. G. viridulum**
    3 Plants not bearing gemmae; leaves with bistratose margins on one or both sides near apex **3. G. mosis**

1. Gymnostomum aeruginosum Sm., Fl. Brit. 3 : 1163 (1804).
G. rupestre Schleich. ex Schwägr., Sp. Musc. Frond., Suppl. 1 : 31 (1811). — Bilewsky (1965) : 359.

Distribution map 34, Figure 34, Plate II : f.

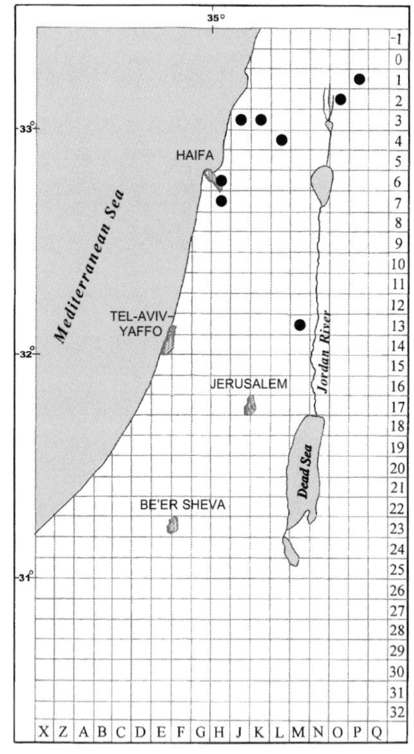

Map 34: *Gymnostomum aeruginosum*

Plants resembling the locally more widespread *Gymnostomum calcareum* but larger, differing in the following characters. Plants 5–7 mm high, glaucous to brown-green. *Leaves* up to 1.5(–2) mm long, more strongly contorted when dry, apex often acute or sometimes narrow-obtuse; costa up to 50(–60) µm wide, cross-section with a ventral and a dorsal stereid band, rarely with dorsal band only; cells in mid-leaf 8–13 µm wide. *Setae* up to 6 mm long; capsules longer and narrower than in *G. calcareum* in diameter, up to 2 mm long including operculum. *Spores* of local plants baculate-gemmate, processes small, solitary or variously joined, not irregularly reticulate as in *G. calcareum* (seen with SEM).

Habitat and Local Distribution. — Plants growing in dense tufts on damp, shaded soils, mainly in mountainous areas in the North. Occasional: Mount Carmel, Upper Galilee, Samaria, and Golan Heights.

General Distribution. — Recorded from Southwest Asia (Turkey, Sinai, Iraq, Iran, and Yemen), around the Mediterranean, Europe (throughout), Macaronesia, North and South Africa, Asia (throughout), and North and South America (confirmed data in Whitehouse & Crundwell 1991).

Gymnostomum calcareum and *G. aeruginosum* are often considered as conspecific (Magill 1981; Zander 1977, 1993), but local plants are easily separated by the characters mentioned above. Sterile plants of *G. aeruginosum* resemble those of *Hymenostylium recurvirostrum* in their habit, but the latter can be distinguished by the short-apiculate leaves with margins recurved, at least on one side, and the operculum, which is often

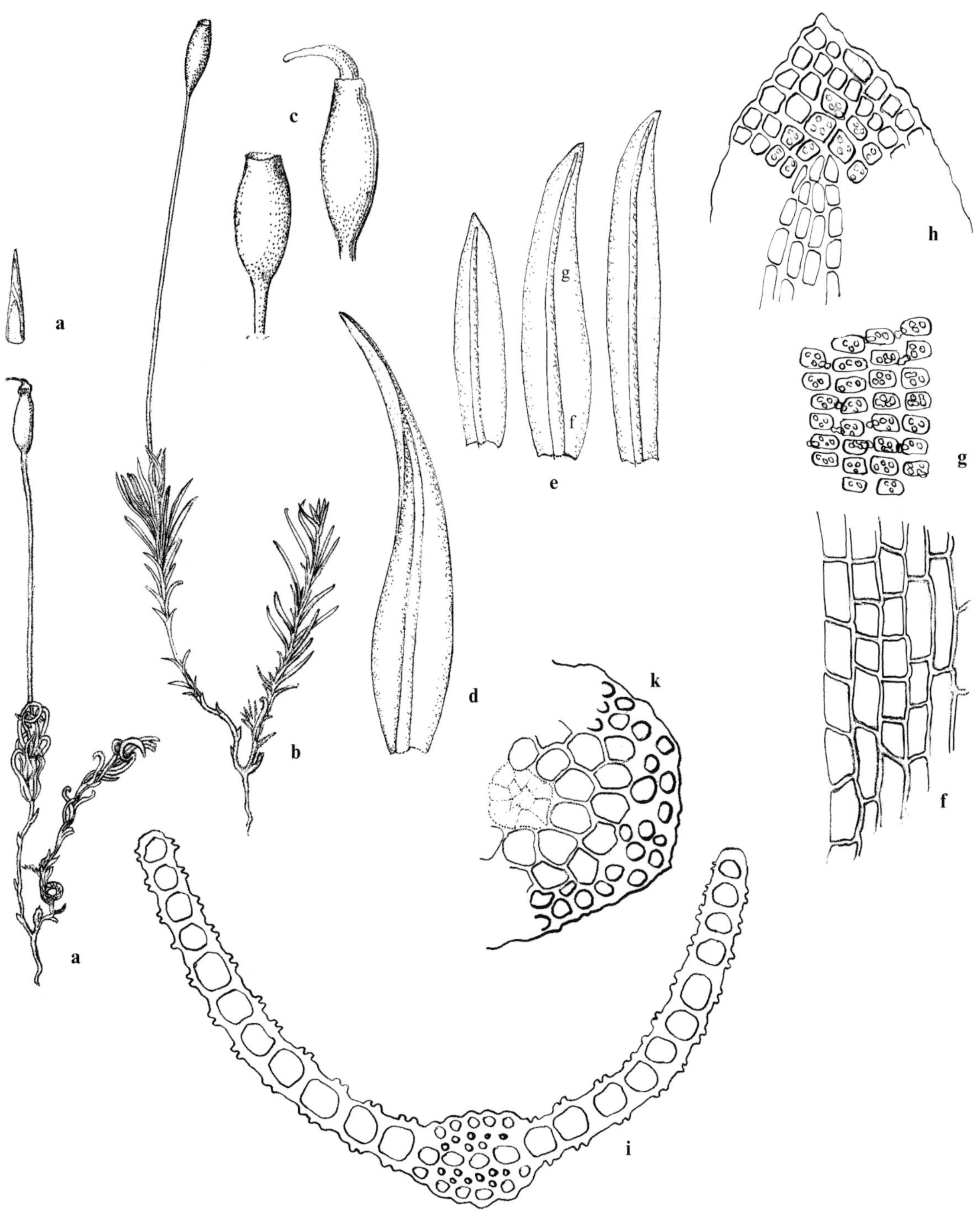

Figure 34. *Gymnostomum aeruginosum*: **a** habit, dry (×10); **b** habit, moist (×10); **c** capsules without and with operculum (×20); **d** perichaetial leaf (×50); **e** leaves (×50); **f** basal cells; **g** mid-leaf cells; **h** leaf apex (all leaf cells ×500); **i** cross-section of leaf (×500); **k** portion of cross-section of stem

not deciduous. In addition, a central strand is usually absent, and the leaf cells of apex do not meet at right angles.

There is some indication that in several species of *Gymnostomum* the development of ventral stereids may be dependent on the age and size of the leaf: in younger and smaller leaves the occurrence of a single stereid band is far more frequent than in larger, more mature leaves. Furthermore, even in the same leaf there may be two stereid bands at the base and only one towards the apex. As the number of stereid bands seems to be a rather plastic character, we tend to include in *G. aeruginosum* also plants with a single stereid band if they do not differ from this species in other characters.

2. Gymnostomum calcareum Nees & Hornsch. *in* Nees, Hornsch. & Sturm, Bryol. Germ. 1 : 153 (1823). [Bilewsky (1965) : 359; Bizot (1945)]

Distribution map 35, Figure 35, Plate II : e.

Plants 1.5–4(–7) mm high, bright green, yellowish-brown below. *Leaves* ± crowded, small, 0.5–0.8 mm long, slightly contorted when dry, ligulate; apex round to obtuse or rarely acute; margins incurved when dry, plane when moist, crenulate-papillose, usually unistratose; costa 27–40 μm wide near base, ending ± below apex, usually with dorsal stereid band only, ventral stereid band rarely present also; cells ± incrassate, densely papillose with simple round papillae; mid-leaf cells 5–8(–9) μm wide; perichaetial leaves up to 1 mm long. *Setae* 3–5 mm; capsules up to 1.5 mm long including operculum, ellipsoid, brown. *Spores* 8–12 μm in diameter, nearly smooth; irregularly reticulate (seen with SEM).

Habitat and Local Distribution. — Plants growing in dense compact tufts on damp, shaded calcareous rocks and soils, frequently in caves. Very common — the most widespread *Gymnostomum* species in the local flora — occurs in nearly all districts: Coastal Galilee, Sharon Plain, Upper Galilee, Mount Carmel, Esdraelon Plain, Mount Gilboa, Samaria, Shefela, Judean Mountains, Judean Desert, Western Negev, Central Negev, Southern Negev, Dan Valley, Bet She'an Valley, and Golan Heights.

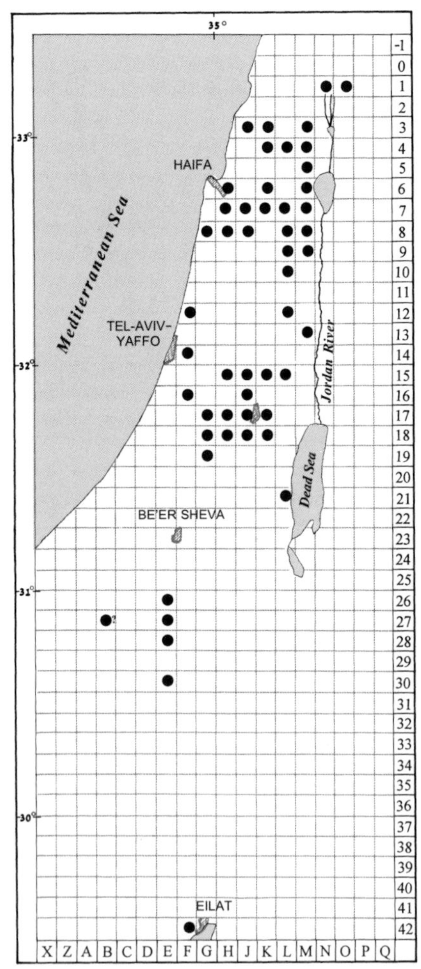

Map 35: *Gymnostomum calcareum*

General Distribution. — A species widely distributed throughout the world. Recorded from Southwest Asia (Cyprus, Turkey, Syria, Lebanon, Jordan, Iraq, Iran, Afghanistan, and Kuwait), around the Mediterranean, Europe (throughout), Macaronesia, North, tropical, and South Africa, north-

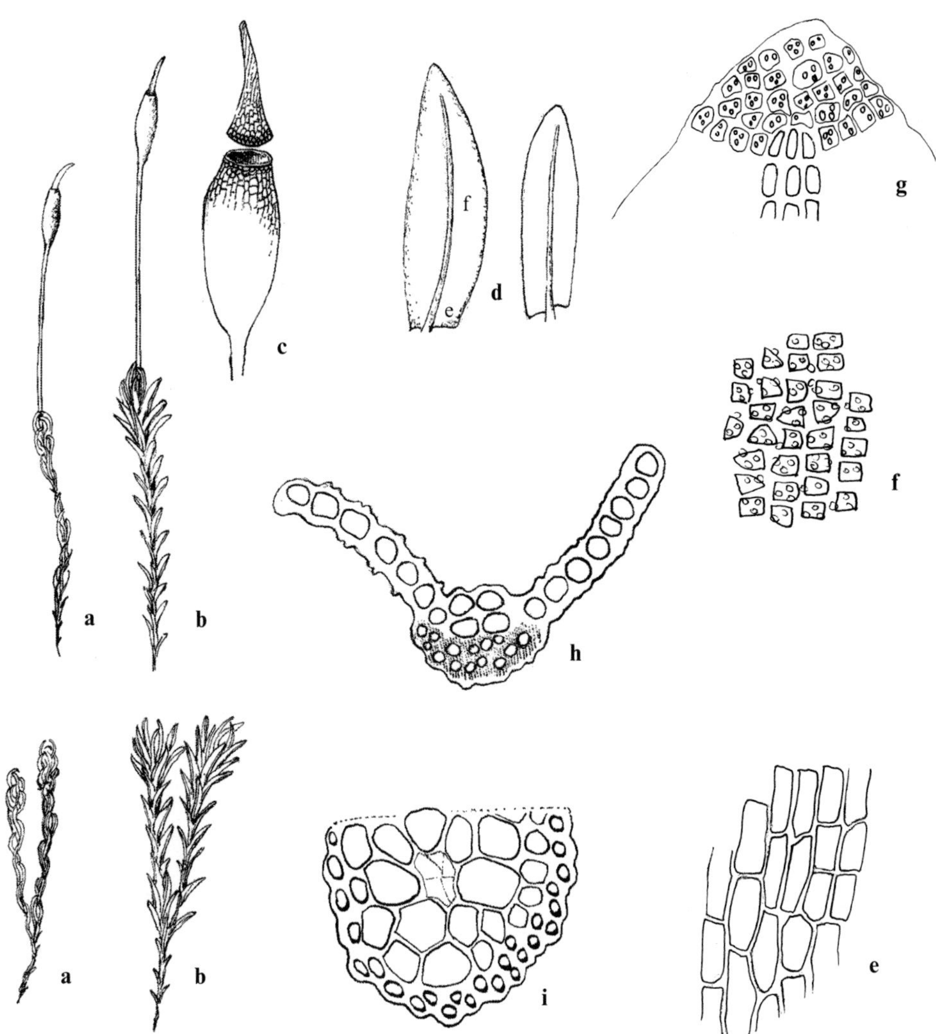

Figure 35. *Gymnostomum calcareum*: **a** habit, dry; plants with and without sporophyte (×10); **b** habit, moist; plants with and without sporophyte (×10); **c** capsule and operculum (×25); **d** leaves (×50); **e** basal cells; **f** mid-leaf cells; **g** leaf apex (all leaf cells ×500); **h** cross-section of leaf (×500); **i** cross-section of stem (×500). (*Markus & Kutiel 77-572-2*)

eastern, eastern, and central Asia, Australia, New Zealand (cf. Whitehouse & Crundwell 1991), and North, Central, and southern South America.

The occurrence of partly bistratose margins near the apex has rarely been observed in some leaves of local plants of *Gymnostomum calcareum*.

Two — apparently sympatric — varieties of *Gymnostomum calcareum* were recorded by Bilewsky (1965): var. *brevifolium* "Schpr.", considered by Whitehouse & Crundwell (1991) as not differing from var. *calcareum*, and var. *tenellum* "Schpr." (*sic*; B.S.G. were the correct authors).

3. Gymnostomum mosis (Lorentz) Jur. et Milde., Verh. K. K. Zool.-Bot. Ges. Wien 20 : 590 (1870).

Trichostomum mosis Lorentz, Phys. Abh. Königl. Akad. Wiss. Berlin 1 : 28, Tab. 3 & 4 (1868); *Gyroweisia mosis* (Lorentz) Paris, Index Bryol. 550 (1896). — Bilewsky & Nachmony (1955); *Barbula mosis* (Lorentz) Hilp., Beih. Bot Centralbl. 50 : 656 (1933).

Distribution map 36, Figure 36.

Plants 1–4 (5) mm high, bright green. *Leaves* appressed at base, ± incurved above, slightly contorted when dry, erectopatent to patent when moist; leaves of sterile plants ca. 0.5 mm long, ovate with rounded apex; upper leaves of fertile plants ca. 1 mm long, greatly varying in size and shape on single plant, ovate-lanceolate with obtuse, apiculate apex; margins plane or one margin slightly recurved above, crenulate, bistratose on one or both sides in upper $\frac{1}{2}$–$\frac{2}{3}$ of lamina; costa strong, nearly equal in width along lamina, 40–50 µm wide, ending below apex, with single dorsal stereid band; mid-leaf and upper cells papillose with up to three simple round papillae per cell, mid-leaf cells 5–9 µm wide. *Setae* 3–4 mm long; capsules up to 1.4 mm long including operculum; operculum conical-rostrate. *Spores* 8–12 µm in diameter; irregularly reticulate.

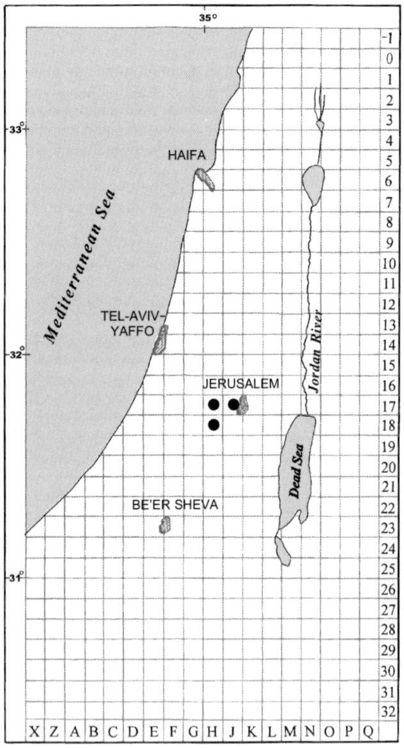

Map 36: *Gymnostomum mosis*

Habitat and Local Distribution. — Plants growing in dense tufts on shaded calcareous rocks. Rare: Shefela bordering with Judean Mountains, and Judean Mountains.

General Distribution. — Records of *Gymnostomum mosis* that were confirmed by Whitehouse & Crundwell (1991) were from Egypt (Sinai), Iran, Afghanistan, and Oman. It was also recorded from southern Spain by Martínez-Sánchez *et al.* (1991).

Gymnostomum mosis was described as *Trichostomum mosis* by Lorentz from a sterile plant collected in Sinai and though the type material seems to have been destroyed,

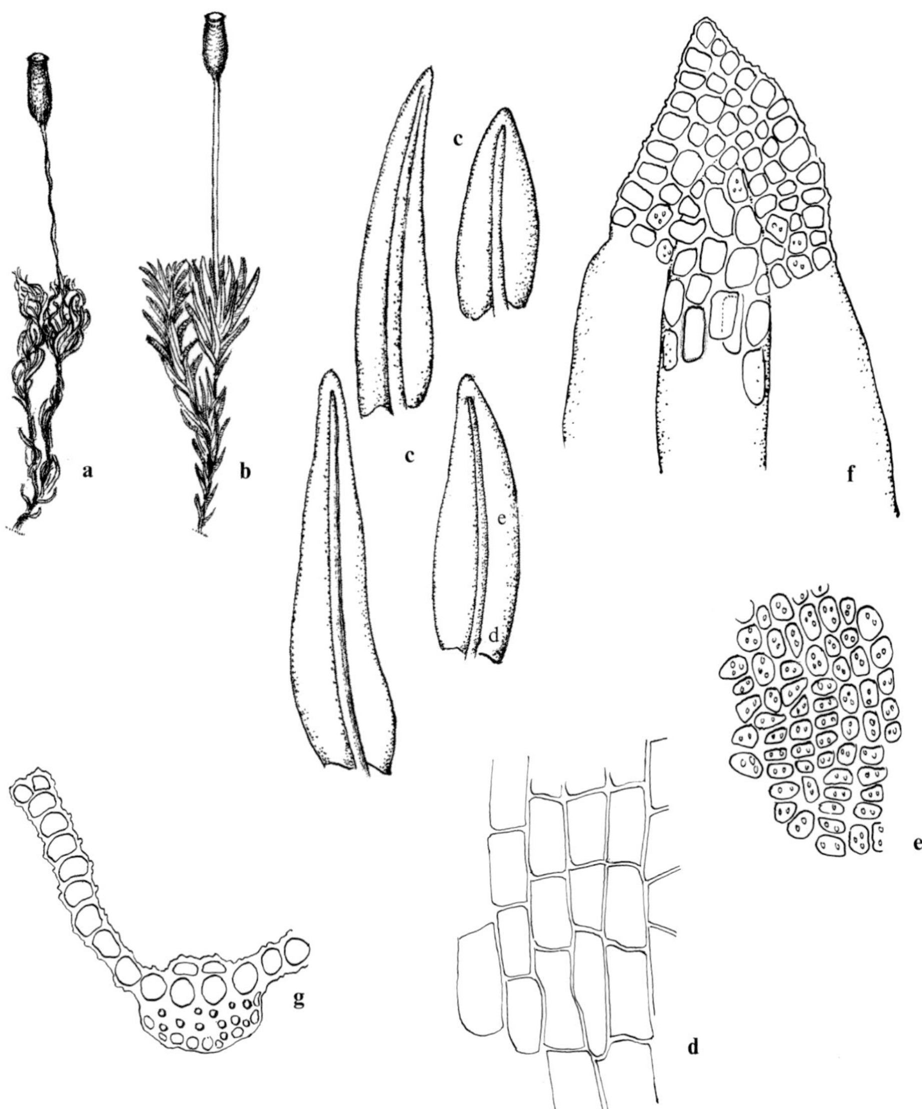

Figure 36. *Gymnostomum mosis*: **a** habit, dry (×10); **b** habit, moist (×10); **c** leaves (×50); **d** basal cells; **e** mid-leaf cells; **f** leaf apex (all leaf cells ×500); **g** cross-section of costa and portion of lamina (×500). (*Kushnir*, 7 Jan. 1941)

his description and illustrations clearly differentiate this species from others in the genus. All later records from Southwest Asia also refer to sterile plants only. Martínez-Sánchez *et al.* (1991) described collections of fertile plants from Almeria, Spain as *G. mosis*, an identification doubted by Whitehouse & Crundwell (1991). One of the three collections made in Israel (Judean Mountains: Jerusalem, *Kushnir*, 4 Mar. 1943), includes fertile plants bearing sporophytes. They agree with Lorentz's plants in the shape and size of their lower leaves, the wide costa, and the bistratose margins. The fertile plants of *G. mosis* from Almeria were described and drawn with a narrower costa (22–24 μm) and more acute upper leaves. In addition, no capsules with large-celled annulus, as depicted in the Spanish plants, were seen in local plants.

Zander (1993) expressed doubts about the differentiation of *Gymnostomum mosis* from *G. viridulum*. However, the absence of gemmae and the bistratose margins are reliable diagnostic characters of *G. mosis*.

A sporophyte-bearing specimen cited by Bilewsky & Nachmony (1955) from Giv'at-Hamore (Esdraelon Plain) as *Gyroweisia mosis* is considered here as *G. calcareum*. All the collections from Netiv HaLamed He (Shefela), previously identified as *G. mosis* (cf. Herrnstadt *et al.* 1982), belong to *G. viridulum*.

Literature. — Lorentz (1868).

4. **Gymnostomum viridulum** Brid., Bryol. Univ. 1 : 66 (1826).

Gyroweisia luisieri Sergio, Bol. Soc. Portug. Ci. Nat. 14 : 82 (1972); *Gymnostomum luisieri* (Sergio) Sergio ex Crundw. J. Bryol. 11 : 603 (1981).

Distribution map 37, Figure 37.

Plants 1–3(–4) mm high, bright green, turning reddish-brown, brownish below. *Leaves* minute, 0.25–0.7 mm long, appressed, ± imbricate, slightly contorted when dry, erectopatent to spreading, sometimes recurved when moist, ± keeled, ovate to ovate-lanceolate, round to obtuse or minutely apiculate; margins plane, crenulate, unistratose; costa 20–30 μm wide near base, ending below apex, with a single dorsal stereid band; cells papillose with simple, round papillae; mid-leaf cells 6–9 μm wide; perichaetial leaves up to 1 mm long. *Gemmae* in leaf axils and on protonemata multicellular, fusiform, ovate or pyriform, sometimes with a long terminal cell, green turning castaneous. *Setae* up to 3 mm long; capsules yellowish-brown, ovoid to ellipsoid; only immature or deoperculate capsules seen in local plants, with theca 0.6–0.8(–1) mm. *Spores* 10–15 μm in diameter, ± papillose; densely baculate, some processes joined (seen with SEM).

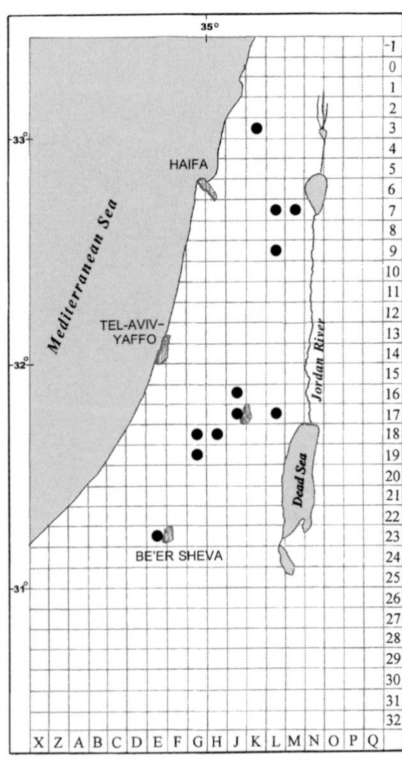

Map 37: *Gymnostomum viridulum*

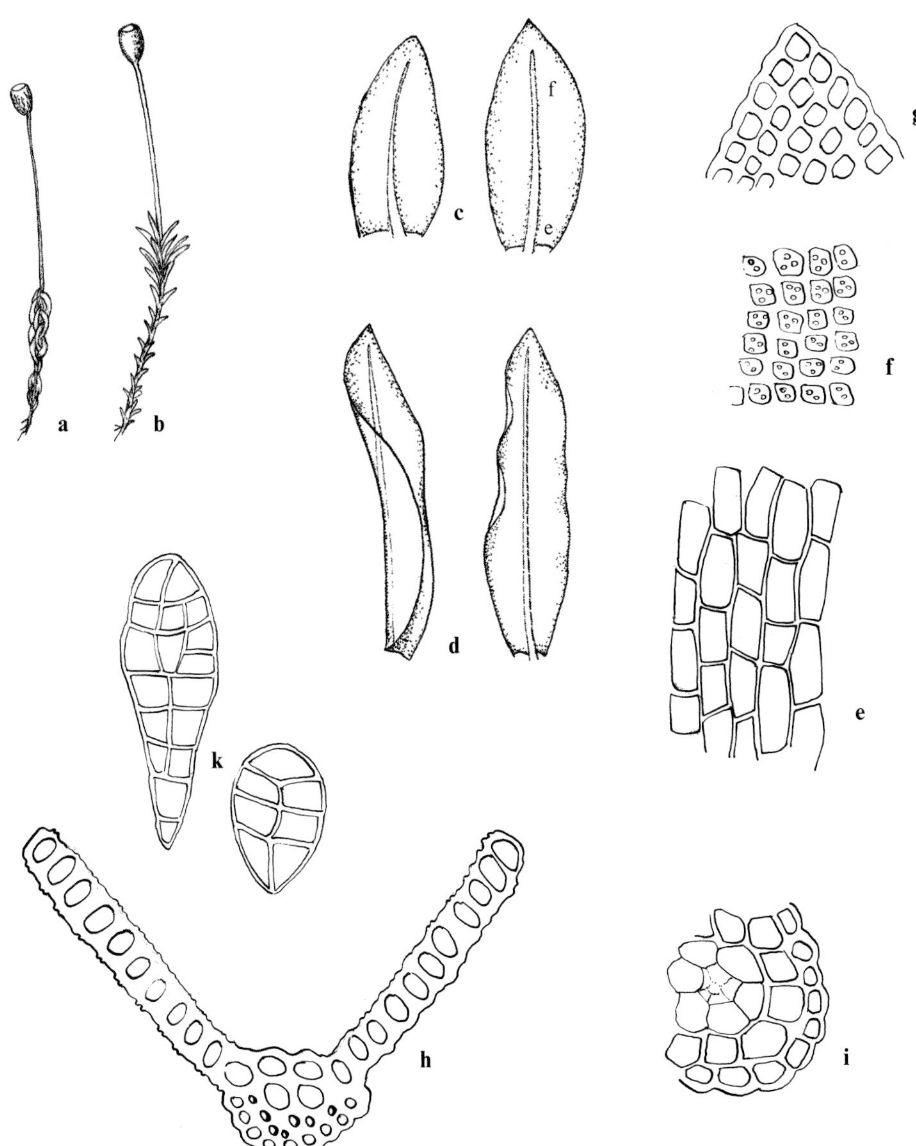

Figure 37. *Gymnostomum viridulum*: **a** habit, dry (×10); **b** habit, moist (×10); **c** lower leaves (×50); **d** upper leaves (×50); **e** basal cells; **f** upper cells; **g** leaf apex (all leaf cells ×500); **h** cross-section of leaf (×500); **i** portion of cross-section of stem (×50); **k** gemmae (×500). (**a**, **b**: *Kushnir*, 1 Sep. 1943; **c**–**k**: *Herrnstadt*, 8 Mar. 1980)

Habitat and Local Distribution. — Plants growing in dense compact tufts on shaded limestone and calcareous rocks, sometimes on vertical walls of caves; base of tuft often encrusted with calcareous deposits, in moist habitats in particular. Occasional to common: Upper Galilee, Lower Galilee, Esdraelon Plain, Shefela, Judean Mountains, Judean Desert, and Northern Negev.

General Distribution. — Recorded from Southwest Asia as *Gymnostomum viridulum* or as *G. luisieri* (Cyprus, Turkey, Lebanon, Sinai, Jordan, Iraq, Iran, and Afghanistan), around the Mediterranean, Europe (widespread, mainly the south), Macaronesia, and tropical and eastern Africa (Somalia) (confirmed data in Whitehouse & Crundwell 1991).

Gymnostomum luisieri is considered here as conspecific with *G. viridulus* (cf. Whitehouse & Crundwell 1991).

Literature. — Sérgio (1984).

11. *Gyroweisia* Schimp.

Plants minute. *Stems* erect, short, mostly simple, sometimes forked, with or without central strand. *Leaves* erect, ± incurved when dry, erectopatent to recurved when moist, linear-lanceolate, ligulate to long-ovate, apex round-obtuse to round-acute; costa usually ending below apex or percurrent, cross-section usually with dorsal stereid band only; basal cells rectangular, smooth; upper cells smaller, round-quadrate, moderately incrassate, papillose; perichaetial leaves differentiated. *Rhizoidal gemmae* often present. *Dioicous. Setae* straight; capsules symmetrical, erect, ovoid to cylindrical; operculum conical to rostrate; peristome teeth 16, rudimentary, short, or absent; annulus of two–three rows of large, often revoluble cells. *Calyptrae* cucullate, smooth. *Spores* smooth to papillose.

A small genus comprising only six species (Zander 1993); two species occur in Israel.

The distinction between sterile plants of *Gyroweisia* and *Gymnostomum* may be sometimes rather difficult.

1 Upper perichaetial leaves reflexed to squarrose when moist; peristome present **1. G. reflexa**

1 Upper perichaetial leaves patent to spreading when moist; peristome absent **2. G. tenuis**

1. Gyroweisia reflexa (Brid.) Schimp., Syn. Musc. Eur., Edition 2 : 39 (1876).
Weissia reflexa Brid., Bryol. Univ. 1 : 355 (1826); *W. tenuis* subsp. *reflexa* (Brid.) Kindb., Eur. N. Amer. Bryin. 2 : 290 (1897).

Distribution map 38 (p. 120), Figure 38 : a–i, Plate II : g.

Plants up to 2 mm high (5 mm with growth of previous year), green, turning brownish. *Stems* with weak central strand, in older plants with a well-defined layer of thick-walled brown cells below the epidermis. *Leaves*: lower leaves erectopatent when moist, 0.5–0.8 mm long, ligulate with rounded apex; upper and perichaetial leaves much larger, up to 1.25 mm long, crowded, reflexed to squarrose when moist, expanded and sheathing at

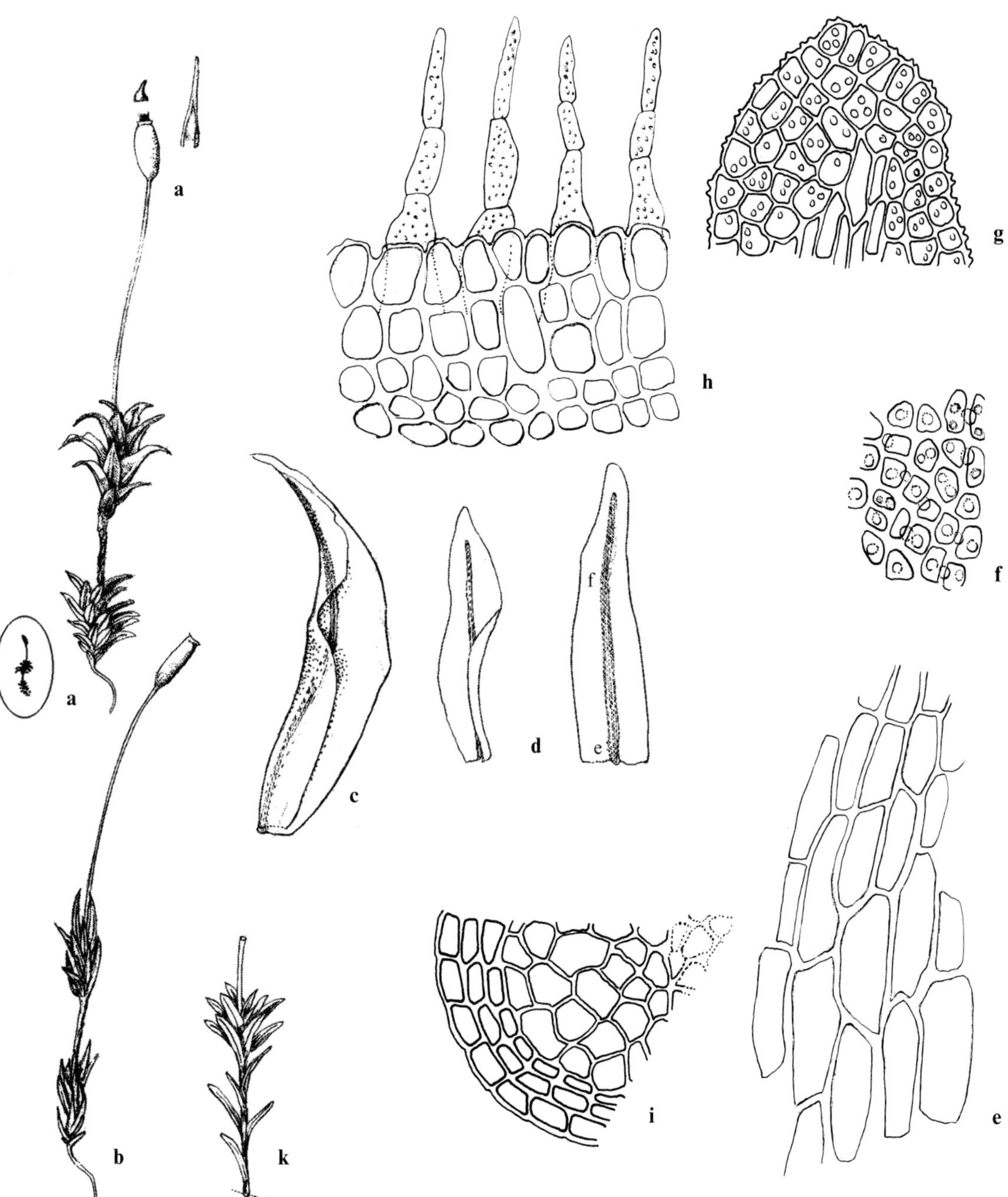

Figure 38. *Gyroweisia reflexa*: **a** habit with detached operculum and calyptra, moist (×10); **b** habit, dry (×10); **c** perichaetial leaf (×50); **d** leaves (×50); **e** basal cells; **f** mid-leaf cells; **g** apical cells (all leaf cells ×500); **h** exothecial cells with peristome teeth (×500); **i** portion of cross-section of stem (×500). *Gyroweisia tenuis*: **k**: habit, moist (×10).
(**a**–**i**: *Herrnstadt*, 28 May 1984; **k**: *Kushnir*, 9 Oct. 1943)

base, acuminate; margins of all leaves plane, papillose-crenulate; costa ending below apex; basal cells rectangular, smooth; mid- and upper leaf cells quadrate, in mid-leaf 6–10 μm wide, with evenly scattered rounded to C-shaped papillae. *Rhizoidal gemmae* not observed in the local collections. *Setae* up to 5 mm long; capsules up to 1.2 mm long including operculum; operculum obliquely rostrate; peristome teeth 16, 80–110 μm long, articulate, papillose, slender and readily breaking. *Spores* up to ca. 13 μm in diameter, papillose; irregularly gemmate to baculate, with variously joined processes (seen with SEM).

Habitat and Local Distribution. — Rare. Plants have been found in only three localities: in the Upper Galilee, Lower Galilee, and Judean Mountains (in 'En Hemed found growing together with *Eucladium verticellatum* and *Barbula tophacea*) in humid habitats, and in Esdraelon Plain (Giv'at HaMore) a single plant (found in a collection of *Gymnostomum calcareum* in Herb. Bilewsky) has been collected from a vertical rock.

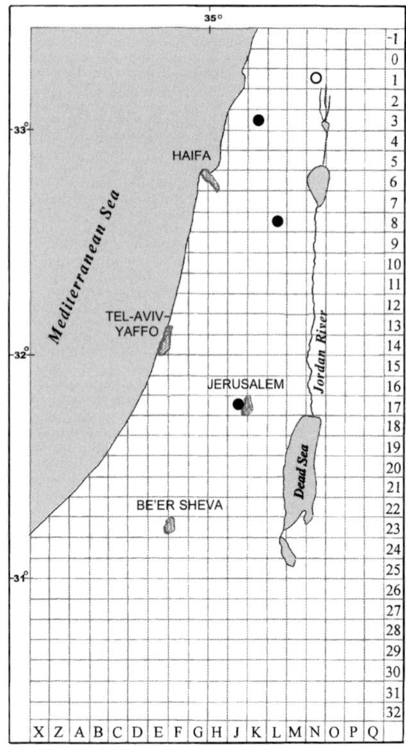

Map 38: ● *Gyroweisia reflexa*
○ *G. tenuis*

General Distribution. — Records are few; recorded from Southwest Asia (Turkey and Jordan), scattered around the Mediterranean, Europe (mainly the south), Macaronesia (in part), North Africa, and North America.

2. Gyroweisia tenuis (Hedw.) Schimp., Syn. Musc. Eur. 2 : 38 (1876).
[Bilewsky (1965) : 360]
Gymnostomum tenue Schrad. ex Hedw., Sp. Musc. Frond. 37 (1801).

Distribution map 38, Figure 38 : k (p. 119).

Plants resemble *Gyroweisia reflexa*, but have perichaetial leaves patent to spreading when moist (not reflexed to squarrose). Peristome not present.

Bilewsky (1965) recorded *Gyroweisia tenuis* from the Central Negev and var. *acutifolia* "Philib.", without details on distribution. The collection from the Central Negev, deposited in Herb. Bilewsky (HUJ), was identified as *Anoectangium handelii* Schiffn. The only collection named *Gyroweisia tenuis* var. *acutifolia* (Esdraelon Plain, Giv'at HaMore) included plants of *Gymnostomum calcareum* and a single plant of *Gyroweisia*. This plant was identified as *G. reflexa* due to its well-developed peristome. A collection from "Hazbani Bridge" (*Kushnir*, 9 Oct. 1943), determined by H. J. O. Reimers as *G. tenuis*, agrees indeed with our concept of this species. Thus, *G. tenuis* was found in Israel, though contrary to previous records, only in the north (in the Dan Valley and not Upper Galilee as erroneously stated by Herrnstadt *et al.* 1991).

General Distribution. — Recorded from Southwest Asia (Turkey, Syria, Jordan, Iraq, and Saudi Arabia), scattered around the Mediterranean, Europe (more in the south), North Africa, eastern Asia (recent record from China), and North America.

12. *Hymenostylium* Brid.

Plants green, sometimes glaucous, brownish below, growing in ± dense tufts. *Stems* erect, branched, central strand usually absent. *Leaves* appressed to ± incurved, sometimes contorted, when dry, erectopatent to spreading and keeled when moist, usually lanceolate to ligulate; apex acute, sometimes obtuse; margins plane or recurved, usually entire; costa strong, ending below apex to excurrent, sometimes with some elongate-rectangular cells on ventral surface, with two stereid bands, usually without ventral epidermis; basal cells rectangular, smooth; upper cells irregularly quadrate to rectangular, rhomboidal or triangular, usually papillose, papillae simple, low; upper leaf cells not meeting at right angle at apex (cf. *Gymnostomum*); perichaetial leaves little differentiated. *Dioicous*. *Setae* straight, up to 1 cm long; capsules symmetrical, erect, ovoid to short rectangular; operculum obliquely rostrate, often still attached to columella at maturity; annulus usually with one row of small persistent cells; peristome absent. *Spores* weakly papillate. *Calyptrae* cucullate, smooth.

Hymenostylium is variously dealt with, even in comparatively recent treatments. Some authors often consider it as part of *Gymnostomum* (e.g., Smith 1978, Crum & Anderson 1981) whereas others as a separate genus (e.g., Corley *et al.* 1981, Casas 1991, Zander *in* Sharp *et al.* 1994). Zander (1993) even accommodates the two genera in two different tribes of the *Merceyoideae*: *Leptodontieae* (*Hymenostylium*) and *Barbuleae* (*Gymnostomum*). The main characters that are usually applied to distinguish between the two genera are the often undeciduous operculum and the absence of a central strand of *Hymenostylium*, as compared to the deciduous operculum and the presence of a central strand in *Gymnostomum*.

A genus comprising 18 species or less (cf. Zander 1993). One species occurs in Israel.

Hymenostylium recurvirostrum (Hedw.) Dixon, Rev. Bryol. Lychénol. 6:96 (1934). [Bilewsky (1974)]

Gymnostomum recurvirostre Hedw., Sp. Musc. Frond. 33 (1801); "*Hymenostylium curvirostre* (Ehrh.) Lindb." (also "var. *scabrum* (Lindb.) Dix."). — Bilewsky (1965): 360.

Distribution map 39, Figure 39.

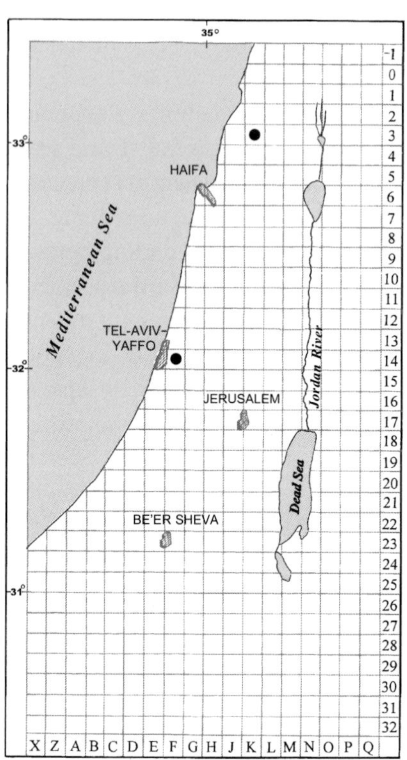

Map 39: *Hymenostylium recurvirostrum*

Plants up to 10–15 mm high, green to yellowish-green. *Stems* usually without, rarely with very weak central strand (in one local collection: Upper Galilee, Wadi Karkara, *Bilewsky*, No. 18 in Herb. Bilewsky). *Leaves* up to 2 mm long, oblong-lanceolate, tapering to acute apex from a wider base; margins recurved on one side, nearly entire; costa ending below apex or percurrent, up to 60(–80) μm wide at base, ventral epidermis absent or little differentiated, ventral stereid band very small, dorsal stereid band much larger; cells ± incrassate, at margin of base sometimes with one or more rows of thin-walled cells; upper cells with two–three papillae per cell; cells in mid-leaf 8–12 μm wide. *Sporophytes* scarce (only few and very young sporophytes could be observed in a single local population).

Habitat and Local Distribution. — Plants growing in tufts on calcareous soils and rocks, cement walls, and sand; in moist habitats often encrusted below. Rare: Sharon Plain and Upper Galilee.

General Distribution. — Recorded from Southwest Asia (scattered records from Turkey, Lebanon, Sinai, Iraq, Iran, Afghanistan, Saudi Arabia, and northern Yemen), around the Mediterranean, Europe (widespread), North, tropical, and South Africa, throughout Asia (except the southeast), and North and South America (except northeastern South America).

13. *Anoectangium* Schwägr.

The affiliation of the taxon named below *Anoectangium handelii* Schiffn. is problematic. The genus *Anoectangium* is usually considered as close to *Gymnostomum* and *Gyroweisia*, differing mainly in the axillary gametangia. Zander (1977, 1993), who placed the genus *Anoectangium* (together with the related ones) in tribe *Barbuleae* of subfamily *Merceyoideae*, transferred *A. handelii* to *Molendoa*, tribe *Hyophileae* of subfamily *Pottioideae*, and included it in the variable species *M. sendtneriana* (B.S.G.) Limpr. (discussed in detail by Geissler 1985). While in both genera the gametangia are lateral, the transfer is based on Zander's concept of *A. handelii* as having the costa anatomy of *Molendoa* — a ventral stereid band in addition to the dorsal one — not the anatomy of the costa of *Anoectangium*, which is characterised by the single,

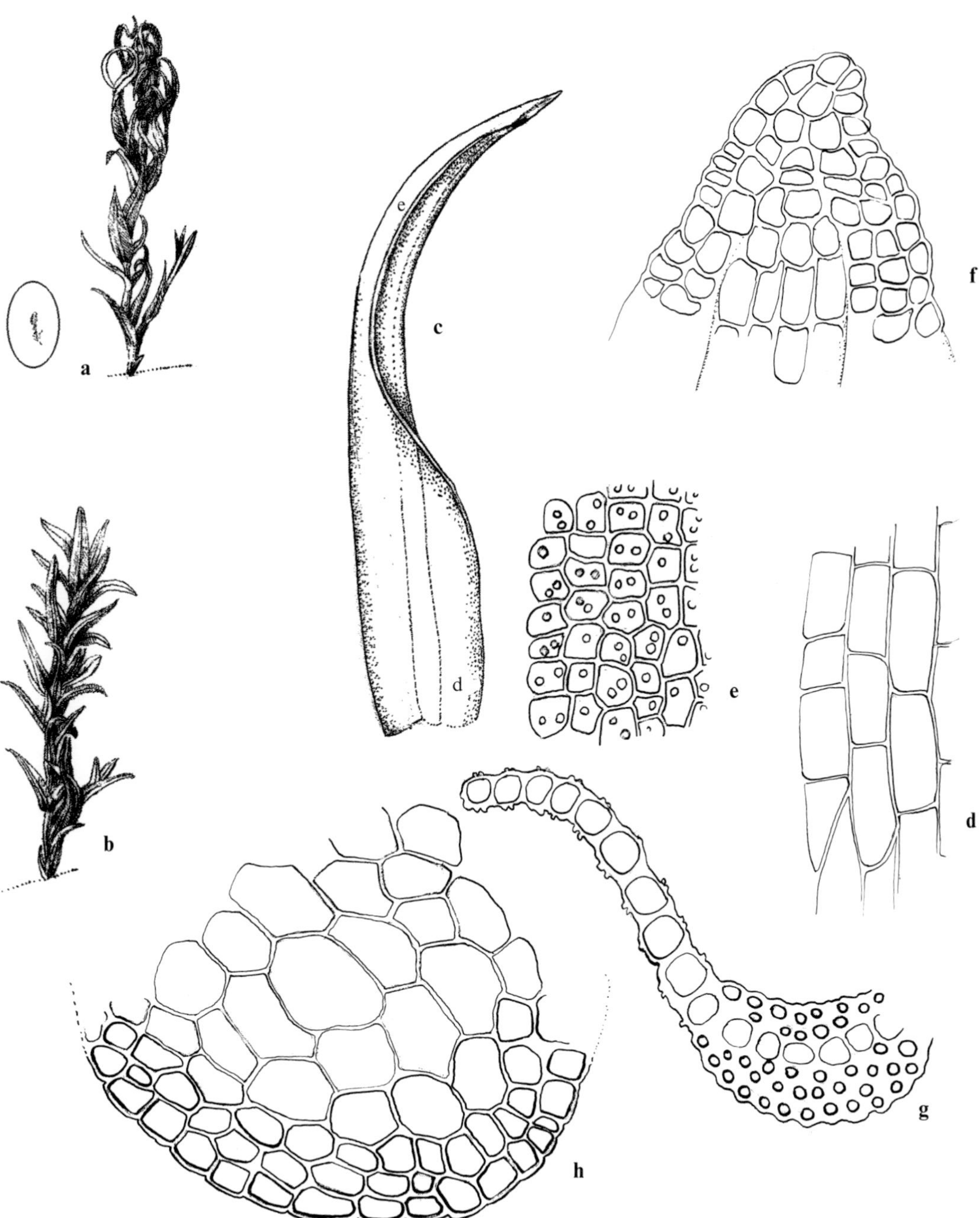

Figure 39. *Hymenostylium recurvirostrum*: **a** habit, dry (×10); **b** habit, moist (×10); **c** leaf (×50); **d** basal cells; **e** upper cells; **f** apical cells (all leaf cells ×500); **g** cross-section of costa and portion of lamina (×500); **h** cross-section of stem (×500). (*Bilewsky*, Feb. 1968, No. 843 in Herb. Bilewsky)

strong stereid band. Neither in the original description and illustration of *Anoectangium handelii* by Schiffner (1913) — who placed the species in the Orthotrichaceae — nor in any later description and illustration (Størmer 1963, Agnew & Vondráček 1975), nor in our own material, have we been able to observe a dorsal and a ventral stereid band. For that reason, we are treating the taxon as *A. handelii* without providing details on the genus, the concept and scope of which remain rather problematic.

Anoectangium handelii Schiffn., Ann. Naturhist. Hofmus. 27:490 (1913).
Molendoa sendtneriana (B.S.G.) Limpr., Laubm. Deutschl. 1:250 (1886).

Distribution map 40, Figure 40.

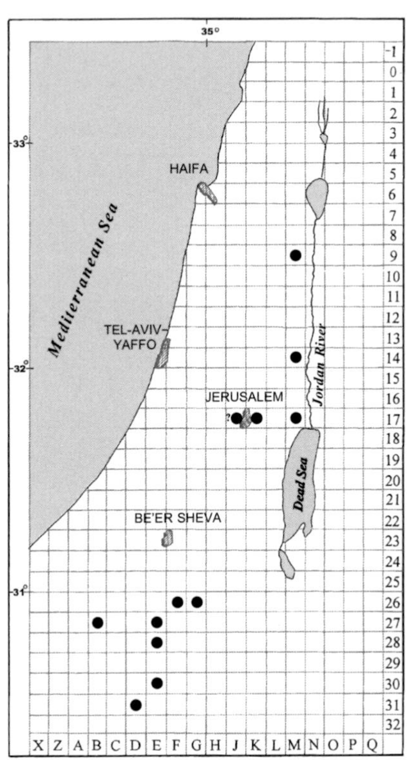

Map 40: *Anoectangium handelii*

Plants up to 3 mm long (in local plants; according to the original description reaching 10 mm), green to olive-green above, brownish below. *Stems* mostly simple, fragile, ± angular, with a central strand of few, small, thin-walled cells. *Leaves* densely spaced, erect to appressed when dry, distinctly carinate, erectopatent when moist, 0.3–0.5 mm long, ovate-lanceolate; apex obtuse to acute; margins plane, minutely crenulate; costa strong, 25–35 μm below, narrowing above and disappearing below apex, cross-section with a single, dorsal, ± hemicircular stereid band, two–four guide cells, and a distinct dorsal epidermis; basal cells few, ± incrassate, smooth, quadrate to rectangular; mid- and upper leaf cells smaller, variable, irregularly round-quadrate, incrassate, papillose with simple or bifid papillae; cells in mid-leaf up to 10 μm wide. *Gemmae* usually present, borne on branched axillary stalks, up to 20 μm across, multicellular, pyriform or elliptical, green, turning brown; perichaetial leaves described as abruptly acuminate from an ovate base, apex acute, and cells as in cauline leaves but without chlorophyll and scarcely papillose. *Archegonia* on short lateral branches. *Sporophyte* description not available.

Habitat and Local Distribution. — Plants growing in very dense tufts on fine-grained calcareous soils and in soil pockets of rocks, mainly in the arid southern districts of Israel. Occasional, more common in the Negev: Samaria, Judean Mountains (?), Judean Desert, Western Negev, Central Negev, and Bet She'an Valley.

General Distribution. — Recorded from Southwest Asia (described from Iraq: western Kurdistan; single records from Turkey, Iran, and Afghanistan), around the eastern Mediterranean, scattered records from Europe (western Russia), and central Asia (Tadzhikistan).

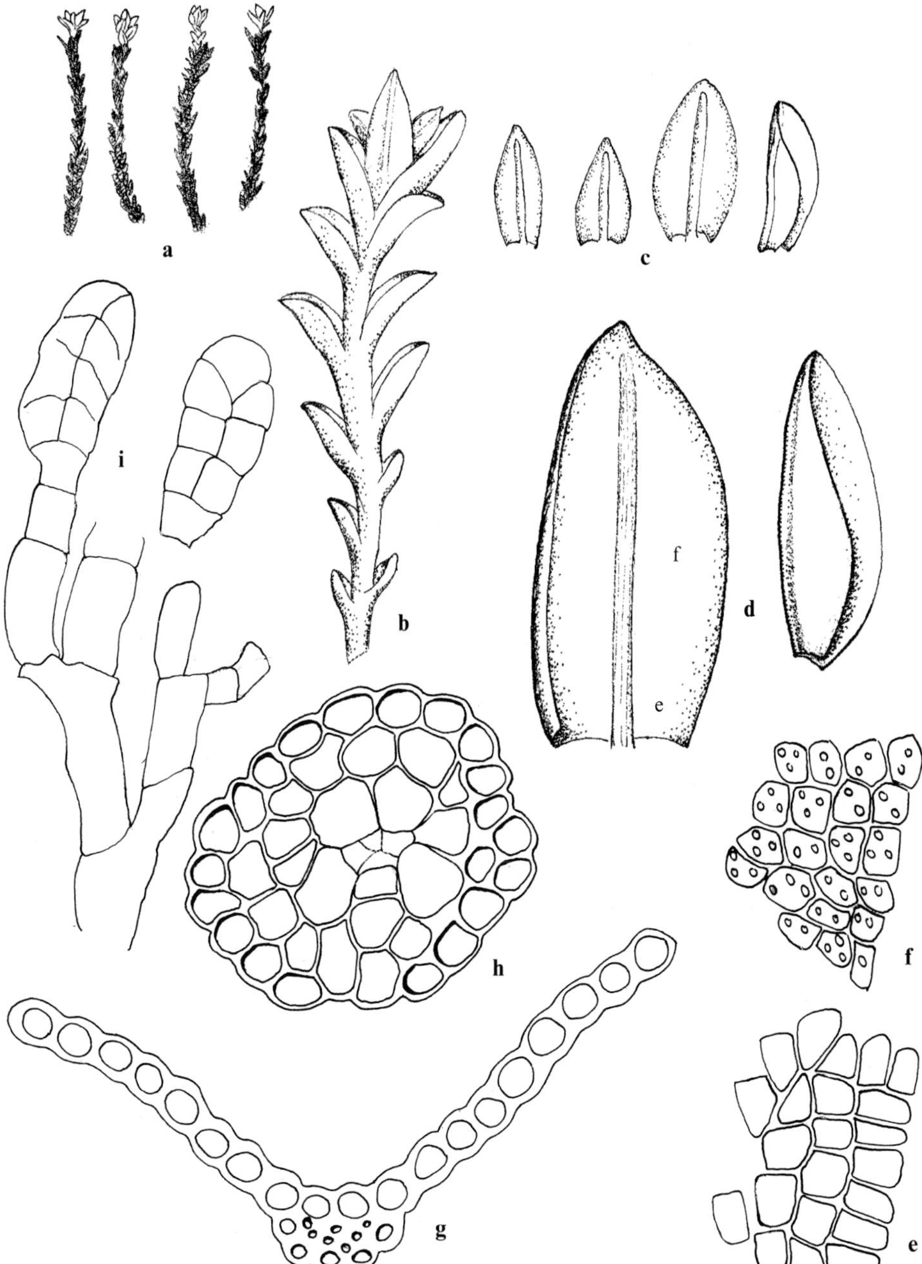

Figure 40. *Anoectangium handelii*: **a** habit of several plants (×10); **b** upper part of stem, moist (×50); **c** leaves (×50); **d** leaves (×125); **e** basal cells; **f** mid-leaf cells, (all leaf cells ×500); **g** cross-section of leaf (×500); **h** cross-section of stem (×500); **i** branched stalks bearing gemmae (×500). (*Markus & Kutiel 77-515-3*)

Plants collected in Israel are most often sterile and usually bear gemmae, but rarely some lateral branches with archegonia have been observed. Agnew & Vondráček (1975) did not find any archegonia in their material from Iraq.

Sterile plants of *Anoectangium handelii* may be separated from plants of *Gymnostomum viridulum*, which they resemble in general habit, by the carinate, densely spaced leaves.

Literature. — Abramova & Abramov (1964).

SUBFAMILY II. POTTIOIDEAE

TRIBE BARBULEAE

14. *Barbula* Hedw. *sensu lato*

Plants green, brown or blackish above, reddish-brown below, growing scattered or in loose or dense tufts, cushions or turfs. *Stems* erect, simple or often branching, usually with a central strand; axillary hairs hyaline throughout or with brownish basal cells. *Leaves* twisted or contorted when dry, spreading when moist, ovate to lanceolate to ligulate, with acute or rounded apex; margins entire, or often papillose, usually recurved, rarely revolute or plane; costa strong (in local species), usually equally wide along leaf, subpercurrent to shortly excurrent, with two stereid bands, the ventral often weak or consisting of substereids, occasionally hydroids present; ventral surface of costa sometimes grooved, with cells distinct or not from adjacent lamina cells; basal cells rectangular, smooth; upper cells quadrate to short-rectangular to hexagonal, papillose to nearly smooth; papillae hollow or solid, C-shaped, bifid or multifid, rarely simple; perichaetial leaves differentiated or not. *Propagula* occasionally present. *Dioicous*. *Setae* elongate; capsules usually erect, symmetrical, ovate to cylindrical; operculum short- to long-conical to rostrate, in local species usually oblique-rostrate; peristome with a low basal membrane, rarely without membrane, with teeth consisting of 32 papillose filaments, straight or spirally twisted. *Spores* more or less distinctly papillose; granulate, verrucate or irregularly reticulate (seen with SEM). *Calyptrae* cucullate, smooth.

A large genus, widespread in all continents, mainly in temperate regions. Zander (1993), who separated *Didymodon* from *Barbula*, still considers *Barbula sensu stricto* as comprising over 200 species.

The species listed below under *Barbula sensu lato* are often variously subdivided between several genera. We prefer here to retain them in one genus because very few uniform differential characters seem to be available for the segregated genera. At least some of these characters seem to represent extreme expressions of continuous infrageneric variations. The often accepted separation of *Didymodon* Hedw. from *Barbula sensu stricto* (in which the majority of local species would be included in that case) by Saito (1975) was based mainly on gametophytic characters, the morphology of axillary hairs, and the shape of the cells on the ventral surface of the costa (which do not seem to be entirely consistent in local material). Other authors (e.g., Crum & Anderson 1981) mainly used the shape and extent of twisting of the peristome teeth for generic distinction. Zander (1978, 1993) generally followed Saito in the concept of the two genera, though when segregating *Didymodon* from *Barbula* pointed out that both genera are heterogeneous and *Didymodon* "remains an apparent potpourri of phyletic lines" (Zander 1993). *Pseudocrossidium* R. S. Williams, often accepted as an independent genus (e.g., Zander 1979, Corley *et al.* 1981), was mainly based by Zander (1993) on the broadly revolute, often thin, leaf margins and the weak or absent ventral stereid band. *Streblotrichum* P. Beauv. is sometimes segregated (Hilpert 1933), mainly because of the long-convolute, sheathing perichaetial leaves. The main characters by which *Hydrogonium* (Müll. Hal.) A. Jaeger is separated from *Barbula* (e.g., Frahm 1995) are the soft and flaccid leaves with plane and smooth margins. Both *Streblotrichum* and *Hydrogonium* were included by Corley *et al.* (1981) and Zander (1993) in separate subgeneric divisions in *Barbula*.

Barbula is one of the largest local genera in the local flora, and is represented by 13 species. Plants can be easily recognised at the generic level and species can often be identified without sporophytes. Nevertheless, in some cases distinction between species may pose considerable difficulties.

1 Leaf margins strongly revolute from base to apex, marginal cells usually thinner-walled than adjacent cells.
  2 Leaves acute, or often obtuse, apiculate, usually with three rows of thin-walled marginal cells **5. B. revoluta**
  2 Leaves acuminate, usually with up to six rows of thin-walled marginal cells **4. B. hornschuchiana**
1 Leaf margins plane or recurved below and ± plane towards apex, marginal cells not thinner-walled than adjacent cells.
  3 Axillary hairs hyaline throughout, leaves oblong-lanceolate, lingulate, or lanceolate-lingulate.
    4 Plants usually more than 20 mm high, often up to 50 mm; leaves soft and flaccid, often encrusted with lime; margins plane, cells smooth or weakly papillose; growing near or under dripping water.
      5 Stems with abundant fusiform axillary gemmae **3b. B. ehrenbergii** var. **gemmipara**
      5 Stems without fusiform axillary gemmae **3a. B. ehrenbergii** var. **ehrenbergii**
    4 Plants less than 20 mm high; leaves not soft and flaccid, not encrusted with lime; margins usually recurved, at least below, cells distinctly papillose; not growing near or under dripping water.
      6 Leaves usually about 1 mm long, weakly recurved at base; perichaetial leaves clearly differentiated, inner perichaetial leaves high-sheathing; setae yellowish to brown-yellow **2. B. convoluta**
      6 Leaves usually more than 2 mm long, recurved up to ⅔ length of leaf; perichaetial leaves little differentiated, inner perichaetial leaves with usually slightly sheathing base; setae reddish **1. B. unguiculata**
  3 Axillary hairs usually with brownish basal cells; leaves lanceolate to ovate-lanceolate, sometimes ovate.
    7 Leaves gradually tapering to long acute apex, costa excurrent, at least in upper leaves **6. B. acuta**

7 Leaves not with long acute apex, costa percurrent or ending below apex.[*]
8 Plants usually of wet habitats, sometimes encrusted with lime; leaves with strongly decurrent base **11. B. tophacea**
8 Plants usually of dry to moist habitats, not encrusted with lime; leaves without decurrent or very slightly decurrent base.
9 Leaf margins bistratose above; gemmae often in axils of upper leaves **9. B. rigidula**
9 Leaf margins unistratose throughout;[**] gemmae absent.
10 Leaves imbricate, not flexuose, twisted to crisped when dry, ovate to ovate-lanceolate, apex rounded to acute; peristome teeth recorded as very short **8. B. imbricata**
10 Leaves loosely imbricate, flexuose-twisted, crisped, lanceolate, tapering to acute, acuminate or subulate apex; peristome teeth long.[***]
11 Cells nearly uniform in shape throughout leaf; cells on ventral surface of costa elongated.
12 Leaves with wide acute apex, usually not recurved when moist; plants rare, brownish-green, growing in moist habitats **10. B. spadicea**
12 Leaves with long-acuminate narrow apex, recurved when moist; plants common, dark green to olive-green, growing in a wide range of drier habitats **7. B. fallax**
11 Cells not uniform in shape throughout leaf; cells on ventral surface of costa not or very slightly different from adjacent cells.
13 Leaves, especially upper leaves, crisped when dry, flexuose when moist, 2–3 mm long; capsules cylindrical **13. B. cylindrica**
13 Leaves not crisped when dry, not flexuose when moist, usually 1–2 mm long; capsules ellipsoid **12. B. vinealis**

According to the generic concept of Zander (1993), only species 1–3 should be included in the genus *Barbula*. According to the differential characters of the sections applied by Zander, *B. unguiculata* should be included in section *Barbula*, *B. convoluta* in section *Convolutae* (syn. *Streblotrichum*), and *B. ehrenbergii* in section *Hydrogonium*. *Barbula hornschuchiana* and *B. revoluta* are considered there as species of *Pseudocrossidium*, and the remaining species, included here in *Barbula sensu lato* (species 6–13), are considered as species of *Didymodon*.

[*] Sometimes shortly excurrent in *B. rigidula*.
[**] Except in some plants of desert populations of *B. imbricata*.
[***] Recorded as short in *B. spadicea* for which no sporophytes have been seen locally.

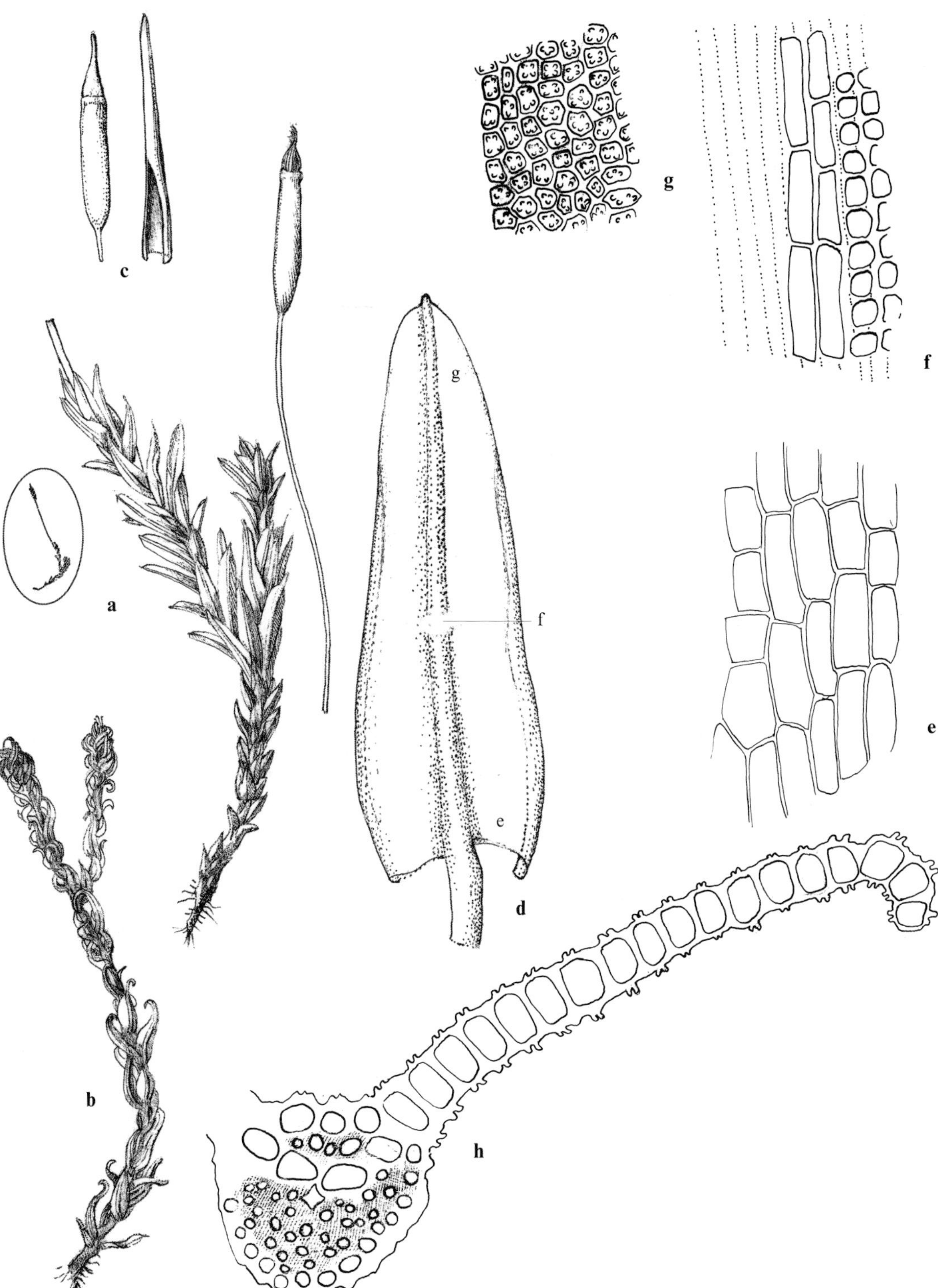

Figure 41. *Barbula unguiculata*: **a** habit with deoperculate capsule, moist (×10); **b** habit, dry (×10); **c** operculate capsule and calyptra (×10); **d** leaf (×50); **e** basal cells; **f** cells on ventral surface of costa at mid-leaf; **g** upper cells (all leaf cells ×500); **h** cross-section of costa and portion of lamina (×500). (*Kushnir*, 5 Feb. 1943)

1. Barbula unguiculata Hedw., Sp. Musc. Frond.: 118 (1801). [Bilewsky (1965): 371; Rabinovitz-Sereni (1931)]
Streblotrichum unguiculatum (Hedw.) Loeske, Stud. Morph. Syst. Laubm. 102 (1910).

Distribution map 41, Figure 41 (p. 129).

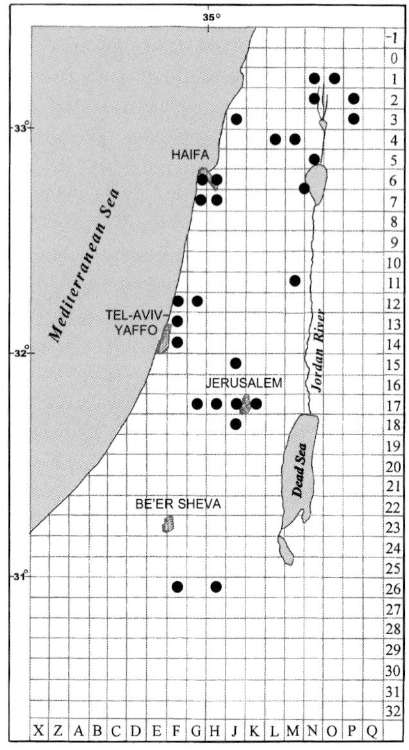

Map 41: *Barbula unguiculata*

Plants usually ca. 10 mm high, yellowish-green to olive-green to brown. *Stem* axillary hairs hyaline throughout. *Leaves* contorted or curved inwards when dry, erectopatent to spreading when moist, 1.5–2.5(–3) mm long (usually longer than 2 mm), oblong-lanceolate or lingulate, obtuse to acute, apiculate; margins usually recurved up to ⅔ length of leaf, papillose; costa green to yellowish-brown, shortly excurrent, with elongate-rectangular cells on ventral surface; leaf cells incrassate; basal cells rectangular, pellucid, upper cells round-hexagonal, 8–10 µm wide, papillose, with crowded C-shaped, bifid papillae, opaque; perichaetial leaves not much differentiated, similar or sometimes larger than other leaves, with only slightly sheathing base. *Setae* reddish; capsules up to 3(–3.5) mm long, including 1 mm long-rostrate operculum, cylindrical to ellipsoid with wide base, light brown; peristome teeth long, spirally twisted. *Spores* nearly smooth, 10–13 µm in diameter.

Habitat and Local Distribution. — Plants growing in tufts on loess, sandy loam, rendzina, or terra rossa, in partly shaded habitats and on river banks. Common: Sharon Plain, Upper Galilee, Lower Galilee, Mount Carmel, Samaria, Shefela, Judean Mountains, Judean Desert, Central Negev, Hula Plain, Upper Jordan Valley, and Golan Heights.

General Distribution. — Recorded from Southwest Asia (Cyprus, Turkey, Lebanon, Syria, Iraq, Iran, and Afghanistan), around the Mediterranean, Europe (throughout), Macaronesia, tropical and North Africa, northeastern, eastern, and central Asia, North and Central America, northwestern and southern South America.

For characters differentiating between *Barbula unguiculata* and *B. convoluta* see the discussion of *B. convoluta* below.

2. Barbula convoluta Hedw., Sp. Musc. Frond. 120 (1801). [Bilewsky (1965): 371]
Streblotrichum convolutum (Hedw.) P. Beauv., Prodr. 89 (1805); *B. convoluta* Hedw. var. *commutata* (Jur.) Husn., Fl. Mousses Nord. Ouest, Edition 2: 164 (1886). — Bilewsky (1965): 371.

Distribution map 42 (p. 132), Figure 42, Plate II: h.

Plants usually ca. 5 mm high, sometimes up to 10 mm, olive-green to brownish. *Stem* axillary hairs hyaline throughout. *Leaves* contorted or curved inwards when dry,

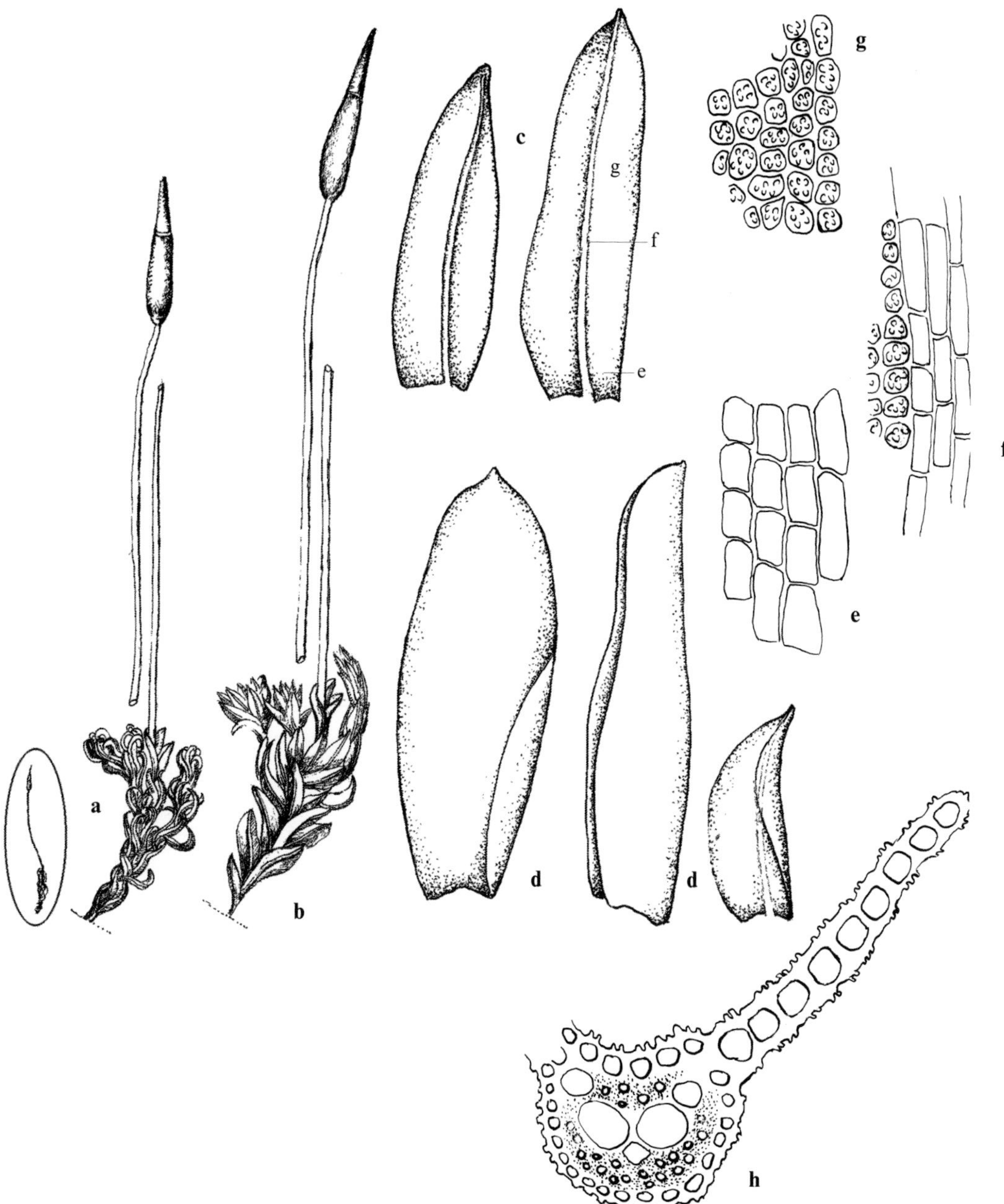

Figure 42. *Barbula convoluta*: **a** habit, dry (×10); **b** habit, moist (×10); **c** leaves (×50); **d** perichaetial leaves (×50); **e** basal cells; **f** cells on ventral surface of costa at mid-leaf; **g** upper cells (all leaf cells ×500); **h** cross-section of costa and portion of lamina (×500). (*D. Zohary*, 28 Mar. 1943)

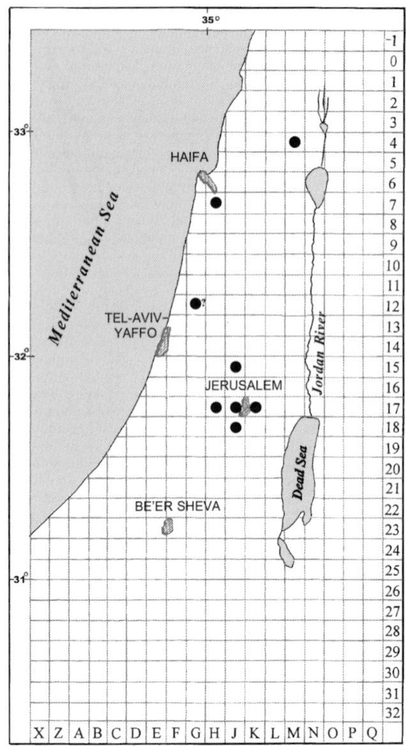

Map 42: *Barbula convoluta*

erectopatent to spreading when moist, about 1 mm long, oblong-lanceolate or lingulate, rounded to acute, rarely acuminate, minutely apiculate; margins usually plane or weakly recurved at base, papillose; costa brownish-green, percurrent or ending just below apex, with elongate cells on ventral surface; basal cells quadrate to short-rectangular, incrassate, upper cells round-quadrate, 6–8 µm wide, papillose with crowded C-shaped papillae, forming compound papillae towards apex, opaque, ± incrassate. *Rhizoidal gemmae* present, brown, spherical, up to 220 µm in diameter; perichaetial leaves clearly differentiated, larger than other leaves, inner pale yellow, convolute, high-sheathing, truncate, rounded or broadly obtuse and apiculate, usually without costa. *Setae* yellow to brown-yellow; capsules up to 3.5(–4) mm long including 1 mm long narrow high-conical operculum, erect or sometimes slightly curved, narrow-ellipsoid to subcylindrical, brown; peristome teeth long, spirally twisted. *Spores* nearly smooth, 8–10 µm in diameter; sparsely gemmate (seen with SEM).

Habitat and Local Distribution. — Plants growing in dense tufts on calcareous rocks, in rock crevices and soil pockets among stones and on calcareous soil, on partly shaded slopes and in maquis. Occasional: Sharon Plain(?), Upper Galilee, Mount Carmel, and Judean Mountains.

General Distribution. — Recorded from Southwest Asia (Cyprus, Turkey, Syria, Lebanon, Iraq, and Iran), around the Mediterranean, Europe (throughout), Macaronesia, tropical and North Africa, northeastern, eastern, and central Asia, Australia, New Zealand, and North and Central America.

Barbula convoluta and *B. unguiculata* have some characters in common. They can be easily distinguished by the high-sheathing perichaetial leaves, the nearly plane leaf margins, and the yellow setae of *B. convoluta*. Conversely, *B. unguiculata* has little differentiated perichaetial leaves, recurved leaf margins up to ⅔ the length, and reddish setae.

3. Barbula ehrenbergii (Lorentz) M. Fleisch., Musci Arch. Indic. Exs. ser. 4: no. 161 (1901).

Trichostomum ehrenbergii Lorentz, Phys. Abh. Königl. Akad. Wiss. Berlin 1: 25, Tab. 465 (1868).

Plants large, usually from 20–50 mm high. *Stems* often with rhizoids along a great part; axillary hairs hyaline throughout. *Leaves* ± remotely spaced, flaccid, erect, flexuose to crisped when dry, erectopatent to patent when moist, 2–4 mm long, lingulate to broadly lanceolate-lingulate, usually rounded-obtuse to broad-acute, sometimes cucullate when moist; margins entire or crenulate, usually plane, sometimes recurved in lower half; costa percurrent or ending shortly below apex, with elongate cells on ventral surface; leaf cells thin-walled to slightly incrassate, pellucid; basal cells laxly elongate-rectangular, mid-leaf cells quadrate to irregularly hexagonal, 10–15 μm wide, smooth to weakly papillose. *Axillary gemmae* present or absent. *Sporophytes* seen in single locality: Upper Galilee (bank of Hazbani river).

Plants of *Barbula ehrenbergii* are often found growing together with *B. tophacea* and *Eucladium verticellatum* in the same wet habitats. They are easily distinguished from the other two species by the paler green colour. *Barbula ehrenbergii* can be differentiated from *B. tophacea* in the larger flaccid leaves, erectopatent not reflexed when moist, the tomentose stems, and the axillary hairs without a brownish basal cell.

3a. **Barbula ehrenbergii** (Lorentz) M. Fleisch. var. **ehrenbergii**

Hydrogonium ehrenbergii (Lorentz) A. Jaeger, Ber. Thätigk. St. Gallischen Naturwiss. Ges. 1877–1878 : 405 (1880). — Bilewsky (1965) : 368 (as "*H. ehrenbergii* Jaeg. et Sauerbr.").

Distribution map 43.

Habit as in the species. Plants green to brown-green, ± yellowish, not bearing any vegetative reproductive organs in local plants (filiform or claviform axillary gemmae seem to occur sometimes elsewhere — see Crum & Anderson 1981).

Habitat and Local Distribution. — Plants growing in tufts in wet habitats, on soil or limestone rocks near or under dripping water, sometimes submerged, usually encrusted with lime. Common near water sources: Upper Galilee, Samaria, Judean Mountains, Central Negev, Southern Negev, Dan Valley, Bet She'an Valley, Lower Jordan Valley, Dead Sea area, and Golan Heights.

General Distribution. — Recorded from Southwest Asia (Cyprus, Turkey, Lebanon, Sinai, Jordan, Iraq, and Iran), around the Mediterranean (scattered records), Europe (mainly in the south), tropical and North Africa, eastern, central and southeastern Asia, Australia, and North and Central America.

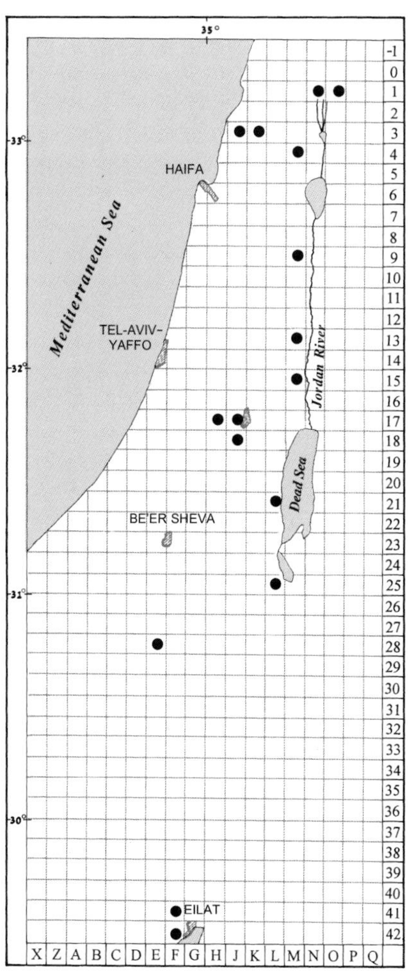

Map 43: *Barbula ehrenbergii* var. *ehrenbergii*

3b. Barbula ehrenbergii (Lorentz) M. Fleisch. var. **gemmipara** Herrnst. & Heyn, Nova Hedwigia 69 : 234 (1991).

Distribution map 44, Figure 43.

Habit as in the species. Plants as var. *ehrenbergii* but with abundant multicellular, fusiform axillary gemmae, which are 27–68 μm long.

Habitat and Local Distribution. — Plants growing in tufts by springs, on wet rocks above water level, or on soil under dripping water. Rare (recorded only from Israel): Upper Galilee, Judean Mountains, and Golan Heights.

The circumscription of this variety on the basis of the shape of gemmae — hence the varietal epithet — has been suggested by Dr. R. H. Zander.

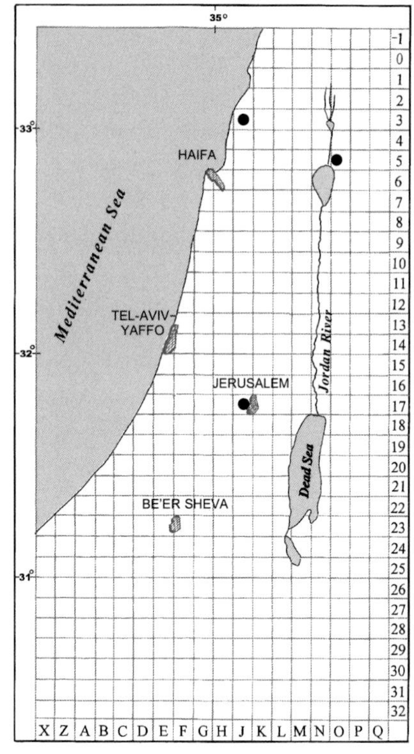

Map 44: *Barbula ehrenbergii* var. *gemmipara*

Figure 43. *Barbula ehrenbergii* var. *gemmipara*: **a** habit, moist (×10); **b** habit, dry (×10); **c** leaves (×50); **d** basal cells; **e** upper cells (all leaf cells ×500); **f** cross-section of costa and portion of lamina (×500); **g** gemmae (×125). (*D. Kaplan*, 22 Jul. 1987)

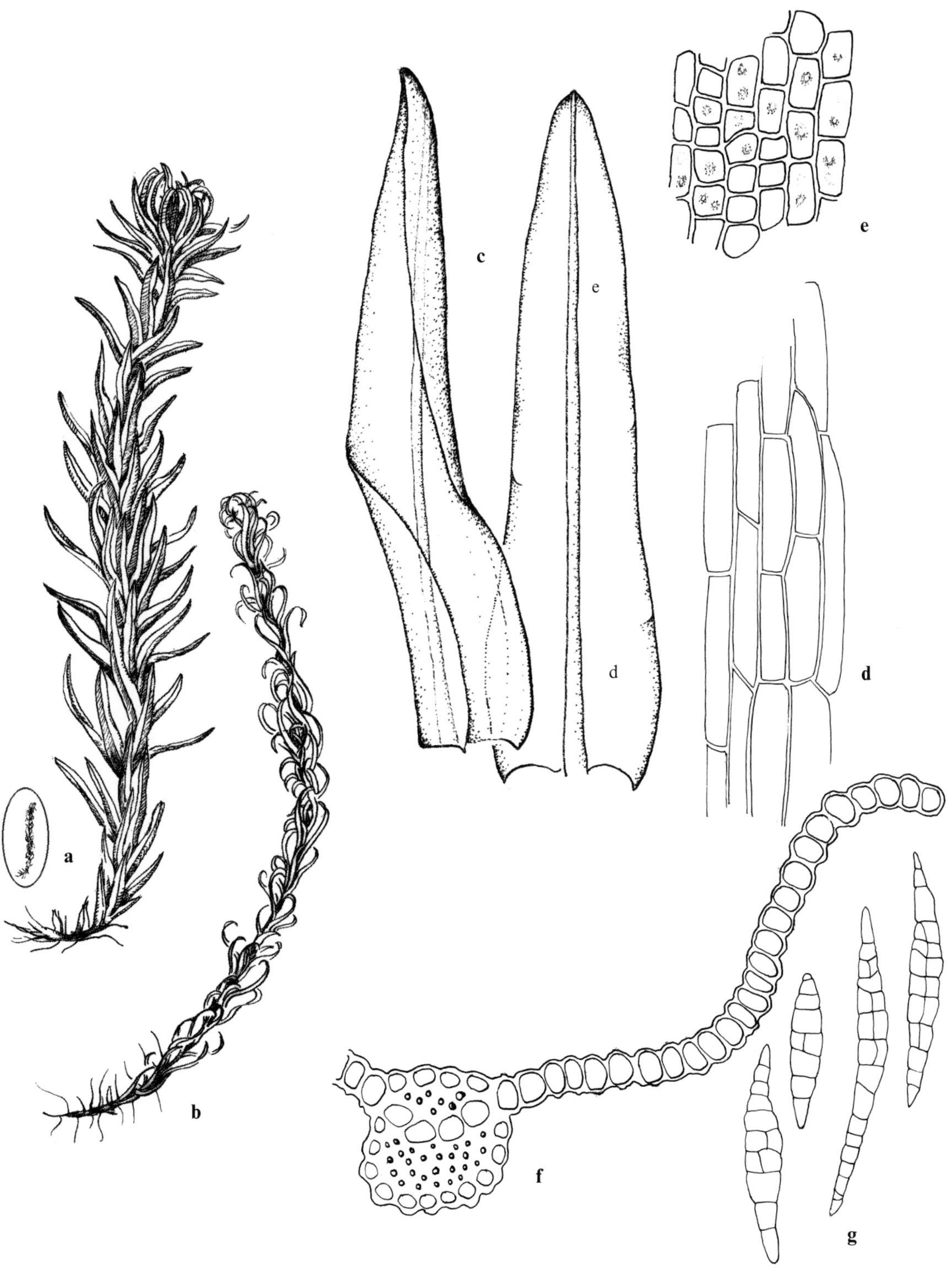
a
b
c
e
d
e
d
f
g

4. Barbula hornschuchiana Schultz, Flora 5 (Syl.) : 35 (1822). [Bilewsky (1965) : 369]
Pseudocrossidium hornschuchianum (Schultz) R. H. Zander, Phytologia 44 : 205 (1979).

Distribution map 45, Figure 44, Plate II : i.

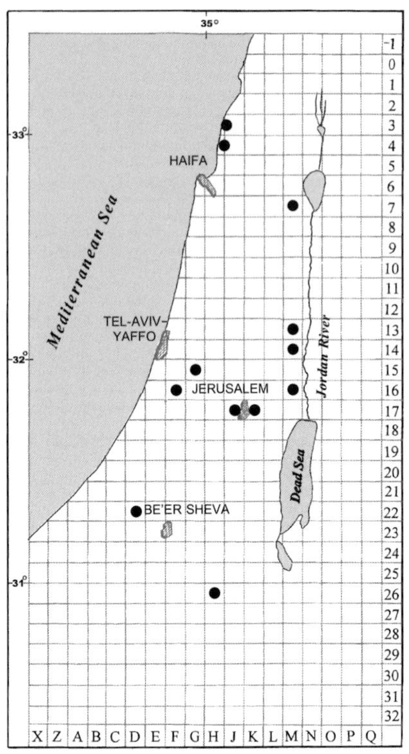

Map 45: *Barbula hornschuchiana*

Plants up to 7.5 mm high, bright green to olive-green, turning yellowish-brown. *Stem* axillary hairs usually hyaline throughout, sometimes lower cells thick-walled. *Leaves* ± twisted, loosely curved inwards when dry, erectopatent when moist, up to 1.25 mm long, lanceolate, acuminate, gradually tapering from near base to apex; margins entire, strongly revolute from base to apex, often to wide costa, enclosing up to six rows of thin-walled marginal cells; costa often slightly grooved in upper part, excurrent, with cells on ventral surface quadrate to short-rectangular; leaf cells (except marginal) incrassate; basal cells quadrate to short-rectangular, upper cells incrassate, irregularly quadrate, usually 10–14 μm wide, papillose, often with three–four round or C-shaped papillae per cell; perichaetial leaves twice as long as other leaves, convolute-sheathing, with weak, excurrent costa, not papillose, pale. *Setae* and capsules as in *B. revoluta*. *Spores* 8–10 μm in diameter; densely gemmate (seen with SEM).

Habitat and Local Distribution. — Plants growing in loose or dense tufts on sandy loam, loess and rendzina, in exposed habitats; usually restricted to the sub-Mediterranean and the Irano-Turanian territories. Occasional to common: Coastal Galilee, Philistean Plain, Lower Galilee, Judean Mountains, Northern Negev, Central Negev, and Lower Jordan Valley.

General Distribution. — Recorded from Southwest Asia (Cyprus, Turkey, Syria, Lebanon, Jordan, and Iraq), around the Mediterranean, Europe (widespread), Macaronesia, North and South Africa, northeastern Asia, and North America.

See the discussion of *Barbula revoluta* for the common characters of *B. hornschuchiana* and *B. revoluta*.

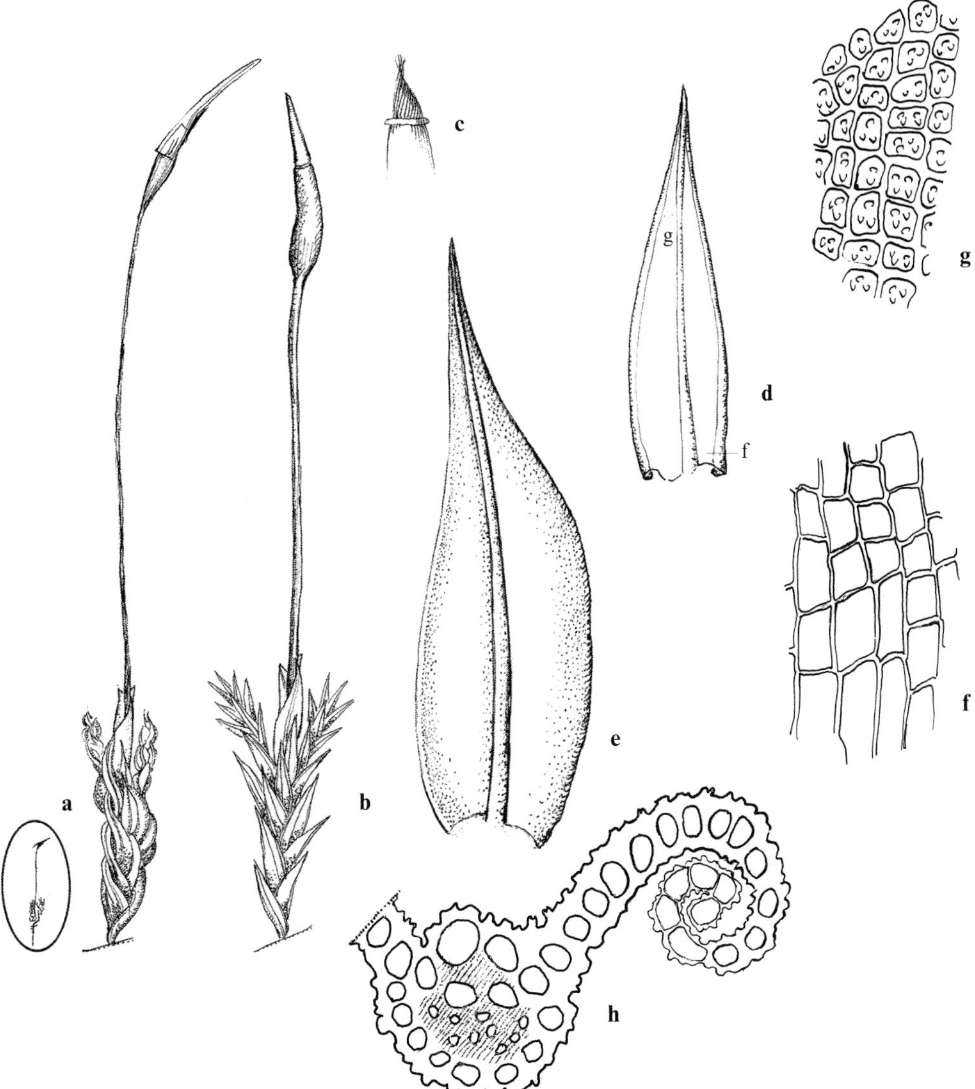

Figure 44. *Barbula hornschuchiana*: **a** habit, dry (×10); **b** habit, moist (×10); **c** peristome (×25); **d** leaf (×50); **e** perichaetial leaf; **f** basal cells; **g** upper cells (all leaf cells ×500); **h** cross-section of costa and portion of lamina (×500). (*Pazy*, *Heyn* & *Herrnstadt*, 21 Feb. 1976)

5. Barbula revoluta Brid. *in* Schrad., J. f. Bot. 1800 : 299 (1801).
[Bilewsky (1965) : 368]
Pseudocrossidium revolutum (Brid. *in* Schrad.) R. H. Zander, Phytologia 44 : 204 (1979).

Distribution map 46, Figure 45.

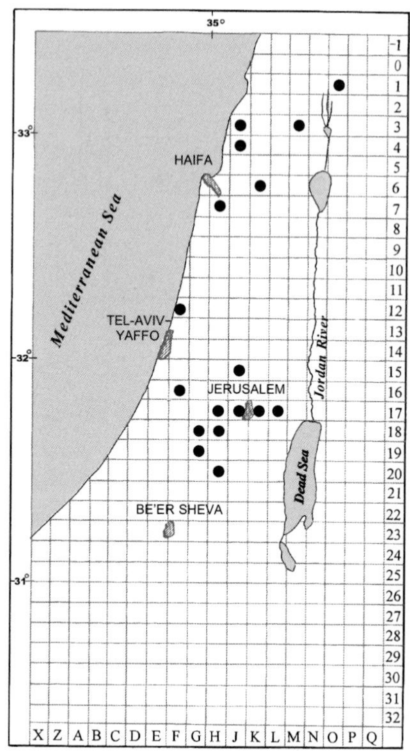

Map 46: *Barbula revoluta*

Plants up to 7 mm long, bright green to olive-green. *Stem* axillary hairs usually hyaline throughout. *Leaves* tightly coiled, curved inwards when dry, erectopatent when moist, up to 1 mm long, oblong-lanceolate to lingulate, acute or obtuse, apiculate; margins entire, strongly revolute from base to apex, often to wide costa, enclosing usually three rows of thin-walled cells; costa excurrent, with cells on ventral surface quadrate to rectangular; leaf cells (except marginal) incrassate; basal cells rectangular, mid- and upper leaf cells round-quadrate, 8–10 μm wide, papillose, with one–four round or C-shaped papillae per cell; perichaetial leaves twice as long as other leaves, convolute-sheathing, with costa weak or absent, not papillose, pale. *Capsules* erect, sometimes slightly curved, up to 3 mm long including ca. 1 mm long operculum, oblong-ellipsoid to subcylindrical, orange-red turning dark brownish-red; peristome teeth long, twisted. *Spores* 10–13 μm in diameter.

Habitat and Local Distribution. — Plants growing in dense tufts or cushions, on sandy loam, rendzina or terra rossa, on soil of rock crevices, on exposed sandstone walls and north-facing limestone rocks. Usually restricted in Israel to the Mediterranean territory. Common: Coastal Galilee, Sharon Plain, Philistean Plain, Upper Galilee, Lower Galilee, Mount Carmel, Shefela, Judean Mountains, Judean Desert, and Golan Heights.

General Distribution. — Recorded from Southwest Asia (Cyprus, Turkey, Lebanon, Syria, Iraq, Iran, and Afghanistan), around the Mediterranean, Europe (widespread), Macaronesia, North and South Africa, central Asia, and North America.

Barbula revoluta, together with *B. hornschuchiana* are often included in the genus *Pseudocrossidium*. They can be distinguished from the other local *Barbula* species mainly by the strongly revolute leaves and the excurrent costa. *Barbula revoluta* differs from *B. hornschuchiana* in the more closely revolute leaves, which are obtuse and apiculate, not acuminate, and the smaller leaf cells. *Barbula revoluta* usually occurs in typically Mediterranean habitats, whereas *B. hornschuchiana* is mainly a plant of sub-Mediterranean and Irano-Turanian territories of Israel.

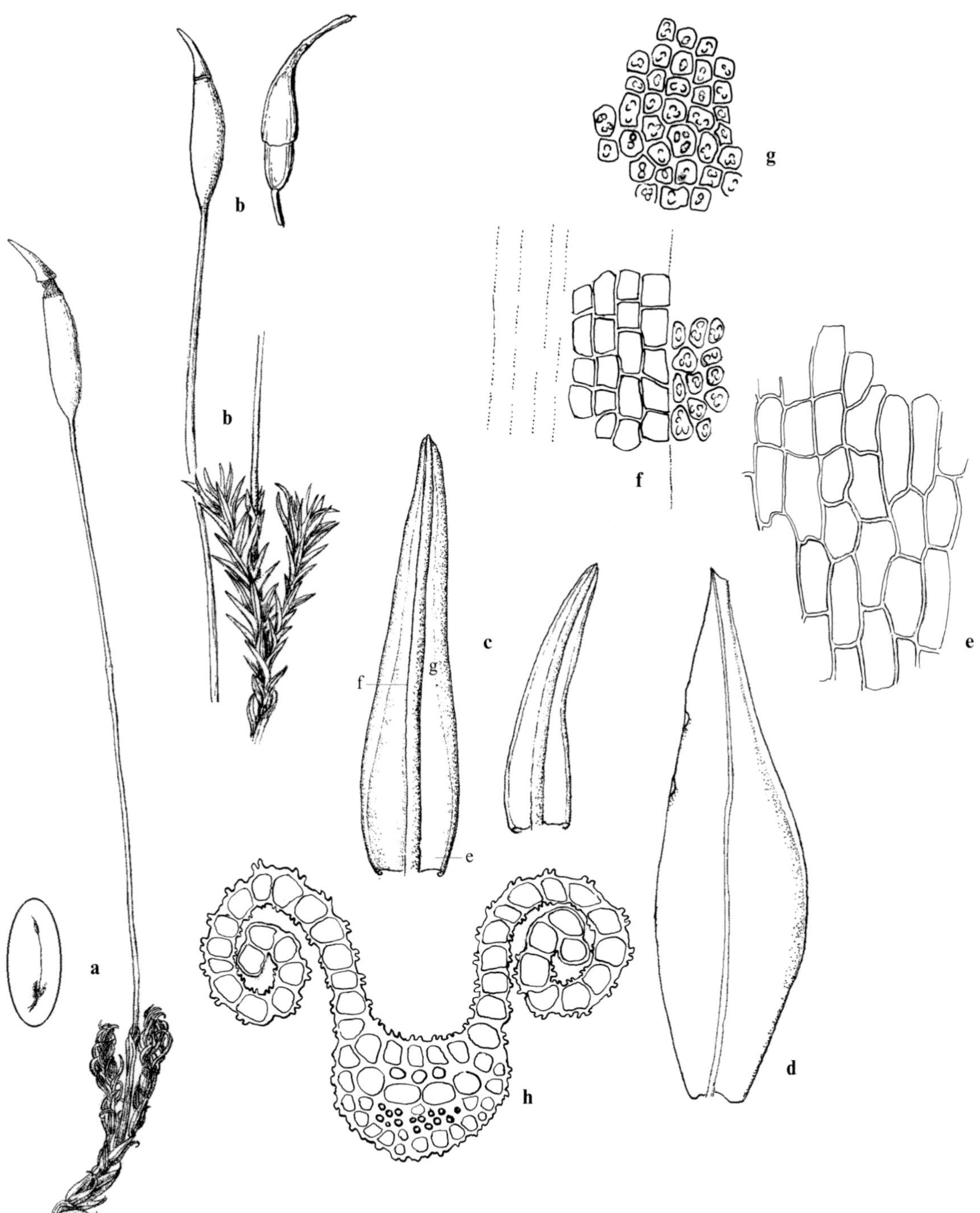

Figure 45. *Barbula revoluta*: **a** habit, dry (×10); **b** habit, moist, and capsule with calyptra (×10); **c** leaves (×50); **d** perichaetial leaf; **e** basal cells; **f** cells on ventral surface of costa; **g** mid-leaf cells (all leaf cells ×500); **h** cross-section of leaf (×500). (*Waisel*, 2 Oct. 1954)

6. Barbula acuta (Brid.) Brid., Mant. Musc. 96 (1819).

Tortula acuta Brid., Spec Musc. 1 : 265 (1806); *B. gracilis* Schwägr., Sp. Musc. Frond., Suppl. 1 : 125 (1811). — Bilewsky (1965) : 369; *Didymodon acutus* (Brid.) K. Saito, J. Hattori Bot. Lab. 39 : 519 (1975); *D. rigidulus* var. *gracilis* (Schleich. ex Hook. & Grev.) R. H. Zander, Cryptog. Bryol. Lichénol. 2 : 393 (1981).

Distribution map 47, Figure 46.

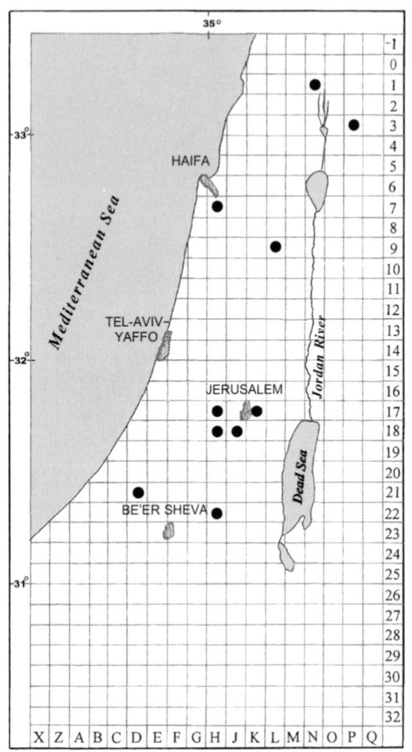

Map 47: *Barbula acuta*

Plants up to 10 mm high, green to yellowish-brown. *Stem* axillary hairs with brownish basal cell. *Leaves* appressed and flexuose when dry, erectopatent when moist, concave, 1–1.5 mm long, ovate-lanceolate, gradually tapering to long acute apex from ovate base; margins entire, recurved in lower half, plane above; costa green to brownish, occupying most of upper part of leaf, excurrent at least in upper leaves, with cells on ventral surface of costa mostly similar in shape to other cells, some slightly larger; leaf cells smooth, incrassate; basal cells short-rectangular, longer towards costa, mid-leaf cells rounded to irregularly hexagonal, 8–10 µm wide; perichaetial leaves longer than other leaves, ca. 2 mm long. *Setae* reddish; capsules up to 2.5 mm long including ca. 1 mm long operculum, ellipsoid, brown to reddish; peristome teeth long, spirally twisted. *Spores* 8–12 µm in diameter.

Habitat and Local Distribution. — Plants growing in tufts on exposed or shaded soil. Occasional: Philistean Plain, Mount Carmel, Mount Gilboa, Judean Mountains, Judean Desert, Dan Valley, and Golan Heights.

General Distribution. — Recorded from Southwest Asia (Cyprus, Turkey, Syria, Lebanon, Sinai, Jordan, Iraq, Afghanistan, Saudi Arabia, and Oman), around the Mediterranean, Europe, (throughout), Macaronesia, North Africa, central Asia, and North and Central America.

Plants of *Barbula acuta* resemble *B. fallax* described below. Yet, they are smaller and their cells on the ventral surface of costa are not distinct from adjacent cells. *Barbula acuta* is considered by Zander (*in* Sharp *et al.* 1994) as part of *Didymodon rigidulus* Hedw., *sensu lato*.

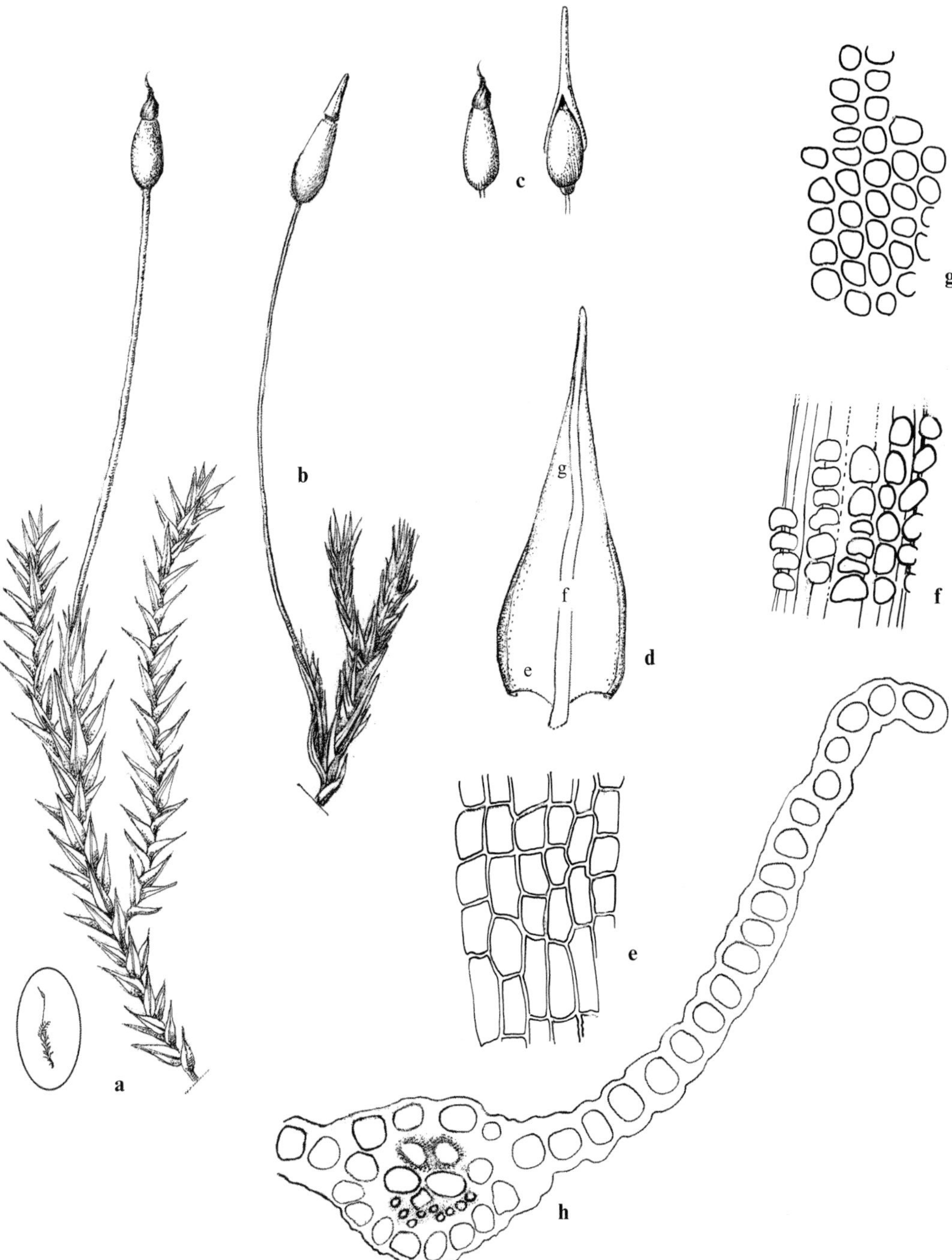

Figure 46. *Barbula acuta*: **a** habit, moist (×10); **b** habit, dry (×10); **c** deoperculate capsule, and capsule with calyptra (×10); **d** leaf (×50); **e** basal cells; **f** cells on ventral surface of costa; **g** upper cells (all leaf cells ×500); **h** cross-section of costa and portion of lamina (×500). (*D. Zohary*, 12 Feb. 1943)

7. Barbula fallax Hedw., Sp. Musc. Frond. 120 (1801). [Bilewsky (1965) : 370]
Didymodon fallax (Hedw.) R. H. Zander, Phytologia 41 : 28 (1978).

Distribution map 48, Figure 47, Plate II : k.

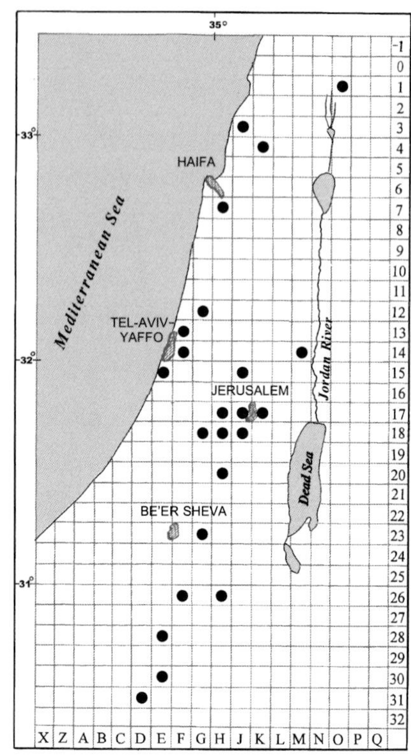

Map 48: *Barbula fallax*

Plants up to 13 mm high, dark green to olive-green, turning yellowish-brown. *Stem* axillary hairs with brownish basal cell. *Leaves* distant, loosely imbricate, erect, slightly twisted when dry, ± keeled in upper part, spreading to widely recurved when moist, 1.0–1.75 mm long, lanceolate, long-acuminate from an ovate base, apex rather narrow, recurved when moist; margins entire, usually recurved in lower half, sometimes to the apex; costa reddish, percurrent, with elongated cells on ventral surface; leaf cells incrassate, nearly uniform throughout leaf; basal cells subquadrate, mid-leaf cells irregularly quadrate to hexagonal, 8–13 μm wide, inconspicuously papillose; perichaetial leaves slightly longer than other leaves. *Setae* yellowish, turning red; capsules up to 3 mm long including ca. 1 mm long operculum, ellipsoid, sometimes with wider lower part, yellowish-brown; peristome teeth about 1 mm long, spirally twisted. *Spores* 12–14 μm in diameter, weakly papillose; gemmate, some processes joined (seen with SEM).

Habitat and Local Distribution. — Plants growing in dense tufts in a wide range of habitats, on shaded walls, limestone rocks, in rock crevices, and on various soils: sandy loam, terra rossa, rendzina, or loess. Common: Sharon Plain, Philistean Plain, Upper Galilee, Mount Carmel, Shefela, Judean Mountains, Central Negev, Lower Jordan Valley, and Golan Heights.

General Distribution. — Recorded from Southwest Asia (Cyprus, Turkey, Syria, Lebanon, Jordan, Iraq, and Afghanistan), around the Mediterranean, Europe (throughout), Macaronesia, North Africa, northeastern, eastern, and central Asia, and North and Central America.

Plants of *Barbula fallax* can be distinguished by the elongate cells on the ventral surface of the costa, the strongly incrassate leaf cells, and the inconspicuous papillosity of the mid-leaf cells.

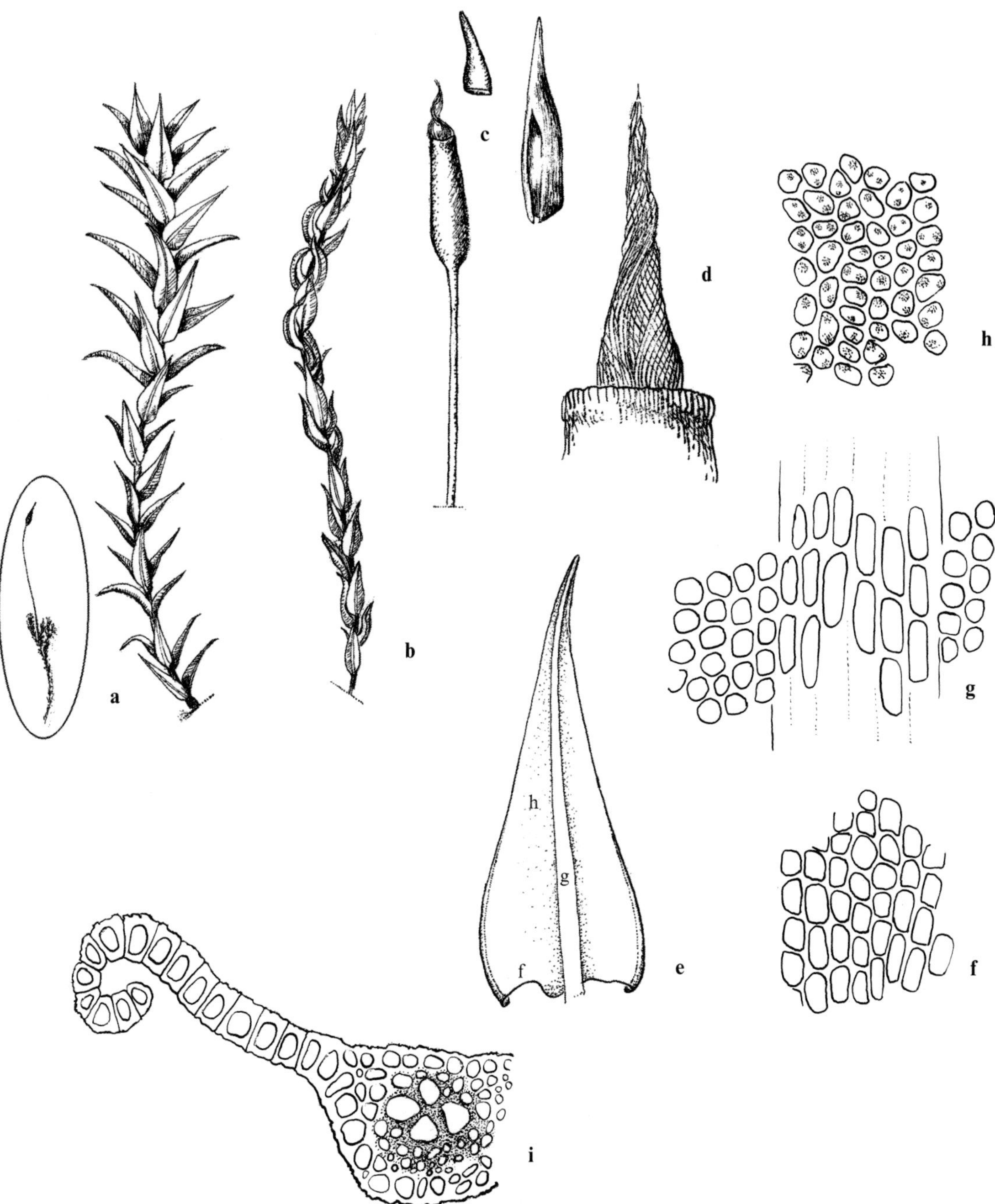

Figure 47. *Barbula fallax*: **a** stem moist (×10); **b** stem dry (×10); **c** seta with deoperculate capsule, operculum and calyptra (×10); **d** peristome (×50); **e** leaf (×50); **f** basal cells; **g** cells on ventral surface of costa **h** mid-leaf cells (all leaf cells ×500); **i** cross-section of costa and portion of lamina (×500). (*D. Zohary*, 12 Feb. 1943)

8. Barbula imbricata Herrnst. & Heyn, Bryologist 94 : 174 (1991).

B. trifaria auct., non (Hedw.) Mitt., J. Linn. Soc. Bot., Suppl. 1 : 36 (1859); *Didymodon luridus* Hornsch. *in* Spreng., Syst. Veg. Fl. Peruv. Chil. 4 : 173 (1827). — Bilewsky (1965) : 367; *D. trifarius* auct., non (Hedw.) Röhl., Deutschl. Fl. (Edition 2), Kryptog. Gew. 3 : 56 (1813); *B. rigidula* (Hedw.) Milde var. *desertorum* J. Froehl., Ann. Naturhist Mus. Wien 63 : 31 (1959); *B. trifaria* var. *desertorum* (J. Froehl.) S. Agnew *in* S. Agnew & Vondr., Feddes Repert. 86 : 366 (1975). — Brullo *et al.* (1991).

Distribution map 49, Figure 48.

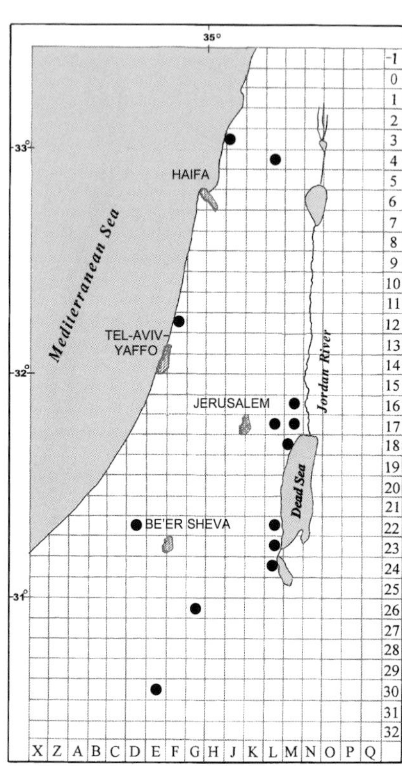

Map 49: *Barbula imbricata*

Plants up to 10 mm high, bright to dark green, turning brownish. *Stems* axillary hairs with brownish basal cell. *Leaves* imbricate, ± appressed, ± curved inwards when dry, patent when moist, ca. 1 mm long, ovate to ovate-lanceolate, apex rounded to acute, base sometimes slightly decurrent; margins entire, recurved below, plane above; bistratose at apex in some plants growing in arid environments; costa nearly round in cross section and 40–55 µm wide near base (in plants from Mediterranean habitats) or ± oval and 70–75 µm wide (in some plants in populations of arid habitats), ending in apex or below it, darker than the rest of leaf; cells on ventral surface of costa not different from adjacent cells; leaf cells usually incrassate; basal cells rectangular, upper cells round-quadrate to hexagonal, 7–9 µm wide, ± incrassate, smooth to weakly papillose, pellucid. *Setae* purplish; capsules ca. 1 mm long, ellipsoid, light brown, without operculum (operculum and peristome not seen in local collections); peristome teeth recorded as very short and straight. *Spores* ca. 12 µm in diameter.

Habitat and Local Distribution. — Plants growing on exposed sandy loam of the coast, in crevices of dolomite rocks and on loess of wadi banks. Occasional: Coastal and Upper Galilee, Sharon Plain, Judean Desert, Northern Negev, Central Negev, Lower Jordean Valley, and Dead Sea area.

General Distribution. — Recorded from Southwest Asia (Cyprus, Turkey, Iran, Syria, Lebanon, Jordan, Iraq, Saudi Arabia, and Oman), around the Mediterranean, Macaronesia, Europe (widespread), tropical and North Africa, eastern and central Asia, and North and Central America.

Plants of *Barbula imbricata* from arid environments may sometimes differ in several characters from the typical plants. They tend to be smaller, have a thicker costa, with cross-section ovate, and sometimes the leaf margin is bistratose near the apex. Froehlich (1959) described such plants as *B. rigidula* var. *desertorum*. Agnew & Vondráček (1975) considered them as *B. trifaria* (= *B. imbricata*) var. *desertorum*. A specimen deposited in BM (Iraq, Khanaqin, *BUH 239*), which was collected and cited

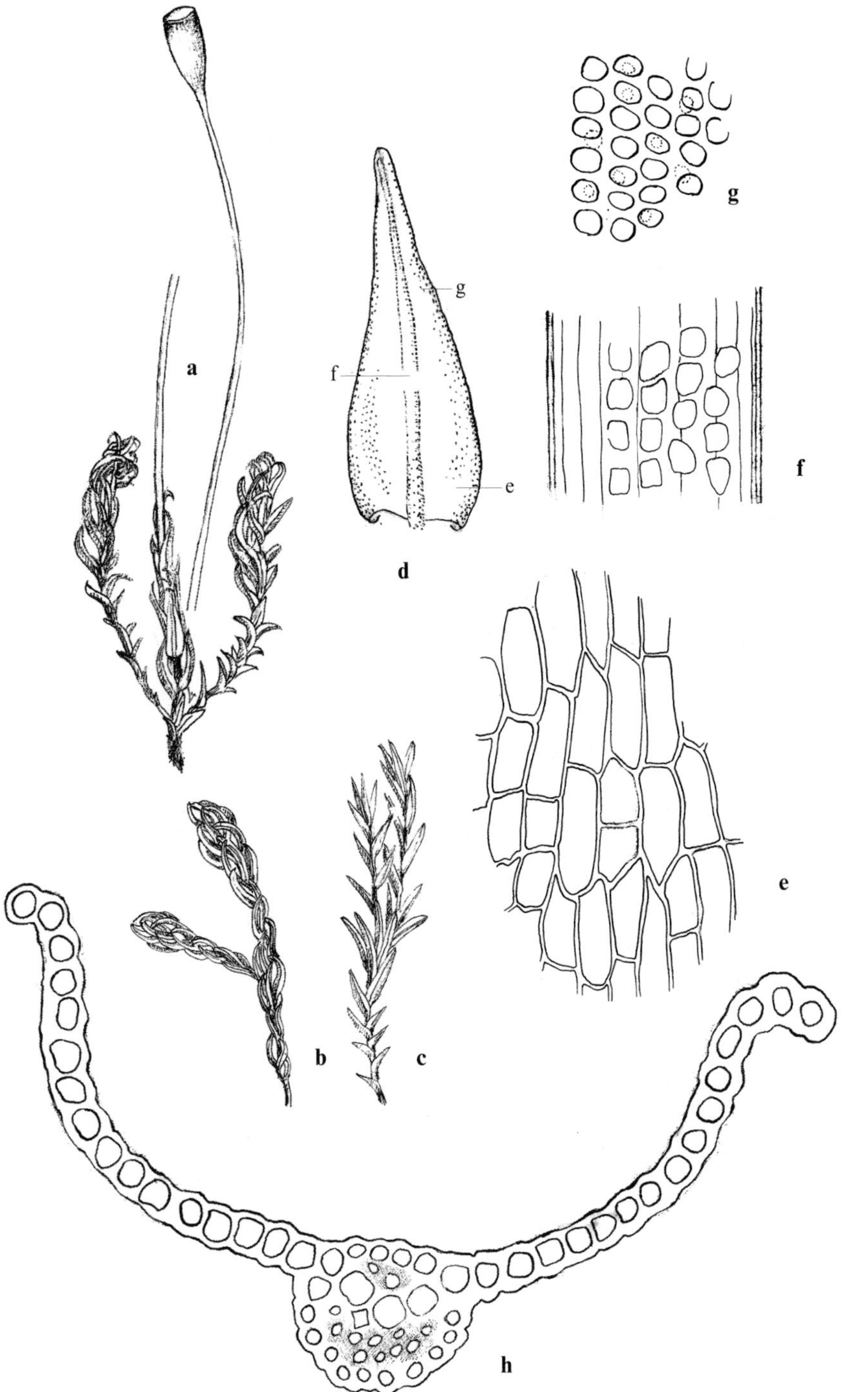

Figure 48. *Barbula imbricata*: **a** habit, ± dry, deoperculate capsule (peristome teeth missing) (×10); **b** plant, dry (×10); **c** plant, moist (×10); **d** leaf (×50); **e** basal cells; **f** cells on ventral surface of costa; **g** upper cells (all leaf cells ×500); **h** cross-section of leaf (×500). (*Rigik*, 27 Aug. 1943)

by them, agrees with some of the local desert specimens. They are not given a formal status in this Flora because the characters considered as typical of var. *desertorum* may be expressed at various degrees and perhaps seem to be correlated with the degree of aridity of the habitat. Similarly, bistratose margins at the leaf apex seem to be a reoccurring character in arid habitats in *Barbula* (e.g., *B. vinealis*).

In dry habitats, *Barbula imbricata* may grow together with *Trichostmopsis aaronis*, which can be distinguished by the contorted leaves with bistratose margins also below the apex and by the frequently bistratose vertical ridges of the lamina. In more mesic habitats, *B. imbricata* may be confused at first sight with *B. tophacea*, but the latter has leaves not imbricate when dry, with strongly decurrent bases, elongated cells on the ventral surface of the costa, and more densely papillose cells.

9. Barbula rigidula (Hedw.) Milde, Bryol. Siles. 118 (1869).
Didymodon rigidulus Hedw., Sp. Musc. Frond. 104 (1801). — Bilewsky (1965): 368.

Distribution map 50, Figure 49.

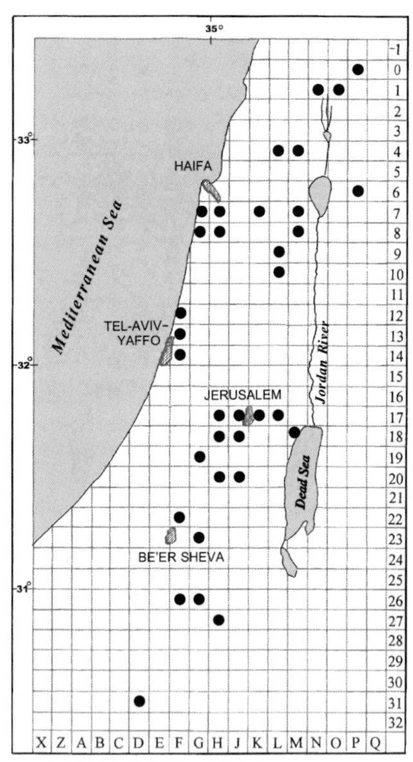

Map 50: *Barbula rigidula*

Plants usually up to 15 mm high, yellowish-green to brownish. *Stem* axillary hairs with brownish basal cell. *Leaves* appressed, sometimes ± twisted when dry, patent to spreading, concave when moist, 1–1.8 mm long, ovate to long-lanceolate, gradually tapering to a subulate, thick apex from a wide base; margins entire, usually narrowly recurved up to $\frac{1}{2}$–$\frac{2}{3}$ the length of leaf, in upper part bistratose, thickened; costa from subpercurrent to shortly excurrent, on ventral surface with quadrate cells hardly differing from adjacent cells, turning reddish-brown; basal cells weakly differentiated, quadrate to rectangular, mid-leaf cells irregularly round-quadrate to hexagonal, incrassate, 8–11 μm wide, weakly to obscurely papillose; papillae, when present, one–two per cell, simple or bifid and C-shaped. *Gemmae* often present in axils of upper leaves, 27–50 μm long, subglobose or rarely ellipsoidal, formed by three–four incrassate brown-walled cells. *Setae* orange red; capsules up to 2.5–3 mm long including 1 mm long operculum, ellipsoid to subcylindrical, reddish-brown; peristome teeth short, nearly straight. *Spores* 8–11 μm in diameter.

Habitat and Local Distribution. — Plants with costa growing in dense tufts or gregarious, often growing with other mosses and hepatics, on calcareous soils, rocks, or on stone fences, often in shaded habitats. Very common: Coastal Carmel, Sharon Plain, Upper Galilee, Lower Galilee, Mount Carmel, Mount Gilboa, Samaria, Shefela, Judean Mountains, Judean Desert, Northern Negev, Central Negev, Dan Valley, Dead Sea area, Golan Heights, and Mount Hermon.

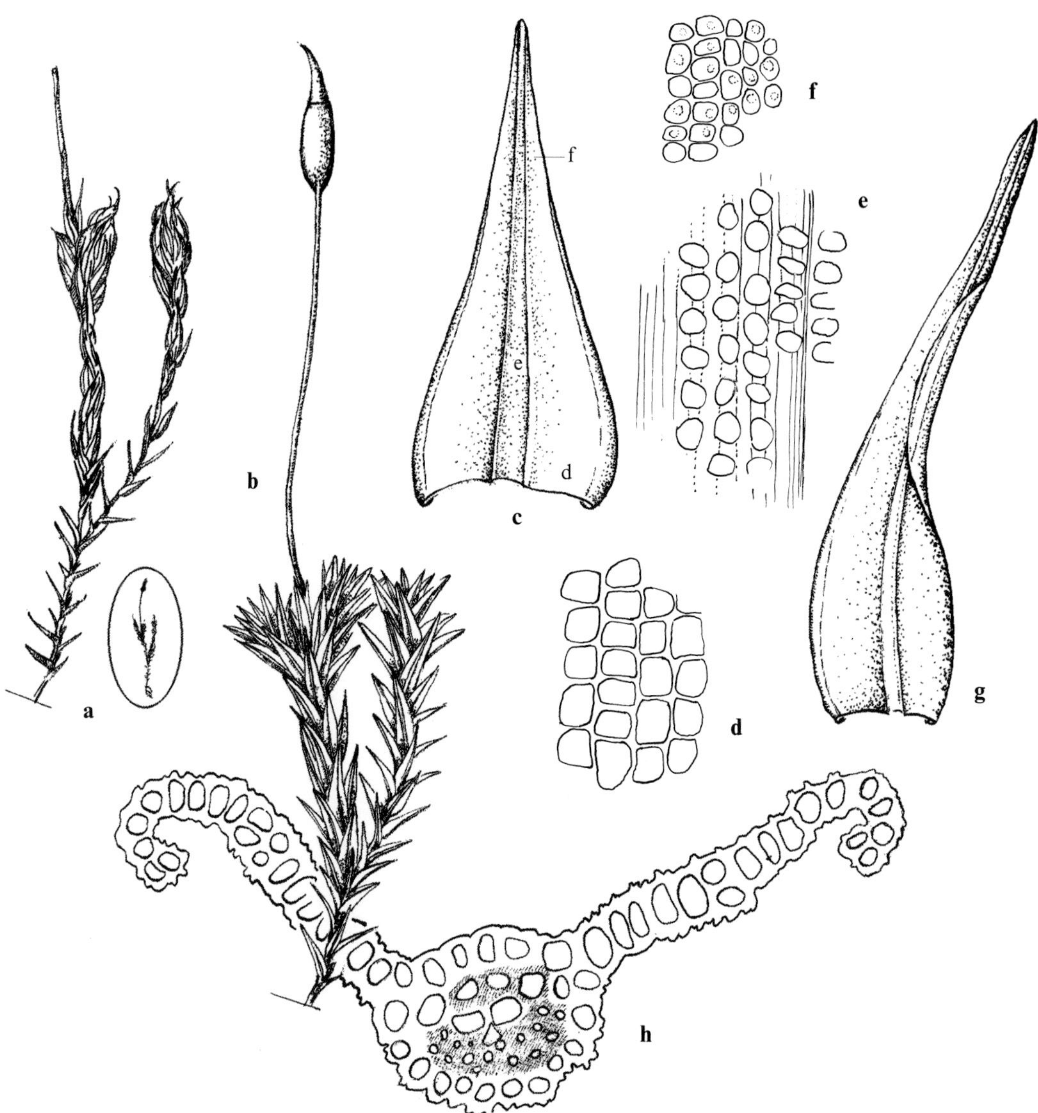

Figure 49. *Barbula rigidula*: **a** habit, dry (×10); **b** habit, moist (×10); **c** leaf (×50); **d** basal cells; **e** cells on ventral surface of costa; **f** upper cells (all leaf cells ×500); **g** leaf (×50); **h** cross-section of leaf (×500). (**a**–**f**: *Reichert*, April 1929; **g**, **h**: *Kushnir*, 5 Feb. 43)

General Distribution. — Recorded from Southwest Asia (Cyprus, Turkey, Syria, Lebanon, and Iraq), around the Mediterranean, Europe (throughout), Macaronesia, tropical and North Africa, northeastern, eastern, and central Asia, North and Central America, and Antarctica.

Barbula rigidula is often treated as *Didymodon rigidulus* Hedw. *sensu lato*, including var. *gracilis* (Schleich. ex Hook. & Grev.) R. H. Zander (= *Barbula acuta*).

Plants of *Barbula rigidula* are not easily distinguished in the field from several other *Barbula* species, in particular *B. fallax*. They can be differentiated from *B. fallax* by the axillary gemmae, the bistratose margins in the upper part of leaves, and the absence of differentiated cells on the ventral surface of the costa.

10. Barbula spadicea (Mitt.) Braithw., Brit. Moss Fl. 1 : 267 (1887).
[Bilewsky (1965) : 370]

Tortula spadicea Mitt., J. Bot. (London) 5 : 326 (1867); *Didymodon spadiceus* (Mitt.) Limpr., Laubm. Deutschl. 1 : 556 (1888).

Distribution map 51, Figure 50.

Plants up to 17 mm high, dull brownish-green. *Stem* axillary hairs with brownish basal cell. *Leaves* distant, appressed, flexuose or slightly twisted when dry, spreading, not or only sometimes slightly recurved when moist, sometimes keeled in upper part, ca. 1.5 mm long, lanceolate, gradually tapering into a wide acute apex from a wide base; margins entire, recurved in lower half; costa percurrent or ending below apex, with at least some elongated cells on ventral surface; leaf cells incrassate, nearly uniform in shape throughout leaf; basal cells subquadrate, mid-leaf cells irregularly quadrate to hexagonal, 8–11 μm wide, inconspicuously papillose. *Sporophytes* not seen in local plants. Recorded as having short and straight peristome teeth.

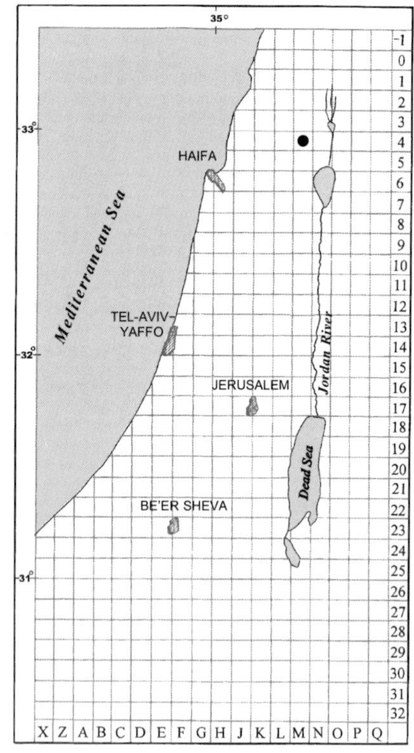

Map 51: *Barbula spadicea*

Habitat and Local Distribution. — Plants growing in dense tufts on rocks near running water. Very rare: Upper Galilee.

General Distribution. — Recorded from Southwest Asia (Cyprus, Turkey, Iran, and Lebanon), the Mediterranean (only few records), Europe (scattered records throughout), and northeastern Asia.

Rechecking of the herbarium material from Israel at HUJ showed the local distribution of *Barbula spadicea* to be more restricted than assumed by Bilewsky (1965) and Herrnstadt *et al.* (1991). So far, only one population could be definitely identified as *B. spadicea*: Upper Galilee, Wadi Tawahin (vicinity of Zefat) near running water

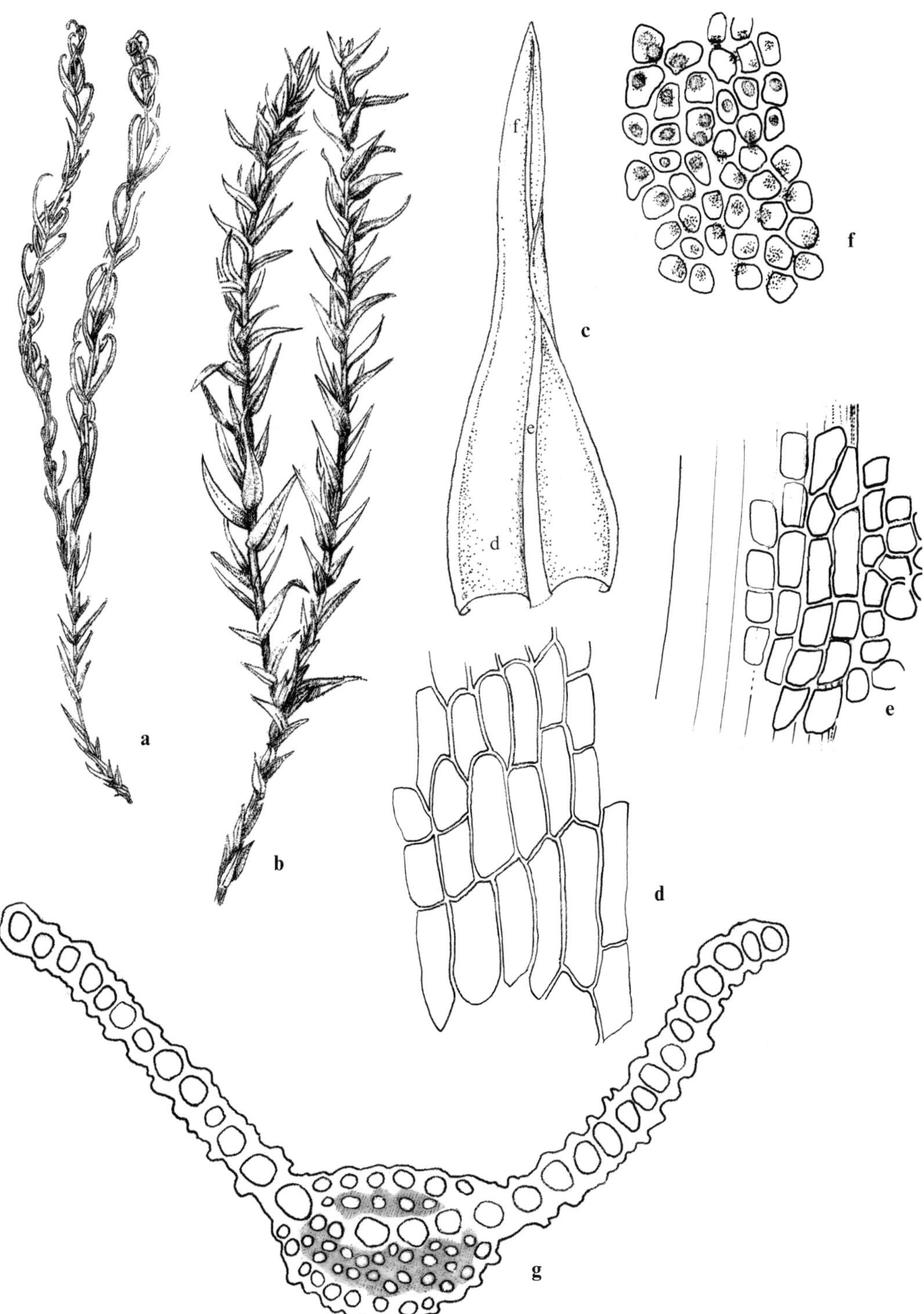

Figure 50. *Barbula spadicea*: **a** habit, dry (×10); **b** habit, moist (×10); **c** leaf (×50); **d** basal cells; **e** cells on ventral surface of costa; **f** upper cells (all leaf cells ×500); **g** cross-section of leaf (×500). (*Bilewsky*, 16 May 1958, No. 41 in Herb. Bilewsky)

(*Bilewsky*, 16 May 1958). Further examination of the collections may reveal other occurrences of *B. spadicea*, which might have been overlooked.

Barbula spadicea seems to be easily confused with luxuriant plants of *B. fallax*. Plants of *B. spadicea* are larger, inhabit, however, habitats that are more humid, and can be distinguished by the less strongly recurved leaves and by the wide-acute — not narrow-acuminate — leaf apex.

11. Barbula tophacea (Brid.) Mitt., J. Linn. Soc. Bot., Suppl. 1 : 35 (1859). [Rabinovitz-Sereni (1931)]

Trichostomum tophaceum Brid., Mant. Musc. 84 (1819); *Didymodon tophaceus* (Brid.) Lisa, Elenc. Musch. 31 (1837). — Bilewsky (1965) : 367 (as *D. tophacaceus* (Brid.) Jur.).

Distribution map 52, Figure 51.

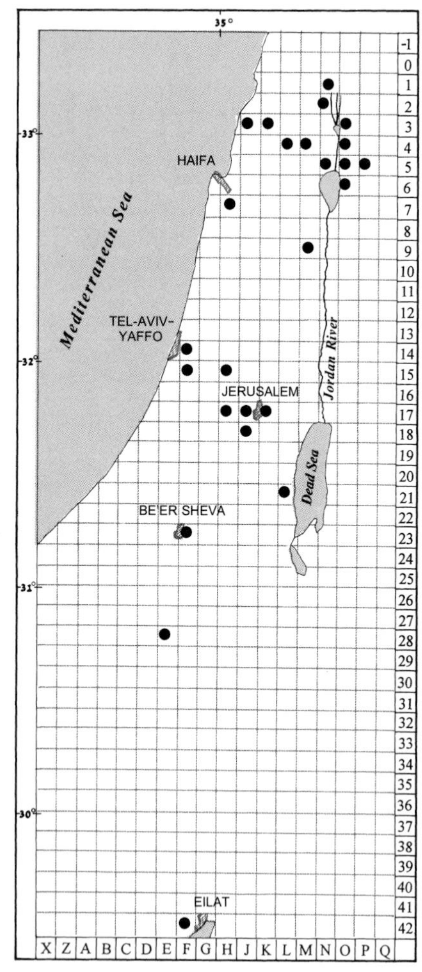

Map 52: *Barbula tophacea*

Plants up to 12(−15) mm high, olive- to yellow-green, brownish at base. *Stems* axillary hairs with brownish basal cell. *Leaves*: lower leaves flexuose, slightly contorted, upper leaves curved inwards when dry, in all leaves patent and recurved when moist, 1–1.5 mm long (lower slightly shorter), ± concave, ovate-lanceolate to oblong-lanceolate, rounded to acute, with strongly decurrent base; margins entire or papillose-crenulate, plane or narrowly recurved, slightly undulate near base; costa ending below apex; cells on ventral surface of costa elongated; leaf cells ± incrassate; basal cells rectangular, upper cells irregularly quadrate-hexagonal, 10–12 μm wide, with one–two rounded papillae per cell, sometimes nearly down to base. *Setae* yellowish-red turning dark red; capsules up to 2.5 mm long including operculum (operculum greatly varying in length, up to ½ of capsule), ellipsoid to cylindrical; peristome teeth very short, 200–250 μm in diameter, erect or weakly twisted. *Spores* 16–18 μm in diameter.

Habitat and Local Distribution. — Plants growing in tufts usually in wet habitats, often under dripping water, sometimes encrusted with lime, on soil or calcareous rocks, on vertical rock surfaces, roadsides and wadi banks. Common, abundant in wet habitats: Sharon Plain, Philistean Plain, Upper Galilee, Lower Galilee, Mount Carmel, Judean Mountains, Northern Negev, Central Negev, Southern Negev, Dan Valley, Hula Plain, Bet She'an Valley, Dead Sea area, and Golan Heights.

General Distribution. — Recorded from Southwest Asia (Cyprus, Turkey, Syria, Lebanon, Sinai, Jordan, Iraq, Iran, and Afghanistan), around the Mediterranean,

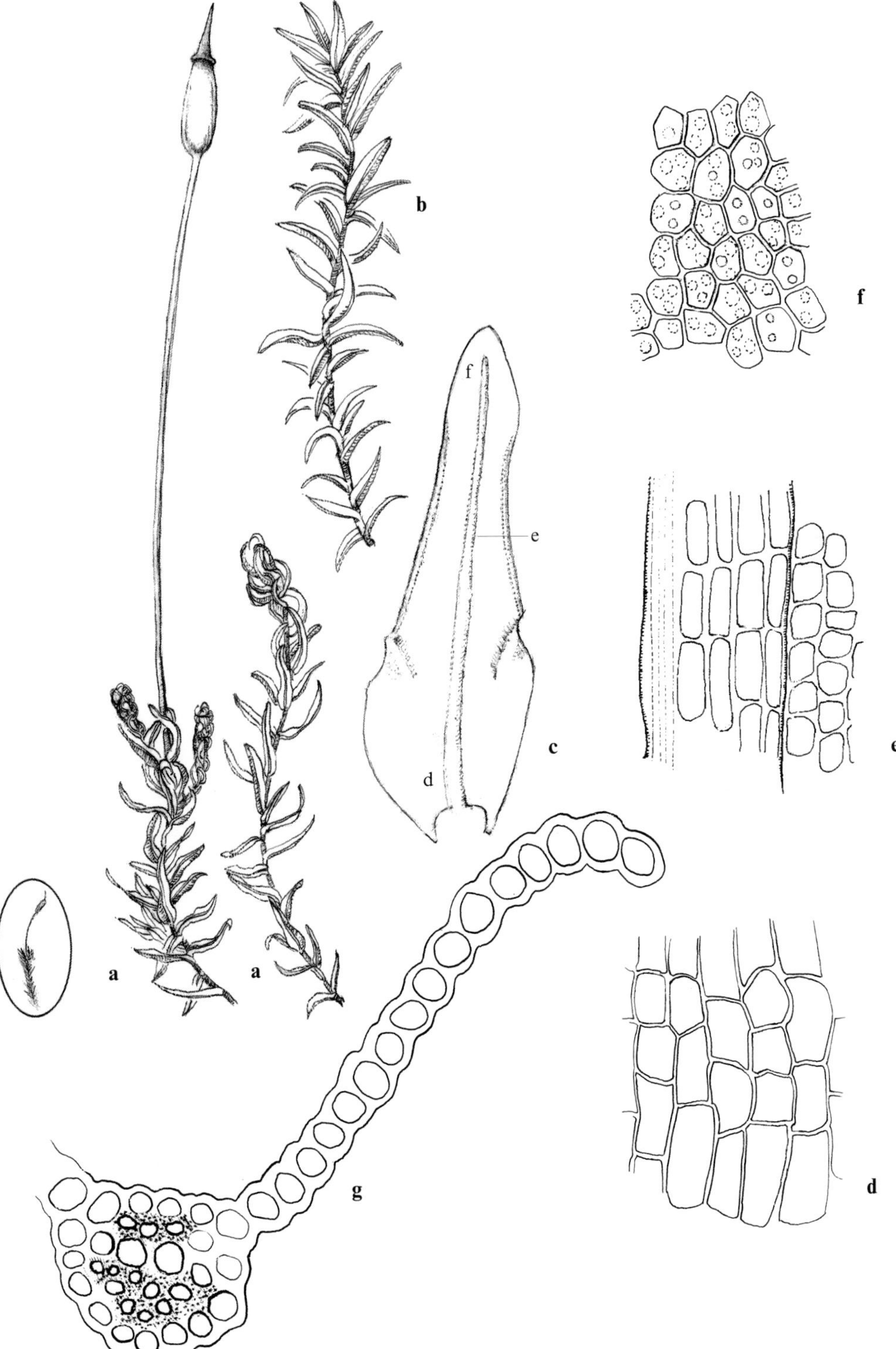

Figure 51. *Barbula tophacea*: **a** habit, dry (×10); **b** habit, moist (×10); **c** leaf (×50); **d** basal cells; **e** cells on ventral surface of costa; **f** upper cells (all leaf cells ×500); **g** cross-section of costa and portion of lamina (×500). (**a**, **b**: *Herrnstadt* & *Crosby 78-11-3*; **c**–**g**: *Herrnstadt*, 28 May 1984)

Europe (widespread throughout), Macaronesia, North and South Africa, northeastern, eastern, and central Asia, and North, Central, and South America.

Plants of *Barbula tophacea* often grow in wet habitats together with *B. ehrenbergii* and *Eucladium verticellatum*, and may occur in ± dryer habitats with *B. imbricata* (see the discussions of these species).

12. Barbula vinealis Brid., Bryol. Univ. 1 : 830 (1827). [Bilewsky (1965) : 371]
Didymodon vinealis (Brid.) R. H. Zander, Phytologia 41 : 25 (1978). — Brullo *et al.* (1991).

Distribution map 53, Figure 52.

Plants up to 10(–15) mm high, olive-green, turning ± dark red. *Stems* usually forked; axillary hairs with brownish basal cell. *Leaves*: upper leaves flexuose, ± twisted, lower leaves imbricate, erect, ± flexuose when dry, all leaves to some degree patent, not flexuose when moist, 1–2(–2.5) mm long, lanceolate, gradually tapering to an acute apex, channelled in upper part; margins entire, recurved from base to near apex; costa percurrent, with quadrate to short-rectangular cells on ventral surface; basal cells irregularly rectangular, mid- and upper leaf cells quadrate to hexagonal, 6–10 μm wide, incrassate, papillose, with simple, round, or bifid papillae; perichaetial leaves slightly longer than other leaves. *Setae* widely varying in length, orange turning dark red; capsules 2–3 mm long including 1 mm long operculum, wide- to narrow-ellipsoid, reddish-brown; peristome teeth long, spirally twisted. *Spores* 9–11 μm in diameter.

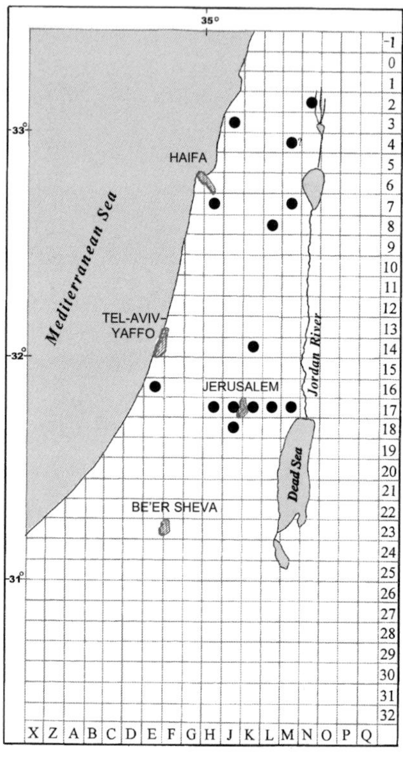

Map 53: *Barbula vinealis*

Habitat and Local Distribution. — Plants growing in tufts on rocks, walls, in rock crevices, and on soil, in shaded and exposed habitats. Occasional to common: Philistean Plain, Upper Galilee, Lower Galilee, Mount Carmel, Samaria, Judean Mountains, Judean Desert, Hula Plain, and Lower Jordan Valley.

General Distribution. — Recorded from Southwest Asia (widespread throughout), around the Mediterranean, Europe (throughout), Macaronesia, North Africa, eastern, central, and Southeast Asia, and North and Central America.

The affinities between *Barbula vinealis* and *B. cylindrica* are presented in the discussion of *B. cylindrica*.

Figure 52. *Barbula vinealis*: **a** habit with detached calyptra, dry (×10); **b** habit moist, capsule with and without operculum (×10); **c** leaf (×50); **d** basal cells; **e** cells on ventral surface of costa; **f** upper cells (all leaf cells ×500); **g** cross-section of leaf (×500). (*Kushnir*, 25 May 1945, No. 3911 in Herb. Bilewsky)

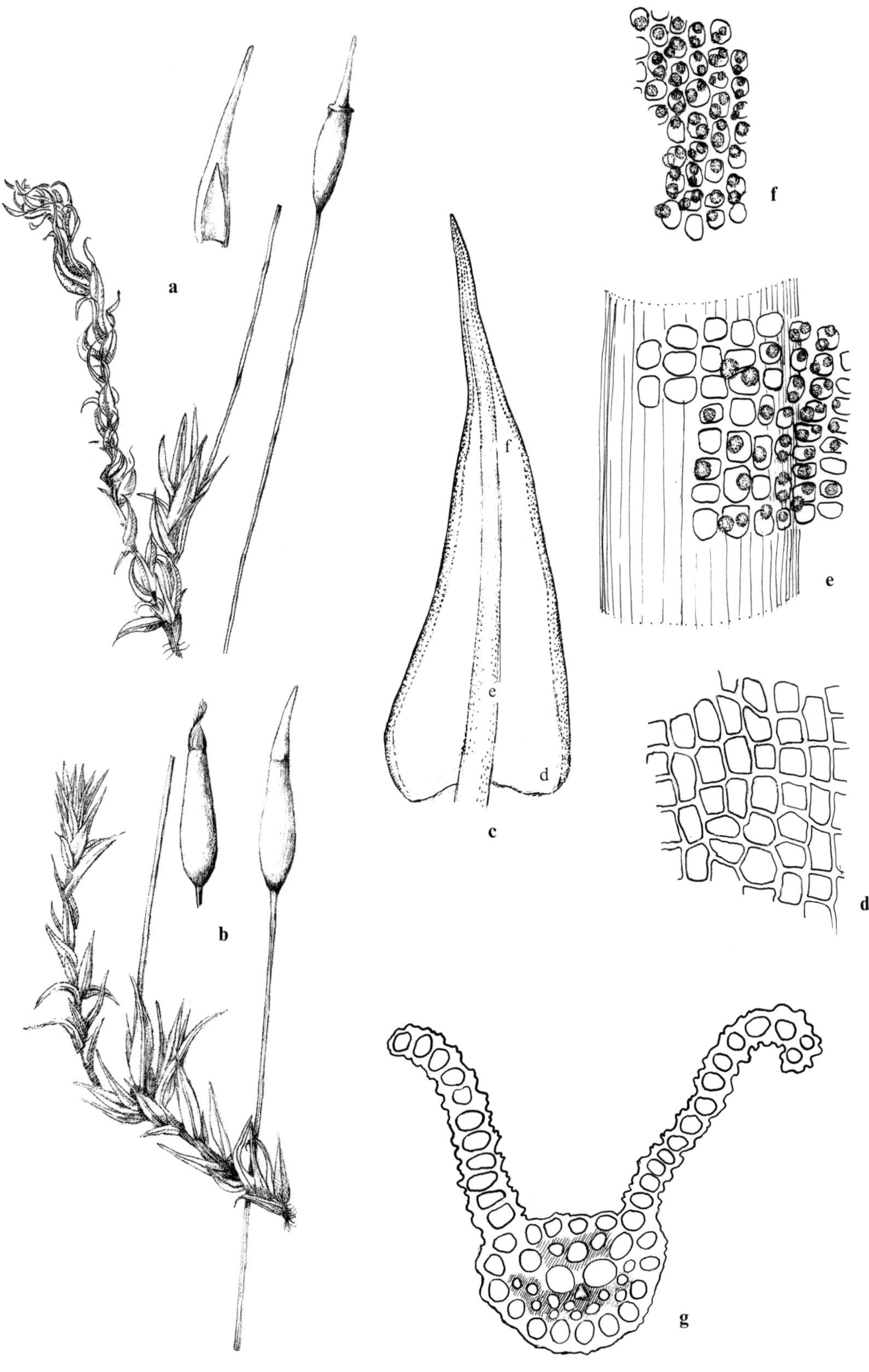
a
b
c
d
e
f
f
e
d
g

13. Barbula cylindrica (Taylor) Schimp. *in* Boulay, Fl. Crypt. Est. Muscin. 430 (1872). [Bilewsky (1965) : 370]

Zygotrichia cylindrica Taylor *in* J. MacKay, Fl. Hibern. 2 : 26 (1836); *B. vinealis* subsp. *cylindrica* (Taylor) Podp., Consp. Musc. Eur. 210 (1954); *Didymodon insulanus* (De Not.) M. O. Hill, J. Bryol. 11 : 599 (1981) [1982].

Distribution map 54, Figure 53.

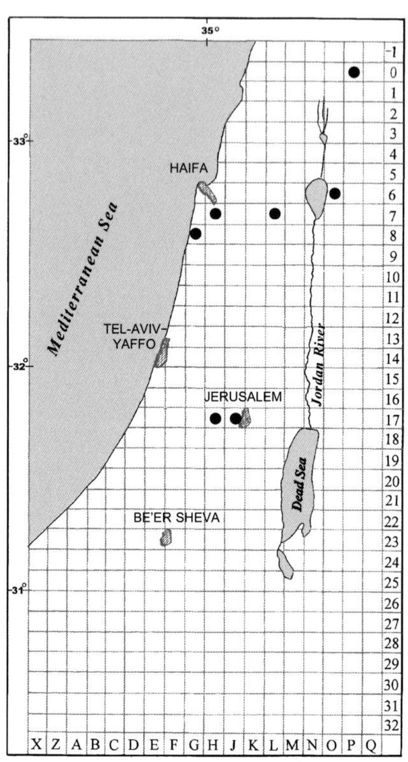

Map 54: *Barbula cylindrica*

Plants up to 15 mm high, olive-green. *Stems* sometimes forked; axillary hairs with brownish basal cell. *Leaves*, mainly upper, curled and crisped when dry, more or less patent, flexuose and sometimes undulate when moist, 2–3 mm long, narrow-lanceolate, gradually tapering to a long-subulate flexuose apex from an ovate base; margins entire, recurved from base to near apex; costa subpercurrent, with cells on ventral surface quadrate to rectangular, only slightly longer than adjacent cells; basal cells irregularly quadrate to rectangular, mid-leaf cells irregularly round-quadrate to hexagonal, 6–9 μm wide, incrassate, papillose, with round papillae; perichaetial leaves slightly longer than other leaves. *Setae* yellowish-red; capsules brownish, cylindrical, ca. 2 mm long without operculum (operculum not seen in local plants); peristome teeth long, spirally twisted. *Spores*: mature spores not seen.

Habitat and Local Distribution. — Plants growing in tufts on calcareous rocks and soil. Rare: Lower Galilee, Mount Carmel, Judean Mountains, Mount Hermon, and Golan Heights.

General Distribution. — Recorded from Southwest Asia (Cyprus, Turkey, Syria, Lebanon, Jordan, and Iraq), around the Mediterranean, Europe (throughout), Macaronesia, North Africa, northeastern and central Asia, and North and Central America.

Barbula cylindrica (Taylor) Schimp. is often considered as a variety of *B. vinealis* (var. *cylindrica* (Taylor) Boulay, or as a synonym of *Didymodon vinealis* (Brid.) R. H. Zander var. *vinealis* (Anderson *et al.* 1990, Zander *in* Sharp *et al.* 1994). Although *B. cylindrica* somewhat resembles *B. vinealis*, plants of *B. cylindrica* are larger, have longer leaves (which are crisped when dry and flexuose when moist) — these two characters are not found in *B. vinealis* — and have cylindrical, not ellipsoid, capsules.

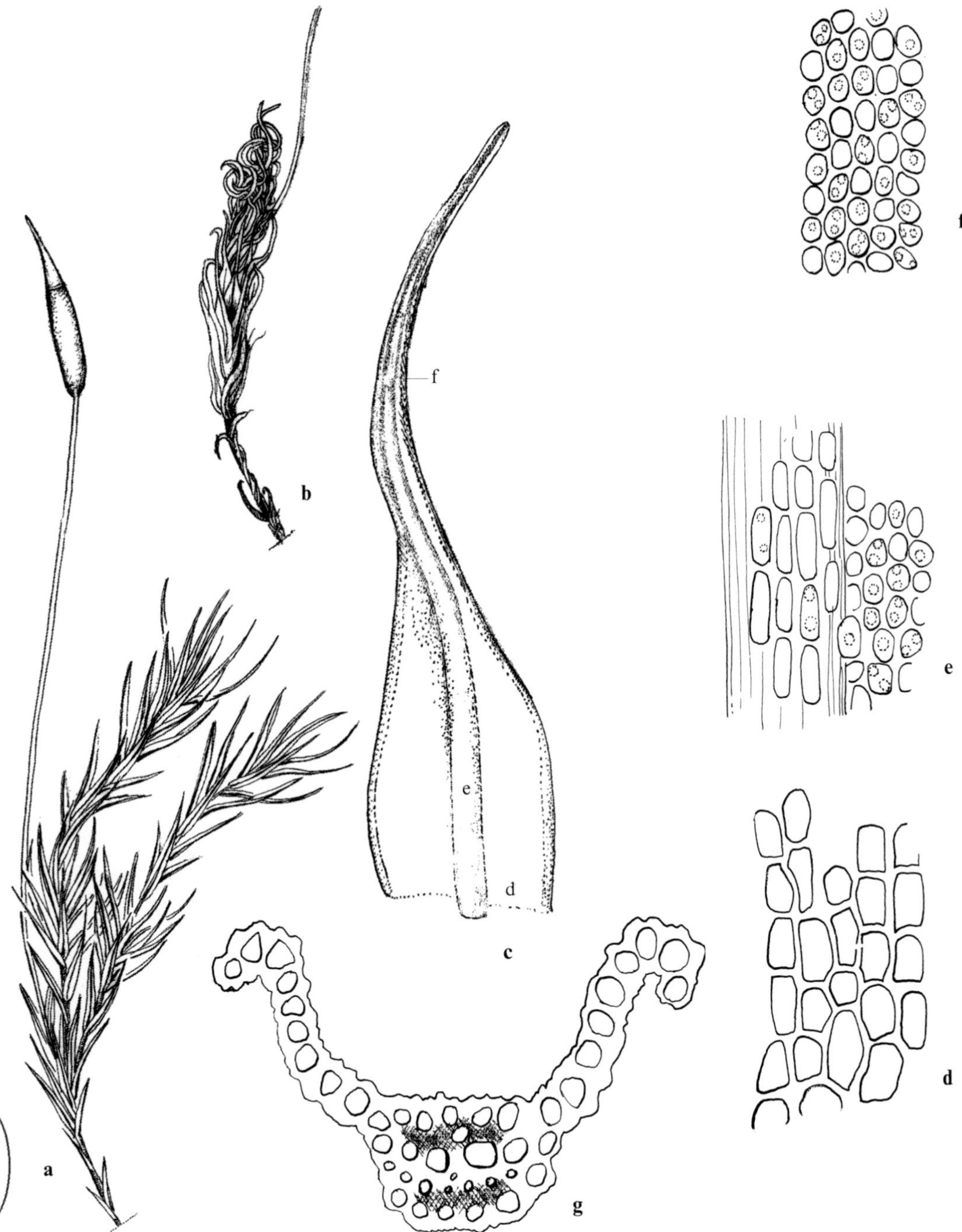

Figure 53. *Barbula cylindrica*: **a** habit, moist (×10); **b** habit, dry (×10); **c** leaf (×50); **d** basal cells; **e** cells on ventral surface of costa; **f** upper cells (all leaf cells ×500); **g** cross-section of leaf (×500). (**a**: *Kushnir*, 12 Mar. 1943; **b**–**g**: *Kushnir*, 15 Dec. 1940, No. 38 in Herb. Bilewsky)

15. *Trichostomopsis* Cardot

Trichostomopsis and its relationships with several genera have been variously treated. Robinson (1970), who revised the genus, considered it as related to *Barbula*. Zander (1978, 1993) included it in *Didymodon*.

Based on the single local species — *Trichostomopsis aaronis* — and data from the literature (e.g., Corley *et al.* 1981, Frahm 1995), *T. aaronis* is treated here as part of a distinct genus. The main diagnostic characters of the genus are the bistratose margins, the bistratose longitudinal patches of the lamina, the single (dorsal) stereid band, and the ± bulging large ventral cells of the costa.

1. Trichostomopsis aaronis (Lorentz) S. Agnew & C. C. Towns., Israel J. Bot. 19:258 (1970).

Trichostomum aaronis Lorentz, Phys. Abh. Königl. Akad. Wiss. Berlin 1:29, Tab. 5 & 6 (1868); *Barbula aaronis* (Lorentz) Hilp., Beih. Bot. Centralbl. 50:656 (1933).

Distribution map 55, Figure 54.

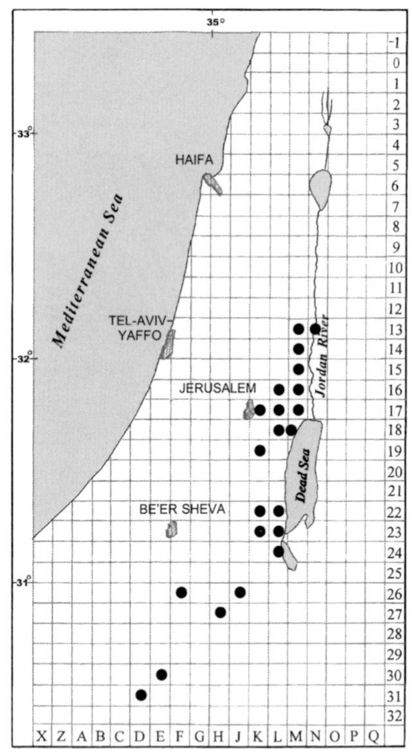

Map 55: *Trichostomopsis aaronis*

Plants up to 5 mm high, green to bluish-green (colour does not fade when dried). *Stems* often branched above. *Leaves* curled inwards and twisted when dry, erectopatent, with an erect base, when moist, 1.0–1.5 mm long, concave, ovate-lanceolate to sublingulate, apex round to short-acuminate, appearing cucullate because of the bistratose margins; margins narrowly recurved, bistratose except at base, entire; costa stout, 60–80 μm wide, width nearly equal throughout leaf, ending shortly below apex, with subquadrate to quadrate cells on ventral surface; in cross-section with a dorsal group of stereids and ± bulging, large ventral cells; basal cells rectangular, achlorophyllous, smooth, pellucid; upper cells incrassate, irregularly round-quadrate or rectangular, mid-leaf cells 8–11(–13.5) μm wide, minutely papillose, densely chlorophyllous; lamina with bistratose longitudinal patches. *Rhizoidal gemmae* have been described in this species, but were not been found in local plants. *Dioicous* (?); some young sporophytes were observed in local material (the only existing record of non-sterile plants).

Habitat and Local Distribution. — Plants, growing in tufts on exposed loess or north-facing crevices of cliffs; a species typical for the arid parts of Israel. Common in arid habitats: Judean Desert, Western Negev, Central Negev, Lower Jordan Valley, and Dead Sea area.

General Distribution. — Recorded around the Mediterranean. Available records are rather scattered and have accumulated mainly during recent years: Southwest

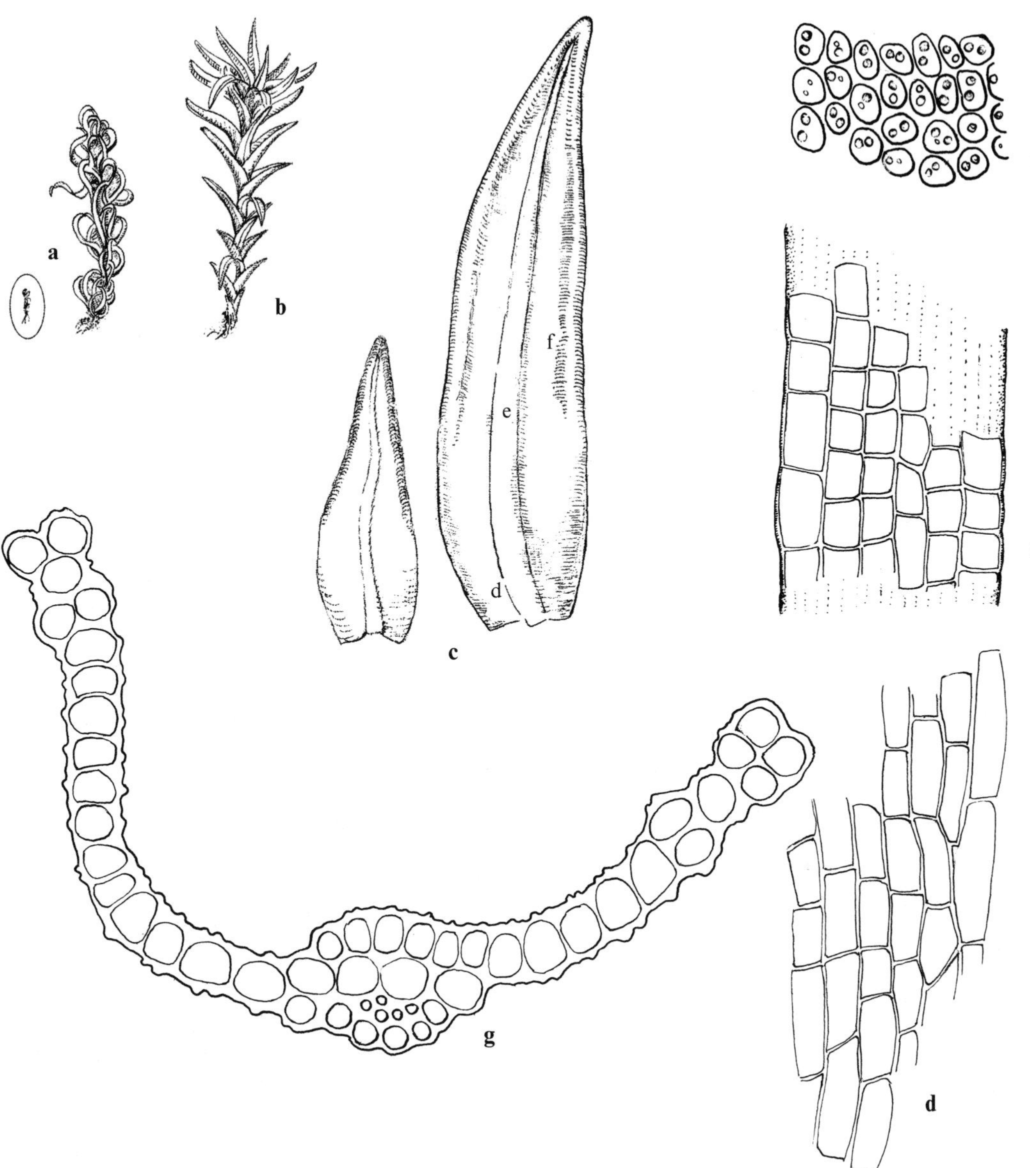

Figure 54. *Trichostomopsis aaronis*: **a** habit, dry (×10); **b** habit, moist (×10); **c** leaves (×50); **d** basal cells; **e** cells on ventral surface of costa; **f** mid-leaf cells (all leaf cells ×500); **g** cross-section of leaf (×500). (**a**–**f**: *Shmida*, 28 Jan. 1981; **g**: *Raven*, 31 Jan. 1975)

Asia (Turkey, Lebanon, Sinai, Jordan, Iraq, Afghanistan, and Saudi Arabia), Europe (Spain and Sicily), North Africa, and central Asia.

Literature. — Robinson (1970).

Trichostomopsis aaronis often grows together with species of *Aloina*, *Barbula*, and *Tortula*. It resembles *Barbula imbricata* in colour and habit, but is easily distinguished by the bistratose margins and longitudinal patches of the lamina, and by the laxer areolation of the basal cells.

Tortula cabulica J. Froehl. and *Barbula bistrata* Rungby are considered as synonyms of *T. aaronis* by Agnew & Townsend (1970) and Agnew & Vondráček (1975). *Tortula hellenica* Schiffn. & Baumgartner is regarded as closely allied with *T. aaronis* by Townsend (1966) and by Corley *et al.* (1981).

Frey & Kürschner (1993) recorded an additional species of *Trichostomopsis* — *T. trivialis* (Müll. Hal.) H. Rob. — from Jordan. This species, so far not observed in the local flora, differs from the local *T. aaronis* in the lanceolate, acuminate to acute leaves and rectangular cells on the ventral surface of the costa.

16. *Leptobarbula* Schimp.

Leptobarbula is a monotypic genus. It is worthwhile to mention that we treat *Leptobarbula* as belonging to the tribe *Barbuleae* while in Zander (1993) it is recognised within the tribe *Pottieae*.

Literature. — Whitehouse & During (1987).

Leptobarbula berica (De Not.) Schimp., Syn. Musc. Eur., Edition 2 : 181 (1876).
[Bilewsky (1965) : 361]
Leptotrichum bericum De Not., Cronac. Briol. Ital. 1 : 14 (1866).

Distribution map 56, Figure 55, Plate II : 1.

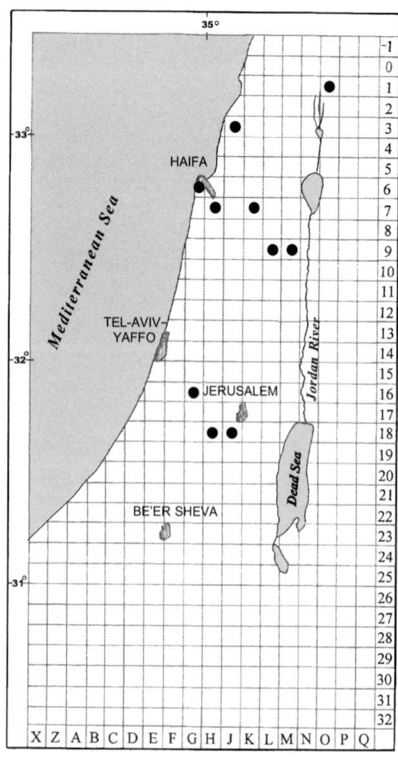

Map 56: *Leptobarbula berica*

Plants small, up to 3 mm high, green to yellowish-green. *Stems* with a central strand. *Leaves*: basal leaves very small, appressed when dry, erectopatent to spreading when moist, upper larger, up to 1 mm long, crisped when dry, spreading and ± recurved when moist, linear-lanceolate, acute or obtuse; margins plane, crenulate; costa strong, percurrent or ending just below apex, cross-section with two stereid bands; basal cells slightly incrassate, rectangular, smooth; upper cells quadrate to round-quadrate, opaque, densely papillose with C-shaped, bifid papillae; mid-leaf cells 8–10 µm wide; perichaetial leaves more than two times longer than other leaves, with a long, convolute sheathing base, abruptly contracted into a long narrow upper part. *Dioicous*.

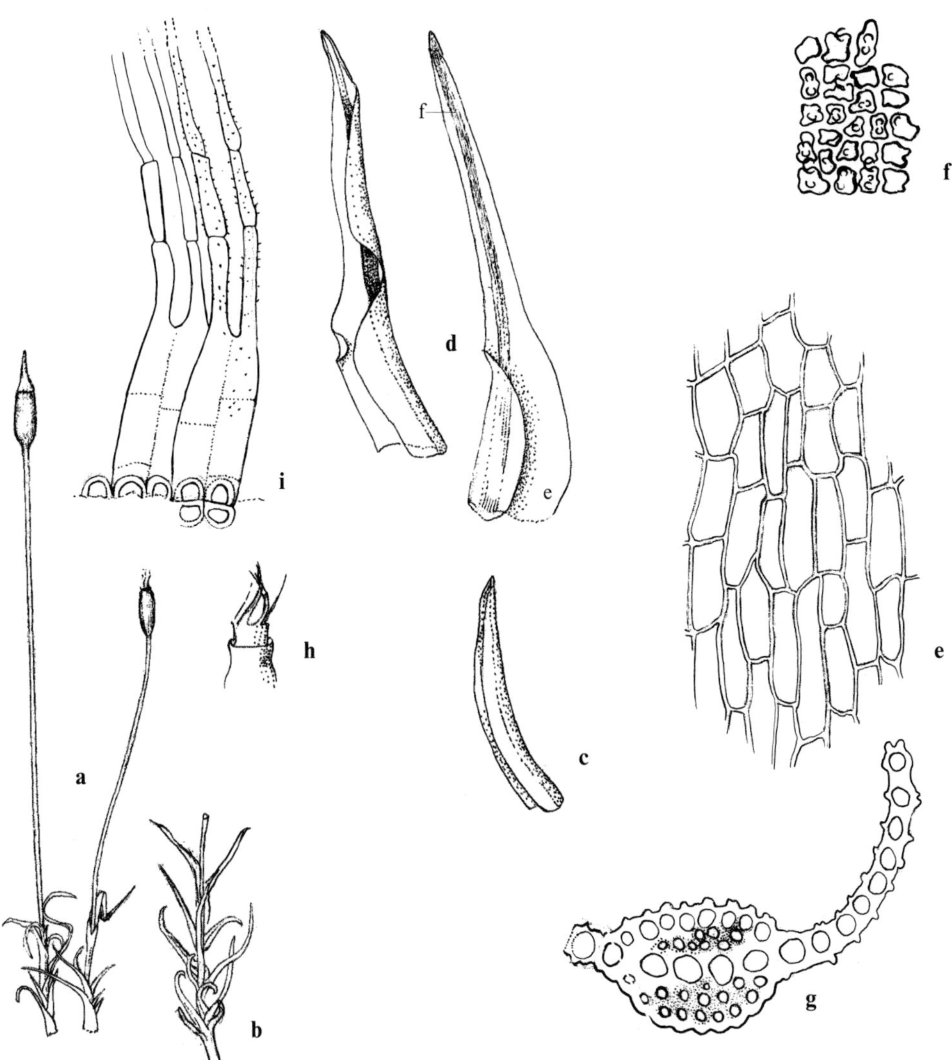

Figure 55. *Leptobarbula berica*: **a** habit, dry (×10); **b** habit, moist (×10); **c** leaf (×50); **d** perichaetial leaves; **e** basal cells; **f** upper cells (all leaf cells ×500); **g** cross-section of costa and portion of lamina (×500); **h** peristome (×25); **i** peristome teeth (×500). (*Kushnir*, 26 Feb. 1944)

Setae much varying in length, 4–9 mm long, yellowish-brown; capsules 0.75–2.25 mm long with operculum, narrow elliptical to cylindrical, brown; operculum conical, long-rostrate; annulus present; peristome with a basal membrane 25–45 μm high, teeth 32, narrow, papillose, arranged in pairs, slightly twisted. *Spores* very small, 7–9.5 μm in diameter, nearly smooth; gemmate and joined into bead-like structures (seen with SEM). *Calyptrae* cucullate.

Habitat and Local Distribution. — Plants growing in small tufts usually in shade, on calcareous rocks and stones. Occasional: Upper Galilee, Lower Galilee, Mount Carmel, Mount Gilboa, Shefela, Judean Mountains, Dan Valley, Bet She'an Valley, and Golan Heights.

General Distribution. — Recorded from Southwest Asia (Turkey and Lebanon), around the Mediterranean (few, scattered records), Europe (few records, mainly from the south), Macaronesia, and North Africa.

Leptobarbula berica resembles small species of *Ditrichum* in the leaf shape — the wide base narrowing into a long narrow upper part, composed mainly of the costa. In cross-section, the leaf resembles *Barbula*, but can be differentiated from *Barbula sensu stricto* by the only slightly-twisted peristome teeth.

SUBFAMILY II. POTTIOIDEAE

TRIBE POTTIEAE

17. *Tortula* Hedw.

(including *Syntrichia* Brid.)

Plants variously green to brownish, growing usually in tufts or cushions. *Stems* erect, simple or forked, usually with a central strand. *Leaves* in upper part of stem often larger and more crowded, appressed and usually twisted when dry, erect to spreading, sometimes recurved when moist, often folded, lingulate, spathulate, or obovate; apex usually acute, obtuse or rounded; margins most often entire (in the majority of local species at least partly recurved), rarely plane, sometimes bordered; costa usually strong, percurrent to excurrent in short or long, often hyaline hair-point, with a single, dorsal stereid band, with or without hydroids; basal cells usually large, long-rectangular, smooth and hyaline, gradually or abruptly delimited from cells above; upper cells round-quadrate to hexagonal, rarely smooth, usually pluripapillose on both surfaces, most often with bifid, branched, sometimes C-shaped papillae, papillae rarely absent; perichaetial leaves only rarely differentiated. *Gemmae* sometimes present (in *T. rigescens* among local species). *Dioicous* or autoicous, rarely synoicous. *Setae* elongate; capsules usually erect, straight or slightly curved, ellipsoid to cylindrical; operculum conical to long-rostrate; peristome usually with a low or high, often tessellated basal membrane and 32 teeth consisting of filiform, papillose filaments, straight or spirally twisted. *Spores* variously papillose; in local species verrucose-gemmate, processes sometimes joined into irregular reticulum (e.g., *T. muralis*), in some species processes densely covered by granules (seen with SEM). *Calyptrae* cucullate, smooth.

A large genus widespread in all continents, mainly in temperate regions. *Tortula* is the largest local genus with 17 species, growing in a great variety of habitats.

The treatment of the genus *Tortula* widely varies in the bryological literature. In particular, *Syntrichia* is either considered a separate genus (e.g., Chen 1941, Agnew & Vondráček 1975, Zander 1989, Ochyra 1992) or retained in *Tortula* (e.g., Smith 1978, Corley *et al.* 1981, Crum & Anderson 1981, Mishler *in* Sharp *et al.* 1994). Zander (1993) segregated *Syntrichia* and several other genera from *Tortula*, but included *Desmatodon*, *Pottia*, and *Phascum* in *Tortula*. The separation of *Syntrichia* from *Tortula* is based on several characters. *Syntrichia* is usually considered (mainly following Chen 1941) as comprising more robust plants than *Tortula*, possessing a high, tessellated basal membrane (Kramer 1980) and long, spirally twisted peristome teeth. Zander (1993) based the segregation of *Syntrichia* from *Tortula* mainly on the KOH reaction of laminal cells, the crescent-shaped stereid band seen in cross-section of the costa, and the lack of dorsal costal epidermal cells.

In Herrnstadt *et al.* (1982, 1991) *Syntrichia* was considered as a genus separate from *Tortula*. Local species can be easily segregated into either genus. However, the differentiation between the genera seems far less clear-cut when they are treated globally, and some of the differential characters may occur in various combinations. A few changes at the specific level have also been made since Herrnstadt *et al.* (1991).

Literature. — Chen (1941), Kramer (1980), Zander (1989).

1 Plants usually large, more than 10 mm high; cells at base of leaf abruptly differentiated from cells above; basal membrane high, spirally tessellated, at least half the length of peristome; dorsal cells of costa not differentiated.

2 Leaves tapering to a short, obtuse or acute apex; costa percurrent or excurrent in a short mucro or apiculus **12. T. inermis**

2 Leaves with round to obtuse apex (rounded or acute in *T. pseudohandelii*); costa excurrent in a long, usually hyaline, hair-point.

3 Leaf lamina at least partly bistratose.

4 Leaves bearing brown, spherical to ellipsoid gemmae **11. T. rigescens**

4 Leaves without gemmae.

5 Costa covered on dorsal side with short, usually simple, papillae; mid-leaf cells 5–7 µm wide; growing in the mountainous areas of northern Israel **9. T. handelii**

5 Costa covered on dorsal side with long, branched, or stellate papillae; mid-leaf cells 8–11 µm wide; growing in the more arid mountainous areas of southern Israel **10. T. pseudohandelii**

3 Leaf lamina unistratose.

6 Leaves constricted at middle or below; margins near middle or up to ½–⅔ of lamina.

7 Plants usually rupestral; leaves with denticulate hair-point **13. T. intermedia**

7 Plants usually arboreal; leaves with nearly smooth hair-point **14. T. laevipila**

8 Leaves ca. 2.5 mm long (not including hair-point), some with border of one–two rows of slightly incrassate cells **14a.** var. **laevipila**

8 Leaves ca. 3.5 mm long (not including hair-point), with distinct border of up to four rows of strongly incrassate cells **14b.** var. **meridionalis**

6 Leaves not or indistinctly constricted at middle, or constricted only on one side; margins recurved nearly up to apex of lamina.

9 Plants dioicous; stems without a central strand; leaves distinctly recurved when moist; margins recurved nearly up to near apex of lamina; hydroids in costa absent **17. T. ruralis**

9 Plants synoicous; stems with a central strand; leaves patent to ± recurved when moist; margins recurved up to ¾ the length of lamina; hydroids in costa present.

10 Mid- and upper leaf cells pluripapillose with three–four irregularly C-shaped papillae **15. T. princeps**

10 Mid- and upper leaf cells with one(–two) large stellate stalked papillae **16. T. echinata**

1 Plants usually small, less than 10 mm long; cells at base of leaf gradually differentiated from cells above, sometimes only little differentiated; basal membrane short, not exceeding ⅓ the length of peristome, often very short; dorsal cells of costa differentiated.

11 Costa thickened in upper part of leaf, with enlarged chlorophyllous cells on ventral surface.

12 Leaves with long hyaline hair-point **2. T. brevissima**
12 Leaves without hyaline hair-point.
13 Leaves short, up to 1 mm long, broad-ovate; peristome teeth once twisted; moist plants forming open rosettes **3. T. fiorii**
13 Leaves long, more than 1 mm long, lanceolate to oblong-ovate; peristome teeth nearly straight; moist plants not forming rosettes **1. T. atrovirens**
11 Costa not thickened in upper part of leaf, without enlarged chlorophyllous cells on ventral surface.
14 Leaf margins bistratose, with elongated cells **6. T. marginata**
14 Leaf margins unistratose, without elongated cells.
15 Leaf pellucid, all cells smooth; margins plane **5. T. cuneifolia**
15 Leaf not pellucid, upper cells papillose; margins at least partly recurved.
16 Leaf margins strongly recurved from base to near apex **7. T. muralis**
17 Leaf cells densely covered by 2.5–5.5 μm high, bi- or trifid papillae or mammillae **7a.** var. **muralis**
17 Leaf cells covered by one–two, 8–13(–15) μm high, simple or bifid mammillae, rarely papillae **7b.** var. **israelis**
16 Leaf margins plane to slightly recurved, sometimes only at the middle.
18 Plants ca. 5 mm long; leaves soft, twisted when dry; capsules slightly curved; peristome teeth almost free from base, basal membrane very short **8. T. vahliana**
18 Plants up to 3 mm long; leaves not soft, scarcely twisted when dry; capsules ± straight; peristome teeth above a basal membrane, basal membrane up to ¼–⅓ the length of the peristome **4. T. canescens**

Of the species listed in the above key, Zander (1993) includes in *Syntrichia*: *Tortula echinata*, *T. handelii*, *T. inermis*, *T. intermedia*, *T. laevipila*, *T. princeps*, *T. pseudohandelii*, *T. rigescens*, and *T. ruralis*.

1. Tortula atrovirens (Sm.) Lindb., Öfvers. Förh. Kongl. Svenska Vetensk.-Akad. 21 : 236 (1864). [Bilewsky (1965) : 381]

Grimmia atrovirens Sm., Engl. Bot. 28 : 2015 (1809); *Desmatodon convolutus* (Brid.) Grout, Moss Fl. N. Amer. 1 : 224 (1939).

Distribution map 57, Figure 56.

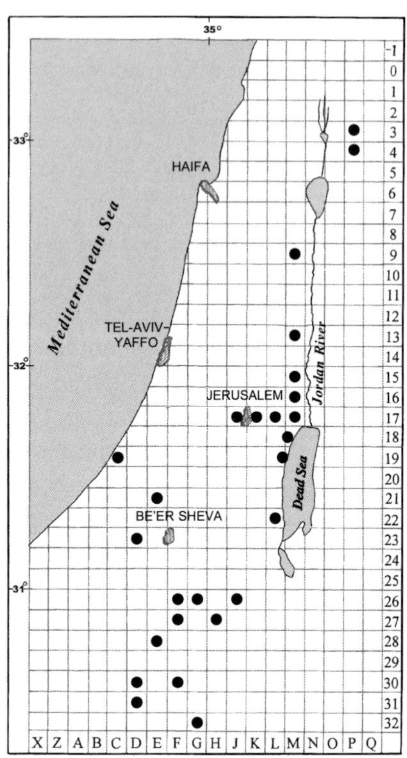

Map 57: *Tortula atrovirens*

Plants small, up to 3 mm high, green to dark green. *Stems* with a central strand. *Leaves* crowded, curved inwards when dry, patent when moist, up to 2 mm long, concave, lanceolate to oblong-ovate, apex acute or obtuse, apiculate; margins entire, usually strongly recurved except near base and apex, turning reddish-brown in upper part; costa strong, thickened and wider in upper part, with enlarged chlorophyllous cells on ventral surface, without papillae on dorsal surface, excurrent in a short, green or reddish mucro, hydroids present; basal cells short-rectangular, smooth; mid-leaf cells irregularly round-quadrate to hexagonal, 10–14 μm wide, densely papillose with C-shaped papillae. *Autoicous. Setae* erect, reddish-yellow; capsules ca. 2 mm long including operculum, ovate-ellipsoid, reddish-brown; operculum ± oblique, conical-ellipsoid, rostrate, often detached together with calyptra; peristome with a short basal membrane, teeth nearly straight. *Spores* 20–24 μm in diameter.

Habitat and Local Distribution. — Plants growing in dense or loose tufts, often on exposed loess, marl, and various desert soils. Common in arid habitats: Philistean Plain, Samaria, Judean Mountains, Judean Desert, Northern Negev, Central Negev, Bet She'an Valley, Lower Jordan Valley, Dead Sea area, and Golan Heights.

General Distribution. — A widespread xerophytic species. Recorded from Southwest Asia (Cyprus Syria, Sinai, Jordan, Iraq, Iran, Afghanistan, Kuwait, and Saudi Arabia), around the Mediterranean, Europe (mainly the south), Macaronesia, Africa (throughout), Asia (except the northeast), Australia, New Zealand, Oceania, and North, Central, and South America (often recorded as *Desmatodon convolutus* — e.g., in North and Central America by Anderson *et al.* (1990) and Zander *in* Sharp *et al.* (1994), respectively).

Tortula atrovirens is sometimes treated within the genus *Desmatodon* (as *D. convolutus*), mainly because of the short, nearly straight peristome teeth. Plants of *T. atrovirens* resemble *T. brevissima* and *T. fiorii*, with which they may share arid habitats, and a similar morphology and anatomy of the costa. The three taxa are often considered as comprising section *Crassinerves* (Milde) Wijk & Margad. (cf. Corley *et al.* 1981). *Tortula atrovirens* can be, however, easily distinguished from *T. brevissima*, which

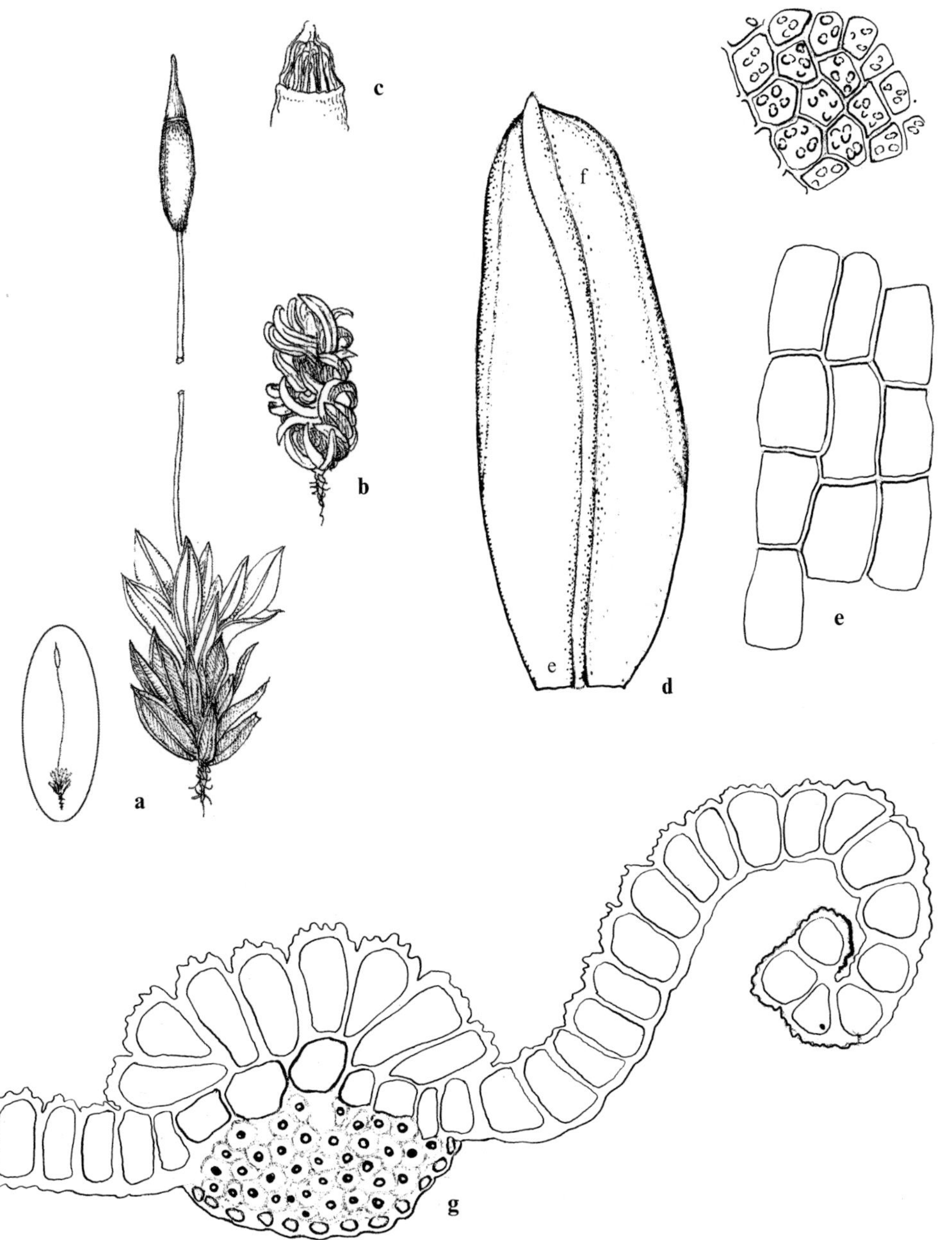

Figure 56. *Tortula atrovirens*: **a** habit, moist (×10); **b** habit, dry (×10); **c** peristome (×25); **d** leaf (×50); **e** basal cells; **f** upper cells (all leaf cells ×500); **g** cross-section of costa and portion of lamina (×500). (*D. Zohary*, 10 Feb. 1943)

has leaves with a hyaline hair-point, and from *T. fiorii*, in which the leaves form rosettes when moist. The enlarged chlorophyllous cells on the ventral surface of the costa seem to enhance photosynthetic activity by increasing the exposed leaf surface and occur in various forms also in other moss genera from the Pottiaceae of arid habitats. An extreme expression of this trend may be seen in the development of photosynthetic filaments (in *Aloina* and *Crossidium*) or lamellae (in *Pterygoneurum*) (see also Frey 1990).

2. Tortula brevissima Schiffn., Ann. K. K. Naturhist. Hofmus. 27:481 (1913). [Bilewsky (1965):383]

Distribution map 58, Figure 57.

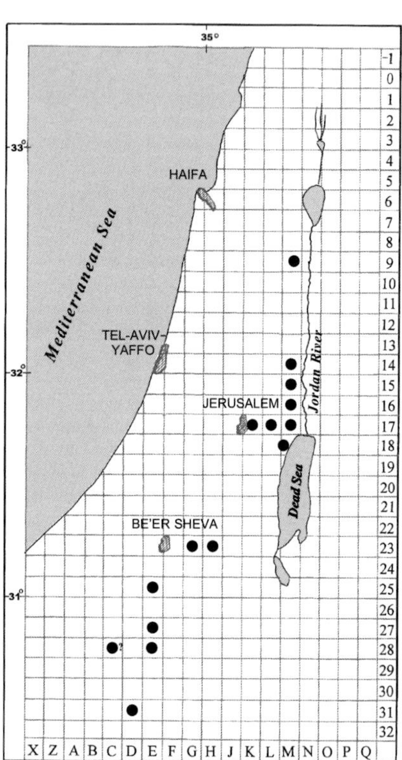

Map 58: *Tortula brevissima*

Plants small, 2–3 mm high, bright green to dull green. *Stems* with or without a central strand. *Leaves* keeled, contorted or twisted when dry, erectopatent when moist, 1–1.5 mm long with hair-point, broad-ovate, apex round to obtuse; margins entire, more or less strongly recurved, except near base and apex; costa strong, slightly thickened and wider in upper part, with enlarged chlorophyllous cells on ventral surface, excurrent in a smooth, hyaline, ± oblique hair-point, half the length of lamina or longer, hydroids present; basal cells subquadrate to rectangular, smooth; mid-leaf cells irregularly quadrate-hexagonal, 11–14 µm wide, densely papillose with C-shaped papillae; perichaetial leaves longer, up to 2.5 mm. *Dioicous*. *Setae* erect, yellowish, ± red below; capsules ca. 2 mm long including operculum, ellipsoid, reddish-brown; operculum ± obliquely long conical-rostrate; peristome with a short basal membrane, teeth twisted. *Spores* 13–15 µm in diameter, nearly smooth.

Habitat and Local Distribution. — Plants growing in tufts, more often gregarious or scattered, on loess, marl, or exposed calcareous soils in rock crevices, rarely on soil on rock. Locally common in arid territories: Judean Desert, Northern, Western and Central Negev, Bet She'an Valley, Lower Jordan Valley, and Dead Sea area.

General Distribution. — Very few records on distribution are available: Southwest Asia (Syria, Jordan, and Iraq), Europe (Germany, France, and Spain), and Ciscaucasia.

Tortula brevissima is related to *T. atrovirens* (see the discussion there).

Leaves of plants growing under more favourable conditions may reach twice the length of those in arid habitats.

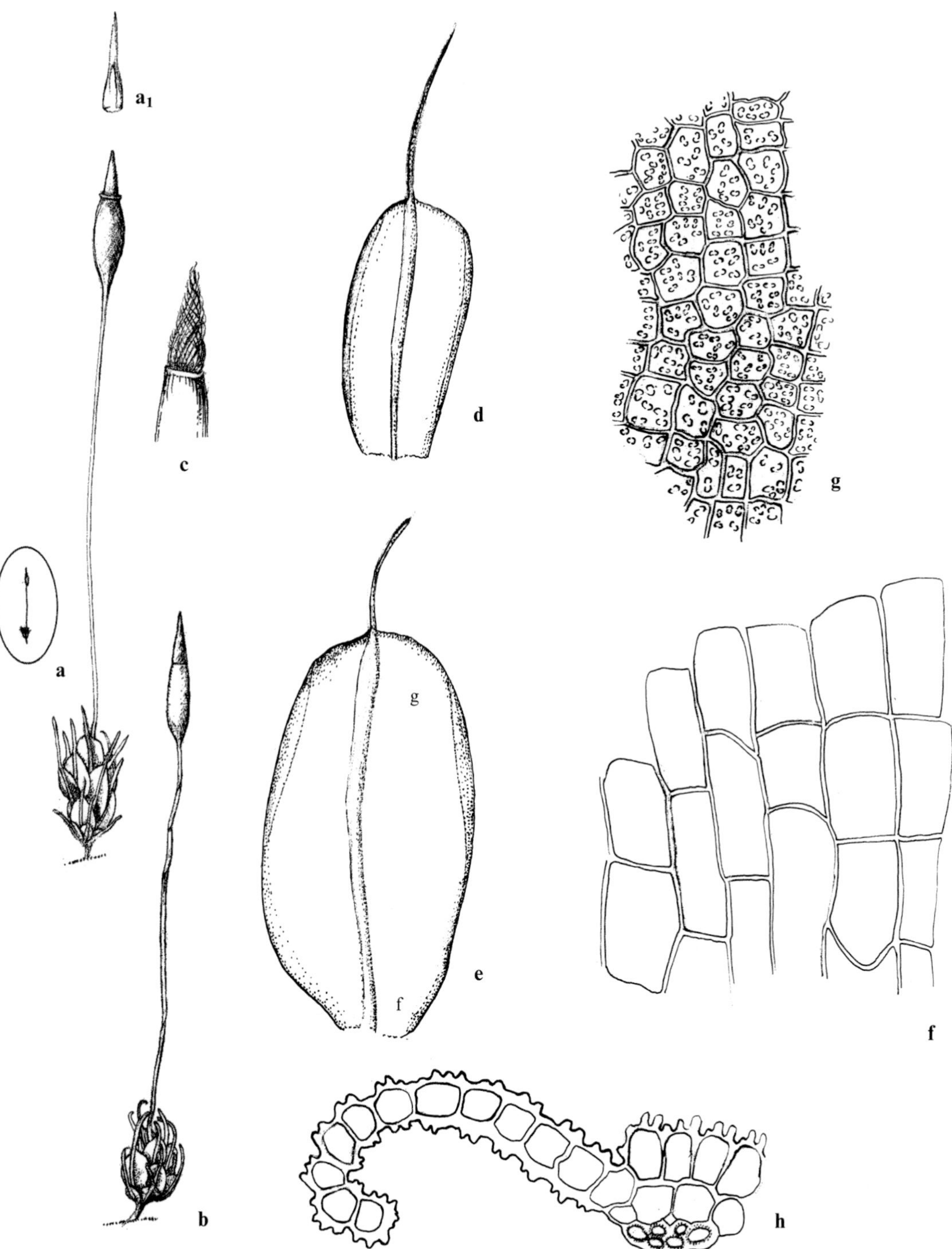

Figure 57. *Tortula brevissima*: **a** habit, moist (×10), $\mathbf{a_1}$ calyptra (×10); **b** habit, dry (×10); **c** peristome (×20); **d** leaf (×50); **e** perichaetial leaf (×50); **f** basal cells; **g** upper cells (all leaf cells ×500); **h** cross-section of costa and portion of lamina (×500). (*Oyserman 79-107-1*)

3. Tortula fiorii (Venturi) G. Roth, Eur. Laubm. 1 : 351 (1904).
Barbula fiorii Venturi, Rev. Bryol. 12:66 (1885); *T. revolvens* (Schimp.) G. Roth var. *obtusata* Reimers, Hedwigia 79 : 284 (1940).

Distribution map 59, Figure 58.

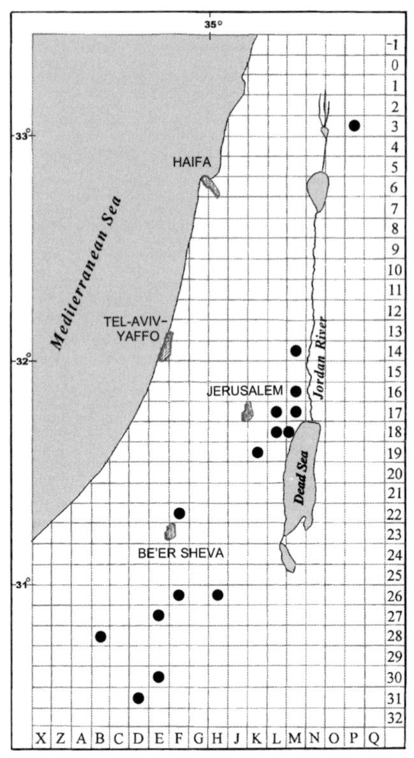

Map 59: *Tortula fiorii*

Plants small, up to 2.5 mm high, bright green to brownish, turning nearly black when dry. *Stems* with a central strand. *Leaves* imbricate, appressed, slightly contorted when dry, erect to patent and forming an open rosette when moist, up to 1 mm long, broad-ovate, apex round to obtuse or round and shortly apiculate; margins entire, strongly recurved up to apex; costa strong, thickened and wider in upper part, ending in apex or below, with enlarged chlorophyllous cells on ventral surface, hydroids present; basal cells rectangular, smooth; mid-leaf cells irregularly quadrate-hexagonal, 7–10 μm wide, papillose with C-shaped papillae; perichaetial leaves up to 1.5 mm long, sometimes twice as long as other leaves. *Autoicous. Setae* erect, reddish; capsules 2–2.5 mm long including operculum, sometimes weakly curved, ovoid-ellipsoid, brown; operculum ± obliquely rostrate; peristome with a very short basal membrane, teeth once twisted. *Spores* 14–20 μm in diameter.

Habitat and Local Distribution. — Plants growing often in dense tufts, on various desert soils (e.g., loess, rendzina), often in soil pockets on rocks. Common in the more arid parts of Israel: Judean Desert, Northern and Western Negev, Central Negev, Lower Jordan Valley, Dead Sea area, and Golan Heights.

General Distribution. — Recorded (often as *Tortula revolvens* var. *obtusata*) from Southwest Asia (Turkey, Jordan, Iran, and Iraq), few records around the Mediterranean, Europe (few records from the south), Macaronesia (in part), and North Africa.

Tortula fiorii is easily distinguished from all the other local species of *Tortula* by the rosette-like appearance of moist plants, the leaves with revolute margins reaching up to rounded apex, and the unexserted costa (see also the discussion of *T. atrovirens*).

Tortula fiorii is often considered as *T. revolvens* var. *obtusata* (e.g., by Corley *et al.* 1981; Frey & Kürschner 1991c). It is treated here at the specific level because it lacks several of the important differential diagnostic characters of *T. revolvens*, i.e., being dioicous, leaves with costa excurrent into a mucro, and nearly smooth lamina cells.

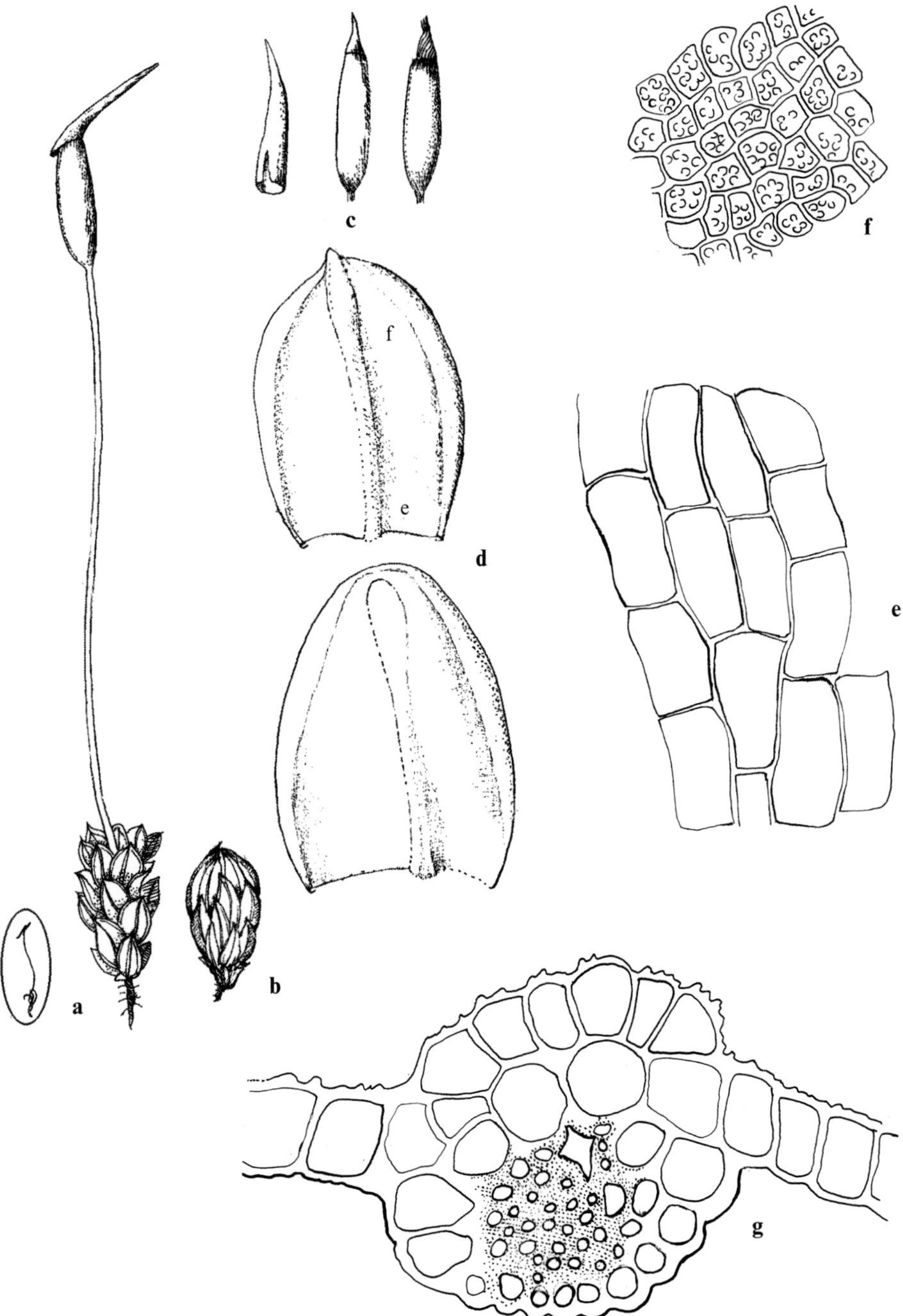

Figure 58. *Tortula fiorii*: **a** habit, moist (×10); **b** habit, dry (×10); **c** calyptra, capsules with and without operculum (×10); **d** leaves (×50); **e** basal cells; **f** upper cells (all leaf cells ×500); **g** cross-section of costa and portions of lamina (×500). (*Danin 79-59-1*)

4. Tortula canescens Mont., Arch. Bot. 1 : 133 (1833). [Bilewsky (1965) : 383]

Distribution map 60, Figure 59.

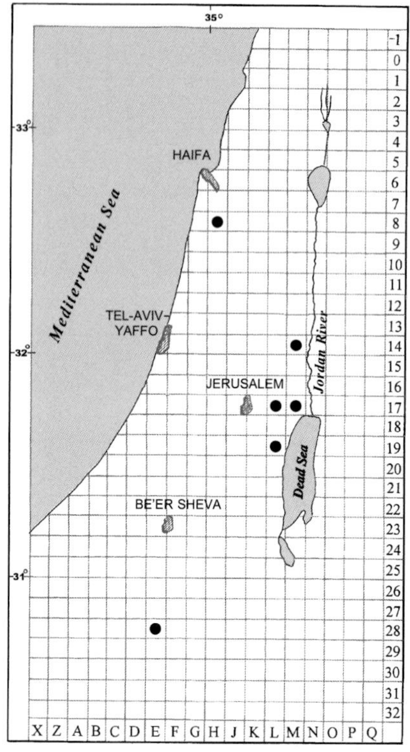

Map 60: *Tortula canescens*

Plants small, up to 3 mm high, yellowish-green. *Stems* with a central strand. *Leaves* scarcely twisted and flexuose when dry, erectopatent to patent when moist, one–two mm long with hair-point, broadly ovate to lingulate-lanceolate, apex round to obtuse; margins slightly recurved near mid-leaf, sometimes up to near apex, with one–two rows of incrassate, usually less papillose cells; costa excurrent in a smooth hyaline hair-point often up to ⅓ the length of lamina or longer, hydroids present; basal cells rectangular; mid-leaf cells irregularly hexagonal, 10–16 µm wide, densely papillose with C-shaped papillae. *Autoicous*. *Setae* erect, yellowish-red; capsules ± straight, 1.5 mm long including operculum, narrow-elliptical to cylindrical, reddish-brown; operculum ± obliquely rostrate; peristome with a basal membrane up to ca. ¼–⅓ the length of peristome, teeth twisted. *Spores* 13–15 µm in diameter.

Habitat and Local Distribution. — Plants growing in tufts, often scattered among other mosses, on horizontal or north-facing chalk, and loess soils and on stones. Occasional: Samaria, Judean Desert, Central Negev, and Lower Jordan Valley.

General Distribution. — Recorded from Southwest Asia only from Turkey, around the Mediterranean (few records), Europe (mainly the south), Macaronesia, and North Africa.

Plants of *Tortula canescens* resemble *T. muralis* except for the only slightly recurved leaf margins and the higher basal membrane. They can be distinguished from plants of *T. vahliana*, with which they may sometimes grow together, by their smaller size and the hyaline hair-point of leaves.

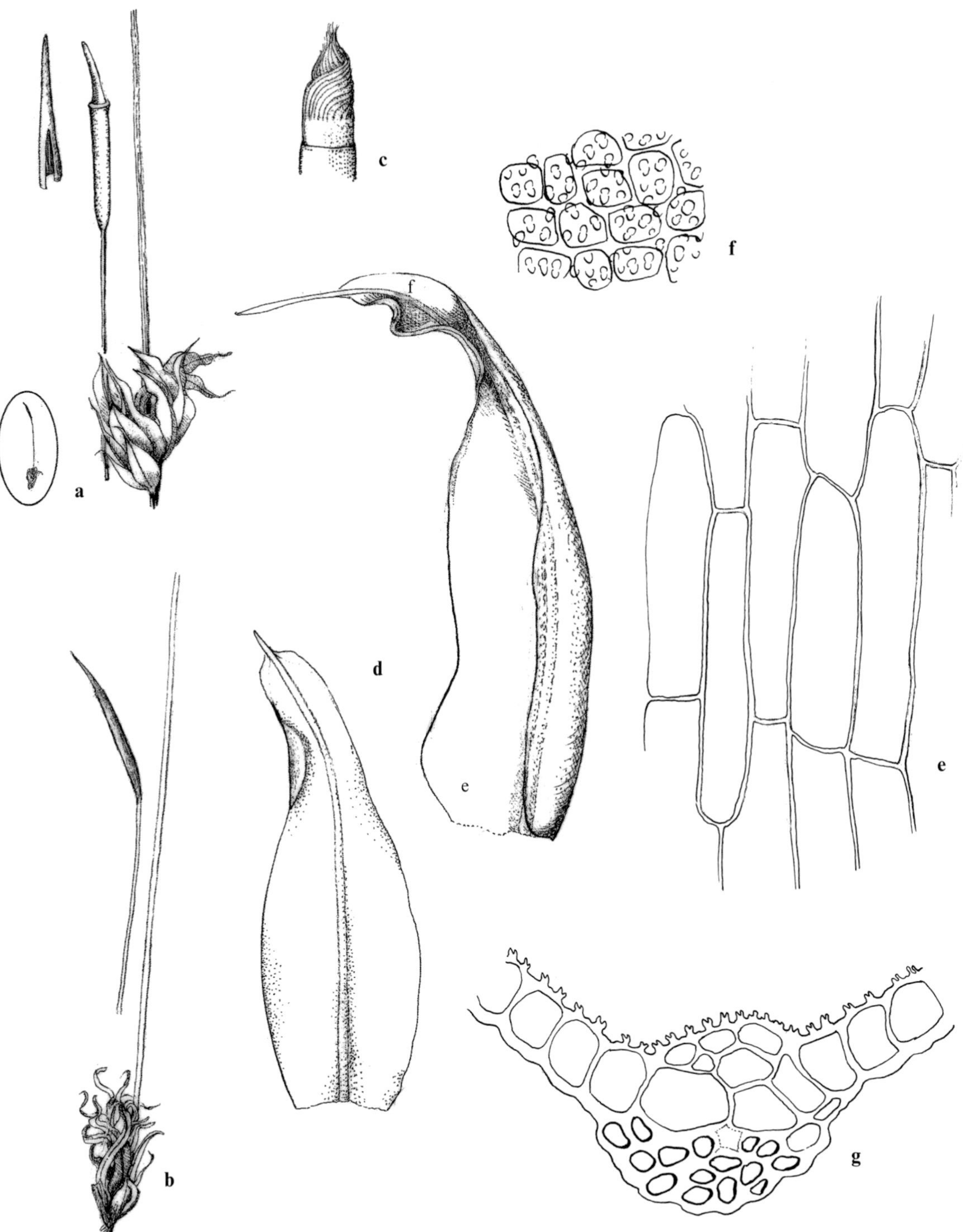

Figure 59. *Tortula canescens*: **a** habit, moist, calyptra (×10); **b** habit, dry (×10); **c** peristome (×25); **d** leaves (×50); **e** basal cells; **f** upper cells (all leaf cells ×500); **g** cross-section of costa and portions of lamina (×500). (*Kushnir*, 21 May 1944)

5. Tortula cuneifolia (With.) Turner, Muscol. Hibern. Spic. 51 (1804).
[Bilewsky (1965) : 381]

Bryum cuneifolium Dicks. ex With., Syst. Arr. Brit. Pl., Edition 4, 3 : 794 (1804); *T. cuneifolia* var. *marginata* M. Fleisch., Atti Congr. Bot. Int. Genova 285 (1893). — Bilewsky (1965) : 381.

Distribution map 61, Figure 60.

Plants usually rather small, 1.5–5 mm high, bright green to yellowish-green. *Stems* with a central strand. *Leaves* concave, thin, soft, lower very small, upper twice as long, crowded, slightly twisted and incurved when dry, erecto-patent to patent, slightly undulate when moist, 1–3 mm long with hair-point (size considerably varying between populations), broadly ovate to spathulate, apex acute, rarely obtuse; margins plane, entire, sometimes partly with one–three rows of slightly incrassate cells; costa excurrent in a hyaline hair-point from very short to ⅓ (½) as long as lamina, hydroids present; all leaf cells thin, smooth, pellucid; basal cells rectangular; mid-leaf cells quadrate-hexagonal, 14–22 µm wide. *Autoicous. Setae* yellowish-red; capsules ± straight, 2.5–3(–4) mm long including operculum, narrow-ellipsoid to cylindrical, reddish-brown; operculum ± obliquely rostrate; peristome with a very short basal membrane. *Spores* 16–19 µm in diameter.

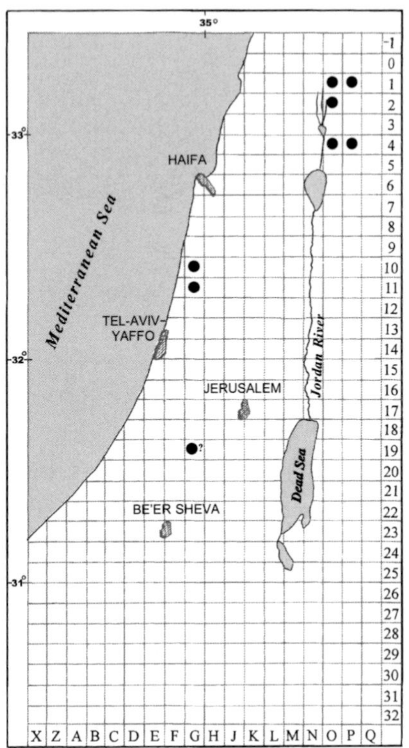

Map 61: *Tortula cuneifolia*

Habitat and Local Distribution. — Plants growing in loose tufts or scattered on exposed sandy loam, terra rossa, basaltic soils, and on rocks. Occasional, locally abundant: Sharon Plain, Shefela (?), and Golan Heights.

General Distribution. — Recorded from Southwest Asia (Cyprus, Turkey, Syria, and Lebanon), around the Mediterranean, Europe (mainly the south), Macaronesia, North Africa, and central Asia.

Tortula kneuckeri Broth. & Geh. (*nom. nud.*) (Geheeb 1903–1904) was described from Sinai ("Dsch. Katherin, 1900–2100 m") as a species belonging to the *T. cuneifolia* group. We were unable to study the type specimen, and it was not possible to decide whether to consider it as a species separate from *T. cuneifolia* on the basis of the description only.

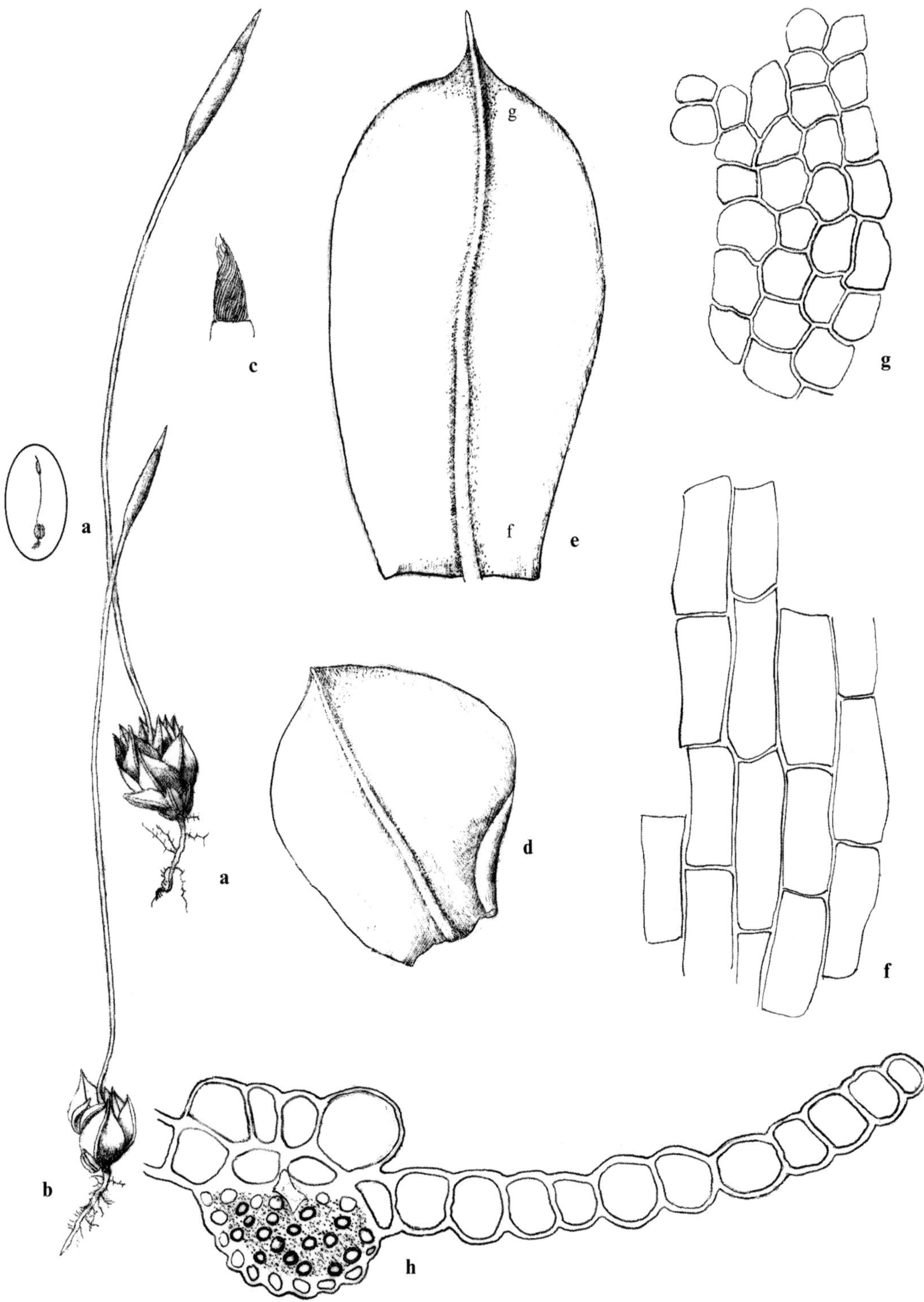

Figure 60. *Tortula cuneifolia*: **a** habit, moist (×10); **b** habit, dry (×10); **c** peristome (×25); **d** lower leaf (×50); **e** upper leaf (×50); **f** basal cells; **g** upper cells (all leaf cells ×500); **h** cross-section of costa and portion of lamina (×500). (*Nachmony*, 2 Mar. 1954)

6. Tortula marginata (B.S.G.) Spruce, London J. Bot. 4 : 192 (1845).
[Bilewsky (1965) : 382]
Barbula marginata B.S.G., Bryol. Eur. 2 : 95 (1843); *Tortula limbata* Lindb., Öfvers. Förh. Kongl. Svenska Vetensk.-Akad. 21 : 238 (1864).

Distribution map 62, Figure 61.

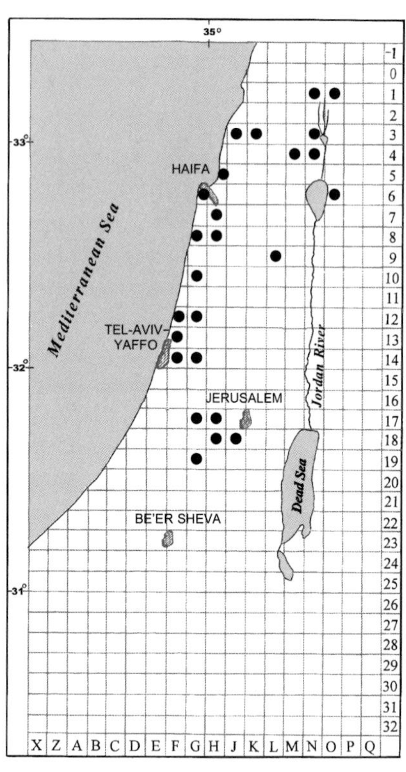

Map 62: *Tortula marginata*

Plants rather small, 2–4(–6) mm high, yellowish-green with yellow costa. *Stems* with a central strand. *Leaves* flexuous to twisted when dry, erectopatent to spreading when moist, 1.5–3(–4) mm long, lingulate to lingulate-spathulate or spathulate, apex acute or obtuse; margins plane, crenulate, with a distinct bistratose border of two–four rows of incrassate, elongate, smooth to weakly papillose cells, extending from base to apex; costa yellow, excurrent in a yellowish mucro, from very short to up to ¼ the length of lamina, hydroids present; basal cells elongate, rectangular; mid-leaf cells irregularly short rectangular or quadrate-hexagonal, 9–14 µm wide, densely papillose with C-shaped papillae. *Dioicous. Setae* erect, orange to red-brown; capsules 2–3 mm long including operculum, cylindrical to elliptical, sometimes slightly curved, reddish-brown; operculum ± obliquely rostrate; peristome with a very short basal membrane. *Spores* 8–10 µm in diameter.

Habitat and Local Distribution. — Plants growing in small, loose or dense tufts, on calcareous or sandy walls and stones. Common: Acco Plain, Coast of Carmel, Sharon Plain, Upper Galilee, Mount Carmel, Mount Gilboa, Shefela, Judean Mountains, Dan Valley, Hula Plain, and Golan Heights.

General Distribution. — Recorded Southwest Asia (Cyprus, Turkey, Lebanon, and Iraq), around the Mediterranean, Europe (mainly the south), Macaronesia, North Africa, and central Asia.

Plants of *Tortula marginata* may be confused with small forms of *T. subulata* because both have leaves with bistratose margins of incrassate cells. However, the peristome teeth of *T. marginata* are almost free, while those of *T. subulata* have a well developed basal membrane. It appears that *T. subulata* is not represented in the local flora, and previous records from the Esdraelon Plain and the Judean Mountains by Bilewsky (1965) and from the Judean Mountains by Herrnstadt *et al.* (1991) proved, after additional studies, to be based on large plants of *T. marginata. Tortula marginata* often shares its habitat with *T. muralis* but can be distinguished by its plane leaf margins and the yellowish, usually short mucro.

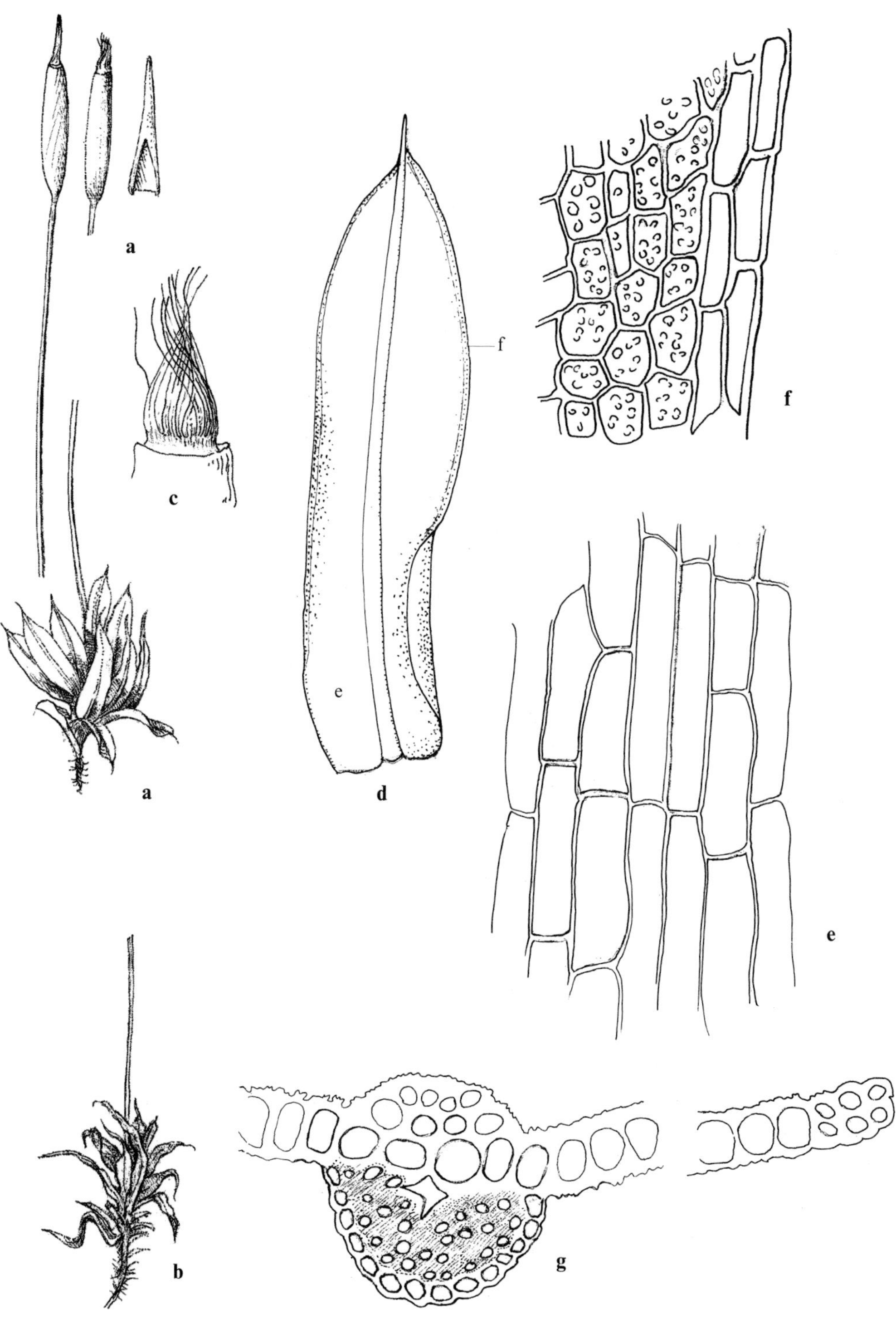

Figure 61. *Tortula marginata*: **a** habit with and without operculum, and calyptra, moist (×10); **b** habit, dry (×10); **c** peristome (×50); **d** leaf (×50); **e** basal cells; **f** marginal upper cells (all leaf cells ×500); **g** cross-section of costa and portion of lamina (×500). (*D. Zohary*, 16 Mar. 1951)

7. Tortula muralis Hedw., Spec. Musc. Frond. 123 (1801). [Bilewsky (1965): 382]
T. aestiva (Hedw.). P. Beauv., Prodr. Aetheogam. 91 (1805).

Plants up to 5(–10) mm long, green when moist, greyish hoary when dry. *Stems* with a central strand. *Leaves* folded and curved or twisted around stem when dry, erect to spreading when moist, up to 2 mm long without hair-point, oblong-lanceolate to lingulate, apex rounded, obtuse or subacute; margins strongly recurved from base to near apex, usually bordered by one–two rows of more incrassate, yellowish, less papillose cells; costa excurrent in a long, ± smooth, usually hyaline hair-point, up to the length of lamina, in lower leaves often excurrent only in a mucro, hydroids present; basal cells rectangular, hyaline; mid-leaf cells irregularly quadrate-hexagonal, incrassate, 8–15 µm wide, densely papillose or mammillose. *Autoicous. Setae* erect, dark red when mature; capsules nearly straight, 2.5–3.5 mm long with operculum, ellipsoid to cylindrical, reddish-brown; operculum ± obliquely rostrate; peristome with very short basal membrane. *Spores* 8–13 µm in diameter; surface rugulate (seen with SEM).

General Distribution. — A cosmopolitan species, growing throughout the temperate regions of the world.

Literature. — Guerra *et al.* (1992), Cano *et al.* (1996).

7a. **Tortula muralis** Hedw. var. **muralis**

Distribution map 63, Figure 62:a–f (p. 179), Plate III:a.

Upper leaf cells (8)10–15 µm wide, usually with about four bi- or trifurcate papillae or mammillae, 2.5–5.5 µm high (seen with SEM).

Habitat and Local Distribution. — Plants growing in small dense tufts or cushions in a great variety of habitats, on walls, stones, calcareous and basaltic rocks, in rock crevices, on soil — loess, chalk, marl, basalt, terra rossa, rendzina, travertine — exposed or partly shaded, sometimes in maquis or forests, rarely on tree trunks. Very common, occurs throughout nearly all districts included in the Flora (except for the far south); the most widespread local species.

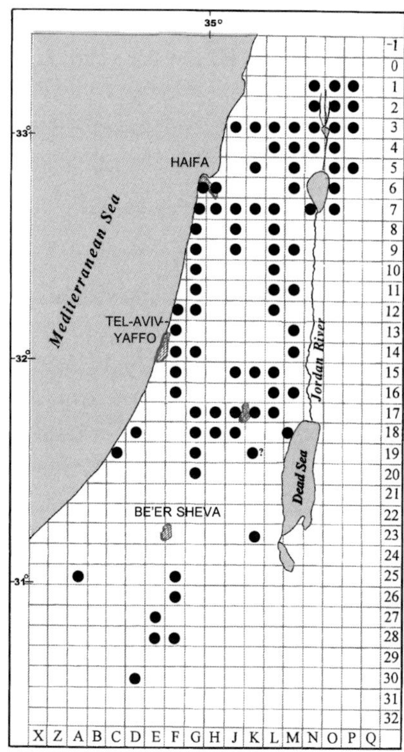

Map 63: *Tortula muralis* var. *muralis*

7b. Tortula muralis Hedw. var. **israelis** (Bizot & F. Bilewsky) Bizot, Rev. Bryol. Lich. 25 : 270 (1956). [Bilewsky (1965) : 382]

T. israelis Bizot & F. Bilewsky, Bull. Res. Counc. Israel Bot. 5D : 51 (1955); *T. muralis* var. *baetica* Casas & R. Oliva, Acta Bot. Malacitana 7 : 104 (1982); *T. baetica* (Casas & R. Oliva) J. Guerra & Ros *in* J. Guerra, Ros & J. S. Carrion, J. Bryol. 17 : 281 (1992).

Distribution map 64, Figure 62 : g, h, Plates II : m, III : b.

Upper leaf cells 6–8 µm wide, usually with one–two simple or bifurcate mammillae or papillae, 8–13(–15) µm high (seen with SEM).

Habitat and Local Distribution. — Plants growing on soft and hard limestone walls and rocks, basalt rocks, facing north and west and in shade of trees. Occasional to common: Mount Carmel, Samaria, Shefela, Judean Mountains, and Golan Heights.

General Distribution. — Recorded in Sicily (Aiello & Dia 2000), Greece (Ros pers. comm.), Cyprus, Turkey, and southern Spain (Cano *et al.* 1996).

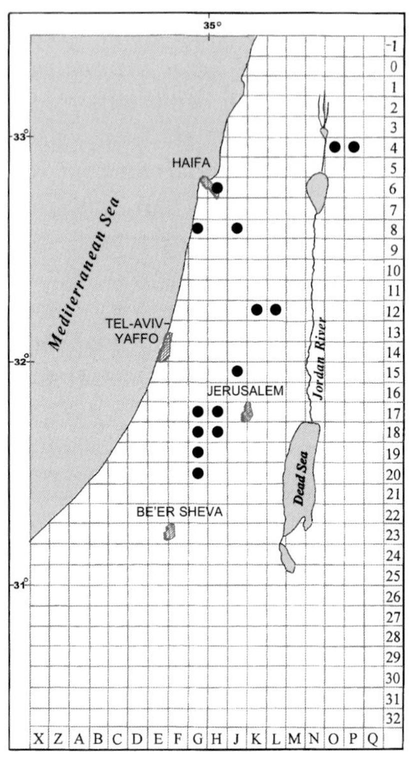

Map 64: *Tortula muralis* var. *israelis*

The infraspecific division of *Tortula muralis* poses many problems because of the continuous variation in the size of plants and leaves, shape of leaves, margin curvature, and the length of the hair-point. Bilewsky (1965) recorded in the local *T. muralis* (in addition to the typical variety) var. *israelis* and var. *aestiva* from the Judean Mountains, and two forms: forma "*incana* Schpr." and forma "*obcordata* Schpr." from "throughout the whole country". Because of the continuous variation of characters in *T. muralis*, it was preferred here to accept at the intraspecific level only var. *muralis* and var. *israelis*. The type specimen of the var. *israelis* (no. 65 in Herb. Bilewsky) shows the mammillae or papillae (in fact usually mammillae) as described by Bizot & Bilewsky in the diagnosis of *T. israelis*. However, the additional characters mentioned there are also part of the continuous variation observed in *T. muralis*. The identification of *T. baetica* as *T. israelis* was proposed by Cano *et al.* (1996).

Figure 62. *Tortula muralis* var. *muralis*: **a** habit and calyptra, dry (×10); **b** habit, capsule without operculum, moist (×10); **c** leaf (×50); **d** basal cells; **e** upper cells (all leaf cells ×500); **f** cross-section of costa and portion of lamina (×500).
Tortula muralis var. *israelis*: **g** upper cells (×500); **h** cross-section of costa and portions of lamina (×500).
(**a–f**: *Markus & Kutiel 77-599-2*; **g, h** *Herrnstadt & Crosby 78-15-3*)

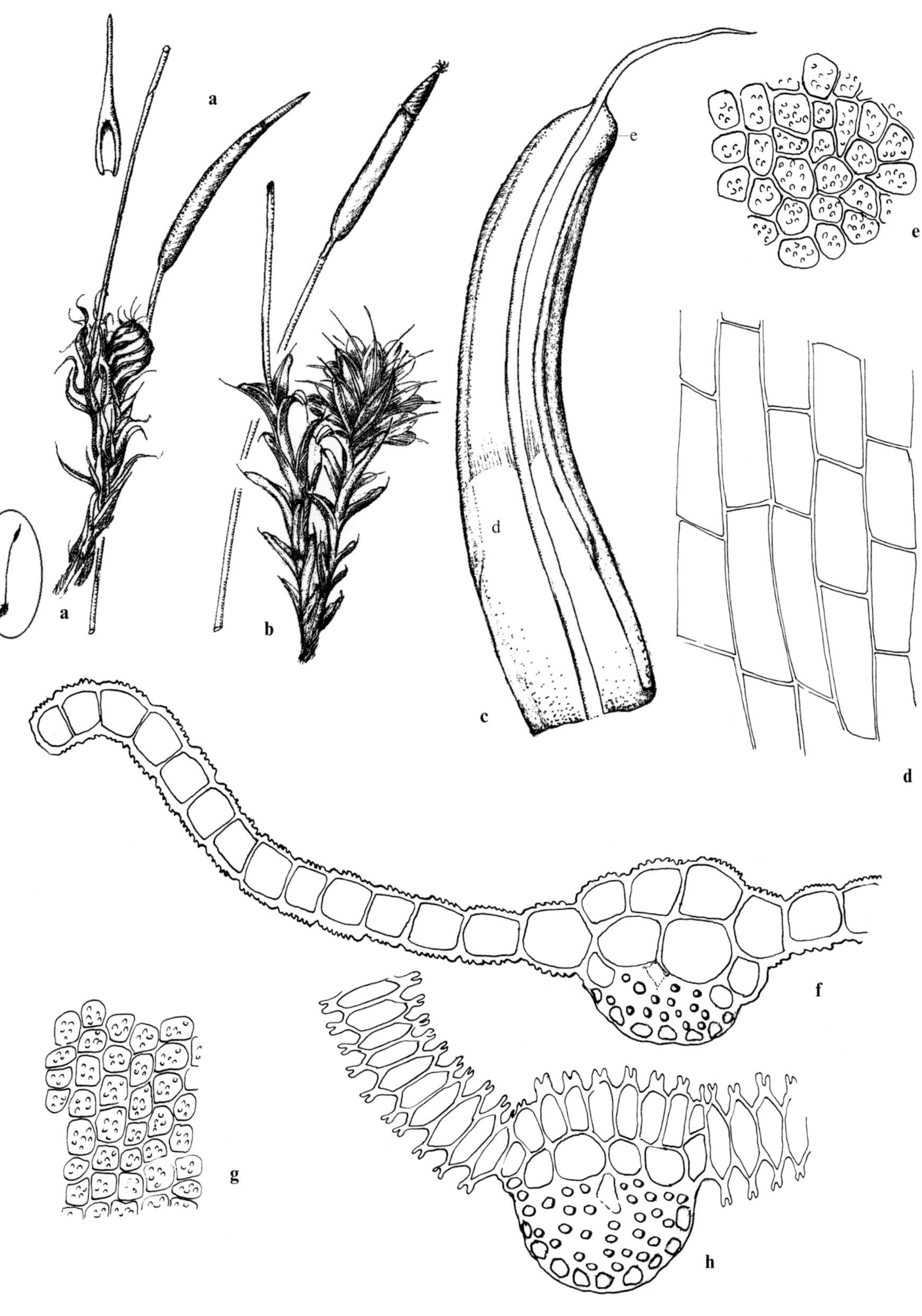
a
e
d
a
b
c
e
d
f
g
h

8. Tortula vahliana (Schultz) Mont., *in* Gay, Fl. Chil. 7 : 153 (1850).
[Bilewsky (1965) : 381]
Barbula vahliana Schultz, Nova Acta Acad. Caes. Leop.-Carol. German. Nat. Cur. 11 : 222 (1823).

Distribution map 65, Figure 63, Plate III : c.

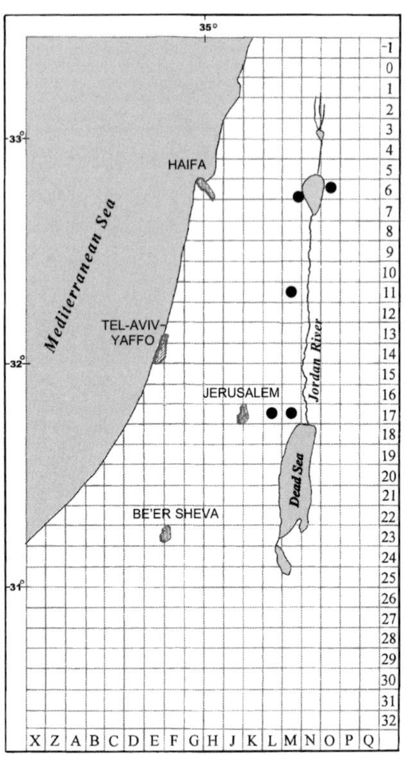

Map 65: *Tortula vahliana*

Plants up to 5 mm long, green to yellowish-green. *Stems* with a central strand. *Leaves* twisted when dry, soft, erect to spreading, with flexuose margins when moist, 3 mm long with hair-point, lingulate to spathulate, slightly constricted below middle; apex obtuse to rounded; margins slightly recurved at mid-leaf or sometimes above, some leaves plane, occasionally leaves with one–two marginal rows of incrassate, less papillose cells; costa excurrent in a smooth, not hyaline point, usually short, sometimes up to ⅓ the length of lamina, hydroids present; basal cells rectangular, hyaline; mid-leaf cells irregularly quadrate, 14–19 μm wide, papillose with irregularly branched papillae. *Autoicous. Setae* erect, yellowish; capsules ca. 3 mm long including operculum, narrow, long-cylindrical, slightly curved, brown; operculum long, ± obliquely rostrate; peristome with a very short basal membrane; teeth twisted. *Spores* 11–15 μm in diameter; gemmate (seen with SEM).

Habitat and Local Distribution. — Plants growing in small tufts or scattered among other mosses, on fine-grained soils, on or between rocks, restricted to the arid parts of Israel, mainly along the Syrian-African Rift Valley. Occasional: Samaria, Judean Desert, Upper Jordan Valley, and Lower Jordan Valley.

General Distribution. — Recorded from Southwest Asia (Cyprus, Turkey, and Iraq), around the Mediterranean, Europe, (mainly the south), Macaronesia, North Africa, and southern South America.

Plants of *Tortula vahliana* can be distinguished from other *Tortula* species by their soft leaves and the long-cylindrical, slightly curved capsules.

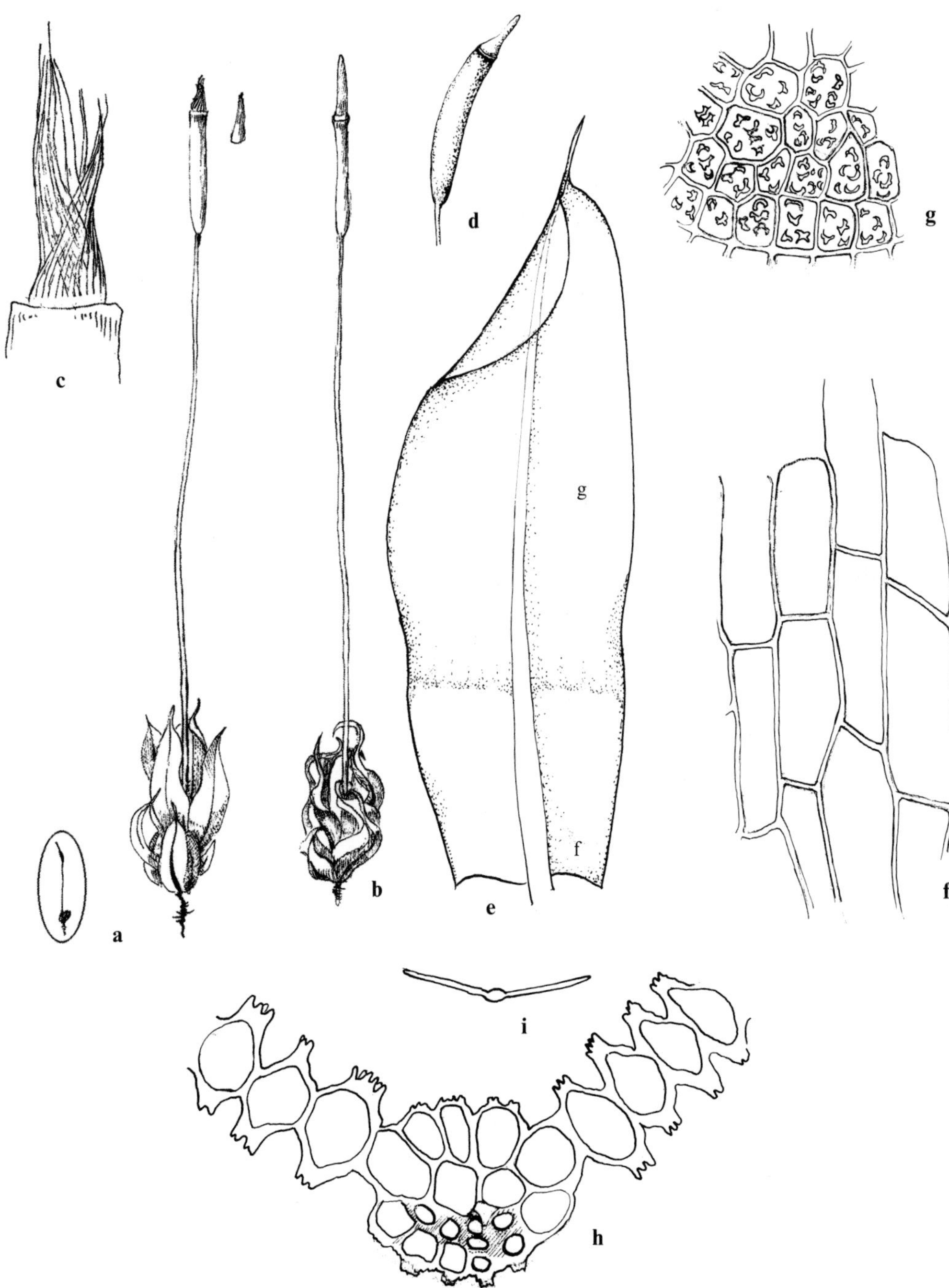

Figure 63. *Tortula vahliana*: **a** habit with detached operculum, moist (×10); **b** habit, dry (×10); **c** peristome (×50); **d** capsule (×10); **e** leaf (×50); **f** basal cells; **g** mid-leaf cells (all leaf cells ×500); **h** cross-section of costa and portions of lamina (×500); **i** schematic cross-section of leaf (×50). (**a**, **b**, **e**–**i**: Tiberias, *Landau* (?); **c**, **d**: *Herrnstadt 80-132-6*)

9. Tortula handelii Schiffn. Ann. K. K. Naturhist. Hofmus. 27:485 (1913).
Syntrichia handelii (Schiffn.) Agnew & Vondr., Feddes Repert. 86:401 (1975). — Herrnstadt *et al.* (1982).

Distribution map 66, Figure 64, Plate III:d.

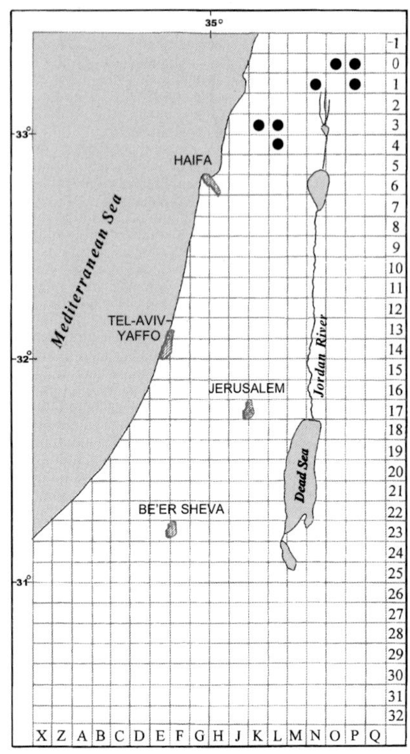

Map 66: *Tortula handelii*

Plants robust, up to 30 mm high, dark green to greenish-brown when dry, yellowish to brownish-green when moist. *Stems* long, branched, without a central strand, lower part buried in soil. *Leaves* folded, appressed and slightly twisted when dry, patent and ± recurved when moist, 2–4 mm long without hair-point, broad-lingulate to spathulate, with a ± wider base, sometimes constricted at middle on one side, apex round to obtuse; margins recurved except near apex; costa excurrent in a densely serrate, oblique, hyaline hair-point, red-brown, hair-point usually about half the length of lamina, on dorsal surface with short, usually simple, papillae, hydroids present or absent; basal cells smooth, rectangular, abruptly differentiated from cells above; mid-leaf cells small, 5–7 µm wide, irregularly quadrate-hexagonal, densely papillose with C-shaped papillae; lamina partly bistratose above middle. *Dioicous*. *Setae* reddish-brown; capsules up to 5 mm long including operculum, cylindrical, weakly curved, dark brown; operculum long, ± obliquely rostrate; peristome with a tessellated basal membrane up to over half of its length, teeth spirally twisted. *Spores* 15–18 µm in diameter; gemmate, processes densely covered by granules (seen with SEM).

Habitat and Local Distribution. — Plants growing on soil on rocks on north- and west-facing slopes and on soil in shade of *Quercus* trees; plants of the mountainous areas of northern Israel. Occasional: Upper Galilee, Dan Valley, Mount Hermon, and Golan Heights.

General Distribution. — Recorded from Southwest Asia (Cyprus, Turkey, Syria, Iraq, Iran, and Afghanistan); otherwise, recorded from southern Europe and western Mediterranean Region.

Plants of *Tortula handelii* resemble *T. pseudohandelii* (for differences see the description of *T. pseudohandelii*).

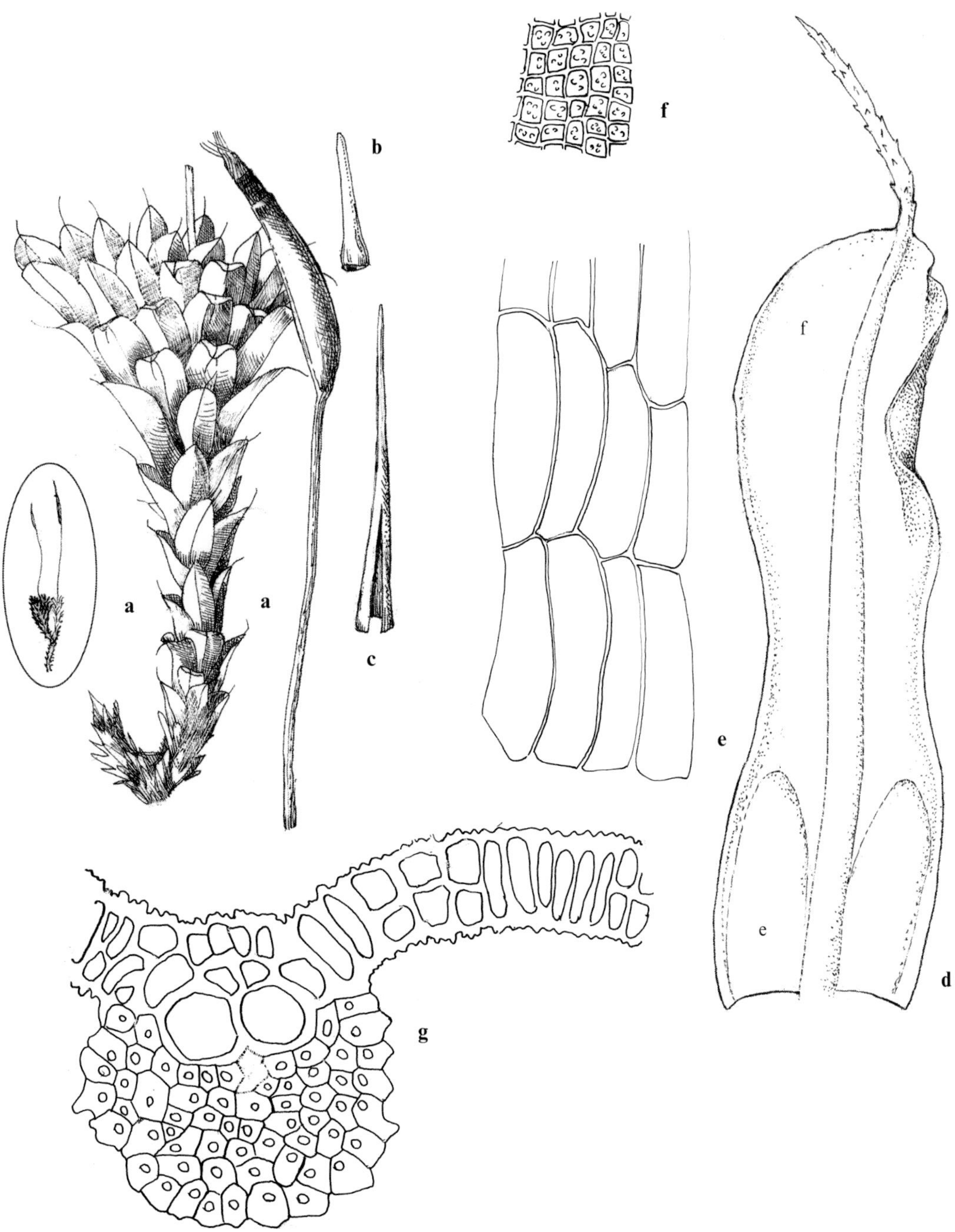

Figure 64. *Tortula handelii*: **a** habit, moist (×10); **b** operculum (×10); **c** calyptra (×10); **d** leaf (×50); **e** basal cells; **f** upper cells (all leaf cells ×500); **g** cross-section of costa and portions of lamina (×500). (*Herrnstadt* & *Crosby 78-48-12*)

10. Tortula pseudohandelii J. Froehl. Ann. Naturhist. Mus. Wien 67: 155 (1964).
Syntrichia pseudohandelii (J. Froehl.) Herrnst. & Heyn, Bryologist 85: 216 (1982).

Distribution map 67, Figure 65.

Plants resembling *Tortula handelii*, but ± smaller, 10–20 mm high, dark green to greyish when dry, green to brownish when moist. *Leaves* smaller, 2–3 mm long without hair-point, patent, not recurved when moist, ligulate with wider base, not constricted at middle, apex rounded or acute, sometimes emarginate, hyaline, narrower than in *T. handelii*; costa excurrent in a longer, slightly twisted hair-point, red-brown, hair-point up to the length of lamina; dorsal surface of costa with long, branched, or stellate papillae; mid-leaf cells larger, 8–11 μm wide; lamina usually bistratose near apex, partly bistratose at middle. *Sporophytes* not seen.

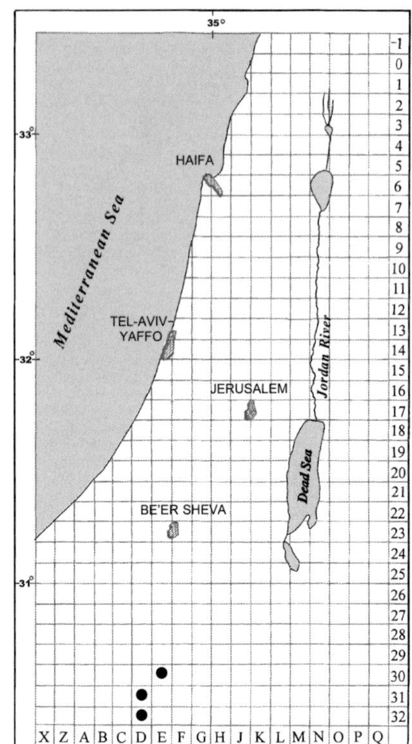

Map 67: *Tortula pseudohandelii*

Habitat and Local Distribution. — Plants growing in dense tufts in loess pockets and at foot of smooth-faced limestone outcrops on north-facing slopes. Rare: so far found only in the arid mountainous parts of Israel, in a few localities in the Central Negev at ca. 900 m altitude.

General Distribution. — Recorded only from Southwest Asia (Turkey, Jordan, Iraq, Iran, and Afghanistan).

The name *Tortula pseudohandelii* is used here with some reservation. Several species have been described as part of a complex related to *T. pseudohandelii* (*T. caninervis* (Mitt.) Broth., *T. desertorum* Broth., *T. pseudodesertorum* J. Froehl.) and have been variously treated in the literature. Agnew & Vondráček (1975) considered the latter two taxa as species of *Syntrichia* separate from *S. pseudohandelii*, though they queried whether *S. pseudohandelii* may not be a long-stemmed form of *S. desertorum* in sheltered habitats. Kramer (1980) treated *T. pseudohandelii* as a separate taxon but considered *T. desertorum* and *T. pseudodesertorum* as synonyms of *T. caninervis*, a species only poorly described up to Kramer's treatment. El-Oqlah *et al.* (1988a) listed both *T. caninervis* (from Amman and Maan) and *T. pseudohandelii* (Amman, Maan, and Karak) from Jordan. However, our study of their cited specimens still left some doubts about the delimitation of these two taxa.

Figure 65. *Tortula pseudohandelii*: **a** stem, moist (×10); **b** stem, dry (×10); **c** leaf (×50); **d** basal cells; **e** mid-leaf cells (all leaf cells ×500); **f** cross-section of costa and portions of lamina (×500). (*Danin 79-60-3*)

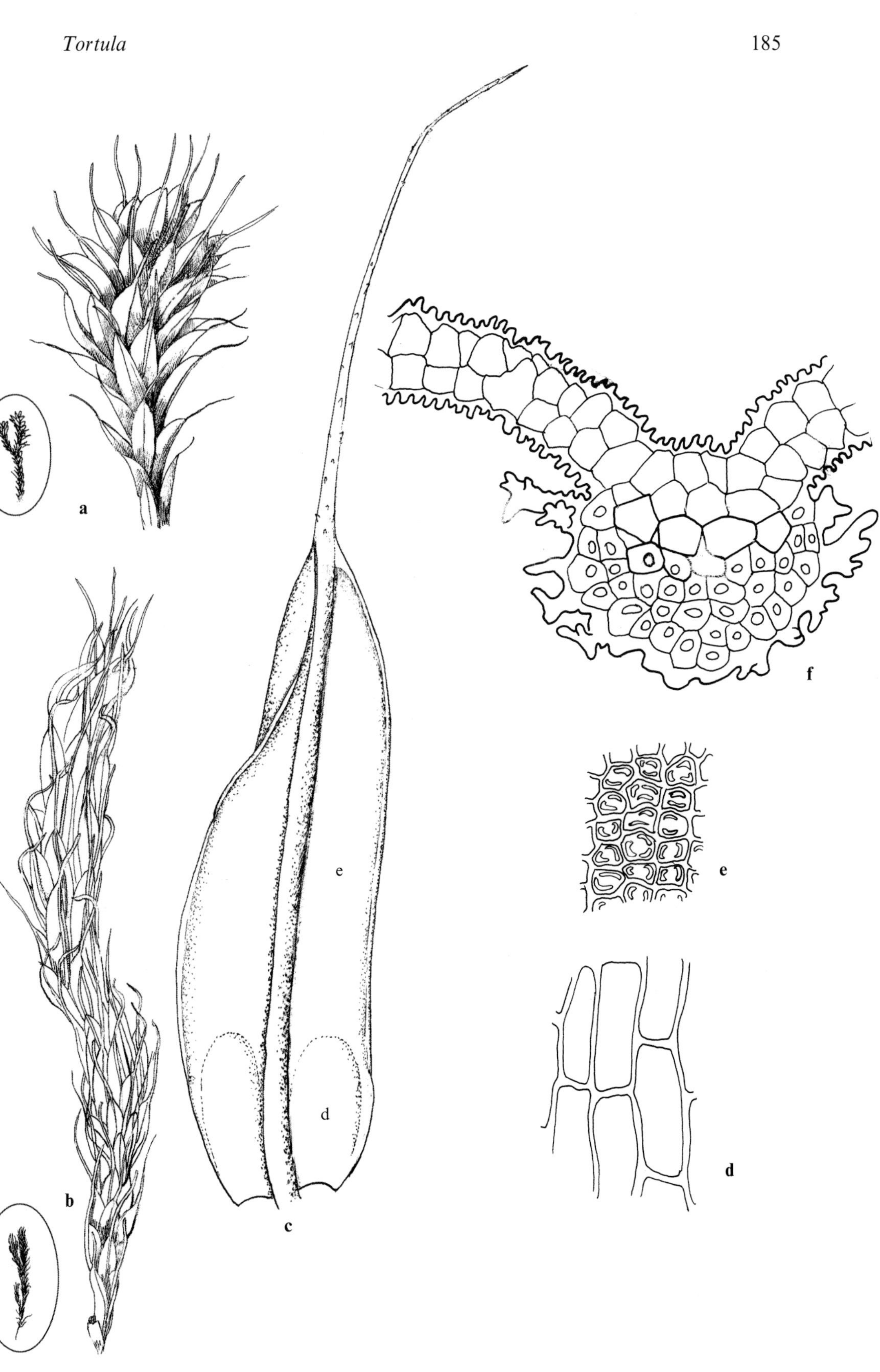

a
b
c
d
e
e
f
d

11. Tortula rigescens Broth. & Geh. Allg. Bot. Z. Syst. 9 : 188 (1903).
Syntrichia rigescens (Broth. & Geh.) Ochyra, Fragm. Florist. Geobot. 37 : 212 (1992).

Distribution map 68, Figure 66.

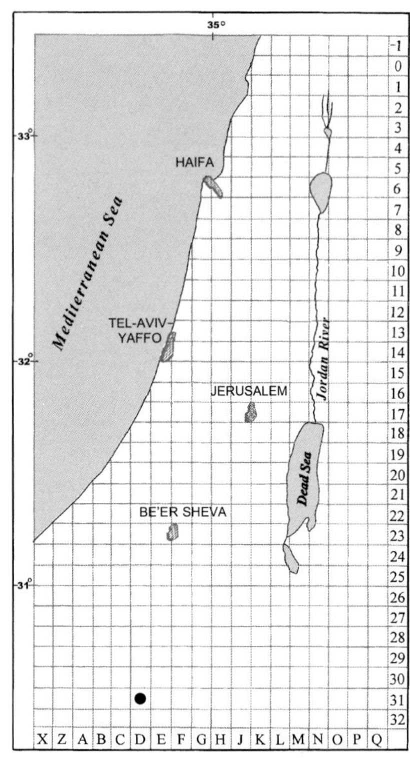

Map 68: *Tortula rigescens*

Plants up to 15 mm high, olive-green turning brownish, upper part readily detached (may serve as dispersal unit). *Stems* without a distinct central strand. *Leaves* appressed, slightly contorted when dry, patent and ± recurved when moist, 1.5–2 mm long without hair-point, oblong-lingulate, apex round; margins crenulate to papillose, recurved from base to near apex, often only on one side; costa strong, excurrent in a long, hyaline, serrate hair-point, brown, hair-point up to ¾ the length of lamina, usually with simple, sometimes with forked or stellate papillae on dorsal surface, hydroids present; basal cells rectangular, abruptly differentiated from cells above; mid-leaf cells irregularly round-hexagonal, 6–11 µm wide, papillose with C-shaped papillae; lamina partly bistratose in upper part; ventral surface of costa with clusters of spherical to ellipsoid four–six-celled gemmae, 30–55 µm wide, brown to dark brown when mature. *Dioicous. Sporophyte* not seen.

Habitat and Local Distribution. — In the single population seen plants growing in tufts or gregarious in loess pockets of north-facing basalt rocks: Central Negev (Qarne Ramon, *Danin*, 25 Oct. 1988).

General Distribution. — Recorded, in addition to the type locality — southern Sinai (on a granite rock) — from Moav and Edom, Jordan (on calcareous rocks) (El-Oqlah *et al.* 1988b).

Plants of *Tortula rigescens* resemble *T. pseudohandelii* in the leaf shape and the partly bistratose lamina, and *T. handelii*, in addition to the latter characters, in the mainly simple papillae on the dorsal surface of costa.

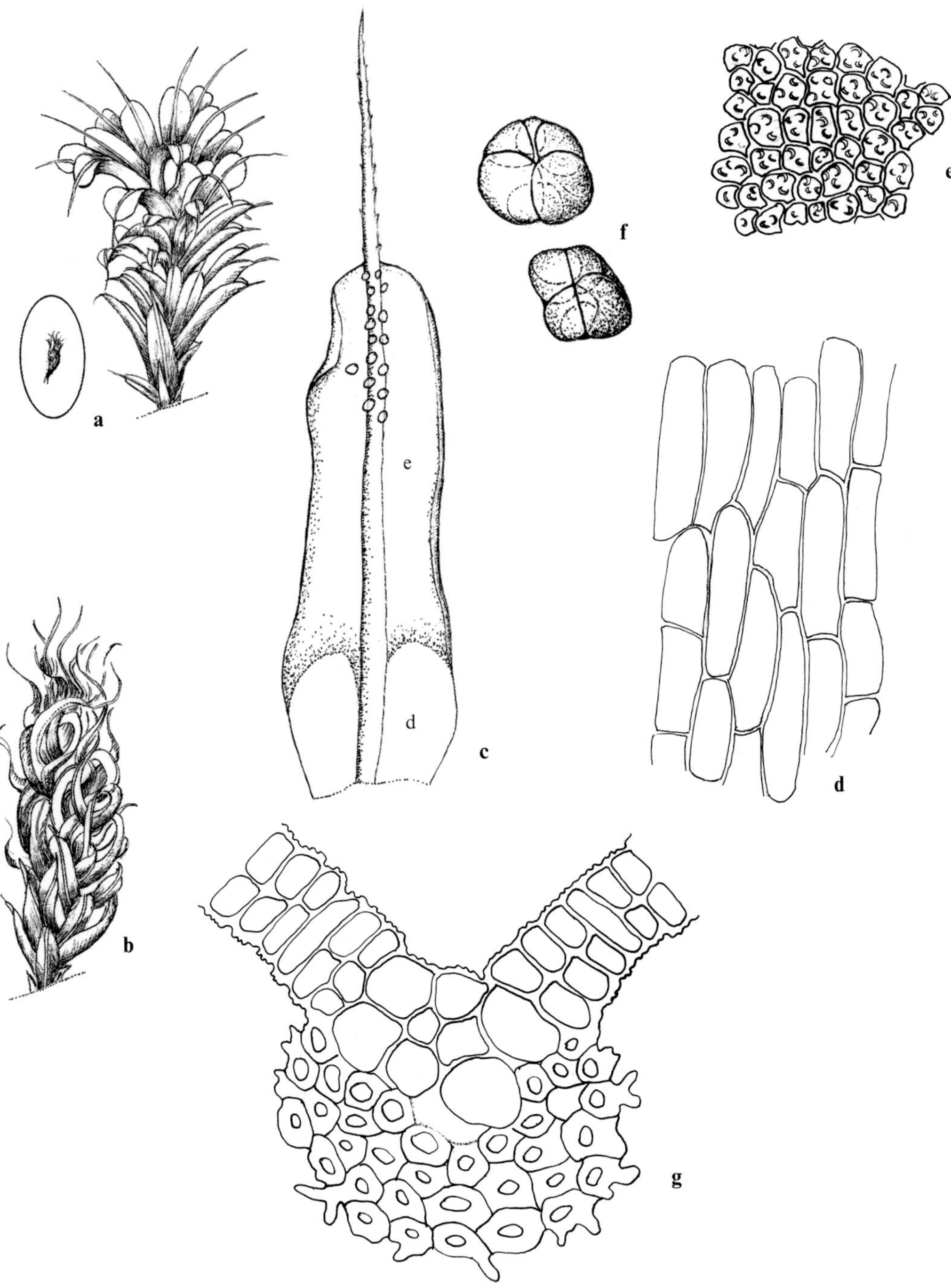

Figure 66. *Tortula rigescens*: **a** stem, moist (×10); **b** stem, dry (×10); **c** leaf (×50); **d** basal cells; **e** mid-leaf cells (all leaf cells ×500); **f** gemmae (×500); **g** cross-section of costa and portions of lamina (×500). (*Danin*, 25 Oct. 1988)

12. Tortula inermis (Brid.) Mont., Arch. Bot. 1 : 136 (1832). [Bilewsky (1965) : 384]
Syntrichia subulata (Hedw.) F. Weber & D. Mohr var. *inermis* Brid., Bryol. Univ. 1 : 581 (1826); *Syntrichia inermis* (Brid.) Bruch *in* Huebener, Muscol. Germ. 335 (1833).

Distribution map 69, Figure 67.

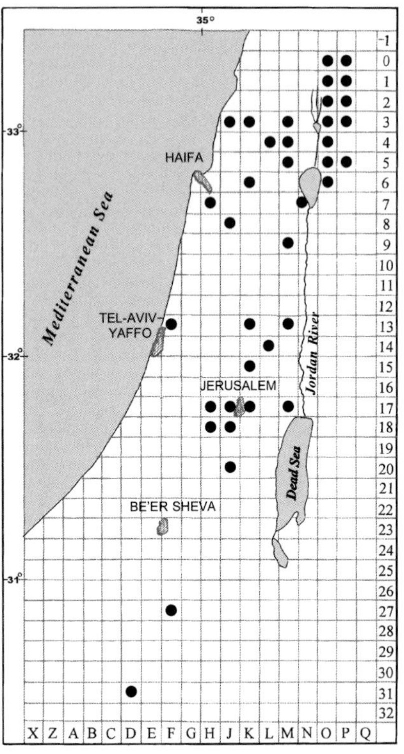

Map 69: *Tortula inermis*

Plants up to 15 mm high, greenish-brown. *Stems* with a central strand. *Leaves* densely spaced, spirally twisted tightly around stem when dry, erectopatent, ± cucullate when moist, 2–4 mm long, lingulate, tapering to a short acute or obtuse, mucronate or apiculate apex; margins narrowly recurved up to near apex; costa very strong, percurrent or excurrent in a short stout mucro, or apiculus, red-brown, hydroids present; basal cells rectangular, smooth, hyaline, narrower towards margin; mid-leaf cells irregularly round-hexagonal, 10–16 μm wide, densely papillose with C-shaped papillae, incrassate. *Autoicous*. *Setae* reddish; capsules erect or sometimes slightly curved, very long, up to 6(–7) mm including operculum, cylindrical, reddish-brown; operculum ± obliquely rostrate; peristome with a tessellated basal membrane up to over half of its length, teeth spirally twisted. *Spores* 13–16 μm in diameter.

Habitat and Local Distribution. — Plants growing in dense tufts on basaltic and calcareous soils, on rocks, stones, walls, and in rock crevices. Very common: Coastal Galilee, Sharon Plain, Upper Galilee, Lower Galilee, Mount Carmel, Mount Gilboa, Samaria, Judean Mountains, Central Negev, Lower Jordan Valley, Mount Hermon, and Golan Heights.

General Distribution. — Recorded from Southwest Asia (nearly throughout), around the Mediterranean, Europe (various parts, mainly the south), Macaronesia (in part), North Africa, northeastern, eastern and central Asia, and North and Central America.

Plants of *Tortula inermis* are easily identified, especially when dry, by their spirally twisted leaves, the red-brown costa, the lack of a hair-point and the recurved margins all along the leaf.

Figure 67. *Tortula inermis*: **a** habit, dry (×10); **b** habit, moist (×10); **c** deoperculate capsule and operculum (×10); **d** leaf (×50); **e** basal cells; **f** mid-leaf cells (all leaf cells ×500); **g** cross-section of costa and portions of lamina (×500). (*Markus & Kutiel 77-602-1*)

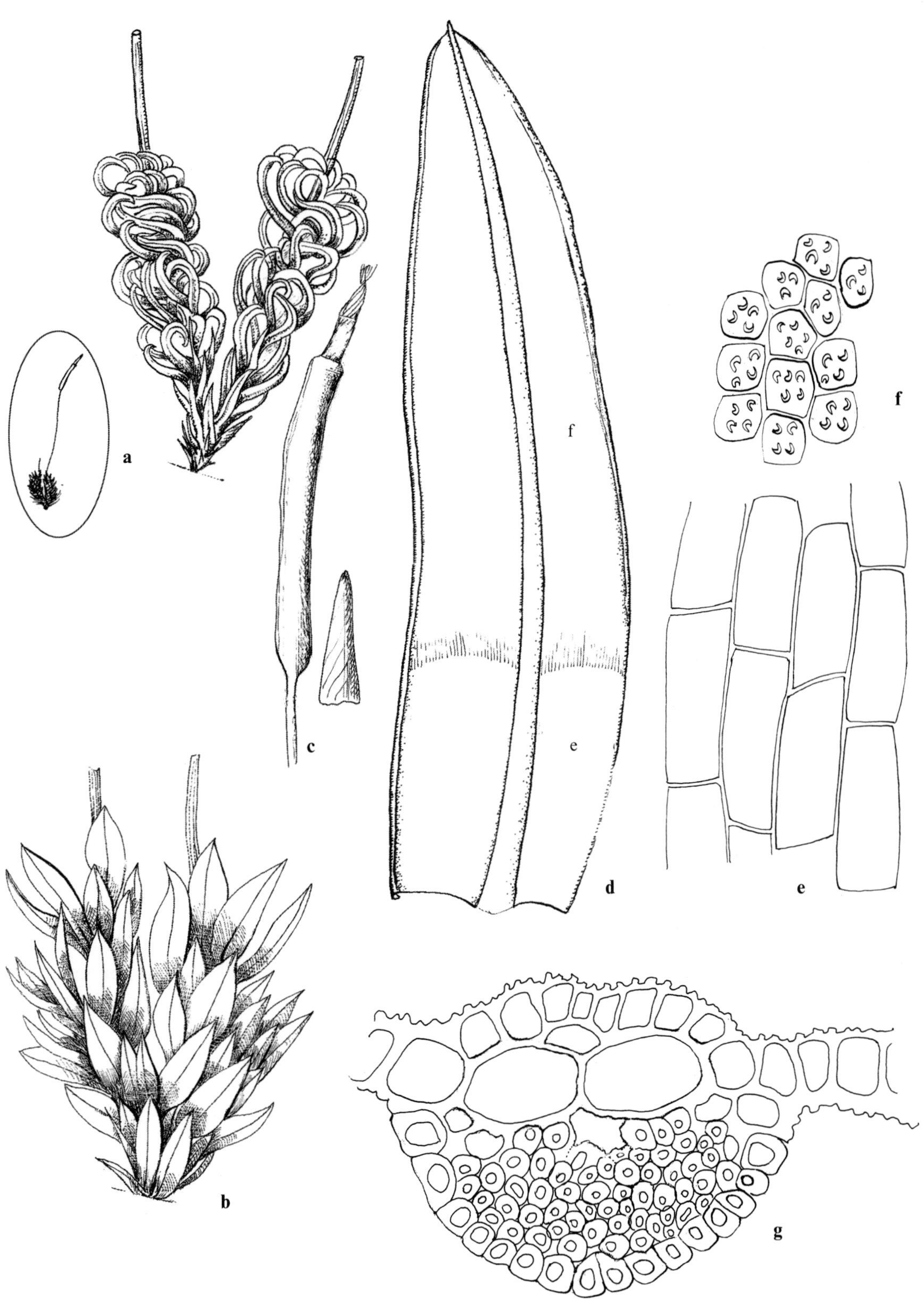
a
f
f
c
e
b
d
e
g

13. Tortula intermedia (Brid.) De Not., Syllab. Musc. 181 (1838). [Bilewsky (1974)]
Syntrichia intermedia Brid., Bryol. Univ. 1:586 (1826); *S. montana* Nees *in* Raab, Flora 2:301 (1819); *T. montana* (Nees) Lindb., Musci Scand. 20 (1879). — Bilewsky (1965):385; *T. ruralis* (Hedw.) P. Gaertn. & Scherb. *in* B. Mey. var. *crinita* De Not., Mem. Reale Accad. Sci. Torino 40:291 (1838).

Distribution map 70, Figure 68.

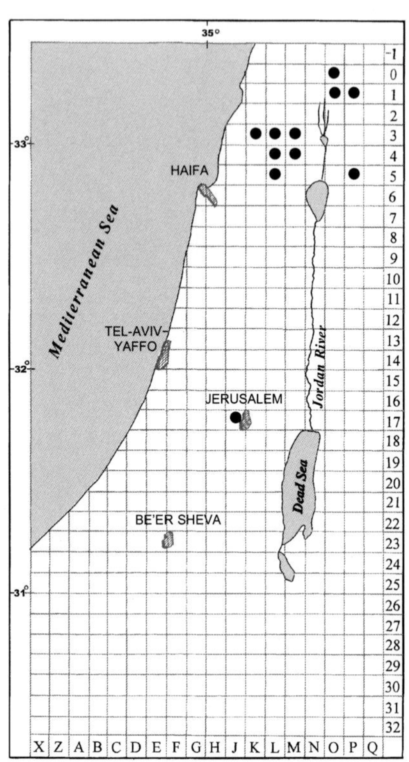

Map 70: *Tortula intermedia*

Plants robust, 10–30(–40) mm high, brownish-green when dry, olive-green when moist. *Stems* forked, without a central strand. *Leaves* more or less crowded towards apex, appressed, flexuose, and incurved when dry, uppermost twisted, erectopatent or ± spreading when moist, ± 3 mm long without hair-point, broad spathulate to lingulate, usually distinctly constricted below middle, round-obtuse to emarginate; margins papillose to crenulate, recurved to ½–⅔ the length of lamina; costa strong, excurrent in a serrulate hyaline hair-point, reddish, hair-point up to ¾ the length of lamina, often oblique, on dorsal surface densely papillose, hydroids present; basal cells rectangular, abruptly differentiated from cells above, towards margins narrower and green; mid-leaf cells shorter, 8–11 µm wide, irregularly hexagonal, densely papillose with C-shaped papillae. *Dioicous*. *Setae* brownish; capsules 5 mm long including operculum, cylindrical, sometimes slightly curved, brown; operculum long, ± obliquely rostrate; peristome with a tessellated basal membrane, up to about half of its length, teeth spirally twisted. *Spores* 13–18 µm in diameter.

Habitat and Local Distribution. — Plants growing in loose tufts on soil on calcareous rocks and on sandstone banks, and in partly shaded habitats in mountainous areas. Occasional: Upper Galilee, Lower Galilee, Judean Mountains, Mount Hermon, and Golan Heights.

General Distribution. — Recorded from Southwest Asia (nearly throughout), around the Mediterranean, Europe (throughout), Macaronesia, North Africa, northeastern and central Asia, and North and Central America.

Tortula intermedia resembles *T. ruralis* but differs in the constriction of the leaves, and the margins which are not recurved up to near the apex. It is treated by some authors (e.g., Anderson *et al.* 1990, Mishler *in* Sharp *et al.* 1994) as part of *T. ruralis*.

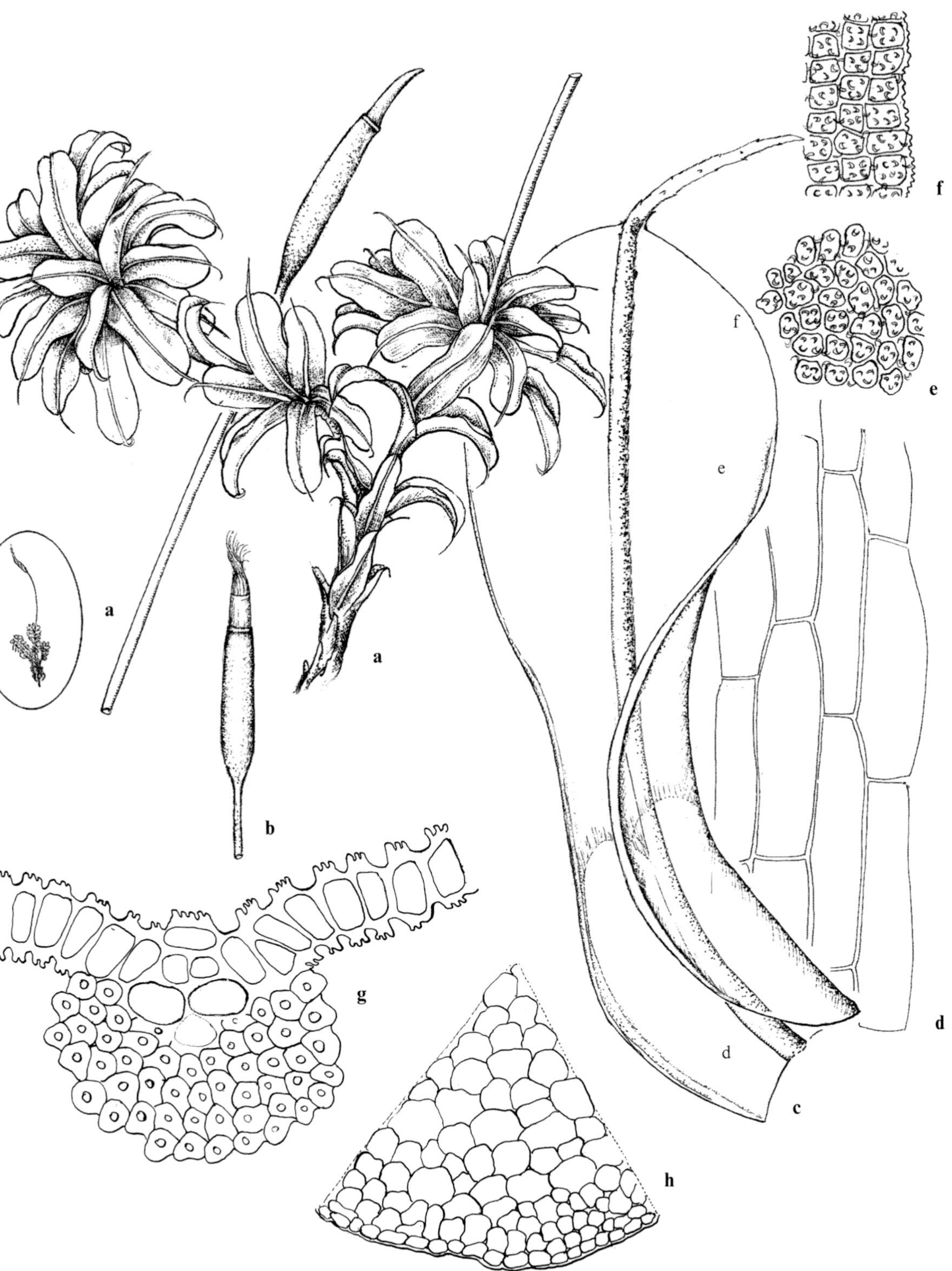

Figure 68. *Tortula intermedia*: **a** habit, moist (×10); **b** deoperculate capsule (×10); **c** leaf (×50); **d** basal cells; **e** upper cells; **f** marginal upper cells (all leaf cells ×500); **g** cross-section of costa and portions of lamina (×500); **h** portion of cross-section of stem (×250). (**a**–**g**: *Herrnstadt*, Apr. 1987; **h**: *Bilewsky*, Apr. 1954, No. 79/80 in Herb. Bilewsky)

14. Tortula laevipila (Brid.) Schwägr., Sp. Musc. Frond. Suppl. 2:66 (1823). [Bilewsky (1965):384]

Syntrichia laevipila Brid., Mant. Musc. 98 (1819).

Distribution map 71.

Plants up to 15(–25) mm high. *Stems* with a central strand. *Leaves* folded and twisted when dry, patent-spreading and ± recurved when moist, spathulate, distinctly constricted below middle, with apex rounded; margins crenulate, narrowly revolute up to mid-leaf, not bordered or bordered by weakly papillose, at least ± incrassate cells, with lumen wider than long; costa strong, excurrent in a nearly smooth hair-point, ¼–⅔ the length of lamina, hydroids present; basal cells rectangular, abruptly differentiated from cells above; mid-leaf cells quadrate to irregularly hexagonal, 8–13 µm wide, densely papillose with C-shaped papillae; lamina partly bistratose. *Setae* yellowish; capsules up to 4.5 mm long including operculum, straight or slightly curved, cylindrical, brown; operculum ± obliquely rostrate; peristome with a tessellated basal membrane about half of its length, teeth spirally twisted. *Spores* 14–16 µm in diameter.

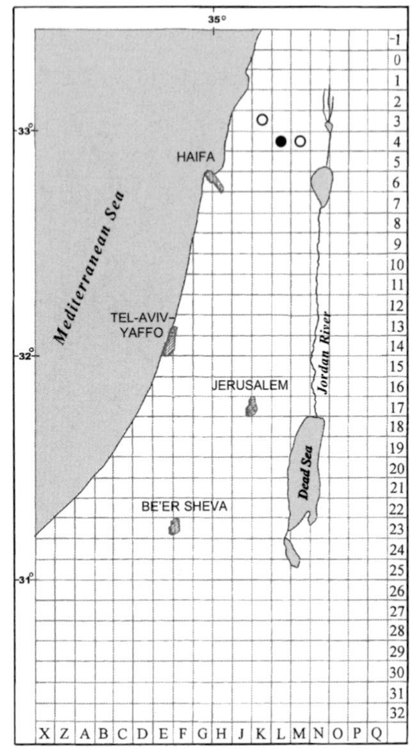

Map 71: ● *Tortula laevipila* var. *laevipila*
○ *T. laevipila* var. *meridionalis*

14a. Tortula laevipila (Brid.) Schwägr. var. **laevipila**

Distribution map 71, Figure 69: a–h.

Plants yellowish-green to dark green. *Leaves* ca. 2.5 mm long not including hair-point; margins sometimes without border or mainly above middle with a border of one–two rows of weakly papillose, slightly incrassate cells, with lumen distinctly wider than long; costa yellow to brown. *Autoicous.*

Habitat and Local Distribution. — Plants growing on the bark of *Olea europaea* L. Found, so far, only in a single locality: Upper Galilee, Hurfeish (Herb. Bilewsky nos. 76 and 739).

General Distribution. — Recorded from Southwest Asia (Turkey, Syria, Lebanon, and Iraq), around the Mediterranean, Europe (throughout), Macaronesia, tropical and North Africa, northeastern and central Asia, Australia, New Zealand, Antarctica, and northern and southern South America.

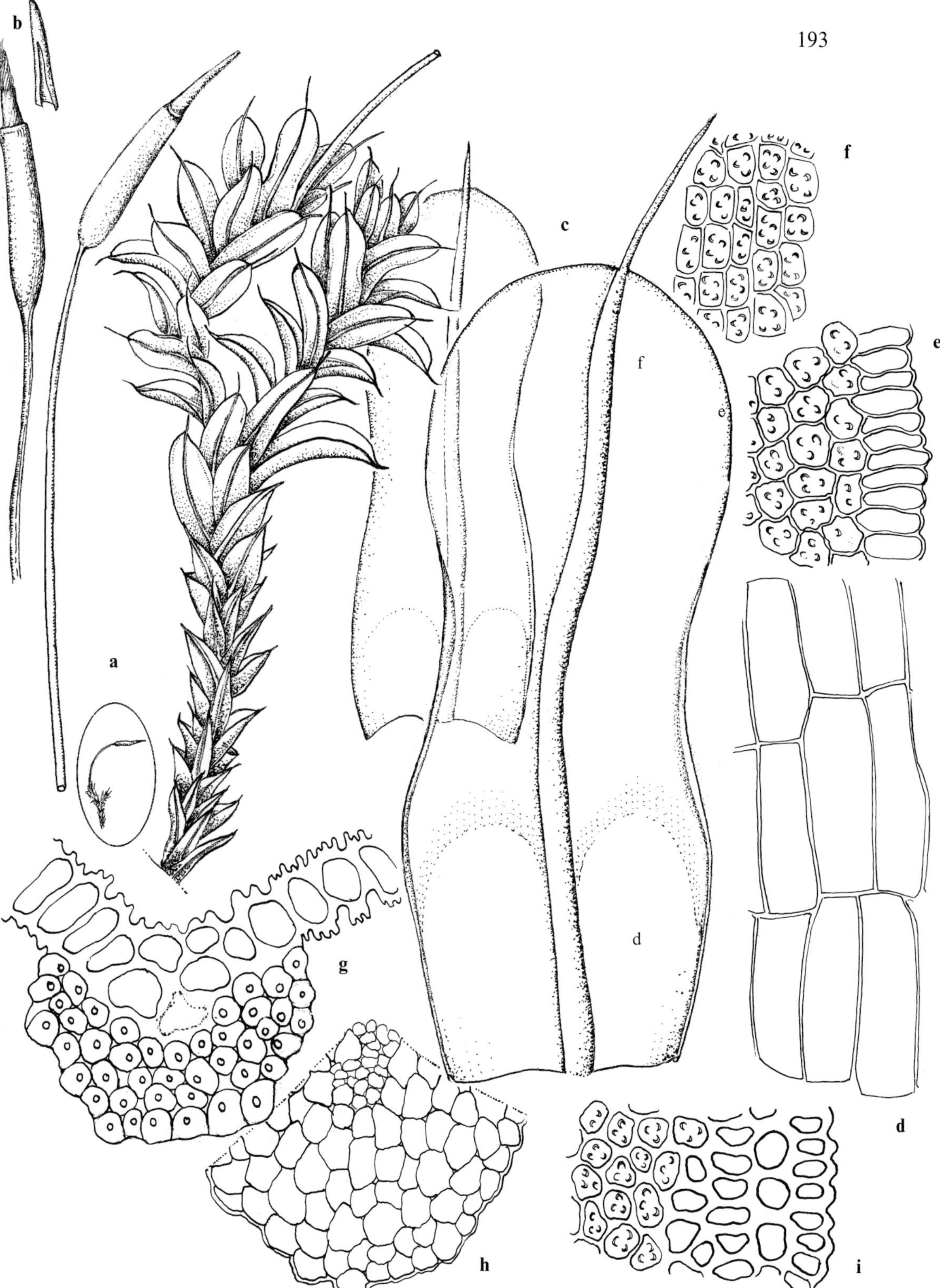

Figure 69. *Tortula laevipila* var. *laevipila*: **a** habit with operculate capsule, moist (×10); **b** seta with deoperculate capsule and calyptra (×10); **c** leaves (×50); **d** basal cells; **e** marginal upper cells; **f** upper cells (all leaf cells ×500); **g** cross-section of costa and portions of lamina (×500); **h** portion of cross-section of stem (×250).

Tortula laevipila var. *meridionalis*: **i** leaf marginal upper cells (×500).

(**a–h**: *Bilewsky*, 16 Apr. 1960, No. 76a in Herb. Bilewsky; **i**: *Kushnir*, "Churfesh", *sine dato*)

14b. Tortula laevipila (Brid.) Schwägr. var. **meridionalis** (Schimp.) Wijk & Margad., Taxon 8 : 75 (1959).

Barbula laevipila (Brid.) Garov. var. *meridionalis* Schimp., Syn. Musc. Eur. 699 (1860); *T. laevipilaeformis* De Not., Musci Ital. 1 : 7 (1862). — Herrnstadt *et al.* (1991) : 176; *T. laevipila* (Brid.) Schwägr. var. *laevipilaeformis* (De Not.) Limpr., Laubm. Deutschl. 1 : 680 (1888). — Bilewsky (1965) : 384.

Distribution map 71 (p. 192), Figure 69 : i (p. 193).

Tortula laevipila var. *meridionalis* differs from var. *laevipila* by the ± longer leaves, ca. 3.5 mm not including hair-point, the well-differentiated border — almost from base to apex — of up to four rows of smooth, strongly incrassate yellowish cells, with lumen only a little wider than long. Recorded as *dioicous*. Sporophytes not seen.

Habitat and Local Distribution. — Plants growing on the bark of *Olea europaea*. Found, so far, in two populations in the Upper Galilee.

General Distribution. — Recorded from Southwest Asia only from Turkey, around the Mediterranean, Europe (mainly the south), Macaronesia (in part), North Africa, and North America.

Two collections deposited in BM — one named *Syntrichia laevipila* and collected by Bridel in 1803 "circa Romam et Naepolim abundad" (perhaps an isotype), and another named *Tortula laevipilaeformis*, collected by Bicchi in Lucca in 1858, and annotated as isotype — fully agree with the concept of the two varieties as accepted here.

Plants of *Tortula laevipila* can be easily differentiated from other local species of *Tortula* by their leaves, which are constricted at their middle and have a nearly smooth hair-point.

15. Tortula princeps De Not, Mem. Reale Accad. Sci. Torino 40 : 288 (1838).

Syntrichia princeps (De Not.) Mitt., J. Linn. Soc. Bot., Suppl. 1 : 39 (1859).

Distribution map 72, Figure 70, Plate III : e.

Plants rather robust, up to 20(–30) mm high, green to brown-green, hoary. *Stems* branched, with a central strand. *Leaves* flexuous, twisted and folded when dry, patent to ± reflexed when moist, up to 3.5 mm long without hair-point, spathulate or broad lingulate, apex round-obtuse; sometimes leaf constricted in the middle on one side; margins entire, recurved up to ¾ the length of lamina; costa with simple, small papillae on dorsal surface, excurrent in a serrate, hyaline, often oblique hair-point, ½–¾ length of lamina, hydroids present; basal cells rectangular, abruptly differentiated from cells above; mid-leaf cells

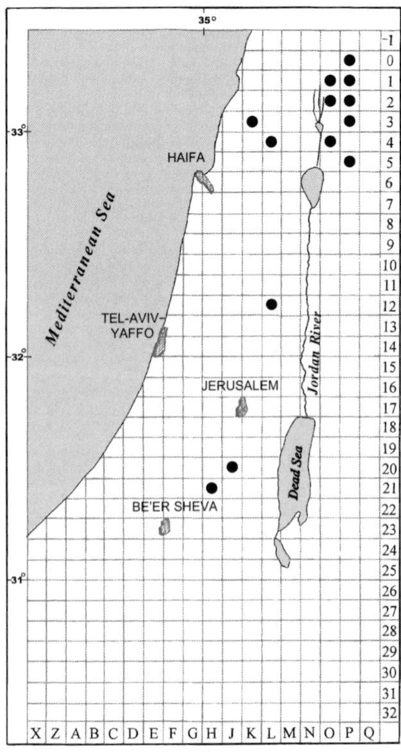

Map 72: *Tortula princeps*

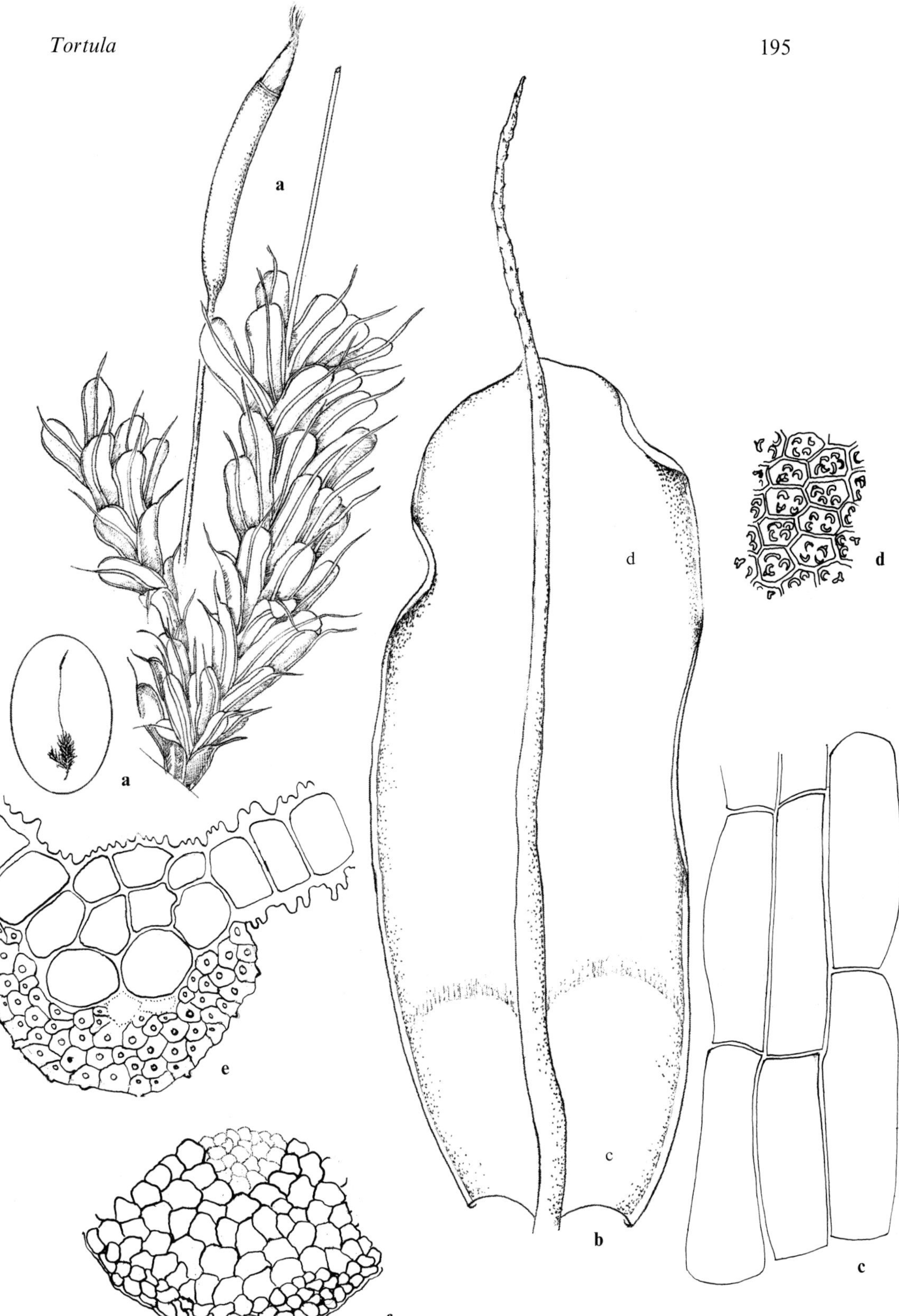

Figure 70. *Tortula princeps*: **a** habit, moist (×10); **b** leaf (×50); **c** basal cells; **d** upper cells (all cells ×500); **e** cross-section of costa and portions of lamina (×500); **f** portion of cross-section of stem (×250). (*Markus* & *Kutiel 77-597-2*)

hexagonal, 10–13 µm wide, with three–four irregularly branched papillae per cell. *Synoicous*. *Setae* reddish-brown; capsules 4–6 mm long, including operculum, erect, longly cylindrical, slightly curved, brown; operculum ± obliquely rostrate; peristome with a tessellated basal membrane, up to ½–⅔ of its length; teeth pale orange, spirally twisted. *Spores* 13–15 µm in diameter; baculate-gemmate (seen with SEM).

Habitat and Local Distribution. — Plants growing in dense tufts among rocks, in soil pockets of rocks, and on tree trunks. Occasional to common: Upper Galilee, Samaria, Judean Mountains, Mount Hermon, and Golan Heights.

General Distribution. — A widespread species; recorded from Southwest Asia (Cyprus, Turkey, Jordan, Iraq, Iran, and Afghanistan), Europe (throughout), Macaronesia (in part), North and South Africa, northeastern, eastern, and central Asia, Australia, New Zealand, and North and Central America.

Tortula echinata is closely related to *T. princeps* (see the discussion of *T. echinata*). Plants of *T. princeps* resemble *T. ruralis*. However, they are synoicous whereas the plants of *T. ruralis* are dioicous. Sterile plants are sometimes difficult to place in either species. The distinguishing vegetative characters used by Kramer (1980) are mainly the distinctly recurved moist leaves, the absence of a central strand in the stem, and hydroids in the costa of *T. ruralis*.

16. Tortula echinata Schiffn., Oesterr. Bot. Z. 65 : 4 (1915).

Syntrichia echinata (Schiffn.) Herrnst. & Ben-Sasson *in* Herrnst., Ben-Sasson & Crosby, Bryologist 85 : 216 (1982); *T. princeps* subsp. *echinata* (Schiffn.) W. A. Kramer, Bryoph. Biblioth. 21 : 86 (1980).

Distribution map 73 (p. 198), Figure 71.

Plants closely resemble *Tortula princeps* in habit and anatomical characters. Therefore, *T. echinata* is often treated as an infraspecific taxon of *T. princeps* (e.g., by Karmer 1980, Frahm 1995). *Tortula echinata* mainly differs from *T. princeps* in the one(–two) large, 8–11 µm high, stalked and stellate papillae in each mid- and upper leaf cell.

Habitat and Local Distribution. — The three populations seen are from similar habitats as occupied by *T. princeps*. Rare: Upper Galilee and Judean Mountains.

General Distribution. — Records are few: from Southwest Asia (Cyprus, Turkey, Lebanon, and Jordan) and Crete.

Plants of one population in the Upper Galilee (Mount Meron, *Herrnstadt* & *Crosby* *78-48-11* — see Figure 71 : g) have larger stellate papillae (13–16 µm high). The plants are very robust, up to 60 mm high, their leaves are flexuose and strongly recurved when moist, and may reach 6 mm in length. In leaf position, they resemble *T. ruralis*, but have a central strand and hydroids in the costa as *T. echinata*. So far, they were not given a distinct taxonomic status.

Figure 71. *Tortula echinata*: **a** habit, moist (×10); **b** leaf (×50); **c** basal cells; **d** upper cells (all leaf cells ×500); **e** cross-section of costa and portions of lamina (×500); **f** portion of cross-section of stem (×250); **g** upper part of stem (×10). (**a**–**f**: *Herrnstadt* & *Crosby* *79-88-19*; **g**: *Herrnstadt* & *Crosby* *78-48-11*)

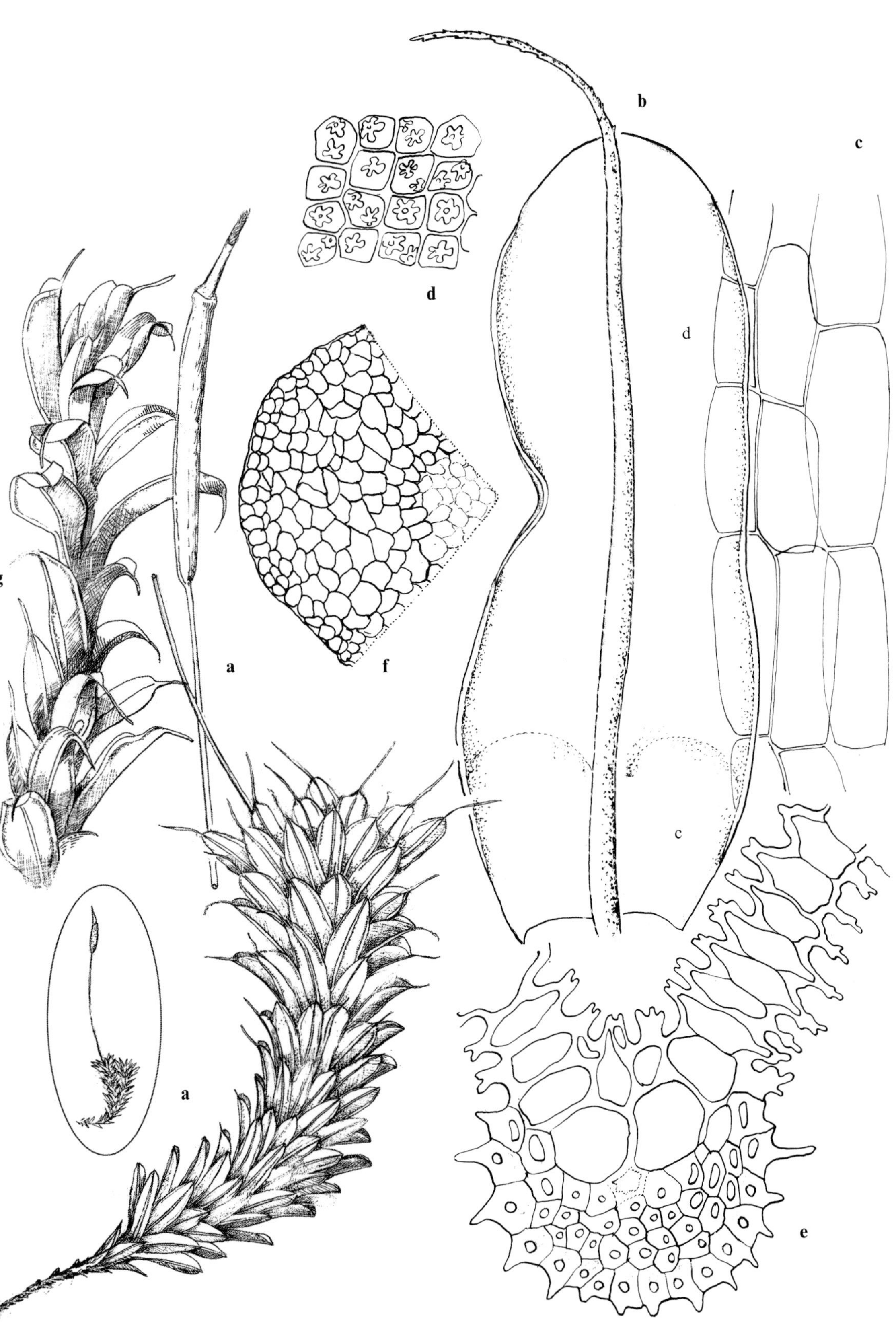
b
c
d
d
f
a
g
a
c
e

A collection identified as *Tortula papillosissima* (Coop.) Broth. was recorded from Israel (Bizot 1954), but later found to have been collected in Cyprus (Bizot 1956).

17. Tortula ruralis (Hedw.) P. Gaertn. *in* B. Mey. & Scherb., Oekon. Fl. Wetterau 3 : 91 (1802). [Bilewsky (1965) : 385 (as "*Tortula ruralis* (L.) Ehrh.")]
Barbula ruralis Hedw., Spec. Musc. Frond. 121 (1801); *Syntrichia ruralis* (Hedw.) F. Weber & D. Mohr, Index Mus. Pl. Crypt. 2 (1803).

Distribution map 74, Figure 72.

Plants robust, up to 25 mm high, brown-green to blackish-green when dry, hoary. *Stems* without a central strand. *Leaves* contorted or twisted around stem and folded when dry, recurved when moist, up to 3 mm long without hair-point, lingulate to spathulate; apex rounded, sometimes emarginate; margins usually strongly recurved, often up to near apex; costa strong, excurrent in a densely serrate to spinulose hyaline hair-point, red-brown, hair-point up to ¾ the length of lamina, rarely longer, few stereids present, hydroids absent; basal cells rectangular, shorter towards margins, smooth, abruptly differentiated from cells above; mid-leaf cells irregularly quadrate-hexagonal, 11–14 μm wide, densely papillose with irregularly branched papillae, sometimes joined in groups. *Dioicous*. *Sporophytes* as in *T. princeps*

Habitat and Local Distribution. — Plants growing in tufts on soil on rocks, in mountainous areas. Occasional: Judean Mountains, Mount Hermon, and Golan Heights.

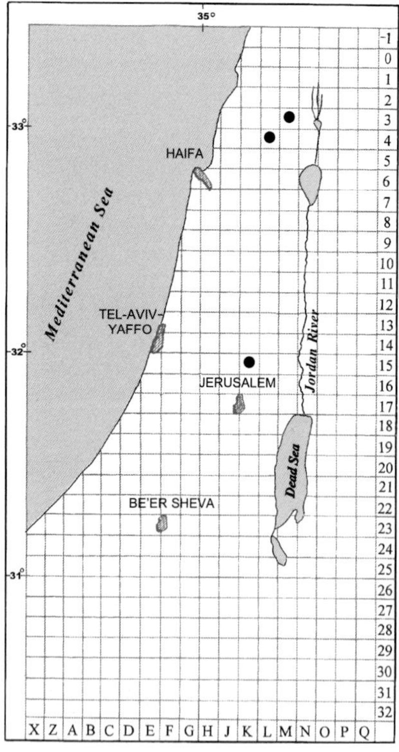

Map 73: *Tortula echinata*

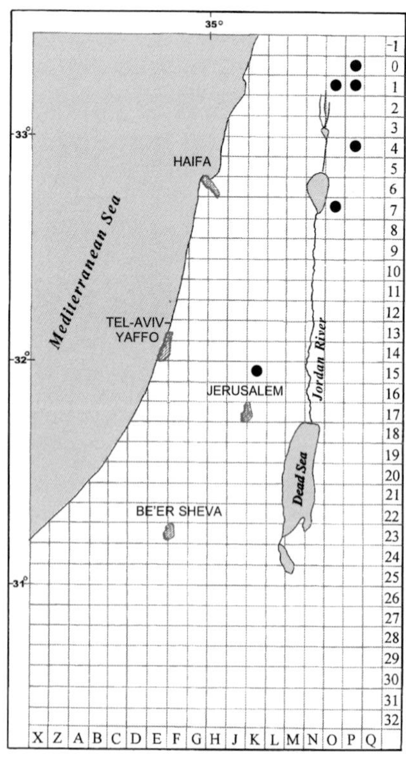

Map 74: *Tortula ruralis*

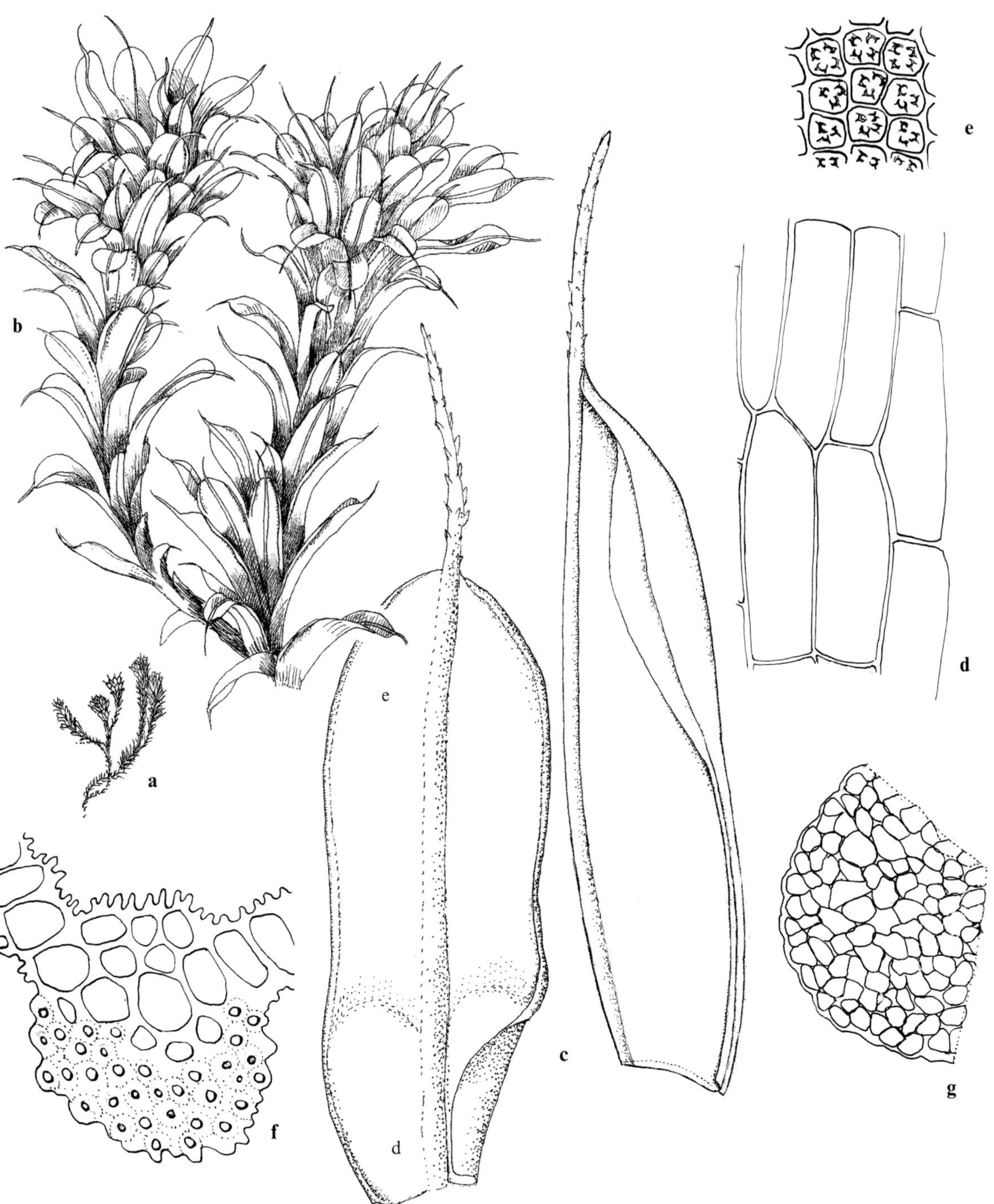

Figure 72. *Tortula ruralis*: **a** habit (×1); **b** habit, moist (×10); **c** leaves (×50); **d** basal cells; **e** upper cells (all leaf cells ×500); **f** cross-section of costa and portions of lamina (×500); **g** portion of cross-section of stem (×125). (*Herrnstadt* & *Crosby 79-97-6*)

General Distribution. — A widespread species; recorded from Southwest Asia (Cyprus, Turkey, Syria, Lebanon, Jordan, Iraq, Iran, Afghanistan, and Saudi Arabia), around the Mediterranean, Macaronesia, North, tropical, and South Africa, northeastern, eastern and central Asia, Australia, Oceania, and north, central, northwestern and southern South America.

Plants of *Tortula ruralis* are easily identified when moist by their distinctly recurved leaves. They differ from some related taxa of section Rurales De Not. (*T. intermedia*, *T. laevipila*, and *T. laevipila* var. *meridionalis*) by the leaves, which are not constricted at the middle, and from some other taxa (e.g., *T. princeps*) by the absence of a central strand and hydroids in the costa (see Kramer 1980).

Tortula ruralis has all the preceding characters in common with *T. norvegica*. Thus, *T. norvegica* is often considered as a variety of *T. ruralis*. Local records of *T. ruralis* subsp. *norvegica* (Web.) Dixon (Bilewsky 1965 : 385 "as var. *norvegica* (Web.) Moenkem".) and *T. norvegica* (Web.) Wahlenb. ex Lindb. (Herrnstadt *et al.* 1991) were based on the misidentification of plants of *T. princeps* De Not. A collection from the Upper Galilee recorded by Bilewsky (1977) as "*T. ruralis* subsp. *calcicola* (Amman) Giac." (= *T. calcicolens* W. Kramer) was not found in Herb. Bilewsky.

18. *Desmatodon* Brid.

Desmatodon is usually considered as a problematic genus, which is not based on clear differential diagnostic characters. Although this view is repeatedly expressed in the literature, authors usually prefer to accept *Desmatodon* as a separate genus comprising 30–40 species and deal with the majority of the traditional species as part of it (e.g., Crum & Anderson 1981, Corley *et al.* 1981, Zander *in* Sharp *et al.* 1994). *Desmatodon* is mostly characterised and differentiated from *Tortula* by sporophytic characters, in particular the nearly straight, 16 deeply cleft peristome teeth vs. the usually twisted 32 filaments of *Tortula*. According to Zander (1993), *Desmatodon* does not deserve a generic status separate from *Tortula* because sporophytic characters (even if constantly found different) "would cut across ... clearly defined generic groupings based on several gametophytic characters". Consequently, he includes *Desmatodon* in *Tortula*.

In the local flora, there are only two species, which are often included in *Desmatodon*: *D. convolutus* and *D. guepinii*. Here, only *D. guepinii* is retained in the genus while *D. convolutus* is included in *Tortula* as *T. atrovirens* (see the discussion above, p. 164).

Desmatodon guepinii B.S.G., Bryol. Eur. 2 : 58 (1843).
Tortula guepinii (B.S.G.) Broth., Nat. Pflanzenfam. 1 : 430 (1902). — Bilewsky (1965) : 383 (as "*Tortula guepini* (Br. Eur.) Limpr".).

Distribution map 75 (p. 202), Figure 73.

Plants yellowish-green when moist, brownish when dry, small, 2–3 mm high. *Leaves* concave, irregularly twisted around stem when dry, erectopatent or patent when moist, up to 1.5 mm with hair-point, concave, obovate or ovate to spathulate, apex

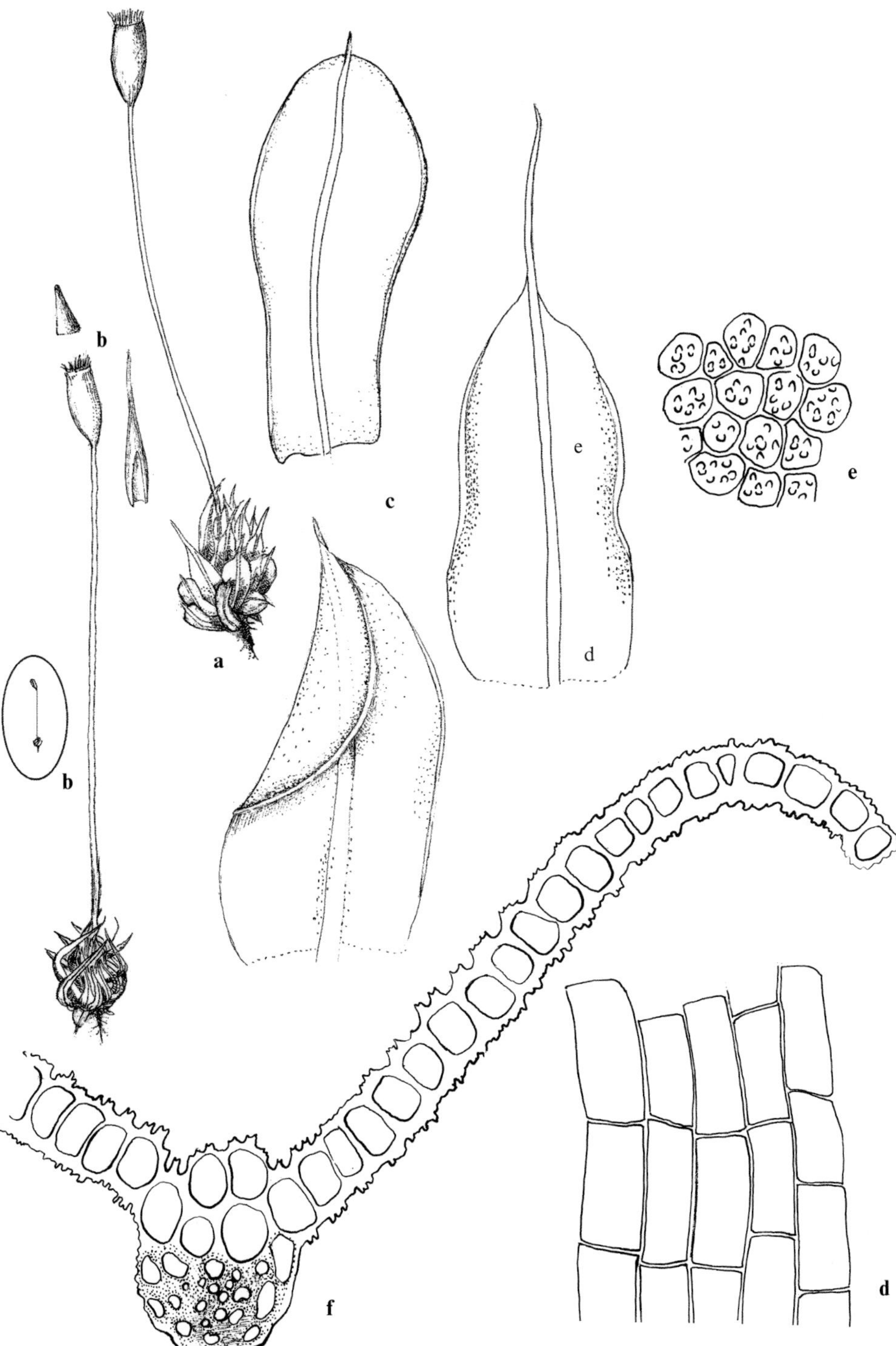

Figure 73. *Desmatodon guepinii*: **a** habit, moist (×10); **b** habit with detached operculum and calyptra, dry (×10); **c** leaves (×50); **d** basal cells; **e** upper cells (all leaf cells ×500); **f** cross-section of costa and portion of lamina (×500). (*Kushnir*, 11 Mar. 1944)

rounded to acute; margins entire, narrowly recurved except near base and apex; costa strong, equally wide throughout, brown excurrent in a green to yellowish-brown or hyaline hair-point, hair-point from short to half the length of lamina; basal cells rectangular, smooth, hyaline; mid- and upper leaf cells irregularly quadrate to hexagonal, 11–17 µm wide, densely papillose with C-shaped papillae. *Autoicous. Setae* erect, yellowish-brown; capsules up to 2 mm long including operculum, ellipsoid; operculum conical-rostrate; peristome teeth 16, from a short basal membrane, deeply cleft, straight. *Spores* 16–19 µm in diameter; nearly smooth, sparsely granulate, granules widely varying in size (seen with SEM).

Habitat and Local Distribution. — Plants gregarious or growing in small tufts, in shaded habitats, on basalt soils and rendzina, in open rock crevices. Rare: Judean Desert, Upper Jordan Valley, and Lower Jordan Valley.

General Distribution. — Records few: Europe (mainly the south), and North and Central America.

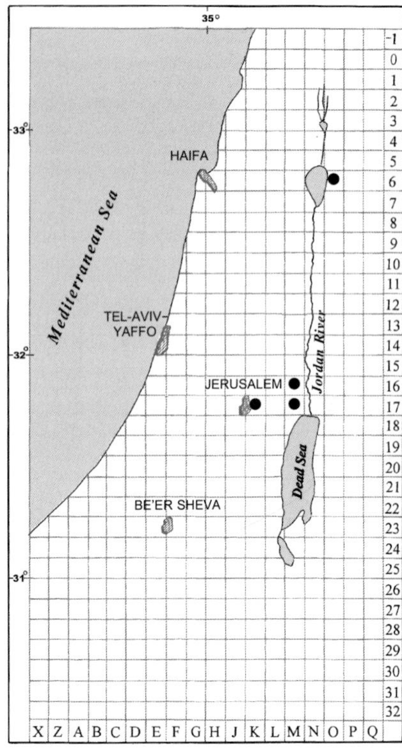

Map 75: *Desmatodon guepinii*

19. *Aloina* Kindb., *nom. cons.*

Plants small, bud-like, fleshy, greenish-brown, growing scattered or gregarious. *Stems* short, often buried in soil, simple, rarely irregularly branched, with or without a central strand. *Leaves* crowded, thick, and rigid, usually incurved when dry, erect to spreading when moist, usually concave above, inflexed above sheathing base, often with distinct shoulders, wide-ellipsoid to lingulate; apex round-obtuse, often cucullate; margins plane, entire to crenulate above, sometimes with a border of several long, thin-walled hyaline cells near base of leaf; costa wide and flat, usually ending in apex or below, rarely excurrent in a hyaline hair-point, usually with a dorsal stereid band, often in several layers; upper part of ventral surface of costa covered by chlorophyllous, simple, or often branched filaments of usually three–five thin-walled cells, each with a conical, apically thickened terminal cell; filaments extending to ventral surface of lamina adjacent to costa; basal cells quadrate to rectangular, smooth, hyaline, thin-walled or slightly thickened at corners and transverse walls; upper cells bistratose near costa, often wider than long, smooth, incrassate especially at corners, with cell lumen rounded; perichaetial leaves sometimes differentiated. *Dioicous. Setae* elongate, straight, (7–)10–15(–20) mm long (in local species), yellowish-red, turning darker when mature; capsules erect or inclined, cylindrical to ellipsoid, symmetrical or slightly curved, yellowish red-brown, turning dark red-brown (in local species); operculum conical-rostrate, ± oblique, up to ⅓ the length of capsule; annulus of one–three rows of cells, usually persistent (revoluble in the local *Aloina rigida*); peristome usually with a low basal membrane (absent among local plants of *A. aloides* var. *aloides*), with 16 teeth, usually cleft into 32 filiform, papillose, twisted segments. *Calyptrae* cucullate, smooth. *Spores* finely papillose.

A small, widespread genus of 12 species (Zander 1993), growing mainly on soil. Three species and one variety occur in the local flora.

Plants of all local species of *Aloina* are olive-green to green, turning dark green to reddish-brown when mature. They are the first to produce sporophytes in the early winter, after the dry season, and can be found bearing sporophytes over a long period (November–April).

Aloina resembles *Crossidium*, in particular in the chlorophyllous filaments on the ventral surface of the leaf (see the discussion of *Crossidium* below, p. 210). This seems a most efficient photosynthetic device, forming an enlarged green surface, which is protected by the inflexed leaf.

Literature. — Delgadillo (1973a, b, 1975; *in* Sharp *et al.* 1994).

1 Leaf apex not cucullate, costa ending in a long hyaline hair-point **2. A. bifrons**
1 Leaf apex cucullate, costa not ending in a hyaline hair-point.
  2 Leaf at base with several rows of thin-walled, elongate hyaline cells, forming a distinct border; annulus revoluble **3. A. rigida**
  2 Leaf at base without a differentiated border; annulus persistent.
    3 Peristome with a distinct basal membrane; spores 13–17 μm in diameter **1b. A. aloides** var. **ambigua**
    3 Peristome without a distinct basal membrane; spores 19–25 μm in diameter **1a. A. aloides** var. **aloides**

1. Aloina aloides (W. D. J. Koch ex Schultz) Kindb., Bih. Kongl. Svenska Vetensk.-Akad. Handl. 7: 136 (1883).

Trichostomum aloides W. D. J. Koch ex Schultz, Nova Acta Caes. Leop. Carol. German. Nat. Cur. 11: 197 (1823).

Plants up to 3 mm high. *Leaves* up to 3 mm long, lingulate to ligulate, with apex cucullate, obtuse and apiculate or acute; margins entire and not differentiated; costa ending below apex or percurrent, with three–six layers of stereids; upper (including upper marginal) cells oblate-oblong, incrassate; upper and mid-leaf cells 10–22 μm wide. *Capsules* up to 3.5(–4.5) mm long including operculum — larger than in the other local species — cylindrical, straight or sometimes slightly curved; annulus of one–two rows of cells, persistent; peristome teeth twisted, with or without a basal membrane. *Spores* densely granulate, granules with rough surface.

1a. Aloina aloides var. **aloides**
["*A. aloides* (Koch) Kindb.", Bilewsky (1965): 378].

Distribution map 76, Figure 74: h.

Peristome without a distinct basal membrane; teeth cleft from base into segments, in dry capsules from nearly straight to incurved when short, twisted if longer. *Spores* 19–25(–28) µm in diameter.

Habitat and Local Distribution. — Plants gregarious or growing scattered among other mosses, exposed or shaded, on soil, stone walls, in rock crevices or at foot of rocks, on basaltic and calcareous soils, loess, rendzina, silt, sandy loam, terra rossa. Common: Sharon Plain, Upper Galilee, Lower Galilee, Mount Carmel, Esdraelon Plain, Samaria, Judean Mountains, Judean Desert, Northern Negev, Central Negev, Dan Valley, Upper Jordan Valley, Lower Jordan Valley, and Golan Heights.

General Distribution. — Recorded from Southwest Asia (Cyprus, Turkey, Syria, and Lebanon), around the Mediterranean, Europe (throughout), Macaronesia, and North Africa.

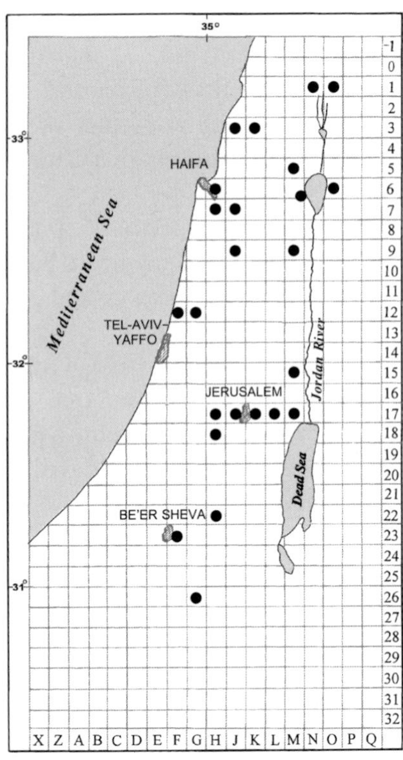

Map 76: *Aloina aloides* var. *aloides*

In some plants marginal cells at the base of the leaf shoulder may be hyaline, as in *Aloina rigida* but, unlike that species, they do not differ in shape and degree of incrassation from adjacent cells and do not form a distinct border (cf. Figure 74: f and Figure 76: f).

1b. Aloina aloides var. **ambigua** (B.S.G.) E. J. Craig *in* Grout, Moss Fl. N. Amer. 1: 214 (1939).

Barbula ambigua B.S.G., Bryol. Eur. 2: 76 (1842); *A. ambigua* (B.S.G.) Limpr., Laubm. Deutschl. 1: 638 (1888). — Bilewsky (1965): 378, Herrnstadt *et al.* (1991).

Distribution map 77, Figure 74: a–g.

Plants similar in habit to var. *aloides*. *Peristome* with a distinct basal membrane; teeth usually long, twisted above when dry. *Spores* 13–17 µm in diameter.

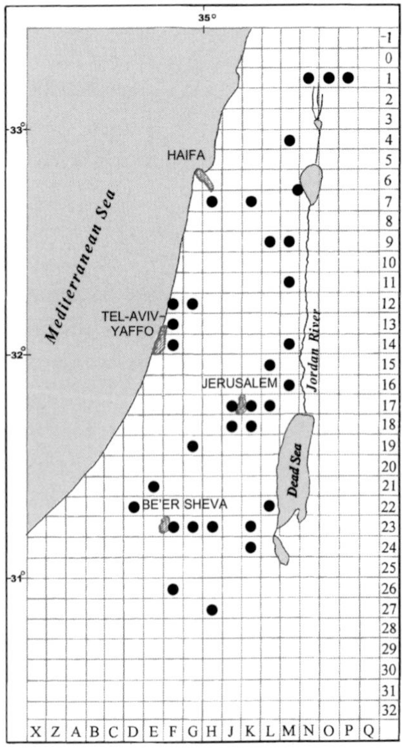

Map 77: *Aloina aloides* var. *ambigua*

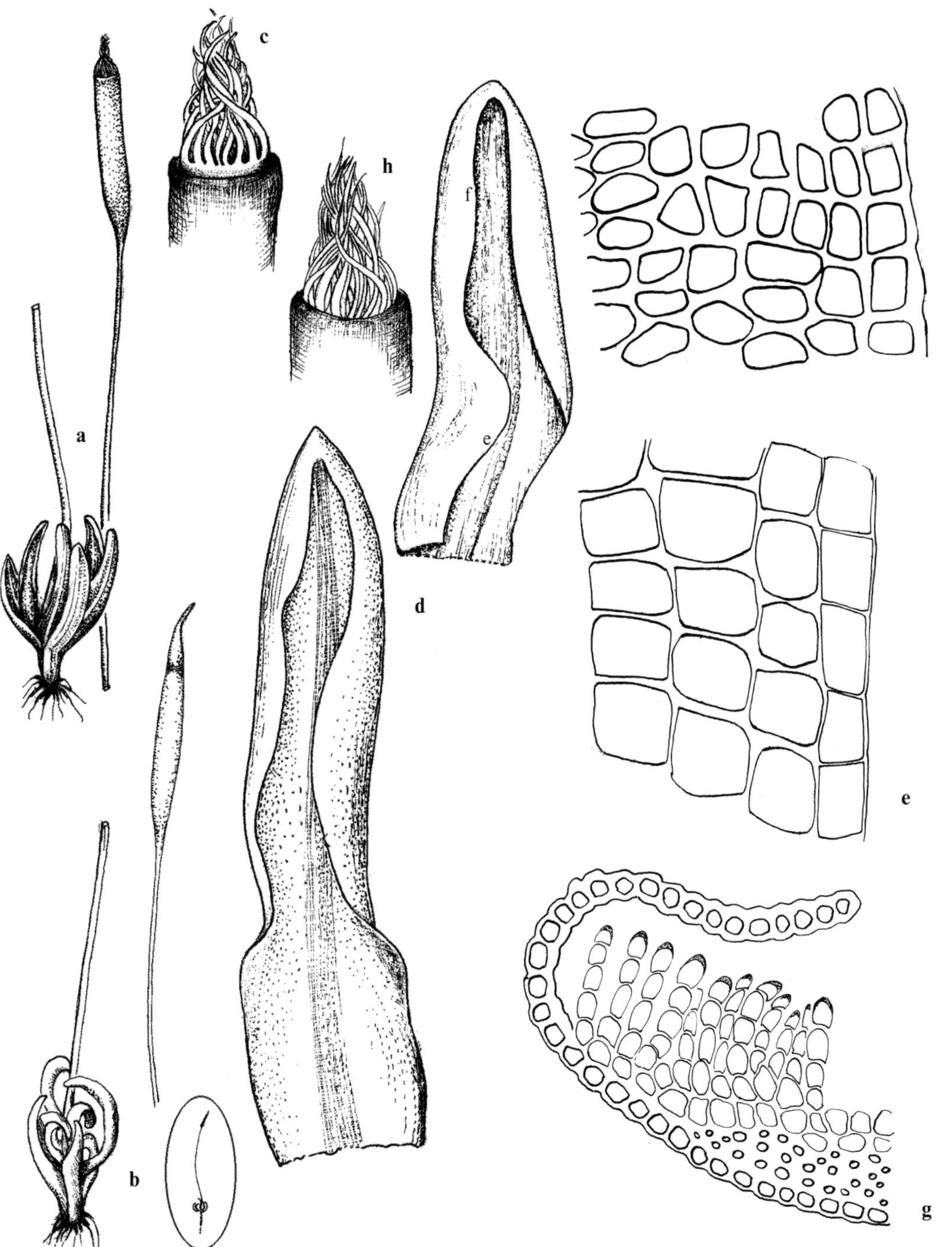

Figure 74. *Aloina aloides* var. *ambigua*: **a** habit, moist (×10); **b** habit, dry (×10); **c** peristome (×50); **d** leaves (×50); **e** marginal cells at leaf shoulder; **f** upper marginal cells (all leaf cells ×500); **g** cross-section of costa and portion of lamina (×250).
Aloina aloides var. *aloides*: **h**: peristome (×50).
(**a**–**g**: *Crosby & Herrnstadt 79-81-12*; **h**: *Crosby & Herrnstadt 79-61-12*)

Habitat and Local Distribution. — Plants growing in similar habitats as var. *aloides*. Very common: Sharon Plain, Upper Galilee, Lower Galilee, Mount Carmel, Mount Gilboa, Samaria, Shefela, Judean Mountains, Judean Desert, Northern Negev, Central Negev, Dan Valley, Bet She'an Valley, Lower Jordan Valley, and Golan Heights.

General Distribution. — Recorded from Southwest Asia (Cyprus, Turkey, Syria, Lebanon, Jordan, Iraq, Iran, and Afghanistan), around the Mediterranean, Europe (throughout), Macaronesia, North Africa, eastern Asia, Australia, and North and Central America.

This taxon is often considered as a separate species (e.g., El-Oqlah *et al.* 1988a, Herrnstadt *et al.* 1991). Here, *Aloina aloides* is treated, following Craig (1939) and Delgadillo (1975), as comprising two varieties: var. *aloides* and var. *ambigua*. The varietal rank is preferred because of the similarity in habit of the two taxa. The characters distinguishing the two varieties are few, and the variation of some of these (e.g., the height of the basal membrane) within a population can be observed.

2. Aloina bifrons (De Not.) Delgad., Bryologist 76 : 273 (1973).
Tortula bifrons De Not., Mem. Reale Accad. Sci. Torino 40 : 305 (1838); *A. pilifera* (De Not) H. A. Crum & Steere, Southw. Naturalist 3 : 119 (1959). — Bilewsky (1974) : 247.

Distribution map 78, Figure 75, Plate III : f.

Plants up to 3 mm long. *Leaves* up to 2 mm long without hair-point, ovate-lingulate to lingulate, often constricted above sheathing base, apex from obtuse to rounded, not cucullate; margins entire, not differentiated; costa weak, ending in a nearly smooth, fragile, hyaline hair-point up to $\frac{1}{2}(-\frac{2}{3})$ the length of lamina, with few stereids; upper cells transversely elongate and irregularly elliptical, incrassate; mid-leaf cells 14–24 μm wide; perichaetial leaves clasping, without involute margins, with hair-point up to length of lamina. *Capsules* up to 3.5 mm including operculum, cylindrical to ovoid-cylindrical; annulus of one–two rows of large, revoluble cells; peristome teeth twisted, with a distinct basal membrane. *Spores* 10–17 μm in diameter; densely gemmate (seen with SEM).

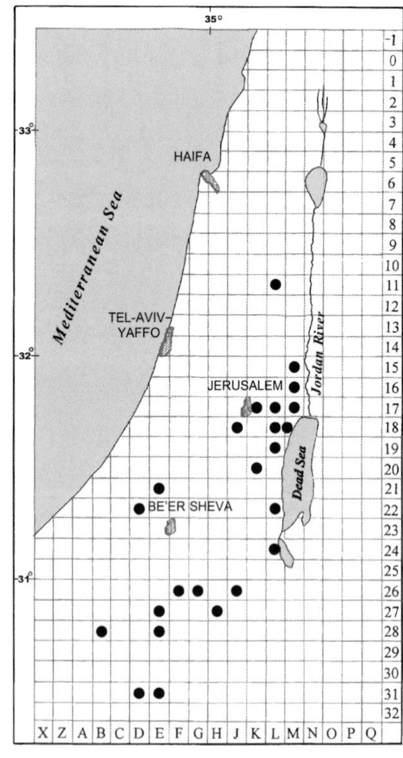

Map 78: *Aloina bifrons*

Figure 75. *Aloina bifrons*: **a** habit, dry (×10); **b** habit (with calyptra) and detached calyptra, moist (×10); **c** peristome (×50); **d** leaf (×50); **e** marginal basal cells; **f** upper cells (all leaf cells ×500); **g** cross-section of costa and portion of lamina (×250). (*Oyserman 79-106-1*)

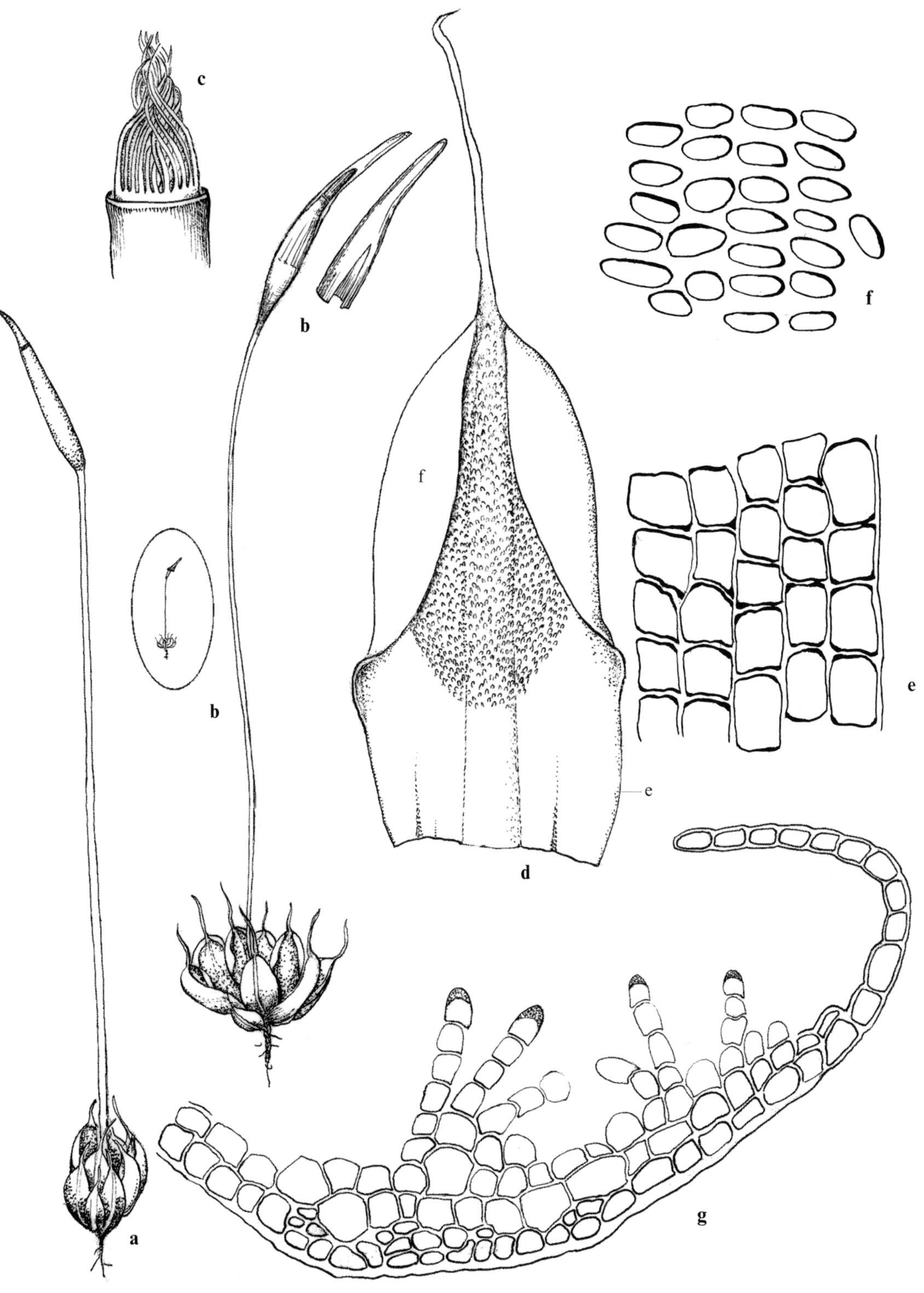

c
b
b
a
f
f
d
e
e
g

Habitat and Local Distribution. — Plants gregarious, growing on calcareous soils, on rocks, among rocks, at foot of limestone outcrops, and on exposed soil, usually confined to the more arid parts of the country. *Aloina bifrons* grows frequently together with *Crossidium crassinervium* var. *laevipilum*. Its growth in very extreme habitats is due to being able to tolerate periodical increase in temperatures and salinity. Common, locally abundant: Samaria, Judean Mountains, Judean Desert, Northern Negev, Western Negev, Central Negev, and Lower Jordan Valley.

General Distribution. — Recorded from Southwest Asia (Turkey, Jordan, and Iraq), Europe (very few), scattered around the Mediterranean, North and South Africa, Australia, New Zealand, North and Central America, and Chile.

3. Aloina rigida (Hedw.) Limpr., Laubm. Deutschl. 1 : 637 (1888). [Bilewsky (1974)]
Barbula rigida Hedw., Sp. Musc. Frond. 115 (1801); *A. stellata* Kindb., Bih. Kongl. Svenska Vetensk.-Akad. Handl. 7 : 137 (1883). — Bilewsky (1965) : 378.

Distribution map 79, Figure 76.

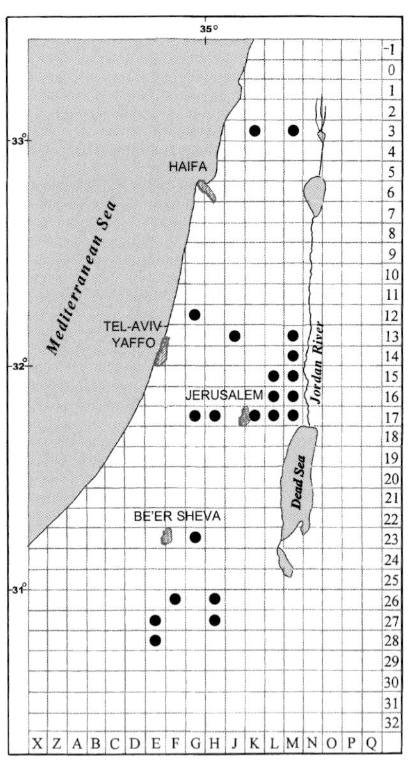

Map 79: *Aloina rigida*

Plants up to 2.5(–3) mm long. *Leaves* up to 2.5 mm long, lingulate, cucullate, apex rounded; margins entire; costa ending in or below apex, cross section with three–six stereid layers; marginal basal cells in several rows, elongate, hyaline, thin walled, often forming a distinct border at the shoulders of the sheathing base; upper cells oblate-oblong, incrassate; mid-leaf cells up to 22 µm wide; perichaetial leaves little differentiated, or sometimes with weakly developed costa and not cucullate apex. *Capsules* up to 3.5 mm long including operculum, narrow-cylindrical; annulus of one–three rows of large revoluble cells; peristome teeth twisted, with a distinct basal membrane. *Spores* 11–18 µm in diameter; densely granulate, granules with rough surface.

Habitat and Local Distribution. — Plants gregarious or growing scattered among other mosses, on exposed calcareous soils, and among rocks. Locally common: Sharon Plain, Upper Galilee, Samaria, Shefela, Judean Desert, Northern Negev, Central Negev, and Lower Jordan Valley.

General Distribution. — Recorded from Southwest Asia (widespread), around the Mediterranean, Europe (throughout), Macaronesia (in part), North Africa, northeastern, eastern, and central Asia, Australia, and North America.

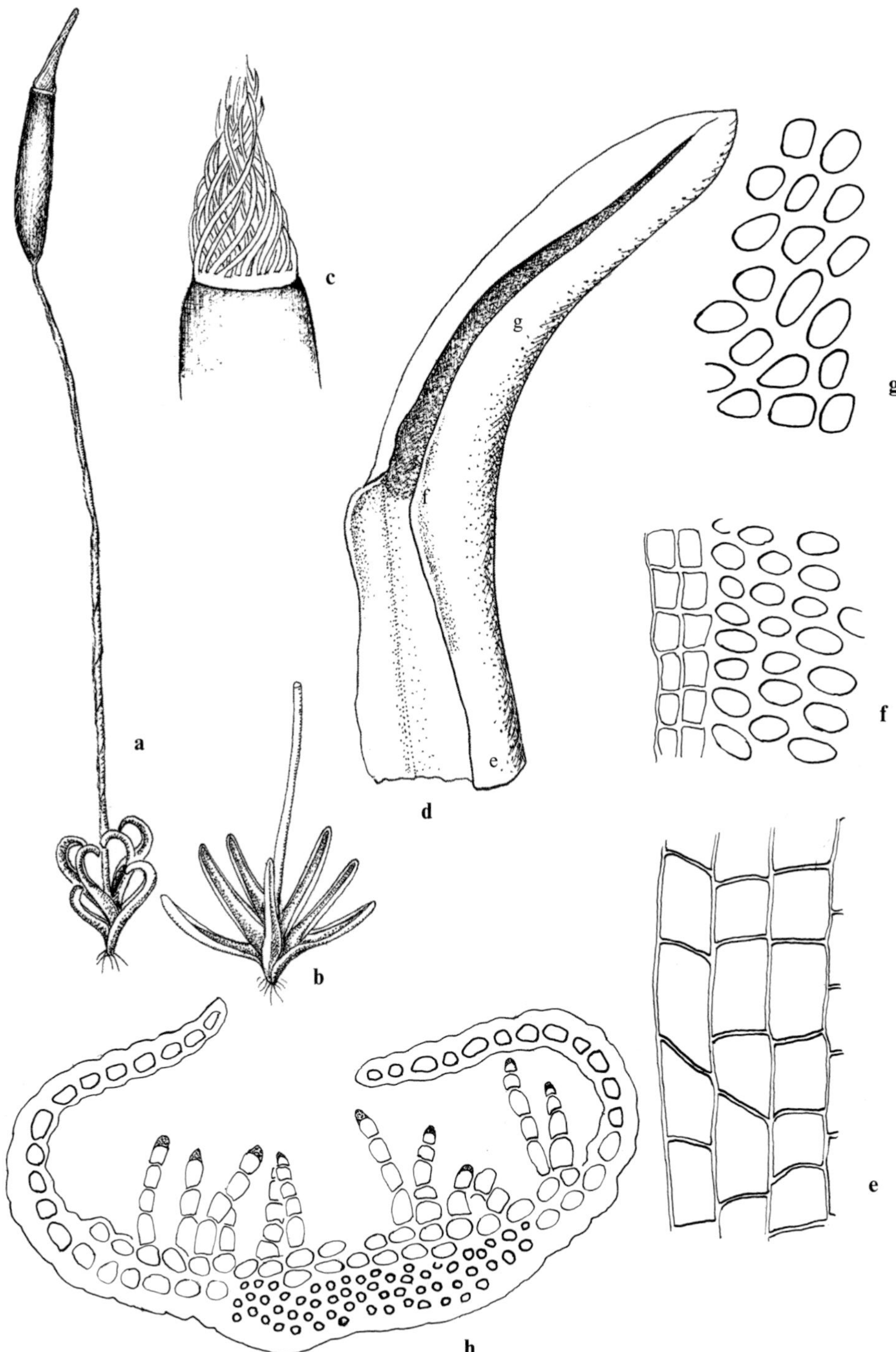

Figure 76. *Aloina rigida*: **a** habit, dry (×10); **b** habit, moist (×10); **c** peristome (×50); **d** leaf (×50); **e** basal cells; **f** marginal cells at leaf shoulder; **g** mid-leaf cells (all leaf cells ×500); **h** cross-section of leaf (×250). (*Bilewsky*, 10 Apr. 1954, No. 60 in Herb. Bilewsky)

Aloina rigida can be distinguished from the other local species by the distinct marginal border at the shoulders of the sheathing base of leaves. Bilewsky (1965) erroneously recorded plants of *A. bifrons* as "*Aloina stellata* (Schreb.) Kindb. var. *pilifera* Moenkem."

20. *Crossidium* Jur.

Plants small, growing in hoary, green or brownish tufts, gregarious or scattered among other moss species. *Stems* erect, short, often buried in soil, simple, rarely branched, with a central strand. *Leaves* small below, larger and more crowded above, appressed-incurved, sometimes ± twisted when dry, erect-spreading to spreading when moist, often concave towards apex, ovate-lanceolate or lingulate, rarely spathulate; apex acute, obtuse, rounded, or emarginate; margins plane or reflexed to recurved mainly in upper half; costa ending in a mucro or a hyaline, nearly smooth hair-point, with a single dorsal stereid band, a small group of hydroids and several layers of thin-walled cells on ventral side of guide cells; on upper part of ventral surface covered by simple or branched chlorophyllous filaments, each of one–twelve thin-walled cells, terminal cell thin-walled or thickened, subspherical to conical, smooth or papillose; basal cells subquadrate to rectangular, smooth, hyaline, thin-walled or occasionally ± incrassate; mid- and upper leaf cells smaller, quadrate to rectangular, rounded-hexagonal to rhomboidal, sometimes wider than long, smooth or papillose, with C-shaped papillae, incrassate to various degrees; perichaetial leaves differentiated or not. *Dioicous* or monoicous. *Setae* elongate; capsules erect, symmetrical to slightly curved, long-ellipsoidal, cylindrical to ovoid-cylindrical; operculum conical to rostrate; peristome with usually low basal membrane, with teeth divided into 32 filiform, papillose segments, nearly straight to twisted. *Spores* smooth to papillose, in local species granulate; granules with rough surface, variously crowded (seen with SEM). *Calyptra* cucullate, smooth.

A small genus comprising 13 (Zander 1993) or 11 species (Cano *et al.* 1993), growing mainly in the arid regions of all continents. Plants can be easily overlooked because of their habit and small size, and new distribution data are often published (e.g., from the Canary Islands by Dirkse & Bouman 1995). Several new species were described in recent years: three species from Southwest Asia (Frey & Kürschner 1984, 1987) and two species from South Africa (Magill 1981).

Three species of *Crossidium* occur in the local flora: *C. aberrans*, *C. crassinervium*, and *C. squamiferum*. *Crossidium crassinervium* and *C. squamiferum* are represented each by two varieties.

Literature. — Delgadillo (1973a, b; 1975).

Plants of *Crossidium* resemble *Aloina* in the chlorophyllous costal filaments, but differ in not having the fleshy habit, infolded leaves, and sheathing leaf base. The filaments in *Crossidium* are confined to the ventral surface of the costa, do not extend to the lamina as in *Aloina*, and sometimes may be papillose, in particular their terminal cell. Some species of both genera may often grow together.

1 Upper leaf cells unipapillose; costal filament cells usually subspherical, papillose **1. C. aberrans**

1 Upper leaf cells smooth; costal filament cells cylindrical to subspherical, not papillose (except terminal cell).

2 Cells strongly incrassate in upper ½(–⅔) of leaf; terminal cell of costal filaments incrassate, with solid papillae.

3 Leaf margins with distinct border of thin-walled cells up to ⅔ length of lamina; setae up to 15 mm long; capsules narrow-cylindrical; peristome teeth up to 1 mm long, teeth twisted **3a. C. squamiferum** var. **squamiferum**

3 Leaf margins without a distinct border; setae up to 9 mm long; capsules wider, elliptical to cylindrical; peristome teeth up to ca. 0.5 mm long, teeth usually not twisted **3b. C. squamiferum** var. **pottioideum**

2 Cells moderately incrassate in upper ½ of leaf; terminal cell of costal filament thin-walled, with hollow papillae.

4 Leaves ovate-lanceolate to lingulate, often twisted when dry, apex obtuse to acute, upper leaves with hyaline hair-point at least as long as lamina **2a. C. crassinervium** var. **crassinervium**

4 Leaves wide-ovate to ovate-lanceolate, not twisted when dry, apex round-obtuse or emarginate, upper leaves with hyaline hair-point usually shorter than lamina **2b. C. crassinervium** var. **laevipilum**

1. Crossidium aberrans Holz & E. B. Bartram, Bryologist 27 : 4 (1924).

Distribution map 80, Figure 77, Plate III : g.

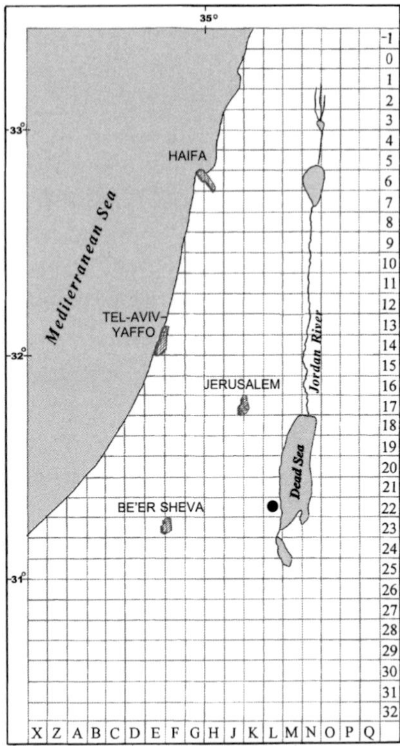

Map 80: *Crossidium aberrans*

Plants resemble *Crossidium crassinervium* in general habit. *Leaves* up to 1 mm long without hair-point, ovate-oblong to lingulate; margins recurved except near base; costa excurrent in a smooth hair-point, from half the length to the length of lamina; costal filaments two–three cells long, cells thin-walled, usually subspherical, papillose; terminal cell subspherical, with two to several hollow papillae; upper leaf cells usually unipapillose. *Sporophytes* as in *C. crassinervium*. *Spores* 11–14 μm; densely granulate-gemmate, processes with rough surface, some joined (seen with SEM).

Habitat and Local Distribution. — The record of *Crossidium aberrans* from Israel (Herrnstadt *et al.* 1991) was based on a single collection (Judean Desert, vicinity of Mezada, *Bilewsky* no. 655 labelled as "*Crossidium chloronotos*"). In the initial phase of the preparation of this Flora, microscopic slides, SEM micrographs, and drawings of a specimen from the collection were made. However, the collection itself has been since displaced and no additional plants of *C. aberrans* have been found to date. The description presented is mainly based on the above documentation of the local material, and is supplemented by data from Delgadillo (1975; *in* Sharp *et al.* 1994).

The habitat in which *Crossidium aberrans* was found (vicinity of Mezada) is similar to that of *C. crassinervium* var. *laevipilum*. For that reason, and due to the similarity in general habit of the two taxa, there is some possibility that plants of *C. aberrans* may have been overlooked in the field and among herbarium collections of *C. crassinervium*.

General Distribution. — Recorded from Southwest Asia only from Jordan and Saudi Arabia, around the Mediterranean (few, scattered records), Europe (Italy and Spain), Canary Islands, North Africa (Algeria), and North and Central America.

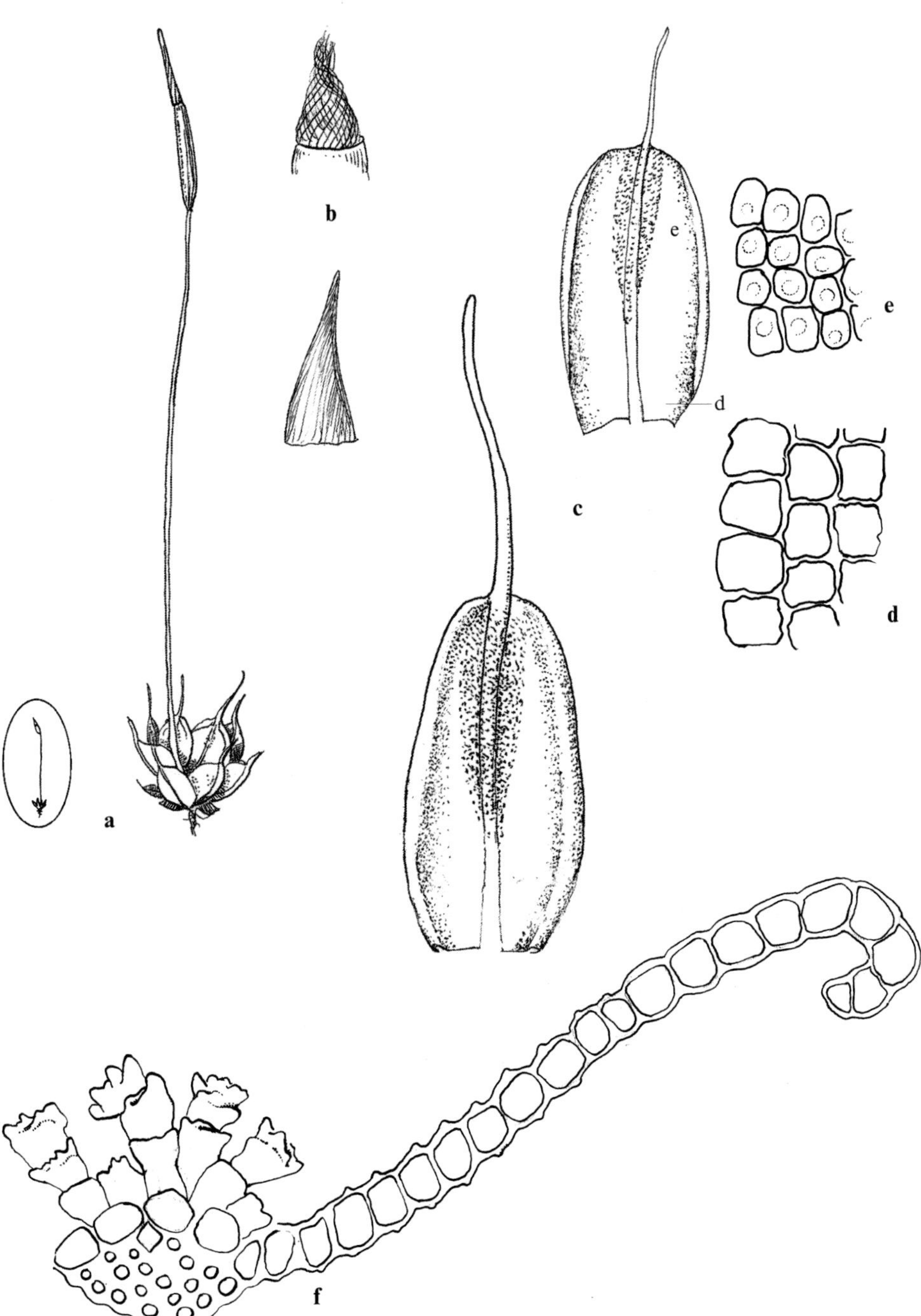

Figure 77. *Crossidium aberrans*: **a** habit, moist (×10); **b** peristome and operculum (×25); **c** leaves (×50); **d** basal cells; **e** upper cells (all leaf cells ×500); **f** cross-section of costa and portion of lamina (×250). (*Bilewsky*, 16 Mar. 1957, No. 655 in Herb. Bilewsky)

2. Crossidium crassinervium (De Not.) Jur. [the epithet has been usually misspelled "*crassinerve*"], Laubm.-Fl. Oesterr.-Ung. 128 (1882).
Tortula crassinervia De Not., Mem. Reale Acad. Sci. Torino 40 : 303 (1838).

Plants up to 3 mm high. *Stems* single or weakly branched. *Leaves* appressed, erect or twisted when dry, slightly spreading when moist, concave above, 0.75–1.25 mm long without hair-point; margins entire or weakly denticulate near apex, recurved to reflexed from near base to near apex; costa excurrent in a mucro or a hyaline hair-point; lower leaves usually mucronate, upper leaves piliferous to various degrees; costal filaments thin-walled, simple or branched, four–ten cells long, smooth except terminal cell; terminal cell cylindrical, conical, or subspherical, with or without two–four irregularly shaped, hollow papillae; basal cells more or less thin-walled; upper cells subquadrate to short-rectangular, moderately incrassate, particularly at angles, smooth, in mid-leaf 9–14 µm wide. *Dioicous* or autoicous. *Setae* up to 12 mm long; capsules ovoid to cylindrical; operculum conical to conical-rostrate, straight to slightly oblique; peristome teeth 350–550 µm long, straight to once twisted. *Spores* 10–17 µm in diameter.

2a. Crossidium crassinervium var. **crassinervium**
C. chloronotos auct., non (Brid.) Limpr., Laubm. Deutschl. 1 : 645 (1888). — Bilewsky (1965) : 377 (as "*C. chloronotos* (Bruch) Jur.").

Distribution map 81, Figure 78, Plate III : h.

Plants green to brownish. *Leaves* often twisted when dry, ovate-lanceolate to lingulate; upper leaves usually with hyaline hair-point, varying in length, at least as long as lamina, apex obtuse to acute. *Setae* up to 12 mm long; capsules up to 3 mm long including operculum; peristome teeth up to 550 µm long, usually twisted. *Spores* densely gemmate, processes with rough surface (seen with SEM).

Habitat and Local Distribution. — Plants growing in tufts of various densities, gregarious, or in the somewhat moister habitats scattered among other moss species; on calcareous, exposed, partly shaded soil, or on soil on rock. Locally common: Upper Galilee, Judean Mountains, Judean Desert, Northern Negev, Central Negev, Dan Valley, Bet She'an Valley, and Lower Jordan Valley.

General Distribution. — This Variety occurs around the Mediterranean. Available records are from Southwest Asia, Europe (mainly the south), Macaronesia, North and South Africa, central Asia (India), and North and Central America.

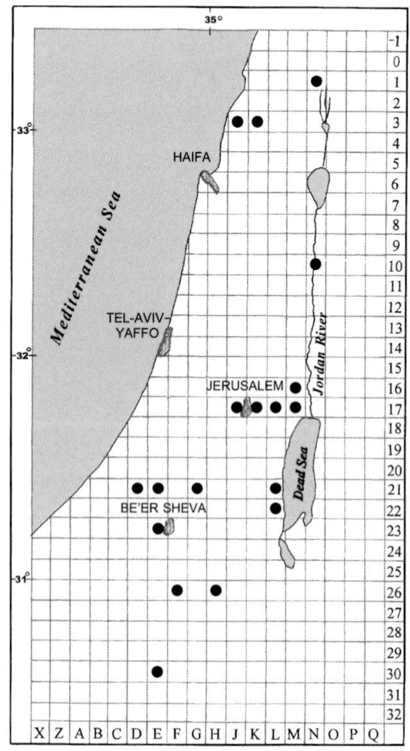

Map 81: *Crossidium crassinervium* var. *crassinervium*

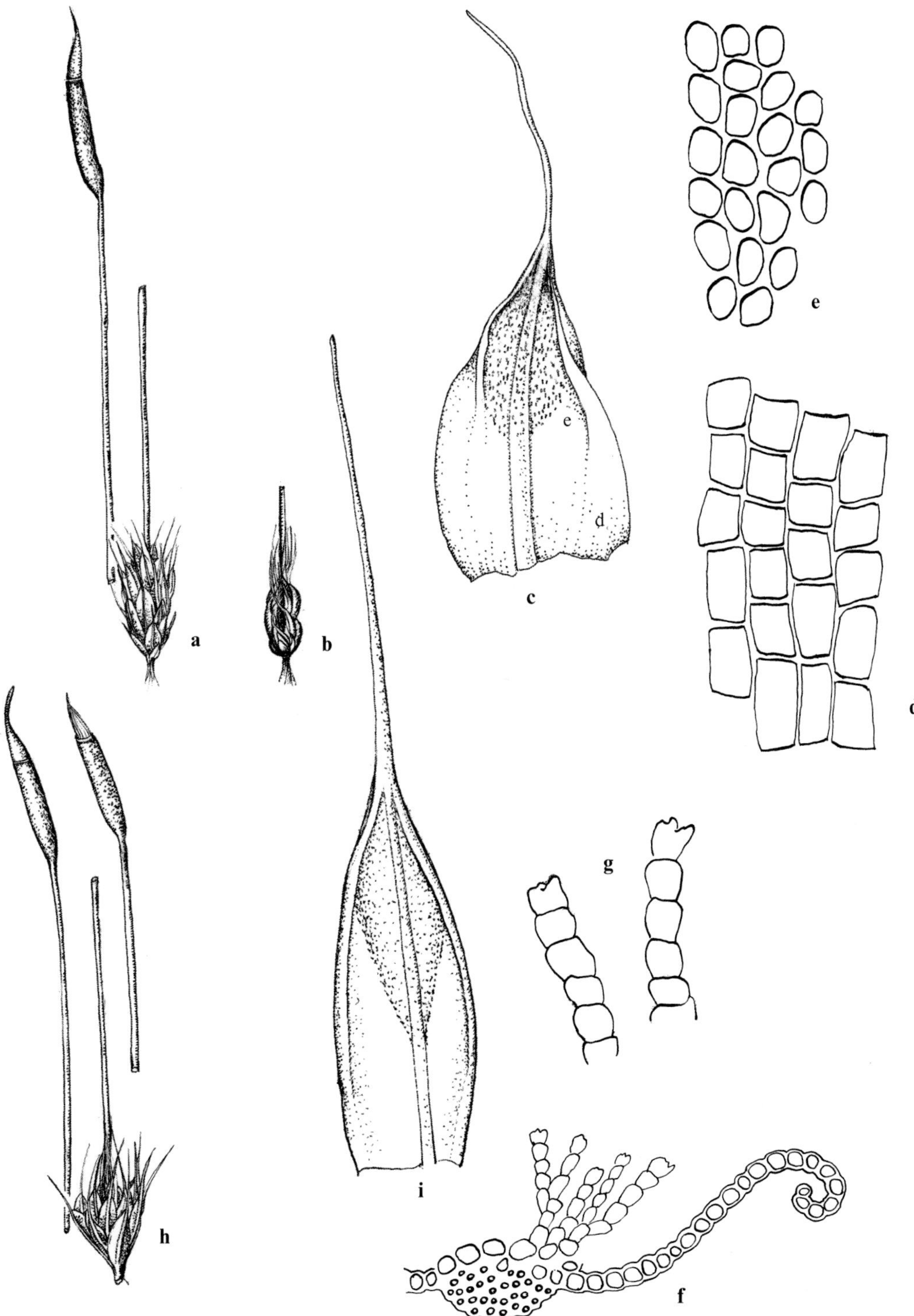

Figure 78. *Crossidium crassinervium* var. *crassinervium*: **a** habit, moist (×10); **b** habit, dry (×10); **c** leaf (×50); **d** basal cells; **e** mid-leaf cells (all leaf cells ×500); **f** cross-section of costa and portion of lamina (×250); **g** costal filaments (×500); **h** habit, moist (×10); **i** leaf (×50). (**a**–**g**: *Kushnir*, 12 Mar. 1943; **h**, **i**: *Herrnstadt* & *Crosby 79-87-13*)

Crossidium chloronotos (Brid.) Limpr. is considered as the valid name for this taxon in Index Muscorum (1959, Vol. I). According to Delgadillo (1975), *C. chloronotos* is a synonym of *C. squamiferum*.

2b. Crossidium crassinervium var. **laevipilum** (Thér. & Trab.) Delgad., Bryologist 78 : 275 (1975).

C. laevipilum Thér & Trab., Bull. Soc. Hist. Nat. Afrique N. 22 : 161 (1931). — Bilewsky (1965) : 377.

Distribution map 82, Figure 79.

Plants yellowish-green turning brownish. *Leaves* imbricate, erect, not twisted when dry, wide-ovate to ovate-lanceolate, lower leaves mucronate, upper leaves with hyaline hair-point varying in length, up to as long as lamina, usually shorter; apex round-obtuse or emarginate; costa turning brown. *Setae* up to 6 mm long; capsules up to 2(–3) mm including operculum; peristome teeth up to 350(–440) μm long, nearly straight.

Habitat and Local Distribution. — Plants growing solitary or gregarious, on fine-grained, exposed calcareous soils, e.g., loess, marl, and clay; in the more arid habitats plants are often bud-shaped and partly embedded in the substrate. Locally common: Judean Mountains, Judean Desert, Northern, Western, and Central Negev, Lower Jordan Valley, and the Dead Sea area.

General Distribution. — According to Frey & Kürschner (1991a), var. *laevipilum* (as *C. laevipilum*) has been identified in Israel, Jordan, and Algeria. Recently, it was identified in Spain (Cano *et al.* 1993).

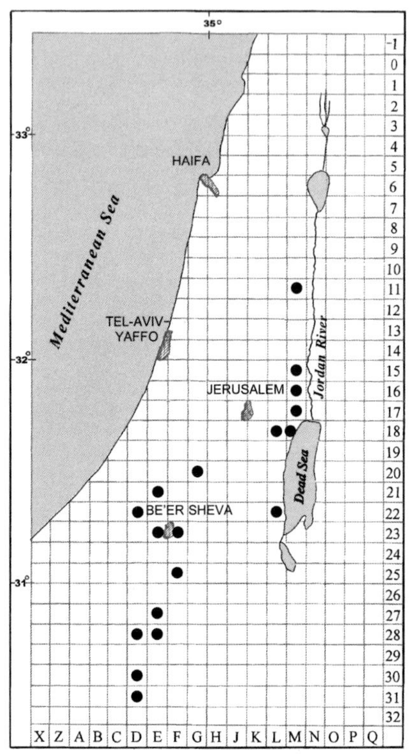

Map 82: *Crossidium crassinervium* var. *laevipilum*

Delgadillo (1975) was the first to reduce *Crossidium laevipilum* to a variety of *C. crassinervium*, though based on very scarce herbarium material. Frey & Kürschner (1991a) reverted to the concept of two well-separable species on the basis of field studies, mainly in Israel and Jordan, and herbarium collections. The main characters differentiating *C. laevipilum* from *C. crassinervium* according to Frey & Kürschner (1991a) are: life form (solitary with "basitone innovations" vs. growth in tufts), leaf apices (emarginate vs. obtuse to acute), and ecological requirements (dry, unstable marl and loess of Saharo-Arabian districts in Israel and Jordan vs. semiarid to sub-Mediterranean habitats).

An unpublished study of populations of the two taxa from various localities and habitats in Israel showed that some transitional forms exist, in particular in transitional habitats (e.g., in the Judean Desert along the transect between the Judean Mountains and the Dead Sea area; compare also Figures 78 : a–c, h, i, and 79 : a–c, h–k). Leaf

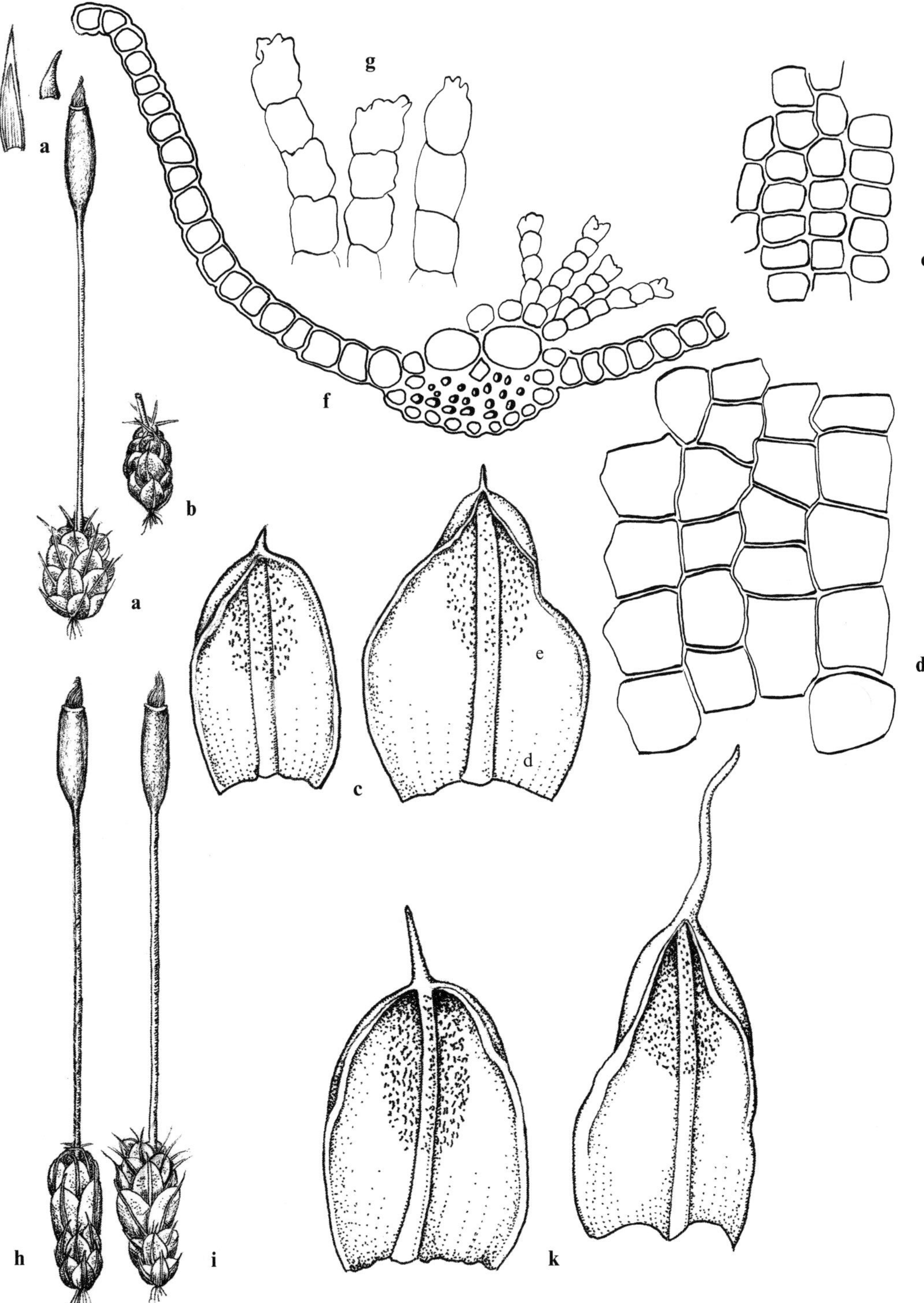

Figure 79. *Crossidium crassinervium* var. *laevipilum*: **a** habit, moist, detached operculum and calyptra (×10); **b** habit, dry (×10); **c** leaves (×50); **d** basal cells; **e** mid-leaf cells (all leaf cells ×500); **f** cross-section of costa and portion of lamina (×250); **g** costal filaments (×500); **h** habit, dry (×10); **i** habit, moist (×10); **k** leaves (×50). (**a**–**g**: *Herrnstadt* & *Danin 79-120a*; **h**–**k**: *Danin*, 18 Feb. 1980)

characters and growth forms may overlap: plants of var. *crassinervium*, which tend to grow in tufts, may occur solitary whereas plants of var. *laevipilum*, which are usually gregarious or solitary, can rarely be found also growing in tufts. Microhabitat conditions may be partly the cause for the density of plants and their pattern of innovation. Those conditions may also be the cause, to some extent, to the degree of twisting of the leaves, which in var. *laevipilum* is related to the partial coverage by soil. For those reasons, it is preferred here to treat the two taxa at the varietal level, following Delgadillo (1975), and not at the species level.

3. Crossidium squamiferum (Viv.) Jur., Laubm.-Fl. Oesterr.-Ung. 127 (1882).
Barbula squamifera Viv., Ann. Bot. 1 : 191 (1804).

Plants up to 6(–7) mm high, green above, brown below, frequently with hoary white appearance resulting from the hyaline hair-points and the nearly hyaline lamina. *Stems* single or branched. *Leaves* slightly imbricate, appressed when dry, erectopatent when moist, concave above, up to 1.5 mm long without hair-point, ovate or ovate-lanceolate, with acute apex; margins plane below, erect to recurved in upper part, entire below, serrulate near apex; costa strong, up to 110 µm wide, excurrent in a long, hyaline, yellowish, weakly serrulate hair-point, up to twice the length of lamina; filaments densely spaced, thin-walled, branched, eight–ten cells long, cylindrical to subspherical, smooth except terminal cell; terminal cell incrassate, usually conical to subspherical, with two–five solid papillae; basal cells ± incrassate; upper cells strongly incrassate, in particular at angles, irregularly quadrate to round-rhomboidal, more elongate towards apex, with lumen often nearly obliterated, sometimes oblique, smooth, in mid-leaf cells 11–27 µm wide. *Autoicous*. *Capsules* brown to dark red. *Spores* 12–21 µm in diameter.

Habitat. — Plants growing in dense tufts (in more arid habitats, tufts less dense, plants sometimes solitary) on calcareous soils: loess, desert rendzina, and marl, in enclaves of limestone rocks, on fossil sandstone and basalt. Both varieties of *Crossidium squamiferum* grow in general in similar habitats and often together, though the less widespread var. *pottioideum* tends to occur only in the more arid habitats of the range of distribution of the species.

3a. Crossidium squamiferum var. **squamiferum**
C. chloronotos (Brid.) Limpr., Laubm. Deutschl. 1 : 645 (1888) (*fide* Delgadillo 1975); — Bilewsky (1965) : 376 as "*C. squamigerum* (Viv.) Jur."

Distribution map 83 (p. 220), Figure 80, Plate III : i.

Leaves with distinctly differentiated margins up to ⅔ the length of lamina, formed by up to seven rows of hyaline, thin-walled, quadrate to rectangular cells, with abrupt transition to the usually strongly incrassate (in the upper ½–⅔ of leaf) round-rhomboidal to oblate cells towards costa. *Setae* up to 15 mm long; capsules narrow-cylindrical, up to 4 mm long including operculum; operculum up to 1.5 mm, conical-rostrate, erect to slightly oblique; peristome teeth 650 µm–1 mm long, twisted. *Spores* densely gemmate to verrucate, processes irregular in shape (seen with SEM).

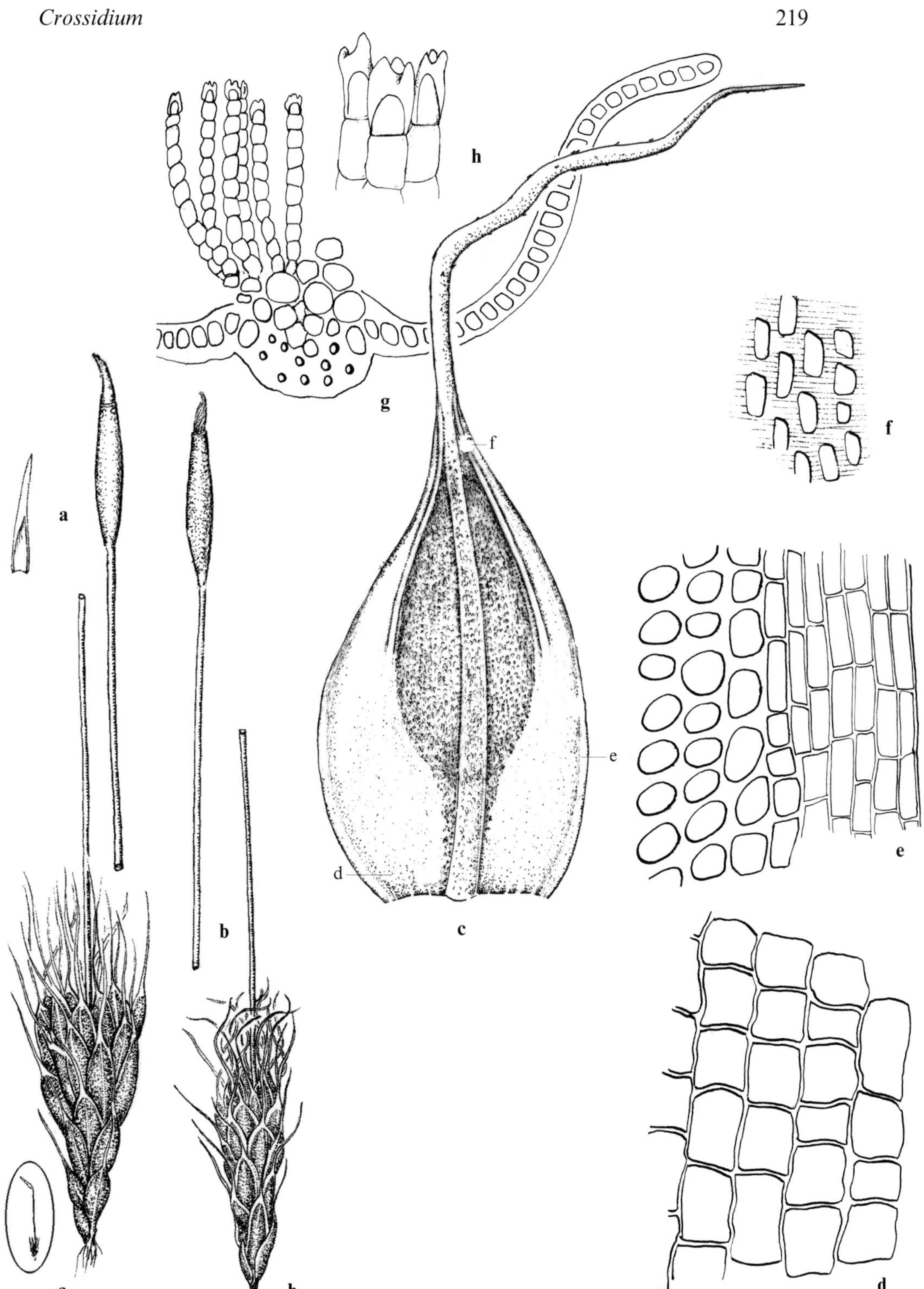

Figure 80. *Crossidium squamiferum* var. *squamiferum*: **a** habit with operculate capsule and calyptra, moist (×10); **b** habit with deoperculate capsule, dry (×10); **c** leaf (×50); **d** basal cells; **e** marginal mid-leaf cells; **f** apical cells (all leaf cells ×500); **g** cross-section of costa with costal filaments and portion of lamina (×250); **h** costal filaments (×500). (*Markus & Kutiel 77-529-1*)

Local Distribution. — Plants of var. *squamiferum* are widespread in the area covered by the Flora, except in the south: Coastal Galilee, Sharon Plain, Philistean Plain, Upper Galilee, Lower Galilee, Mount Carmel, Samaria, Shefela, Judean Mountains, Judean Desert, Northern Negev, Central Negev, Dan Valley, Bet She'an Valley, Lower Jordan Valley, and Golan Heights; very common.

General Distribution. — Recorded from Southwest Asia (throughout), around the Mediterranean, Europe (more often from the south), Macaronesia, North Africa, eastern and central Asia, and North America.

3b. Crossidium squamiferum var. **pottioideum** (De Not.) Mönk., Laubm. Eur. IV: 315 (1927).

Tortula squamifera (Viv.) De Not var. *pottioidea* De Not., Musci Ital. 22 (1862); *C. griseum* (Jur.) Jur., Laubm.-Fl. Oesterr.-Ung. 127 (1882). — Bilewsky (1965): 377 (as "C. *griseum* (Vent.) Jur."); *C. squamiferum* var. *griseum* (Jur.) G. Roth, Eur. Laubm. 1: 332 (1904).

Distribution map 84, Figure 81.

Leaves usually without differentiated margins; cells towards costa more incrassate (in the upper ½–⅔ of leaf), round-rhomboidal, oblate. *Setae* shorter than in var. *squamiferum*, up to 9 mm long; capsules wider, elliptical to cylindrical, up to 3 mm long including operculum; operculum conical, not oblique; peristome teeth up to 550 μm long, nearly straight.

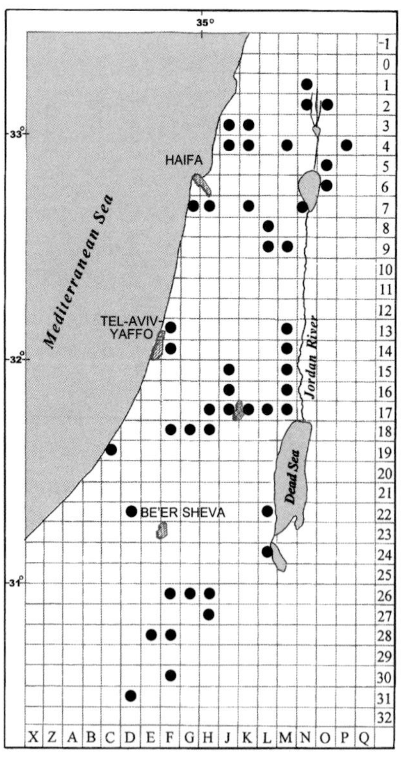

Map 83: *Crossidium squamiferum* var. *squamiferum*

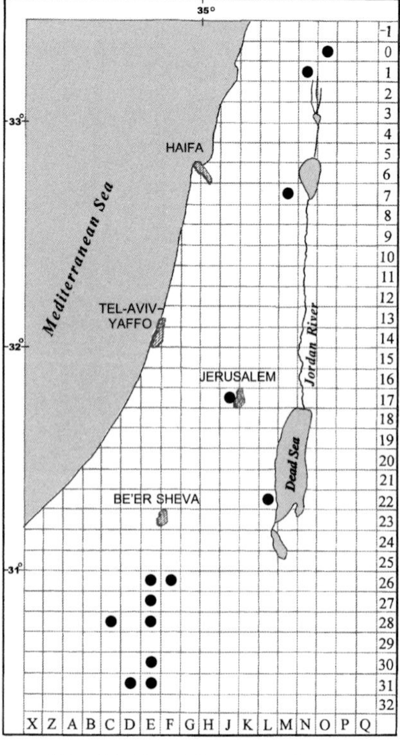

Map 84: *Crossidium squamiferum* var. *pottioideum*

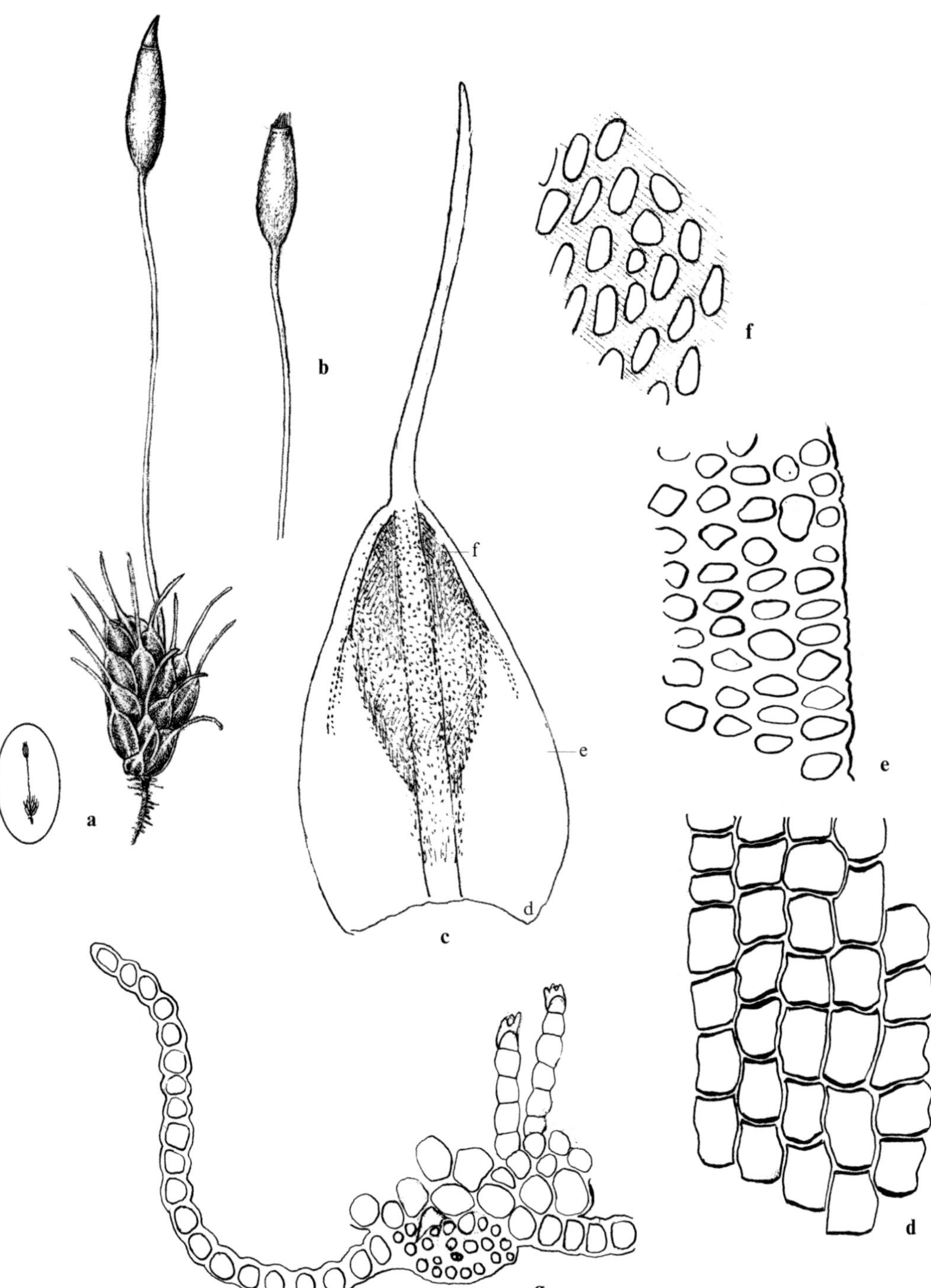

Figure 81. *Crossidium squamiferum* var. *pottioideum*: **a** habit, moist (×10); **b** seta with deoperculate capsule (×10); **c** leaf (×50); **d** basal cells; **e** marginal mid-leaf cells; **f** upper cells (all leaf cells ×500); **g** cross-section of costa and portion of lamina (×250). (**a**–**f**: *Herrnstadt* & *Crosby 78-30-5*; **g**: *D. Zohary*, 28 Mar. 1943)

Local Distribution. — Plants of var. *pottioideum* grow in Upper Galilee, Lower Galilee, Judean Mountains, Judean Desert, Western and Central Negev, and Mount Hermon; occasional, more frequent in the Western and Central Negev.

General Distribution. — Recorded from Southwest Asia (throughout), around the Mediterranean, Europe (scattered, mainly in the south), Macaronesia (in part), North Africa, eastern and central Asia, and North and Central America.

In one locality in the Upper Galilee (*Herrnstadt & Crosby 78-30-5*) both varieties of *Crossidium squamiferum* grow together, and some plants displayed various combinations of characters.

21. *Pterygoneurum* Jur.

Plants small, often bulbiform, growing in loose or dense tufts. *Stems* with a central strand. *Leaves* appressed when dry, weakly spreading when moist, concave, broad-ovate to oblong-ovate; margins plane or erect above; costa with two–four chlorophyllous longitudinally inserted lamellae on ventral surface, usually of upper half of lamina, excurrent (in local species) in a long, hyaline hair-point, sometimes lower leaves cuspidate; basal cells subquadrate to rectangular, smooth; mid- and upper leaf cells quadrate to short-rectangular, usually smooth, rarely papillose at back. *Autoicous. Setae* straight, short, or long; capsules immersed or exserted, ± erect, symmetrical; operculum rostrate; annulus and peristome usually absent. *Calyptrae* cucullate or mitrate, smooth.

A small genus comprising 12 species (Zander 1993), growing on most continents, usually in dry climates.

1 Lamellae on ventral side of costa starting from the base, on lower half of lamina, bearing branched filaments in upper half **1. P. crossidioides**
1 Lamellae on ventral side of costa only in upper half, not bearing filaments.
  2 Hyaline hair-point of leaf serrulate; capsules immersed; spores up to 40 μm in diameter, covered with sparsely short papillae; calyptrae mitrate **3. P. subsessile**
  2 Hyaline hair-point of leaf nearly smooth; capsules shortly exserted; spores up to 30 μm in diameter, more densely covered with longer papillae; calyptrae cucullate **2. P. ovatum**

1. Pterygoneurum crossidioides W. Frey, Herrnst. & Kurschner, Nova Hedwigia 50 : 239 (1990).

Distribution map 85 (p. 224), Figure 82.

Plants up to 2.5 mm high, yellowish-green, turning brown yellow when dry. *Stems* often dichotomously branched. *Leaves* strongly imbricate, evenly spaced along stem, up to 1.75 mm long with broadly ovate hair-point; apex nearly cucullate, abruptly narrowed into a hyaline hair-point, hair-point ⅔ the length of lamina to equal; margins plane, entire, except for weak serrulation near apex; costa strong, excurrent into hair-point, bearing on ventral side two–three lamellae, starting from the base, on lower half of lamina; in upper half of lamina branched filaments develop from both sides of the lamellae, increasingly branching and proliferating towards leaf

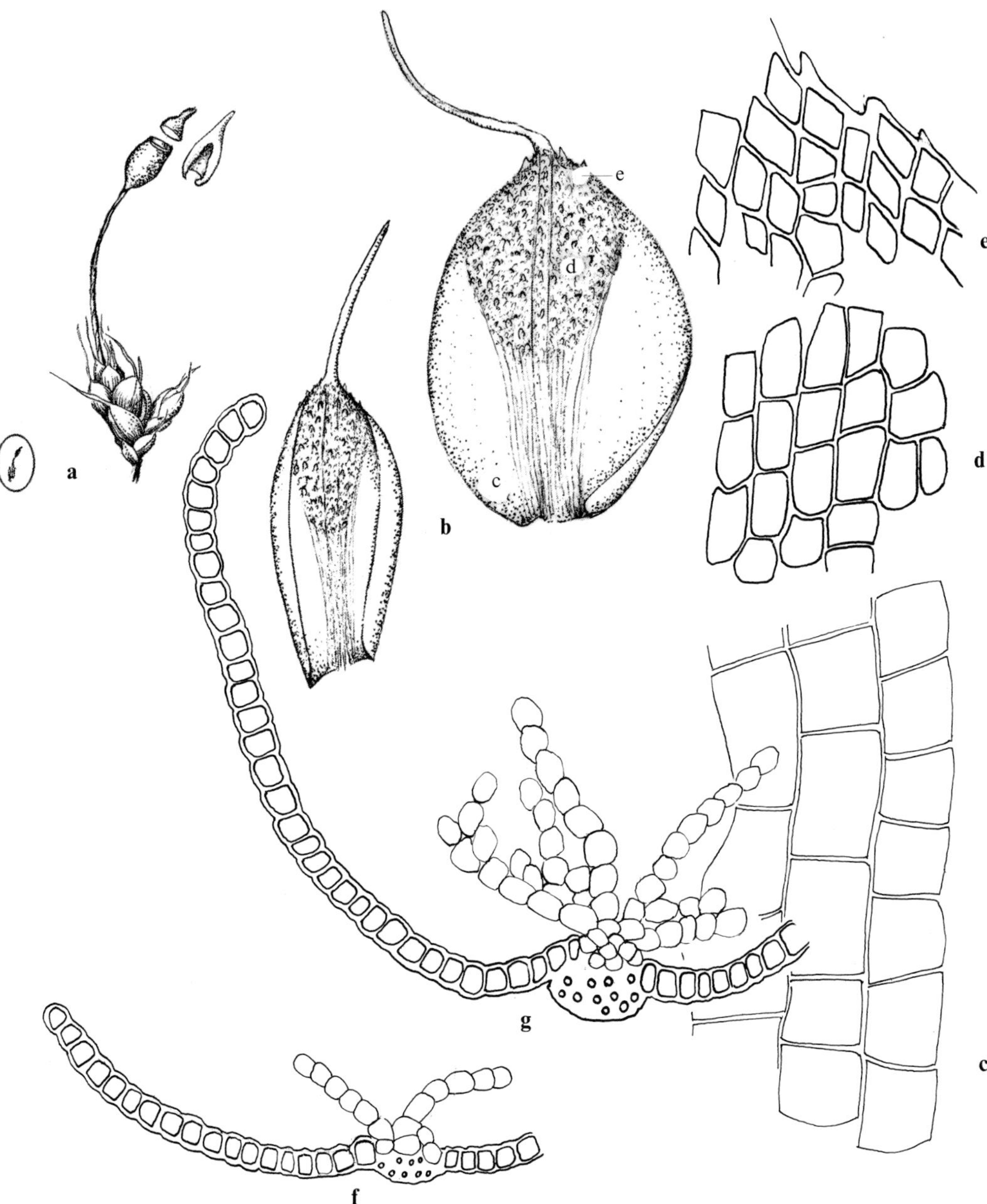

Figure 82. *Pterygoneurum crossidioides*: **a** habit, detached operculum and calyptra, moist (×10); **b** leaves (×50); **c** basal cells; **d** upper cells; **e** marginal apical cells (all leaf cells ×500); **f** cross-section of costa and portion of lamina at lower part of leaf (×250); **g** cross-section of costa and portion of lamina at upper part of leaf (×250). (*Frey*, *Herrnstadt* & *Kürschner*, 29 Apr. 1989 — holotype)

apex; lamina cells smooth, chlorophyllous except at hyaline apex; upper cells ± incrassate towards apex, apical cells irregularly quadrate to hexagonal, hyaline; mid-leaf cells 10–14 µm wide. *Autoicous*. *Setae* erect, strongly twisted, ca. 2.5 mm long; capsules 1.0–1.25 mm long including operculum, subglobose, brownish; operculum straight to ± obliquely rostrate; peristome absent. *Spores* (20–)24–31(–35) µm in diameter; verruculose. *Calyptrae* cucullate.

Habitat and Local Distribution. — Plants growing in loose tufts on exposed soil, on loess, deeply embedded except for leaf tips. Rare: Dead Sea area, found only in three localities.

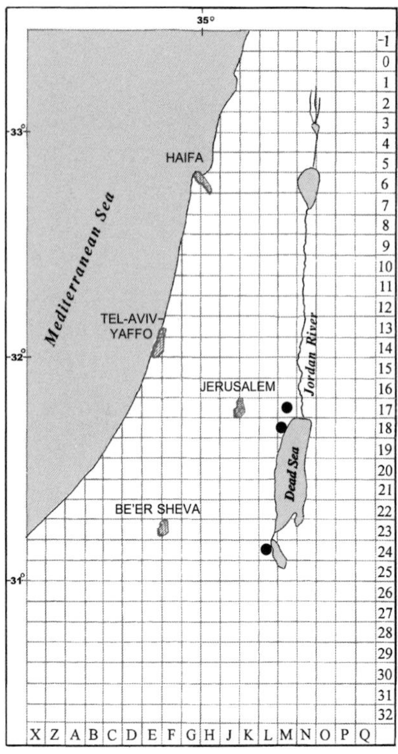

Map 85: *Pterygoneurum crossidioides*

Pterygoneurum crossidioides is characterised by the well developed lamellae (typical for the genus), together with photosynthetic filaments (typical for *Crossidium*). It occupies an intermediate position between *Pterygoneurum* and *Crossidium*, but the lack of central strand and peristome indicate some closer affiliation with *Pterygoneurum* (see Frey *et al.* 1990).

Recently, a closely related species, *Pterygoneurum compactum*, was described from Spain (Cano *et al.* 1994). The main differences between the two species (*P. crossidioides* and *P. compactum*) are the papillose apical cells of the filaments and the papillose lamina cells of *P. compactum* in contrast to the smooth apical filament cells and lamina cells of *P. crossidioides*.

2. Pterygoneurum ovatum (Hedw.) Dixon, Rev. Bryol. Lichénol. 6 : 96 (1934). [Bilewsky (1965) : 376]

Gymnostomum ovatum Hedw., Sp. Musc. Frond. 31 (1801); *P. cavifolium* Jur., Laubm.-Fl. Oesterr.-Ung. 95, 96 (1882). — Bilewsky & Nachmony (1955).

Distribution map 86 (p. 226), Figure 83, Plate III : k.

Plants 1–2 mm yellowish-green to brown. *Leaves* imbricate when dry, erect-spreading when moist, strongly concave, 1.5–2.0 mm long with hair-point, hair-point usually about as long as lamina, ovate to oblong-ovate; basal leaves shorter, 0.5–1 mm long, cuspidate; margins entire, plane to erect above; costa strong, excurrent into hyaline or yellowish hair-point, smooth or slightly serrulate at base, bearing on upper half of ventral side two–four lamellae; some upper lamina cells very weakly papillose, ± incrassate; mid-leaf cells 10–19 µm wide. *Autoicous*. *Setae* erect, 1.5–3 mm long, yellowish-brown; capsules shortly exserted, much varying in size, usually 1–2 mm long including operculum, ovoid to wide ellipsoid, dark purplish-brown, sulcate when dry; operculum obliquely rostrate; peristome absent. *Spores* up to 30(–40) µm in diameter

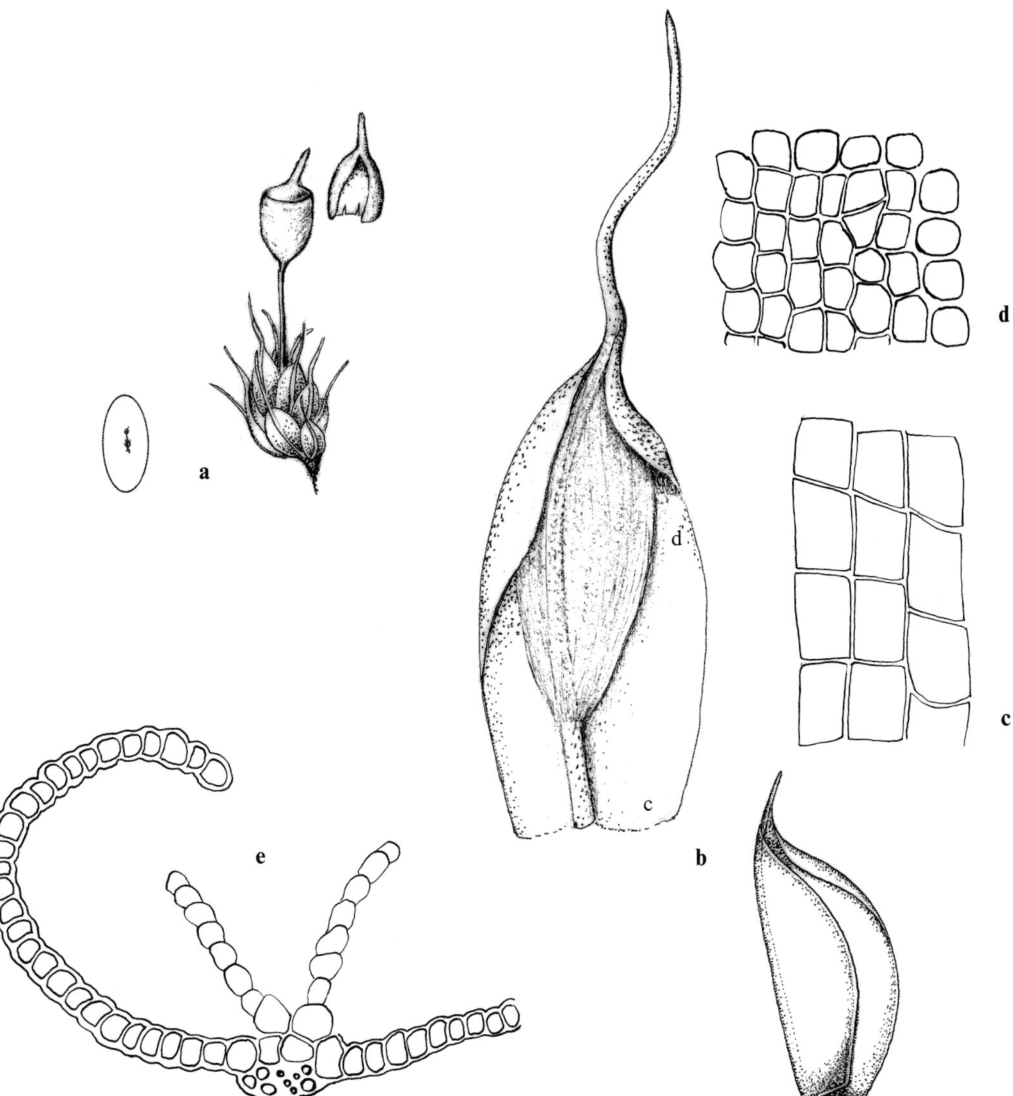

Figure 83. *Pterygoneurum ovatum*: **a** habit, moist, detached calyptra (×10); **b** leaves (×50); **c** basal cells; **d** upper cells (all leaf cells ×500); **e** cross-section of costa and portion of lamina (×250). (*Nachmony*, 9 Jan. 1954)

densely papillose; baculate, surface of baculae irregularly granulate (seen with SEM). *Calyptrae* cucullate.

Habitat and Local Distribution. — Plants growing in tufts on exposed calcareous soils and loess, in the desert parts of Israel and in the mountains of some of the Mediterranean districts, sometimes in association with *Poa bulbosa* L. Occasional to common: Judean Mountains, Judean Desert, Northern Negev, Central Negev, Lower Jordean Valley, Dead Sea area, Mount Hermon, and Golan Heights.

General Distribution. — Recorded from Southwest Asia (Cyprus, Turkey, Syria, Jordan, Iraq, and Kuwait), around the Mediterranean, Europe (throughout), Macaronesia, North Africa, northeastern and central Asia, Australia, and North and Central America.

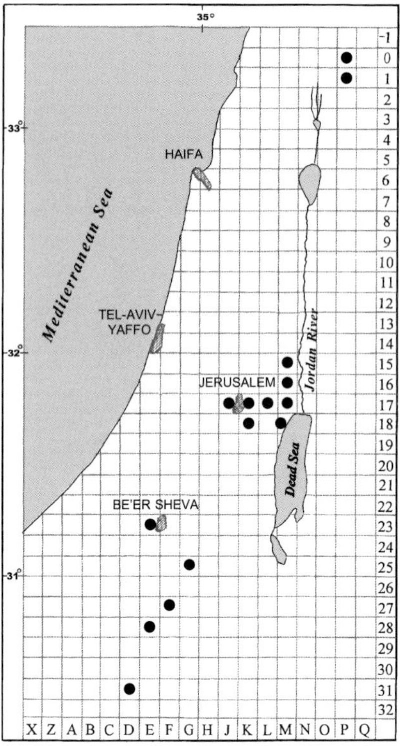

Map 86: *Pterygoneurum ovatum*

Plants of *Pterygoneurum ovatum* resemble in habit *Crossidium crassinervium* and some species of *Pottia*, with which they may also share the habitat. The smaller spores and the exserted capsules of *Pterygoneurum ovatum* may serve as distinguishing characters of this species from *P. subsessile*. The local pattern of distribution of *P. ovatum* is noteworthy. In the southern part of Israel it extends from the Judean Mountains into the arid, semidesert to desert south, and in the northern part it occurs on the Golan Heights and Mount Hermon. This pattern was also described by Shmida (1980) for some higher plants and is often considered as being of relictic origin.

3. Pterygoneurum subsessile (Brid.) Jur., Laubm.-Fl. Oesterr.-Ung. 96 (1882). [Herrnstadt *et al.* (1982)]

Gymnostomum subsessile Brid., Muscol. Recent., Suppl. 1 : 36 (1806).

Distribution map 87 (p. 228), Figure 84.

Plants up to 4 mm high green, hoary, turning brown-green when dry. *Leaves* imbricate when dry, erect when moist, strongly concave, 1.5–2.5 mm long with hair-point, hair-point the length of lamina to twice as long, broad-ovate, abruptly narrowing into hair-point; margins entire below, conspicuously serrulate near base of hair-point, plane; costa strong, excurrent into a long, serrulate hair-point, bearing on upper part of ventral side two(–four) lamellae; upper lamina cells smooth to ± papillose, near apex irregularly hexagonal, strongly incrassate, hyaline; mid-leaf cells 10–14 μm wide. *Autoicous. Setae* very short, up to 1.25 mm long, yellowish-brown; capsules immersed, ca. 1 mm long including operculum, subhemispheric, brown, irregularly wrinkled when

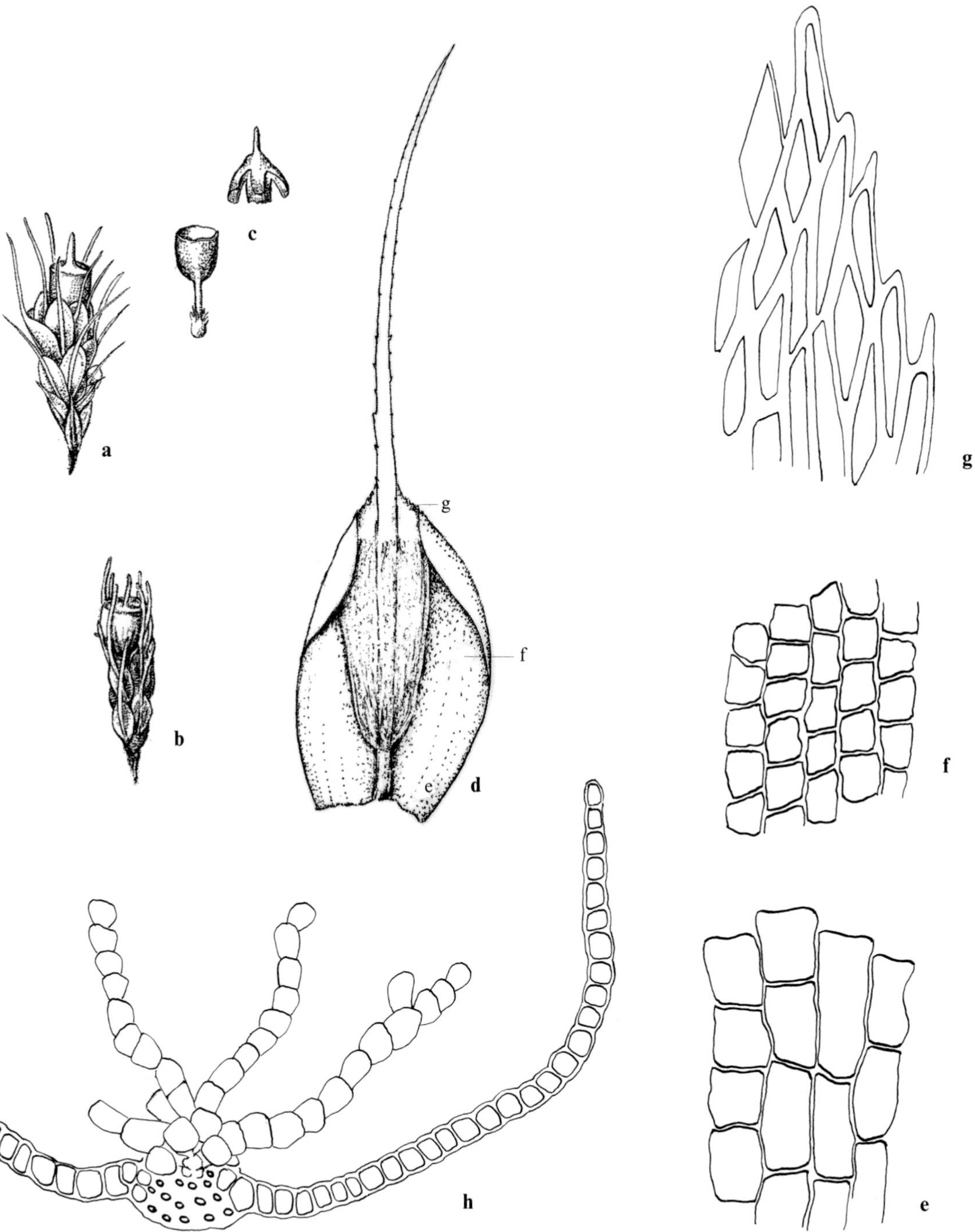

Figure 84. *Pterygoneurum subsessile*: **a** habit, moist (×10); **b** habit, dry (×10); **c** detached sporophyte with deoperculate capsule, and calyptra (×10); **d** leaf (×50); **e** basal cells; **f** mid-leaf cells; **g** apical cells including marginal cells (all leaf cells ×500); **h** cross-section of costa and portion of lamina (×250). (*Danin*, 5 Jun. 1980)

dry; operculum straight-rostrate; peristome absent. *Spores* up to 40 µm in diameter; sparsely short-papillose, baculate (bacula shorter and less densely spaced than in *P. ovatum*), rarely with some round bodies scattered on surface of bacula. *Calyptrae* mitrate, dehiscing together with operculum.

Habitat and Local Distribution. — Plants growing in tufts on soil, often accumulating soil grains among the long hair-points, thereby forming small, rounded, easily recognisable soil humps in desert habitats. Occasional: Judean Mountains, Judean Desert, Northern Negev, Western Negev, Central Negev, and Dead Sea area.

General Distribution. — The only previous record of *Pterygoneurum subsessile* from Southwest Asia is from Syria (Schiffner 1913), scattered records from around the Mediterranean; recorded from Europe, North Africa, northeastern and central Asia, and North, Central, and southern South America.

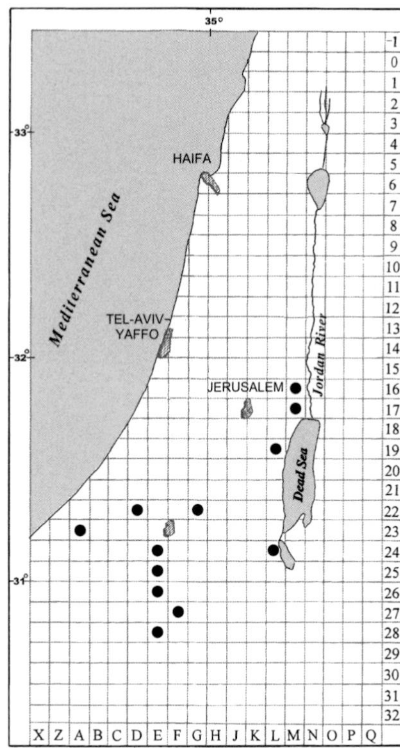

Map 87: *Pterygoneurum subsessile*

22. *Pottia* (Rchb.) Ehrh. ex Fürnr.

Plants small, terrestrial, scattered, gregarious or growing in loose tufts. *Stems* erect, simple, or rarely branched, with central strand. *Leaves* erect to spreading, ovate-lanceolate to elliptical, apiculate, cuspidate or with a yellowish-green point; plane or recurved, sometimes crenulate above; costa often strong, percurrent or excurrent, with only one dorsal stereid band, with bulging cells on ventral surface; basal cells rectangular, smooth; mid- and upper cells subquadrate to hexagonal, usually pluripapillose, with round, hollow or C-shaped papillae on one or both surfaces, rarely smooth; perichaetial leaves not differentiated. *Autoicous*, paroicous, or synoicous. *Setae* elongate; capsules exserted, symmetrical, erect, or slightly inclined, ovoid or cylindrical, often distinctly varying in size within populations, stegocarpous, very rarely cleistocarpous (*Pottia recta* among local species); operculum conical or rostrate; peristome with 16 papillose, entire, or irregularly divided, articulated, sometimes bi- or trifid teeth, or rudimentary to various degrees, or absent. *Spores* widely varying in shape and surface sculpturing. *Calyptrae* cucullate, smooth or papillose.

Spore shape, size, and surface texture in particular (especially as seen with SEM) are often considered as good delimiting characters of the local species.

A cosmopolitan genus comprising ca. 35 species. Eight terrestrial species occur in the local flora, all growing in exposed habitats, often two or more species together. Problems in the delimitation of *Pottia* from *Phascum* are discussed in the treatment of *Phascum*.

Among the local taxa, those considered below as *Pottia commutata*, *P. davalliana*, *P. mutica*, and *P. starckeana* are showing great resemblance in gametophytic characters. Accordingly, they have been treated in the literature in different ways, and have been united at various taxonomic levels and under different names (compare, for instance, treatments of Chamberlain 1978, Corley *et al.* 1981, Carrión *et al.* 1993). The species concept adopted here is to a large extent based on sporophytic characters (shape of capsules and peristome development), and on spore shape and texture. In local plants, a fair correlation between certain peristome structures and spore morphology could usually be observed (contrary to Carrión *et al.* 1993). Zander (1993) considers *Pottia*, together with *Phascum*, as part of *Tortula*.

Data on geographic distribution of species found in literature and cited in part below should be taken with caution because of the varying species concepts among authors of floristic accounts.

Literature. — Warnstorf (1916), Casas (1991).

1 Capsules cleistocarpous **8. P. recta**
1 Capsules stegocarpous.
  2 Peristome absent or reduced to basal membrane, teeth absent.
    3 Mid- and upper leaf cells papillose, operculum short-conical **5. P. davalliana**
    3 Mid- and upper leaf cells smooth to weakly papillose, operculum oblique-rostrate.
      4 Plants with bud-shaped propagula in leaf axils **7. P. gemmifera**
      4 Plants without bud-shaped propagula in leaf axils **6. P. intermedia**
  2 Peristome present, teeth with (one–)two–ten articulations.
    5 Operculum oblique-rostrate, peristome teeth long and narrow, with six–ten articulations **1. P. lanceolata**
    5 Operculum conical, peristome teeth short, broad, and truncate, with one–four articulations.
      6 Spores with large, hollow warts, not spinulose **2. P. starckeana**
      6 Spores without hollow warts, at least ± spinulose.
        7 Peristome teeth usually with three–four articulations **3. P. commutata**
        7 Peristome teeth with only one–two articulations **4. P. mutica**

1. Pottia lanceolata (Hedw.) C. Müll. Hal., Syn. Musc. Frond. 1 : 548 (1849).
Encalypta lanceolata Hedw., Sp. Musc. Frond. 63 (1801); *Anacalypta lanceolata* (Hedw.) Nees & Hornsch., Bryol. Germ. 1 : 141 (1831); *Tortula lanceola* R. H. Zander, Genera of the Pottiaceae 223 (1993).

Distribution map 88, Figure 85, Plate III : l, m.

Plants up to 2.5 mm high, yellowish-green. *Leaves* up to 1.5 mm long, ovate-lanceolate to oblong-lanceolate, cuspidate; margins entire, usually broadly recurved from base to apex or near to apex; costa very stout, in particular in upper part, excurrent in a 100–300 µm long stout point; mid- and upper cells quadrate, 11–17 µm wide, densely papillose. *Autoicous. Setae* 5–6 mm long; capsules up to 2 mm long including operculum, ovoid to narrow-ellipsoid, brown; annulus present; operculum ± obliquely rostrate; peristome teeth long and narrow, with six–ten articulations, irregularly shaped, usually longitudinally cleft and often perforated. *Spores* up to 25 µm in diameter; finely and densely papillate, densely gemmate (seen with SEM). *Calyptrae* smooth.

Habitat and Local Distribution. — Plants growing in tufts on dry soil in shade of rocks. Rare to occasional: Judean Mountains, Judean Desert, and Dead Sea area.

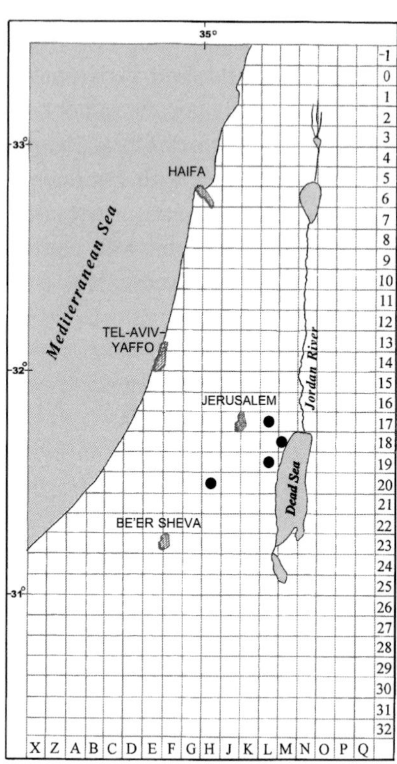

Map 88: *Pottia lanceolata*

General Distribution. Recorded from Southwest Asia (Turkey, Syria, Iraq, and Kuwait), around the Mediterranean, Europe (throughout), Macaronesia, North Africa, eastern and central Asia, and North America.

Pottia lanceolata is easily distinguished from the other local *Pottia* species by the well developed peristome teeth, the rostrate operculum, and the long-excurrent costa of the leaf.

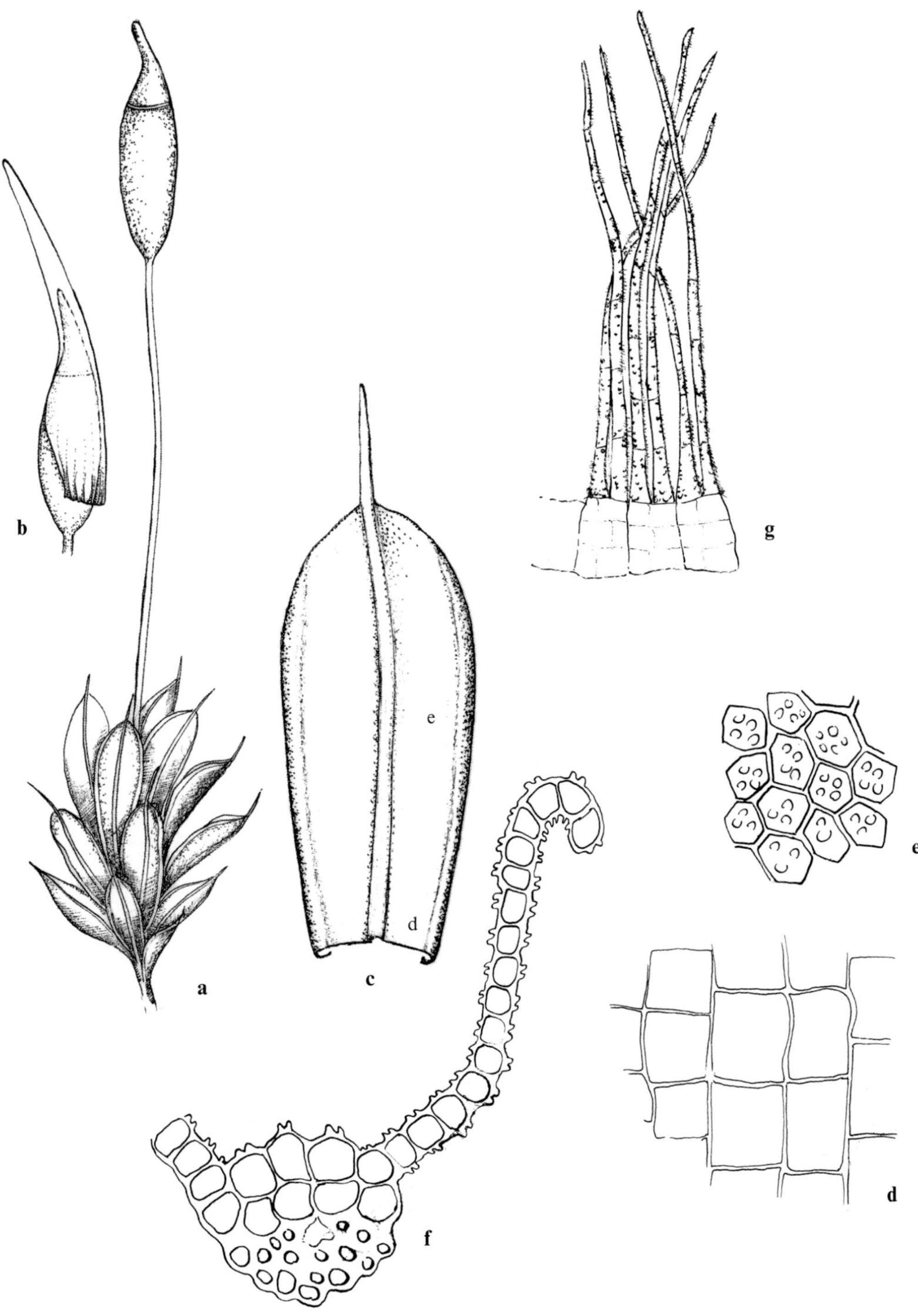

Figure 85. *Pottia lanceolata*: **a** habit, moist (×20); **b** capsule with calyptra (×20); **c** leaf (×50); **d** basal cells; **e** mid-leaf cells (all leaf cells ×500); **f** cross-section of costa and portion of lamina (×250); **g** portion of peristome (×250). (*Kushnir*, 4 Apr. 1943)

2. Pottia starckeana (Hedw.) Müll. Hal., Syn. Musc. Frond. 1 : 547 (1849). [Bilewsky (1965) : 374]

Weissia starckeana Hedw., Sp. Musc. Frond. 65 (1801); *Anacalypta starckeana* (Hedw.) Fürnr., Flora Erg. 12 : 25, 26 (1829); *Microbryum starckeanum* (Hedw.) R. H. Zander, Genera of the Pottiaceae 240 (1993).

Distribution map 89, Figure 86, Plate IV : a, b.

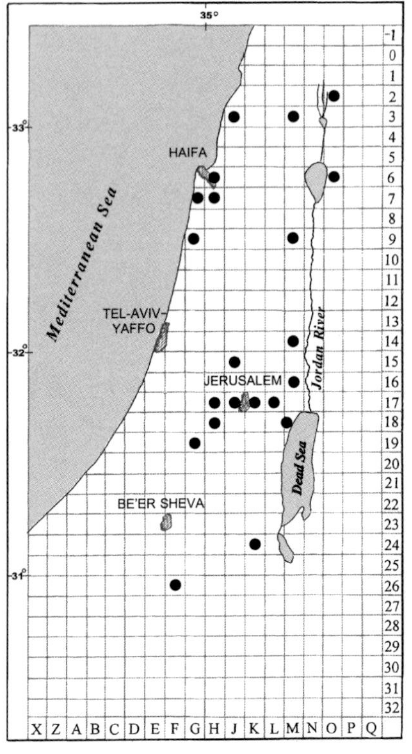

Map 89: *Pottia starckeana*

Plants up to 2 mm high, green to olive-green. *Leaves* only slightly contorted when dry, erectopatent when moist, up to 2 mm long: ovate-lanceolate, ± abruptly narrowing into cuspidate apex, margins recurved; costa excurrent as a cuspidate point, turning brown; mid- and upper leaf cells quadrate to rectangular, 10–12 µm wide, papillose, ± incrassate. *Autoicous. Setae* yellow, 1–4 mm long; capsules erect, 0.75–1 mm long including operculum, broad-ellipsoid, widest at its middle, constricted at mouth when operculum detached, brown; cells below mouth thick-walled, differentiated (as in *P. commutata*); operculum long-conical; peristome well developed, teeth short, broad, truncate, papillose, usually with up to four articulations. *Spores* ca. 24 µm in diameter; with large, hollow warts, surface porose (seen with SEM). *Calyptrae* papillose.

Habitat and Local Distribution. — Plants growing in dense tufts, on exposed soils of various types. Common: Coastal Galilee, Acco Plain, Coast of Carmel, Sharon Plain, Upper Galilee, Mount Carmel, Shefela, Judean Mountains, Judean Desert, Central Negev, Bet She'an Valley, Lower Jordan Valley, and Golan Heights.

General Distribution. — Recorded from Southwest Asia (Cyprus, Turkey, Syria, Lebanon, Jordan, Iraq, and Kuwait), around the Mediterranean, Europe (nearly throughout), Macaronesia, North Africa, Australia, New Zealand, and North and Central America.

Literature. — Chamberlain (1969), Carrión *et al.* (1993).

Pottia starckeana resembles *P. commutata* but the leaves in *P. starckeana* are somewhat more abruptly narrowing into a cuspidate point.

In two populations (Sharon Plain and Judean Mountains), plants were found to have only a rudimentary peristome, thus resembling *Pottia mutica*, but their irregularly shaped spores were typical of *P. starckeana*. They may be identical to *P. starckeana* subsp. *starckeana* var. *brachyodus* Müll. Hal. (cf. Chamberlain 1969). The type specimen of *P. affinis* Hook. & Taylor (BM), was found by us to be a plant possessing a

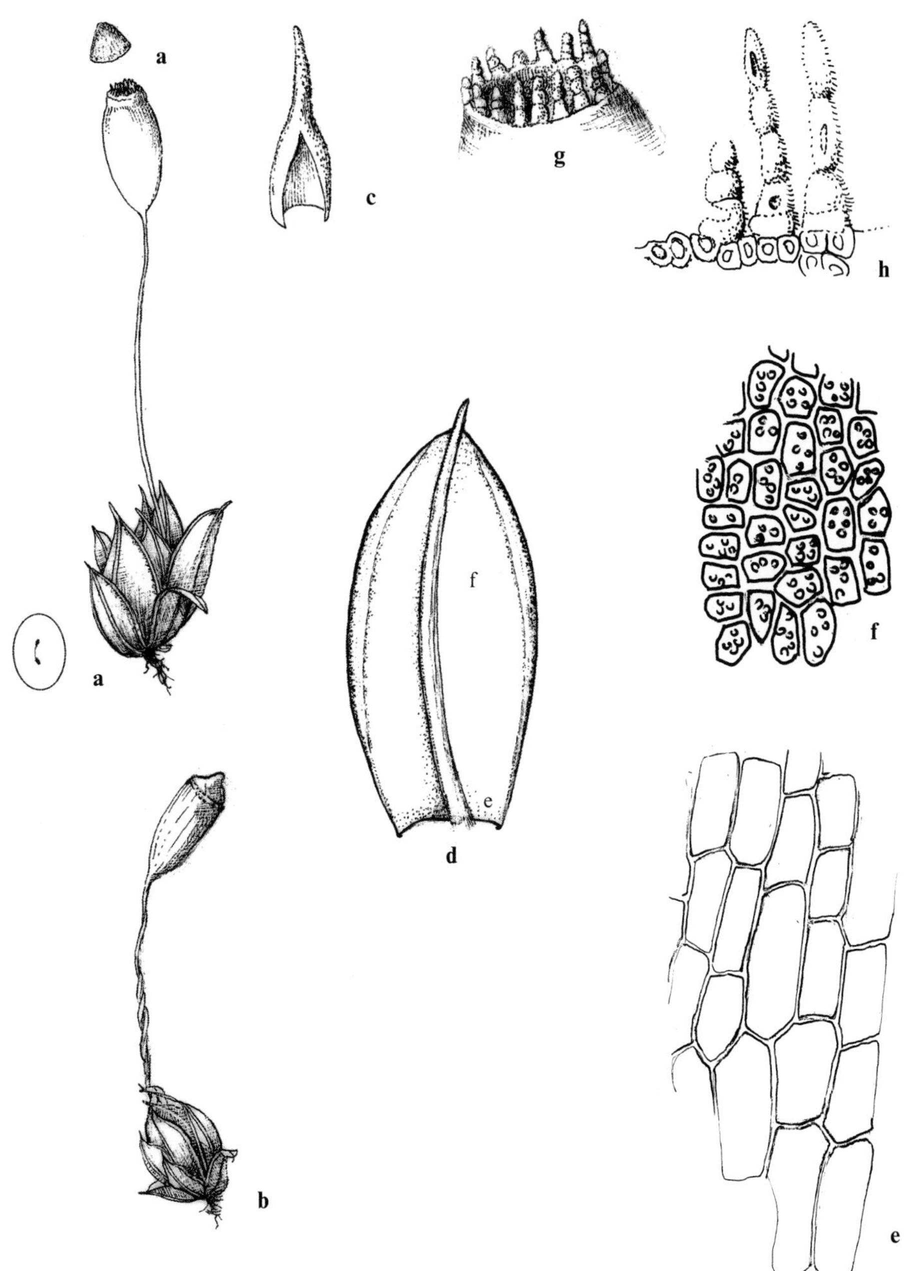

Figure 86. *Pottia starckeana*: **a** habit with detached operculum, moist (×20); **b** habit, dry (×20); **c** calyptra (×20); **d** leaf (×50); **e** basal cells; **f** upper cells (all leaf cells ×500); **g** peristome (×100); **h** peristome teeth (×500). (*Herrnstadt* & *Crosby 78-26-3*)

rudimentary peristome and spores with hollow warts similar to those of *P. starckeana* (see the discussion of *P. mutica*).

Bilewsky (1965) listed in addition to the species two varieties: "var. *brachyoda* Wild." with very short and truncate peristome, and "var. *leiostoma* Corb." without a peristome. Such plants were neither found by us in the field nor among Bilewsky's collections.

The identity of *Weissia starckeana* Hedw., the taxon on which *Pottia starckeana* is based, is not entirely clear. The taxon was first described and depicted by Hedwig (1791–1792, pp. 83–84, Tab. 34, Figure 13). Hedwig's plant conforms with the usually accepted concept of *P. starckeana*, except for the important diagnostic characters of the spores: no spines and an irregularly wavy outline. The spores of Hedwig's W. starckeana are described as spiny and are drawn as having a regular outline. Thus, we cannot exclude the possibility that the nomenclature of this species may change after typification of the original material.

3. Pottia commutata Limpr., Laubm. Deutschl. 1 : 537 (1888). [Bilewsky (1965) : 375]
P. davalliana (Sm. ex W. F. M. Drake) C. E. O. Jensen subsp. *commutata* (Limpr.) Podp., Consp. Musc. Eur. 226 (1954); *Microbryum davallianum* (Sm.) R. H. Zander var. *commutatum* (Limpr.) R. H. Zander, Genera of the Pottiaceae 240 (1993).

Distribution map 90, Figure 87, Plate IV : c, d.

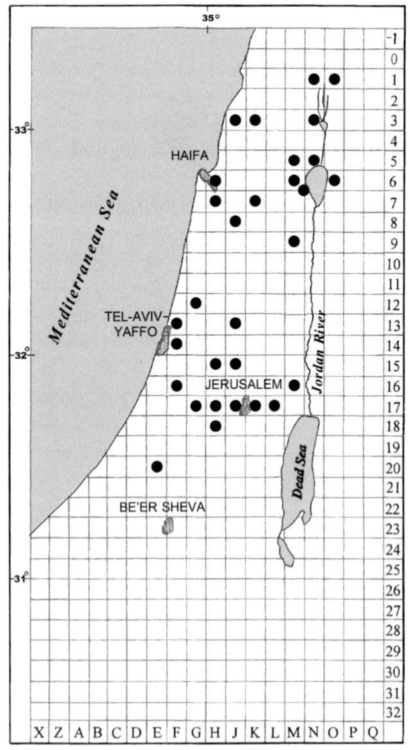

Map 90: *Pottia commutata*

Plants 1–2 mm long, bright green. *Leaves* slightly contorted when dry, erect-spreading when moist, 0.6–1.2 mm long, ovate-lanceolate, cuspidate; margins recurved, entire to papillose, often papillose-crenulate towards apex; costa excurrent into a cuspidate point 10–100(–130) μm long; mid-leaf and upper cells quadrate, sometimes rectangular, 10–14 μm wide, papillose, ± incrassate. *Autoicous*. *Setae* up to 3(–3.5) mm long; capsules slightly inclined, usually less than 1 mm long including operculum, wide-ellipsoid to wide-cylindrical, usually ± narrowed at mouth, yellowish-brown to reddish-brown; cells below mouth in two rows, thick-walled, differentiated; operculum conical; peristome teeth ± short, broad, truncate, papillose, with (two–)three–four (–five) articulations. *Spores* up to 30 μm in diameter; spinulose with perforate surface and spinulose processes (seen with SEM). *Calyptrae* papillose, often detached together with operculum.

Habitat and Local distribution. Plants growing in dense tufts, on dry to somewhat moist, exposed soils of various types. Very common: Coastal Galilee, Sharon Plain, Philistean Plain, Upper Galilee, Lower Galilee, Mount Carmel, Mount Gilboa, Samaria, Shefela, Judean Mountains, Judean Desert,

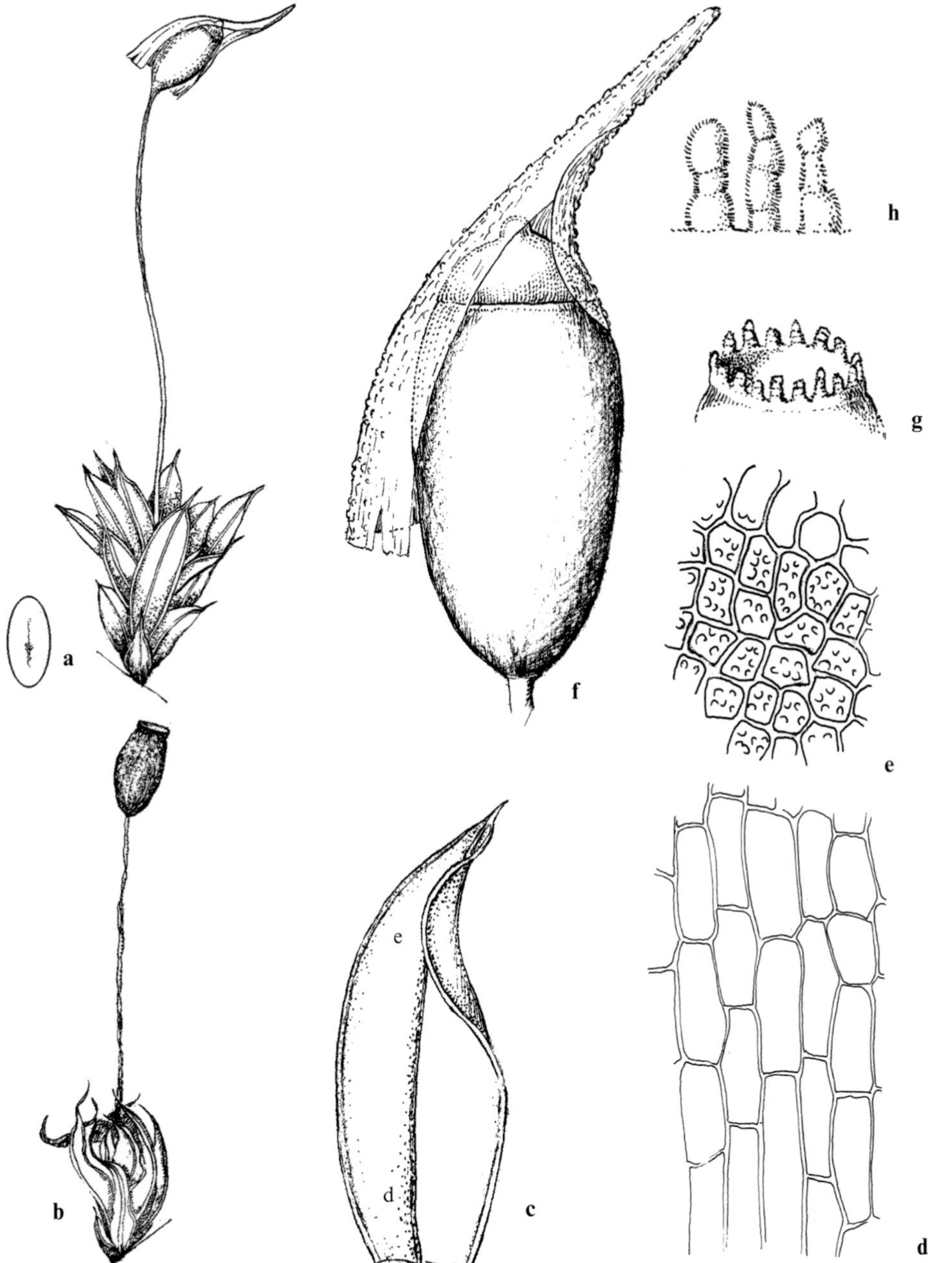

Figure 87. *Pottia commutata*: **a** habit, moist (×20); **b** habit, capsule with detached operculum, dry (×20); **c** leaf (×50); **d** basal cells; **e** upper cells (all leaf cells ×500); **f** capsule with calyptra (×100); **g** peristome (×100); **h** peristome teeth (×500). (*Nachmony*, 6 Feb. 1955)

Northern Negev, Dan Valley, Hula Plain, Bet She'an Valley, Mount Hermon, and Golan Heights.

General Distribution. Records of *Pottia commutata* are few, possibly due to various existing species concepts: recorded from Southwest Asia (Turkey, Syria, Jordan, and Iran), around the Mediterranean (few scattered records), Europe (few records), and North Africa.

Pottia commutata is often found growing together with either *P. davalliana*, *P. mutica*, or *P. starckeana*.

4. Pottia mutica Venturi *in* De Not., Atti R. Univ. Genova 1 : 592 (1869).
[Bilewsky (1965) : 375 (as "var. *gymnostoma* Corb.")]
Microbryum starckeanum (Hedw.) R. H. Zander var. *brachyodus* (B.S.G.) R. H. Zander, Genera of the Pottiaceae 240 (1993).

Distribution map 91, Figure 88, Plate IV : e.

Plants brownish-green, turning brownish-red, similar to *Pottia commutata*, but sometimes with longer, up to 1.5 mm long, and wider leaves. *Autoicous. Sporophyte* resembling *P. commutata* in shape, but setae longer, up to 3 mm long. *Capsules* nearly erect, larger, up to 1.25 mm long including operculum, peristome teeth shorter, with only one–two articulations. *Spores* 24–27(–30) μm in diameter ± triangular, at least slightly spinulose; with short spinulose processes (seen with SEM).

Habitat and Local Distribution. — Plants gregarious or growing in mats on exposed open ground on various soils. Occasional, most often together with other *Pottia* species: Philistean Plain, Upper Galilee, Mount Carmel, Judean Mountains, Judean Desert, Bet She'an Valley, and Lower Jordan Valley.

General Distribution. Records are few (often as *Pottia davalliana* var. *brachyodus*); recorded from Southwest Asia (Turkey, Syria, and Iraq), around the Mediterranean, Europe (mainly from the south), Macaronesia (in part), and North Africa.

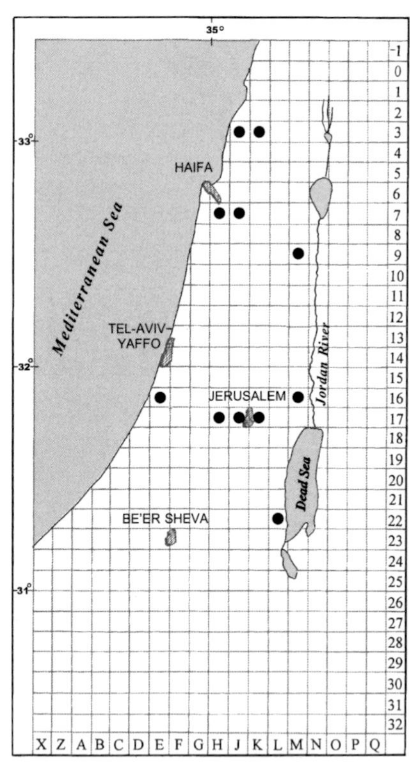

Map 91: *Pottia mutica*

Pottia mutica seems a somewhat doubtful species. Perhaps it is based on various hybrids between *Pottia* species growing together. This could give some explanation to the fact that spores of *P. mutica* tend to vary in size, shape, and texture: some spores may be triangular and the degree of spininess also varies. The short peristome occupies an intermediate position between the gymnostomous *P. davalliana* and the partly developed peristome of *P. commutata* and *P. starckeana*.

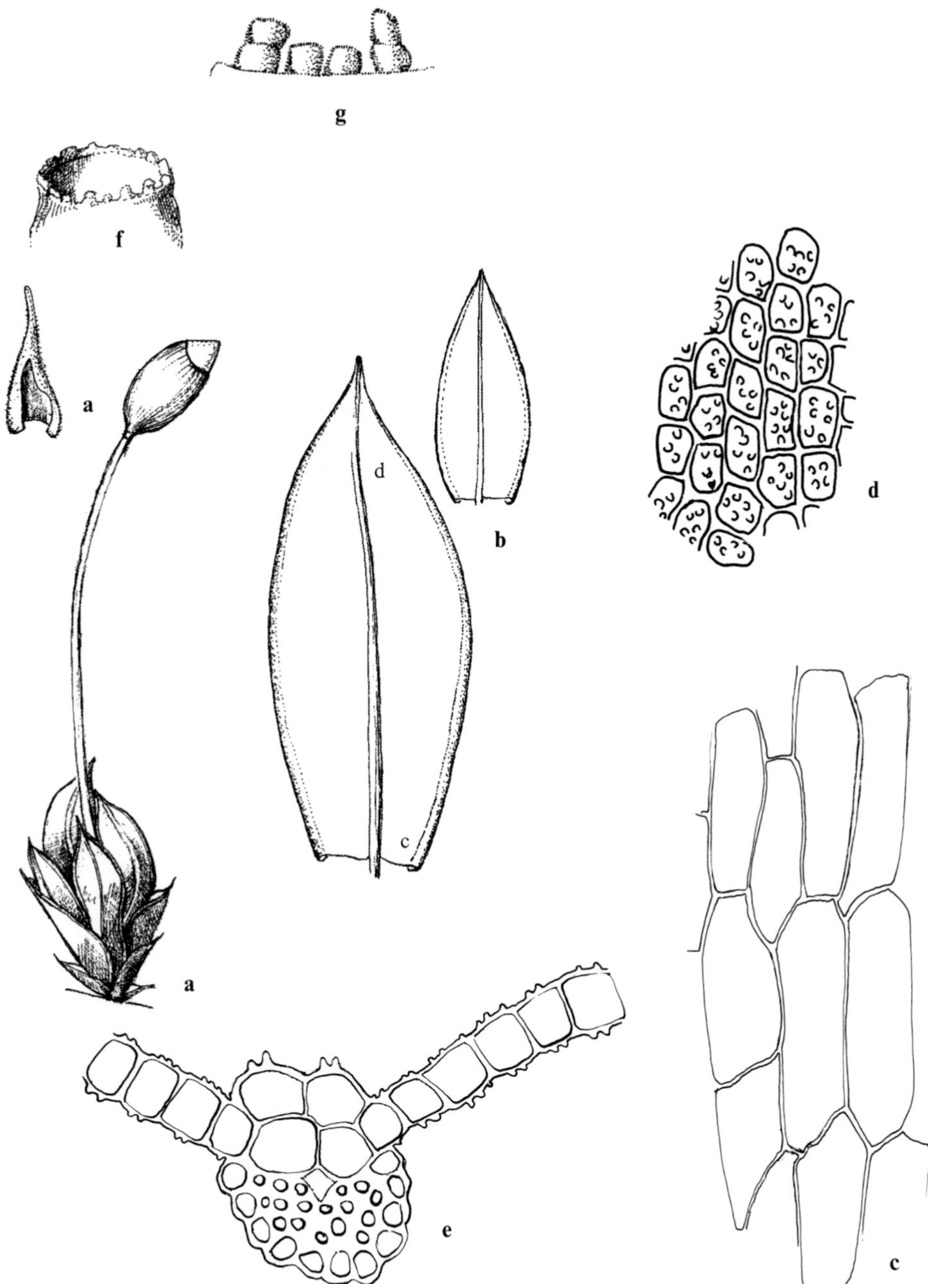

Figure 88. *Pottia mutica*: **a** habit and calyptra, dry (×20); **b** leaves (×50); **c** basal cells; **d** upper cells (all leaf cells ×500); **e** cross-section of costa and portions of lamina (×500); **f** peristome (×100); **g** peristome teeth (×500). (*Kushnir* & *Ginzburg*, 23 Jan. 1943)

Herrnstadt *et al.* (1991) assumed "*affinis*" to be the earliest epithet for this taxon, based on *Weissia affinis* Hook. & Taylor (1818, p. 44, Tab. 14), and proposed the new combination *Pottia affinis* (Hook. & Taylor) Herrnst. & Heyn, with *P. mutica* as its synonym. However, a later study of the type specimens of both taxa in BM established that: (a) the type specimen of *W. affinis* (Ireland, Dublin, Phoenix Park, *Taylor*) is not identical with that of *P. mutica* Venturi (Italy: Martignano ed a Cognola, presso Trento, *Venturi*, 2 Mar. 1868); (b) the type specimen of *W. affinis* has ± bilateral spores with warty outline, resembling *P. starckeana*, whereas the type specimen of *P. mutica* has spiny spores with a subspherical outline; (c) the type illustration of *Weissia affinis* Hook. & Taylor (1818, p. 44, Tab. 14.) shows a peristome with up to four articulations, thus resembling *P. starckeana* and not *P. mutica*, which typically has only one–two articulations. Therefore, the valid name of the taxon described here is *P. mutica* Venturi, and *P. affinis* should be considered as a synonym of *P. starckeana* subsp. *starckeana* var. *brachyodus* (cf. Chamberlain 1969).

5. Pottia davalliana (Sm. in W. F. N. Drake) C. E. O. Jensen, Danmarks Mosser 2 : 342 (1923).

Gymnostomum davallianum Sm. in W. F. N. Drake, Ann. Bot. 1 : 577 (1805); *P. minutula* (Schleich. ex Schwägr.) Fürnr. ex Hampe, Flora 20 : 287 (1837); — Bilewsky (1965) : 374 (as var. *conica* Schl.); *P. rufescens* (Schultz) Fürnr. ex Warnst., Krypt.-Fl. Brandenburg Laubm. 2 : 209 (1904); *P. starkeana* (Hedw.) Müll. Hal. subsp. *minutula* (Schleich. ex Schwägr.) D. F. Chamb., Notes Roy. Bot. Gard. Edinburgh 29 : 403 (1969); *Microbryum davallianum* (Sm.) R. H. Zander, Genera of the Pottiaceae 240 (1993).

Distribution map 92, Figure 89, Plate IV : f, g.

Plants minute, up to 1.5 mm long, bright green, turning brownish-red; gametophytes resembling *Pottia commutata*, but leaves sometimes ± elliptical. *Paroicous*. *Setae* up to 2.5(–3) mm long; capsules up to 1 mm long including operculum, shortly ovoid to cylindrical, turbinate, wide-mouthed and empty when dry, brownish; operculum short-conical; peristome absent. *Spores* 25–27 μm in diameter, ± spinulose; surface perforate with short, spinulose processes, processes irregular in shape (seen with SEM). *Calyptrae* slightly papillose.

Habitat and Local Distribution. — Plants gregarious, scattered or growing in loose tufts, on moist open ground on banks of wadis or ditches. Common: Coastal Galilee, Coast of Carmel, Sharon Plain, Upper Galilee, Mount Gilboa, Samaria, Judean Mountains, Judean Desert, Central Negev, Bet She'an Valley, and Lower Jordan Valley.

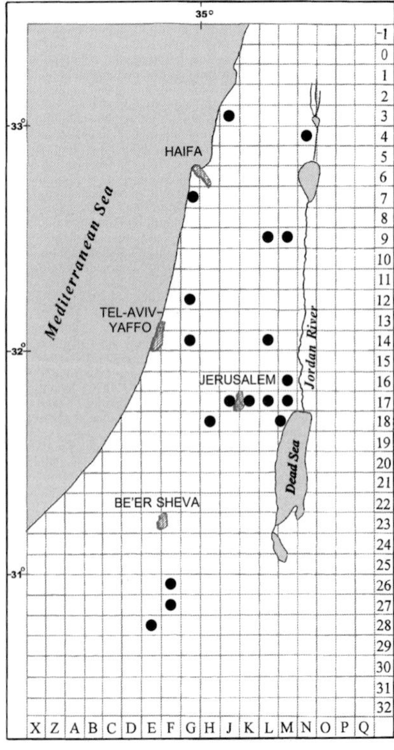

Map 92: *Pottia davalliana*

General Distribution. — Recorded from Southwest Asia (Cyprus, Turkey, Jordan,

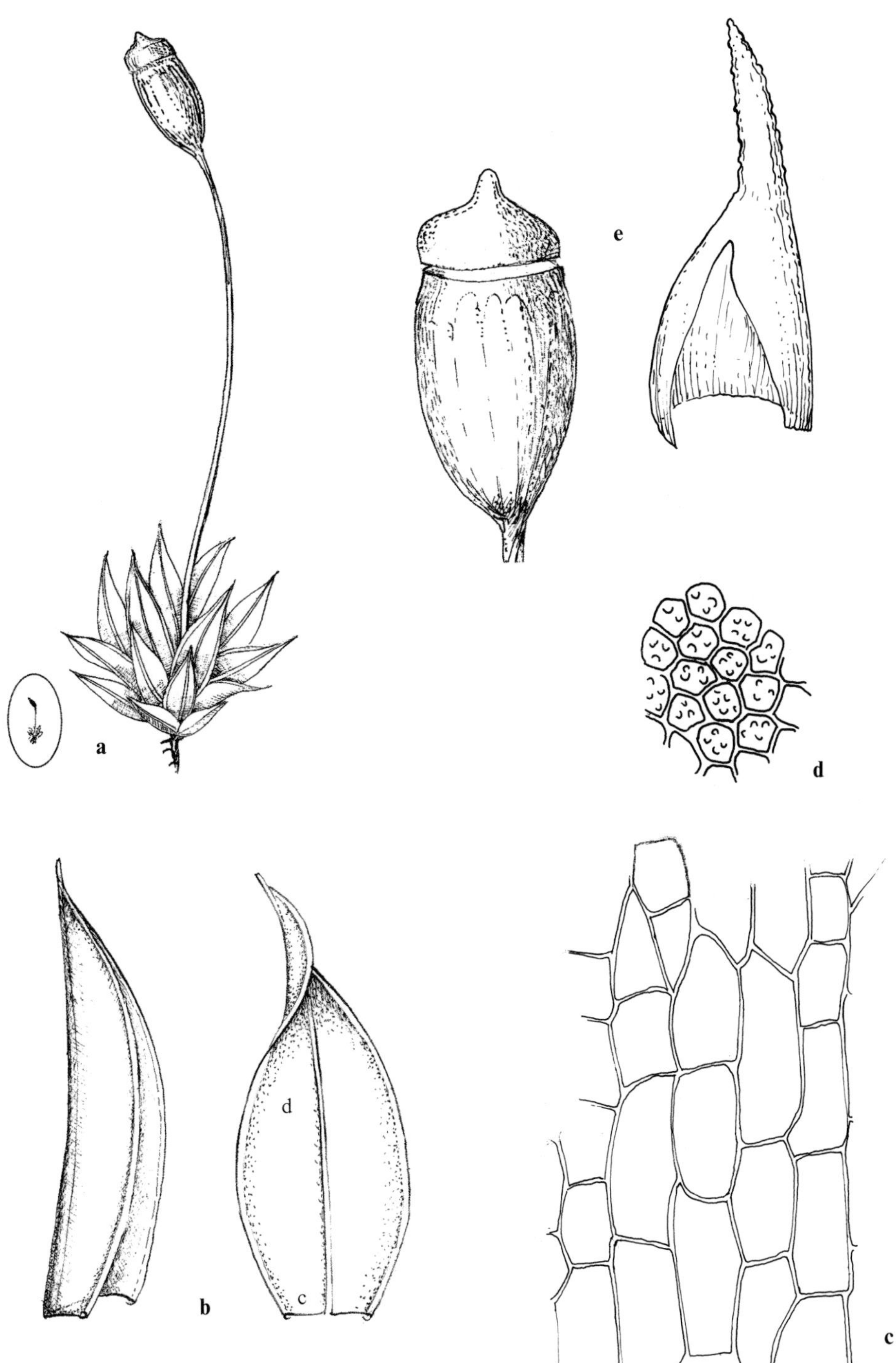

Figure 89. *Pottia davalliana*: **a** habit, moist (×20); **b** leaves (×50); **c** basal cells; **d** upper cells (all leaf cells ×500); **e** capsule and calyptra (×50). (Wadi Qilt, *Kushnir*, *sine dato*)

Iraq, Iran, and Kuwait), around the Mediterranean, Europe (throughout), Macaronesia, North Africa, Australia, and North America.

Pottia davalliana is easily distinguished from *P. commutata*, *P. mutica*, and *P. starckeana* by the elliptical leaves and the capsules, which are widest at the mouth and have a short-conical operculum.

Chamberlain (1978) expressed doubts about the identity of *Gymnostomum davallianum* and proposed to use the epithet *minutulum* instead. He included the taxon as subsp. *minutula* (= *P. davalliana*) in *P. starckeana*, together with other subspecies, including subsp. *conica*. Bilewsky (1965) listed "var. *conica* Schl." under *P. minutula*. The concept of that variety does not seem clear enough.

6. Pottia intermedia (Turner) Fürnr., Flora 12, 2 (Erg. 2): 13 (1829).
Gymnostomum intermedium Turner., Muscol. Hibern. Spic. 7 (1804); *P. truncata* var. *major* (F. Weber & D. Mohr) B.S.G., Bryol. Eur. 2: 37 (1843); *P. lanceolata* (Hedw.) Müll. Hal. var. *gymnostoma* Schimp., Syn. Musc. Eur., Edition 2: 158 (1876); *Tortula modica* R. H. Zander, Genera of the Pottiaceae 226 (1993).

Distribution map 93, Figure 90, Plate V: a, b.

Plants 2.5–4 mm high, yellowish-green, turning brown. *Leaves* 2–2.5(–3) mm long, oblong-lanceolate to oblong-spathulate; margins plane or narrowly recurved, entire or crenulate towards apex; costa excurrent in a yellowish-green cuspidate point 100–650 μm long, much varying in length between leaves of single plants; all leaf cells thin-walled, mid- and upper leaf cells irregularly subquadrate, 11–16 μm wide, smooth or weakly papillose. *Autoicous*. *Setae* 5–7 mm long; capsules ca. 2 mm long including operculum, ovoid to subcylindrical, ± narrowed at mouth; with ± three rows of small, incrassate, differentiated cells below mouth; operculum obliquely rostrate; annulus present; peristome absent or reduced to a basal membrane. *Spores* up to 30 μm in diameter; weakly papillose, irregularly reticulate (seen with SEM). *Calyptrae* smooth.

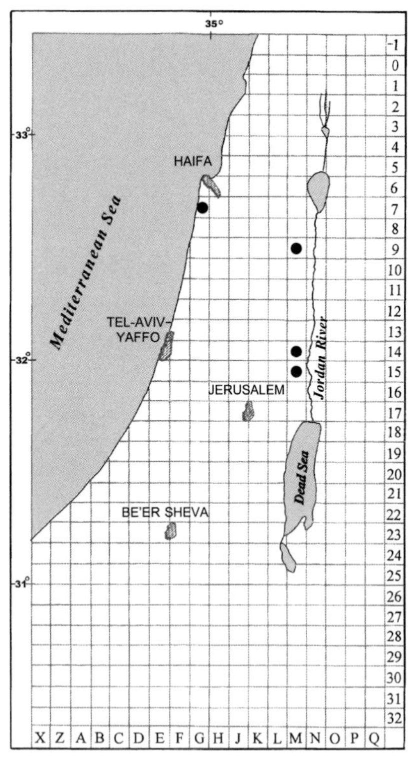

Map 93: *Pottia intermedia*

Habitat and Local Distribution. — Plants growing scattered or in tufts on calcareous soils and on walls. Occasional: Coast of Carmel, Mount Gilboa, Bet She'an Valley, and Lower Jordan Valley.

General Distribution. — Recorded from Southwest Asia (Turkey and Jordan), around Mediterranean (scattered records), Europe (throughout), North Africa, eastern Asia, Australia, and North and Central America.

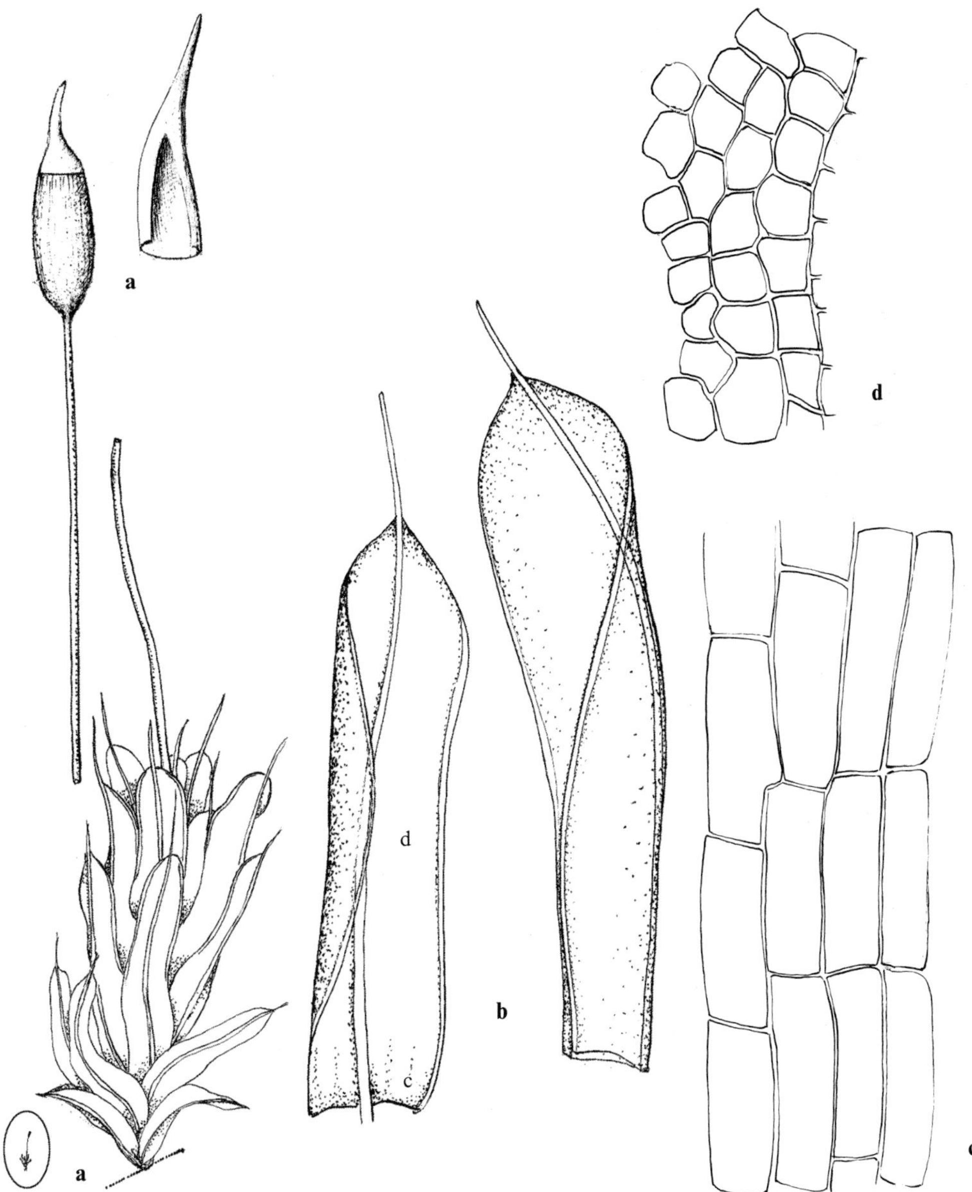

Figure 90. *Pottia intermedia*: **a** habit, moist, and calyptra (×20); **b** leaves (×50); **c** basal cells; **d** mid-leaf cells (all leaf cells ×500). (*Kushnir*, 30 Mar. 1945)

The original diagnosis and the illustrations of *Gymnostomum intermedium* (the basionym of *Pottia intermedia*) by Turner (Muscol. Hibern. Spic., p. 7; Tab. 1 : a–c. 1804) refer to a plant with cuspidate leaves. In many floras the concept of this species is somewhat broader and also includes plants possessing leaves with a long-excurrent costa. In some publications (e.g., Warnstorf 1916, Crum & Anderson 1981), the taxon is treated as *P. truncata* var. *major*.

Among the few local populations found so far, those from Mount Gilboa and from the Lower Jordan Valley tend to have more spathulate leaves with plane margins and longer excurrent costa. Those plants are identical, perhaps, with *Pottia pallida* Lindb. var. *longicuspis* Warnst., syn. *P. venusta* Jur. (see Warnstorf 1916). Further studies are needed.

Plants of *Pottia intermedia* resemble *P. lanceolata*, but can be easily distinguished by the absence of even rudimentary peristome teeth and by the different spore texture.

7. Pottia gemmifera Herrnst. & Heyn, Nova Hedwigia 69 : 232 (1999).

Distribution map 94, Figure 91, Plate V : c.

Plants 2.5–4 mm long, yellowish-green. *Stems* usually simple, rarely branched. *Leaves* crowded, up to 3.5 mm long, oblong to oblong-spathulate, cuspidate; margins plane or slightly revolute in upper part, entire or weakly crenulate near apex; costa stout, excurrent 100–165(–300) µm long; mid- and upper leaf cells hexagonal to irregularly rectangular, 15–25 µm wide, smooth. *Propagula* present in axils of at least some leaves, large, 1.2–1.5 mm long, bud-shaped, with leaves having costa excurrent into long hair-point. *Autoicous*. *Setae* 6–7 mm long; capsules greatly varying in size, 0.75–1.5 mm long including operculum, obloid, gradually narrowing towards base, dark red-brown; in two–three rows of incrassate cells below mouth; deoperculate capsules widest at mouth; operculum obliquely beaked; annulus present; peristome absent. *Spores* 27–30 µm in diameter, irregularly granulate; verrucate, verrucae irregularly fused, with granulate surface (seen with SEM). *Calyptrae* smooth.

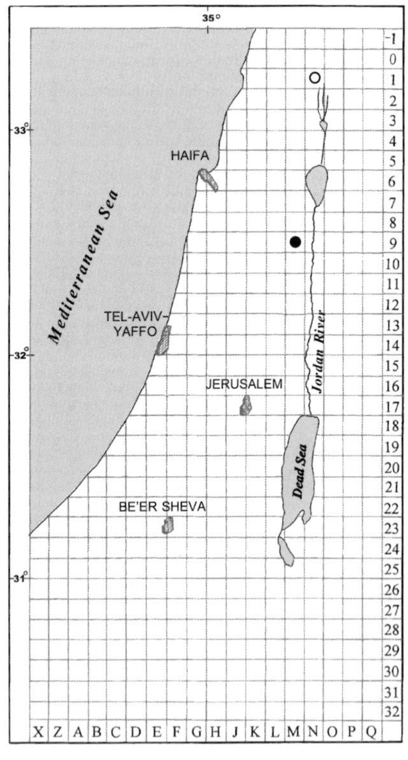

Map 94: ● *Pottia gemmifera*
○ *P. recta*

Habitat and Local Distribution. — Plants growing in dense tufts. Found in a single locality: Bet She'an Valley, Gan-Hashlosha.

Pottia gemmifera most resembles *P. intermedia* among the local *Pottia* species (a population of *P. intermedia* was found growing near the locality of *P. gemmifera* listed above). The species can be characterised, in addition to the presence of propagules, by the crowded leaves with plane margins, large and smooth lamina cells, and the

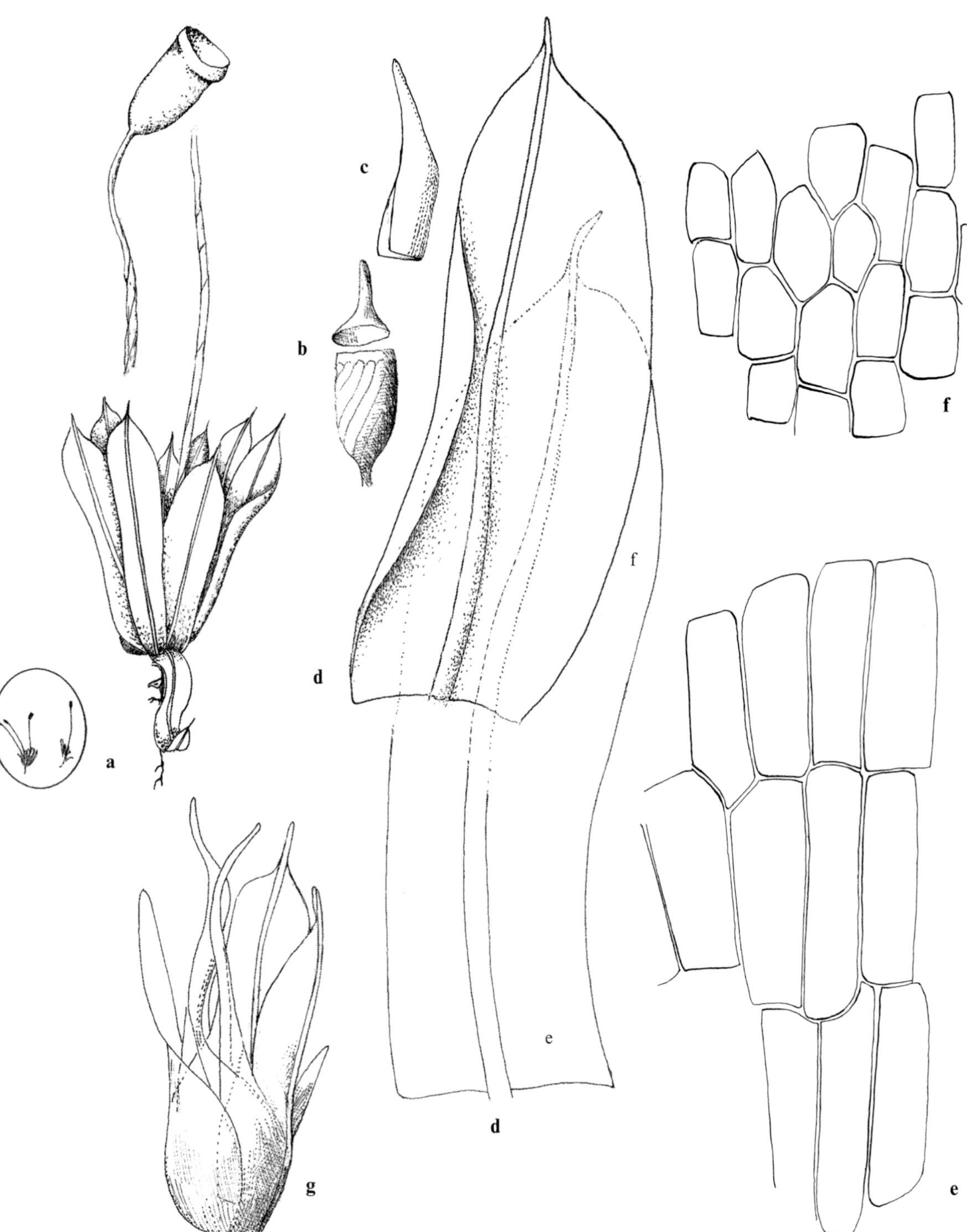

Figure 91. *Pottia gemmifera*: **a** habit, moist (×20); **b** capsule with detached operculum (×20); **c** calyptra (×20); **d** leaves (×50); **e** basal cells; **f** upper cells (all leaf cells ×500); **g** propagulum (×50). (Gan Hashlosha, Herb. Bilewsky No. 94a)

widened mouth of the deoperculate capsules. Propagules in *Pottia* have been recorded, so far, only in *P. propagulifera* Herzog from Sardinia. However the propagules of that species were described as having various shapes and as always being attached to the costa.

8. Pottia recta (With.) Mitt., Ann. Mag. Nat. Hist. ser. 2, 8 : 311 (1851).
Phascum rectum With., Syst. Arr. Britt. Pl., Edition 4 : 771 (1801); *Pottiella recta* (With.) Gams, Krypt. Fl. Mitteleur., Edition 2, 1 : 101 (1948); *Microbryum rectum* (With.) R. H. Zander, Genera of the Pottiaceae 240 (1993).

Distribution map 94 (p. 242), Figure 92, Plate V : d.

Plants minute, 1(–1.25) mm high, green to olive-green. *Leaves* up to 1 mm long, ovate-lanceolate, cuspidate; margins entire, recurved almost to apex; costa stout, reddish; mid- and upper leaf cells sub-quadrate, 8–13(–14) μm wide, densely papillose. *Paroicous*. *Setae* up to 1 mm long; capsules 0.5–1 mm long, dark red, cleistocarpous, subspherical to broad ellipsoid; operculum persistent, beaked; peristome absent. *Spores* ca. 24 μm in diameter, spinulose; long-spinulose (seen with SEM). *Calyptrae* rough.

Habitat and Local Distribution. — Plants growing on open, exposed, dark-brown soil. Plants of *Pottia recta* were found, so far, only in a single locality in Israel (Dan Valley: Horeshat Tal) scattered among other mosses, including *P. commutata*. These very small plants may have been overlooked in other localities.

General Distribution. — Records are few: recorded from Southwest Asia (Turkey and Iraq), from few localities around the Mediterranean, Europe (few), North Africa, and southern Asia.

23. *Phascum* Hedw.

Plants minute to small, terrestrial, solitary, gregarious or growing in tufts. *Stems* erect, simple, or branched, without central strand. *Leaves* obovate, ovate to lanceolate, acute to acuminate; margins plane or recurved, usually entire; costa excurrent, with a single, dorsal stereid band; basal cells thin-walled, rectangular, lax, smooth, upper cells sub-quadrate, hexagonal or rhomboidal, often papillose, with C-shaped papillae on one or both surfaces. *Monoicous*. *Setae* short, straight or curved; capsules immersed among leaves to shortly exserted, sometimes in pairs, globose to ellipsoid, apiculate to beaked; cleistocarpous, operculum and peristome not formed. *Calyptrae* cucullate, rarely mitriform.

A small genus comprising about 20 species. Four species occur in the local flora (one consisting of five varieties; two are perhaps endemic). The genus may be more wide-spread than recorded as plants of *Phascum* are easily overlooked because of their short life cycle, small size, and gregarious growth, often in ephemeral habitats.

The delimitation of *Phascum* from *Acaulon* and *Pottia* is difficult (see the discussions of *Acaulon* and *Phascum floerkeanum* below). According to Chamberlain (1978) the separation of *Phascum* and *Pottia* is "somewhat arbitrary and it might be more satisfactory taxonomically if all the species were placed in only 1 genus". While cleistocarpous species of *Pottia* could perhaps be better placed in *Phascum*, two recently described species of *Phascum* with longer setae than assumed as typical for that

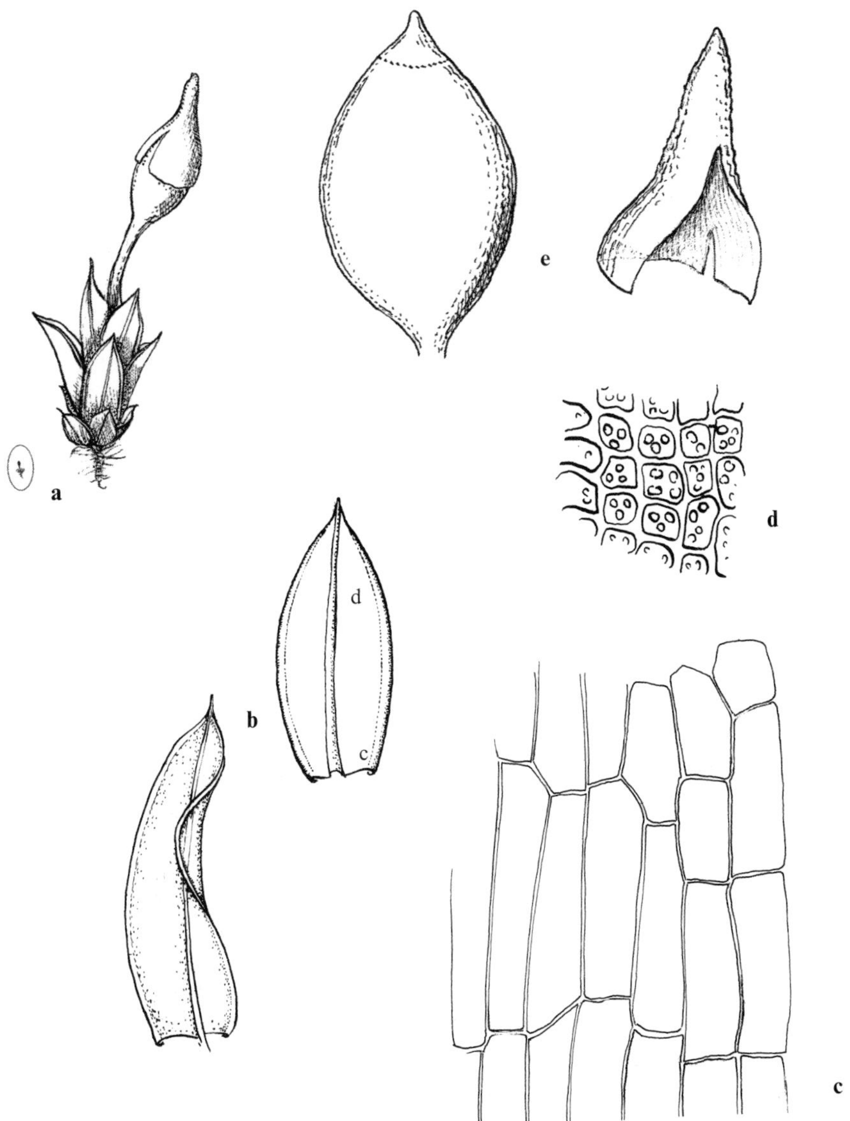

Figure 92. *Pottia recta*: **a** habit, moist (×20); **b** leaves (×50); **c** basal cells; **d** upper cells (all leaf cells ×500); **e** capsule and calyptra (×75). (*Markus & Kutiel 77-555-2*)

genus — *P. longipes* J. Guerra, J. J. Martínez & R. M. Ros (Guerra *et al.* 1990) and *P. galilaeum* Herrnst. & Heyn (Herrnstadt *et al.* 1991, Herrnstadt & Heyn 1993 — could perhaps be accommodated in *Pottia*. Zander (1993) includes *Phascum*, together with *Pottia*, in the genus *Tortula*.

Literature. — Zander (1989), Carrión *et al.* (1990), Guerra *et al.* (1991).

1 Plants minute, up to 1.5 mm high; leaves not exceeding 1 mm in length; spores ± smooth.
  2 Leaf margins recurved; setae strongly curved up to 1 mm long; capsules ovoid, laterally exserted; calyptrae cucullate **3. P. curvicolle**
  2 Leaf margins plane, rarely narrowly recurved in upper part; setae straight, very short, up to 0.2 mm long; capsules globose, immersed among leaves; calyptra mitriform or subcucullate **4. P. floerkeanum**
1 Plants larger, 2–6 (7) mm high; leaves at least 2.5 mm long; spores distinctly papillose.
  3 Setae at least 1 mm long; capsules ± exserted, subglobose to ovoid, obliquely beaked **2. P. galilaeum**
  3 Setae up to 0.5 mm long; capsules immersed among leaves, globose, apiculate **1. P. cuspidatum**
    4 Setae strongly curved **1c. P. cuspidatum** var. **arcuatum**
    4 Setae straight.
      5 Plants 4–6 mm high; stems unbranched in lower part, branched above **1d. P. cuspidatum** var. **schreberianum**
      5 Plants up to 4 mm high; stems unbranched or branched from base.
        6 Leaves distinctly bordered by one–two(–three) rows of smooth, rectangular, incrassate cells **1e. P. cuspidatum** var. **marginatum**
        6 Leaves not bordered.
          7 Leaves with costa excurrent in a short yellowish-green hair-point, hair-point up to 0.25 mm long **1a. P. cuspidatum** var. **cuspidatum**
          7 Leaves with costa excurrent in a long, often hyaline hair-point, hair-point up to 1 mm long **1b. P. cuspidatum** var. **piliferum**

1. Phascum cuspidatum Hedw., Sp. Musc. Frond. 22 (1801).
Tortula atherodes R. H. Zander, Genera of the Pottiaceae 222 (1993).

Plants up to 6 mm high, green to yellowish-brown, often branched at base or above. *Leaves*: upper leaves up to 2.75 mm long without point, keeled, ± contorted or twisted when dry, erectopatent to spreading when moist, ovate to lanceolate, acute or acuminate; margins usually recurved except at apex, entire; costa strong, excurrent into a short point or with hyaline hair-point 0.1–1.5(–2) mm long; upper cells subquadrate to irregularly hexagonal, sometimes slightly incrassate, smooth or papillose, 10–19 µm wide. *Autoicous* or paroicous. *Setae* short, up to 1 mm long, straight or curved (sometimes two capsules terminating plant); capsules immersed among leaves, up to 0.5 mm, ellipsoid to globose, light brown, with a short, obtuse apiculus. *Spores* 27–40 µm in diameter, strongly papillose. *Calyptrae* small, cucullate.

A most variable species in which many infraspecific taxa have been recognised. Five distinct varieties are accepted in the local flora. Two varieties were only recently described (Herrnstadt *et al.* 1991).

The protologue gives a clear description of a typical *Phascum cuspidatum*. However, the references cited there seem to refer to taxa with various other character combinations. For instance, the description of von Schreber (1770) of "*Phascum (cuspidatum)*" and the figures cited there (Tab. 1, Figs. 1 and 2) refer to a plant with a distinct stem whereas the Linnean synonym "*Phascum* (*acaulon*)" refers to a stemless plant. Dickson (1801) based his description of *P. schreberianum* (here accepted as a variety of *P. cuspidatum*) on the same figures of von Schreber.

1a. Phascum cuspidatum Hedw. var. **cuspidatum**
P. acaulon With., Syst. Arr. Brit. Pl., Edition 4, 3 : 768 (1801).

Distribution map 95, Figure 93.

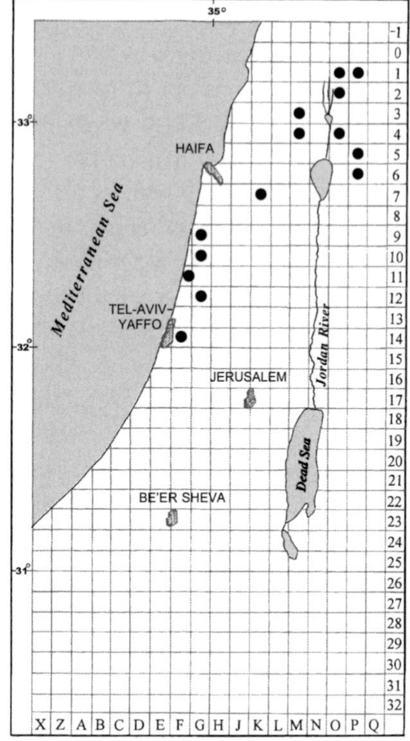

Map 95: *Phascum cuspidatum* var. *cuspidatum*

Plants 2–3 mm high green to yellowish. *Stems* simple or sometimes branched from base. *Leaves* crowded, up to 2.5(–2.75) mm long without point, costa excurrent in a short yellowish-green hair-point up to 0.25 mm long; cells smooth or weakly papillose above. *Capsules* on a short straight seta.

Habitat and Local Distribution. — Plants gregarious or growing in tufts on exposed soils of various types, sometimes in very moist habitats of the Mediterranean district. Occasional, locally common: Sharon Plain, Philistean Plain, Upper Galilee, Lower Galilee, and Golan Heights.

General Distribution. — Recorded from Southwest Asia (Turkey, Iraq, and Iran), around the Mediterranean, Europe (widespread), Macaronesia, North Africa, northeastern, eastern, and central Asia, and North, Central, and northwestern South America.

Plants of the typical variety from Israel seem to have nearly smooth leaf cells, contrary to the papillose cells described from plants of this taxon from Europe.

Figure 93. *Phascum cuspidatum* var. *cuspidatum*: **a** habit, moist (×20); **b** habit, dry (×20); **c** sporophyte and calyptra (×20); **d** leaves (×50); **e** basal cells; **f** upper cells (all leaf cells ×500); **g** cross-section of costa and portion of lamina (×500). (*Crosby & Herrnstadt 79-93-3*)

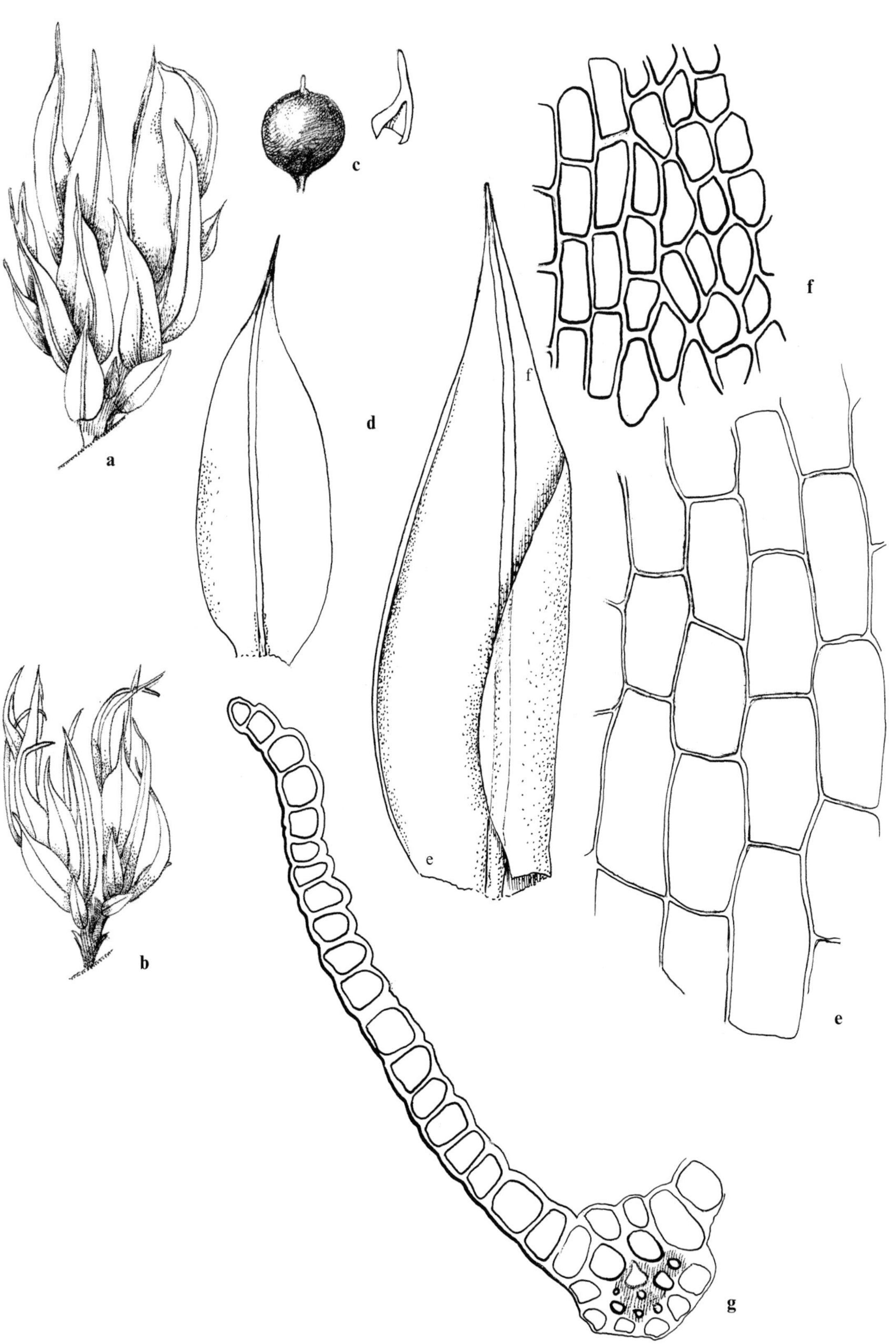
a
b
c
d
e
f
g
e
f

1b. Phascum cuspidatum Hedw. var. **piliferum** (Hedw.) Hook. & Taylor, Muscol. Brit. 8 (1818). [Bilewsky (1965): 372]
P. piliferum Hedw., Sp. Musc. Frond. 20 (1801).

Distribution map 96, Figure 94: a–f.

Plants green or often yellowish-brown; generally resembling var. *cuspidatum* except for the following characters. *Leaves* with costa excurrent in a long, often hyaline hair-point, hair-point up to 1 mm long; upper cells usually densely papillose.

Habitat and Local Distribution. — Plants gregarious or growing in tufts on various, usually exposed, soils; more common than var. *cuspidatum* and extending also to more arid areas of the Mediterranean and the Irano-Turanian districts. Occasional: Sharon Plain, Lower Galilee, Mount Gilboa, Judean Mountains, Judean Desert, Bet She'an Valley, and Golan Heights.

General Distribution. — Recorded from Southwest Asia (Turkey, Syria, and Lebanon), around the Mediterranean, Europe (mainly the south), North Africa, northeast Asia, and North America.

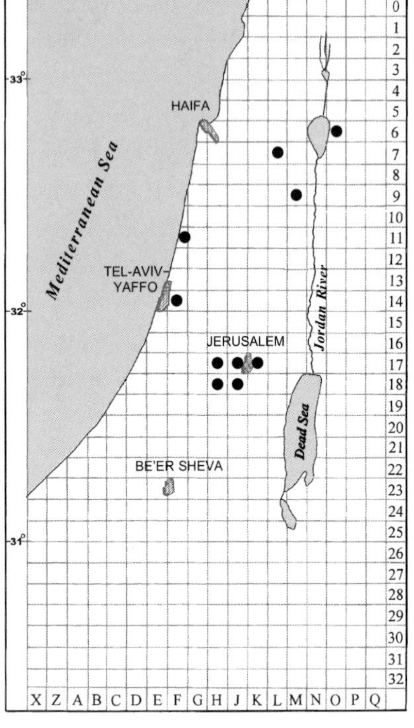

Map 96: *Phascum cuspidatum* var. *piliferum*

The delimitation of var. *piliferum* from the typical variety, which is mainly based on the length of the hair-point, is somewhat arbitrary and is often not given a separate status (e.g., Anderson *et al.* 1990).

1c. Phascum cuspidatum Hedw. var. **arcuatum** Herrnst. & Heyn, Bryologist 94: 175 (1991).

Distribution map 97 (p. 252), Figure 94: g–i, Plate V: e.

Plants green, turning yellowish-green to light brown when mature; resembling var. *piliferum* except for the following characters. *Leaves* with long hyaline hair-point, up to 1.5(–2) mm long. *Setae* curved, up to 1 mm long; capsules inclined, shorter than setae. *Spores* baculate or clavate, processes mostly with lobed apex (seen with SEM).

Habitat and Local Distribution. — Plants growing in dense tufts on exposed or partly shaded terra rossa and basaltic soils. Occasional, locally common (so far, this variety was recorded only from the local flora): Upper Galilee, Mount Gilboa (?), Judean Mountains, and Golan Heights.

Plants of var. *Phascum cuspidatum* var. *arcuatum* resemble var. *curvisetum* (Dicks.) Nees & Hornsch. in the curved seta. However, they differ from Dickson's (1801,

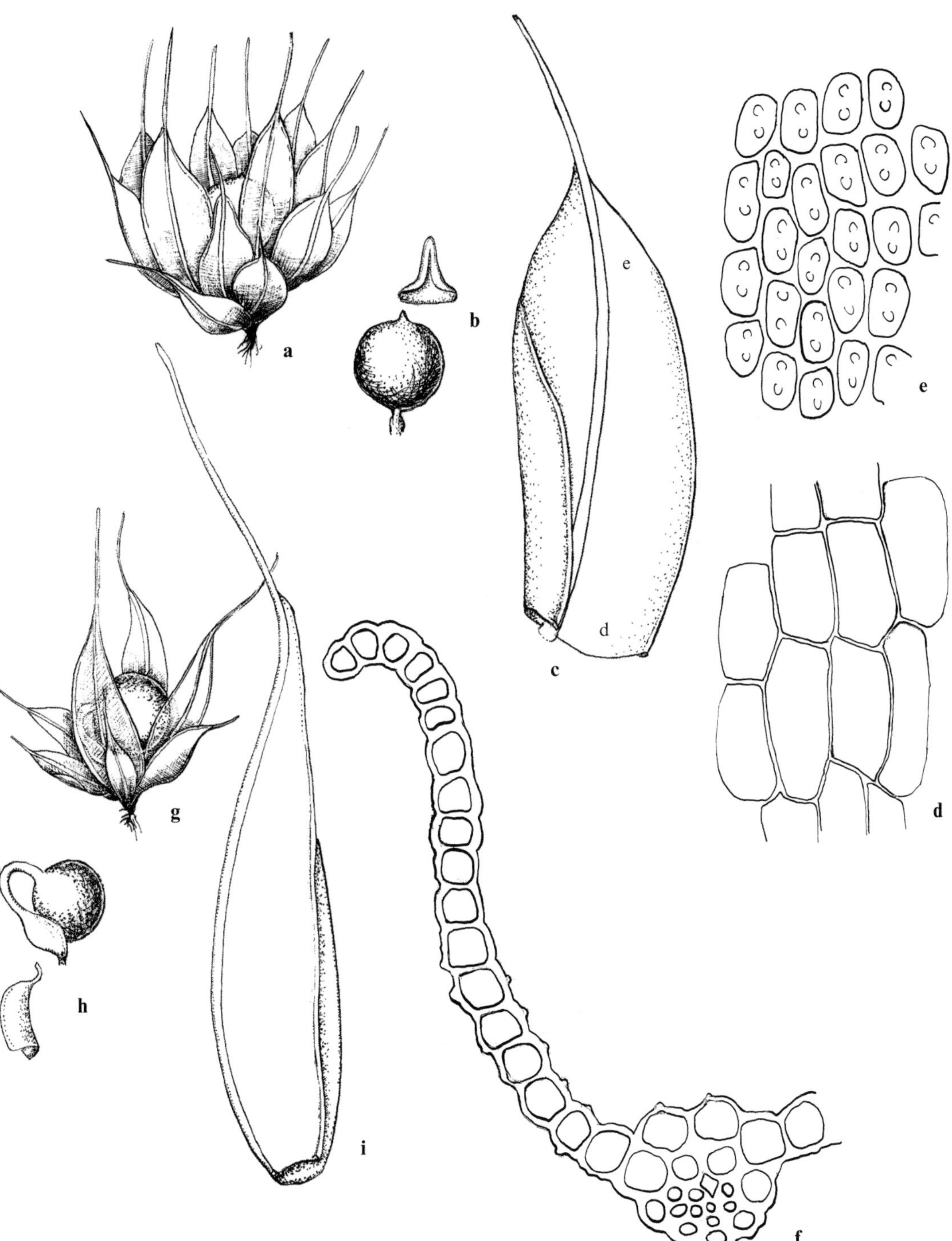

Figure 94. *Phascum cuspidatum* var. *piliferum*: **a** habit, moist (×20); **b** sporophyte and calyptra (×20); **c** leaf (×50); **d** basal cells; **e** upper cells (all leaf cells ×500); **f** cross-section of costa and portion of lamina (×500).
Phascum cuspidatum var. *arcuatum*: **g** habit, moist (×20); **h** sporophyte and calyptra (×20); **i** leaf (×50).
(**a**–**f**: *Herrnstadt* & *Crosby 78-27-1*; **g**–**i**: *Herrnstadt* & *Crosby 78-48-3*)

p. 3 and Tab. 10, Figure 4) original description and drawing of var. *curvisetum* by the inclined capsule, which is shorter than the curved seta, and the long-piliferous leaves. Plants of var. *curvisetum* have longer, upward pointing capsule borne on a cygneous seta, and short-cuspidate leaves. Although *P. cuspidatum* var. *curvisetum* was recorded by Bilewsky (1965), no plants with that name could be found in his herbarium.

1d. Phascum cuspidatum Hedw. var. **schreberianum** (Dicks.) Brid. Sp. Musc. Frond. 1:9 (1806).

P. schreberianum Dicks., Pl. Crypt. Brit. 4:2 (1801).

Distribution map 98, Figure 95.

Plants yellowish green, resembling var. *cuspidatum* except for the following characters. Plants up to 6 mm high. *Stems* simple below, branched above into two–three branches (a character not observed in other varieties of *P. cuspidatum*); rarely producing rhizoidal gemmae.

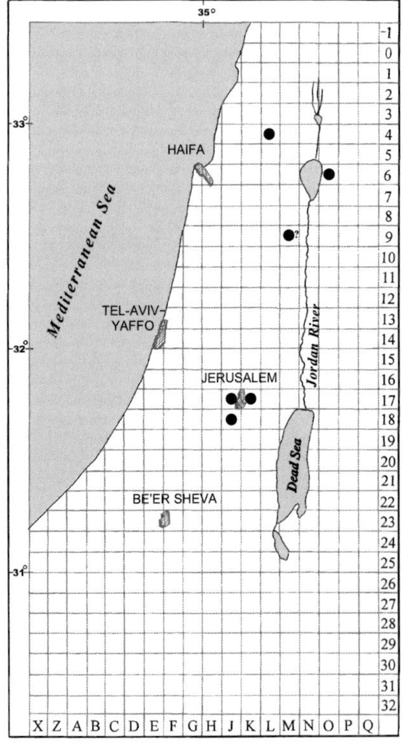

Map 97: *Phascum cuspidatum* var. *arcuatum*

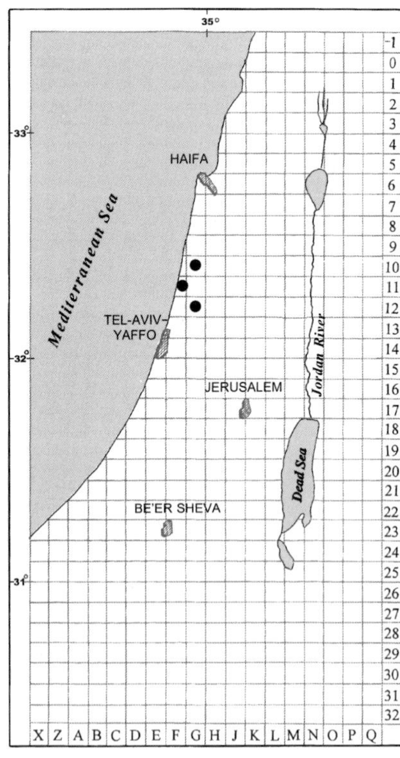

Map 98: *Phascum cuspidatum* var. *schreberianum*

Figure 95. *Phascum cuspidatum* var. *schreberianum*: **a** habit, moist (×20); **b** habit, dry (×20); **c** sporophyte and calyptra (×20); **d** leaves (×50); **e** basal cells; **f** upper cells (all leaf cells ×500); **g** cross-section of costa and portion of lamina (×500); **h** rhizoidal gemma (×150). (*Kushnir*, 21 Mar. 1943)

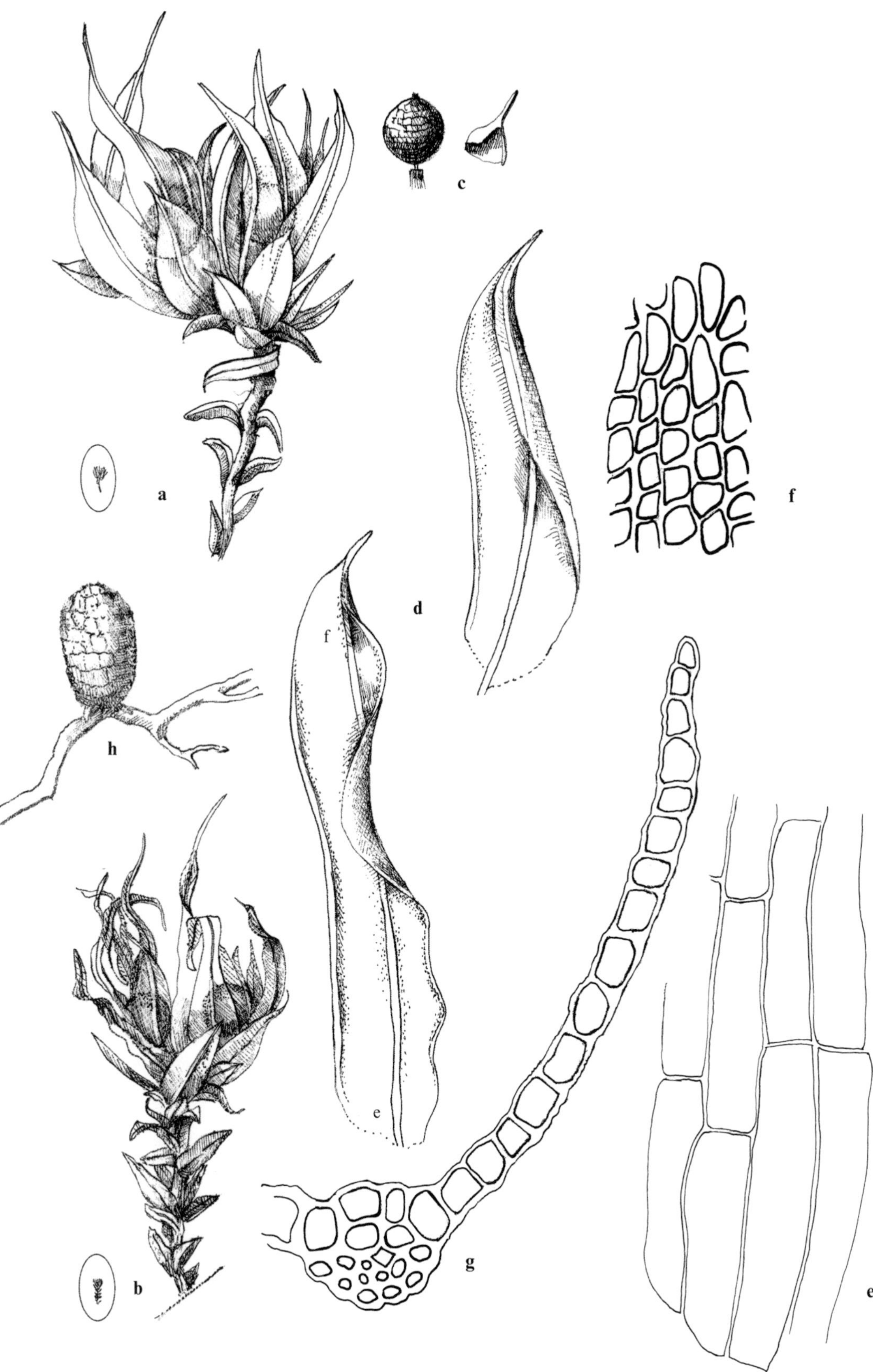
c
a
d
f
f
h
e
g
b
e

Habitat and Local Distribution. — Plants growing in tufts on exposed sandy loam. Rare: Sharon Plain (in three localities).

General Distribution. — Recorded from Europe and Australia.

Rhizoidal gemmae have been observed on some rhizoids in the two populations of this variety (Figure 95: h), in particular in sterile plants, and have not been previously reported in this variety. In similar habitats on sandy substrates, rhizoidal gemmae have often been observed also in species of other genera (*Bryum*, *Ditrichum*, *Dicranella*, and others).

The validity of this variety is often doubted (e.g., by Anderson *et al.* 1990).

1e. Phascum cuspidatum Hedw. var. **marginatum** Herrnst. & Heyn, Bryologist 94: 175 (1991).

Distribution map 99, Figure 96.

Plants bright green, resembling var. *piliferum* except for the following characters. *Leaves* with shorter hyaline hair-point, up to 0.5 mm long, cells densely papillose except for one–two(–three) rows of smooth, incrassate marginal cells forming a distinct border. *Setae* seldom very slightly curved; capsules ± laterally exserted.

Habitat and Local Distribution. — Plants growing in small groups or scattered among plants of *Bryum dunense*, *Pottia davalliana*, and *Acaulon muticum* on exposed sandy loam. Found in a single locality: Sharon Plain.

Phascum cuspidatum var. *marginatum* should perhaps be considered as a species separate from *P. cuspidatum* mainly because of the distinctly bordered leaves. However, some additional populations should be studied before the elevation of the rank is considered.

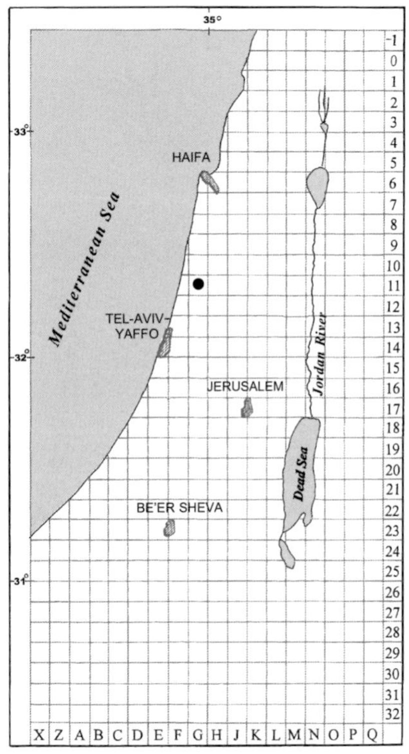

Map 99: *Phascum cuspidatum* var. *marginatum*

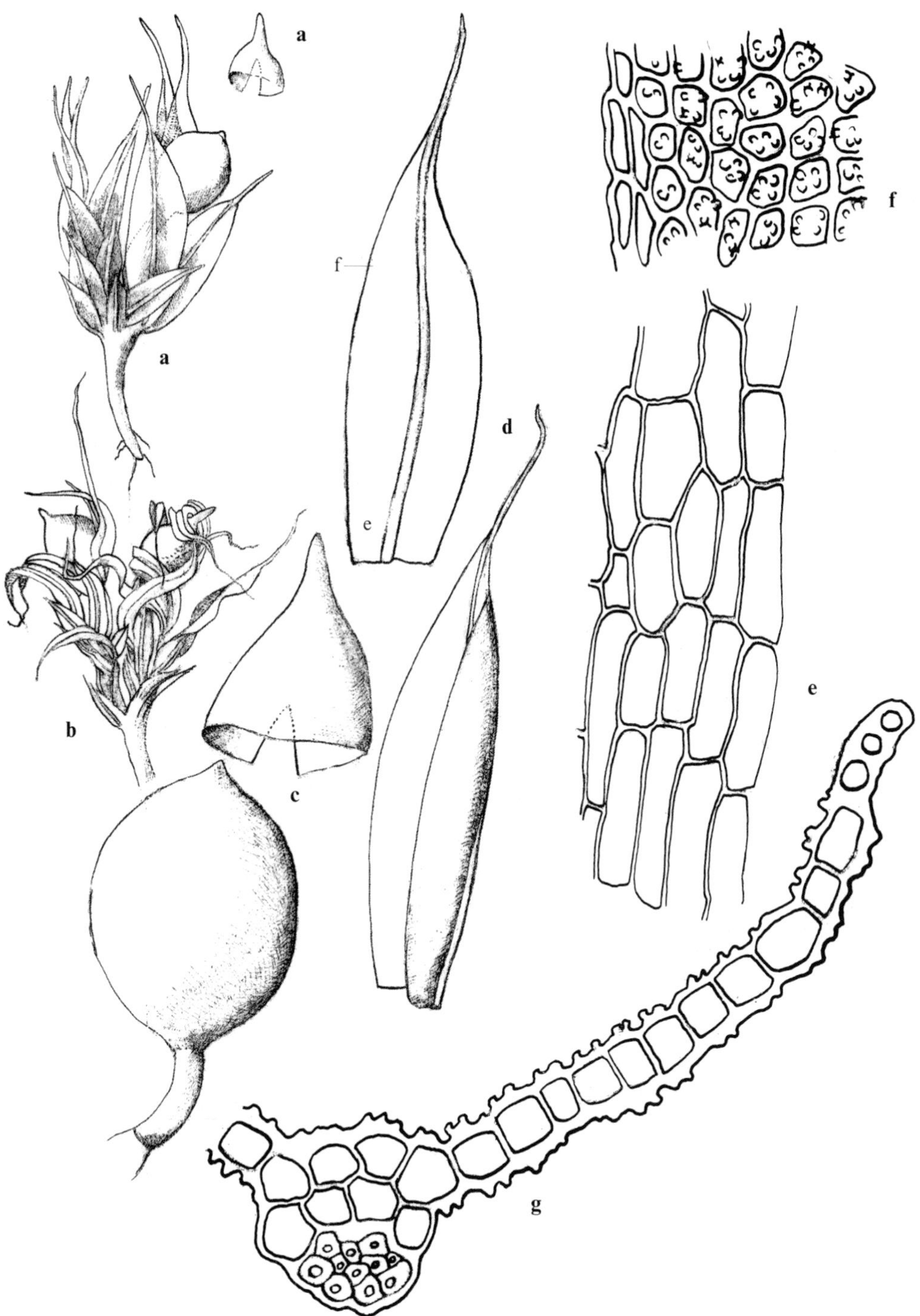

Figure 96. *Phascum cuspidatum* var. *marginatum*: **a** habit, moist, and calyptra (×20); **b** habit, dry (×20); **c** sporophyte and calyptra (×50); **d** leaves (×50); **e** basal cells; **f** marginal upper cells (all leaf cells ×500); **g** cross-section of costa and portion of lamina (×500). (*Herrnstadt* & *Boaz*, 29 Jan. 1986 – holotype)

2. Phascum galilaeum Herrnst. & Heyn, Bryologist 94 : 175 (1991).

Distribution map 100, Figure 97, Plate V : f.

Plants 3–6(–7) mm high (at times terminated with two sporophytes), sometimes branched at base, yellowish-green turning brownish when dry. *Leaves* up to 3 mm long, 2.5–4 times longer than wide, concave, slightly contorted when dry, erectopatent with tips ± pointing inwards when moist, lanceolate; lower leaves smaller, 1.5–2 mm long; margins plane, slightly recurved at base, entire; costa strong, excurrent in a short stout point, 0.16–0.22 mm long; upper cells much smaller than basal cells, quadrate to short-rectangular, 13–19 × 19–32 µm, smooth. *Autoicous. Setae* up to 1.5 mm long, straight; capsules ± exserted, as long as or slightly shorter than setae, subglobose to ovoid, with a ± oblique conical beak, brown; operculum not differentiated nor rudimentary (see the discussion below). *Spores* 32–38 µm in diameter, strongly papillose; baculate or clavate, processes often with lobed apex (seen with SEM). *Calyptrae* smooth, cucullate.

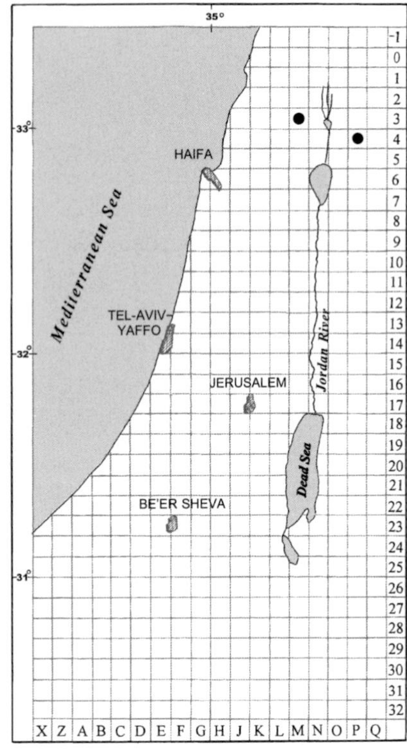

Map 100: *Phascum galilaeum*

Habitat and Local Distribution. — Plants growing solitary, gregarious, or in tufts, on very moist dark soils of basaltic origin (found growing together with *Phascum cuspidatum* var. *cuspidatum*). Very rare (known only as an endemic species): Upper Galilee and Golan Heights.

The above description is based on a study of a larger number of plants (though from the same two populations) and is somewhat modified compared to the original diagnosis of the species (Herrnstadt & Heyn 1993).

Phascum galilaeum resembles *P. cuspidatum* in gametophytic characters, but differs in the longer setae and the subglobose obliquely beaked capsules.

Phascum galilaeum resembles the cleistocarpous *Pottia bryoides* (Dicks.) Smith (which has not been found locally) in the exserted capsules, but the capsules of *P. bryoides* are borne on a much longer seta and are recorded as having a peristome (at least rudimentary) attached to the inner side of the operculum (e.g., by Nyholm 1956, Smith 1978). In *P. galilaeum* there is no peristome. However, in some of the capsules two–three rows of cells at the base of the beak may differ in shape from the other cells. *Phascum galilaeum* and *Pottia bryoides* form perhaps a link between the genera *Phascum* and *Pottia*, together with *Phascum longipes* J. Guerra, J. J. Martínez & R. M. Ros described from Spain (Guerra *et al.* 1990).

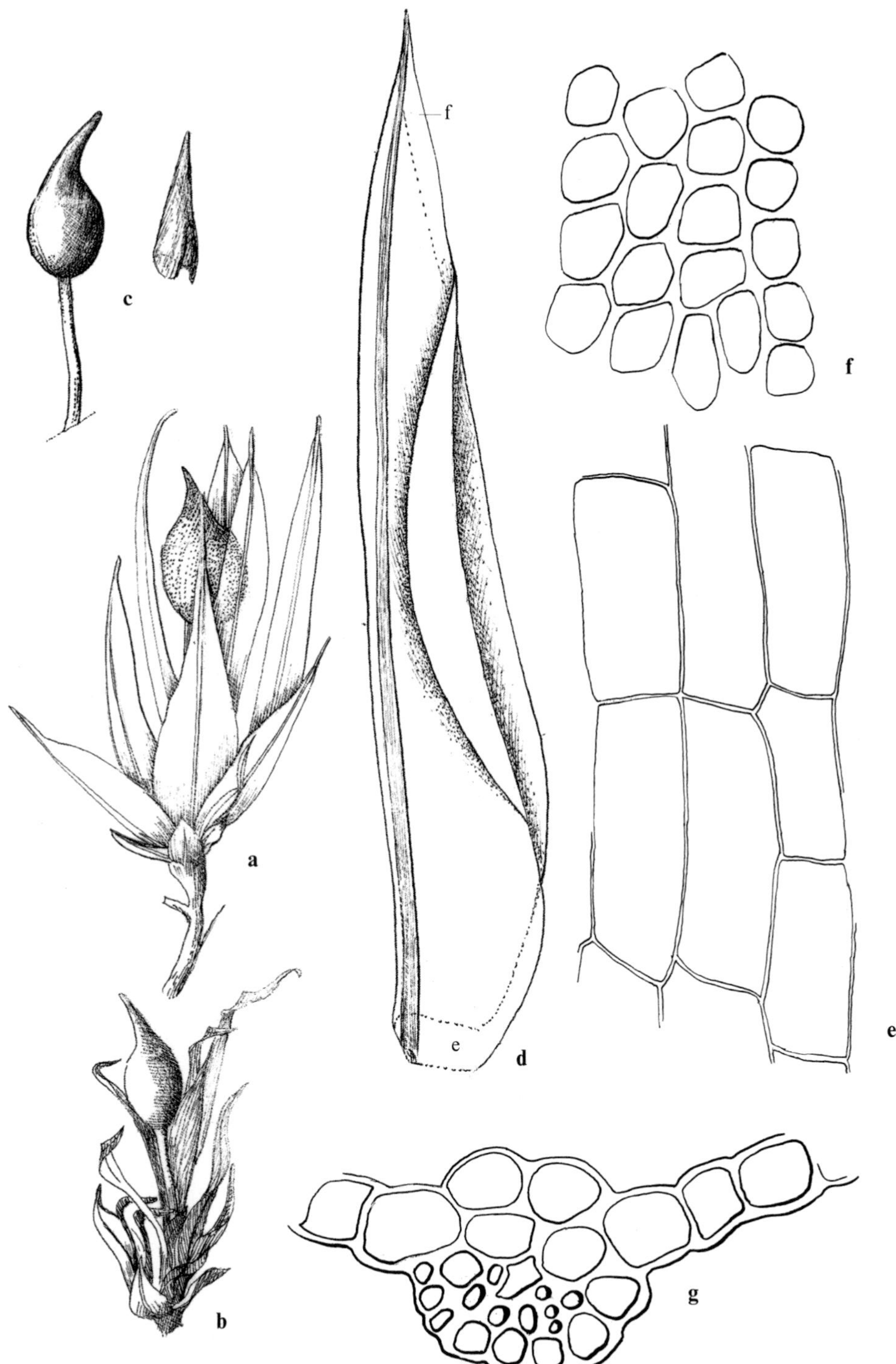

Figure 97. *Phascum galilaeum*: **a** habit, moist (×20); **b** habit, dry (×20); **c** sporophyte and calyptra (×50); **d** leaf (×50); **e** basal cells; **f** upper cells (all leaf cells ×500); **g** cross-section of costa and portions of lamina (×500). (*Herrnstadt 79-93-1* – holotype)

3. Phascum curvicolle Ehrh. ex Hedw., Sp. Musc. Frond. 21 (1801).
[Bilewsky (1965):373 (as "*Phascum curvicollum*" Ehrh.)]
Pottiella curvicollis (Hedw.) Gams, Krypt. Fl. Mitteleur., Edition 2, 1:101 (1948); *Microbryum curvicolle* (Hedw.) R. H. Zander, Genera of the Pottiaceae 240 (1993).

Distribution map 101, Figure 98, Plate VI:a.

Plants minute, 0.5–1.5 mm high, yellowish-brown. *Leaves.* Upper leaves up to 1 mm long, erectopatent when moist, bent at their tips when dry, lanceolate, acuminate; lower leaves smaller, ovate-lanceolate, acuminate; margins entire or crenulate above, recurved; costa excurrent in a stout point, 0.1–0.25 mm long; upper cells irregularly rounded to quadrate, 8–13 μm wide, densely papillose. *Paroicous. Setae* 0.5–1.0 mm long, arcuate or cygneous; capsules horizontal to pendulous, laterally exserted, 0.4–0.8 mm long, ovoid, with an oblique obtuse apiculus, brown. *Spores* about 27 μm in diameter, smooth; nearly smooth to slightly granulate (seen with SEM). *Calyptrae* cucullate.

Habitat and Local Distribution. — Plants gregarious, growing on exposed soil. Rare: Upper Galilee, Judean Mountains, and Golan Heights.

General Distribution. — Recorded from Southwest Asia (Turkey), around the Mediterranean (few scattered records, including Minorca), Europe (few and scattered reliable records), and North Africa.

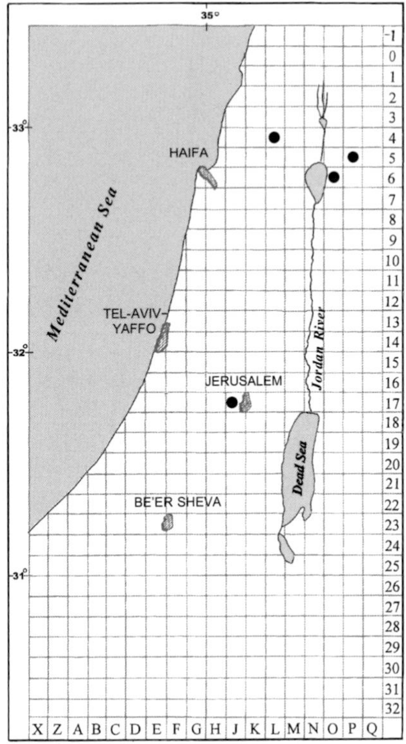

Map 101: *Phascum curvicolle*

Sterile plants of *Phascum curvicolle* may be easily mistaken for those of some species of *Pottia*.

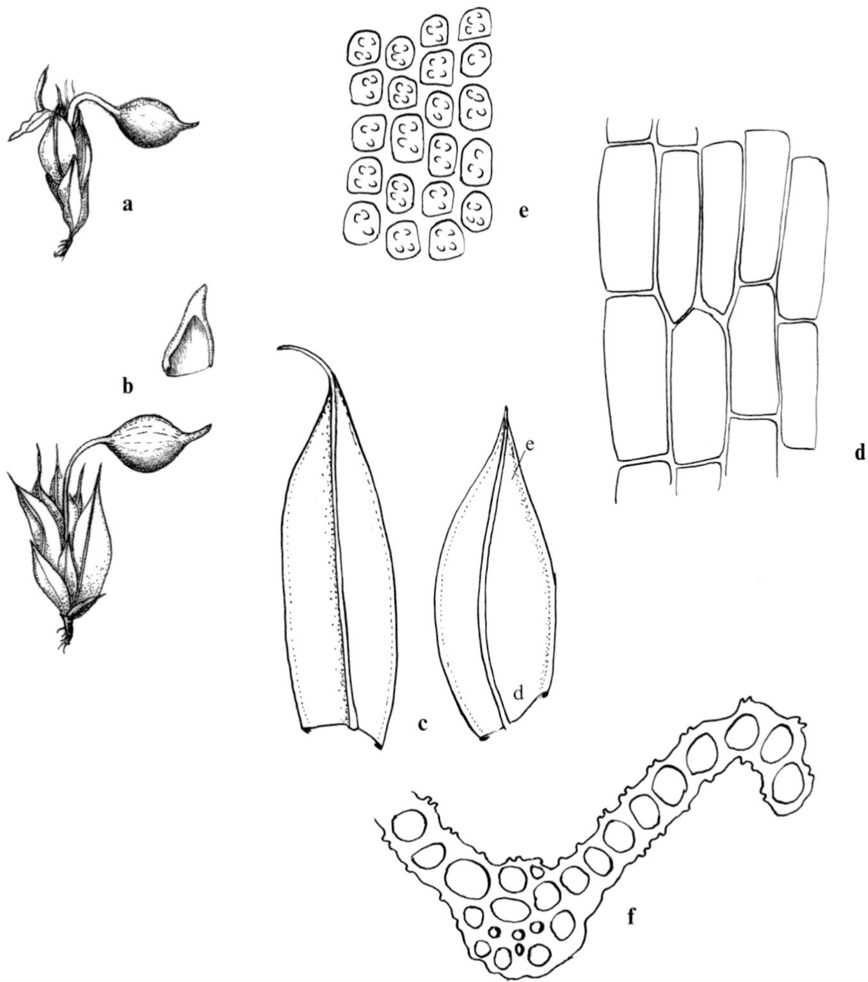

Figure 98. *Phascum curvicolle*: **a** habit, dry (×20); **b** habit and calyptra, moist (×20); **c** leaves (×50); **d** basal cells; **e** upper cells (all leaf cells ×500); **f** cross-section of costa and portion of lamina (×500). (Jerusalem, *Kushnir*, *sine dato*, No. 47 in Herb. Bilewsky)

4. Phascum floerkeanum F. Weber & D. Mohr, Bot. Taschenb. 70:45 (1807).
Acaulon floerkeanum (F. Weber & D. Mohr) Müll. Hal., Bot. Zeitung 5:99 (1847); *Microbryum floerkeanum* (F. Weber & D. Mohr) Schimp., Syn. Musc. Eur. 11 (1860).

Distribution map 102, Figure 99, Plate VI: b.

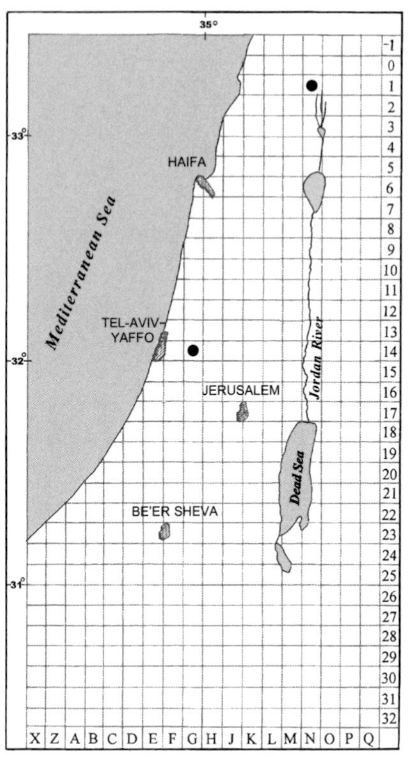

Map 102: *Phascum floerkeanum*

Plants minute, 0.7–1.5 mm high, reddish-brown. *Leaves* up to 1 mm long, erectopatent when moist, ± contorted when dry, concave, ovate, acuminate; margins plane, sometimes narrowly recurved in upper part, papillate from below mid-leaf upwards weakly bordered, except at base and apex, by narrow-rectangular cells, more thick-walled and less papillose than other cells of lamina; costa turning reddish-brown, very strong above, excurrent in a stout point, 0.15–0.17(–0.27) mm long; upper cells quadrate to rhomboidal, 10–15 µm wide, papillose at back. *Paroicous. Setae* very short, 0.1–0.2 mm, stout, straight; capsules immersed, 0.3–0.5 mm, globose, with a minute blunt apiculus, brown. *Spores* 19–24 µm in diameter, nearly smooth; densely granulate, granules sometimes joint in groups (seen with SEM). *Calyptrae* mitriform with entire margins or subcucullate.

Habitat and Local Distribution. — Plants gregarious or growing in small groups; so far found only in two localities on bare calcareous soil. Very rare: Sharon Plain and Upper Galilee.

General Distribution. — Records few: Southwest Asia (Turkey), scattered around the Mediterranean (mainly the centre and north), North Africa, and North America.

Phascum floerkeanum is easily distinguished from other *Phascum* species in the local flora by the reddish-brown colour and the concave leaves with a dark red excurrent costa. This species is often considered as occupying an intermediate position between *Acaulon* and *Phascum* (e.g., Mönkemeyer 1927, Scott & Stone 1976, Magill 1981). For that reason, the genus *Microbryum* was established by Schimper in order to accommodate *P. floerkeanum. Microbryum* is sometimes maintained on subgeneric level and was reinstated more recently as a genus by Zander (1989) including several species.

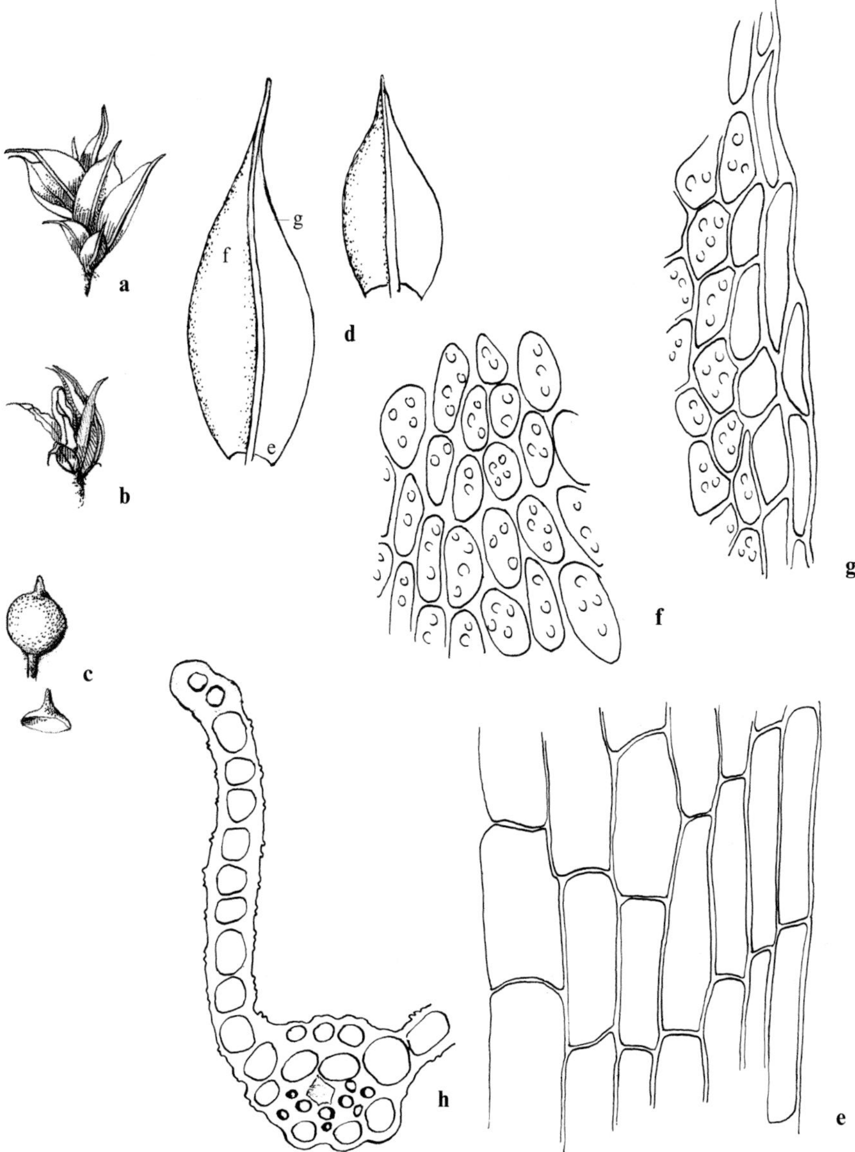

Figure 99. *Phascum floerkeanum*: **a** habit, moist (×20); **b** habit, dry (×20); **c** sporophyte and calyptra (×20); **d** leaves (×50); **e** basal cells; **f** upper cells; **g** marginal upper cells (all leaf cells ×500); **h** cross-section of costa and portion of lamina (×500). (*Herrnstadt* & *Crosby 78-30-1*)

24. *Acaulon* Müll. Hal.

Plants minute, bud-like, terrestrial, gregarious or scattered. *Stems* very short, simple or forked, without a central strand. *Leaves* few, crowded, imbricate, usually appressed when dry, erect to spreading when moist, concave, sometimes keeled, usually broadly oblong-ovate, acute or acuminate, cuspidate to piliferous; margins plane or slightly recurved, entire or irregularly dentate above; costa subpercurrent to percurrent or excurrent, with a single dorsal stereid band; basal cells rectangular, smooth; upper cells quadrate, hexagonal or rhomboidal, usually smooth. *Dioicous* or monoicous. *Setae* very short, straight or curved; capsules immersed, cleistocarpous, without apiculus or sometimes with minute apiculus; peristome not formed. *Spores* papillose to spinulose. *Calyptrae* minute, conical.

A small genus of ca. 15 species. In the local flora, three species occur in ephemeral habitats, often growing together with other terrestrial Pottiaceae, especially *Phascum* and *Pottia*.

Acaulon and *Phascum* are two closely related genera. Their delimitation often poses problems (as in the floras of Australia and South Africa — see Magill 1981 and Stone 1989 respectively) and consequently, generic concepts differ. For example, subgen. *Acaulonopsis* I. G. Stone, described as a subgen. of *Phascum* (Stone 1989), was included by Zander (1993) in *Acaulon*. The main diagnostic generic characters adopted here are for *Acaulon*: the budlike habit of the dioicous, rarely monoicous, plants, the usually plane leaf margins, the percurrent or often shortly excurrent costa, smooth lamina cells, capsules without apiculus or with minute apiculus, and tiny conical calyptrae. Conversely, plants of *Phascum* are considered as monoicous, have usually recurved leaf margins, excurrent costa ending in a cuspidate point, usually papillose lamina cells, capsules with longer apiculus, and small cucullate calyptrae.

1 Plants triangular when viewed from above; leaves strongly keeled (at least three uppermost); setae arcuate **3. A. triquetrum**
1 Plants rounded when viewed from above; leaves not keeled; setae straight.
  2 Plants up to 1.5 mm; leaves broad-ovate to broad-elliptic, margins ± dentate near apex; costa subpercurrent to shortly excurrent; mature capsules concealed among leaves **1. A. muticum**
  2 Plants up to 2.5(–3) mm; leaves ovate-lanceolate to oblong-lanceolate, margins entire; costa excurrent in a long apiculus; mature capsules not or only partly concealed among leaves **2. A. longifolium**

1. Acaulon muticum (Hedw.) Müll. Hal., Bot. Zeitung (Berlin) 5:99 (1847).
Phascum muticum Hedw., Sp. Musc. Frond. 23 (1801).

Distribution map 103 (p. 264), Figure 100, Plate VI:c, d.

Plants minute, up to 1.5 mm high, yellowish-brown. *Leaves* 0.5–1.5 mm long, strongly concave, not keeled, not convolute when dry, broad-ovate to broad-elliptic, acute or obtuse; margins plane, entire below, ± dentate or rarely incised near apex; costa subpercurrent to shortly excurrent into apiculus ca. 110 µm long; basal cells rectangular, four–seven times as long as wide; upper cells subquadrate to rhomboidal, one–three

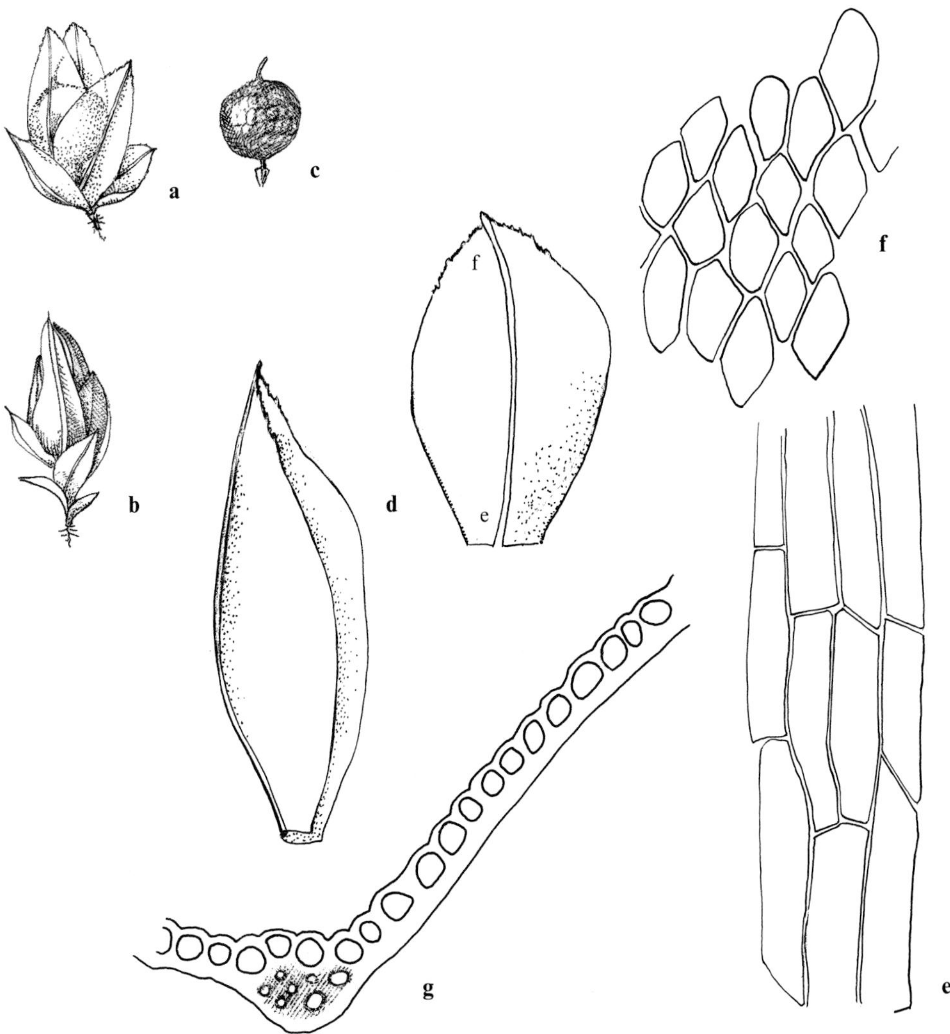

Figure 100. *Acaulon muticum*: **a** habit, moist (×20); **b** habit, dry (×20); **c** sporophyte (×20); **d** leaves (×50); **e** basal cells; **f** upper cells (all leaf cells ×500); **g** cross-section of costa and portions of lamina (×500). (*Kushnir*, 21 Mar. 1943 – mixed sample with *Acaulon longifolium*)

times as long as wide, short-rhomboidal towards upper margins and apex; mid-leaf cells 13–20 µm wide. *Dioicous. Setae* minute, straight; capsules ± 0.5 mm long, globose, minutely apiculate, red-brown, concealed among leaves when mature. *Spores* 27–35 µm in diameter; baculate, bacula profusely and irregularly lobed at apex (seen with SEM) (cf. Herrnstadt & Heyn 1989).

Habitat and Local Distribution. — Plants gregarious or scattered, growing on exposed or shaded wet sand or sandy loam, in one locality found together with *Acaulon longifolium.* Very rare: Sharon Plain.

General Distribution. — Recorded from Southwest Asia (Turkey), around the Mediterranean, Europe (throughout), Macaronesia, North Africa, Madagascar, northern Asia, and North America.

Bilewsky (1965, p. 372) lists *Acaulon muticum* var. *cuspidatum* (= *A. triquetrum*) as the only local taxon of *Acaulon.*

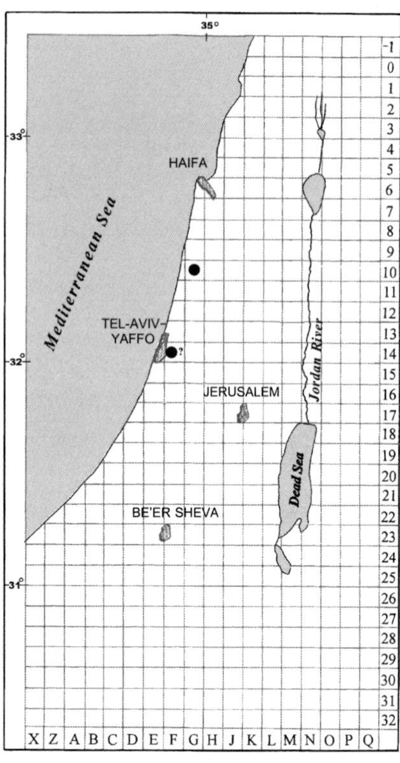

Map 103: *Acaulon muticum*

2. Acaulon longifolium Herrnst. & Heyn, Nova Hedwigia 69 : 229 (1991).

Distribution map 104, Figure 101, Plate VI : e, f.

Plants up to 2.5(–3) mm high, yellowish-green. *Stems* often forked. *Leaves* up to 2 mm long, concave, not keeled, ± convolute when dry, erect-spreading when dry and moist, ovate-lanceolate to oblong-lanceolate, long-acute; margins plane, entire; costa excurrent into a long apiculus, up to 275 µm long; in cross-section with large thin-walled ventral cells and hydroids; cells often with thicker dorsal than ventral walls; basal cells rectangular, three times as long as wide; upper cells irregularly quadrate, smaller, rectangular near margins, marginal cells up to 1.5 times as long as wide, mid-leaf cells 13–21 µm wide. *Dioicous. Setae* thin, minute, straight; capsules 0.3–0.5 mm long, globose, minutely apiculate, brown, not or only partly concealed among leaves when mature. *Spores* 22–26 µm in diameter, spinulose; baculate to spinulose (seen with SEM).

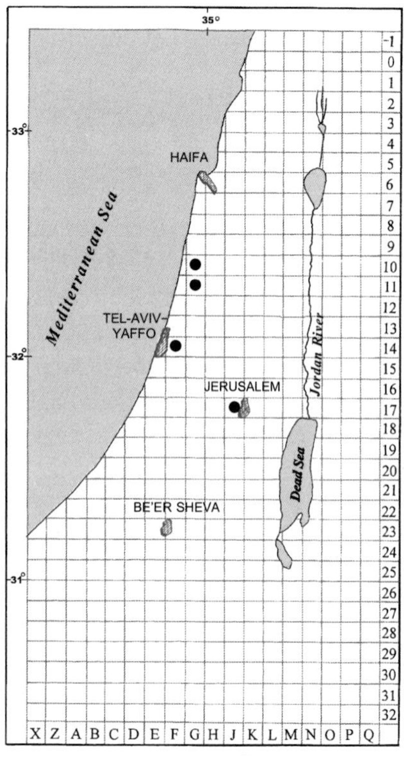

Map 104: *Acaulon longifolium*

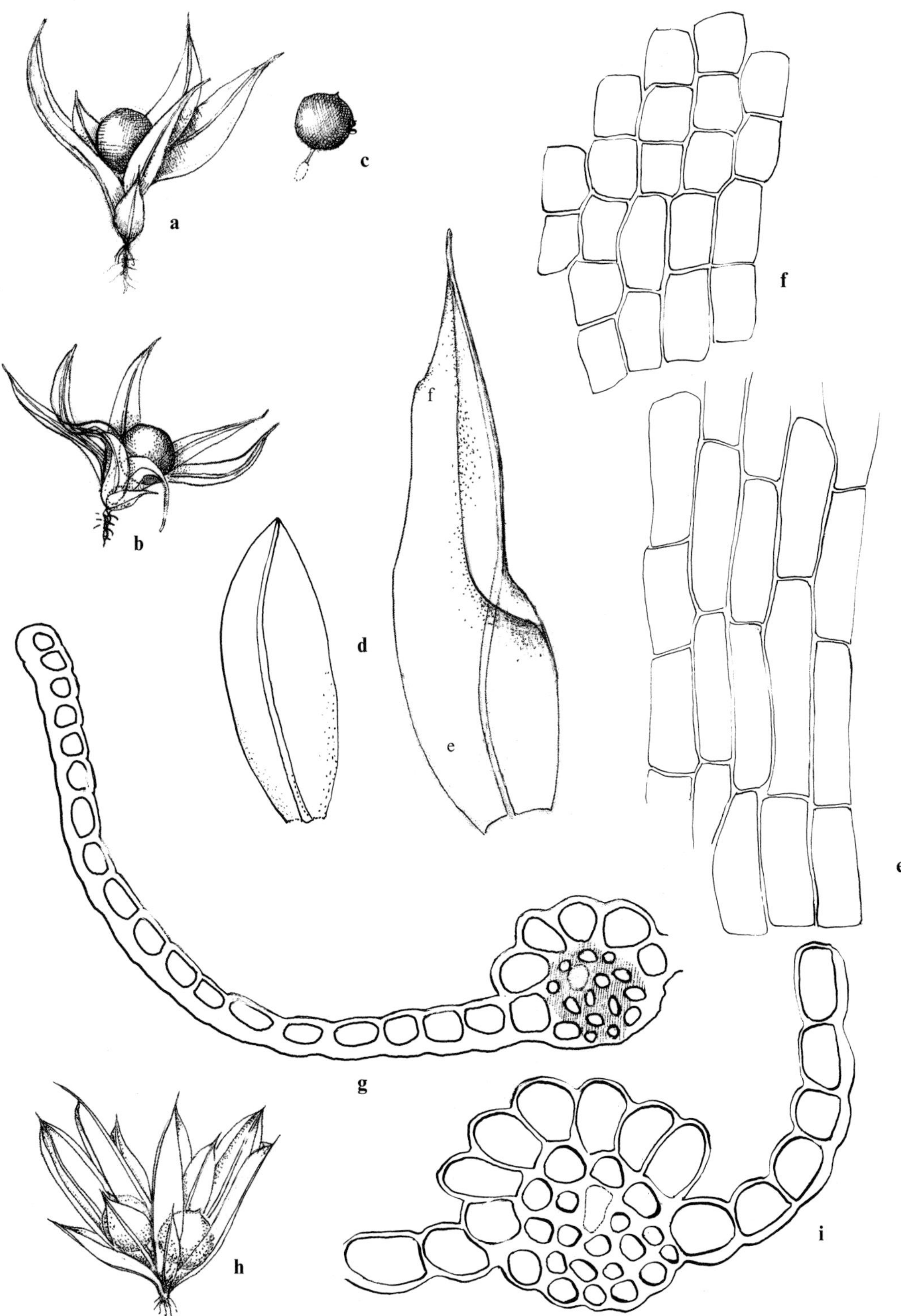

Figure 101. *Acaulon longifolium*: **a, b** habit, moist (×20); **c** sporophyte (×20); **d** leaves (×50); **e** basal cells; **f** upper cells (all leaf cells ×500); **g** cross-section of costa and portion of lamina (×500); **h** habit (×10); **i** cross-section of costa and portion of lamina. (**a**–**g**: *Kushnir*, 21 Mar. 1943 – mixed sample with *Acaulon muticum*; **h**, **i**: *Herrnstadt*, 15 Feb. 1992)

Habitat and Local Distribution. — Plants gregarious or growing in tufts on exposed, wet sandy loam or alluvial soils, in shade, also found growing together with *Acaulon muticum* and *Astomum crispum.* Rare to occasional: Sharon Plain, Philistean Plain, and Judean Mountains.

The specimen from the Judean Mountains, Jerusalem (Mar. 1943, *Zohary*, no. 45 in Herb. Bilewsky) was determined by Bilewsky as "*Acaulon muticum* var. *cuspidatum* Amann."

No records of this taxon from outside Israel have been recorded so far.

Acaulon longifolium was previously recorded from Israel as *Acaulon mediterraneum* (Herrnstadt *et al.* 1982) and as *A. minus* (Hook. & Taylor) A. Jaeger (Herrnstadt *et al.* 1991). A comparison between those species is provided in Herrnstadt & Heyn (1999).

Plants from the four collections examined differ from local *Acaulon muticum* in their larger size and larger, erect-spreading leaves with entire margins and excurrent costa, and mature capsules, which are not or only partly concealed by the leaves. In addition, the spores are spiny, and not bearing lobed bacula.

Spinulose spores have also been described in another species of *Acaulon*: *A. mediterraneum* Limpr. (Laubm. Deutschl. 1:180. 1885). According to the original description, plants of *A. mediterraneum* are smaller than of *A. muticum*, possess untoothed leaves, not entirely covering the immersed capsules, and spinulose spores (Limpricht considered *A. muticum* as having finely verrucose spores). Corley *et al.* (1981) and Hill (1982) included *A. mediterraneum* together with another species — *A. minus* — in *A. muticum*. Other authors, e.g., Sérgio (1972), treat *A. mediterraneum* as a variety of *A. muticum*, a concept also adopted by Casas *et al.* (1990) and Casas (1991).

There is no doubt that *Acaulon mediterraneum* and *A. longifolium* are two different taxa, though they both have spinulose spores. Some of the characters of *A. longifolium* resemble those of at least some species of *Phascum* subgen. *Acaulonopsis* (Stone 1989) in the larger plant size, somewhat spreading leaves, long-excurrent costa with large, thin-walled ventral cells and hydroids, and the lamina cells that often have thicker dorsal than ventral walls. In other characters, e.g., the plane margins and the smooth lamina cells, *A. longifolium* cannot be affiliated with subgen. *Acaulonopsis*.

3. Acaulon triquetrum (Spruce) Müll. Hal., Bot. Zeitung (Berlin) 5:100 (1847).
[Bilewsky (1965) as "*Acaulon muticum* (Schreb.) C. M. var. *cuspidatum* Schpr."]
Phascum triquetrum Spruce, London J. Bot. 4:189 (1845).

Distribution map 105 (p. 268), Figure 102, Plate VI:g, h.

Plants minute, up to 1.5 mm high, yellowish-brown. *Leaves* up to 1.5 mm long, concave, three uppermost leaves most strongly keeled, making plants appear triangular when viewed from above, ovate, acuminate; lower leaves ca. 0.5 mm long, acuminate; margins plane or slightly recurved in upper part, entire or weakly dentate near apex, apex usually reflexed when dry; costa strong, excurrent into a ± reflexed apiculus, 110–250(–450) μm long, in lower leaves only shortly excurrent; basal cells rectangular, up to five times as long as wide; upper cells subquadrate to rhomboidal, one–three times as long as wide, mid-leaf cells 13–22 μm wide. *Dioicous. Setae* up to 0.5 mm long, ca. ⅔ the length of capsule, thin and fragile, arcuate or cygneous; capsules

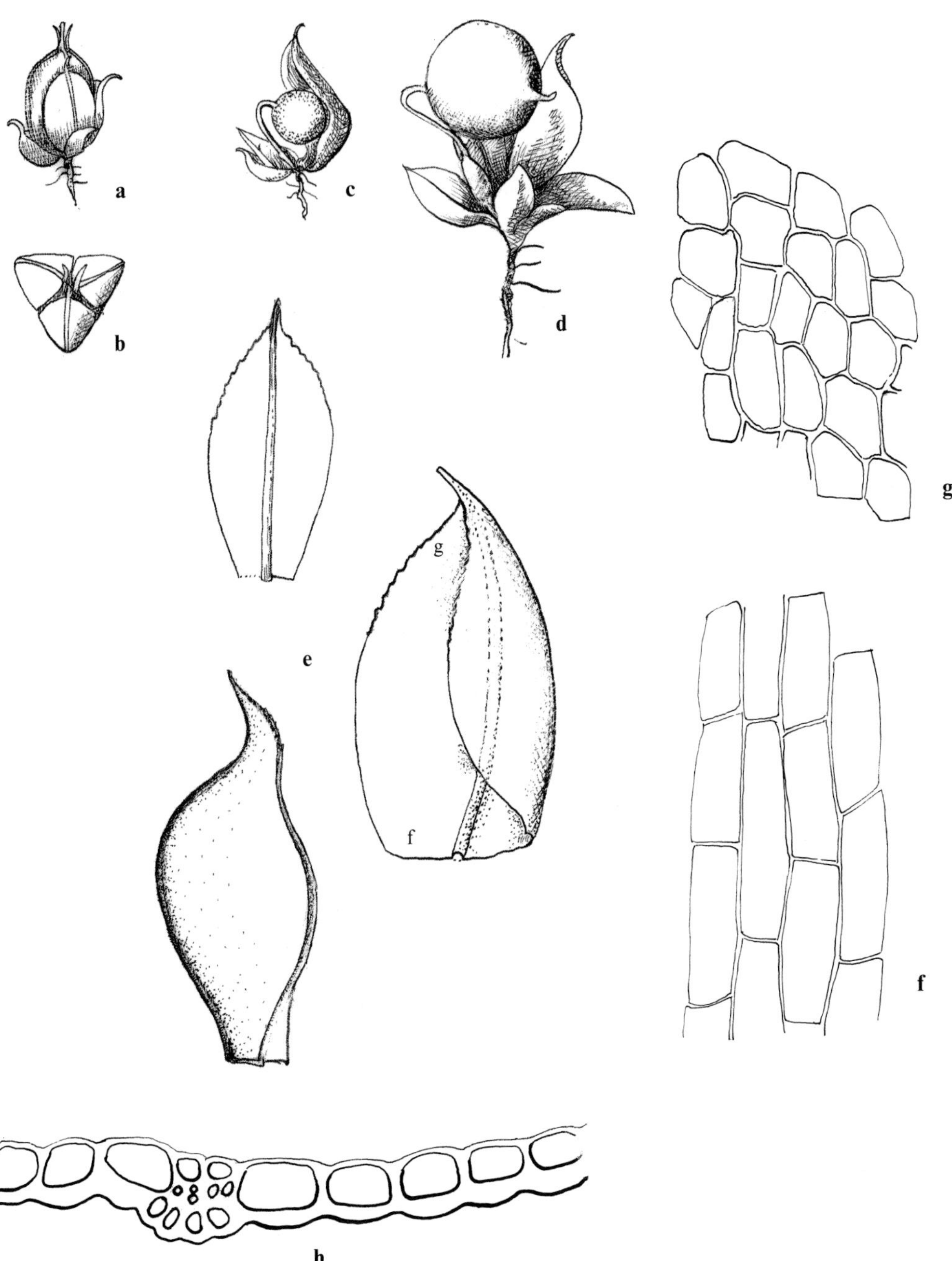

Figure 102. *Acaulon triquetrum*: **a** habit, dry (×20); **b** habit as seen from above, dry (×20); **c** habit, moist (×20); **d** habit, moist (×40); **e** leaves (×50); **f** basal cells; **g** upper cells (all leaf cells ×500); **h** cross-section of costa and portions of lamina (×500). (*Heyn & Herrnstadt 80-136-1*)

immersed, ca. 0.75 mm long, globose, minutely apiculate. *Spores* ± 32 µm in diameter; baculate, baculae large, some lobed at apex (seen with SEM).

Habitat and Local Distribution. — Plants gregarious or scattered, growing on various kinds of exposed soils, mainly under conditions of high temperatures and low rainfall. Occasional (the most common species of *Acaulon* in the local flora): Judean Mountains, Judean Desert, Northern Negev, Central Negev, Bet She'an Valley, and Lower Jordan Valley.

General Distribution. — Recorded from Southwest Asia (Turkey and Iraq), around the Mediterranean (few records), Europe, Macaronesia (in part), North Africa, Australia and North America.

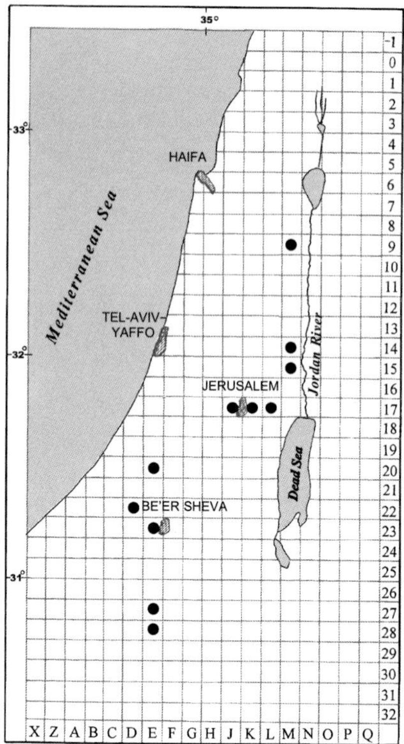

Map 105: *Acaulon triquetrum*

Acaulon triquetrum is easily distinguished from the other local species of the genus by the triangular shape of the plant as seen from above, and by the arcuate to cygneous seta.

Bilewsky (1965) recorded this taxon (as "*Acaulon muticum* (Schreb.) C. M. var. *cuspidatum* Schpr.") from two districts: Samaria and the Judean Mountains. The plants of the single collection bearing that name from the Judean Mountains (Jerusalem) have been determined as *Acaulon longifolium.*

CINCLIDOTACEAE

Plants differ from those of the Pottiaceae in being robust, aquatic, and cladocarpous. *Leaves* with strongly thickened leaf margins, formed by several cell layers.

The family, comprising the single genus *Cinclidotus*, is often considered as part of the Pottiaceae. Zander (1993) suggested that *Cinclidotus* could be considered as a taxon bridging between the Pottiaceae and the Grimmiaceae.

Cinclidotus P. Beauv.

Plants cladocarpous, dark green to blackish, robust, usually aquatic, rarely growing in periodically aquatic habitats (*Cinclidotus mucronatus*), usually attached to calcareous rocks and boulders by smooth, thick, reddish-brown rhizoids. *Stems* fairly rigid, dichotomously or fastigiately branched, without a central strand, with cells becoming incrassate towards the outside. *Leaves* erect or flexuose when dry, erectopatent to spreading when moist, linear-lanceolate to lingulate, apex acute or obtuse; margins usually plane (sometimes recurved in *C. mucronatus*), thickened by 2–16 layers of marginal cells; costa strong, ending in apex or shortly excurrent (long-excurrent in *C. pachyloma*), in cross section with two stereid bands; cells incrassate; basal cells rectangular, usually restricted to few cell layers only; mid- and upper leaf cells irregularly quadrate to hexagonal, ± more elongate towards margins; old leaves sometimes eroded to costa. *Dioicous* usually. *Setae* short or long; capsules immersed or exserted; peristome single, consisting of sixteen, usually filiform, teeth.

A small genus comprising ten, mainly Mediterranean species. Five species occur in the local flora, of which four share the same habitats in the northern part of Israel.

1 Plants periodically submerged in running water; upper leaf cells distinctly papillose **1. C. mucronatus**

1 Plants always submerged in running water; upper leaf cells smooth or weakly papillose.

  2 Plants up to 60 cm high; leaves erect to distinctly secund, costa at base about ⅓ the width of leaf **4. C. aquaticus**

  2 Plants up to 8(–10) cm high; leaves weakly secund or not secund when dry, costa at base less than ⅓ the width of leaf.

    3 Leaf margins with eight–sixteen layers of cells **5. C. pachyloma**

    3 Leaf margins with less than six layers of cells.

      4 Leaves crowded along stem, ± rigid and concave, not flexuose when dry, oblong lanceolate to lingulate; margins with two(–three) layers of cells **3. C. riparius**

      4 Leaves laxly spaced along stem, soft, not concave, flexuose when dry, oblong-lanceolate to narrow-lanceolate; margins usually with four layers of cells **2. C. fontinaloides**

1. Cinclidotus mucronatus (Brid.) Guim., Cat. Descr. Briol. Port. 57 (1919).
Barbula mucronata Brid., Muscol. Recent., Suppl. 1:268 (1806); *Dialytrichia mucronata* (Brid.) Broth., Nat. Pflanzenfam. 1:412 (1902).

Distribution map 106, Figure 103.

Plants up to 5 cm high, dark green. *Stems* sparsely branched, often repeatedly dichotomously branched. *Leaves* about 3 mm long, flexuose when dry, erectopatent when moist, oblong-lanceolate to broadly lingulate, obtuse, shortly mucronate; margins plane or recurved usually on one side, papillose-crenulate, with two–three layers of cells; costa less than ⅓ the width of leaf, excurrent into a mucro; cells ± incrassate; basal cells subquadrate to rectangular, smooth; upper cells irregularly hexagonal, distinctly papillose; mid-leaf cells 8–12 µm wide. *Dioicous*; archegonia on terminal branches, antheridia at apex of short side-branches. *Sporophytes* not seen in local populations.

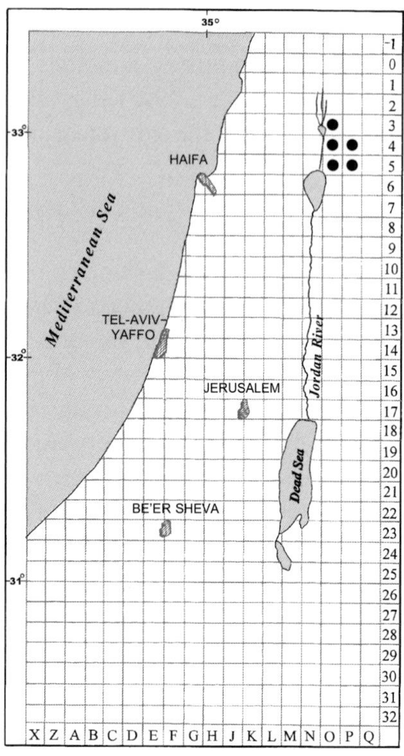

Map 106: *Cinclidotus mucronatus*

Habitat and Local Distribution. — Plants growing in tufts attached to basalt rocks, periodically submerged in running water in the rainy season and exposed in the summer, restricted to wadis. Occasional: southern Golan Heights.

General Distribution. — Recorded from Southwest Asia (Turkey and Lebanon), around the Mediterranean, Europe (mainly the south), Macaronesia, North Africa, and eastern Asia.

Leaves of submerged parts are sometimes eroded to costa. In one population (Golan Heights: Nahal Samakh) stems are tomentose and the rhizoids bear abundantly brown gemmae (Figure 103:h).

The flexuous leaves of dry plants of *Cinclidotus mucronatus* resemble those of *C. fontinaloides* (Hedw.) P. Beauv. When moist, the differences in the leaves can be more easily observed: a mucronate — not acute — apex, papillose upper cells, and thinner margins. Plants of *C. mucronatus* are the only ones among local *Cinclidotus* that grow in periodically dry habitats.

Figure 103. *Cinclidotus mucronatus*: **a** habit (×1); **b** part of habit, moist (×10); **c** part of habit, dry (×10); **d** leaves (×50); **e** basal cells; **f** upper cells (all leaf cells ×500); **g** cross-section of costa and portion of lamina (×500); **h** rhizoidal gemma (×150). (*D. Kaplan*, 11 May 1987)

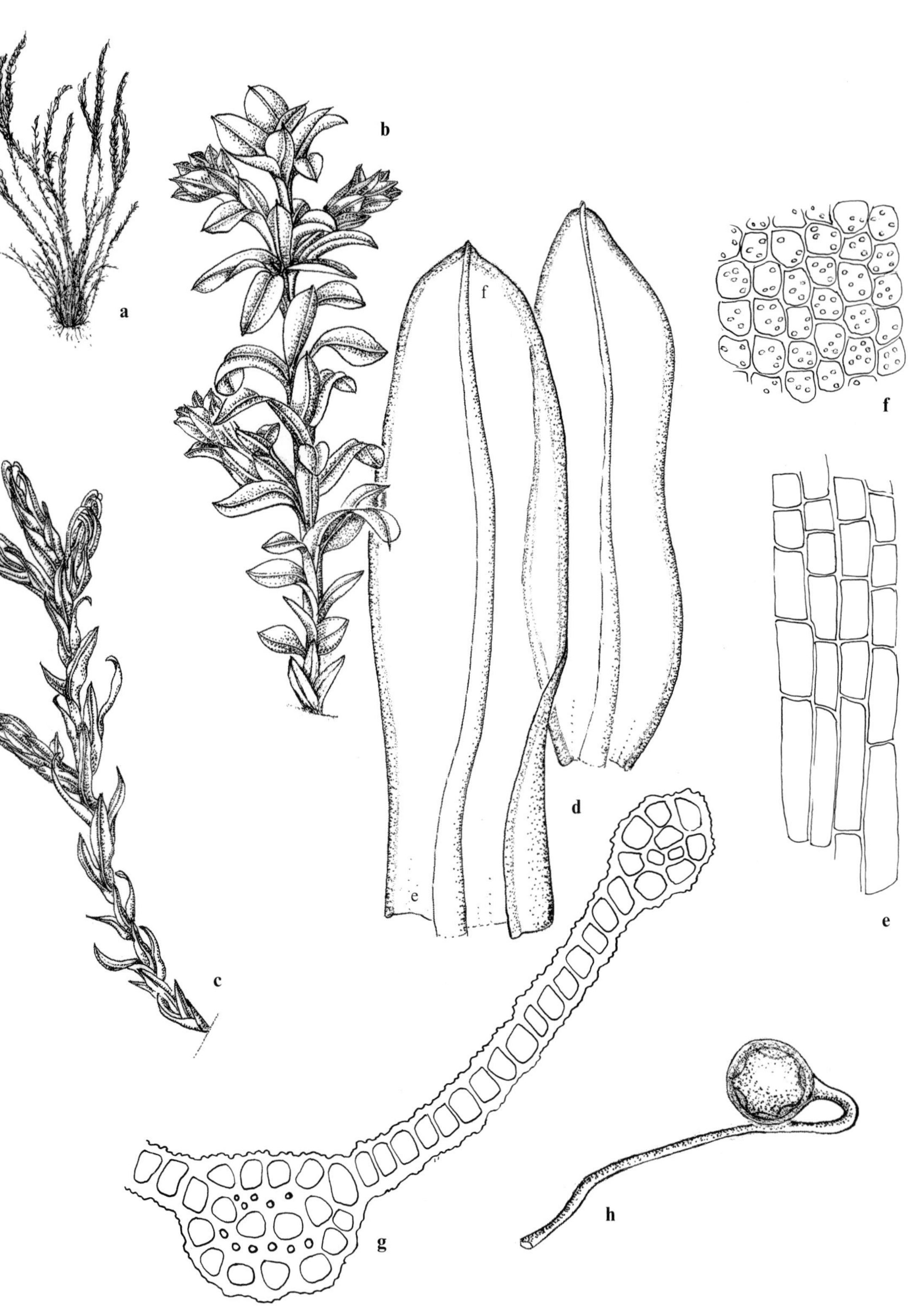
b
a
f
d
f
e
e
c
g
h

2. Cinclidotus fontinaloides (Hedw.) P. Beauv., Prodr. 52 (1805).
Trichostomum fontinaloides Hedw., Sp. Musc. Frond. 114 (1801).

Distribution map 107, Figure 104.

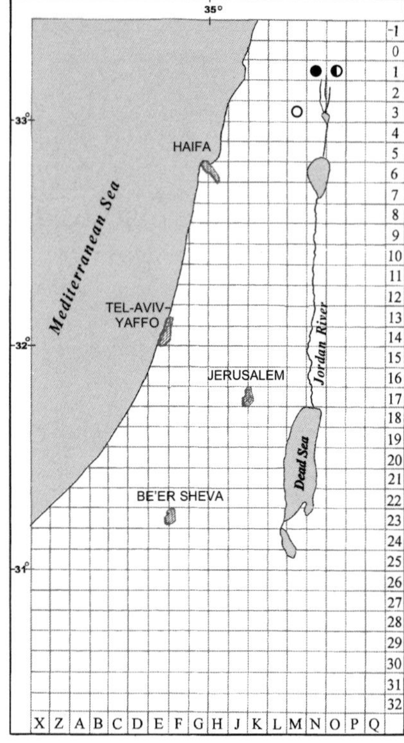

Map 107: ● *Cinclidotus fontinaloides*
○ *C. riparius*
◐ *C. fontinaloides* + *C. riparius*

Plants 4–8 cm high, green in young parts, blackish-green below. *Stems* densely and fastigiately branched from near base. *Leaves* laxly spaced along stem, 2.5–3.5 mm long, soft, flexuose when dry, patent to spreading when moist, oblong-lanceolate to narrow-lanceolate, acute; margins plane, usually with four layers of cells, entire or irregularly denticulate near apex; costa less than ⅓ width of leaf, percurrent or shortly excurrent; basal cells rectangular, smooth, restricted to extreme base, followed by irregularly quadrate cells; upper cells hexagonal to irregularly quadrate, smooth or weakly papillose; mid-leaf cells 10–16 µm wide. *Capsules* sessile, immersed (a single capsule observed in one population — Dan Valley: Ein el Bard).

Habitat and Local Distribution. — Plants submerged in running water, attached to rocks and boulders. Very rare, locally abundant: Dan Valley (Dan River) and Golan Heights (Banyas River).

General Distribution. — Recorded from Southwest Asia (Lebanon, Iraq, and Iran), scattered around the Mediterranean, Europe (many records throughout), Macaronesia, tropical and North Africa, and northeastern, eastern, and central Asia.

Cinclidotus fontinaloides is a most variable species. It differs from *C. riparius*, which it may resemble, in the laxly spaced, flexuose, soft, lanceolate, acute leaves. Sterile plants of one population (Dan Valley: Nukheila Nature Reserve) cannot be definitely identified as a compact form of either *C. fontinaloides* or as *C. riparius*. The leaf margins of *C. fontinaloides*, recorded as having up to six layers of cells, were found to have only up to four layers in local plants.

Figure 104. *Cinclidotus fontinaloides*: **a** habit (×1); **b** habit of upper part of plant, dry (×10); **c** part of habit, moist (×10); **d** leaf (×50); **e** basal cells; **f** mid-leaf cells, (all leaf cells ×500); **g** cross-section of costa and portion of lamina (×500). (*Herrnstadt* & *Hefez*, 18 Jun. 1987)

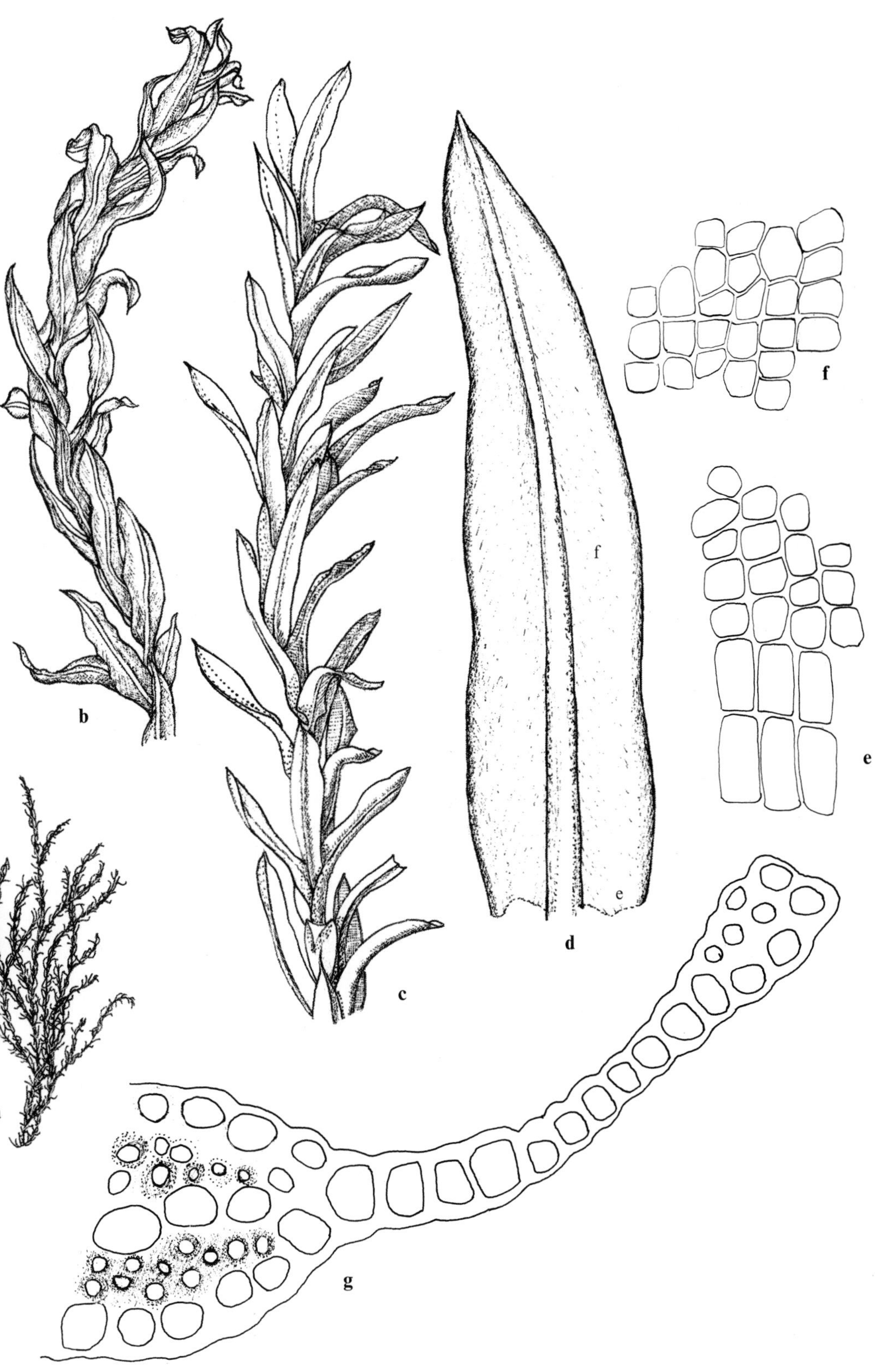
b
c
d
e
f
g
f
e

3. Cinclidotus riparius (Brid.) Arn, Mém. Soc. Linn. Paris 7 : 247 (1827).
Gymnostomum riparium Host ex Brid., J. Bot. (Schrader) 1800 : 274 (1801); *C. nigricans* (Brid.) Wijk & Margad., Buxbaumia 1 : 51 (1947).

Distribution map 107 (p. 272), Figure 105.

Plants up to 8(–10) cm high, green to dark green. *Stems* densely, fastigiately branched. *Leaves* crowded along stem, up to 3 mm long, erect, sometimes slightly contorted towards apex when dry, erectopatent to patent when moist, oblong-lanceolate to lingulate, ± rigid and concave; apex obtuse and apiculate or sometimes leaves abruptly narrowing into acute apex; margins plane, with two(–three) layers of cells; costa at base less than ⅓ width of leaf, ending in apex or shortly excurrent; basal cells short-rectangular, smooth, restricted to few layers at base; upper cells hexagonal to irregularly quadrate, smooth, or rarely weakly papillose; mid-leaf cells 9–12 µm wide. *Dioicous*; archegonia usually acrogynous. *Sporophytes* not seen in local populations.

Habitat and Local Distribution. — Plants immersed in running water in strong currents, growing in tufts attached to rocks and boulders. Rare, locally abundant: Upper Galilee and Golan Heights.

General Distribution. — Recorded from Southwest Asia (Turkey, Lebanon, and Iraq), around the Mediterranean, Europe (mainly the centre and south), North Africa, and central Asia.

Plants of *Cinclidotus riparius* may resemble *C. fontinaloides* (see the discussion there). *Cinclidotus danubicus* was recorded from Israel by Bilewsky (1970, 1977), but the collections on which the record was based are typical for *C. riparius*. *Cinclidotus danubicus* was described by Schiffner and Baumgartner (Oesterr. Bot. Z. 56 : 154. 1906) from plants without sporophytes, growing together with *C. riparius*. Although those authors pointed out the resemblance of their plants to *C. riparius*, several differential characters were considered by them as typical for *C. danubicus*: leaves gradually narrowing, acuminate, with thinner and narrower margins, cells usually larger, less incrassate outer cells of the costa, and archegonia in lateral position not acrogynous. One local collection (Dan Valley: Nukheila, *D. Kaplan*, 4 Nov. 1990) shows some of the characters attributed to *C. danubicus*: smaller and narrower leaves and wider (12–15 µm wide) mid-leaf cells (Figure 105 : h–m; "danubicus type"). It is difficult to decide whether *C. danubicus* may not merely represent part of the range of variability of *C. riparius*. Therefore, it is treated here as part of that species.

Figure 105. *Cinclidotus riparius*: **a** habit (×1); **b** part of habit, dry (×10); **c** part of habit, moist (×10); **d** leaf (×50); **e** basal cells; **f** upper cells; **g** cross-section of costa and portion of lamina (×500); **h** habit (×1); **i** part of habit, dry (×10); **k** leaf (×50); **l** basal cells; **m** upper cells (all leaf cells ×500). (**a**–**g**: Wadi Hindaj, *Bilewsky*, *sine dato*, No. 81 in Herb. Bilewsky; **h**–**m**: "danubicus type", *D. Kaplan*, 4 Nov. 1990)

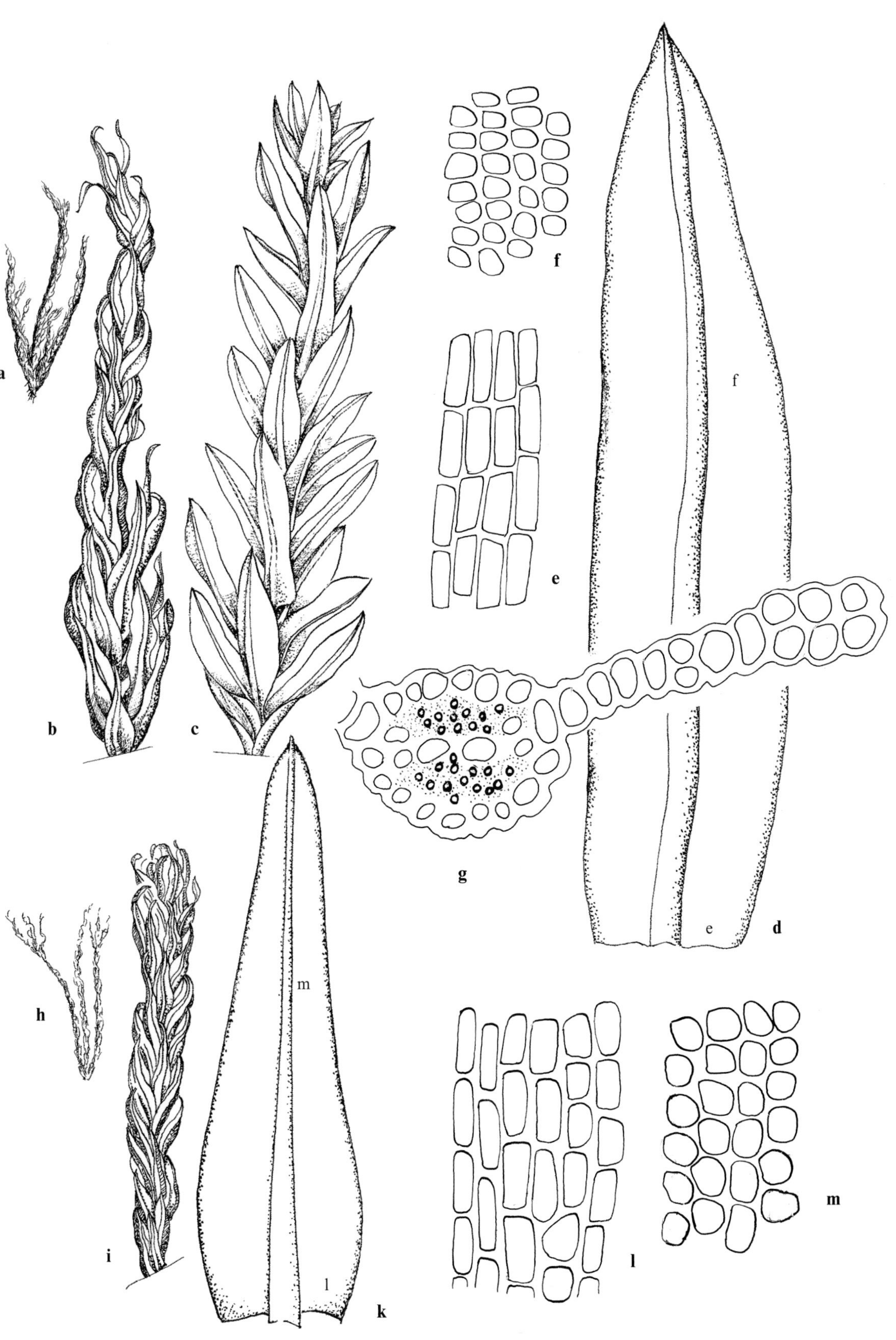
a
b
c
f
e
f
g
e
d
h
m
i
l
k
l
m

4. Cinclidotus aquaticus (Hedw.) B.S.G., Bryol Eur. 3 : 170 (1842).
Anictangium aquaticum Hedw., Sp. Musc. Frond. 41 (1801).

Distribution map 108, Figure 106.

Plants very tall, up to 60 cm high, dark green. *Stems* densely branched at base, dichotomously above. *Leaves* 3–4 mm long, appressed, imbricate, erect to distinctly secund when dry, erectopatent when moist, linear-lanceolate, tapering to acuminate apex; margins plane, with two(–three) layers of cells; costa strong, at base about ⅓ width of leaf, excurrent in a subula; cells incrassate, smooth, subquadrate, nearly uniform in size throughout leaf, sometimes rectangular towards margins; mid-leaf cells 8–11 μm wide. *Dioicous*; gametangia on short side branches. *Sporophytes* not seen in local populations.

Habitat and Local Distribution. — Plants submerged in running water, attached to calcareous rocks and boulders. Very rare, locally abundant: Dan Valley (in the Dan River) and Golan Heights (in the Banyas River).

General Distribution. — Records are few: Southwest Asia (Turkey and Lebanon), sparsely scattered around the Mediterranean, Europe, North Africa, and central Asia.

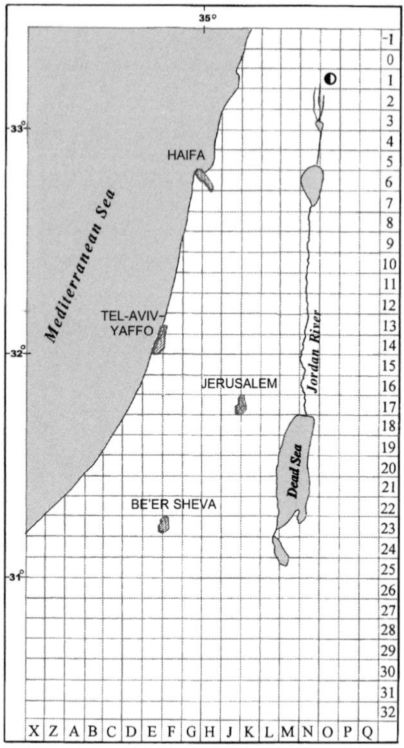

Map 108: ◐ *Cinclidotus aquaticus* + *C. pachyloma*

Plants of this species are the largest in the genus and among local mosses in general.

In one population on the Golan Heights (Banyas, *Kushnir*, 4 Oct. 1942), paraphyllia-like structures were observed among the perichaetial leaves, they are one to several cells wide, ± incrassate, sometimes branched (Figure 106 : i).

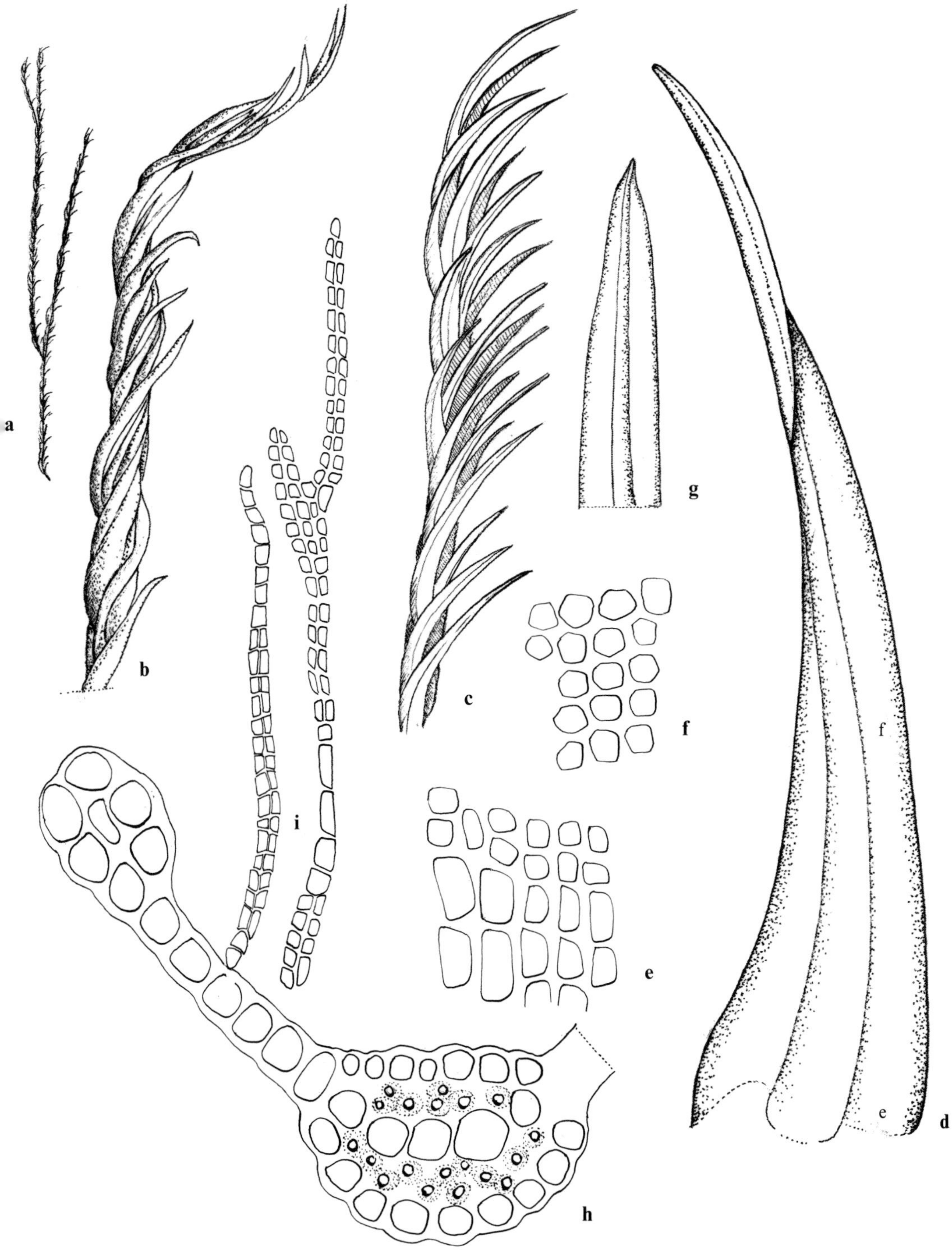

Figure 106. *Cinclidotus aquaticus*: **a** habit (×1); **b** habit of upper part of plant, dry (×10); **c** part of habit, moist (×10); **d** leaf (×50); **e** basal cells; **f** mid-leaf cells (all leaf cells ×500); **g** leaf apex (×50); **h** cross-section of costa and portion of lamina (×500); **i** paraphyllia-like structures (×125).
(**a**–**h**: *Lipkin*, 24 Jul. 1979; **i**: *Kushnir*, 4 Oct. 1942)

5. Cinclidotus pachyloma E. S. Salmon, Rev. Bryol 27 : 29 (1900).
C. nyholmiae Çetin, J. Bryol. 15 : 269 (1988).

Distribution map 108 (p. 276), Figure 107.

Plants up to 5 cm high, dark bluish-green. *Stems* densely and fastigiately branched. *Leaves* up to 4 mm long (including the long cuspidate point), slightly contorted when dry, patent when moist, oblong-lanceolate, obtuse, cuspidate; margins plane, entire, strongly thickened with eight-sixteen layers of incrassate cells; outer layers green, colour of layers gradually decreasing towards inner layers; costa very strong, at base up to about ⅓ width of leaf, long excurrent as a long cuspidate point; cells incrassate, smooth; basal cells quadrate to rectangular, upper cells irregularly quadrate to hexagonal, in mid-leaf 8–11 μm wide. *Gametangia* and sporophytes not seen in local populations.

Habitat and Local Distribution. — Plants growing in dense tufts, immersed in strong water currents, attached to calcareous rocks and boulders along the Dan River. Very rare, locally abundant: Dan Valley.

General Distribution. — Recorded from very few localities: Turkey (as *Cinclidotus nyholmiae*), Syria, and Lebanon (Zahle — type locality).

Cinclidotus nyholmiae Çetin, (J. Bryol. 15 : 269. 1988) seems to agree with *C. pachyloma* in all but small quantitative characters, which may fit into the range of variation of *C. pachyloma. Cinclidotus pachylomoides* was described by Bizot on the basis of two specimens, one from Lebanon and the other from Syria (Bizot *et al.* 1952). In spite of the similar epithets, the species differs in many characters from *C. pachyloma*, and therefore should not be confused. The characteristic spiny appearance of *C. pachyloma* is due to the erosion of the leaf lamina, leaving the costa and the thickened margins intact. The erosion is due to epiphytic algae (unpublished observations). The snail *Bythinia phialis* (identified by Y. Heller) is often found associated with *C. pachyloma.*

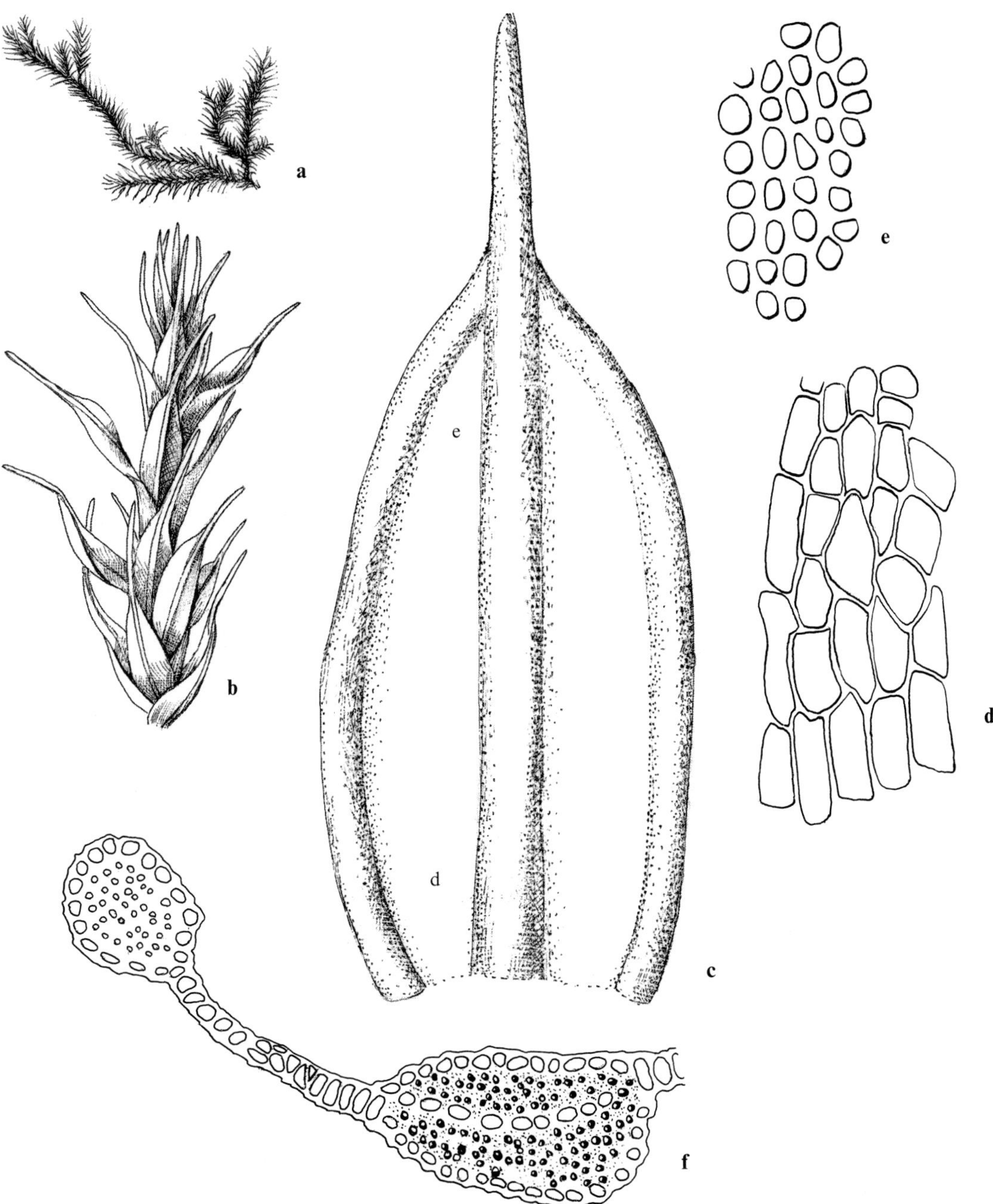

Figure 107. *Cinclidotus pachyloma*: **a** habit (×1); **b** stem, dry (×10); **c** leaf (×50); **d** basal cells; **e** upper cells (all leaf cells ×500); **f** cross-section of costa and portion of lamina (×150). (*Lipkin*, 23 Jul. 1979)

GRIMMIALES

GRIMMIACEAE

Plants acrocarpous, sometimes cladocarpous, perennial, forming usually dark, often hoary, tufts or cushions. *Stems* simple or more often branched. *Leaves* usually strongly hygroscopic, mostly lanceolate, often piliferous; lamina unistratose, partly or throughout bistratose to multistratose; margins usually entire, plane or recurved (often on one side only); costa single, ending below apex, percurrent or excurrent; cells usually smooth, often with incrassate, sinuate, or nodulose walls. *Setae* terminal or lateral, very short to elongate, straight or curved (cygneous); capsules immersed or exserted, symmetrical or bulging to various degrees on one side of base, smooth, or striate; columella attached to operculum or not; operculum convex to conical, mammillate or rostrate; peristome usually present, single, consisting of 16 teeth, entire or variously perforated and cleft. *Calyptrae* mitrate or cucullate.

Habitat. — Plants growing mainly on rocks.

Three genera are represented in the local flora: *Schistidium*, *Grimmia*, and *Racomitrium*.

1 Plants acrocarpous; leaves with lateral walls of basal cells straight to weakly sinuate; stems with central strand; peristome teeth entire, perforated, or cleft down to the middle or above.
 2 Columella attached to operculum, both falling off together **1. Schistidium**
 2 Columella not attached to operculum, persistent **2. Grimmia**
1 Plants cladocarpous; leaves with lateral walls of basal cells distinctly sinuate to nodulose; stems without central strand; peristome teeth cleft nearly to base into two or three divisions **3. Racomitrium**

1. *Schistidium* Brid.

Plants acrocarpous. *Stems* decumbent or erect, often branched, with central strand. *Leaves* usually carinate, ovate to lanceolate; basal cells usually elongate-rectangular to quadrate, with straight or weakly sinuate lateral walls; upper cells irregularly quadrate or rounded, often bistratose especially near apex, sometimes bistratose in patches. *Autoicous. Setae* short, straight; capsules ± entirely immersed, usually smooth; columella attached to and falling off with operculum; peristome teeth entire or irregularly perforated and cleft to their middle or above. *Calyptrae* small, mitrate, or cucullate.

A genus comprising ca. 20 species, with world-wide distribution, often considered as part of *Grimmia* (e.g., Bilewsky 1965, Crum *in* Sharp *et al.* 1994). One species occurs in the local flora.

Literature. — Bremer (1980).

Figure 108. *Schistidium apocarpum*: **a** habit, moist (×10); **b** mouth of capsule with detached operculum (×15); **c** calyptra (×15); **d** leaves (×50); **e** basal cells; **f** mid-leaf cells; **g** upper cells (all leaf cells ×500); **h** variation of leaf apices (×125); **i** cross-section of leaf (×500). (*Herrnstadt & Crosby 79-98-4*)

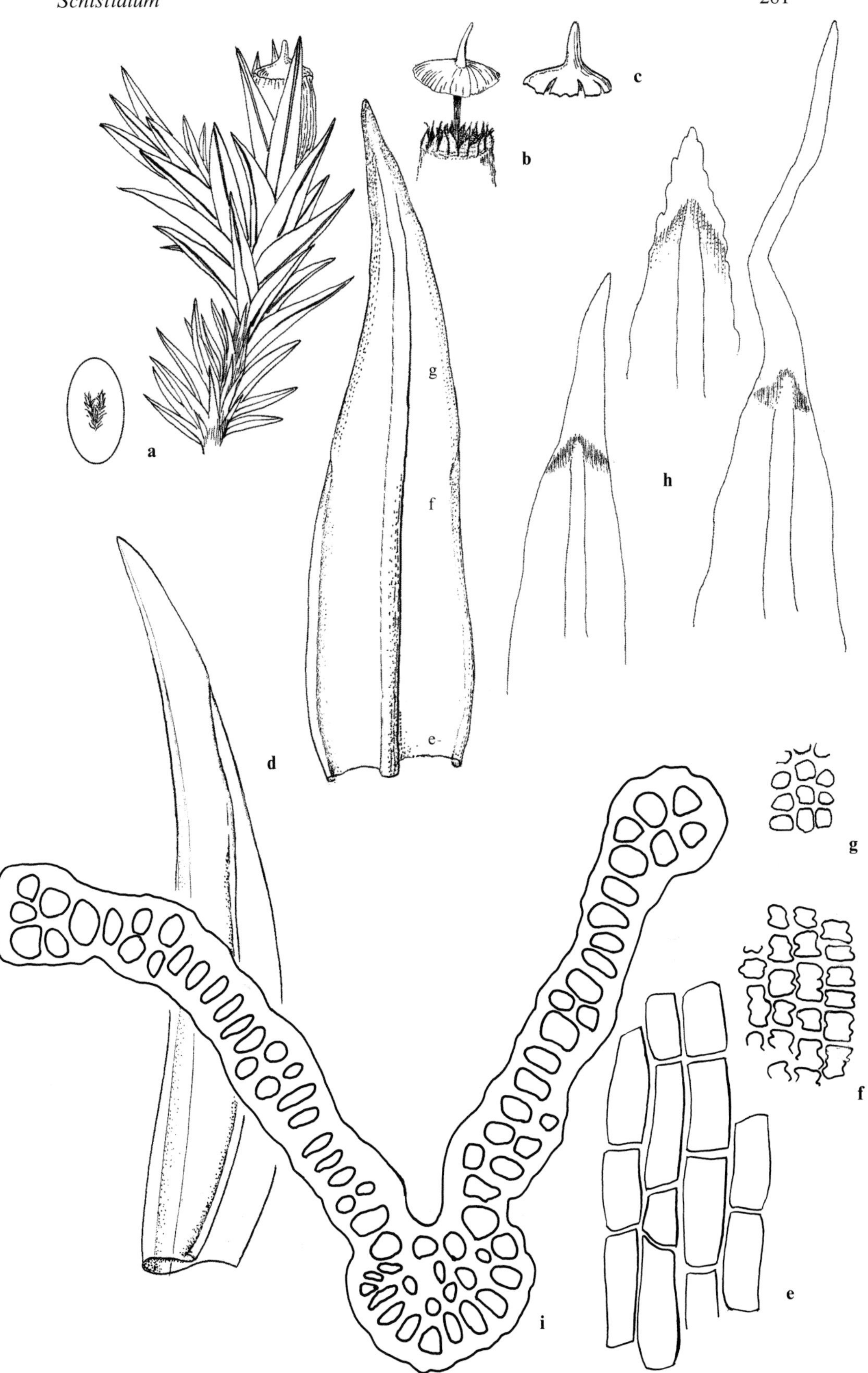
a
b
c
d
e
f
g
h
i
g
f
e-
g
f
e

Schistidium apocarpum (Hedw.) B.S.G., Bryol. Eur. 3:99 (1845).

Grimmia apocarpa Hedw., Sp. Musc. Frond. 76 (1801). — Bilewsky (1965): 389; *G. conferta* Funck, Deutschl. Moose 18 (1820); *G. atrofusca* Schimp., Syn. Musc. Eur., Edition 2:240 (1876); *S. atrofuscum* (Schimp.) Limpr., Laubm. Deutschl. 1:713 (1889); *G. singarensis* Schiffn., Ann. Naturhist. Hofmus. 27:487 (1913).

Distribution map 109, Figure 108 (p. 281), Plate VII: a.

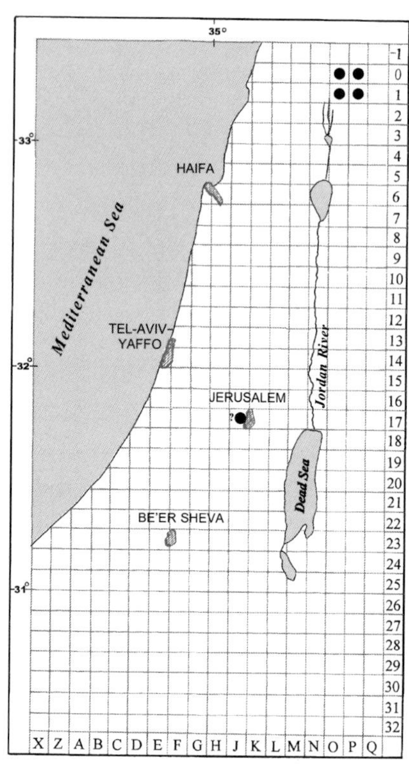

Map 109: *Schistidium apocarpum*

Plants up to 20 mm high, much branched, green to brownish. *Leaves* imbricate and erect when dry, erectopatent to spreading or ± recurved when moist, carinate, 2–2.5 mm long, ovate-lanceolate to lanceolate, acuminate, rarely muticous, often tapering into a ± denticulate hyaline hair-point, varying in shape and size; margins entire, usually recurved on one or both sides, bistratose above; costa percurrent or disappearing at base of hair-point; cells unistratose, with some bistratose patches, thick-walled; basal cells quadrate at margins, rectangular near costa, not or weakly sinuate; upper cells irregularly rounded-quadrate, mid-leaf cells rectangular to quadrate, weakly sinuate, 6–8 μm wide. *Setae* short, straight, ca. 0.6 mm long; capsules up to 2 mm long including operculum, immersed, broadly ellipsoid to obloid, widest at mouth when dry and empty; peristome red, finely papillose, perforated above; operculum rostrate, attached to the columella, both falling off together. *Spores* 8–14 μm in diameter, slightly papillose; densely granulate-gemmate (seen with SEM).

Habitat and Local Distribution. — Plants growing in tufts on north-facing calcareous rocks and rock enclaves on mountains of 700–1200 m altitude. Occasional: Judean Mountains (?), Mount Hermon, and Golan Heights. The records of this species from the Upper Galilee and Lower Jordan Valley (Bilewsky 1965) could not be verified.

General Distribution. — A widespread species throughout the temperate regions of the world.

Schistidium apocarpum is often treated as *Grimmia apocarpa*. When treated as a member of *Schistidium* or *Grimmia* it is subdivided into many infraspecific taxa. The epithets *atrofusca* and *conferta* have been used either for species of *Schistidium* or *Grimmia*, or for infraspecific taxa of either *S. apocarpum* or *G. apocarpa*. In this treatment of the species, no separate status was assigned to them, as the diagnostic characters recorded for each seem to occur in local plants in various combinations.

2. *Grimmia* Hedw.

Plants acrocarpous, usually small. *Stems* mostly erect, often branched, with central strand. *Leaves* carinate, ovate to lanceolate; cells usually smooth, basal cells elongate-rectangular to ± quadrate, with straight or weakly sinuate lateral walls; upper cells ± irregularly quadrate or rounded, often bistratose especially near apex, sometimes bistratose in patches. *Autoicous* or dioicous. *Setae* straight or curved; capsules smooth or striate, exserted, emergent, or immersed; columella not attached to operculum; peristome usually present, teeth entire or irregularly perforated and cleft down to the middle or above.

A large genus of up to 90 species (Greven pers. comm.) or 242 species (Magill 1981) with world-wide distribution. Eight species occur in the local flora.

Literature. — Greven (1995).

1 All leaves without hair-point; capsules distinctly bulging on one side of base **2. G. pitardii**

1 At least upper leaves with hair-point; capsules not or slightly bulging on one side of base.

  2 Leaves unistratose throughout.

    3 Perichaetial leaves with conspicuous hyaline cells at base; capsules distinctly striate when moist; calyptrae mitrate **4. G. mesopotamica**

    3 Perichaetial leaves with hardly conspicuous hyaline cells at base; capsules faintly striate or smooth when moist; calyptrae cucullate.

      4 Setae very short, ca. 1 mm long, slightly curved to sigmoid when moist; capsules immersed to shortly emergent, slightly bulging on one side of base **3. G. crinita**

      4 Setae longer, ca. 2 mm long, curved when moist; capsules emergent, not bulging on one side of base **6. G. orbicularis**

  2 Leaves at least partly bistratose.

    5 Leaves concave, not carinate, with unistratose margins and bistratose lamina **5. G. laevigata**

    5 Leaves carinate, with bistratose margins, lamina usually unistratose, sometimes with bistratose patches.

      6 Capsules exserted; peristome present; annulus revoluble.

        7 Plants in compact, hoary cushions, dark green to blackish above; at least upper leaves with hair-point the length of lamina or sometimes longer; autoicous **7. G. pulvinata**

        7 Plants in loose, not or slightly hoary tufts, yellowish-green above; leaves with shorter hair-point, usually not exceeding ¾ the length of lamina; dioicous **8. G. trichophylla**

      6 Capsules immersed; peristome absent; annulus persistent, not revoluble **1. G. anodon**

1. Grimmia anodon B.S.G., Bryol. Eur. 3 : 110, Tab. 236 (1845). [Bilewsky (1965) : 389]
Schistidium anodon (B.S.G.) Loeske, Laubm. Eur. 1 : 49 (1913).

Distribution map 110, Figure 109, Plate VII : b.

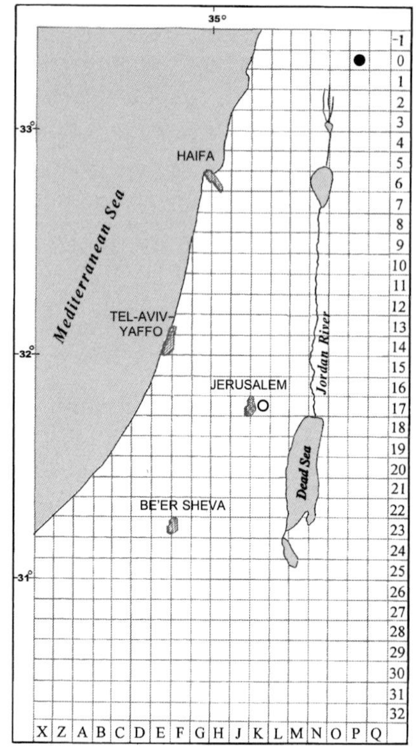

Map 110: ● *Grimmia anodon*
○ *G. pitardii*

Plants up to 6 mm high, dark green. *Leaves* imbricate when dry, patent when moist, concave to carinate above, oblong-lanceolate, 1–1.5(–2) mm long not including hair-point; lower leaves muticous to short-awned, upper leaves ending in a smooth to denticulate hyaline hair-point, up to the length of lamina; margins plane, bistratose above; costa subpercurrent to excurrent; lamina unistratose with bistratose patches, bistratose in upper part; basal cells rectangular, hyaline; mid- and upper cells irregularly quadrate, incrassate, with weakly sinuate walls, in mid-leaf 8–10 µm wide; perichaetial leaves wider than upper leaves, especially at base, up to 2.5 mm long without hair-point (hair-point up to the length of lamina), ovate-lanceolate. *Autoicous. Setae* curved, much shorter than capsules; capsules immersed, up to 1 mm long including operculum, slightly bulging on one side of base, subglobose, nearly smooth, wide-mouthed when empty; annulus narrow, persistent; operculum almost flat or obscurely mammillate; peristome absent. *Spores* 8–10 µm in diameter; minutely baculate, some baculae fused into bead-like structures (seen with SEM). *Calyptrae* mitrate.

Habitat and Local Distribution. — Plants growing in hoary cushions in shaded rock crevices: Mount Hermon; rare. Only a single population from Mount Hermon was examined by us. Although *Grimmia anodon* was recorded by Bilewsky (1965) from the Central Negev and Upper Galilee, no specimens from those two districts were found in his herbarium. Furthermore, those districts were not mentioned again in Bilewsky's later publications.

General Distribution. — Recorded from Southwest Asia (Turkey, Lebanon, Jordan, Sinai, Saudi Arabia, Iraq, Iran, and Afghanistan), around the Mediterranean, Europe (throughout), Macaronesia (in part), North Africa, north and central Asia, and North and Central America.

Literature. — Ochyra (1989)

Grimmia anodon is easily distinguished by the capsules that are immersed, subglobose, borne on very short arcuate setae and lack a peristome.

Grimmia anodon var. *sinaitica* Renauld & Cardot was described from Sinai: "Djebel Senah" (Bull. Herb. Boissier 2 : 33. 1894) as differing from the typical European plants by several characters, e.g., the larger size of various organs of the plants and the shorter and denticulate hair-point, possessing a wider base.

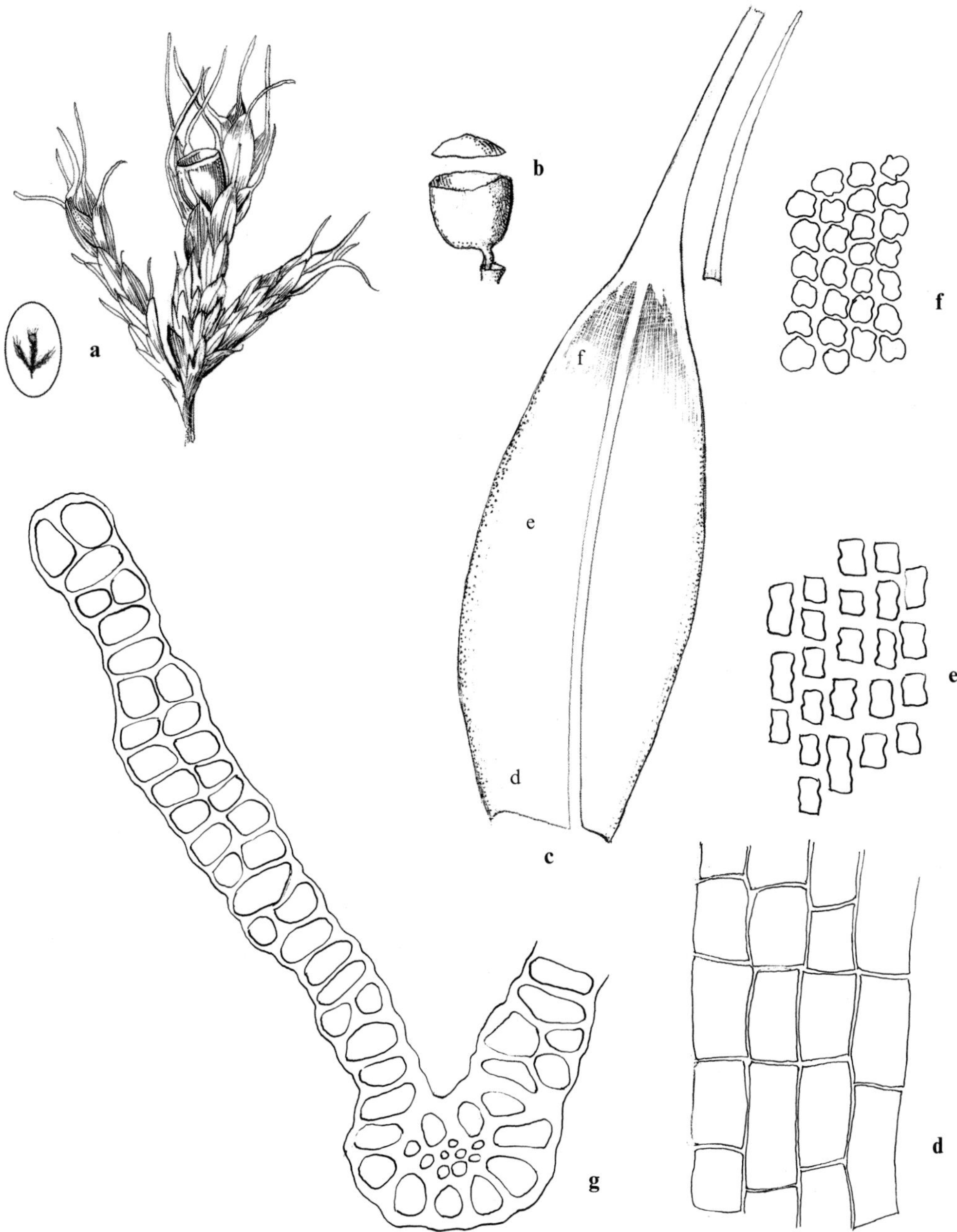

Figure 109. *Grimmia anodon*: **a** habit, moist (×10); **b** sporophyte and detached operculum (×15); **c** leaf (×50); **d** basal cells; **e** mid-leaf cells; **f** upper cells (all leaf cells ×500); **g** cross-section of costa and portion of lamina (×500). (*Herrnstadt 79-96-3*)

2. Grimmia pitardii Corb., Bull. Soc. Bot. France 56: LVI (1909).
G. gibbosa Agnew, J. Bryol. 7: 339 (1973).

Distribution map 110 (p. 284), Figure 110, Plate VII: c.

Plants up to 6 mm high. *Leaves* erect and twisted when dry, patent when moist, 0.5–1.5 mm long, ± carinate, usually lingulate, obtuse, always without hair-point, lower leaves brown, upper leaves olive-green; margins plane; costa wide, ending below apex or percurrent; cells unistratose; basal cells rectangular, hyaline; upper cells irregularly rounded, pellucid, in mid-leaf 5–8 µm wide; perichaetial leaves 2 mm long, ovate-lanceolate, acute, with costa percurrent or shortly excurrent. *Paroicous. Setae* short, less than 1 mm long, curved, twisted; capsules emergent, distinctly bulging on one side of base, up to 1 mm long including operculum, ovoid-obloid, weakly striate; annulus of two rows of cells; operculum with a rostrate beak; peristome of narrow, papillose, sometimes perforated erect teeth, cream-coloured, reflexed in dry and empty capsules. *Spores* ca.13 µm in diameter, slightly papillose; baculate, bacula with rough surface (seen with SEM). *Calyptrae* mitrate, uniformly lobed, covering operculum only.

Habitat and Local Distribution. — Plants small growing in crevices of calcareous rocks. So far, found only in a single locality: Judean Mountains.

General Distribution. — Recorded from Southwest Asia (Cyprus, Turkey, and Iraq), around the Mediterranean, Europe (few records, all from the south) Macaronesia (in part), and North Africa.

Grimmia pitardii differs from the other local *Grimmia* species in the lack of hair-points in all leaves, the distinctly asymmetric capsules, and the twisted setae.

Plants of this inconspicuous moss may have been overlooked by collectors.

In Maier (1998), *Grimmia pitardii* Corb. ("fide Herrnstadt *et al.* 1982") was transferred back to *Grimmia gibbosa* S. Agnew on the basis of a comparison of the types. *Grimmia pitardii* as described by Corbier is included in the genus *Campylostelium* B.S.G. (Ptychomitriaceae), thus creating a new combination: *C. pitardii* (Corb.) E. Maier.

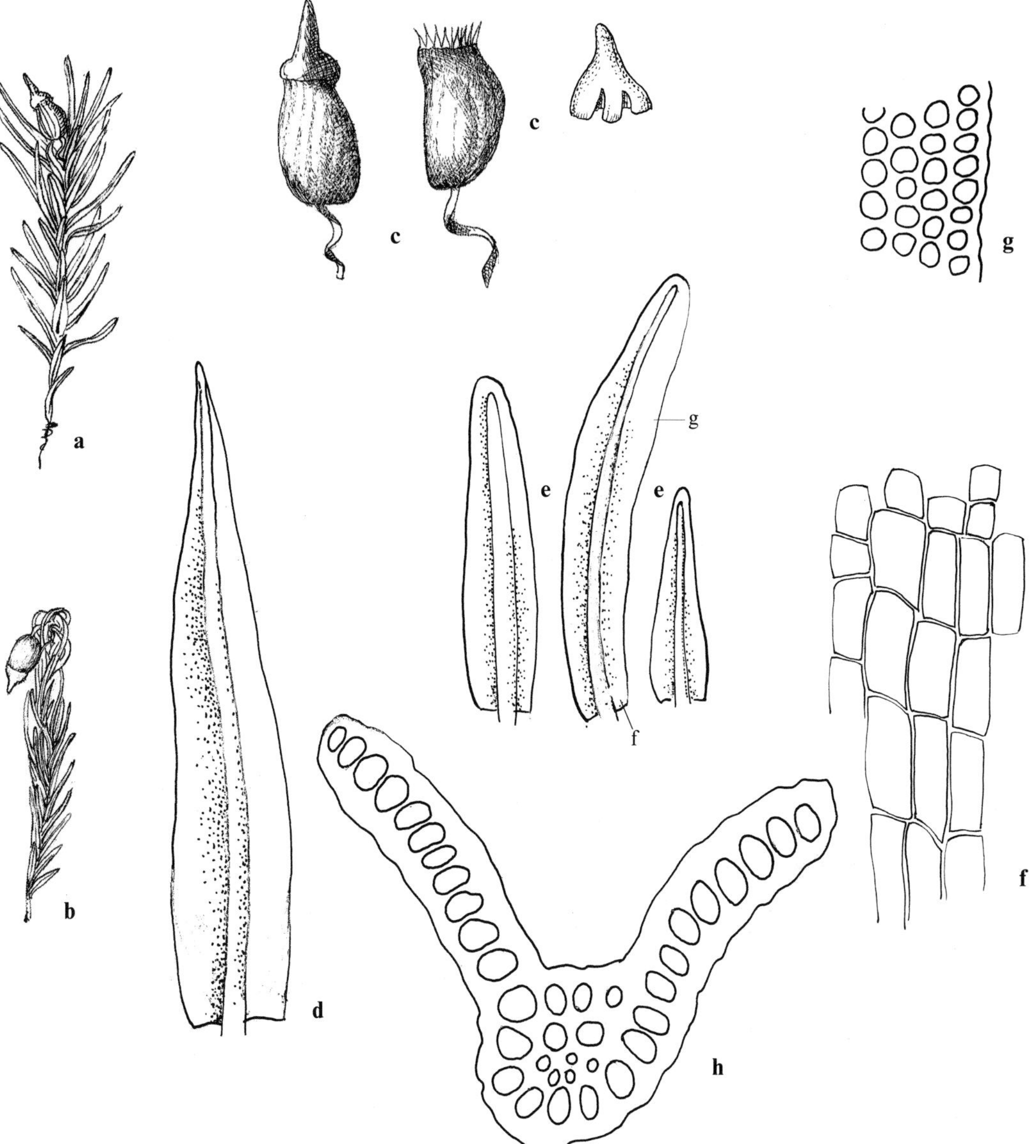

Figure 110. *Grimmia pitardii*: **a** habit, moist (×10); **b** habit, dry (×10); **c** capsules with and without operculum and calyptra (×25); **d** perichaetial leaf (×50); **e** leaves (×50); **f** basal cells; **g** marginal upper cells (all leaf cells ×500); **h** cross-section of leaf (×500). (*D. Zohary*, 23 Feb. 1943)

3. Grimmia crinita Brid., Sp. Musc. Frond. 1:95 (1806). [Bilewsky (1965):388]
G. sinaica B.S.G., Bryol. Eur. 3:113 (1845); *G. crinita* Brid. var. *libani* Bizot, Rev. Bryol. Lichénol. 13:51 (1942). — Bilewsky (1965):388.

Distribution map 111, Figure 111, Plate VII:d.

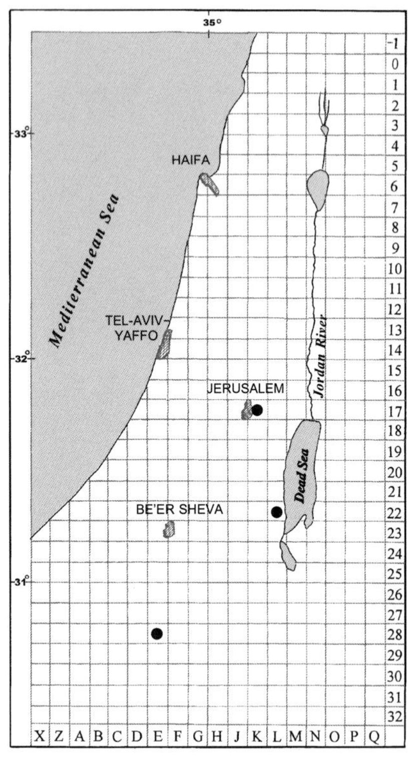

Map 111: *Grimmia crinita*

Plants up to 10 mm high, greyish. *Leaves* imbricate when dry, not much changing when moist, concave; lower leaves up to 1.5 mm long including short hair-point (hair-point not exceeding half the length of lamina), obovate, obtuse; upper leaves and perichaetial leaves up to 2.5 mm long (including hair-point); hair-point as long or longer than lamina, abruptly narrowing into nearly smooth, hyaline hair-point; margins plane; costa subpercurrent; cells unistratose; basal cells rectangular, hyaline; upper cells irregularly quadrate to short-rectangular, incrassate, with slightly sinuate walls, in mid-leaf 9–11 µm wide. *Paroicous* (see also Guerra *et al.* 1993). *Setae* slightly curved to sigmoid, very short, ca. 1 mm long, nearly covered by vaginula; capsules immersed to shortly emergent, 1.2–1.3 mm long including operculum, slightly bulging on one side of base, ovoid, smooth or faintly striate when dry; operculum mammillate; annulus broad, consisting of two–three rows of inflated cells, reaching up to ½ of peristome teeth, persistent; peristome teeth dark red, perforated, cleft above. *Spores* ca. 12 µm in diameter, weakly papillose; densely clavate, clavae small (seen with SEM). *Calyptrae* cucullate, reaching just below operculum.

Habitat and Local Distribution. — Plants growing in low grey tufts on loess and calcareous rocks and stones. Rare: Judean Mountains, Judean Desert, and Central Negev.

General Distribution. — Recorded from Southwest Asia (Turkey, Syria, Lebanon, Sinai, Jordan, Iran, Iraq, Afghanistan, and Saudi Arabia), sparsely around the Mediterranean, Europe (few records), Macaronesia (in part), North Africa, and northeastern and eastern Asia.

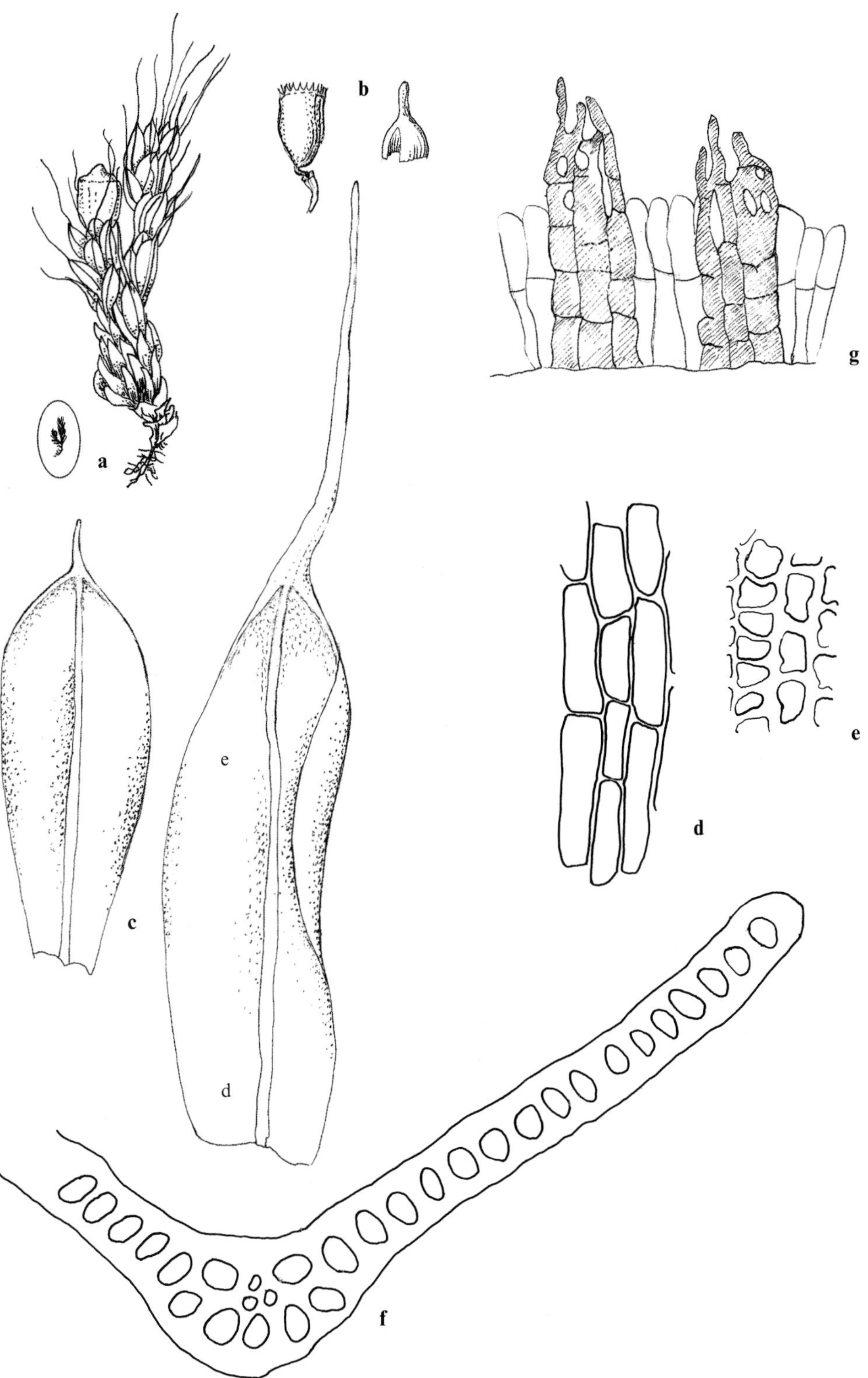

Figure 111. *Grimmia crinita*: **a** habit, dry (×10); **b** deoperculate capsule and calyptra (×10); **c** leaves (×50); **d** basal cells; **e** upper cells (all leaf cells ×500); **f** cross-section of costa and portion of lamina (×500); **g** peristome teeth and part of annulus (×250). (*Kushnir*, 15 Dec. 1942)

4. Grimmia mesopotamica Schiffn., Ann. Naturhist. Hofmus. 27:488 (1913).

Distribution map 112, Figure 112, Plate VII: e.

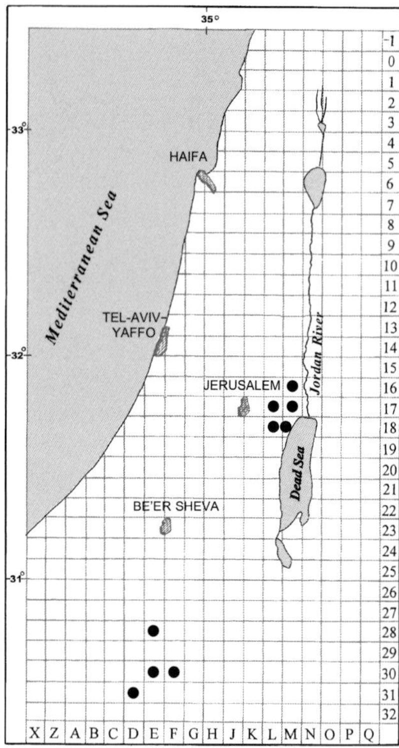

Map 112: *Grimmia mesopotamica*

Plants up to 6 mm high, greyish. *Leaves* imbricate, appressed when dry, ± recurved when moist, strongly carinate, ovate-oblong with wide hyaline apex, ending in a nearly smooth hyaline hair-point; leaves up to 2 mm long including hair-point (hair-point shorter than lamina); margins recurved almost to apex; costa subpercurrent; cells unistratose; basal cells rectangular, hyaline; upper cells irregularly quadrate, slightly sinuate, incrassate, in mid-leaf 9–13 μm wide; perichaetial leaves up to 3.5 mm long including hair-point (hair-point as long as lamina or longer), concave, obovate, with plane margins and large patches of hyaline cells at base, often with one–three rows of thin-walled cells ascending along margins up to ⅔ the length of lamina. *Autoicous. Setae* 0.75–1 mm long, straight to slightly curved, half its length covered by the vaginula; capsules shortly emergent, erect, ca. 1.5 mm long including operculum, wide-cylindrical, very slightly bulging on one side at base, striate even when moist; operculum obtusely conical; annulus broad, consisting of two–three rows of inflated cells, up to half the length of peristome teeth, persistent; peristome teeth dark red, papillose, perforated and cleft in upper part into three. *Spores* 14–18 μm in diameter, slightly papillose, the largest among local species; foveolate, with irregularly joined small clavae, clavae with rough surface (seen with SEM). *Calyptrae* mitrate, long-beaked, reaching just below operculum.

Habitat and Local Distribution. — Plants growing in hoary dense cushions, mainly on loess and calcareous substrates. Occasional, locally abundant (cf. Frey *et al.* 1990): Judean Desert, Central Negev, Lower Jordan Valley, and Dead Sea area.

General Distribution. — *Grimmia mesopotamica*, previously considered an endemic species of the semideserts of Southwest Asia (Syria, Israel, Jordan, and Iraq) and Turkmenia, was recently recorded also from Spain (Guerra *et al.* 1993).

Grimmia mesopotamica was described by Schiffner from two localities in Mesopotamia on gypsum-rich desert soils and calcareous marl rocks. Greven (1995) considers it conspecific with *G. capillata* De Not. from Sardinia, though the description of *G. capillata*, as provided by Greven, neither fully agrees with the original diagnosis of *G. mesopotamica* nor with observations in local material. It is therefore preferred here, before further studies can be carried out, to retain *G. mesopotamica* as a separate species. *Grimmia mesopotamica* most closely resembles *G. crinita* among the other local *Grimmia* species. The main distinguishing characters of *G. mesopotamica*, as pointed

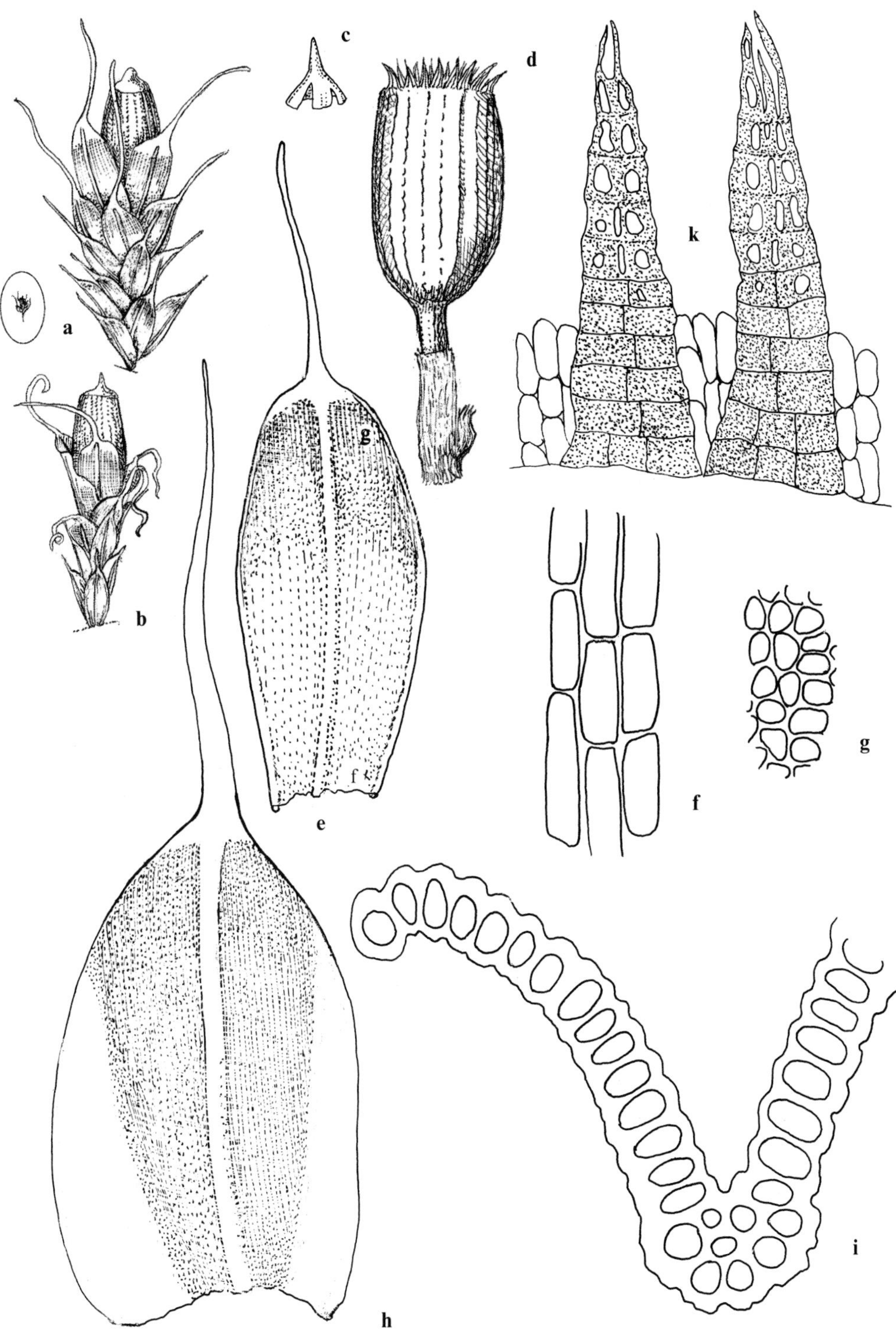

Figure 112. *Grimmia mesopotamica*: **a** habit, moist (×10); **b** habit, dry (×10); **c** calyptra (×10); **d** sporophyte with deoperculate capsule (×25); **e** leaf (×50); **f** basal cells; **g** upper cells (all leaf cells ×500); **h** perichaetial leaf (×50); **i** cross-section of costa and portion of lamina (×500); **k** peristome teeth and part of annulus (×250). (*Danin*, 1 Mar. 1980)

out by Schiffner, were: the different arrangement of "inflorescences", leaves with recurved — not plane — margins, mostly straight — not curved — setae, and the less clearly observable bulging base of the capsule.

5. Grimmia laevigata (Brid.) Brid., Bryol. Univ. 1 : 183 (1826). [Bilewsky (1974)]
Campylopus laevigatus Brid., Muscol. Recent., Suppl. 4 : 76 (1819); *G. campestris* Burch. ex Hook., Musci Exot. 2, Pl. 129 (1819). — Bilewsky (1965) : 387; *G. leucophaea* Grev., Mem. Wern. Nat. Hist. Soc. 4 : 87, Tab. 6 (1822).

Distribution map 113, Figure 113, Plate VII : f.

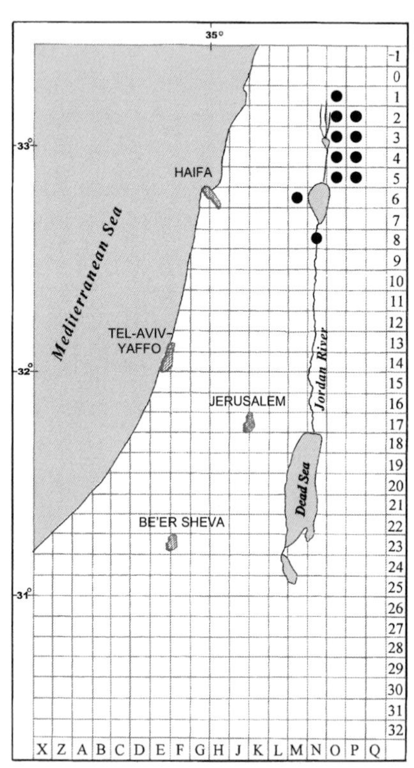

Map 113: *Grimmia laevigata*

Plants up to 10 mm high, dark green to blackish. *Leaves* imbricate, densely appressed, erectopatent when dry, spreading when moist, concave, ovate-lanceolate; lower leaves with or without hair-point, 1–1.5 mm long; upper leaves up to 2.5–3 mm long not including hair-point; hair-point hyaline, emerging from hyaline leaf apex, ± denticulate, up to length of lamina; margins plane, unistratose; costa slender, wide at base, narrower and obscure above, percurrent; cells bistratose except at leaf base; basal cells quadrate to short-rectangular, pellucid; upper cells incrassate, not sinuate, towards the margin wider than long, with ovate lumen, narrower than basal cells, irregularly round-quadrate, opaque, in mid-leaf ca. 8 µm wide; cells on ventral side of costa elongated; perichaetial leaves similar to upper leaves or larger. *Dioicous. Setae* ca. 2 mm, erect; capsules emergent to exserted, ca. 2 mm long including operculum, oblong-cylindrical, smooth; operculum with a short straight rostrum; annulus deciduous; peristome teeth short, dark brown. *Spores* ca. 11 µm in diameter, nearly smooth; densely clavate, clavae with rough surface, forming reticulum (seen with SEM). *Calyptrae* mitrate.

Habitat and Local Distribution. — Plants growing in readily disintegrating, hoary tufts, mainly on basaltic, more rarely on calcareous rocks. Occasional, locally common: Lower Galilee, Upper Jordan Valley, Mount Hermon, and Golan Heights. The record of Bilewsky (1965) of *Grimmia campestris* (= *G. laevigata*) also from the Judean Mountains and Judean Desert could not be verified.

General Distribution. — Recorded from Southwest Asia (Cyprus, Turkey, Syria, Lebanon, Sinai, Jordan, Iraq, Iran, and Afghanistan), around the Mediterranean, Europe (widely throughout), Macaronesia, Africa (throughout), northeastern, eastern, and central Asia, Australia, New Zealand, and North and Central America: a nearly cosmopolitan species.

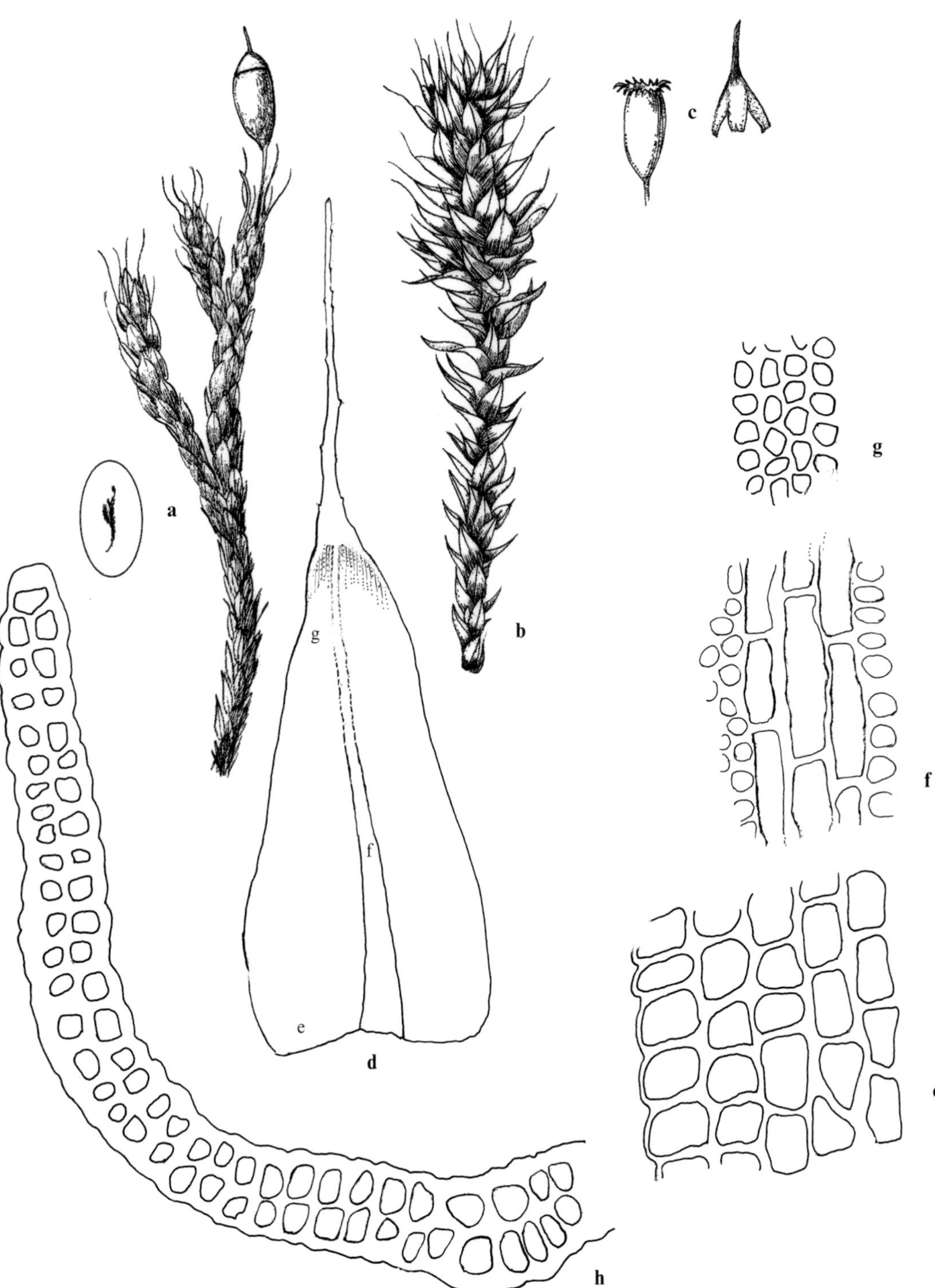

Figure 113. *Grimmia laevigata*: **a** habit, dry (×10); **b** habit, moist (×10); **c** deoperculate capsule and calyptra (×10); **d** leaf (×50); **e** basal cells; **f** cells on ventral surface of costa; **g** upper cells (all leaf cells ×500); **h** cross-section of costa and portion of lamina (×500). (*Herrnstadt & Crosby 78-28-2*)

Grimmia laevigata can be characterised by the densely imbricate concave leaves, and the readily disintegrating tufts.

6. Grimmia orbicularis Bruch ex Wilson, Engl. Bot., Suppl. 4 : Pl. 2288 (1844). [Nachmony (1961)]
G. orbicularis var. *persica* Schiffn., Oesterr. Bot. Z. 47 : 129 (1897).

Distribution map 114, Figure 114.

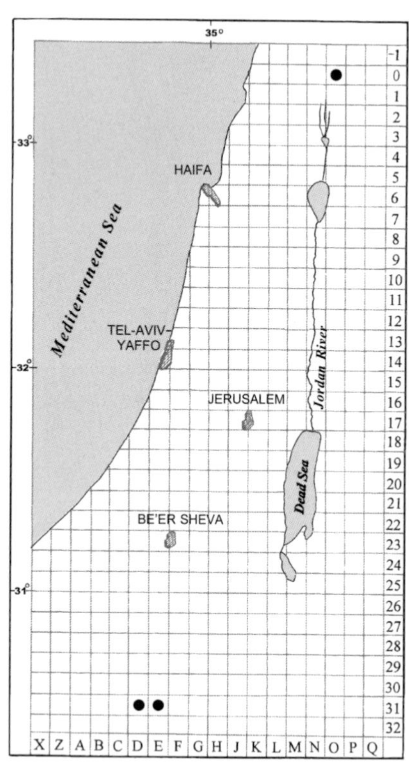

Map 114: *Grimmia orbicularis*

Plants up to 10 mm high, dull green-brown. *Leaves* imbricate when dry, erectopatent when moist, up to 1–1.5 mm long not including hair-point, some lower leaves muticous with rounded apex, others oblong-lanceolate, carinate above, usually abruptly narrowed into a long, ± denticulate, readily broken hyaline hair-point, 0.5–1(–2) the length of lamina; margins plane or recurved, sometimes on one side only, unistratose; costa percurrent to excurrent; cells unistratose; basal cells elongate, rectangular, shorter and hyaline towards margins, ± sinuate; upper cells incrassate, irregularly quadrate to rounded, distinctly sinuate, in mid-leaf 9–12 µm wide; perichaetial leaves up to 3 mm long (up to 5 mm with hair-point). *Autoicous. Setae* ca. 2 mm long, erect when dry, curved when moist; capsules emergent, ca. 1 mm long not including operculum, ovoid to subglobular, sometimes only slightly longer than wide, faintly striate when mature; operculum mammillate, annulus deciduous; peristome teeth red below, pale above, cleft in upper part into three. *Spores* ca. 13 µm in diameter, weakly papillose; resembling spores of *Grimmia crinita*, but less densely baculate (seen with SEM). *Calyptrae* cucullate.

Habitat and Local Distribution. — Plants growing in hoary cushions on dolomite rocks, in rock fissures, and on loess. Rare: highlands of Central Negev and Mount Hermon.

General Distribution. — Recorded from Southwest Asia (Cyprus, Turkey, Syria, Lebanon, southern Sinai, Jordan, Iran, Iraq, and Afghanistan), around the Mediterranean, Europe (throughout), Macaronesia (in part), North Africa, eastern Asia, Australia, and North, Central, and South America.

Grimmia orbicularis superficially resembles *G. pulvinata*, but can be distinguished by the duller green colour, the leaves, widest at mid-leaf rather than at base with unistratose margins, the mammillate operculum and cucullate calyptrae.

Greven (1995) considered *Grimmia orbicularis* var. *persica* Schiffn. as a separate taxon, characterised mainly by the extremely long and secund perichaetial leaves and the short setae of the capsules. The small number of local collections of *G. orbicularis*

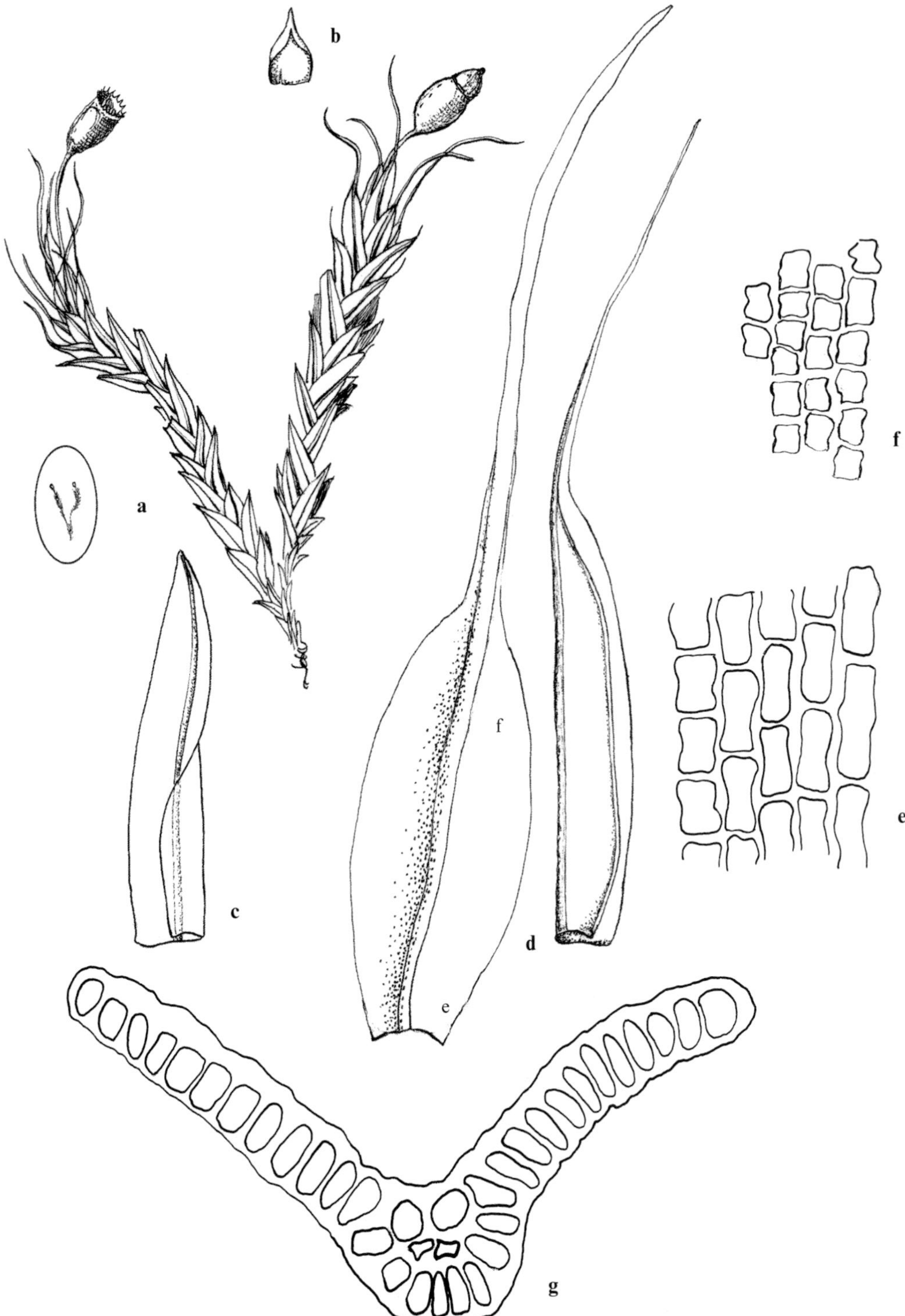

Figure 114. *Grimmia orbicularis*: **a** habit with deoperculate and operculate capsules (×10); **b** cucullate calyptra (×10); **c** lower leaf (×50); **d** upper leaves (×50); **e** basal cells; **f** upper cells (all leaf cells ×500); **g** cross-section of leaf (×500). (*Kushnir*, 9 Oct. 1943)

has not been adequate for determination whether the above characters are always correlated (cf. also Agnew & Vondráček 1975). Hence, it is preferred here not to separate var. *persica* from the typical *G. orbicularis*.

7. Grimmia pulvinata (Hedw.) Sm. ex Sm. & J. C. Sowerby, Engl. Bot. 24: Tab. 1728 (1867). [Bilewsky (1965): 388]
Fissidens pulvinatus Timm. ex Hedw., Sp. Musc. Frond. 158 (1801).

Distribution map 115, Figure 115, Plate VII: g.

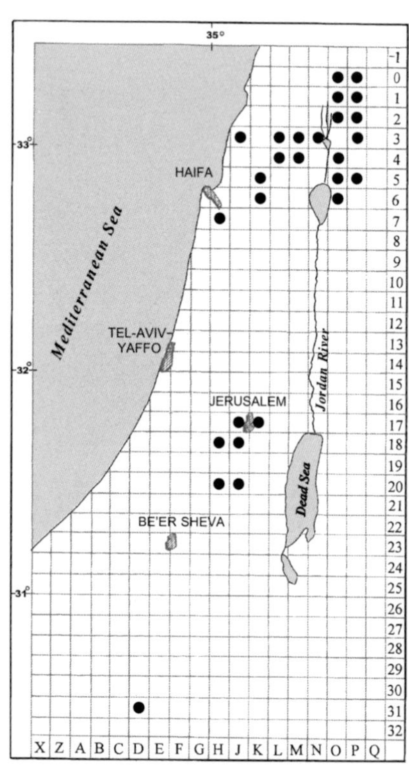

Map 115: *Grimmia pulvinata*

Plants up to 15 mm high, usually dark green to blackish throughout. *Leaves* imbricate, erect, when dry, erectopatent to spreading when moist, carinate, up to 2 mm long not including hair-point, oblong-lanceolate, abruptly narrowed into denticulate hyaline hair-point, 0.5–1(–2) times the length of lamina; margins recurved, usually on both sides, bistratose above; costa disappearing in base of hyaline hair-point; cells unistratose; basal cells short-rectangular, much longer towards costa, one–two marginal rows ± hyaline; mid-leaf cells rounded-quadrate to rectangular, incrassate, pellucid, ± sinuate, in mid-leaf 8–13 μm wide, more rounded towards apex; perichaetial leaves up to 2.5 mm long not including hair-point. *Autoicous. Setae* 2–3 mm long, curved to cygneous when moist, erect and twisted when dry; capsules exserted, ca. 2 mm long including operculum, oblong, brown, conspicuously plicate when dry; operculum with a long straight or oblique beak; annulus revoluble; peristome teeth red-brown, straight. *Spores* ca. 10 μm, with partly smooth and partly ± papillose surface (seen with light microscope); papillose surface baculate, bacula joined in groups, with roughened surface (seen with SEM). *Calyptrae* mitrate.

Habitat and Local Distribution. — Plants growing in dense, rounded, hoary compact cushions on calcareous rocks, stones, and fences, rarely on basaltic rocks. Common; the most widespread local *Grimmia* species: Upper Galilee, Lower Galilee, Mount Carmel, Judean Mountains, Central Negev, Mount Hermon, and Golan Heights.

General Distribution. — A widely distributed species; recorded from Southwest Asia (Cyprus, Turkey, Syria, Lebanon, Jordan, Iraq, Iran, Afghanistan, and Saudi Arabia), around the Mediterranean, Europe (throughout), Macaronesia, Africa (throughout), Asia (except the southeast), New Zealand, and North and Central America.

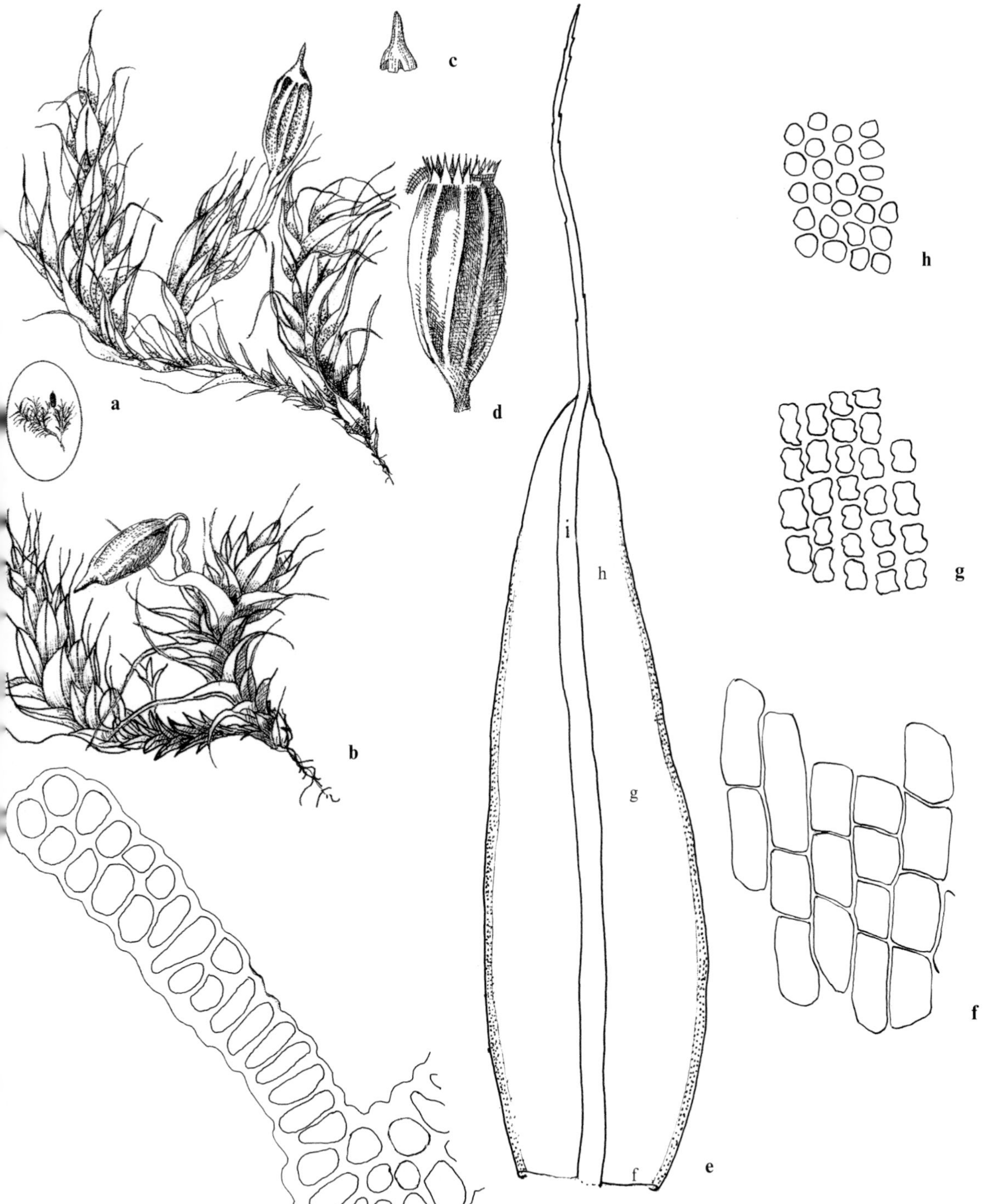

Figure 115. *Grimmia pulvinata*: **a** habit, dry (×10); **b** habit, ± moist (×10); **c** mitrate calyptra (×10); **d** deoperculate capsule (×20); **e** leaf (×50); **f** basal cells; **g** mid-leaf cells; **h** upper cells (all leaf cells ×500); **i** cross-section of costa and portion of lamina at upper part of leaf (×500). (*Markus* & *Kutiel 77-595-4*)

8. Grimmia trichophylla Grev., Fl. Edin. 235 (1824).

Distribution map 116, Figure 116, Plate VII : h.

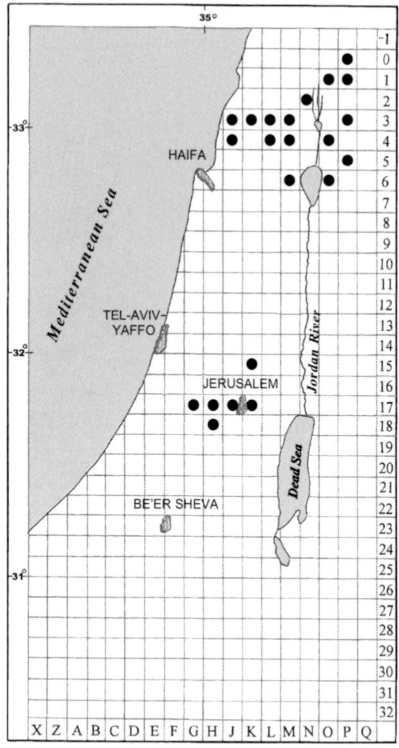

Map 116: *Grimmia trichophylla*

Plants up to 20 mm high, yellowish-green above, blackish below. *Leaves* imbricate, loosely appressed, slightly twisted when dry, spreading to squarrose-recurved when moist, up to 2.5 mm long not including hair-point, carinate, lanceolate to narrow-lanceolate, gradually tapering into a nearly smooth hyaline hair-point, in upper leaves up to ¾ times as long as lamina, in lower leaves shorter; margins recurved on one or both sides, thickened and bistratose above, with one–two hyaline marginal cell rows; costa percurrent; cells usually unistratose; basal cells elongate-rectangular, with sinuate walls; upper cells irregularly quadrate, more incrassate and less sinuate, becoming rounded toward apex, in mid-leaf 8–11 µm wide. *Costal gemmae* sometimes present; perichaetial leaves like upper stem leaves. *Dioicous. Setae* up to 3 mm long, curved to cygneous when moist, erect and twisted when dry; capsules exserted, ca. 2 mm long including operculum, oblong, ribbed when dry; operculum with a long oblique beak; annulus revoluble; peristome dark red, paler above. *Spores* ca. 12 µm in diameter, nearly smooth; surface densely baculate-clavate, clave often jointed (seen with SEM). *Calyptrae* mitrate.

Habitat and Local Distribution. — Plants growing in lax, not or slightly hoary, readily disintegrating tufts, usually on thin soil layers on rocks. Common, having a very similar distribution pattern to *Grimmia pulvinata* though less widespread: Coastal Galilee, Upper Galilee, Lower Galilee, Shefela, Judean Mountains, Mount Hermon, and Golan Heights.

General Distribution. — Recorded from Southwest Asia (Turkey, Lebanon, and Iraq), around the Mediterranean, Europe (throughout), Macaronesia, North Africa, Australia, New Zealand, Oceania, and North, Central, and South America.

Plants of *Grimmia trichophylla* closely resemble *G. pulvinata*, but form less hoary, looser and more readily disintegrating cushions. They are dioicous (not autoicous) and in local plants only rarely bear capsules.

Bilewsky & Nachmony (1955) and Bilewsky (1974) recorded *Grimmia trichophylla* var. *meridionalis* (Müll. Hal.) Loeske as the only representative of the species in Israel. The variety is characterised in literature as growing in dense cushions and having shorter leaves with longer hair-points. Because of the gradual variation of these characters in *G. trichophylla*, the variety does not seem to merit a separate status.

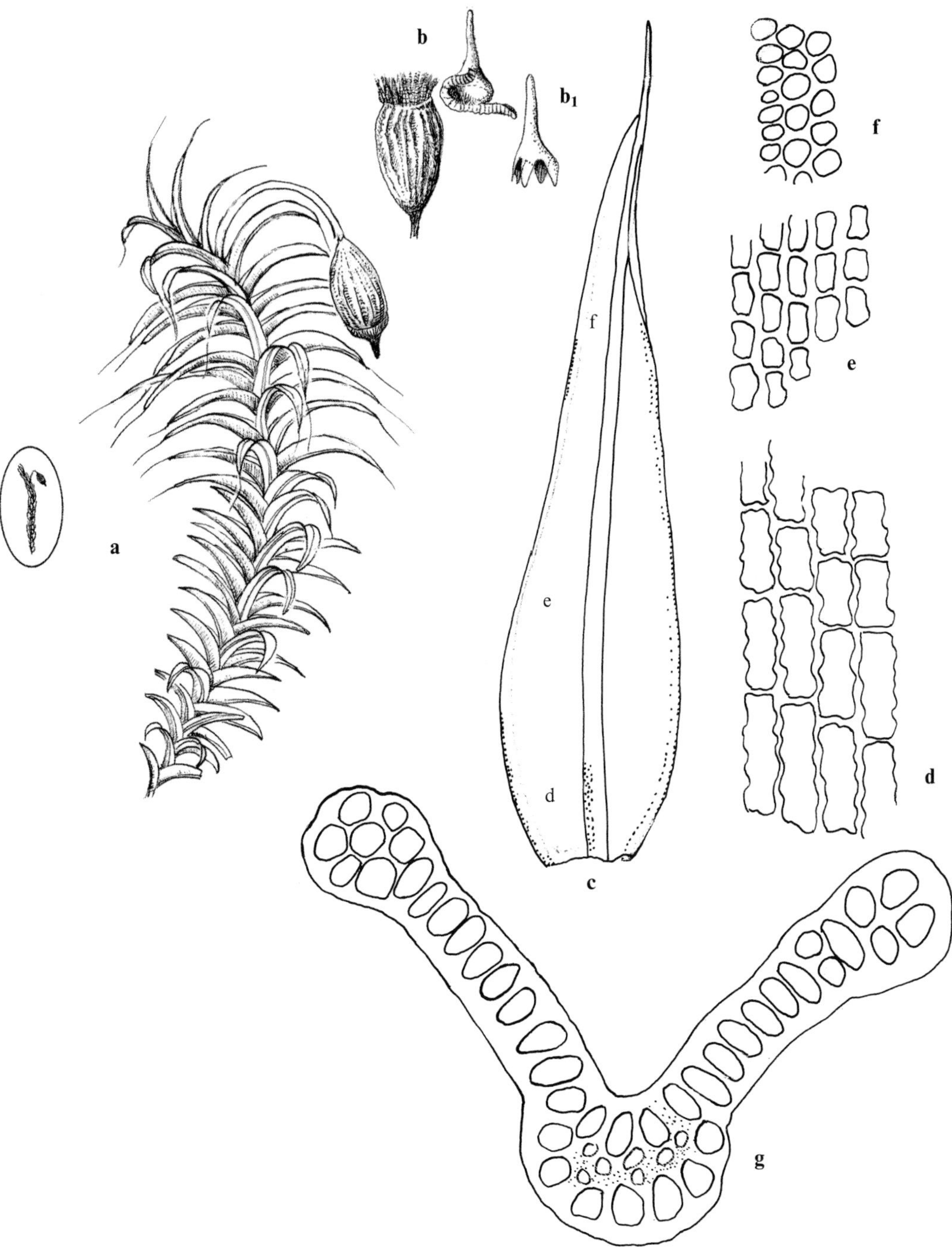

Figure 116. *Grimmia trichophylla*: **a** habit, moist (×10); **b** deoperculate capsule, operculum with detached annulus, $\mathbf{b_1}$ calyptra (×12); **c** leaf (×50); **d** basal cells; **e** mid-leaf cells; **f** upper cells (all leaf cells ×500); **g** cross-section of leaf (×500). (*Herrnstadt* & *Crosby 79-71-11b*)

3. *Racomitrium* Brid.

Plants cladocarpous. *Stems* irregularly branched, erect or prostrate, without central strand. *Leaves* lingulate to narrow-lanceolate; lamina unistratose or bistratose at margins and towards apex; cells smooth, thick-walled, distinctly sinuate to nodulose, in particular from above mid-leaf towards base; basal cells irregularly rectangular; mid- and upper leaf cells irregularly quadrate to narrowly rectangular. *Dioicous*; antheridia and archegonia borne on lateral branches. *Capsules* symmetrical, smooth, exserted on usually straight setae; annulus well differentiated; peristome teeth cleft nearly to base into two(–three) divisions. *Calyptrae* mitrate.

A large genus of ca. 80 species with world-wide distribution. One species occurs in the local flora.

Racomitrium heterostichum (Hedw.) Brid., Mant. Musc. 79 (1819).
Trichostomum heterostichum Hedw., Sp. Musc. Frond. 109 (1801).

Distribution map 117, Figure 117.

Plants up to 30 mm high, yellowish to brownish-green. *Leaves* imbricate, ± appressed, slightly curved and with twisted tips when dry, spreading with ± recurved tips when moist; 1.5–2.5 mm long, lanceolate, obtuse or acuminate, muticous or with a slightly denticulate hair-point varying in size from ⅕ to half the length of lamina; margins entire, usually bistratose above, often recurved on one side of leaf; costa channelled, percurrent or disappearing at base of hair-point; cells in mid-leaf 7–10 μm wide, shorter and wider towards margins. *Setae* 3–5 mm long, twisted when dry, erect, or flexuose or ± curved when moist; only immature capsule seen; operculum long-rostrate.

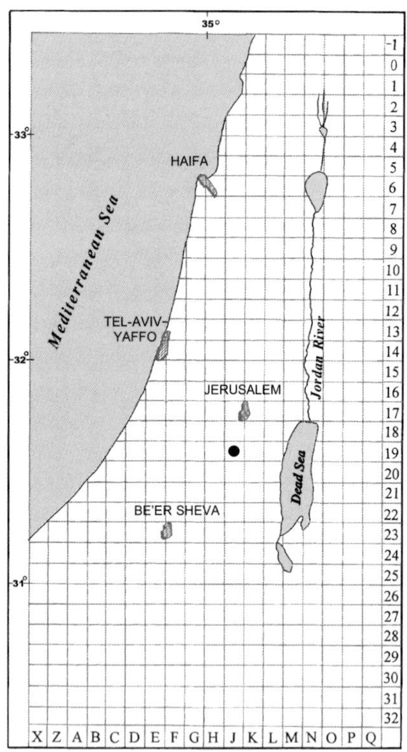

Map 117: *Racomitrium heterostichum*

Habitat and Local Distribution. — Plants growing on calcareous rocks (usually recorded as plants of acidic rocks): found so far in a single locality in the Judean Mountains.

General Distribution. — Recorded from Southwest Asia (Turkey, Syria, and Lebanon), around the Mediterranean, Europe (throughout), Macaronesia, North and South Africa, northeastern, eastern, and central Asia, Tasmania, New Zealand, and central and southern South America.

Racomitrium heterostichum can be easily recognised by the lateral inflorescences and the sinuate to nodulose leaf cells. Sterile plants may have been overlooked and considered as sterile *Grimmia* species.

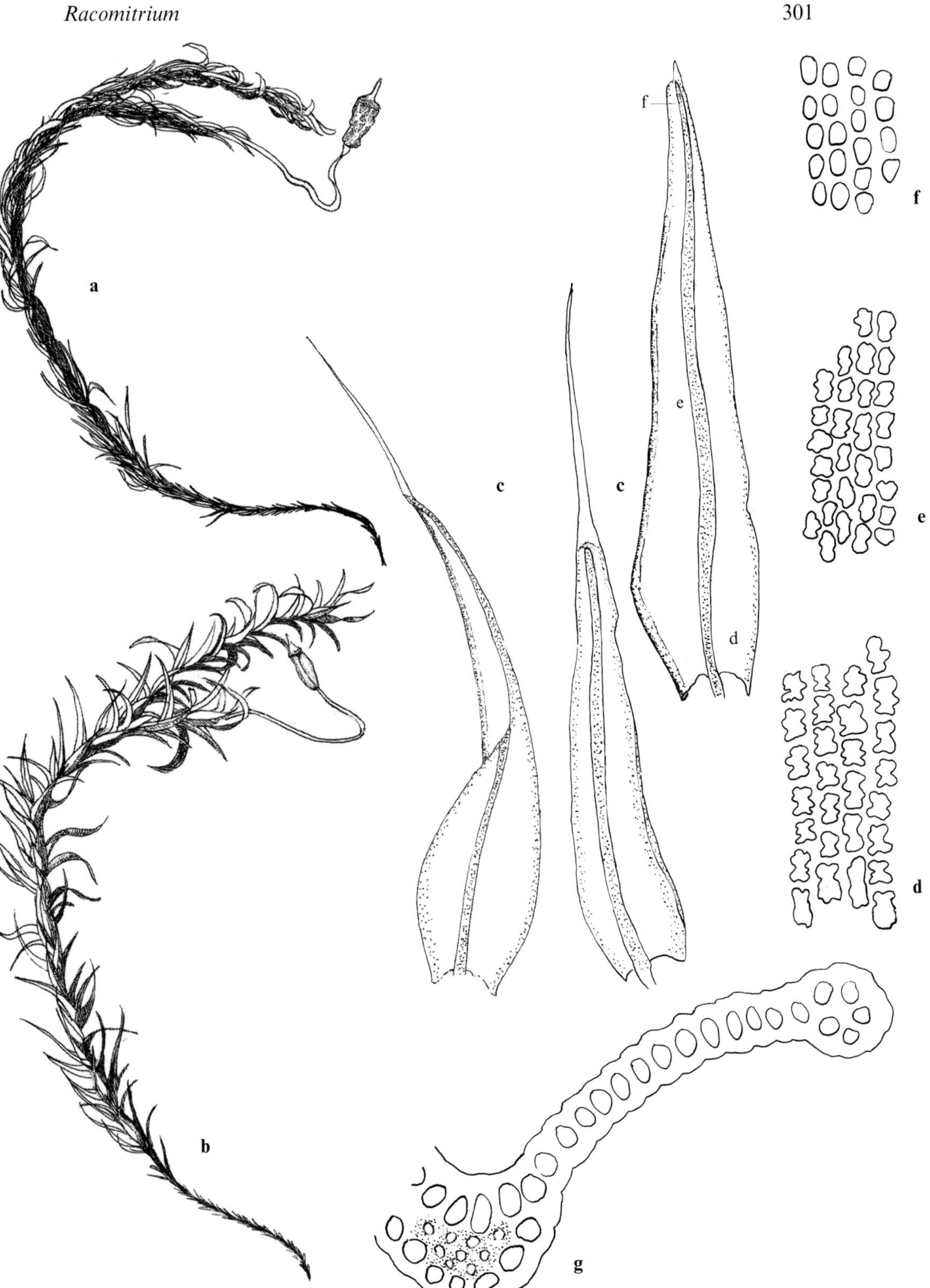

Figure 117. *Racomitrium heterostichum*: **a** habit, dry (×10); **b** habit with immature capsule, moist (×10); **c** leaves (×50); **d** basal cells; **e** mid-leaf cells; **f** apical cells (all leaf cells ×500); **g** cross-section of costa and portion of lamina (×500). (*Kushnir*, 6 Feb. 1943)

ENCALYPTALES

ENCALYPTACEAE

A small family comprising only two genera: *Encalypta* Hedw. (with ca. 35 species) and *Bryobrittonia* R. S. Williams (with only two species). *Bryobrittonia* does not occur in the area of this Flora and differs from *Encalypta* mainly by being dioicous and having non-papillose leaf cells.

For family characters see, the description of *Encalypta* below.

Encalypta Hedw.

Plants up to 5(–10) mm high, growing in tufts. *Stems* erect, branched, central strand often absent (present in the two local species). *Leaves* ± contorted to folded when dry, ± concave, erectopatent, spreading to reflexed when moist, lingulate, spathulate to oblong-obovate; apex obtuse or acute, with or without apiculus, sometimes piliferous; margins plane or recurved below, sometimes erect or inflexed near apex; costa ending just below apex to excurrent, strongly prominent at back, with a single, dorsal stereid band; basal cells smooth, hyaline, oblong-rectangular, with incrassate yellowish transverse walls marginal rows often narrower; upper cells irregularly quadrate to hexagonal, obscure, incrassate, densely papillose, in particular towards apex, with C-shaped branched papillae on both surfaces. *Autoicous*. *Setae* erect to flexuose; capsules exserted, erect and usually symmetrical, cylindrical to ellipsoid, smooth, striate, or furrowed, with or without ribs, ± constricted below mouth when dry; peristome single, double or absent; operculum long-rostrate. *Spores* papillose, in the local taxa bipolar. *Calyptrae* large, extending below capsule, conic-cylindrical, mitrate, long-rostrate, entire, erose, fringed, or lobed at base, scabrous at rostrum (in local plants).

The family and the genus are circumboreal, with 18 species in *Encalypta* (Horton 1988). Two species occur in the local flora.

Literature. — Vitt & Hamilton (1974), Horton (1983, 1988, *in* Sharp *et al.* 1994), Nyholm (1995).

1 Setae up to 7(–10) mm long; capsules smooth to striate, cylindrical; peristome usually absent; costa gradually narrowing in upper part, usually ending below apiculate apex **1. E. vulgaris**

1 Setae up to 4 mm long; capsules furrowed and ribbed, ellipsoid and widest towards base; peristome, at least rudimentary, often comprising pale fragile teeth; costa not narrowing in upper part, percurrent or excurrent **2. E. rhaptocarpa**

1. Encalypta vulgaris Hedw., Sp. Musc. Frond. 60 (1801). [Bilewsky (1965) : 356]

Distribution map 118 (p. 304), Figure 118, Plate VIII : b, d.

Plants up to 1(–1.5) cm high, dull green or yellowish-green, brown below. *Stems* often forked. *Leaves* 2–4 mm long, incurved and folded when dry, ± keeled, sometimes slightly constricted at middle, muticous; margins plane, ± undulate, sometimes erect near apex when dry; costa strong, narrowing in upper part, usually ending below apex, not extending into the short apiculus, at back usually distinctly papillose, turning

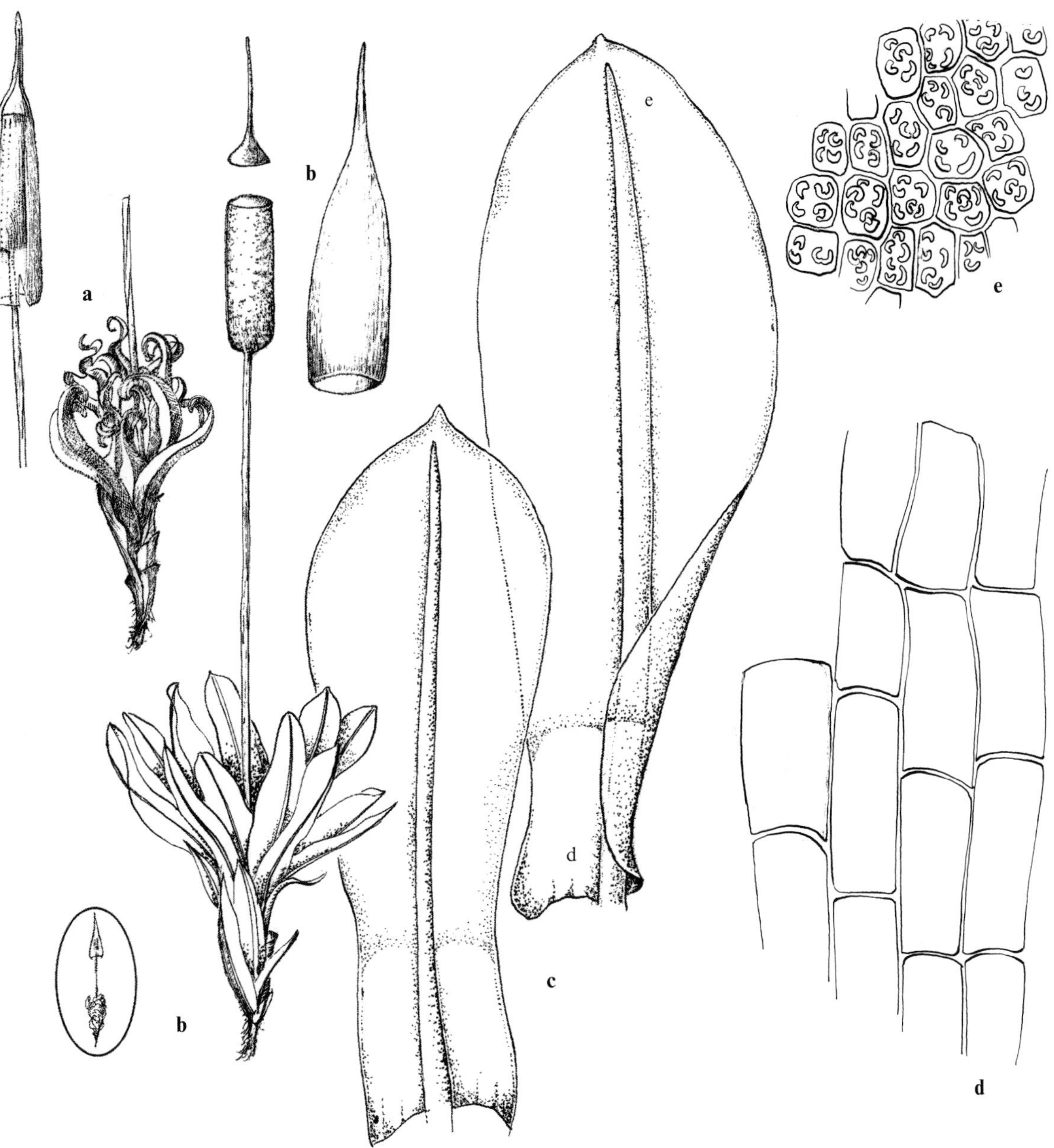

Figure 118. *Encalypta vulgaris*: **a** habit with calyptra, dry (×10); **b** habit, moist, capsule with detached operculum and calyptra (×10); **c** leaves (×50); **d** basal cells; **e** upper cells (all leaf cells ×500). (**a, b**: *Herrnstadt* & *Crosby 78-45-12*; **c**–**e**: *Herrnstadt* & *Crosby 79-61-1*)

yellowish-brown; basal cells rectangular, ± hyaline, distinct from opaque upper cells, sometimes forming V-shaped border line; marginal basal cells rectangular-linear, ± incrassate, often yellowish; upper cells hexagonal, in mid-leaf 15–21 μm wide, with four–five papillae per cell. *Autoicous. Setae* up to 7(–10) mm long, turning red when mature; capsules (without operculum) up to 3 mm long, cylindrical, yellowish, gradually turning reddish near mouth, smooth to striate; preperistome and peristome usually absent, sometimes a low white membrane present; operculum up to half the length of capsule, long-rostrate from a convex base. *Spores* up to 40 μm in diameter, markedly bipolar; proximal surface with weak radial ridges (less marked than in *Encalypta rhaptocarpa*), distal surface with very large, densely gemmate projections (surface appears much more rough than in *E. rhaptocarpa*) (seen with SEM). *Calyptrae* up to 5 mm long, with rostrum up to 2 mm long, scabrous, entire to erose, sometimes contracted and incurved at base.

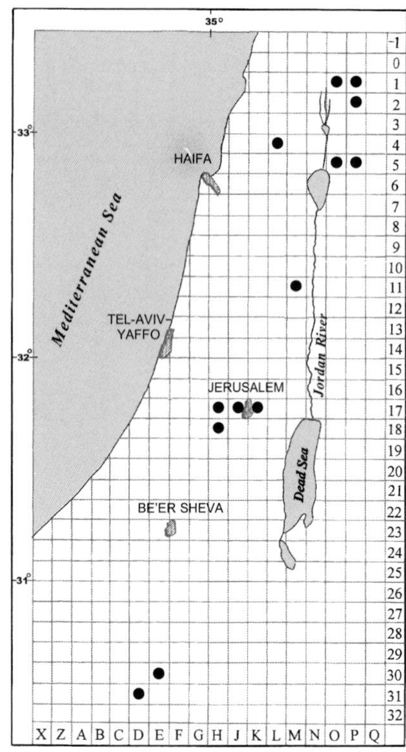

Map 118: *Encalypta vulgaris*

Habitat and Local Distribution. — Plants growing in tufts, often together with other mosses, on soil of calcareous or basaltic rock outcrops and in rock crevices. Occasional to common: Upper Galilee, Samaria, Judean Mountains, Central Negev, Mount Hermon, and Golan Heights.

General Distribution. — Widely recorded from Southwest Asia (Cyprus, Turkey, Syria, Lebanon, Sinai, Jordan, Iraq, Iran, Afghanistan, Saudi Arabia, and Oman), around the Mediterranean, Europe (throughout), Macaronesia, Africa (throughout, but not recorded from tropical Africa), Australia, New Zealand, and North and Central America. Records may also include *Encalypta rhaptocarpa.*

Sterile plants of *Encalypta vulgaris* may resemble *Tortula inermis* in their general appearance. The recently described *E. ovatifolia* Nyholm (1995) seems to resemble *E. vulgaris* but the leaf costa is "excurrent in elongate, usually finely dentate ± yellow point" as opposed to *E. vulgaris* in which the costa usually ends below apex.

2. Encalypta rhaptocarpa Schwägr., Sp. Musc., Suppl. 1 : 56 (1811) (= "*rhabdocarpa*", corr. Fürar., Flora 12: 16. 1829).

Distribution map 119 (p. 306), Figure 119, Plate VIII : a, c.

Encalypta rhaptocarpa closely resembles the locally more widespread *E. vulgaris.* The differential characters are mainly, but not exclusively, sporophytic.

Plants dark green to yellowish-green. *Costa* strong throughout, less distinctly papillose at back than in *Encalypta vulgaris*, percurrent or excurrent into cuspidate apex,

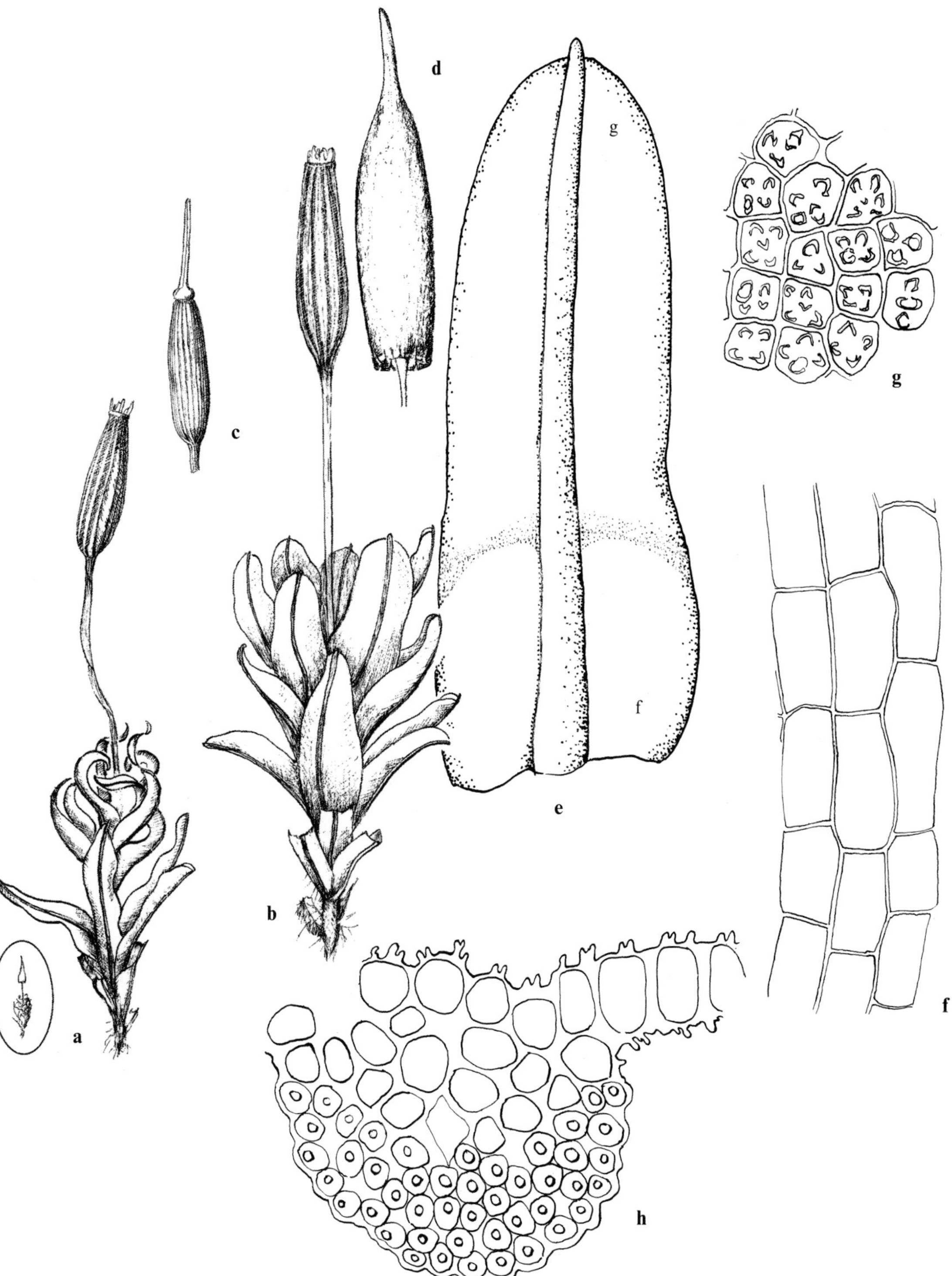

Figure 119. *Encalypta rhaptocarpa*: **a** habit with deoperculate capsule, dry (×10); **b** habit, moist (×10); **c** capsule with operculum (×10); **d** capsule with calyptra (×10); **e** leaf (×50); **f** basal cells; **g** upper cells (all leaf cells ×500); **h** cross-section of costa and portion of lamina (×500). (*Danin & Liston*, 6 Apr. 1983)

turning dark reddish-brown. *Setae* shorter, up to 4 mm long; capsules (without operculum) up to 3 mm long, ± ellipsoid, often widest towards base, slightly constricted below mouth, yellowish-brown, furrowed, with dark reddish-brown conspicuous ribs; peristome usually present, single, often incomplete, inserted ± below mouth; teeth short, up to 120 μm long, pale, fragile; preperistome absent. *Spores* markedly bipolar, up to 50 μm in diameter, generally resembling spores of *E. vulgaris*; proximal surface with distinct radial ridges, distal surface with very large, closely spaced, ± smooth, warty, clavate projections (seen with SEM). *Calyptrae* usually smaller, with rostrum up to 1.5 mm long, scabrous, slightly erose to sometimes lobed at base. *Antheridia* found either in a bud-like arrangement below the perichaetia, as usually described in the genus, or in leaf-axils beside or below them.

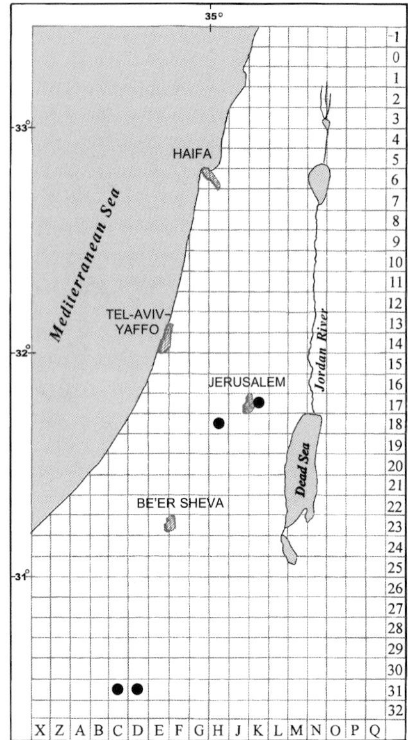

Map 119: *Encalypta rhaptocarpa*

Habitat and Local Distribution. — Plants growing in tufts on exposed calcareous rock outcrops and in soil pockets in shaded habitats, often on northern exposure and at foot of rocks. Rare to occasional, locally abundant in the Negev highlands up to 900 m, in some localities growing together with *Encalypta vulgaris*: Judean Mountains and Central Negev.

The important role of *Encalypta rhaptocarpa* (as "*Encalypta* sp."), with a few other mosses, in trapping airborne dust in the desert of the Negev Highlands was discussed by Danin & Ganor (1991). In a collection from the Judean Mountains (near Jerusalem), plants have capsules with less prominent ribs than in populations of the Negev Highlands.

General Distribution. — Recorded from Southwest Asia (Turkey, Lebanon, Jordan, Iraq, Iran, and Afghanistan), around the Mediterranean (scattered records), Europe (few records from Mediterranean countries), North Africa, northeastern, eastern, and central Asia, Oceania, and North and Central America. The considerable variation of specific concepts in *Encalypta* makes it difficult to evaluate existing records.

Encalypta rhaptocarpa, as circumscribed here, includes the plants previously reported from Israel as *E. intermedia* Jur. and, with some doubts, as *E. rhabdocarpa* (Herrnstadt *et al.* 1991). *Encalypta rhaptocarpa, E. intermedia,* and *E. vulgaris* are part of a complex of polymorphic taxa in which there are difficulties in the delimitation of individual species (Horton 1983, *in* Sharp *et al.* 1994). The leaves of all local plants of *E. rhaptocarpa* resemble *E. intermedia* in the unpiliferous leaves, but differ in the longer setae, the darker colour of the ribs of the capsules, and the presence of a peristome, though sometimes incomplete. *Encalypta buxbaumoidea* T. Cao, C. H. Gao & X. L. Bai (Cao & Gao 1990) somewhat resembles *E. rhaptocarpa* in the shape of the capsules and probably belongs to the same complex.

FUNARIALES

GIGASPERMACEAE

Plants often biennial, sometimes perennial, small, with prostrate, subterranean fleshy rhizomes, bearing short erect branches; central strand absent or weak. *Leaves* concave, larger in upper part of stem, broadly ovate to broadly obovate or elliptic; apex round and apiculate or acuminate; margins plane, entire; costa present or absent; cells lax, thin-walled, smooth. *Paraphyses* present, filiform. *Autoicous* or dioicous. *Capsules* often immersed, sometimes exserted, erect, globose, urceolate to short-cylindrical, without a differentiated neck; stomata with two guard cells, ± imperfectly divided; operculum sometimes not differentiated; peristome absent. *Spores* moderate to very large. *Calyptrae* minute, mitrate or cucullate, and fugacious.

A family comprising six small, often monotypic, terricolous genera, mainly restricted to the Southern Hemisphere. In the local flora, the family is represented by *Gigaspermum mouretii* Corb.

Gigaspermum Lindb.

A small genus, with two (or three) species. *Gigaspermum mouretii* Corb. from the Northern Hemisphere (in the Mediterranean Region) and *G. repens* (Hook.) Lindb. from the Southern Hemisphere seem to be very closely related (Herrnstadt *et al.* 1980). For generic characters, see the descriptions of the family above and *G. mouretii* below.

Literature. — Fife (1980), Herrnstadt *et al.* (1980).

Gigaspermum mouretii Corb., Rev. Bryol. 40 : 10 (1913).

Distribution map 120 (p. 308), Figure 120 (p. 309), Plate VIII : e–g (micrographs of *G. repens*; see the discussion below)

Plants whitish (due to the scarious leaf tissue and the lack of chlorophyll, except at leaf base). *Branches* 2–3 mm long (up to 4 mm at reproductive stage), club-shaped. *Leaves* at base of stem small and scattered, in upper part longer, 0.5–1 mm long, crowded, forming a bud-like, nearly circular in shape, upper leaves 0.5–1 mm long, with a long hair-point; margins plane, entire; costa absent; cells incrassate, chlorophyllous only at base; basal cells rectangular, upper cells irregularly hexagonal, mid-leaf cells 14–22 μm wide; perichaetial leaves more than twice the length of stem leaves, similar in shape, up to 2.5 mm long, with a long, twisted, hair-point. *Paroicous. Capsules* almost sessile, ± concealed between perichaetial leaves, up to 1 mm long including operculum, wide-urceolate, strongly wrinkled when dry; operculum wide, slightly convex, mammillate; peristome absent. *Spores* very large, 100–140 μm in diameter, usually angular, with rough surface; gemmate, processes irregularly deposited on surface (seen with SEM). *Calyptrae* minute, campanulate.

Habitat and Local Distribution. — Plants growing in dense, low tufts, on thin soil layers over limestone and basaltic rocks, in exposed habitats, mainly along the Syrian-African Rift Valley. Occasional: Judean Desert, Upper Jordan Valley, Lower Jordan Valley, Dead Sea area, and Golan Heights.

General Distribution. — Recorded from Portugal, Spain, Morocco, and Cyprus (Blockeel personal communication)

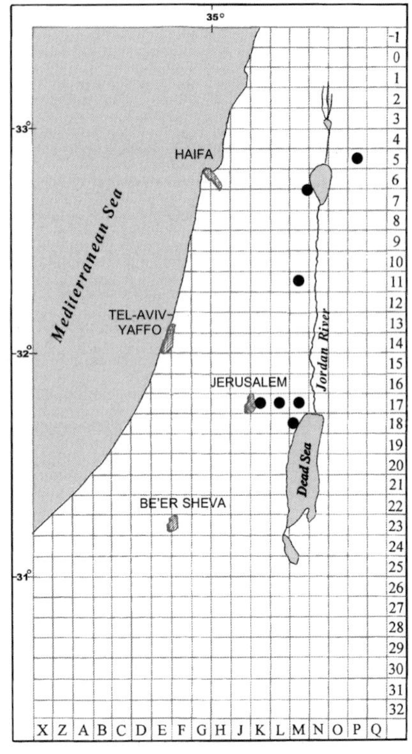

Map 120: *Gigaspermum mouretii*

Gigaspermum mouretii and *G. repens* seem to differ very little. Their separation is based on a single character: paroicous vs. autoicous condition. It is still doubtful whether they should be treated as separate species (Herrnstadt *et al.* 1980)

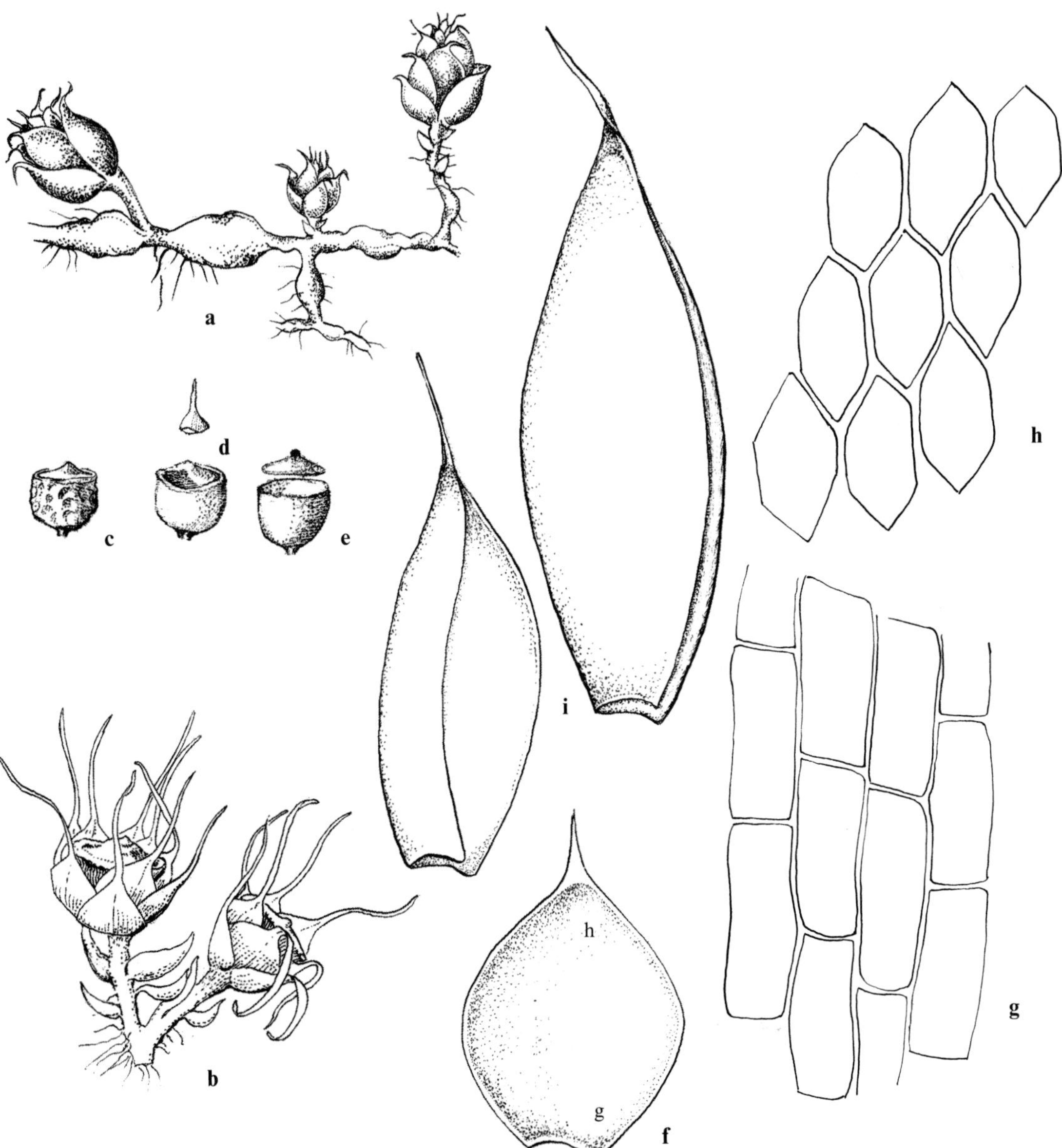

Figure 120. *Gigaspermum mouretii*: **a** habit, sterile plant (×10); **b** habit, plant with sporophytes (×10); **c** capsule, dry; **d** capsule, moist with detached calyptra; **e** capsule with detached operculum (all capsules ×10); **f** leaf (×50); **g** basal cells; **h** upper cells (all leaf cells ×500); **i** perichaetial leaves (×50). (*Herrnstadt 80-65-3*)

FUNARIACEAE

Plants annual or biennial. *Stems* short, erect, usually simple or shortly branching once; central strand usually present. *Leaves* larger and more crowded in upper part of stem, contorted when dry, erect to spreading when moist, soft, plane or concave, broad; margins plane or erect, entire, crenulate or serrulate to serrate above; costa ending below apex to excurrent; cells large, lax, thin-walled, smooth, rectangular at base, from oblong-hexagonal to hexagonal or rectangular above, narrower at margins; perigonial branch basal, short; perichaetial innovation much longer, forming the main stem; paraphyses present within perigonia, usually clavate, absent within perichaetia. *Autoicous* (in local plants). *Capsules* immersed to long-exserted, erect or inclined, symmetrical or asymmetrical, globose to pyriform, often with distinct neck; stomata confined to neck, each with single round guard cell; annulus differentiated or not; operculum from nearly plane to conical, sometimes rostrate; peristome absent, rudimentary, single or double (erect and symmetrical capsules usually with reduced peristomes; inclined and asymmetrical capsules with well developed, often double peristomes); exostome teeth 16, papillose-striate; endostome segments opposite teeth, without basal membrane and cilia. *Spores* often subspherical. *Calyptrae* cucullate or mitrate, inflated or small, and smooth.

A widespread cosmopolitan family, comprising according to different authors up to 16 genera. Four genera are represented in the local flora.

The Funariaceae are characterised by soft, laxly areolated leaves, clavate paraphyses, and stomata with a single round guard cell.

Literature. — Loeske (1929), Fife (1985).

1 Calyptrae four-angled, covering the whole capsule, persistent until maturity **2. Pyramidula**

1 Calyptrae not angled, covering at most the upper half of capsule, not persistent until maturity.

- 2 Calyptrae not inflated at base, mitrate; capsule urceolate or pyriform, neck short, exothecial cells irregularly hexagonal **1. Physcomitrium**
- 2 Calyptrae inflated at base, cucullate or mitrate (sometimes unequally lobed); capsule pyriform, neck long, at least half as long as theca, exothecial cells oblong-hexagonal to linear.
  - 3 Capsules erect*, symmetrical or rarely slightly asymmetrical; peristome absent, rudimentary, or single, very rarely double, inserted well below mouth, not joined at tips **4. Entosthodon**
  - 3 Capsules inclined to pendent, asymmetrical; peristome double, inserted ± below mouth, joined at tips **3. Funaria**

* Inclined in *Entosthodon curvisetus*.

1. *Physcomitrium* (Brid.) Brid.

Plants small. *Leaves* ovate-lanceolate to oblong-lanceolate; costa ending just below apex to shortly excurrent; cells laxly arranged, thin walled, upper cells oblong-hexagonal, basal cells rectangular. *Autoicous. Setae* erect; capsules erect, urceolate or pyriform, usually with a wide mouth when mature and empty, neck short; exothecial cells irregularly hexagonal, shorter than in *Entosthodon* and *Funaria*; operculum convex, sometimes mammillate to rostellate; annulus differentiated; peristome absent. *Calyptrae* not inflated, mitrate, three–four-lobed, long-rostrate.

A cosmopolitan genus comprising over 80 species. One species occurs in the local flora.

Aneuploidy and polyploidy (Crum & Anderson 1981) as well as hybrids between species are known.

Physcomitrium resembles *Entosthodon* but its leaves are more lanceolate with longer costa ending near apex to shortly excurrent, the capsules are urceolate, and the calyptrae are always mitrate.

Physcomitrium eurystomum Sendtn., Denkschr. Bayer. Bot. Ges. Regensburg 3 : 142 (1841).

P. acuminatum B.S.G., Bryol. Eur. 3 : 247 (1841).

Distribution map 121, Figure 121 (p. 313), Plate VIII : h, i.

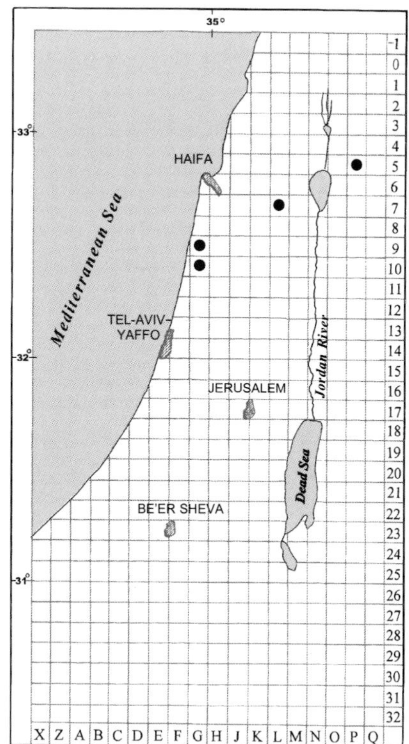

Map 121: *Physcomitrium eurystomum*

Plants up to 3.5 mm, bright green when wet, yellow when dry. *Leaves* slightly contorted, up to 2.5 mm long, ovate-lanceolate, acuminate; margins plane, bluntly serrate in upper part, one row of marginal cells longer and sometimes narrower; costa ending below apex to percurrent, sometimes shortly excurrent; mid-leaf cells 20–22 µm wide. *Setae* 2–4(–5) mm long; capsules up to 1.5(–2) mm long, urceolate, yellowish brown; operculum convex, rostellate; annulus of one row of small quadrate cells, persistent. *Spores* 25–27 µm in diameter, papillose to indistinctly spinulose; baculate, baculae varying in density and with granules on surface (seen with SEM).

Habitat and Local Distribution. — Plants growing gregariously or in patches on soil turning impermeable when wet. Rare: Sharon Plain, Lower Galilee, and Golan Heights.

General Distribution. — Recorded from Southwest Asia only from Turkey, few records from around the Mediterranean, Europe (scattered records throughout), North, tropical, and South Africa, and Asia (scattered records throughout, except from the southeast).

Literature. — Ducker & Warburg (1961).

Physcomitrium eurystomum is regarded by Ducker & Warburg (1961) as an intermediate between *P. pyriforme* and *P. sphaericum* (C. F. Ludw.) Fürnr. and as a variety of *P. sphaericum* by Husnot (1884–1894).

Contrary to European floras (Smith 1978, Frahm 1995), which describe the operculum of *Physcomitrium eurystomum* as bluntly conical, mammillate, or short-apiculate, only plants with convex, rostellate opercula have been seen in the local flora. In this character, plants resemble *P. pyriforme*, but otherwise have urceolate capsules typical for *P. eurystomum*. In addition, we could not verify the differential character — spiny spores — proposed by Fife (1985) as characteristic of the genus *Physcomitrium*. Spores seen with a light microscope may appear somewhat spinulose, but spores of local plants seen with SEM are distinctly baculate.

The previous record of *Physcomitrium* aff. *pyriforme* (Herrnstadt *et al.* 1982, 1991) was based on a single misidentified collection. Hence, that species does not occur in the local flora.

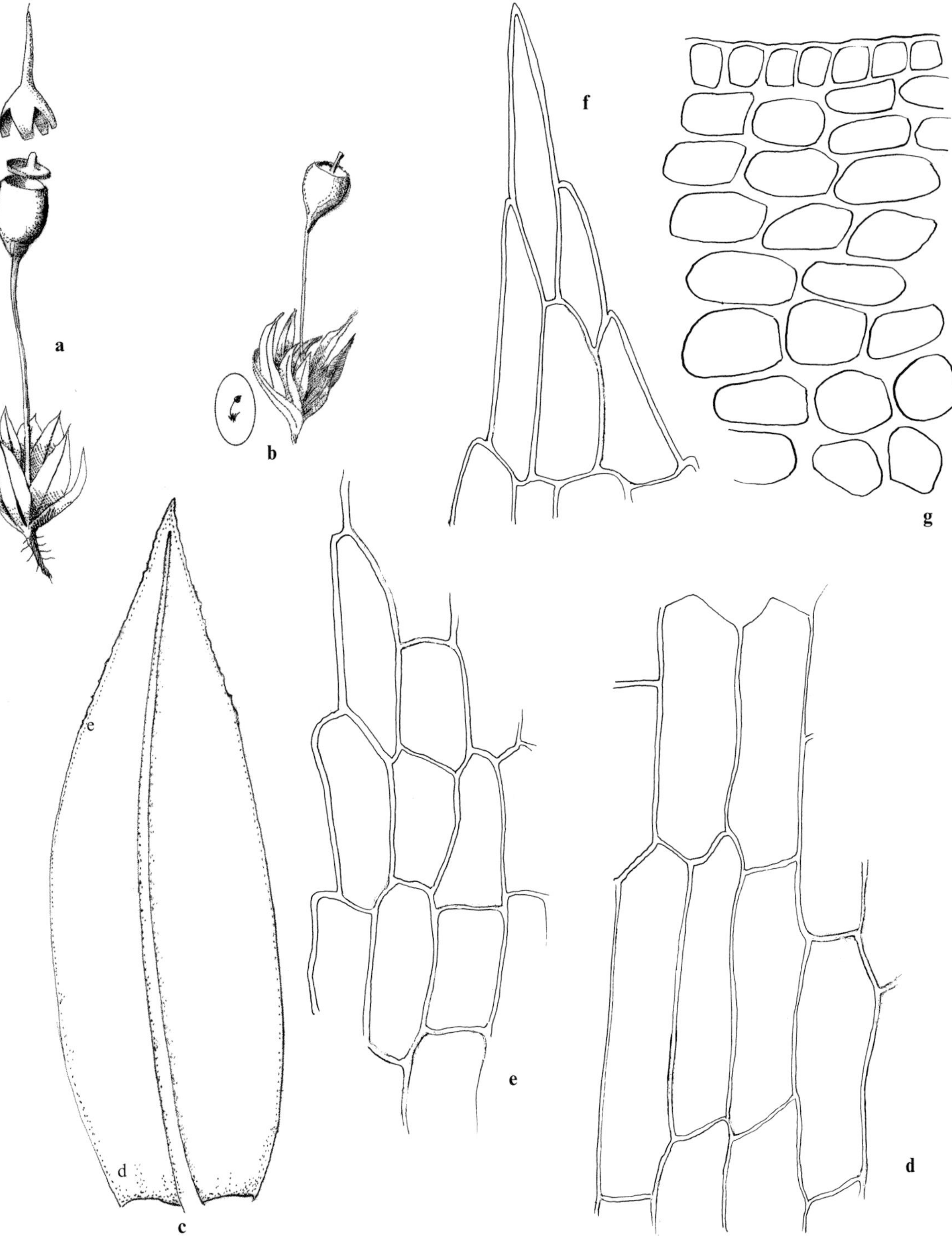

Figure 121. *Physcomitrium eurystomum*: **a** habit, moist, with detached operculum and calyptra (×10); **b** habit, dry (×10); **c** leaf (×50); **d** basal cells; **e** marginal upper cells; **f** apical cells (all leaf cells ×500); **g** exothecial cells (×500). (**a**: *Herrnstadt* & *Crosby 79-82-3*; **b**–**g**: *Kushnir*, 19 Mar. 1944)

2. *Pyramidula* Brid.

A small genus of two species, of which one — *Pyramidula tetragona*, a rare species in some parts of the Northern Hemisphere — has been found in the Golan Heights. The other species — *P. algeriensis* (recorded from two localities in Spain in addition to North Africa) — is mainly differentiated by the longer excurrent costa and spherical capsules (see Frahm 1995)

For generic characters, see the description of *Pyramidula tetragona*.

Plants of *Pyramidula* resemble *Physcomitrium*, but differ in the shorter setae, narrow mouth of the capsules, and four-angled calyptrae (hence the generic name), which enclose the whole capsule until maturity.

Pyramidula tetragona (Brid.) Brid., Mant. Musc. 20 (1819).
Gymnostomum tetragonum Brid., Sp. Musc. Frond. 1 : 270 (1806); *Physcomitrium tetragonum* (Brid.) Fürnr. *in* Hampe, Flora 20 : 285 (1837).

Distribution map 122, Figure 122.

Plants annual, small, up to 5(–6) mm high, bright green, with purplish rhizoids. *Stems* simple or forked. *Leaves* in bud-like clusters, concave, suberect when wet, ± contorted and folded, with twisted yellowish tips when dry; upper leaves 1.5–2.5 mm long, oblong-ovate, long-acuminate; margins plane, entire; costa excurrent; basal cells rectangular, upper cells ± hexagonal to short rectangular 27–40 μm wide in mid-leaf. *Autoicous*. *Setae* straight, short, 1.5–2 mm long, yellow; capsules slightly exserted, erect, symmetric, subglobose, reddish-brown when mature, finely wrinkled and empty when dry, exothecial cells with verruculose cuticle; operculum small, conical, mammillate; annulus and peristome absent. *Spores* (no mature spores seen in local plants) up to 70 μm in diameter, nearly smooth. *Calyptrae* 2 mm long, beaked, inflated, four-angled, covering the whole capsule until maturity, opening by longitudinal slit.

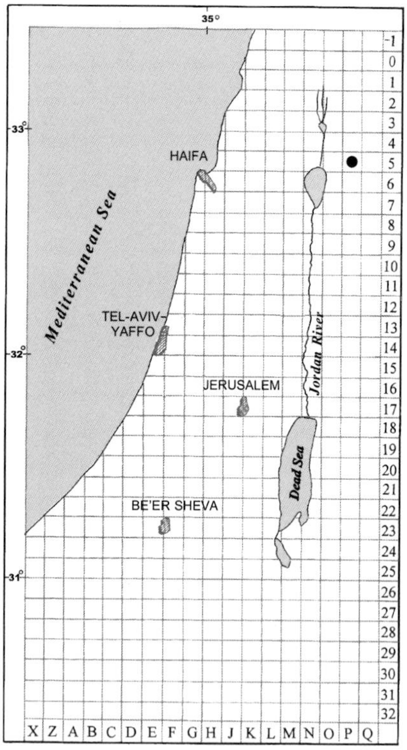

Map 122: *Pyramidula tetragona*

Habitat and Local Distribution. — Plants gregarious or loosely caespitose; so far found in a single locality on an exposed shallow basaltic soil layer on basaltic rock: Golan Heights (southern branch of Wadi Daliyyot).

General Distribution. — Recorded from around the Mediterranean (few records), Europe (few and scattered records), North Africa, northern Asia, and North America.

Local plants are larger than recorded from other parts of the area of distribution of *Pyramidula tetragona* (cf. Husnot 1884–1894, Grout 1935, Crum & Anderson 1981).

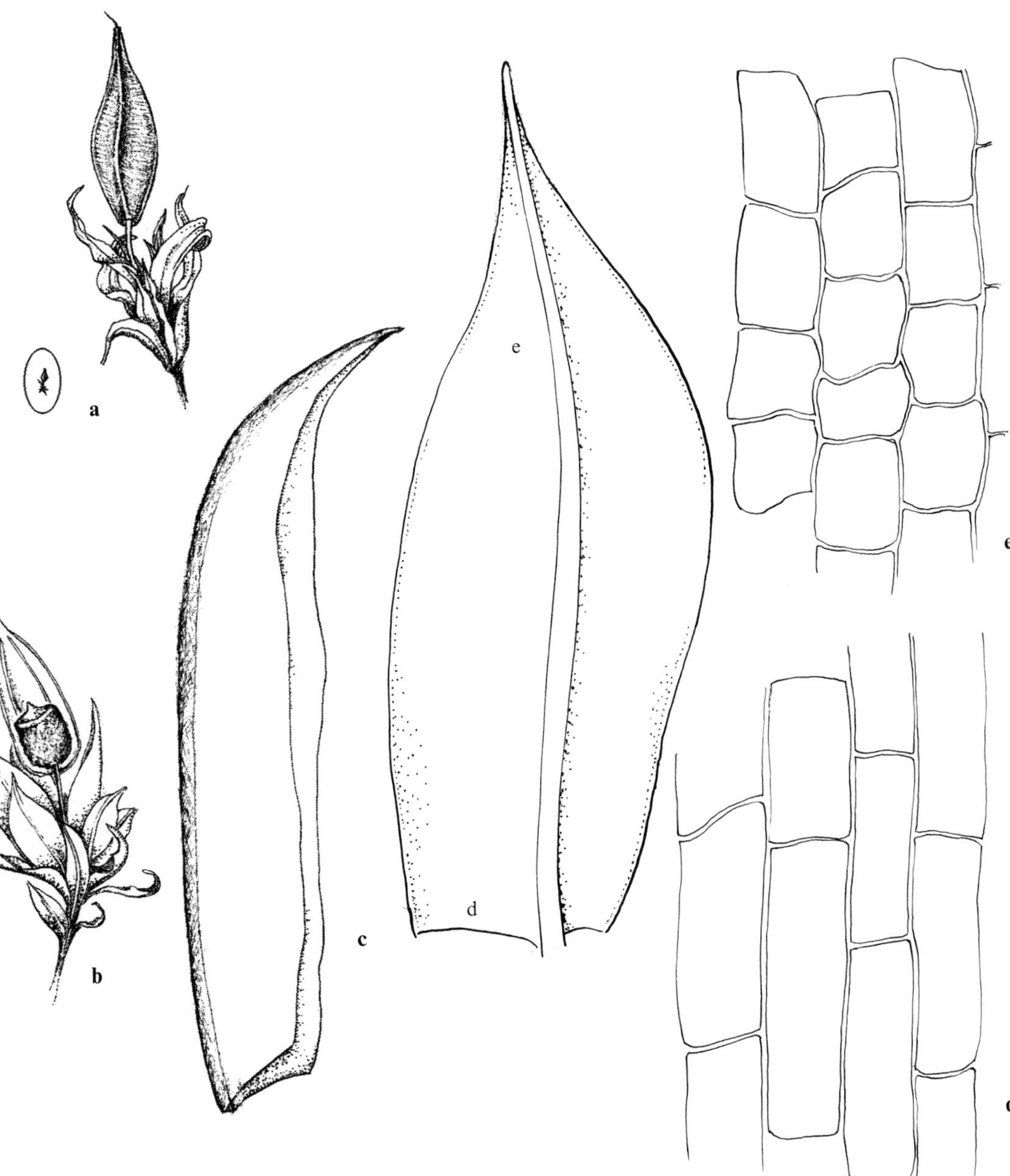

Figure 122. *Pyramidula tetragona*: **a** habit with calyptra, dry (×10); **b** habit and calyptra (in longitudinal section), moist (×10); **c** leaves (×50); **d** basal cells; **e** upper cells (all leaf cells ×500). (*Herrnstadt* & *Crosby 79-77-5*)

3. *Funaria* Hedw.

Plants small to medium-sized. *Leaves* oblong-ovate to broadly obovate-spathulate, rounded to acute or acuminate; margins plane, entire, or bluntly serrulate, rarely serrate, above, sometimes bordered; costa ending below apex to percurrent, more rarely excurrent; upper cells oblong-hexagonal. *Setae* elongate, usually slender, erect and flexuose to twisted and irregularly curved; basal cells long-rectangular; capsules exserted, inclined to pendent, asymmetrical, at least ± curved, pyriform, with distinct, long neck; when dry, theca sometimes sulcate, neck often wrinkled; often constricted below mouth when empty, exothecial cells oblong-hexagonal to linear; operculum convex to convex-conic, with cells in obliquely radial rows; annulus not differentiated or differentiated and revoluble; peristome well developed, double, inserted ± below mouth, curved-sigmoid, joined at tips. *Calyptrae* inflated at base, long-rostrate, cucullate.

A cosmopolitan genus of ca. 150 species. Four species occur in the local flora.

Literature. — Crundwell & Nyholm (1974).

The genus is often considered as including also *Entosthodon* (for differential characters see the description of *Entosthodon*). Differences in spore texture on generic level, as proposed by Fife (1985), could not be verified by SEM studies of local plants. The identification of *Funaria* species is mainly based on sporophytic characters, often making the determination of sterile plants impossible.

Funaria handelii, described by Schiffner (1913) from Turkey, has been also recorded from additional parts of Southwest Asia (sometimes as *Entosthodon*). It seems to resemble *E. durieui* in the unbordered leaves and to differ in the rudimentary (instead of very short) peristome teeth and the somewhat inclined capsules. Various degrees of inclination of capsules have been observed even within local populations of *E. durieui*, so some doubts remain as to the affiliation of *F. handelii* with *Funaria*, and even as to its separation from *E. durieui*.

1 Setae flexuose and twisted when dry; capsules up to 3.5 mm long including operculum; mature capsules sulcate when moist; annulus well differentiated and revoluble **4. F. hygrometrica**

1 Setae not flexuose and twisted when dry; capsules up to 2.5 mm long including operculum; mature capsules not sulcate when moist; annulus absent.

  2 Leaves with terminal cell usually less than 130 µm long; operculum convex, not mammillate **1. F convexa**

  2 Leaves with terminal cell usually longer than 130 µm; operculum conical-mammillate.

    3 Margins distinctly serrate; terminal cell usually 200–300 µm long **2. F. muhlenbergii**

    3 Margins entire to crenulate; terminal cell usually 130–180 µm long **3. F. pulchella**

Funaria convexa Spruce, *F. muhlenbergii* Turner, and *F. pulchella* H. Philib. form a group of species ("*F. dentata*" *sensu* Loeske 1929) quite similar in habit, with capsules having a well developed peristome and no distinct annulus. They have often been considered, mainly following Loeske (1929), as a single species: *F. dentata* Crome (Samml. Deut. Laubm. 2:26. 1806). The identity and delimitation of those three species was revised by Crundwell & Nyholm (1974), whose approach we are accepting here. They mainly distinguish between the species by the shape of the leaves and the operculum, and by the length of the terminal cell of the leaves (see the key above). SEM studies of spores of the three species in local material showed that all have differentiated proximal and distal parts and similar surface texture (Plate IX: a–c). Some intermediate forms between *F. convexa* and *F. pulchella* seem to occur among local plants.

Bilewsky (1965) accepted in this group two species: *Funaria convexa* and *F. dentata*, the latter name referring to *F. muhlenbergii*. He recorded under *F. dentata*, in addition to the typical variety, also "var. *mediterranea* (Lindb.) Lske". (which is a synonym of *F. muhlenbergii*) and "var. *patula* Br. Eur.", though the single collection in his herbarium named *F. dentata* var. *patula* is a misidentified specimen of *Entosthodon attenuatus*.

1. Funaria convexa Spruce, Musci Pyren. no. 149 (1847). [Bilewsky (1965): 393]
F. calcarea Wahlenb. var. *convexa* (Spruce) Husn., Muscol. Gall. 217 (1888); *F. pustulosa* Zodda, Malpighia 24: 270 (1910); — Bilewsky (1965): 393 (as "*F. dentata* Cr. var. *convexa* (Spr.) Lske.").

Distribution map 123, Figure 123, Plate IX: a, b.

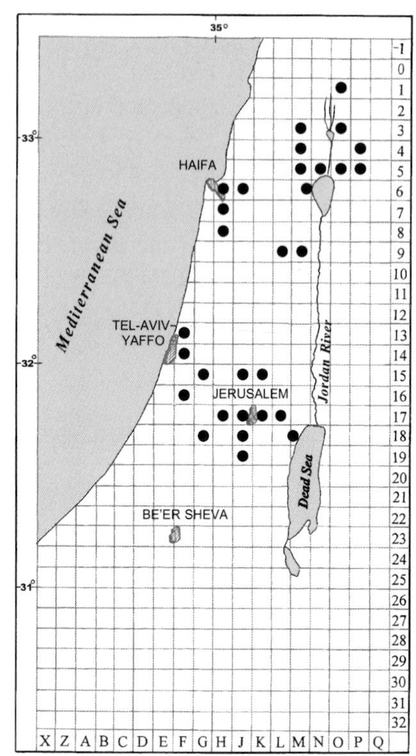

Map 123: *Funaria convexa*

Plants 5–10 mm high, pale yellowish-green. *Leaves*: lower leaves spreading, upper more erect; leaves up to 4 mm long, obovate to spathulate, obtuse, abruptly narrowing to a short apiculus, 215–230 μm long; terminal cell 90–130 μm; margins crenulate to ± serrulate above, marginal cells hardly differing from adjacent cells, sometimes yellowish; costa greatly varying in length, even in single plants, ending well below apex to percurrent, turning red with age; mid-leaf cells 20–36 μm wide. *Setae* 4–11 mm long, straight, yellowish-red; capsules inclined to horizontal, asymmetrical, ca. 2 mm long including operculum, yellowish-brown, constricted below red-coloured mouth, neck sulcate when dry, half the length of theca; operculum from nearly plane to convex, with orange-red border contrasting with theca. *Spores* 22–25 μm in diameter, coarsely papillose; surface uneven, foveolate, covered by insulated, compound, sometimes fused gemmae (seen with SEM).

Habitat and Local Distribution. — Plants growing in loose tufts in calcareous and basaltic rock crevices and on shaded or exposed soil. Common: Sharon Plain, Philistean Plain, Upper Galilee, Lower Galilee, Mount Carmel, Mount Gilboa, Samaria, Shefela, Judean Mountains, Judean Desert, Dan Valley, Dead Sea area, and Golan Heights.

General Distribution. — Recorded from Southwest Asia (Cyprus, Turkey, and Lebanon), around the Mediterranean, Macaronesia, Europe (the south in particular), tropical and North Africa, and northwestern South America.

Figure 123. *Funaria convexa*: **a** habit, moist (×20); **b** habit with deoperculate capsule, dry (×10); **c** capsule with operculum (×10); **d** calyptra (×10); **e** exostome and endostome teeth (×125); **f** leaf (×50); **g** basal cells; **h** upper cells; **i** apical cells (all leaf cells ×500). (*Herrnstadt & Crosby 78-28-5*)

c
e
b
d
h
b
a
g
f
i
h
g

2. Funaria muhlenbergii Turner, Ann. Bot. 2 : 198. (1805).

F. calcarea Wahlenb., Vet. Ak. Nya. Handl. 27 : 137 (1806); *F. dentata* Crome, Samml. Deut. Laubm. 2 : 26 (1806). — Bilewsky (1965) : 393, Rabinovitz-Sereni (1931); *F. mediterranea* Lindb., Öfvers. Foerh. Kongl. Svenska Vetensk.-Akad. 20 : 399 (1863). — Rabinovitz-Sereni (1931), Bizot (1945); *F. dentata* Crome var. *mediterranea* (Lindb.) Limpr., Laubm. Deutschl. 2 : 197 (1891). — Bilewsky (1965) : 394 (as "var. *mediterranea* (Lindb.) Lske.").

Distribution map 124, Figure 124.

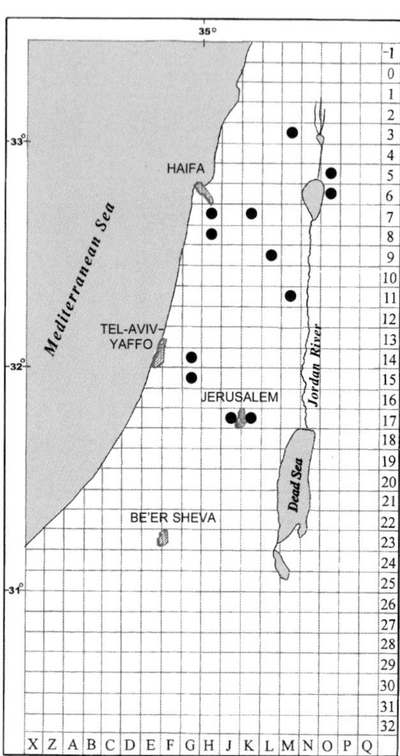

Map 124: *Funaria muhlenbergii*

Plants 5(–10) mm high, pale yellowish-green. *Leaves*: lower leaves spreading, upper more erect; leaves up to 3 mm long, obovate to oblong-lanceolate, ± abruptly narrowing into a long acumen up to 450 µm long; terminal cell 200–300(–450) µm long; margins distinctly serrate in upper ⅔ of leaf, strongest at base of acumen, marginal cells longer and narrower than adjacent cells; costa usually ending well below apex, sometimes percurrent; cells in mid-leaf 21–35 µm wide. *Setae* ca. 8 mm long, straight, yellowish-red; capsules inclined to horizontal, asymmetrical, up to 2.5 mm long including operculum, yellowish-brown, smooth, neck sulcate when dry, half the length of theca; operculum conical, mammillate, yellowish orange, without contrasting margin; peristome double, well-developed. *Spores* 25–30 µm in diameter, resembling spores of *E. convexa* as seen with light microscope and SEM.

Habitat and Local Distribution. — Plants growing in loose tufts on exposed soil on north-facing slopes and in partly shaded habitats, on soil among calcareous rocks. Occasional: Upper Galilee, Lower Galilee, Mount Carmel, Mount Gilboa, Samaria, Shefela, Judean Mountains, and Golan Heights.

General Distribution. — Recorded from Southwest Asia (widespread in Cyprus, Turkey, Syria, Lebanon, Jordan, Iraq, Iran, and Saudi Arabia), around the Mediterranean, Europe (throughout), Macaronesia, North Africa, eastern Asia, and North and Central America.

Sterile plants of *Funaria muhlenbergii* may resemble those of *Pyramidula tetragona* but can be distinguished by the serrate upper margins and the shorter, not excurrent costa of the leaves. *Funaria muhlenbergii* is treated as *Entosthodon muhlenbergii* by Fife

---

Figure 124. *Funaria muhlenbergii*: **a** habit, moist (×10); **b** habit, dry (×10); **c** calyptra (×10); **d** exostome and endostome teeth (×125); **e** leaf (×50); **f** basal cells; **g** marginal upper cells; **h** apical cells (all leaf cells ×500). (*D. Zohary*, 4 Feb. 1943)

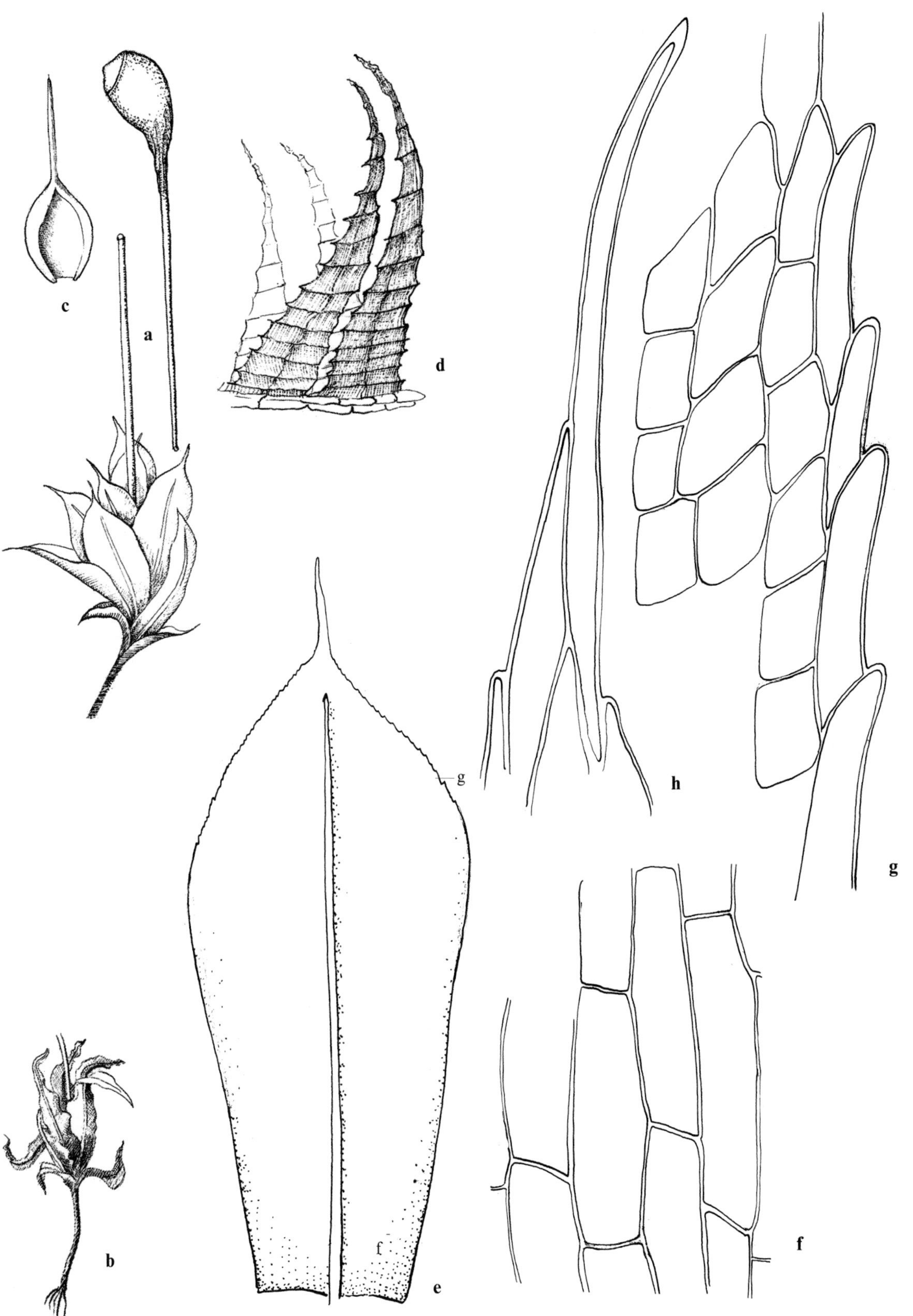
c
a
d
h
g
g
f
e
b
f

(1985), apparently because of the absence of an annulus, considered by him as the main character differentiating between the two genera. If this concept is consistently applied, *F. convexa* and *F. pulchella* have to be similarly treated. Bilewsky (1965) recorded this species as *F. dentata* and *F. dentata* var. *mediterranea*. The specimens bearing those names represent parts of the range of variability of *F. muhlenbergii* (see also the discussion following the key to *Funaria*).

3. Funaria pulchella H. Philib., Rev. Bryol. 11 : 41 (1884).

Distribution map 125, Figure 125, Plate IX : c.

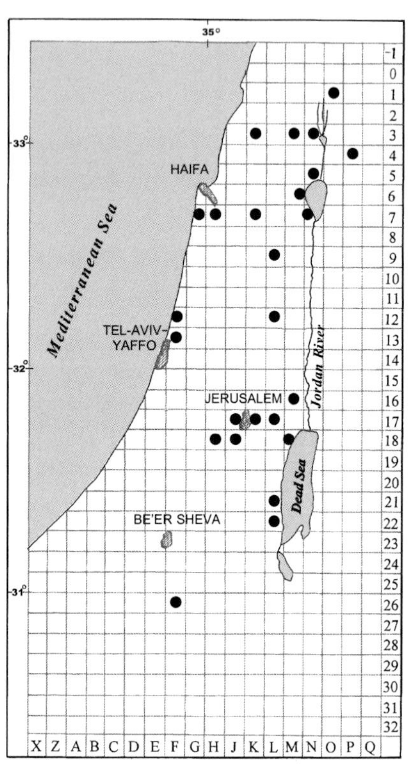

Map 125: *Funaria pulchella*

Plants up to 5(–10) mm high, pale to yellow-green. *Leaves* resembling *F. muhlenbergii* but ± wider, obovate to ovate, acuminate, ± gradually narrowing into medium-sized acumen 220–350 μm long; terminal cell 130–180 μm long; margins entire to weakly crenulate, marginal cells hardly different from adjacent cells; mid-leaf cells 20–35 μm wide. *Setae* and capsules like of *F. muhlenbergii*, but operculum yellowish with reddish contrasting margin. *Spores* ca. 25 μm in diameter; resembling spores of *F. convexa* as seen with light microscope and SEM.

Habitat and Local Distribution. — Plants growing in tufts on sandstone, limestone, and basalt soil on rocks, in open habitats, in rock crevices, on walls, and on north-facing banks of streams. Common: Coast of Carmel, Sharon Plain, Upper Galilee, Lower Galilee, Mount Carmel, Mount Gilboa, Samaria, Judean Mountains, Judean Desert, Central Negev, Hula Plain, Upper and Lower Jordan Valley, Dead Sea area, and Golan Heights.

General Distribution. — Recorded from Southwest Asia (Cyprus, Turkey, Jordan, Saudi Arabia, and Oman), around the Mediterranean, Europe (throughout), Macaronesia, North Africa (Morocco, Algeria, and Egypt), northeastern and central Asia, and North America.

Funaria pulchella is more common in the local flora than *F. muhlenbergii*. It can be distinguished mainly by the nearly entire — not serrate — margins, the shorter apex, and terminal cell of the leaves. It was recorded in the local flora only in 1991 (Brullo *et al.* 1991, Herrnstadt *et al.* 1991).

Figure 125. *Funaria pulchella*: **a** habit, dry (×10); **b** habit, moist (×10); **c** operculate and deoperculate mouth of capsules (×10); **d** capsule with partly open peristome (×10); **e** calyptra (×10); **f** exostome and endostome teeth (×125); **g** leaf (×50); **h** basal cells; **i** upper cells; **k** apical cells (all leaf cells ×500). (*Danin & Herrnstadt 80-132-4*)

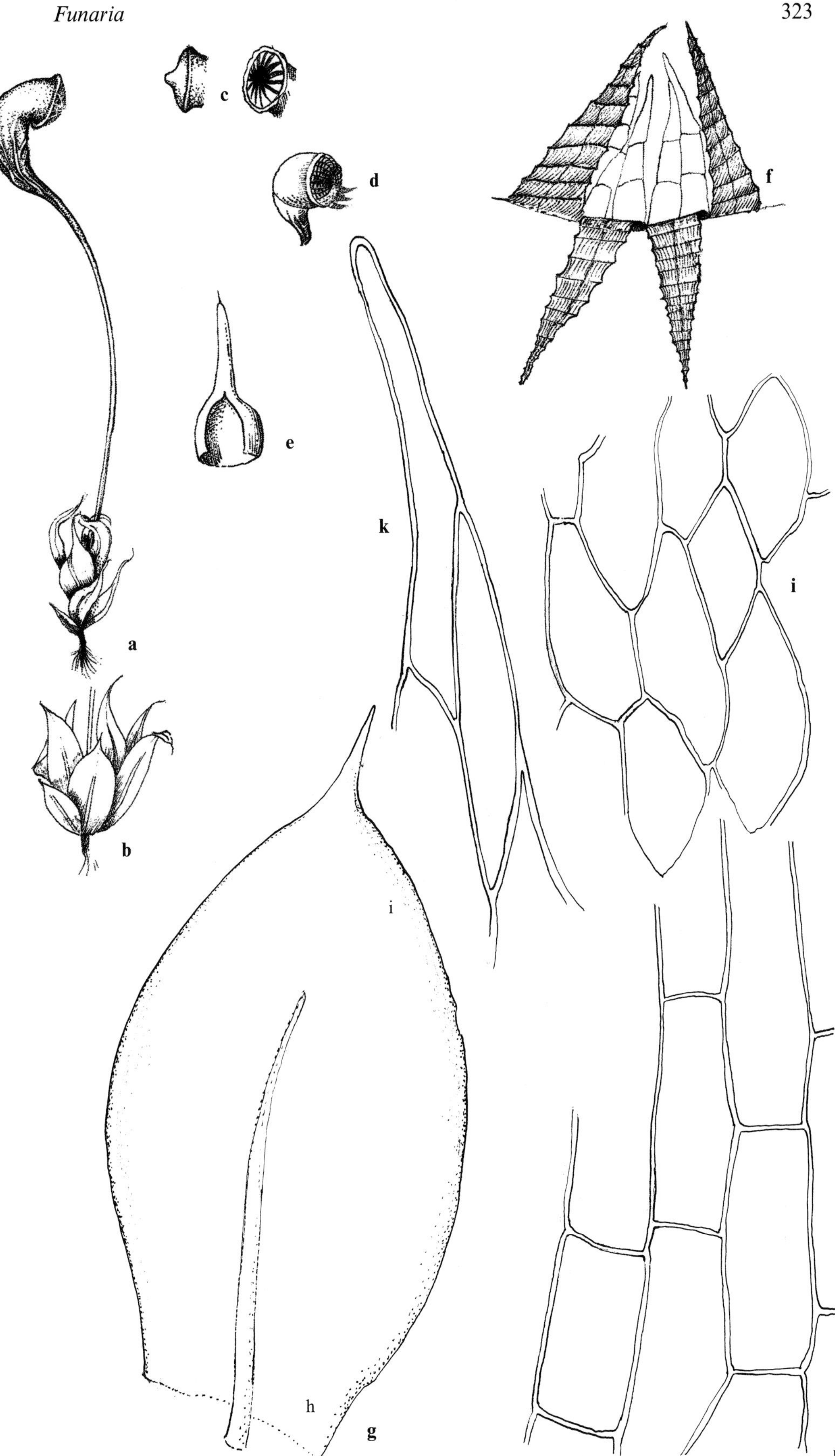
c
d
f
e
k
i
a
b
i
h
g
h

4. Funaria hygrometrica Hedw., Sp. Musc. Frond. 172 (1801). [Bilewsky (1965): 394]

Distribution map 126, Figure 126, Plate IX: d.

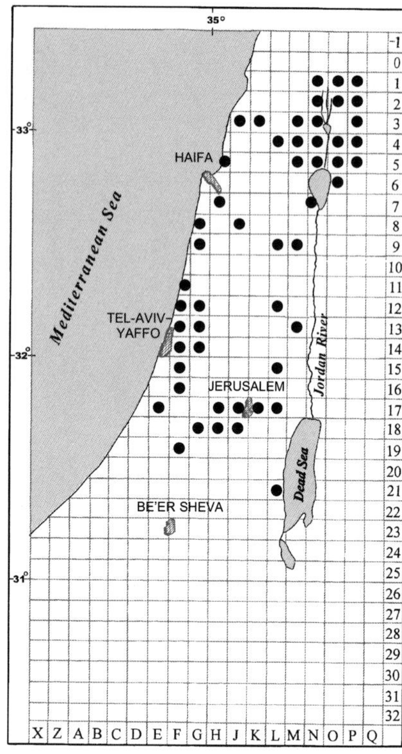

Map 126: *Funaria hygrometrica*

Plants 4–20 mm high, pale green to yellowish-green. *Stems* branched near base. *Leaves* erectopatent to spreading below, erect above, deeply concave, 2–4(–5) mm long, ovate, oblong-ovate to obovate, acute to short-acuminate; margins entire to weakly serrulate; costa subpercurrent to short-excurrent; cells in mid-leaf 27–40 μm wide. *Setae* much varying in length, up to 50 mm long, slender, hygroscopic, flexuous and twisted when dry; capsules asymmetrical, inclined and distinctly curved or suberect and slightly curved, 2–3.5 mm long including operculum, pyriform, yellowish to reddish-brown; mature (in particular empty) capsules sulcate when dry and moist; mouth oblique, neck short, distinct; operculum plane to convex, sometimes mammillate, with red border contrasting in colour; annulus well differentiated, wide, red, revoluble. *Spores* 14–20 μm in diameter, densely papillose; densely baculate, baculae small with rough surface (seen with SEM).

Habitat and Local Distribution. — Plants growing in loose or dense tufts or gregarious, mainly on soil in disturbed habitats, often following fire, also in rock crevices and soil on rocks and walls. Very common: growing throughout Israel, except in the most extreme arid conditions (Negev and Arava valley).

General Distribution. — A cosmopolitan weed.

Literature. — Weitz & Heyn (1981).

The record of *Funaria hygrometrica* var. *calvescens* (Schwägr.) Mont. (Ann. Sci. Nat. Bot. 2, 12: 54. 1839) by Bilewsky (1970) is based on a single specimen from the Sharon Plain. It does not seem to have characters that fall outside the range of variation of the typical variety of the species. The intra-specific variation of *F. hygrometrica* was detailed in Weitz & Heyn (1981).

Figure 126. *Funaria hygrometrica*: **a** habit, dry (×10); **b** habit with base of seta, moist (×10); **c** calyptra (×10); **d** peristome with partly detached annulus (×10); **e** closed peristome (×20); **f** exostome and endostome teeth (×125); **g** leaves (×50); **h** basal cells; **i** upper cells (all leaf cells ×500). (*Weitz*, 31 Mar. 1971)

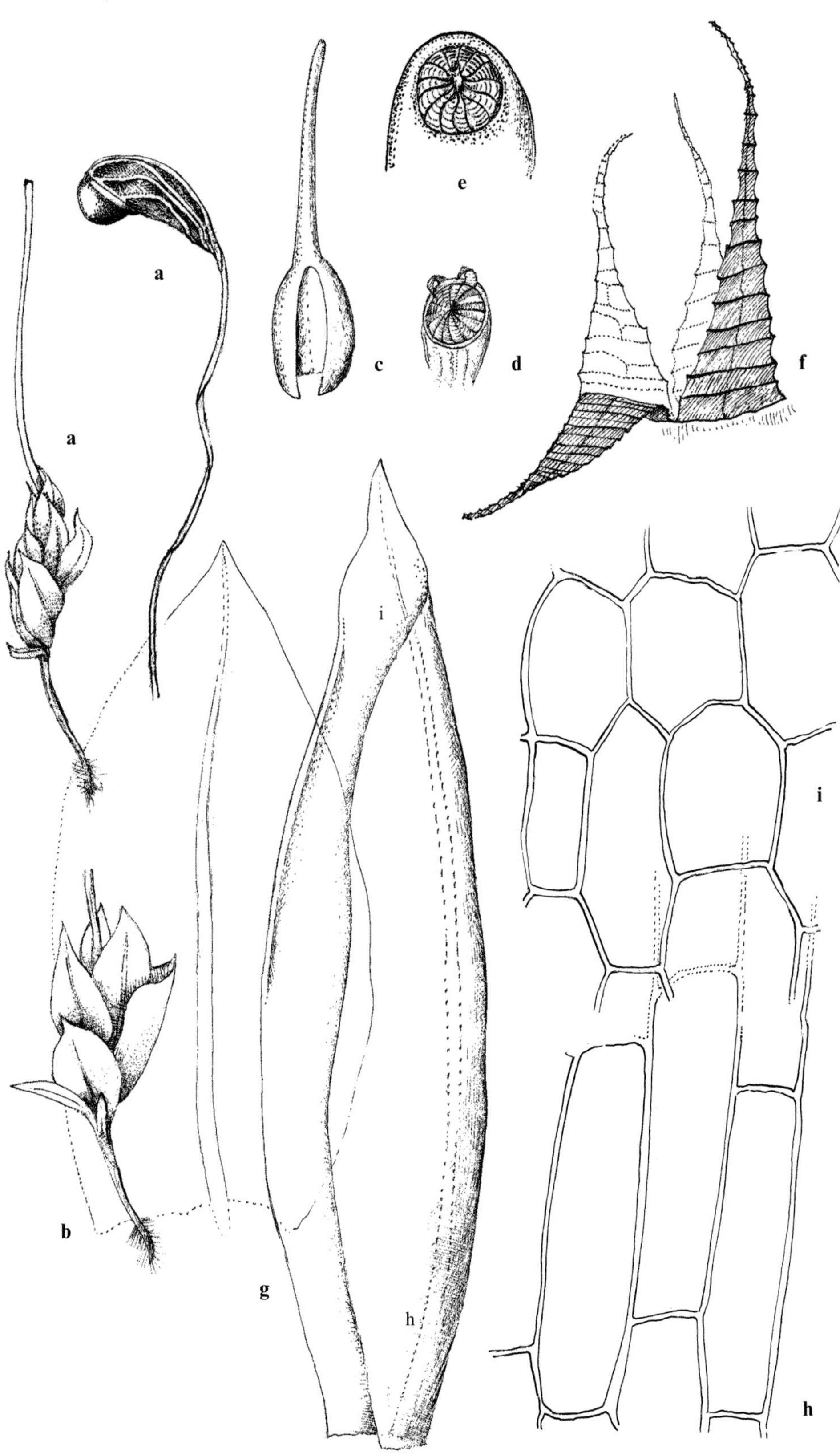
a
a
c
d
e
f
b
g
h
i
i
h

4. *Entosthodon* Schwägr.

Plants small, resembling *Funaria* (in which the genus is sometimes included) but differ in several sporophytic characters. *Setae* straight (regularly curved in *Entosthodon curvisetus*); capsules exserted, erect, symmetrical, rarely slightly asymmetrical; operculum convex to convex-conic, rarely nearly plane, with cells only rarely in obliquely radial rows; annulus usually not differentiated; peristome absent, rudimentary, or single, very rarely double, inserted well below mouth, ± oblique, not curved-sigmoid, not joined at tips. *Calyptrae* inflated at base, rostrate, cucullate with entire or lobed base, sometimes mitrate.

A cosmopolitan genus of ca. 100 species. Six species occur in the local flora.

Entosthodon angustifolius Jur. & Milde recorded from Jordan (El-Oqlah *et al.* 1988a) has not been found so far in the local flora.

In the local flora, the majority of *Entosthodon* species are less widespread than species of *Funaria*, though they may be locally abundant (as *E. attenuatus* and *E. curvisetus*), and species of both genera may grow in mixed populations. Some of the local species seem to tolerate quite extreme xeric habitat conditions.

Entosthodon is often considered as part of *Funaria*; the differential characters are presented above. Identification of species of *Entosthodon* is mainly based on sporophytic characters, which makes determination of sterile plants often rather difficult.

Entosthodon was initially described as having cucullate calyptrae. It seems that mitrate calyptrae were first mentioned by Casares-Gil & Beltrán (1912) as characteristic of *E. physcomitrioides* Casares-Gil & Beltrán (for the nomenclatural history of this taxon see the discussion of *E. durieui*). Boros (1925) described *E. hungaricus* (as *Funaria hungarica*) as having "Calyptra cucullata, haud raro irregulariter bi- et trilobata" and depicted a deeply lobed calyptra. Vondráček & Hadač *in* Vondráček (1965) described a new genus — *Steppomitra* — to accommodate "*Funaria hungarica*" and *S. hadacii* Vondr. They considered the new genus different from *Funaria* and *Entosthodon* mainly in the mitrate — not cucullate — calyptrae, and from *Physcomitrium* in the longer rostrum of the operculum and the longer exothecial cells (oblong-hexagonal to linear, like in *Entosthodon* and *Funaria*). Sometimes, it appears difficult to treat the calyptrae of *Entosthodon* as strictly cucullate or mitrate. This may be due, at least in part, to their inflated base, which may appear lobed when torn. However, in other cases calyptrae were described as having one of the following character states: (a) unequal lobes (in "*Funaria hungarica*"); (b) a mitrate, regularly lobed calyptrae; or (c) calyptrae with much deeper slit (in "*Funaria pallescens* var. *mitratus*"; Casas de Puig 1979). Similar character states were observed by us in some local populations of *E. durieui*.

1 Setae cygneous; capsules curved, ± asymmetrical **6. E. curvisetus**
1 Setae straight; capsules erect, symmetrical.
  2 Capsules round-pyriform, with neck shorter than theca; peristome absent or rudimentary.
    3 Leaf margins not bordered, in upper half distinctly serrate **3. E. fascicularis**
    3 Leaf margins distinctly bordered (usually with one row of elongated marginal cells), entire to weakly serrate **4. E. obtusus**
  2 Capsules oblong-pyriform, with neck equal to theca or longer; peristome present or rudimentary.
    4 Leaf margins with a narrow border, usually of one row of cells; peristome teeth up to 320 μm long; calyptrae cucullate, entire or sometimes with short lobes at base **1. E. attenuatus**
    4 Leaf margins not bordered; peristome teeth rudimentary, not exceeding 55 μm in length; calyptrae equally (mitrate) or unequally lobed.
      5 Operculum plane-convex; costa usually reaching up to about mid-leaf **2. E. durieui**
      5 Operculum rostrate; costa longer, up to ¾ length of leaf **5. E. hungaricus**

1. Entosthodon attenuatus (Dicks.) Bryhn, Kongel. Norske Vidensk. Selsk. Skr. (Trondheim) 1908 (8) : 25 (1908).

Bryum attenuatum Dicks., Pl. Crypt. Brit., Fasc. 4 : 10 (1801); *E. templetonii* (Sm.) Schwägr., Sp. Musc. Frond., Suppl. 21 : 44 (1823); *E. commutatus* Durieu & Mont., Ann. Sci. Nat. Bot. 3, 12 : 317 (1849).

Distribution map 127, Figure 127, Plate IX : e.

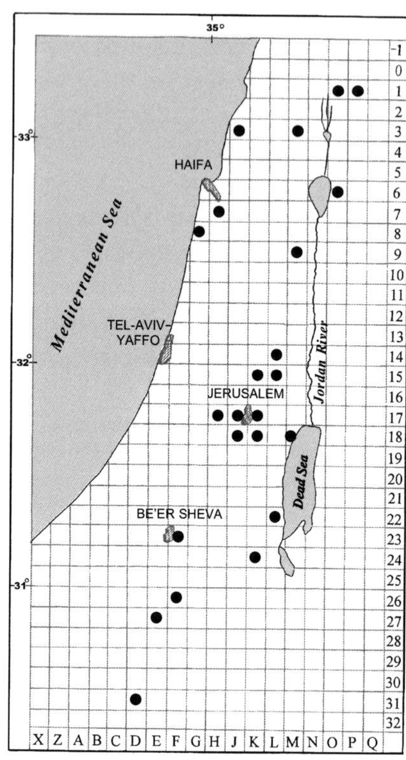

Map 127: *Entosthodon attenuatus*

Plants up to 3.5 mm high, bright green, turning pale yellow when dry. *Leaves* up to 3 mm long, ovate, oblong, or obovate-spathulate, acute to acuminate; margins bordered, crenulate or weakly serrulate, teeth blunt; costa ending in or below apex; basal cells rectangular, upper cells shorter, irregularly hexagonal to rectangular, mid-leaf cells 22–35 μm wide; one row of marginal cells longer, usually ± narrower than adjacent cells, sometimes yellow, forming the border. *Setae* pale yellow to reddish-yellow, straight, about 10 mm long; capsules erect, symmetrical (sometimes both symmetrical and ± asymmetrical capsules within single populations), greatly varying in size, up to 3.5 mm long including operculum, oblong-pyriform, with neck the length of theca or longer, turning reddish-brown at maturity; operculum plane-convex; peristome single, rarely with rudimentary endostome, teeth usually 200–320 μm long, often fugacious. *Spores* 24–30 μm in diameter, papillose; baculate, baculae on foveolate, uneven surface, sometimes fused (seen with SEM). *Calyptrae* cucullate, sometimes with short lobes at base.

Habitat and Local Distribution. — Plants growing usually in small tufts or scattered on calcareous basalt or loess soils, in soil pockets among rocks and stones, in shade of rock outcrops or at foot of rocks and at cave entrances; often found in dry areas under conditions of additional water supply (e.g., Negev Highlands); sometimes growing together with *Funaria pulchella*, *F. convexa*, and *Gigaspermum mouretii.* Common (*Entosthodon attenuatus* is the most widespread local *Entosthodon* species): Upper Galilee, Mount Carmel, Mount Gilboa, Samaria, Judean Mountains, Judean Desert, Northern Negev, Central Negev, Dead Sea area, Mount Hermon, and Golan Heights.

General Distribution. — Recorded from Southwest Asia (throughout), around the Mediterranean, Europe (mainly the south), Macaronesia, North Africa, northeastern and eastern Asia, and North America.

Literature. — Fife (1987).

Entosthodon attenuatus has the longest peristome teeth among all local species of the genus. Although *E. attenuatus* closely resembles *E. durieui* in habit, it can be

Figure 127. *Entosthodon attenuatus*: **a** habit, moist (×10); **b** habit, dry (×10); **c** detached operculum (×10); **d** calyptra (×10); **e** part of mouth of capsule with peristome teeth (×125); **f** leaf (×50); **g** basal cells; **h** marginal mid-leaf cells; **i** upper cells (all leaf cells ×500). (*Weitz 1390*)

distinguished from that species, in addition to the long peristome teeth, by the darker capsules and the bordered leaves with longer costa. In dried plants the colour and shape of the border may be obliterated and identification of sterile herbarium material is often rather difficult.

Fife (1985) described the spores of *Entosthodon attenuatus* as "nearly smooth" (SEM observations) and "without appreciable ornamentation" (light microscope observations). This does not agree with Loeske (1929) and with the distinctly baculate-insulate spores observed in local plants (Plate IX: e).

According to Fife (1987), who studied Dickson's type specimen of *Bryum attenuatum*, it "matches the protologue well". However, the protologue includes the sentence: "Bryum capsulis erectis nudis", which refers to capsules lacking teeth. As peristome teeth are often lost in dry specimens, it may be correct to assume that Dickson's description was based on such material. We have examined a specimen deposited in the Dickson collection (BM), which has bordered leaves. However, for obvious reasons, the single sporophyte of that sample could not be studied. In spite of the difficulties pointed out, *E. attenuatus* seems the name that should be adopted for the *Entosthodon* species with bordered leaves and well developed peristome teeth.

Loeske (1929) pointed out the difficulties of identification and delimitation of several species that seem related to *Entosthodon attenuatus*, as well as to "*E. pallescens*" (see the discussion of *E. durieui*).

2. Entosthodon durieui Mont., Ann. Sci. Nat. Bot., ser. 3, 11: 33 (1849).

E. pallescens Jur. *in* Unger & Kotschy, Ins. Cypern 170 (1865); *E. mustaphae* Trab., Atlas Fl. Algerie, Fasc. 1: 12, Pl. 7, Figs. 1–9 (1886); *Funaria pallescens* (Jur.) Lindb. *in* Broth., Trudy Imp. S. Peterburgsk. Bot. Sada 10: 562 (1888). — Bilewsky (1965): 392; *E. physcomitrioides* Casares-Gil & Beltrán, Bol. Soc. Esp. Hist. Nat. 12: 375 (1912); *E. pallescens* Jur. var. *mitratus* Casares-Gil, Enum. Distr. Géogr. Musc. Penins. Iberica 110 (1915); *E. mitratus* (Casares-Gil) Loeske, Repert. Spec. Nov. Regni Veg. Sonderbeih. 3: 55 (1929), *nom. inval.*; *Funaria pallescens* (Jur.) Lindb. *in* Broth. var. *mitratus* (Casares-Gil) Wijk & Margad., Taxon 9: 50 (1960).

Distribution map 128 (p. 332), Figure 128, Plate IX: f.

Plants resembling *Entosthodon attenuatus* in habit, bright green, turning pale yellow when dry. *Leaves* differing in the shorter costa, usually reaching up only to about mid-leaf; margins entire to crenulate, not bordered. *Capsules* erect, symmetrical, up to 2 mm long including operculum, oblong-pyriform, turning yellowish-red at maturity, operculum plane-convex; peristome rudimentary, teeth very short, not exceeding 55 µm, truncate, one–three articulated. *Calyptrae* equally lobed (mitrate) or unequally lobed (with one deeper slit). *Spores* only up to 27 µm in diameter, resembling in surface texture *E. attenuatus*; sparsely verrucate on uneven, foveolate surface, verrucae with rough apexes (seen with SEM).

Figure 128. *Entosthodon durieui*: **a** habit, moist (×10); **b** habit, dry (×10); **c** deoperculate capsule (×10); **d** leaf (×50); **e** basal cells; **f** marginal upper cells; **g** apical cells (all leaf cells ×500). (*Bilewsky*, April 1954, No. 89 in Herb. Bilewsky)

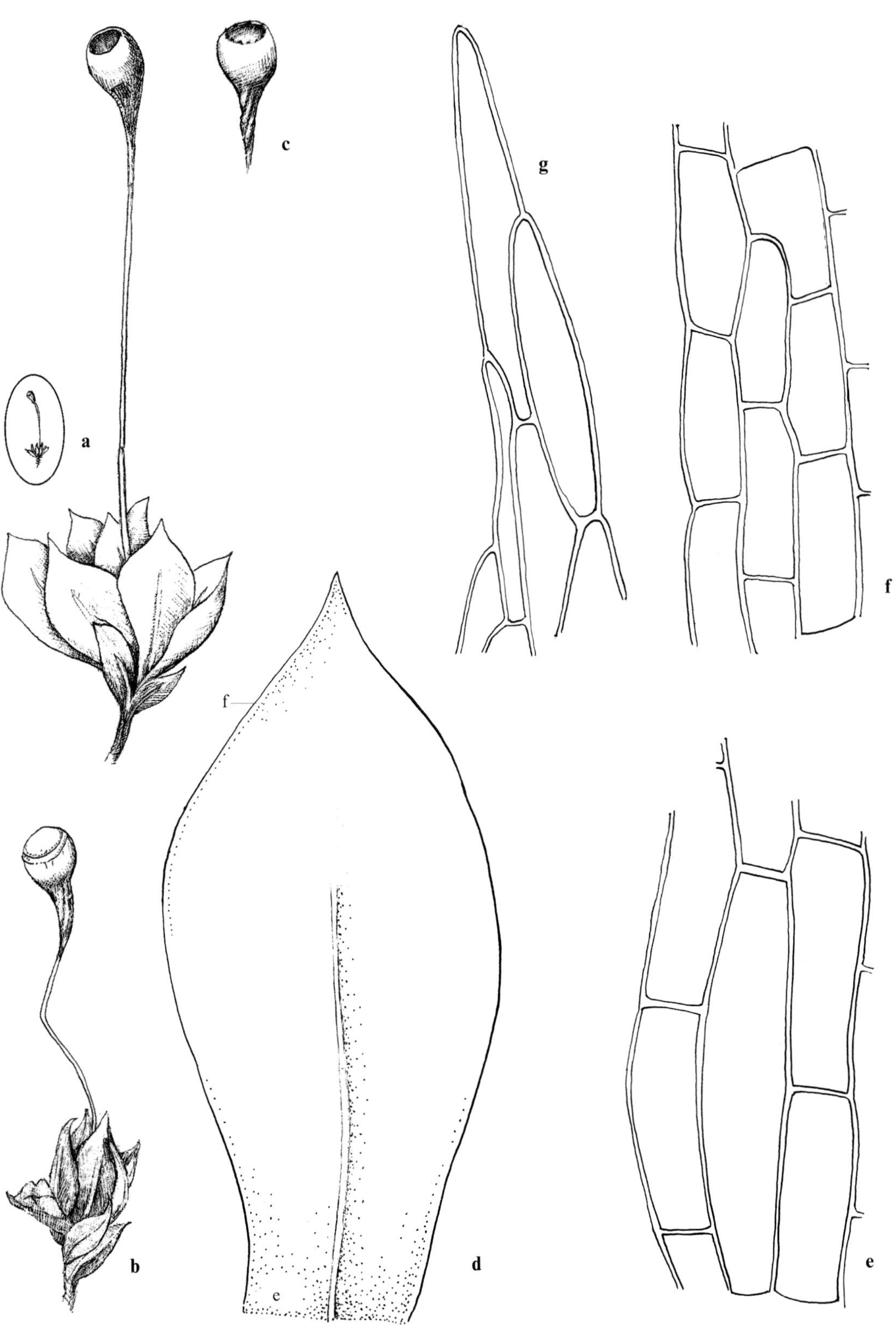

a
b
c
d
e
f
g
f
e

Habitat and Local Distribution. — Plants growing in small tufts on calcareous soils and rocks, in shade of rock outcrops, at foot of rocks, and at cave entrances; in dry areas under conditions of additional water supply. Occasional and common: Upper Galilee, Lower Galilee, Mount Carmel, Samaria, Shefela, Judean Mountains, Judean Desert, Central Negev, Mount Hermon, and Golan Heights.

General Distribution. — Recorded (when considered as including *Entosthodon pallescens*) from Southwest Asia (Cyprus, Turkey, Sinai, and Iran), few localities around the Mediterranean, Europe (scattered records from the south), Macaronesia, and North Africa.

Literature. — Casas de Puig (1979), Brugués (1998).

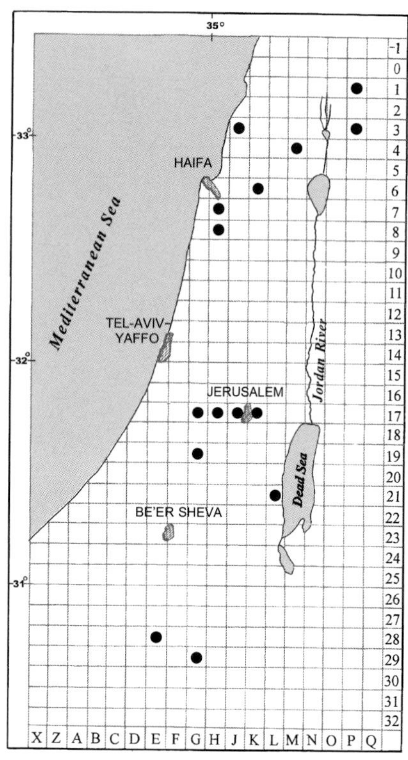

Map 128: *Entosthodon durieui*

Two species: *Entosthodon commutatus* Durieu & Mont. (Ann. Sci. Nat. Bot., ser. 3, 12:317. 1849) and *E. mustaphae* Trab. (Corley *et al.* 1981), in addition to *E. durieui*, were described from Algeria, variously treated, and often affiliated with either *E. attenuatus* or "*E. pallescens*." Corley *et al.* (1981) considered *E. commutatus* and *E. mustaphae* as synonymous of *E. durieui*. *Entosthodon mustaphae* is usually included in *E. durieui* (e.g., Bizot 1945), often at the varietal rank (e.g., Bilewsky 1965, Casas 1991). *Entosthodon commutatus* is treated either as a synonym of *E. durieui* (Frey & Kürschner 1991c), or as a separate species (El-Oqlah *et al.* 1988a). Among those three taxa, only one, *E. commutatus*, was described by Durieu and Montagne as having long peristome teeth, whereas originally *E. durieui* and *E. mustaphae* were described and depicted as having short peristome teeth (with up to three articulations). An original specimen of Montagne, named by him *E. durieui*, which has been seen by us in the British Museum (Algeria, Montagne, BM), has short — though perhaps broken — peristome teeth and unbordered leaves. Considering all, *E. commutatus* should be either regarded as synonymous with *E. attenuatus* (which has, however, bordered leaves) or may be a separate species. *Entosthodon mustaphae* clearly seems synonymous with *E. durieui*.

Entosthodon pallescens Jur., which was described from Cyprus, is often treated as a species of the eastern Mediterranean separate from *E. durieui* (Bilewsky 1965, cf. also Frey & Kürschner 1991c). In Herrnstadt *et al.* (1991), the name *E. pallescens* was regarded as a separate species whereas *E. durieui*, *E. mustaphae*, and *E. commutatus* were regarded as synonyms of *E. attenuatus*. The original description of *E. pallescens* does not include the operculum and calyptra (which were not seen by the author). In all other characters it seems identical with *E. durieui*. If the two species are indeed similar, *E. durieui* has priority and should be the correct name for this taxon. However, the possibility that *E. pallescens* may be identical with *E. durieui* in all characters, except for the shape of operculum and calyptra, still cannot be entirely excluded.

Entosthodon mitratus (= *E. physcomitrioides* according to Casares-Gil & Baltrán 1912) is mainly separated from *E. pallescens* or *E. durieui* by the mitrate calyptra, which is assumed to be cucullate in the other two species. Since the occurrence of both typical mitrate and irregularly lobed calyptrae have been observed in *Funaria pallescens* var. *mitratus* (Casas de Puig 1979), as seen by us also in *E. durieui*, the separation of *E. mitratus* does not seem valid. In this treatment, *E. pallescens* and *E. mitratus* are considered as synonyms of *E. durieui* (cf. Brugués 1998).

The first local record of *Entosthodon durieui* (Mount Carmel: "Zichron Yaacov") was published by Bizot (1945) as *Funaria durieui* Mont. This combination is in fact invalid as *Funaria durieui* was previously used by Schimper (*in* Bescherelle, Cat. Mous. Alg.: 23. 1882) for another taxon. Bilewsky (1965) recorded *Funaria durieui* (Mont.) Broth. and Bilewsky & Nachmony (1955) recorded *F. durieui* (Mont.) Broth. var. *mustaphae* (Trab.) Bizot from the Upper Galilee, Sharon Plain, and Judean Desert. The occurrence of *E. durieui* in the Sharon Plain has not been verified.

3. Entosthodon fascicularis (Hedw.) Müll. Hal., Syn. Musc. Frond. 1 : 120 (1848). *Gymnostomum fasciculare* Hedw., Sp. Musc. Frond. 38 (1801); *Funaria fascicularis* (Hedw.) Lindb., Öfvers. Förh. Kongl. Svenska Vetensk.-Akad. 21 : 597 (1865).

Distribution map 129, Figure 129.

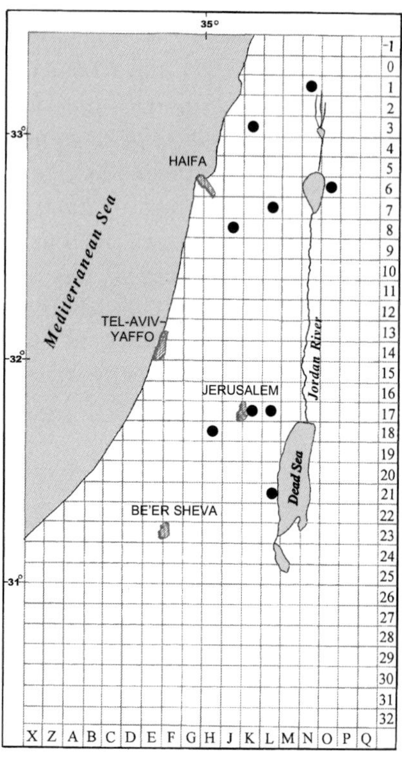

Map 129: *Entosthodon fascicularis*

Plants up to 5 mm high, bright green, turning yellow when dry. *Leaves* up to 2.5 mm long, concave, obovate to lanceolate, acute to acuminate; margins serrate in upper half; costa ending from slightly above mid-leaf to shortly below apex; basal cells rectangular, upper cells smaller, irregularly rectangular to hexagonal, in mid-leaf 25–35 µm wide; sometimes some cells of marginal row longer and narrower than adjacent cells, but not forming a distinct border. *Setae* 5–8 mm long, straight, reddish-yellow; capsules erect, symmetrical to very slightly asymmetrical, 1–3 mm long including operculum, round-pyriform, neck shorter than theca, constricted at mouth, yellowish, turning yellowish-red brown at maturity; operculum flat to convex, not mammillate; peristome rudimentary, fugacious. *Spores* 27–30 µm in diameter, papillose (as in *Entosthodon attenuatus*). *Calyptra* cucullate, distinctly lobed at base.

Habitat and Local Distribution. — Plants growing in patches or scattered among other bryophytes, on calcareous soils (usually rich in calcium-carbonate). Occasional: Upper Galilee, Lower Galilee, Samaria, Judean Mountains, Judean Desert, and Golan Heights.

General Distribution. — Recorded from Southwest Asia (Cyprus, Lebanon, and Jordan), around the Mediterranean, Europe (throughout), and North Africa.

The separation of *Entosthodon fascicularis* from *E. obtusus* is rather difficult. The capsules in both species are round-pyriform and have a short neck. The main differential characters used in the literature (e.g., Loeske 1929) are the serrate unbordered leaves of *E. fascicularis* and its larger capsules as compared with the nearly entire, bordered leaves of *E. obtusus* and the smaller capsules.

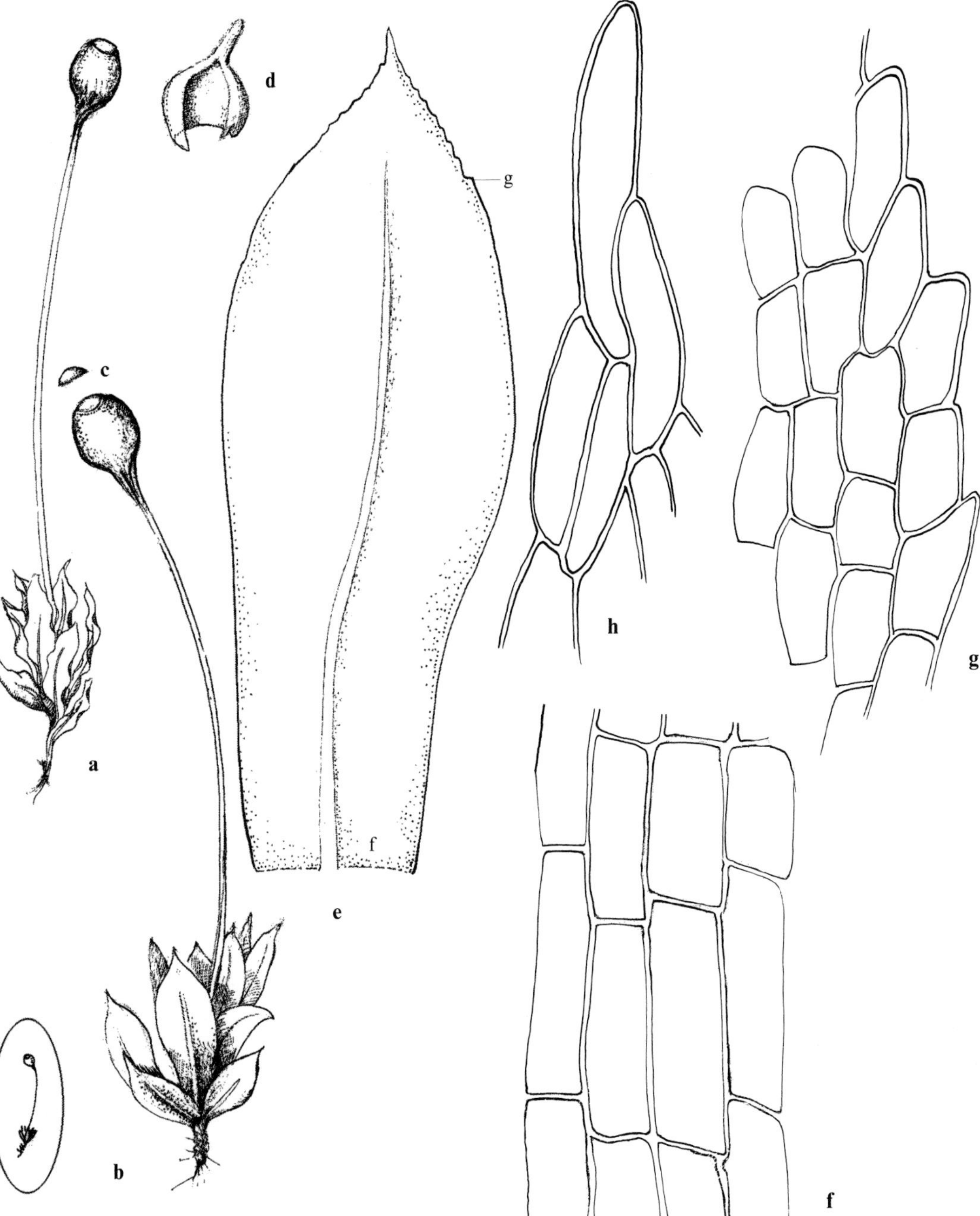

Figure 129. *Entosthodon fascicularis*: **a** habit, dry (×10); **b** habit, moist (×10); **c** detached operculum (×10); **d** calyptra (×10); **e** leaf (×50); **f** basal cells; **g** marginal upper cells; **h** apical cells (all leaf cells ×500). (*Pazy* & *Herrnstadt 81-178-1*)

4. Entosthodon obtusus (Hedw.) Lindb., Öfvers. Förh. Kongl. Svenska Vetensk. Akad. 21 : 221 (1865).

Gymnostomum obtusum Hedw., Sp. Musc. Frond. 34 (1801); *E. ericetorum* (De Not.) Müll. Hal., Syn. Musc. Frond. 1 : 122 (1848); *Funaria obtusa* (Hedw.) Lindb., not. Sällsk., Fauna Fl. Fenn. Förh. 11 : 65 (1870). — Bilewsky (1977).

Distribution map 130, Figure 130.

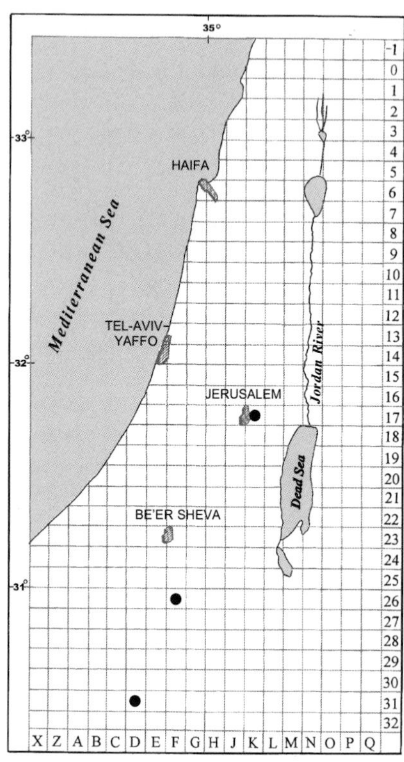

Map 130: *Entosthodon obtusus*

Plants up to 5 mm high, bright yellowish-green, closely resembling *Entosthodon fascicularis*, but differing in the following characters. *Leaves* entire to weakly serrate, with one–two rows of elongated, yellowish, marginal cells, forming a distinct border. *Capsules* light brown to dark reddish-brown when mature. *Spores* with texture like *E. fascicularis*, ± larger, up to 32 μm in diameter.

Habitat and Local Distribution. — Plants growing in patches or scattered, on soil at the base of a stone wall and on soil among north-facing calcareous rocks. Rare: Judean Mountains and Central Negev (two localities).

General Distribution. — Recorded from Southwest Asia only from Turkey, few records from around the Mediterranean, Europe (throughout), Macaronesia, and North Africa.

Entosthodon obtusus is found in Europe and North Africa on peaty or loamy, wet acid soils, very different from the alkaline substrate on which the local populations of this species grow. In our flora the plants considered by us as *E. obtusus* have the same habitat requirements as *E. fascicularis*.

Bilewsky (1977) recorded "*Funaria obtusa* var. *notarisii* (Schimp.) Pavl." (= var. *notarisii* (Schimp.) Latzel) from one locality in the Esdraelon Plain ("dry parts of the banks of the Sachne river") and considered this variety as the only representative of *Entosthodon obtusus* in Israel. The specimen cited by Bilewsky was not found in his herbarium.

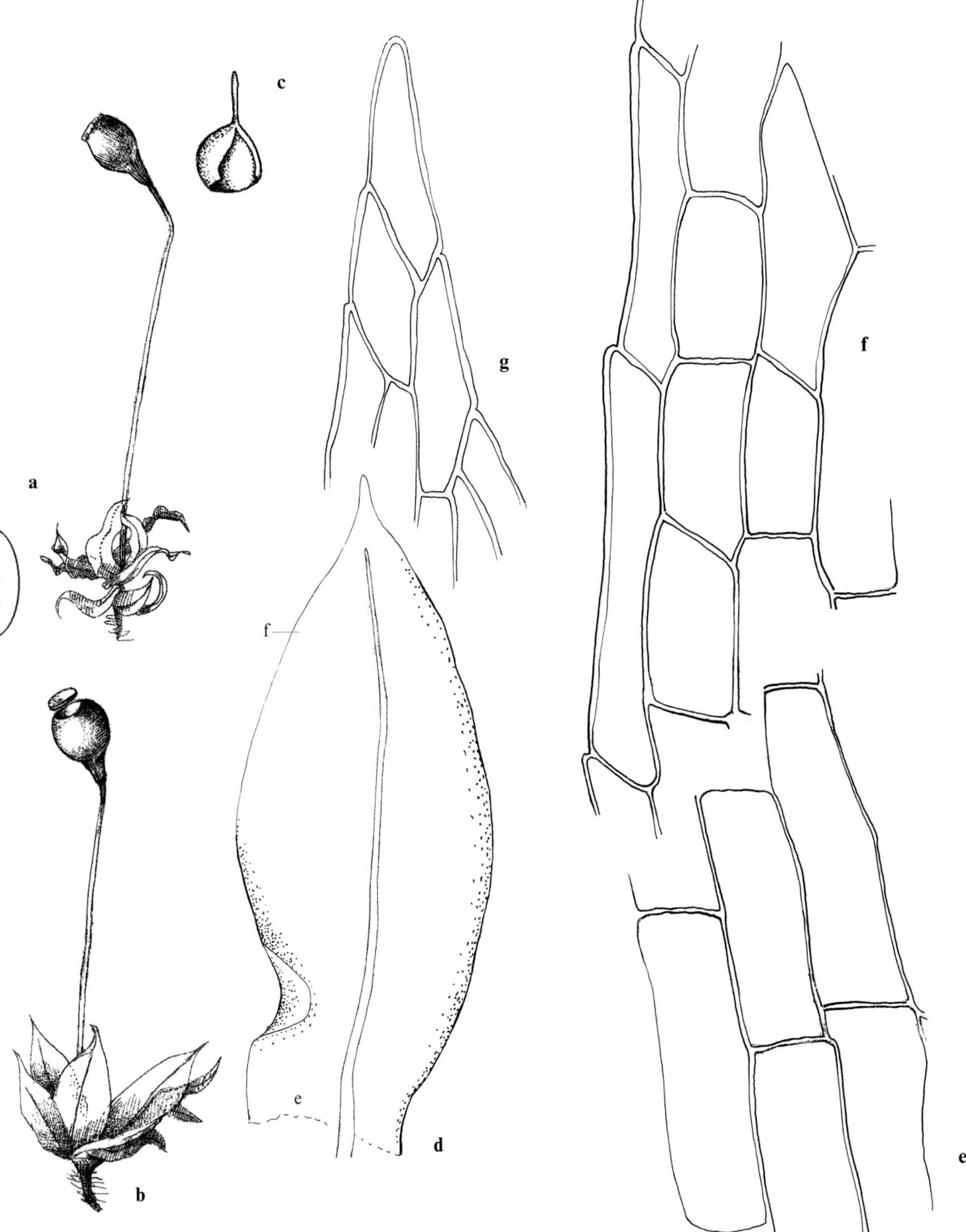

Figure 130. *Entosthodon obtusus*: **a** habit, dry (×10); **b** habit, moist, capsule with detached operculum (×10); **c** calyptra (×10); **d** leaf (×50); **e** basal cells; **f** marginal upper cells; **g** apical cells (all leaf cells ×500). (*Danin* & *Liston*, 25 Mar. 1983)

5. Entosthodon hungaricus (Boros) Loeske, Repert. Spec. Nov. Regni Veg. Sonderbeih. 3 : 115 (1929).

Funaria hungarica Boros, Magyar Bot. Lapok 23 : 73 (1925); *Steppomitra hungarica* (Boros) Vondr., Bull. Soc. Amis Sci. Lett. Poznan, sér. D, Sci. Biol. 6 : 118 (1965).

Distribution map 131, Figure 131, Plate X : a.

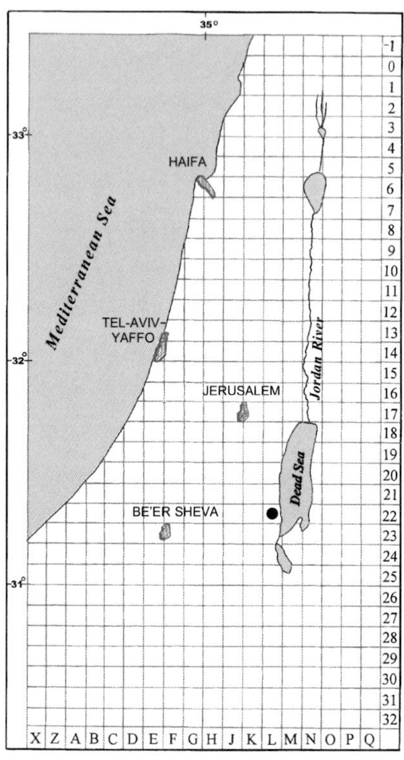

Map 131: *Entosthodon hungaricus*

Plants up to 3 mm high, bright green, turning yellow when dry. *Leaves* ca. 2 mm long, broad-lanceolate to obovate, acuminate; margins entire; costa up to $\frac{3}{4}$ the length of leaf; basal cells, rectangular, upper cells ± smaller, irregularly oblong to hexagonal, mid-leaf cells 21–27 µm wide. *Setae* 6–8 mm long, straight, yellowish; capsules erect, symmetrical to very slightly asymmetrical, ca. 2 mm long including operculum, clavate or pyriform, neck up to or equal to theca, reddish-brown at maturity; operculum with flat base narrowing into a rostrum; peristome rudimentary, fugacious, absent in mature capsules. *Spores* 25–35 µm in diameter, papillose; sparsely covered by insulated, compound gemmae, sometimes nearly fused, distal and proximal ends clearly differentiated (seen with SEM). *Calyptra* mitrate with up to four deep lobes.

Habitat and Local Distribution. — Plants growing on wet sand on wadi banks. Found only in two collections of Herb. Bilewsky from the same locality: Dead Sea area (Nahal Tse'elim, near Massada); one named "*Funaria mediterranea*", and the other "*F. dentata*".

General Distribution. — Recorded from few localities from western, eastern, and central Europe; perhaps more widespread but overlooked or recorded as *Physcomitrium*.

Literature. — Vondráček (1965), Casas & Brugués (1980).

The problem of the cucullate versus mitrate calyptrae of species of *Entosthodon* was discussed above in the discussion of the genus. Boros (1925) described *Funaria hungarica* as having cucullate, but irregularly lobed calyptrae, whereas Vondráček & Hadač *in* Vondráček (1965) included this species in a new genus *Steppomitra* (with *S. hadacii* Vondr.), mainly characterised by the mitriform — not cucullate — calyptrae. Plants of *E. hungaricus* from Israel well agree with the description and illustration of the type by Boros (1925) and with the illustration of a specimen collected by García in Monegros, Spain (Casas & Brugués 1980). However, the opercula of local plants seem to have a longer beak than in the Hungarian and Spanish plants.

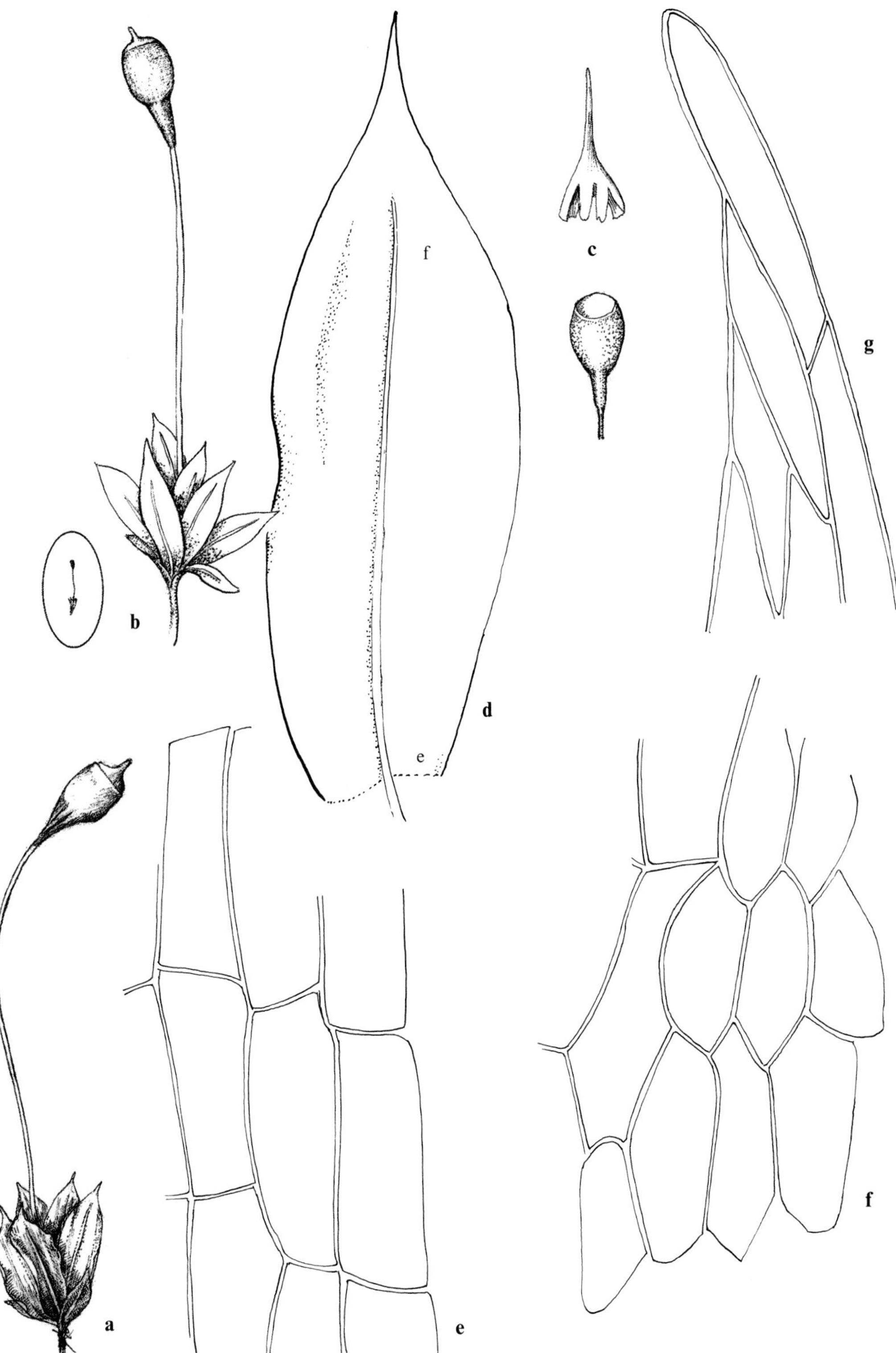

Figure 131. *Entosthodon hungaricus*: **a** habit, dry (×10); **b** habit, moist (×10); **c** deoperculate capsule and calyptra (×10); **d** leaf (×50); **e** basal cells; **f** upper cells; **g** apical cells (all leaf cells ×500). (*Bilewsky*, 16 Mar. 1957, No. 668 in Herb. Bilewsky)

6. Entosthodon curvisetus (Schwägr.) Müll. Hal., Syn. Musc. Frond. 1 : 12 (1848).
Gymnostomum curvisetum Schwägr., Sp. Musc. Frond., Suppl. 2 : 17. 105 (1823); *Funaria curviseta* (Schwägr.) Milde, Bryol. Siles. 196 (1869). — Bilewsky (1965) : 393, Rabinovitz-Sereni (1931); *Funariella curviseta* (Schwägr.) Sérgio, Orsis 3 : 10 (1988).

Distribution map 132, Figure 132, Plate X : b, c.

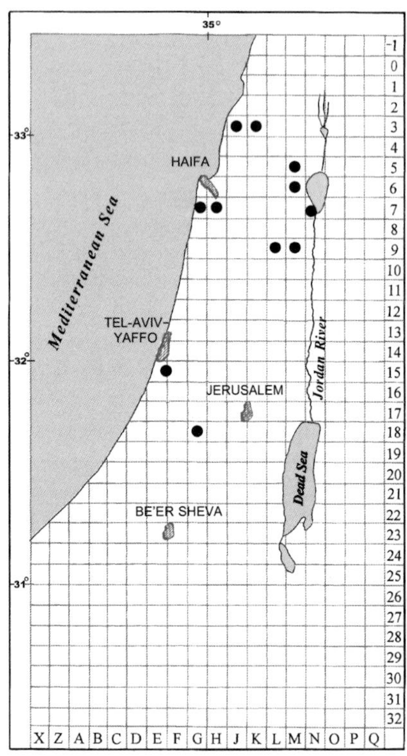

Map 132: *Entosthodon curvisetus*

Plants up to 4 mm high, bright green, turning yellowish when dry. *Leaves* up to 3 mm long, obovate, sometimes lanceolate, acute, with a short point; margins serrate in upper ⅓ of leaf; costa ending below apex; mid- and upper cells irregularly rectangular-hexagonal, much shorter than rectangular basal cells, in mid-leaf 25–29 μm wide; marginal cells ± longer than others. *Setae* short, as long as capsules, cygneous; capsules often emergent or just projecting beyond upper leaves, curved, ± asymmetrical, ca. 2 mm long including operculum, pyriform, turning red brown at maturity, neck up to or equal to theca, mouth wide open; operculum plane-convex, minutely mammillate, with cells in obliquely radial rows; peristome absent. *Spores* 24–27 μm in diameter, reticulately papillose; reticulum on foveolate surface composed of baculae-verrucae lobed at apex (seen with SEM). *Calyptrae* cucullate.

Habitat and Local Distribution. — Plants growing usually in small tufts in soil pockets among calcareous rocks. Occasional: Coast of Carmel, Philistean Plain, Upper Galilee, Lower Galilee, Mount Carmel, Mount Gilboa, Shefela, and Upper Jordan Valley.

General Distribution. — Recorded from Southwest Asia (Cyprus, Turkey, and Lebanon), around the Mediterranean, southern Europe, Macaronesia, and North Africa.

Literature. — Sérgio (1988).

Entosthodon curvisetus can be easily distinguished from all the local species of the genus by the often emergent capsules with curved, cygneous seta.

Sérgio (1988) proposed the new monotypic genus *Funariella* to accommodate *Entosthodon curvisetus*, based on the combination of the following characters: reticulate spore surface, short setae, spirally arranged operculum cells, absence of peristome, and chromosome number n = 5.

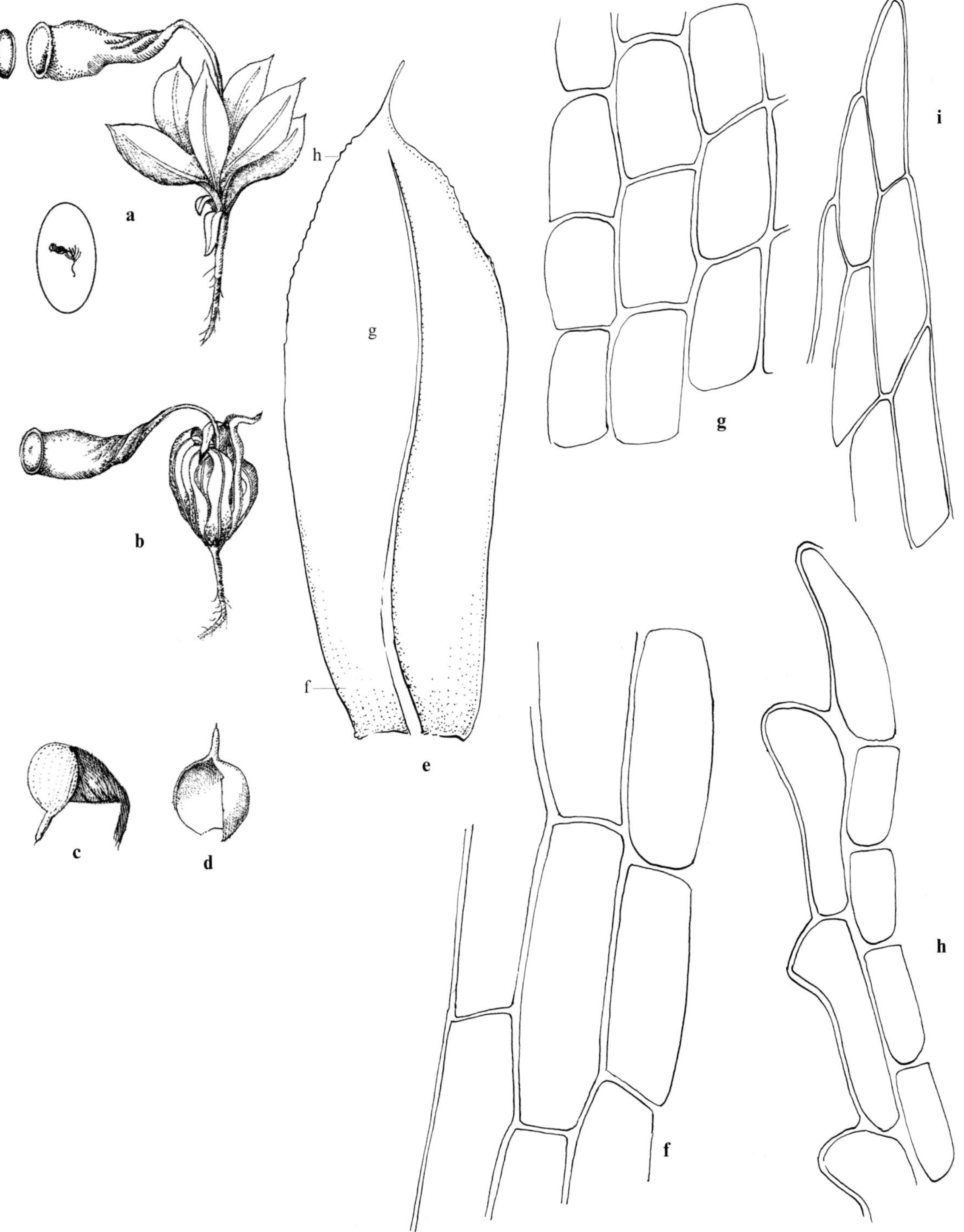

Figure 132. *Entosthodon curvisetus*: **a** habit, moist, capsule with detached operculum (×10); **b** habit, dry (×10); **c** capsule with calyptra (×10); **d** calyptra (×10); **e** leaf (×50); **f** basal cells; **g** upper cells; **h** marginal upper cells; **i** apical cells (all leaf cells ×500). (*Naftolsky*, 2 Mar. 1927)

EPHEMERACEAE

Plants minute, bud-like, annual, ephemeral, terrestrial, usually with persistent protonema. *Stems* short, erect, simple, with few leaves. *Leaves*: lower leaves small, upper leaves larger, forming a rosette, linear-lanceolate, oblong, or ovate-lanceolate; margins entire or serrate; costa present, often weak or absent; cells usually lax, smooth, or sometimes prorate; basal cells rectangular, mid- and upper cells ± hexagonal, rhomboidal to oblong. *Capsules* on short seta or sessile, immersed, subglobose, thin-walled, apiculate or not, indehiscent or dehiscing along a line near middle, stomata, if present, with two guard cells; peristome absent. *Spores* large, reniform. *Calyptrae* usually campanulate.

A small family with only two genera, represented in Israel by the genus *Ephemerum*.

Ephemerum Hampe

Characters as described for the family. *Capsules* indehiscent, with stomata and apiculus. *Calyptrae* campanulate.

A genus comprising ca. 28 species. Three species occur in the local flora.

Literature. — Risse (1996).

1 Leaves without distinct costa; margins strongly serrate; cells in mid-leaf 16–22 μm wide; spores surrounded by hyaline membrane **3. E. serratum** var. **minutissimum**

1 Leaves with distinct costa; margins usually irregularly serrulate; cells in mid-leaf 8–13 μm wide; spores without hyaline membrane.

  2 Dioicous; upper leaves recurved when moist; capsule with oblique apiculus; spores up to 55 μm in diameter; calyptrae campanulate-cucullate **1. E. recurvifolium**

  2 Autoicous; upper leaves erectopatent when moist; capsule with straight apiculus; spores up to 100 μm in diameter; calyptrae campanulate-mitrate **2. E. sessile**

1. Ephemerum recurvifolium (Dicks.) Boulter, Fl. Crypt. Est. Muscin. 694 (1872). [Bilewsky (1965) : 390]

Distribution map 133 (p. 344), Figure 133.

Plants 1.5–2 mm high, pale green, turning yellowish-brown when dry. *Leaves* tortuose when dry, some concave; lower leaves erectopatent, upper leaves recurved when moist, up to 1.75 mm long, linear-lanceolate, gradually tapering into acuminate apex; margins plane, irregularly serrulate with few coarse teeth near apex; costa distinct, strong, excurrent into short acumen; basal cells rectangular; cells in mid-leaf 8–13 μm wide. *Dioicous. Setae* short; capsules yellowish brown, ca. 0.5 mm, with oblique apiculus. *Spores* nearly smooth, 40–55 μm in diameter. *Calyptrae* campanulate-cucullate, covering about ⅓ of capsule.

Habitat and Local Distribution. — Plants growing in tufts on protonemal mats on calcareous soil, exposed or partly shaded, also on dry soil after flooding, found growing together with *Pottia* spp. Rare: Coast of Carmel, Judean Mountains (unverified record of Bilewsky 1965), Hula Plain, and Bet She'an Valley.

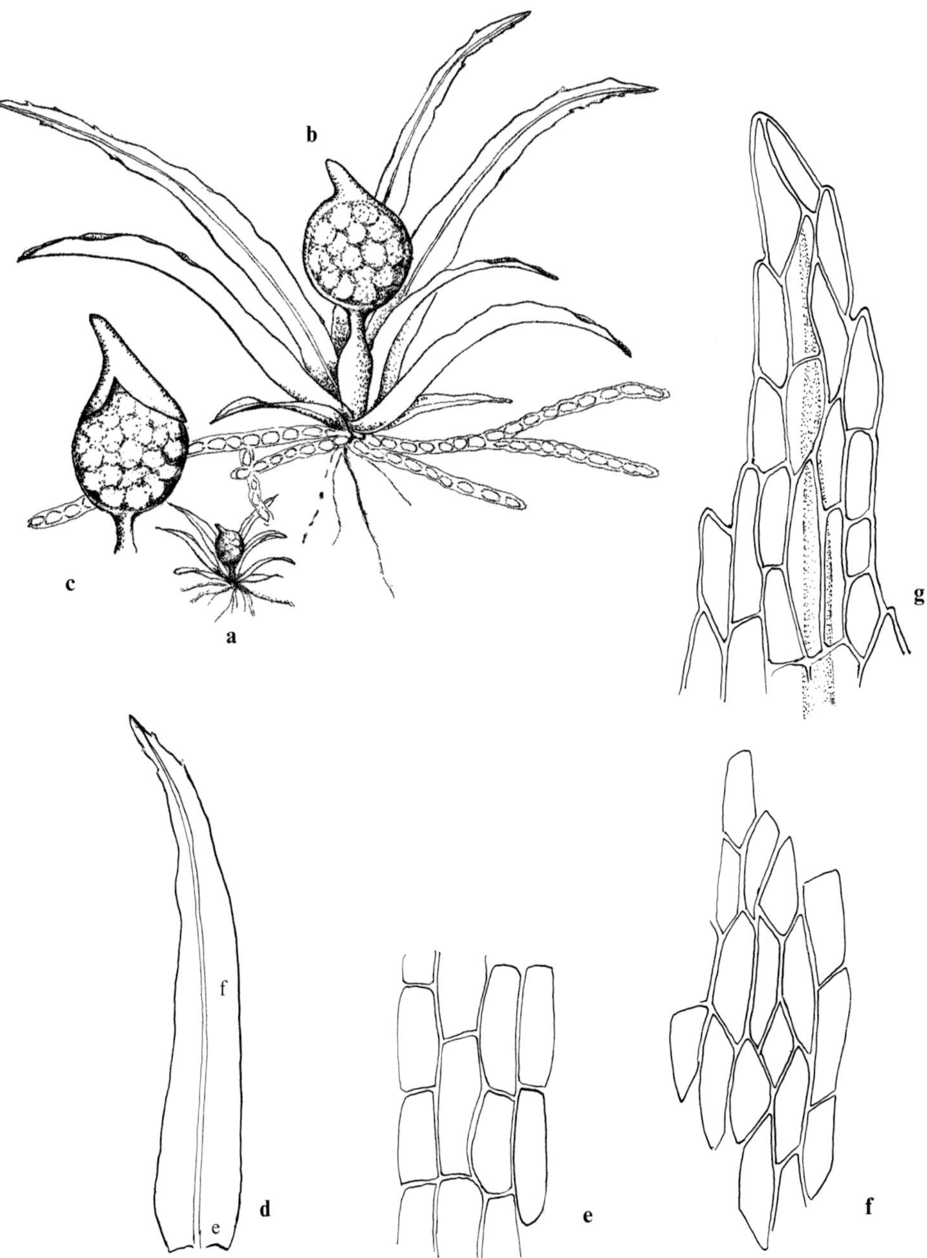

Figure 133. *Ephemerum recurvifolium*: **a** habit, moist (×10); **b** habit, moist (×50); **c** capsule with calyptra (×50); **d** leaf (×50); **e** basal cells; **f** mid-leaf cells; **g** apical cells (all leaf cells ×500). (*Kushnir*, Feb. 1944, No. 87 in Herb. Bilewsky)

General Distribution. — Recorded from few localities outside Europe, Southwest Asia (Turkey), around the Mediterranean (very scarce records), Europe (various parts), North Africa, and northwestern South America.

2. Ephemerum sessile (Bruch & Schimp.) Müll. Hal., Syn. Musc. Frond. 1 : 33 (1848). [Bilewsky (1965) : 391 ("var. *kushniri* Bizot")]

Distribution map 134, Figure 134, Plate X : d.

Plants 1.75–2 mm high, pale green, turning yellowish-brown when dry. *Leaves* erecto-patent when moist; upper leaves up to 1.75 mm long, narrow-lanceolate, gradually tapering into long acuminate apex, ending in one long cell; margins plane, bluntly and irregularly serrulate towards apex; costa distinct, excurrent into acumen; basal cells rectangular; cells in mid-leaf narrower, 8–12 μm wide. Recorded as *autoicous*. *Capsule* up to 0.5 mm long, yellowish-brown, with short, straight apiculus. *Spores* exceptionally large, up to 100 μm in diameter, verrucate; with smooth verrucae of various sizes (seen with SEM). *Calyptrae* wide-conical, short, covering only the apiculus, mitriform.

Habitat and Local Distribution. — Plants growing scattered on persistent protonemal mats; found on compressed hardened sand in a single locality: Sharon Plain (Pardes Hanna).

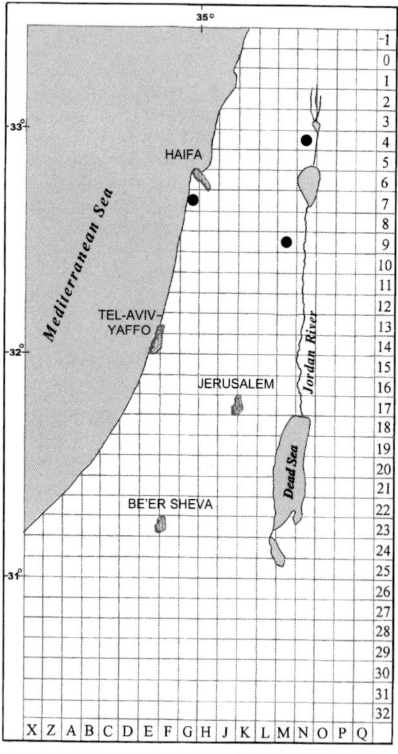

Map 133: *Ephemerum recurvifolium*

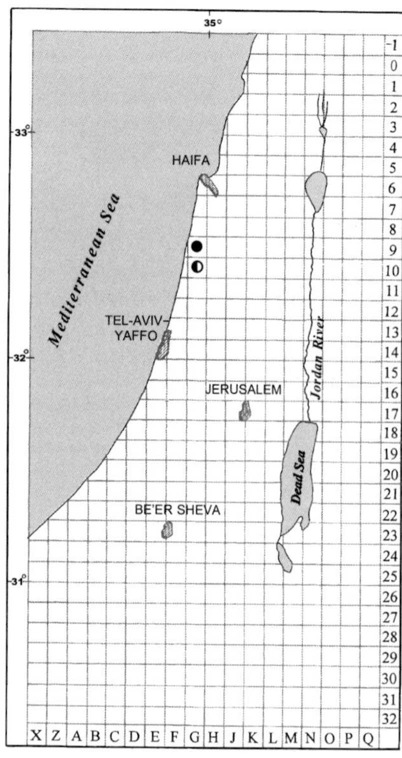

Map 134: ◐ *Ephemerum sessile* + *E. serratum* var. *minutissimum*
● *E. serratum* var. *minutissimum*

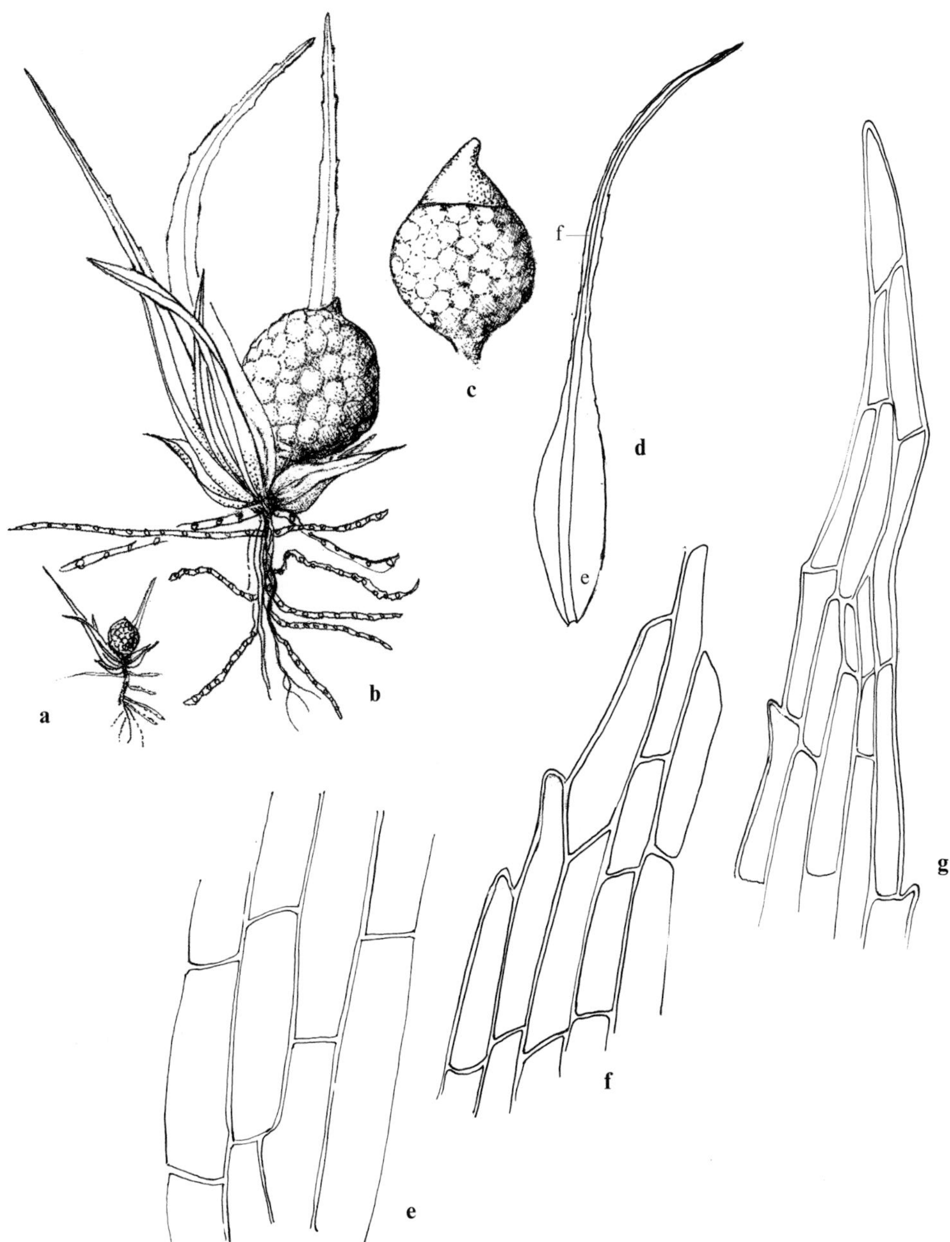

Figure 134. *Ephemerum sessile*: **a** habit, moist (×10); **b** habit, moist (×50); **c** capsule and calyptra (×50); **d** leaf (×50); **e** basal cells; **f** marginal upper cells; **g** apical cells (all leaf cells ×500). (*Kushnir*, 21 Mar. 1943, No. 88 in Herb. Bilewsky)

General Distribution. — Recorded from few localities: Southwest Asia only from Turkey, scattered records from around the Mediterranean, Europe (mainly the south), Macaronesia (single record), and North Africa. Previous records from South Africa do not apply to this species (see Magill 1987).

Ephemerum sessile var. *kushniri*, based on the collection mentioned above from "env. of Pardess Hanna", was described by Bizot in Bilewsky & Nachmony (1955, p. 54). According to Bizot, the characters differentiating this variety from the typical *E. sessile* are "capsules smaller, spores 90–100 μm (not as in type 50–80 μm)". Because of the lack of available information on the range of variability of *E. sessile* and the scarcity of local material, the species is not treated here at the varietal rank.

3. Ephemerum serratum (Hedw.) Hampe, Flora 20:285 (1837).
Phascum serratum Schreb. ex Hedw. Sp. Musc. Frond. 23 (1801).

Plants up to 1.7(–2) mm high, yellowish-green to yellowish-brown when dry. *Leaves* erectopatent when moist, upper leaves 1–1.5 mm long, lanceolate to narrow-lanceolate, gradually tapering into acuminate apex, ending in one long cell; margins plane, irregularly strongly serrate from half the length of leaf to apex, with upper teeth longer and ± recurved; costa absent; basal cells rectangular, wide; upper cells irregularly oblong to rhomboidal, in mid-leaf 16–22 μm wide. *Dioicous. Setae* short; capsules 0.5–0.75 mm long, reddish-brown, apiculate. *Calyptrae* campanulate-mitriform, covering about ⅓ of capsule.

3a. Ephemerum serratum (Hedw.) Hampe var. **minutissimum** (Lindb.) Grout, Moss Fl. N. Amer. 2:68 (1935).
E. minutissimum Lindb., not. Sällsk., Fauna Fl. Fenn. Förh. 13:411 (1874). — Bilewsky (1965):390.

Distribution map 134 (p. 344), Figure 135, Plate X:e, f.

Plants up to 1.5 mm high. *Leaves* up to 1.25 mm long, narrow-lanceolate, margins with recurved teeth. *Spores* 65–75 μm in diameter, papillose, surrounded by hyaline membrane; with elongated, ± baculate processes, partly or completely covered by membrane (seen with SEM).

Habitat and Local Distribution. — Plants growing in tufts on protonemal mat on sand, at margins of swamps, and on sandy loam, several populations in shade of *Thymelaea hirsuta* (L.) Endl. Rare, locally abundant: Sharon Plain (in two localities; erroneously recorded as Shomron in Bilewsky 1965).

General Distribution. — Recorded from Southwest Asia only from Turkey, from the northern Mediterranean (perhaps the scattered records of *Ephemerum serratum* from other parts of the Mediterranean partly apply to this variety), Europe (few records), and North America.

The characters differentiating *Ephemerum serratum* var. *minutissimum* from the typical variety are the smaller size, the narrower lanceolate leaves with recurved (vs. erect) marginal teeth, and the smaller spores, which are finely (not coarsely) papillose. The spores of var. *minutissimum* are usually described as being surrounded by hyaline membrane (e.g., Smith 1978, Frahm 1995, Risse 1996). In the few local collections, the spores are surrounded by hyaline membrane (Plate X:e, f) and have lanceolate leaves with somewhat recurved teeth.

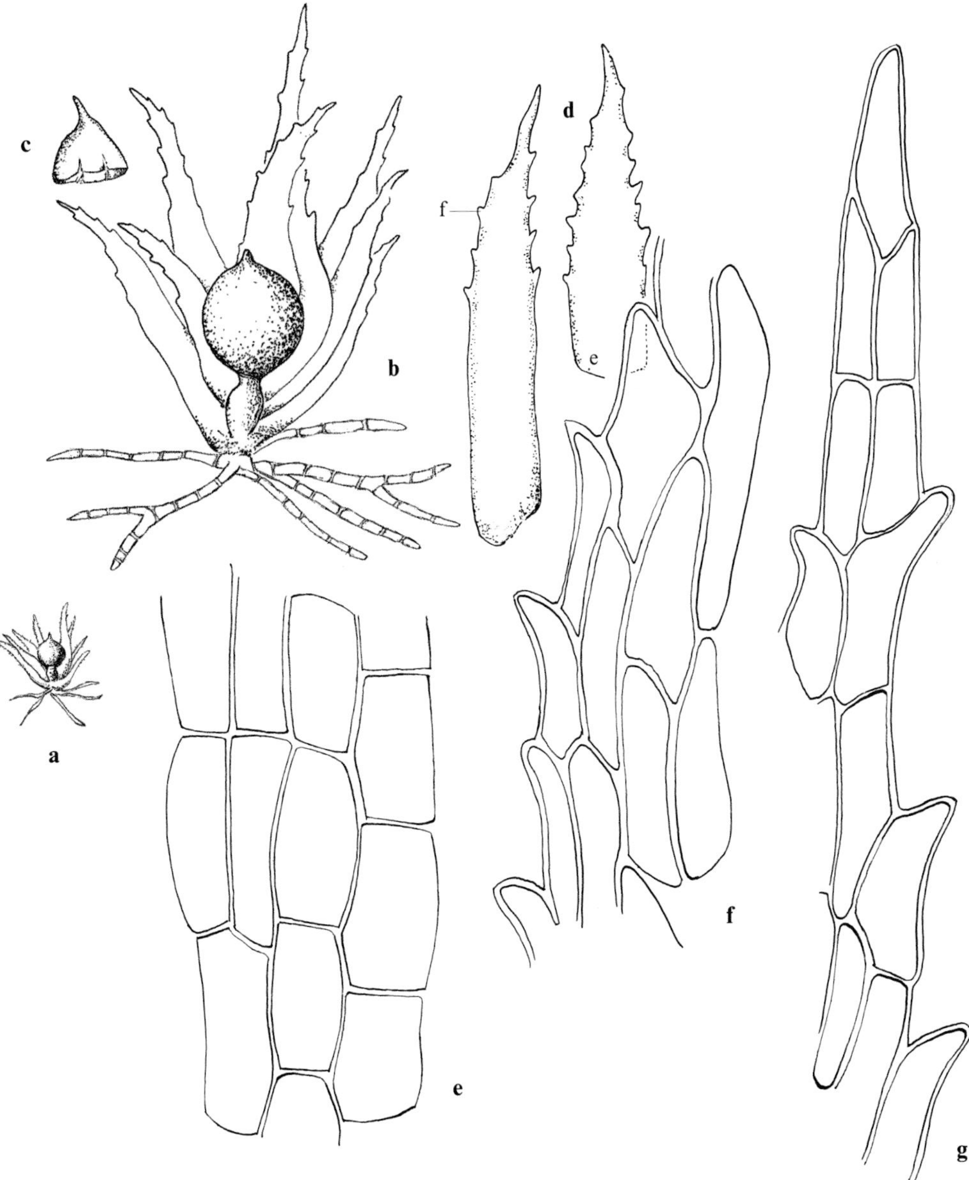

Figure 135. *Ephemerum serratum* var. *minutissimum*: **a** habit, moist (×10); **b** habit, moist (×50); **c** calyptra (×50); **d** leaves (×50); **e** basal cells; **f** marginal upper cells; **g** apical cells (all leaf cells ×500). (*Crosby* & *Herrnstadt 79-82-1*)

Bryales

BRYACEAE

Plants acrocarpous, medium to ± robust, green or yellowish, usually growing in tufts, sometimes radiculose or tomentose. *Stems* erect, simple, or branched by subfloral innovations, usually with central strand. *Leaves* often larger and more crowded above, sometimes forming comal tufts, usually erect to erect-spreading when dry, changing little when moist, with or without border; costa usually strong, often excurrent, with a single dorsal stereid band; cells large, smooth; basal cells rectangular to subquadrate; upper cells most often rhomboidal-hexagonal to rhomboidal; perichaetial leaves little differentiated; perigonia with filiform paraphyses. *Setae* elongate; capsules mostly inclined to pendulous, usually symmetrical, ovoid, pyriform, subglobose or subcylindrical, with distinct neck most often wrinkled when dry; stomata superficial, confined to neck; operculum convex to conical, mostly mammillate, rarely rostrate; annulus frequently differentiated, usually revoluble; peristome usually double, exostome teeth 16, papillose; endostome with yellow or hyaline segments alternating with teeth, often with basal membrane, usually with cilia. *Calyptrae* cucullate.

A worldwide family comprising according to different concepts up to 20 genera. Three genera are represented in the local flora.

The Bryaceae are most often characterised by elongate, inclined to pendulose capsules with distinct neck and double peristome. The inter- and intraclassification of the genera of the family is mainly based on sporophytic characters. Consequently, sterile material may be difficult to identify. It is possible that some sterile collections, considered here as belonging to *Bryum*, include plants of related genera, e.g., *Brachymenium* (some sterile plants from the Dead Sea area were identified by Brullo *et al.* (1991), with some uncertainty, as *Brachymenium exile* (Dozy & Molk.) Bosch & Lacout.).

1 Leaves (except some lower) linear-subulate; costa occupying most of width of upper part of leaf, apex subulate **2. Leptobryum**

1 Leaves ovate, ovate-lanceolate, narrow-lanceolate to oblong; costa not occupying most of width of upper part of leaf, apex not subulate.

  2 Leaves not bordered; costa percurrent or ending below apex, never excurrent; most upper cells elongate, at least four times longer than wide; endostome cilia simple or nodulose, not appendiculate **1. Pohlia**

  2 Leaves often bordered by one to several rows of narrower cells; costa often excurrent, sometimes percurrent, rarely ending below apex; upper cells less elongate, usually less than four times longer than wide; endostome cilia usually appendiculate **3. Bryum**

1. *Pohlia* Hedw.

The genus is represented in the local flora by only two species of subgen. *Mniobryum* (e.g., as in Shaw 1981) or section *Mniobryum* (e.g., as in Corley *et al.* 1981; Nyholm 1993). Accordingly, the generic description applies to that part of the genus. *Mniobryum* is often considered as a genus separate from *Pohlia*. *Pohlia* is mainly characterised by thin-walled leaf cells and short pyriform to ovoid capsules having short exothecial cells with sinuous walls, often with immersed stomata. Plants of *Mniobryum* tend to grow on fine-grained soils, sometimes in disturbed habitats.

Plants gregarious or growing in tufts among other mosses. *Stems* erect, simple, or branched by subperichaetial innovations. *Leaves* ovate, ovate-lanceolate to narrow-lanceolate; margins plane to recurved, not bordered, sometimes serrulate to serrate towards apex; costa percurrent or ending below apex; cells elongate, linear to rhomboidal. *Rhizoidal gemmae* sometimes present. *Dioicous*. *Setae* varying in size, often sigmoid; capsules pendulose to cernuous, wide-mouthed when empty, brown when mature, neck short; annulus not differentiated; peristome double, endostome with basal membrane, segments perforated, cilia simple or nodulose. *Spores* finely to coarsely papillose. *Sporophytes* infrequent in local material.

Pohlia is a cosmopolitan genus comprising about 150 species. Two species occur in the local flora, usually growing in wet habitats.

Literature. — Shaw (1981), Nordhorn-Richter (1982), Nyholm (1993).

1 Plants up to 10(–15) mm high, not glaucous, readily absorbing water on surface when dry; leaf margins distinctly serrulate towards apex; perichaetial leaves narrow-lanceolate **1. P. melanodon**

1 Plants up to 15–25(–30) mm high, often glaucous, slowly absorbing water on surface when dry; leaf margins usually ± entire to serrulate towards apex; perichaetial leaves wider, ovate-lanceolate to lanceolate **2. P. wahlenbergii**

1. Pohlia melanodon (Brid.) A. J. Shaw, Bryologist 84 : 506 (1981).

Bryum melanodon Brid., Bryol. Univ. 1 : 845 (1827); *Pohlia carnea* (Schimp.) Lindb., Musci Scand. 17 (1879). — Bilewsky (1965) : 396; *Mniobryum delicatulum* (Hedw.) Dixon, Rev. Bryol. Lichénol. 6 : 107 (1934).

Distribution map 135, Figure 136.

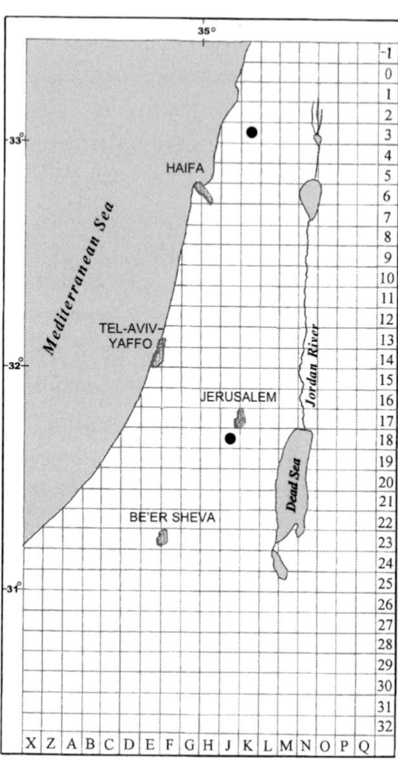

Map 135: *Pohlia melanodon*

Plants up to 10(−15) mm high, pale to yellowish-green when dry, not glaucous. *Stems* reddish. *Leaves* slightly contorted when dry, erectopatent to patent when moist, up to 1.5 mm long, ovate to lanceolate, acute to acuminate, usually decurrent, bases reddish; margins plane, distinctly serrulate towards apex; costa ending usually below apex; basal cells irregularly rectangular, not differing in size from mid-leaf cells; mid- and upper leaf cells ± thin-walled, rhomboidal, 13–26 μm wide, narrower at margins; apical cells slightly shorter; perichaetial leaves crowded, narrow-lanceolate, up to 2.5 mm long. *Setae* up to 10 mm long, often sigmoid; capsules pendulous, 1–1.5 mm long including operculum, short-pyriform, slightly contracted at mouth when dry and empty.

Habitat Local Distribution. — Plants growing in small, loose tufts, scattered among other mosses, often among plants of *Pohlia wahlenbergii*, on wet soil, mainly on rocks near running water; dry plants readily absorb water on surface. Rare, locally abundant: Upper Galilee and Judean Mountains.

General Distribution. — Recorded from Southwest Asia (Cyprus, Turkey, Lebanon, and Iraq), around the Mediterranean, Europe (throughout), Macaronesia, North Africa, northeastern, eastern, and central Asia, and North America.

Figure 136. *Pohlia melanodon*: **a** habit, moist (×10); **b** leaves (×50); **c** basal cells; **d** marginal mid-leaf cells; **e** apical cells (all leaf cells ×500); **f** cross-section of costa and portion of lamina (×500). (*Bilewsky*, Mar. 1953, No. 99 in Herb. Bilewsky)

2. Pohlia wahlenbergii (F. Weber & D. Mohr) A. L. Andrews, *in* Grout, Moss Fl. N. Amer. 2:203 (1935).

Hypnum wahlenbergii F. Weber & D. Mohr, Bot. Taschenb. 280 (1807); *Mniobryum albicans* (Wahlenb.) Limpr., Laubm. Deutschl. 2:272 (1892). — Nachmony (1961); *M. wahlenbergii* (F. Weber & D. Mohr) Jenn., Man. Mosses W. Pennsylvania 146 (1913).

Distribution map 136, Figure 137, Plate XI: a.

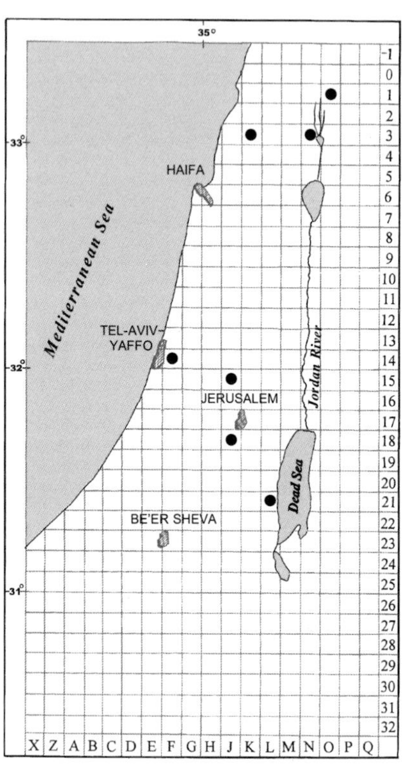

Map 136: *Pohlia wahlenbergii*

Plants 15–25(–30) mm high, whitish-green to dull green, often glaucous. *Stem* turning reddish with age. *Leaves* shrunken when dry, erectopatent to patent when moist, up to 2.5 mm long, ovate to lanceolate, acute to acuminate, usually decurrent; perichaetial leaves longer, ovate-lanceolate to lanceolate, gradually tapering towards apex; margins plane, ± entire to serrulate towards apex; costa percurrent or ending below apex, base turning reddish with age; cells ± thin-walled, elongate, rhomboidal, mid-leaf cells 11–22(–24) μm wide, narrower and more elongate towards margins; apical cells slightly shorter. *Setae* long, 10–25 mm, often sigmoid when dry; capsules pendulous, sometimes cernuous when moist, about 1 mm long, short-pyriform, nearly globose and wide-mouthed when empty. *Spores* densely gemmate, processes small, with rough surface (seen with SEM).

Habitat and Local Distribution. — Plants growing in loose tufts on wet calcareous soils, on rocks or stones near running water, sometimes partly submerged; dry plants slowly absorb water on surface. Occasional: Sharon Plain, Upper Galilee, Judean Mountains, Judean Desert, Dan Valley, and Hula Plain.

General Distribution. — Recorded from Southwest Asia (Cyprus, Turkey, Syria, Lebanon, Iraq, Iran, and Afghanistan), around the Mediterranean, Europe (widespread throughout), Macaronesia, North Africa, Asia (throughout, except the southeast), and North, Central, and South America.

In Herrnstadt *et al.* (1991), *Pohlia wahlenbergii* was treated as including two varieties: the typical variety and var. *calcareum* (F. Weber & D. Mohr) A. L. Andrews. The

---

Figure 137. *Pohlia wahlenbergii*: **a** habit, moist, deoperculate capsule (×10); **b** habit with base of seta, dry (×10); **c** part of young sporophyte with calyptra (×10); **d** young capsule with operculum (×10); **e** leaf (×50); **f** marginal basal cells; **g** upper cells; **h** apical cells (all leaf cells ×500); **i** perichaetial leaf (×50); **k** cross-section of leaf (×500). (*Kushnir*, 5 Feb. 1943)

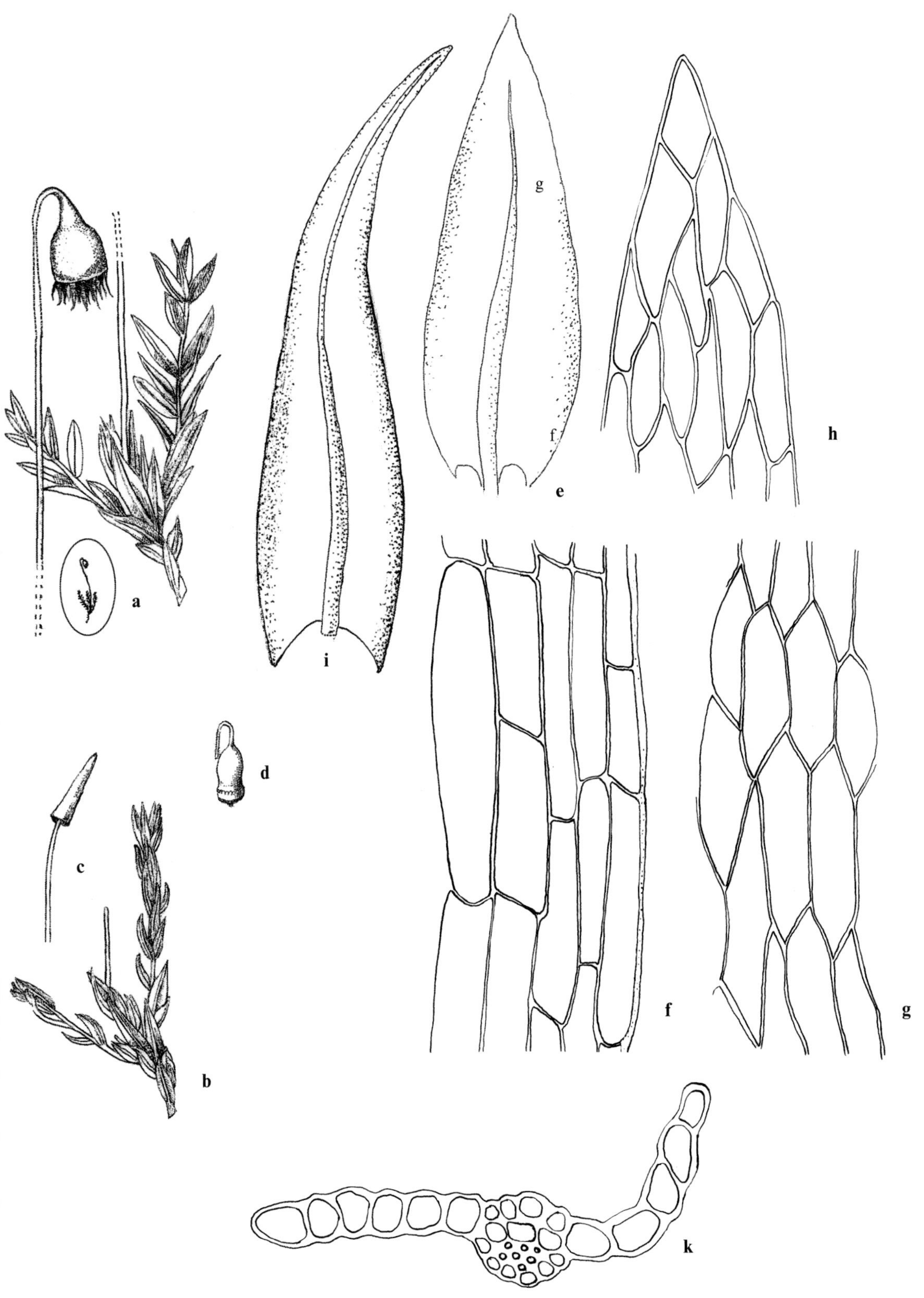
g
f
e
h
i
a
d
c
b
f
g
k

concept of var. *calcareum* differs in various treatments of *Pohlia*. Variety *calcareum* is often considered as being smaller, dull green — not pale green and glaucous like the "typical variety" — with narrower leaves and leaf cells. Although each of these differentiating characters may sometimes be observed in plants of *P. wahlenbergii*, they do not seem to be correlated. For that reason, it was refrained here from separating the species into two formal varieties. In two samples (Upper Galilee: Wadi Karkara, *Kushnir*, 27 Aug. 1943 (with *P. melanodon*); Sharon Plain: Yarkon River bank, no. 355 in Herb. Bilewsky) plants have a combination of several of the characters often attributed to var. *calcareum*.

In Bilewsky (1965, p. 396), both *Mniobryum latifolium* Schiffn. and *Pohlia commutata* are recorded from the Sharon Plain. They were found to be based on a single collection (Yarkon River, Seven Mills, Herb. Bilewsky no. 355). The plants of this collection were, however, identified as *P. wahlenbergii*.

In two collections from the Judean Desert, Wadi Qilt (*Kushnir*, 28 Mar. 1943 (?); and *Farstey*, Jun. 1988) the plants, which grow on clay on the bank of running water, approach in a number of characters *Mniobryum latifolium* [≡ *M. wahlenbergii* (F. Weber & D. Mohr) Jenn. var. *latifolium* (Schiffn.) Wijk. & Margad.]. The most notable characters are the concave, broad-ovate, muticous lower leaves, and short-apiculate upper leaves. No further specimens possessing those characters were found in the local flora. In addition, the habitat, in which those plants were found, extremely differs from Schiffner's type locality of this taxon: a wet high-mountain region. Therefore, we are not including that variety in this work.

2. *Leptobryum* (B.S.G.) Wilson

Plants slender, growing in loose or dense tufts. *Stems* usually simple, with large central strand. *Leaves* distant below, comose above, linear-subulate; costa wide, subpercurrent to shortly excurrent, occupying most of the width of subula; margins plane, entire or ± serrulate at apex. *Setae* long, flexuose-curved; capsules inclined to pendulous, pyriform, contracted at the mouth; neck long, narrow; annulus revoluble; peristome double; endostome with well developed basal membrane, keeled segments, and appendiculate cilia; operculum convex-conical, mammillate. *Rhizoidal gemmae* present.

A small cosmopolitan genus comprising ca. five species. One species occurs in the local flora.

Leptobryum pyriforme (Hedw.) Wilson, Bryol. Brit. 219 (1855).
Webera pyriformis Hedw., Sp. Musc. Frond. 169 (1801).

Distribution map 137 (p. 356), Figure 138.

Plants up to 15 mm high, green to yellowish-green. *Stems* slender, rarely branching, dark brownish-red. *Leaves* flexuose when dry, widely spreading and ± flexuose when moist, distantly spaced below, comose and much larger above; lower leaves 1–1.25 mm long, lanceolate, upper leaves 3.5–4.5 mm long, linear-subulate with a wider base; margins plane, ± serrulate near apex, not bordered; costa wide, occupying about half of leaf base and most of the width of subula, subpercurrent in lower, excurrent in upper leaves; basal cells long-rectangular; mid- and upper leaf cells smaller, long-rectangular to linear, long-rhomboidal towards apex, in mid-leaf 5–8 μm wide. *Rhizoidal*

Figure 138. *Leptobryum pyriforme*: **a** habit, moist (×10); **b** habit, dry (×10); **c** deoperculate capsule (×10); **d** part of exostome and endostome teeth (×125); **e** leaf (×50); **f** basal cells; **g** upper cells; **h** apical cells (all leaf cells ×500); **i** rhizoidal gemma (×50). (*D. Kaplan*, 17 Aug. 1982)

gemmae abundant, dark brown-red, ellipsoid, 130–185 µm long. *Synoicous*. *Setae* 10–15 mm long; capsules up to 2 mm long, brown; neck well developed, sulcate, half the length of capsule; peristome teeth yellowish. *Spores* small, nearly smooth, 13–15 µm in diameter.

Habitat and Local Distribution. — Plants growing in loose tufts; found in one locality near dripping water together with *Adiantum capillus-veneris* L. Very rare: so far found only in two localities on the Golan Heights in depressions of volcanic ash.

General Distribution. — *Leptobryum pyriforme* is a widespread species of the temperate regions. Recorded from Southwest Asia (Cyprus, Turkey, Lebanon, Iran, Afghanistan, and Kuwait), around the Mediterranean, Europe (widespread throughout), Macaronesia, Africa (throughout except tropical Africa), Asia (throughout), Australia, New Zealand, Oceania, and America (throughout).

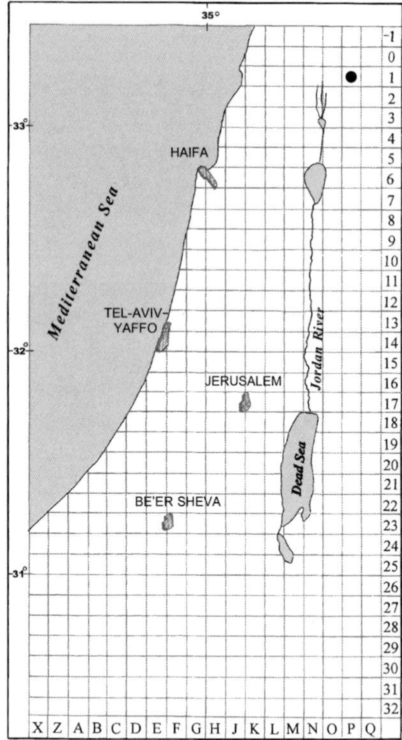

Map 137: *Leptobryum pyriforme*

The linear-subulate leaves with long-linear cells of *Leptobryum pyriforme* resemble those of *Dicranella* species and are unusual in the Bryaceae. However, the pyriform capsules with bryoid peristome do not leave any doubt about the affiliation of this species with the family.

3. *Bryum* Hedw.

Plants growing in dense tufts or forming turfs. *Stems* erect, usually branched by subfloral innovations. *Leaves* on stem distant below, forming comal tufts above, ovate to lanceolate to elliptical, with or without border; margins plane to recurved, entire to denticulate; costa usually strong, ending below apex to longly excurrent, often with hydroids; leaves often bordered by one to several rows of narrower cells; upper cells less than four times longer than wide. *Axillary bulbils* or rhizoidal gemmae often present. *Dioicous*, synoicous, or autoicous. *Setae* widely varying in length within species, usually reddish; capsules usually inclined to pendulous, rarely erect; operculum conical, often mammillate (apiculate in *B. cellulare*); peristome double; endostome with hyaline basal membrane, perforated segments in local species usually with appendiculate cilia (cilia absent in *B. cellulare*, nodulose in *B. intermedium*). *Spores* from nearly smooth to finely papillose; gemmate (seen with SEM) in local species, except in *B. cellulare* where processes are larger and with more distinctly granulate distal surface.

A large cosmopolitan genus in which up to 800 species are accepted; considered most difficult taxonomically and accordingly treatments vary greatly. Locally, *Bryum* is the largest genus comprising 19, mainly terrestrial, species. Some species, recorded as widely distributed in Southwest Asia (for example *B. syriaca* Lorentz), have not been found so far in Israel. The possibility that some additional species of *Bryum* occur in the local flora can not be ruled out.

Literature. — Ochi (1980, 1981).

Bryum includes several groups of closely related species. Some are often considered as "species groups", "species complexes", or "species aggregates". Among the local species, the following species complexes are recognised here: *B. bicolor* complex (species 3–6) and *B. erythrocarpum* complex (species 7–9).

1 Plants whitish to silvery pale green; leaves hyaline in upper part **2. B. argenteum**

1 Plants variously green, not whitish to silvery green; leaves not hyaline in upper part.

2 Capsules erect to inclined; neck at least half the length of capsule; endostome without cilia **1. B. cellulare**

2 Capsules horizontal to pendulous; neck shorter, not exceeding half the length of capsule, sometimes not distinct; endostome with cilia.

3 Leaves obtuse to acute; costa subpercurrent to percurrent.

4 Stem red; costa reddish-brown; upper leaf cells slightly incrassate, 20–27×50–75 µm; no axillary bulbils present **11. B. muehlenbeckii**

4 Stem and costa without red pigmentation; upper leaf cells distinctly incrassate, 10–14×40–55 µm; axillary bulbils abundant **10. B. gemmiparum**

3 Leaves acute to acuminate; costa usually excurrent.

5 Leaves with very stout two–three-stratose border confluent with costa at apex **12. B. donianum**

5 Leaves with unistratose border (of up to three rows of cells) or not bordered, not confluent with costa at apex.

6 Leaf margins distinctly serrulate to serrate above; often comal tufts of two successive years present **15. B. canariense**

6 Leaf margins entire or finely serrulate above; comal tufts, if present, usually from a single year only.
  7 Leaves widest above middle, spirally twisted around stem when dry.
    8 Plants dioicous; leaves strongly spirally twisted around stem when dry; costa excurrent in a long cuspidate point or piliferous **13. B. capillare**
    8 Plants synoicous; leaves slightly twisted around stem when dry; costa excurrent in shorter point, cuspidate **14. B. torquescens**
  7 Leaves widest at or below middle, not twisted or slightly twisted around stem when dry.
    9 Plants synoicous or dioicous; without rhizoidal gemmae; leaves distinctly or obscurely bordered.
      10 Leaves obscurely bordered; cilia of endostome nodulose **19. B. intermedium**
      10 Leaves distinctly bordered; cilia of endostome appendiculate **18. B. pallescens**
    9 Plants dioicous; usually at least some plants with rhizoidal gemmae; leaves unbordered or obscurely bordered.
      11 Plants without axillary bulbils; rhizoidal gemmae always present, abundant, large, up to 250 μm in diameter.
        12 Mature capsules yellowish-brown to reddish-brown.
          13 Leaves 0.5–1 mm long, imbricate, strongly concave, broadly ovate to ovate-oblong or obovate; margins plane; capsules ovoid **17. B. kunzei**
          13 Leaves 1.2–2 mm long, slightly imbricate, concave, ovate-oblong to ovate-lanceolate; margins recurved; capsules oblong-pyriform **16. B. caespiticium**
        12 Mature capsules dark red.
          14 Rhizoids bright violet to deep-violet or purple **8. B. ruderale**
          14 Rhizoids paler, not violet or purple.
            15 Rhizoidal gemmae yellowish-brown, not contrasting with colour of rhizoids; rhizoids coarsely papillose; mid-leaf cells up to 13 μm wide **7. B. radiculosum**
            15 Rhizoidal gemmae scarlet to brick-red, contrasting with colour of rhizoids; rhizoids finely papillose; mid-leaf cells up to 19 μm wide **9. B. subapiculatum**
      11 Plants always with axillary bulbils; rhizoidal gemmae often absent, when present few and smaller than 250 μm in diameter.
        16 Axillary bulbils yellowish-green, up to five per axil, with rudimentary or indistinct leaf primordia **6. B. gemmilucens**

16 Axillary bulbils green, one–two(–three) per axil, with distinct leaf primordia.

17 Costa of upper leaves longly excurrent, 100–190 μm long; bulbils in leaf axils up to 550 μm long **5. B. dunense**

17 Costa of upper leaves ending below apex to shortly excurrent; bulbils smaller, up to 400 μm long.

18 Setae and capsules brownish-red at maturity; capsule neck very thick, strongly wrinkled when dry **4. B. versicolor**

18 Setae and capsules purplish to dark red at maturity; capsules neck not thick, wrinkled when dry **3. B. bicolor**

1. Bryum cellulare Hook. *in* Schwägr., Spec. Musc., Suppl. 3 : 214a (1827).

Bryum splachnoides (Harv.) Müll. Hal., Syn. 1 : 291 (1848). — Bilewsky (1965) : 399, Reimers (1957); *Brachymenium cellulare* (Hook.) A. Jaeger, Ber. St. Gall. Naturw. Ges., 1873–74 : 111 (1875); *Mielichhoferia coppeyii* Cardot *in* Copp., Bull. Soc. Sci. Nancy, ser. 3, 10 : 25 (1909).

Distribution map 138, Figure 139 (p. 361), Plate XI : b.

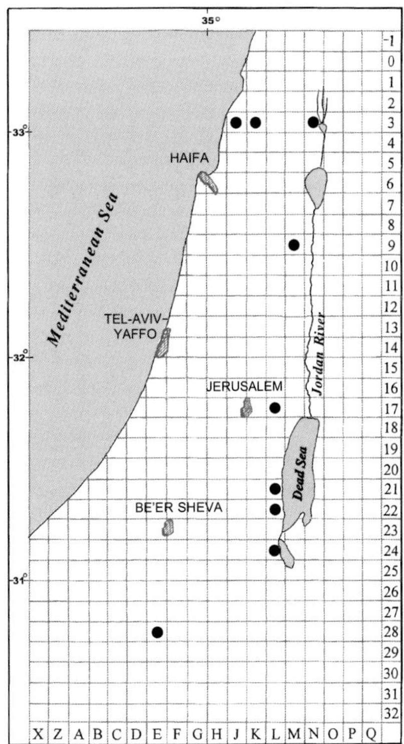

Map 138: *Bryum cellulare*

Plants 5–20 mm high, green above, reddish below. *Stems* often tomentose, red to brown. *Leaves* ± imbricate, equally distributed along the stem, strongly concave, 1–1.7(–2) mm long, ovate to ovate-lanceolate, acute, or short apiculate; margins entire, plane or slightly recurved, not bordered; costa narrow, percurrent to shortly excurrent, turning reddish; cells large and thin-walled; basal cells rectangular, upper cells elongate, rhomboidal to hexagonal, in mid-leaf 16–22 μm wide; perichaetial leaves ± larger than lower leaves. *Rhizoids* dark red, papillose; rhizoidal gemmae present, up to 175 μm in diameter. *Dioicous. Setae* 15–20(–25) mm long; capsules erect to inclined, up to 3–4 mm long including operculum, oblong to pyriform, yellowish-brown to reddish; neck long, narrowing gradually into seta, at least half the length of capsule; operculum conical, apiculate; annulus readily detached; endostome segments linear with elongate perforations; cilia absent. *Spores* 21–27 μm in diameter; with ± baculate projections having a distinctly granulate distal surface (seen with SEM).

Habitat and Local Distribution. — Plants growing in dense tufts on wet calcareous rocks and on clay, near streams and waterfalls, under dripping water or constant spray. Lower parts of plants from previous years undergo crustation by accumulating calcareous material. Often grows together with *Barbula ehrenbergii* and *Eucladium verticillatum.* Occasional: Upper Galilee, Judean Desert, Central Negev, Hula Plain, Bet She'an Valley, and Dead Sea area.

General Distribution. — Mainly a tropical species of the Southern Hemisphere; recorded from Southwest Asia (Cyprus, Turkey, Lebanon, Sinai, Yemen, Oman, and Afghanistan), from scattered localities around the Mediterranean, southern Europe (few records), Macaronesia, Africa (throughout), eastern, central, and Southeast Asia, and Central America.

In southern Europe, as in Israel, *Bryum cellulare* is mainly restricted to warm and humid habitats, which seem to be the only enclaves where this species is able to survive in the Northern Hemisphere. Coppey (1909), who described this species as *Mielichhoferia coppeyi* Cardot from his Greek material, considered some of its characters as primitive and the species a relict of an ancient tropical flora.

Bryum cellulare occupies a unique position in the genus because of its nearly erect capsule. The long narrow neck — a very distinct character among local plants — is generally not mentioned in the literature (e.g., Magill 1987, Ochi *in* Sharp *et al.* 1994).

Figure 139. *Bryum cellulare*: **a** habit, moist (×10); **b** habit, dry (×10); **c** detached operculum (×10); **d** exostome and endostome teeth (×125); **e** leaves (×50); **f** basal cells; **g** upper cells (all leaf cells ×500). (*Herrnstadt 85-11*)

a
c
b
d
e
g
a
b
f
g
f

2. Bryum argenteum Hedw., Sp. Musc. Frond. 181 (1801). [Bilewsky (1965): 401]

Distribution map 139, Figure 140.

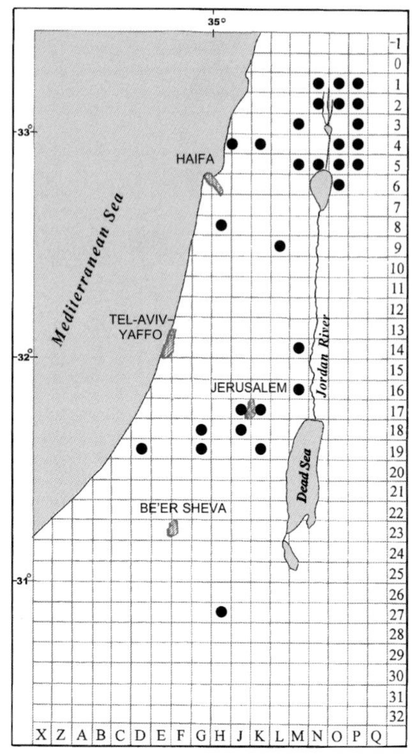

Map 139: *Bryum argenteum*

Plants up to 12 mm high, whitish to silvery pale green. *Stems* sometimes reddish, slender, fragile, julaceous. *Leaves* imbricate when dry, hardly changing when wet, distant below, crowded above, 0.5–1.2 mm long, concave, broadly ovate, acuminate to long-apiculate, hyaline in upper part; margins entire, plane or slightly recurved below, not bordered; costa not strong, sometimes reddish at base, ending well below apex or in acumen; basal cells sometimes reddish, rectangular to quadrate, extending to about ⅓ the length of leaf along the margins; mid- and upper leaf cells hexagonal, rhomboid-hexagonal to narrowly rhomboidal, sometimes very incrassate, mid-leaf cells 11–16 μm wide. *Rhizoids* papillose; rhizoidal gemmae absent; axillary bulbils rare, with rhizoids at their base. *Sporophytes* seen in single locality Golan Height (Birkat Ram).

Habitat and Local Distribution. — Plants growing in loose or dense tufts, forming turfs, or solitary mixed with other mosses, on exposed calcareous or basaltic soils, sometimes also in rock crevices; in dry habitats and disturbed places. Common: Coastal Galilee, Philistean Plain, Upper Galilee, Lower Galilee, Mount Gilboa, Samaria, Shefela, Judean Mountains, Judean Desert, Central Negev, Dan Valley, Hula Plain, Lower Jordan Valley, and Golan Heights.

General Distribution. — A cosmopolitan species, often growing as a weed.

Bryum argenteum can be easily recognised among all other local *Bryum* species by the whitish-silvery colour resulting from the partly hyaline leaves. One collection (Central Negev: "Wadi Jerka", no. 113 in Herb. Bilewsky) was recorded in Bilewsky (1970) as "*B. argenteum* var. *majus* Schwaegr." and in Bilewsky (1977) as *B. argenteum* "var. *lanatum* (P. Beauv.) Hamp.". Variety *lanatum* is characterised by leaves with a larger hyaline apical area and a long-excurrent costa. Such plants are not considered here as a separate taxon (cf. Nyholm 1954–1969, Crum & Anderson 1981) because of the continuous variation of those characters in the species.

Figure 140. *Bryum argenteum*: **a** habit, moist (×10); **b** habit, dry (×10); **c** leaves (×50); **d** basal cells; **e** mid-leaf cells; **f** upper cells (all leaf cells ×500); **g** leaf (×50). (**a**–**f**: *Herrnstadt & Crosby 79-77-8*; **g**: *Herrnstadt, Liston & Boaz 83-284-5*)

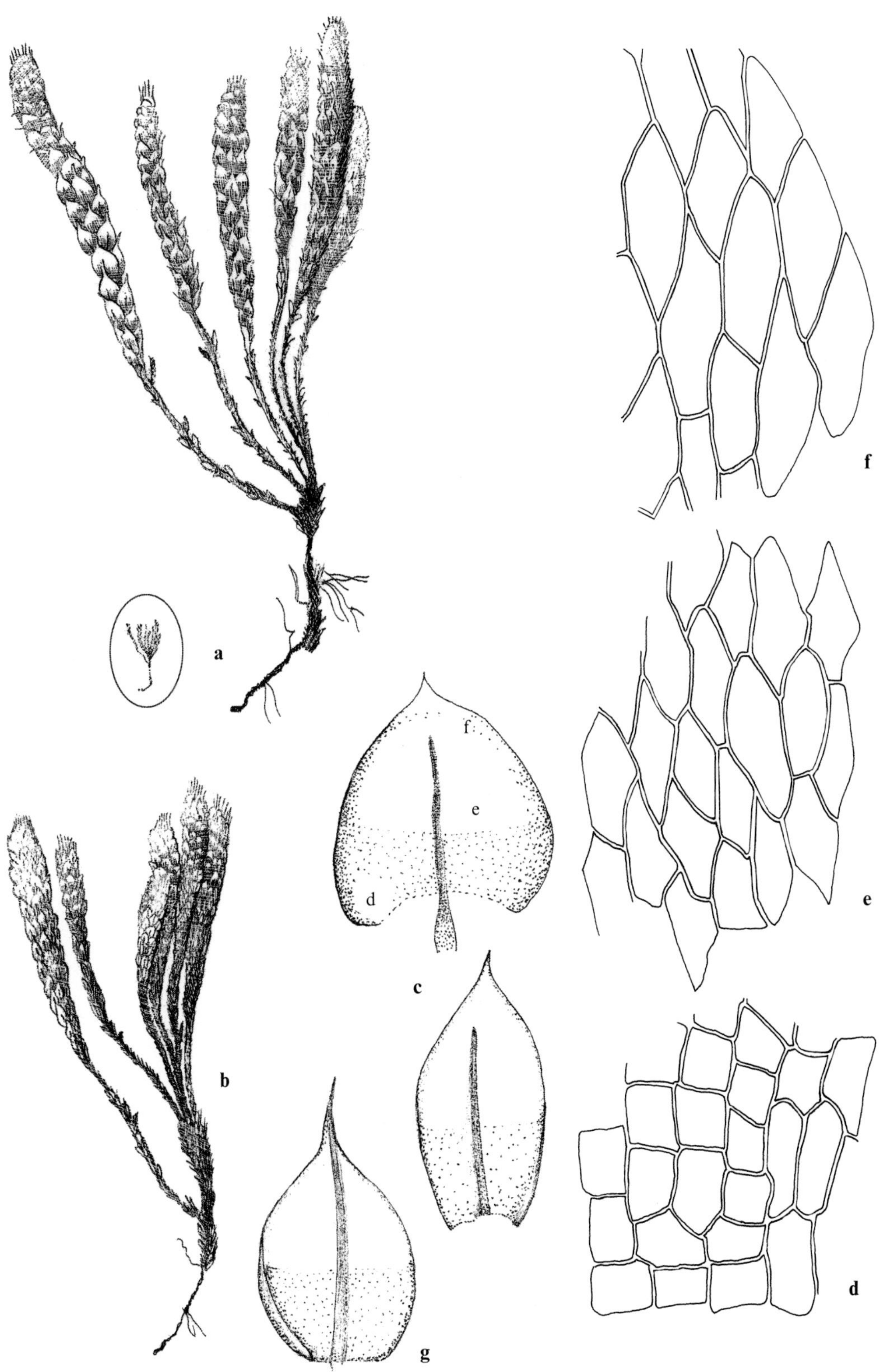
a
b
c
d
e
f
g
f
e
d

Bryum bicolor complex: Species 3–6

Bryum bicolor is treated in recent literature either *sensu lato* as one large variable species (cf. Magill 1987) or *sensu stricto* as part of a complex of several related species (Wilczek & Demaret 1976a, Smith & Whitehouse 1978). Although some doubts remain as to whether each of the species of the complex deserves specific status, it was preferred here to subdivide the plants of the complex growing in the local flora among four species: *B. bicolor*, *B. dunense*, *B. versicolor*, and *B. gemmilucens*. Whereas *B. gemmilucens* can be easily identified as a separate species by its axillary bulbils with only rudimentary primordia, the delimitation of the other three species may be more difficult. *Bryum bicolor* is usually characterised by ovate to ovate-lanceolate leaves with costa ending below apex to shortly excurrent, as compared with *B. dunense* comprising plants with ovate to lanceolate leaves with long-excurrent costa. The two species bear similarly shaped axillary bulbils, somewhat larger in *B. dunense*. Leaf shape and the degree of excurrence of the costa vary within populations and among populations of both species, thereby often making it difficult to place a single plant in one of the two. *Bryum bicolor* and *B. dunense* are, perhaps, not orthospecies but represent two extreme ends of a continuous range of variation. In the leaf characters, *B. versicolor* occupies an intermediate position. However, it can be distinguished by the thick and wrinkled neck of the capsule.

Literature. — Wilczek & Demaret (1976a, b), Smith & Whitehouse (1978).

3. Bryum bicolor Dicks., Pl. Crypt. Brit., Fasc. 4 : 16 (1801). [Bilewsky (1965) : 401; Rabinovitz-Sereni (1931); Bizot (1945)]
B. atropurpureum B.S.G., Bryol. Eur. 4 : 143 (1839).

Distribution map 140, Figure 141 : a–f (p. 367).

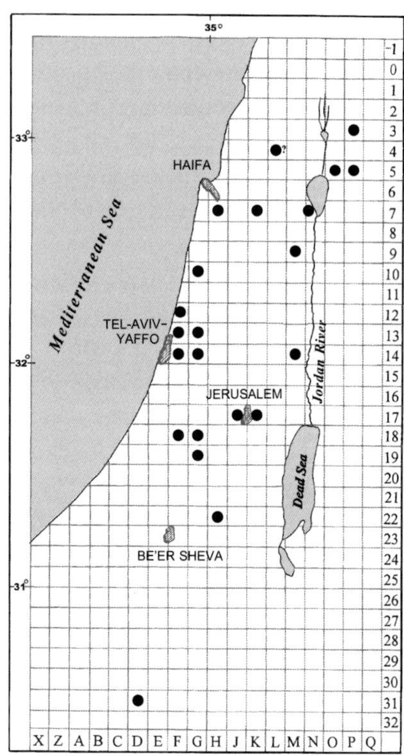

Map 140: *Bryum bicolor*

Plants up to 5 mm high, small, green to pale green. *Stems* green above, red below. *Leaves* imbricate, erect and scarcely contorted when dry, erect to erectopatent when moist, 0.75–1.25 mm long, concave, ovate to ovate-lanceolate, acuminate (acumen varying in length among different plants within populations); margins ± recurved, entire or obscurely serrulate at apex, not bordered; costa ending below apex to shortly excurrent, turning red in old leaves; basal cells irregularly quadrate to rectangular, not incrassate, upper cells more incrassate, rhomboid-hexagonal, in mid-leaf up to 14 µm wide; marginal cells usually narrower, sometimes ± longer than adjacent cells but not forming a distinct border. *Axillary bulbils* present, usually solitary, up to 400 µm long, with leaf primordia usually about half of bulbil length, green. *Rhizoids* usually weakly papillose, yellowish-brown to brown; rhizoidal gemmae very rare. *Dioicous. Setae* up to 12 mm long, purplish-red at maturity; capsules pendulous, greatly varying in size, usually 1–2 mm long with operculum, ovoid or wide-ellipsoid; neck about ¼ the length of capsule, wrinkled when dry, not thick, purplish-red to dark red; operculum conical with a short apiculus; endostome with appendiculate cilia. *Spores* 8–12 µm in diameter.

Habitat and Local Distribution. — Plants terrestrial, growing on calcareous soils (mainly sand, sandy loam, rendzina, and loess). Common: Sharon Plain, Upper Galilee (?), Lower Galilee, Mount Carmel, Shefela, Judean Mountains, Central Negev, Upper Jordan Valley, Bet She'an Valley, Lower Jordan Valley, and Golan Heights.

General Distribution. — Records of *Bryum bicolor* often include all the species of the *B. bicolor* complex (cf. Frey & Kürschner 1991c). Recorded from Southwest Asia (usually not individual species of the complex), around the Mediterranean, Europe (throughout), Macaronesia, North and South Africa, Asia (throughout), Australia, New Zealand, and North and Central America.

Ochi (1980, *in* Sharp *et al.* 1994) contended that the valid name of this taxon is the earlier name *Bryum dichotomum* Hedw. (Sp. Musc. Frond 183. 1801).

4. Bryum versicolor A. Braun ex B. & S., Bryol. Eur. 4: 145 (1839). [Nachmony (1961)]

Distribution map 141, Figure 141: g.

Plants resembling *Bryum bicolor* in gametophytic characters, except for the costa, which is excurrent and not ending below apex. *Setae* strongly curved; capsules smaller than in *B. bicolor*, with very thick and strongly wrinkled neck; setae and capsules brownish-red at maturity.

Habitat and Local Distribution. — Terrestrial plants, growing on terra rossa and loess of wadis and on fine limestone silt of road sides. Occasional: Sharon Plain, Upper Galilee, Shefela, Judean Mountains, Judean Desert, Western Negev (?), Lower Jordan Valley, and Golan Heights.

General Distribution. — Definite records of *Bryum versicolor* as a separate species seem to be available so far only from western, central, and southern Europe.

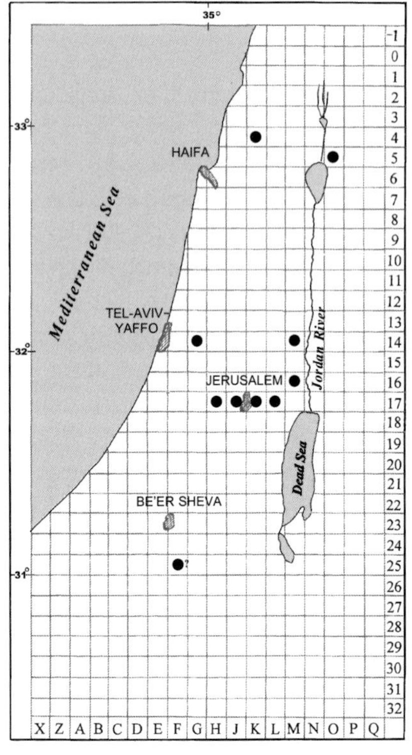

Map 141: *Bryum versicolor*

Figure 141. *Bryum bicolor*: **a** habit, moist (×10); **b** habit, dry (×10); **c** leaves (×50); **d** basal cells; **e** upper cells (all leaf cells ×500); **f** axillary bulbil (×50).
Bryum versicolor: **g** capsule (×10).
(**a**–**f**: *D. Zohary*, 11 Feb. 1943; **g**: *Kushnir*, 29 Feb. 1943)

5. Bryum dunense A. J. E. Sm. & H. Whitehouse, J. Bryol. 10 : 41 (1978).

Distribution map 142, Figure 142.

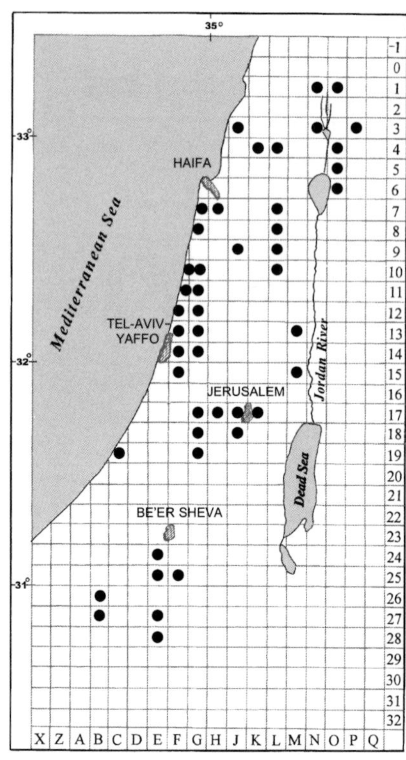

Map 142: *Bryum dunense*

Plants green, turning yellowish-green, rarely brown when dry; resembling *Bryum bicolor* except for the following characters. *Leaves* ovate to lanceolate, widest below middle, long-acuminate; costa excurrent into a smooth or slightly serrulate leaf point, 100–190 µm long. *Axillary bulbils* mainly in upper leaves, usually solitary, sometimes up to three per axil, up to 550 µm long, green. *Rhizoidal gemmae* at end of rhizoids sometimes present, especially on sandy substrates, tuber-like, reddish.

Habitat and Local Distribution. — Plants terrestrial, growing on exposed calcareous soils (e.g., loess, sandy soil) or dunes. Very common: locally the most common species of the complex; previously recorded as *B. bicolor*: Coastal Galilee, Coast of Carmel, Sharon Plain, Philistean Plain, Upper Galilee, Lower Galilee, Mount Carmel, Mount Gilboa, Samaria, Shefela, Judean Mountains, Western Negev, Central Negev, Dan Valley, Lower Jordan Valley, and Golan Heights.

General Distribution. — Records few and scattered, most probably included in part in *Bryum bicolor* (see the comment in General Distribution there): Southwest Asia, northern and eastern Mediterranean, Europe (few, mainly from the north and south), and Macaronesia.

Bryum dunense may resemble small forms of *B. caespiticium* in habit (see the discussion below, p. 392).

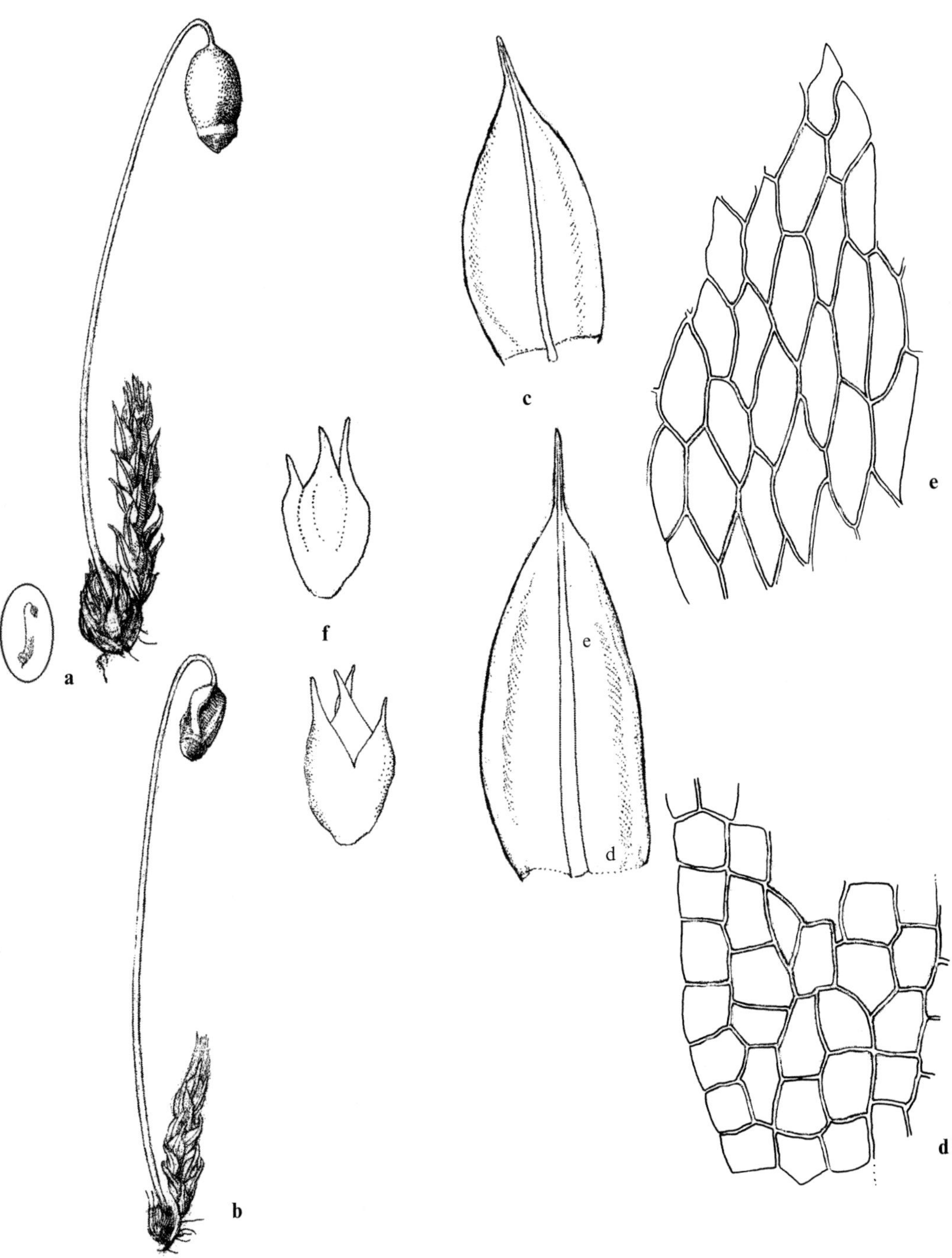

Figure 142. *Bryum dunense*: **a** habit, moist (×10); **b** habit, dry (×10); **c** leaves (×50); **d** basal cells; **e** upper cells (all leaf cells ×500); **f** axillary bulbils (×50). (*D. Zohary*, 10 Feb. 1943)

6. Bryum gemmilucens R. Wilczek & Demaret , Bull. Jard. Bot. Natl. Belgique 46 : 537 (1976).

Distribution map 143, Figure 143.

Plants up to 3 mm high, yellowish-green, resembling *Bryum bicolor*. *Stems* yellowish-green above, orange-red below. *Leaves* ovate to broadly ovate. *Axillary bulbils* at young stage up to five per axil, about 165 µm long, ellipsoid, obovoid or pyriform, with leaf primordia rudimentary and restricted to apex or indistinct, with rounded base, often with a stalk of uniseriate cells, yellowish-green. *Sporophytes* as in *B. bicolor*, abundantly produced in single population seen so far (no previous records of sporophytes seem to be available). *Setae* and capsules brownish-red at maturity.

Habitat and Local Distribution. — Plants growing in tufts on basalt soil in the single locality found. Very rare: Golan Heights.

General Distribution. — Records are few and scattered, most probably included in part in *Bryum bicolor* (see the comment in General Distribution there): Southwest Asia, Europe (southern Britain, Belgium, France, Portugal, and Hungary), Canary Islands, and California.

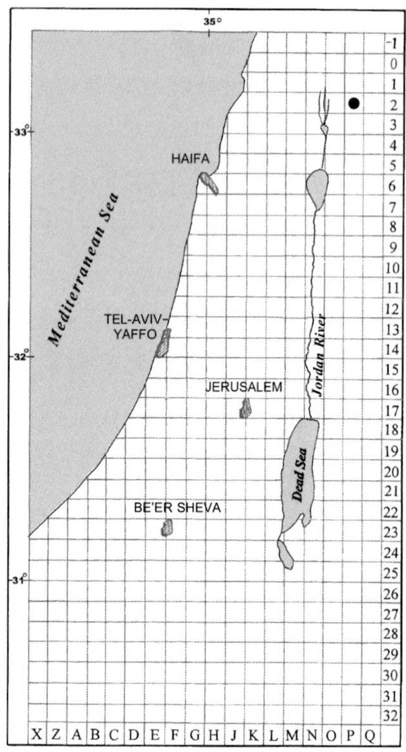

Map 143: *Bryum gemmilucens*

The length of the costa varies in the investigated population (*Herrnstadt & Crosby 79-72-3*) and is often excurrent. In the local species of the *Bryum bicolor* complex, leaves with excurrent costa are more frequent than in European material, and may perhaps be correlated with the more xeric conditions prevailing in the eastern Mediterranean (Richardson 1981).

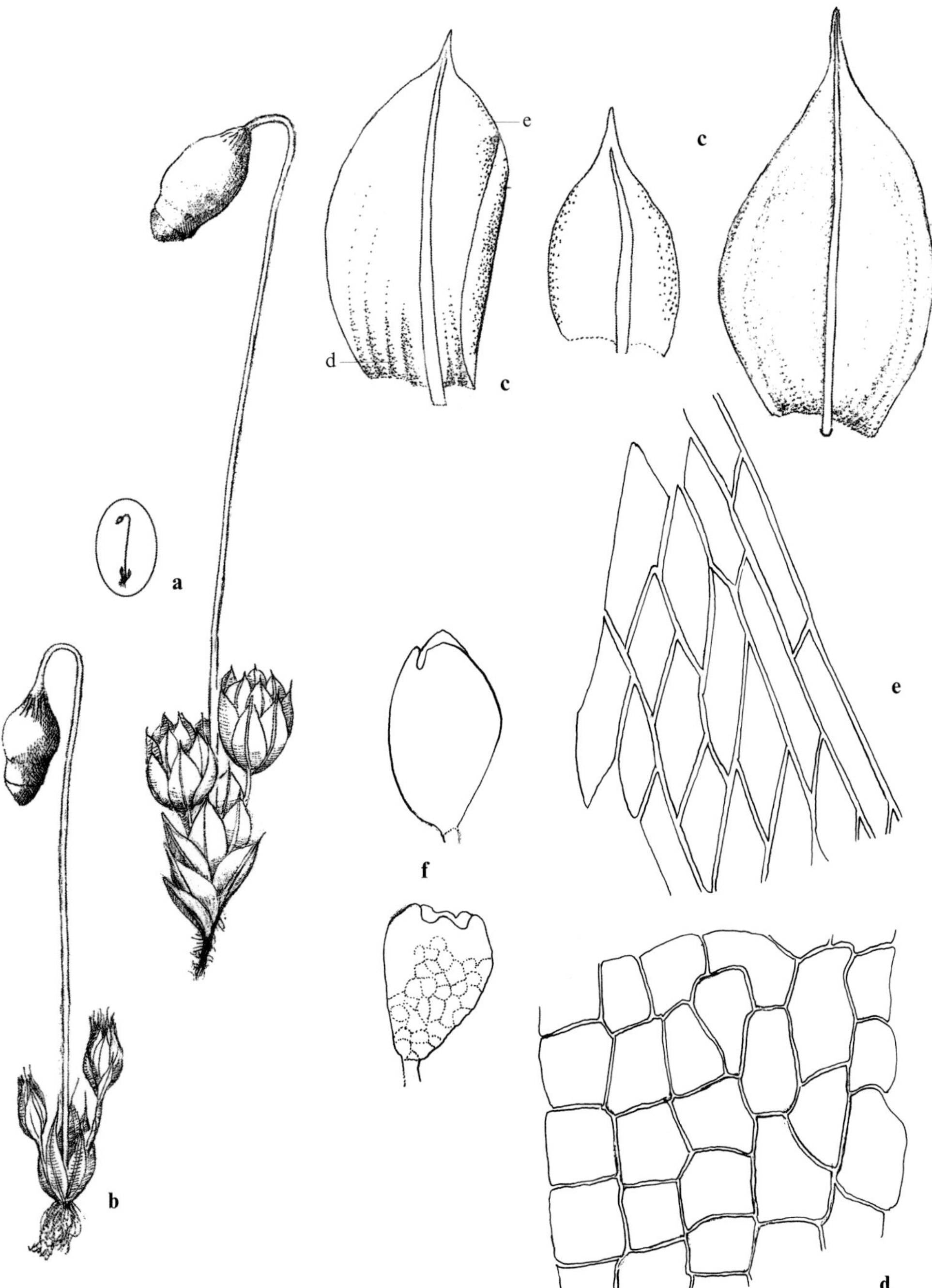

Figure 143. *Bryum gemmilucens*: **a** habit, moist (×10); **b** habit, dry (×10); **c** leaves (×50); **d** basal cells; **e** marginal upper cells (all leaf cells ×500); **f** axillary bulbils (×50). (*Herrnstadt & Crosby 79-72-3*)

Bryum erythrocarpum complex: Species 7–9

In Crundwell & Nyholm (1964), nine species are included in the "*Bryum erythrocarpum* complex" (*B. micro-erythrocarpum* Müll. Hal. & Kindb. is considered there as the valid name of *B. erythrocarpum*). The name of the complex is also retained here although *B. subapiculatum* is used as the valid name of *B. erythrocarpum*. The three local species of the complex can be identified by the following characters.

Plants up to 10 mm high, tinged with red. *Leaves* erect to erect-spreading, ovate-lanceolate to oblong lanceolate, acuminate; margins plane to slightly recurved in upper leaves, serrulate towards apex, not or inconspicuously bordered; costa strong, excurrent. *Rhizoids* densely papillose, always bearing rhizoidal gemmae. *Dioicous*. *Setae* 10–30 mm long, red; capsules pendulose, cylindrical to ovoid-pyriform, usually dark red; endostome with appendiculate cilia.

Literature. — Wilczek & Demaret (1974).

Sterile plants of this complex superficially resemble plants of *Bryum bicolor* but can be distinguished by the coarse rhizoids and the lack of axillary bulbils.

7. Bryum radiculosum Brid., Sp. Musc. 3 : 18 (1817).

B. murorum (Schimp.) Berk., Handb. Brit. Mosses 320 (1863). — Bilewsky (1974); *B. murale* Wilson ex Hunt, Mem. Lit. Soc. Manchester, ser. 3, 3 : 239 (1868). — Bilewsky (1965) : 402, Rabinovitz-Sereni (1931), Bizot (1945); *B. atrovirens* Vill. ex Brid. var. *radiculosum* (Brid.) Wijk & Margad., Taxon 8 : 72 (1959).

Distribution map 144, Figure 144.

Plants up to 10 mm high, bright green. *Leaves* up to 2 mm long, sometimes slightly twisted when dry; margins recurved or not in upper leaves; costa strong, yellowish, long-excurrent; basal cells quadrate to rectangular; upper cells hexagonal to broad-linear, mid-leaf cells 10–13 µm wide; marginal cells usually narrower and longer. *Rhizoids* yellowish-brown, coarsely papillose; rhizoidal gemmae up to 180(–220) µm in diameter, spherical, with peripheral cells not protuberant or slightly so, yellowish-brown. *Capsules* varying in shape and size, up to 4 mm long including operculum; neck to ⅓ the length of capsule; exostome teeth united at base up to level of mouth. *Spores* about 11 µm in diameter.

Habitat and Local Distribution. — Plants growing in tufts, on stones, limestone rocks, chalk and calcareous soils. Occasional to common: Sharon Plain, Upper Galilee, Lower Galilee, Mount Carmel, Mount Gilboa, Samaria, Shefela, Judean Desert, Northern Negev, Dan Valley, and Upper Jordan Valley.

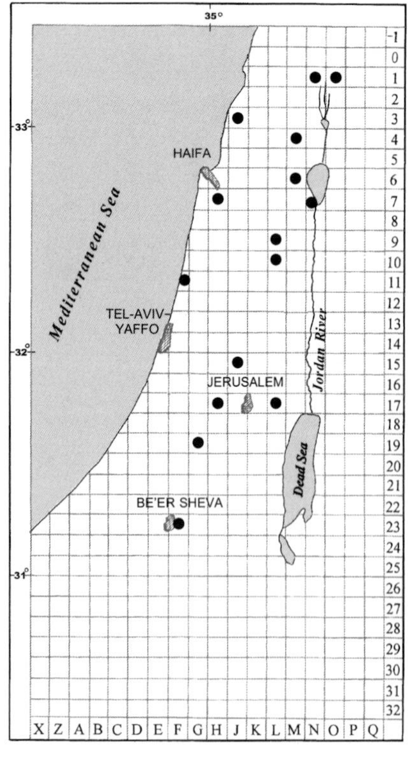

Map 144: *Bryum radiculosum*

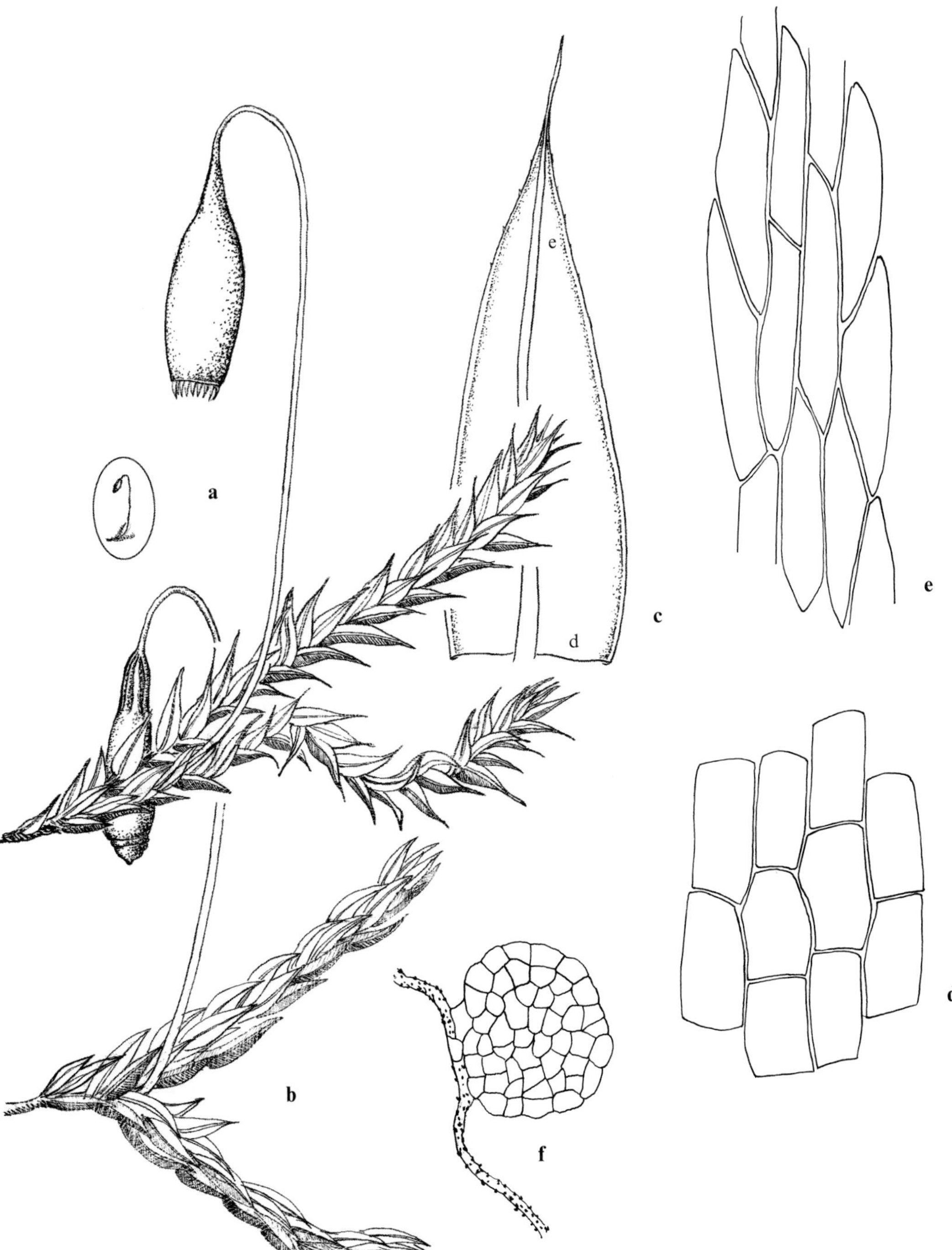

Figure 144. *Bryum radiculosum*: **a** habit, moist (×10); **b** habit, dry (×10); **c** leaf (×50); **d** basal cells; **e** upper cells (all leaf cells ×500); **f** rhizoidal gemma (×125). (*Landau*, 4 Apr. 1937)

General Distribution. — Recorded from Southwest Asia (Cyprus, Turkey, and Lebanon), around the Mediterranean, Europe (throughout), Macaronesia, North and South Africa, eastern and central Asia, North America (recorded by Crum & Anderson (1981) as representing several taxa of the "erythrocarpum complex"), and Central America.

Bryum radiculosum is the most widespread local species of the *B. erythrocarpum* complex. It can be distinguished from the common species of the *B. bicolor* complex, in addition to the distinguishing characters of the two complexes listed above, by the longer and narrower leaf cells.

8. Bryum ruderale Crundw. & Nyholm, Bot. Not. 116:95 (1963).

Distribution map 145, Figure 145.

Plants up to 8 mm high (slightly smaller than the other two species of the complex), bright green. *Stems* usually red. *Leaves* about 1 mm long; costa strong, excurrent, red at leaf base; cells incrassate; basal cells quadrate to short-rectangular, upper cells elongate, hexagonal, mid-leaf cells 8–12 μm wide. *Rhizoids* bright violet to deep-violet or purple, papillose; mature rhizoidal gemmae 110–180 μm in diameter, spherical, peripheral cells protuberant, usually contrasting in colour from rhizoids, bright purple or sometimes orange. *Capsules* about 2 mm long including operculum; neck about ⅓ the length of capsule; exostome teeth united at base to level of mouth or above. *Spores* 11(–14) μm in diameter.

Habitat and Local Distribution. — Plants growing in loose tufts on layers of rendzina. Occasional, but perhaps has been overlooked: Sharon Plain, Philistean Plain, Mount Carmel, and Judean Mountains.

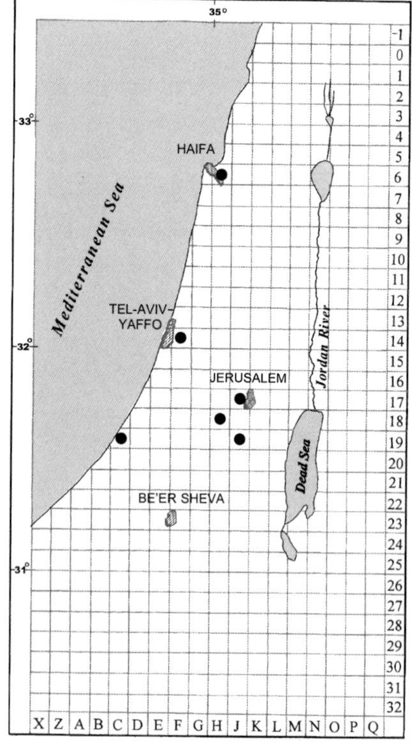

Map 145: *Bryum ruderale*

General Distribution. — Records are few. As *Bryum ruderale* was the latest species described in the "*B. erythrocarpum* complex", previous records of the other species of the complex may include it. Recorded from Southwest Asia (Turkey, Sinai, and Saudi Arabia), few records from the northern Mediterranean, Europe (scattered data throughout), Macaronesia, and North America (see the comment in General Distribution of *B. radiculosum*).

Bryum ruderale can usually be distinguished by the unique colour of the rhizoids and the rhizoidal gemmae. Contrary to records from other parts of its distribution, all local plants bear sporophytes. In some samples, tuber-like structures at the end of rhizoids can be observed, resembling similar structures in *B. dunense*. Those may perhaps be

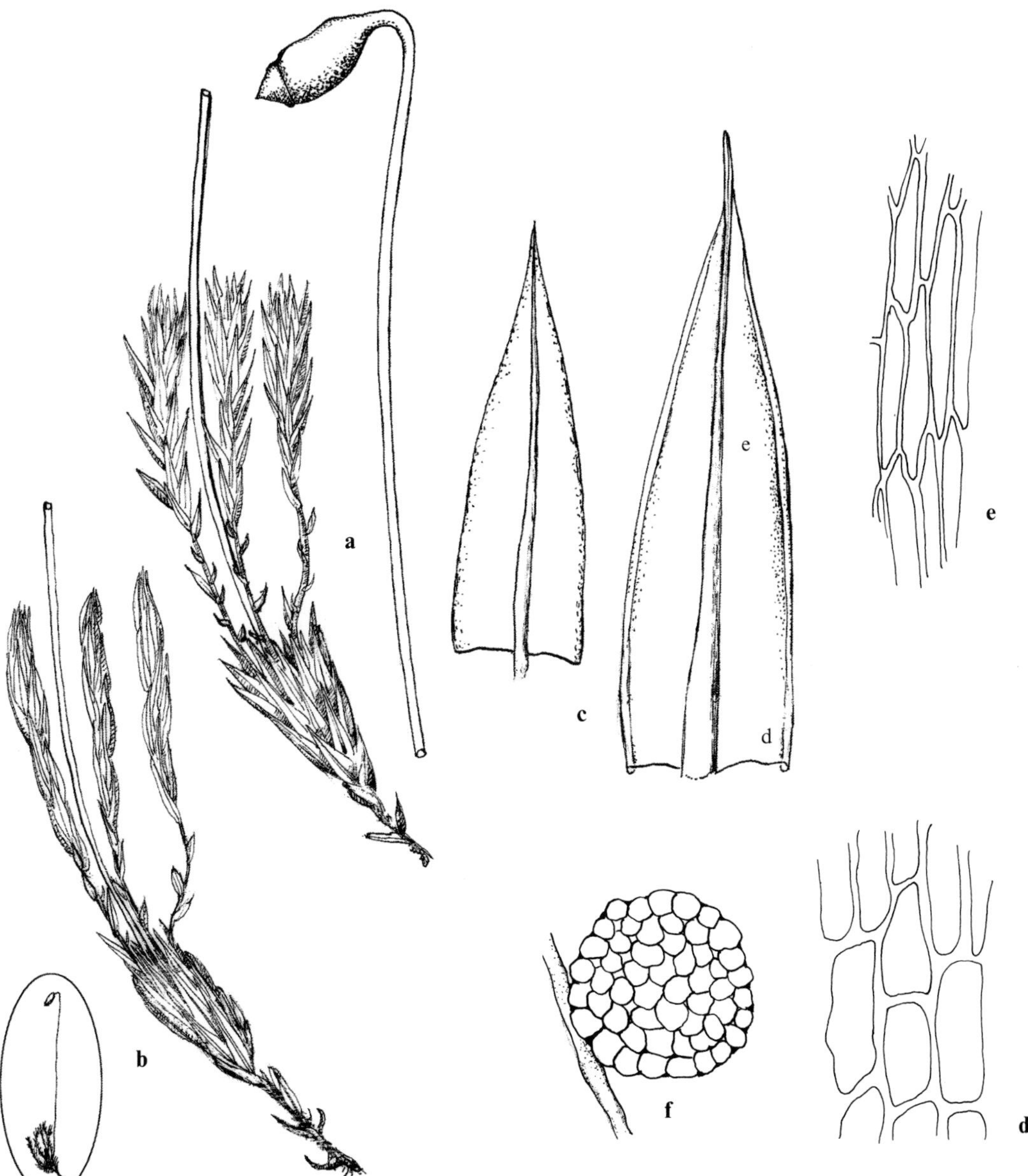

Figure 145. *Bryum ruderale*: **a** habit with sporophyte, moist (×10); **b** habit with only lower part of seta, dry (×10); **c** leaves (×50); **d** basal cells; **e** upper cells (all leaf cells ×500); **f** rhizoidal gemma (×125). (*Kushnir*, Khirbet Karin, *sine dato*)

regarded as initial stages in the development of rhizoidal gemmae (see Wilczek & Demaret 1974).

9. Bryum subapiculatum Hampe, Viddensk. Meddel. Dansk. Naturhist. Foren. Kjøbenhavn, ser. 3, 4 : 51 (1872).

B. erythrocarpum Schwägr., Sp. Musc. Frond., Suppl. 1 : 100 (1816); *B. erythrocarpum* "forma A" Correns, Verm. Laubm. 181 (1899). — Bilewsky (1965) : 401, Bizot (1945); *B. micro-erythrocarpum* Müll. Hal. & Kindb. *in* Macoun, Cat. Canad. Pl. Musci 6 : 124 (1892).

Distribution map 146, Figure 146, Plate XI : c.

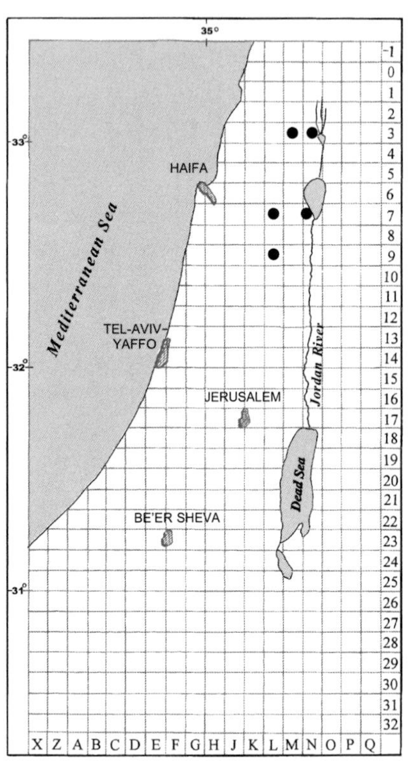

Map 146: *Bryum subapiculatum*

Plants up to 10 mm high, bright green. *Stems* reddish. *Leaves* up to 2 mm long, inconspicuously bordered; margins plane; costa strong, reddish, from short-excurrent to ± long-excurrent; basal cells rectangular; upper cells rhomboidal, elongate, greatly varying in size, mid-leaf cells 10–19 μm wide, marginal cells narrower, longer and more incrassate. *Rhizoids* brownish, finely papillose; rhizoidal gemmae up to 200(–260) μm in diameter, spherical, with slightly protuberant cells, scarlet to brick-red, contrasting with the lighter colour of the rhizoids. *Capsules* up to 3 mm long including operculum; neck up to ¼ the length of capsule; exostome teeth free at base. *Spores* 10–13 μm in diameter; densely covered by very small, rough gemmate-baculate processes (seen with SEM).

Habitat and Local Distribution. — Plants growing in maquis on calcareous rocks. Occasional: Upper Galilee, Lower Galilee, and Mount Gilboa.

General Distribution. — Records of *Bryum subapiculatum* are few; the northern Mediterranean Region, Macaronesia (the Canary Islands), Europe (scattered data from throughout), tropical and South Africa, New Zealand, North America (see the comment in General Distribution of *B. radiculosum*), and Central America.

Bryum atrovirens Brid. is often accepted as the valid name for this taxon (e.g., Bilewsky 1974, Corley *et al.* 1981).

Bryum subapiculatum can be distinguished from the other local species of the complex by the large, red gemmae. It differs from *B. radiculosum* also in the wider mid-leaf cells, the longer reddish costa and the less strongly papillose rhizoids. In Europe, *B. radiculosum* is considered as calcicolous and *B. subapiculatum* as preferring mildly acidic habitats; in the local flora both grow on calcareous substrates.

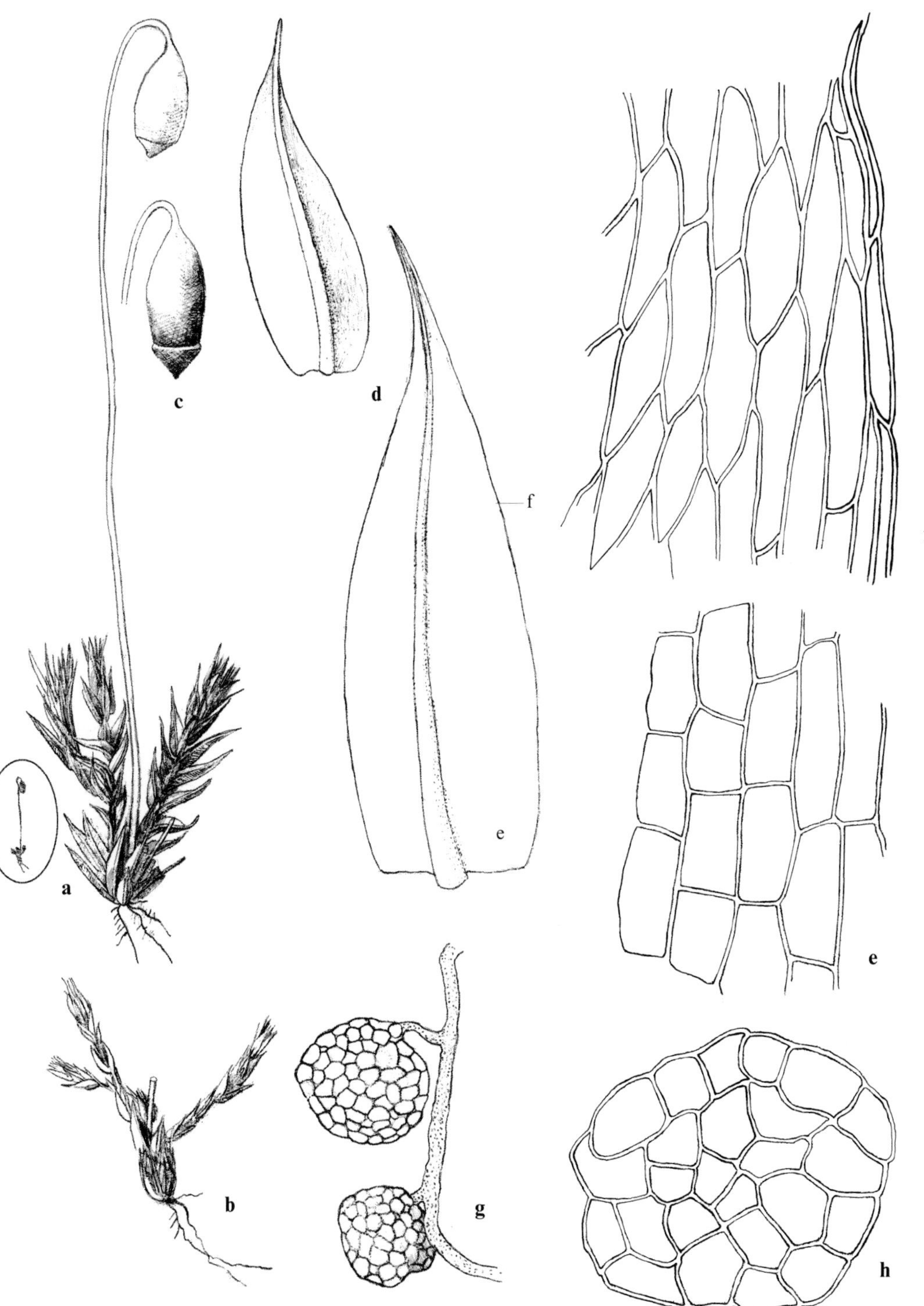

Figure 146. *Bryum subapiculatum*: **a** habit with sporophyte, moist (×10); **b** habit without sporophyte, dry (×10); **c** mature capsule (×10); **d** leaves (×50); **e** basal cells; **f** marginal upper cells (all leaf cells ×500); **g** rhizoid with young gemmae (×125); **h** mature rhizoidal gemma (×250). (**a**, **b**, **d**–**h**: *Grizi*, 16 Apr. 1954; **c**: *Kushnir*, 26 Jun. 1943)

10. Bryum gemmiparum De Not., Cronac. Briol. Ital. 1:26 (1866).
B. alpinum Huds. ex With. subsp. *gemmiparum* (De Not.) Kindb., Eur. N. Amer. Bryin. 2:356 (1897).

Distribution map 147, Figure 147.

Plants up to 20 mm high, green-brown below, turning brownish when dry. *Stems* ± tomentose. *Leaves* erect or slightly contorted when dry, imbricate, erectopatent when moist, concave, up to 1.5 mm long, ovate-lanceolate, usually obtuse or obtuse to acute; margins plane or slightly recurved below, entire, not bordered; costa strong, usually percurrent; cells gradually narrower and longer towards margins; basal cells ± quadrate, mid- and upper leaf cells rhomboidal to hexagonal, incrassate, in mid-leaf 40–55 µm long, 10–14 µm wide. *Axillary bulbils* bright green, sessile, in groups in upper leaves, about 700 µm long; bulbils in lower leaves larger and shoot-like, with rhizoids at base. *Rhizoids* brown, papillose, bearing rhizoidal gemmae, sometimes with terminal tuber-like structures. *Dioicous. Sporophytes* seem very rare in local populations; capsules pendulous, narrowly ellipsoid to nearly cylindrical, symmetrical; operculum mammillate.

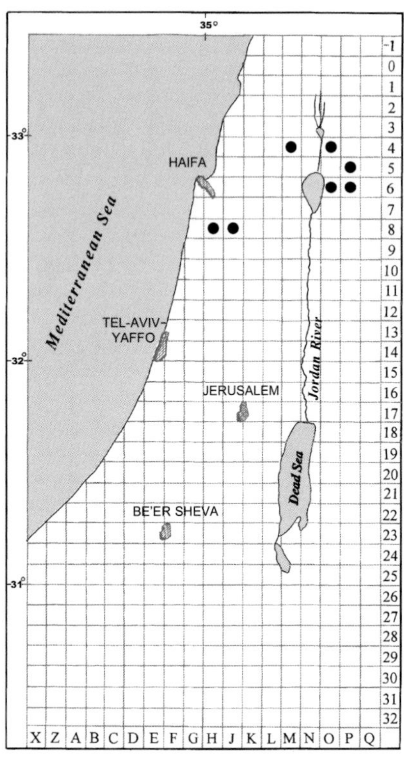

Map 147: *Bryum gemmiparum*

Habitat and Local Distribution. — Plants growing in very dense tufts on wet basalt soil on banks of creeks or under dripping water. Occasional: Upper Galilee, Samaria, and Golan Heights.

General Distribution. — Recorded from Southwest Asia (Cyprus, Turkey, and Sinai), around the Mediterranean, Europe (throughout, many records from the south), North Africa, central Asia (India and China — Yunnan), North America (recorded without gemmae; see Crum & Anderson 1981), and southern South America.

Bryum gemmiparum can be recognised by the ovate-lanceolate, concave leaves, with usually obtuse apex and percurrent costa. It has various brood bodies: abundant bulbil-like axillary gemmae (mainly in upper leaves), axillary shoot-like propagules (in lower leaves), and rhizoidal gemmae. In the occurrence of axillary bulbils and tuber-like structures terminating the rhizoids, it resembles species of the *B. bicolor* complex.

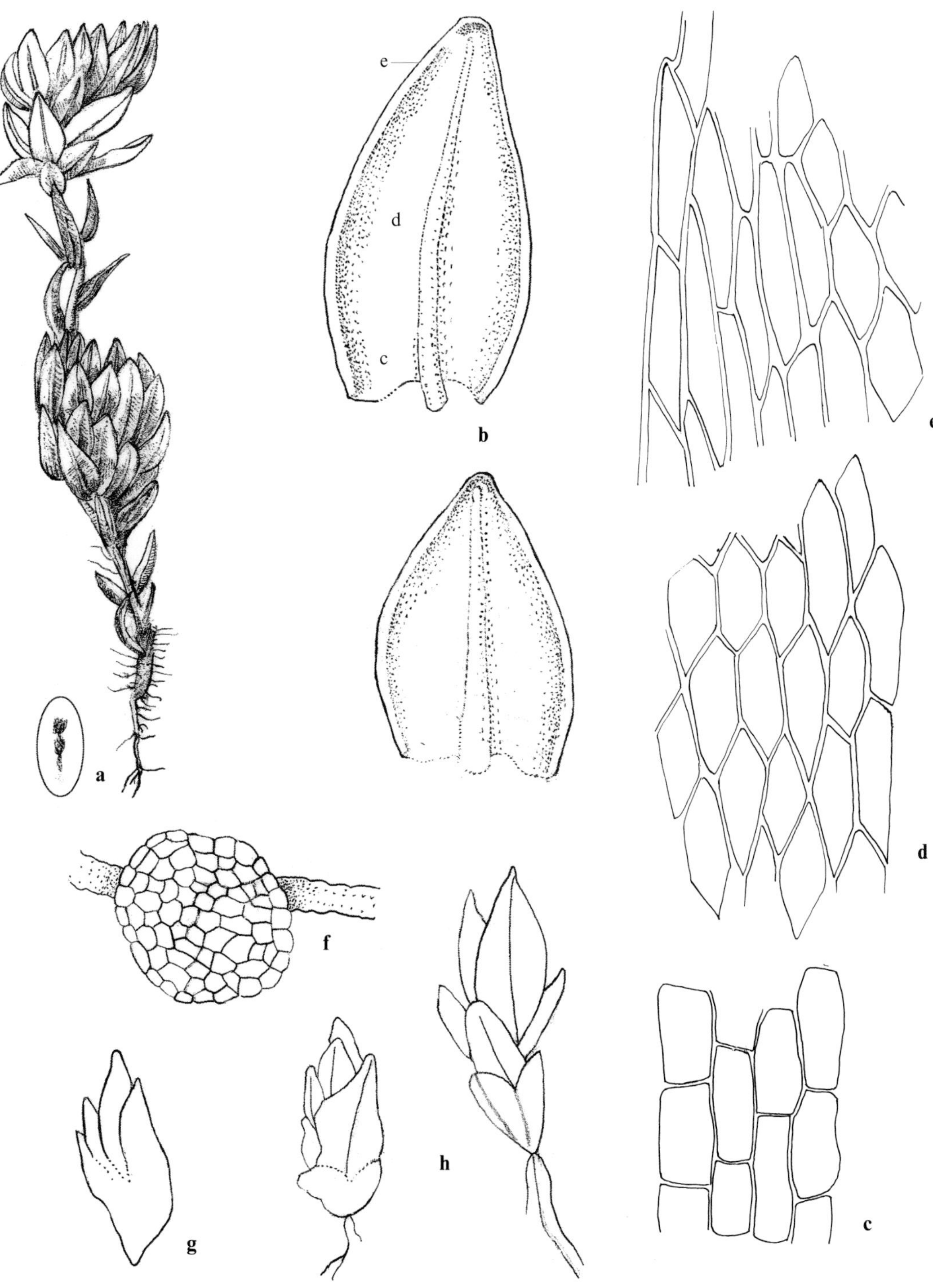

Figure 147. *Bryum gemmiparum*: **a** habit, moist (×10); **b** leaves (×50); **c** basal cells; **d** mid-leaf cells; **e** marginal upper cells (all leaf cells ×500); **f** rhizoidal gemma (×125); **g** axillary bulbil of upper leaves (×50); **h** axillary bulbils of lower leaves (×50). (*Herrnstadt*, 9 May 1984)

11. Bryum muehlenbeckii B.S.G., Bryol. Eur. 4:163. (1846).

Distribution map 148, Figure 148.

Plants 15–30(–40) mm high, dull green to reddish. *Stems* reddish. *Leaves* erect, imbricate when dry, partly erectopatent when moist, strongly concave, carinate, 1–1.5 mm long, oblong-ovate, obtuse to acute; margins plane, not bordered; costa subpercurrent, sometimes percurrent, reddish-brown; cells large, with slightly incrassate reddish-brown walls; basal cells few, rectangular, upper cells rhomboidal, in mid-leaf 50–75 μm long, 20–27 μm wide. *Rhizoids* finely papillose, yellowish-brown; rhizoidal gemmae sometimes present, spherical, about 170 μm in diameter. Only sterile plants seen. Recorded as *dioicous* with capsules as in *Bryum gemmiparum*.

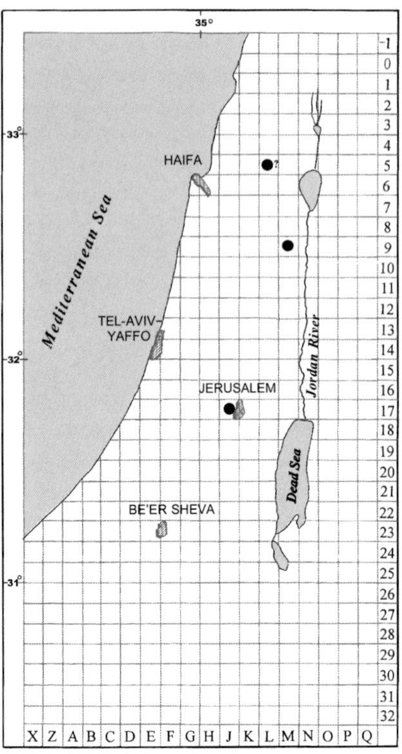

Map 148: *Bryum muehlenbeckii*

Habitat and Local Distribution. — Plants growing in tufts in or near water, on calcareous rocks, on terra rossa, and among soft and hard limestone rocks. Rare: Upper Galilee (?), Judean Mountains, and Bet She'an Valley.

Local habitat data do not agree with other published data noting *Bryum muehlenbeckii* as a calcifuge ("Kalkmeidend") plant (Mönkemeyer 1927) or as growing on acidic rocks (in northern Scotland; Smith 1978). However, according to Crum & Anderson (1981) it can be found "at least sometimes on limestone" in North America.

General Distribution. — Recorded from Southwest Asia (Lebanon, Iran, and Saudi Arabia), scattered from around the Mediterranean (except in the southern areas), Europe (throughout), Macaronesia, Australia, and North and Central America.

Sterile plants can be easily identified by a combination of leaf characters: strongly concave leaves, subpercurrent or sometimes percurrent costa, and large reddish-brown cells.

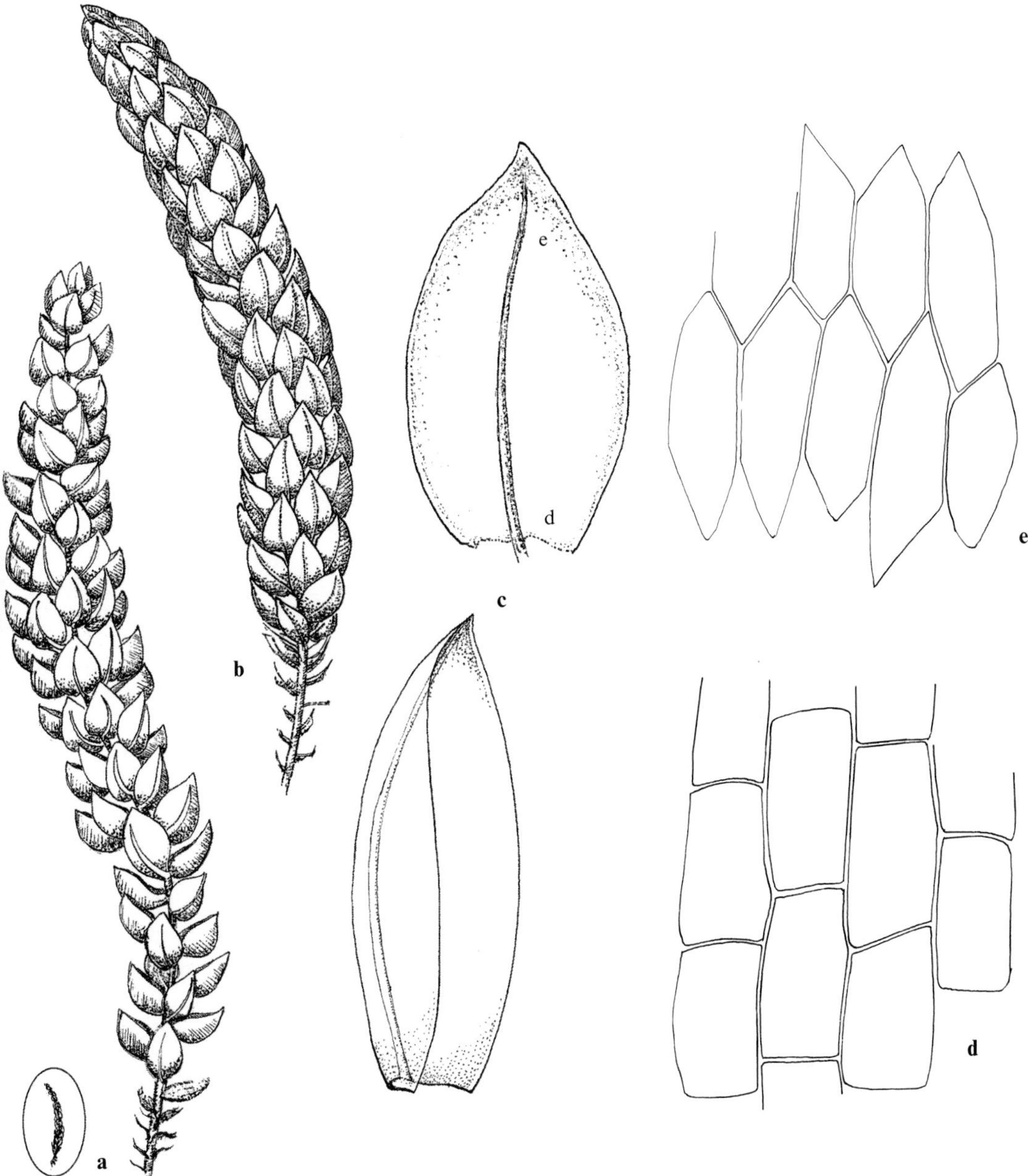

Figure 148. *Bryum muehlenbeckii*: **a** habit, moist (×10); **b** habit, dry (×10); **c** leaves (×50); **d** basal cells; **e** upper cells (all leaf cells ×500). (*Weitz*, 29 Mar. 1971)

12. Bryum donianum Grev., Trans. Linn. Soc. London 15 : 345 (1827). [Bilewsky (1965) : 403]

Distribution map 149, Figure 149.

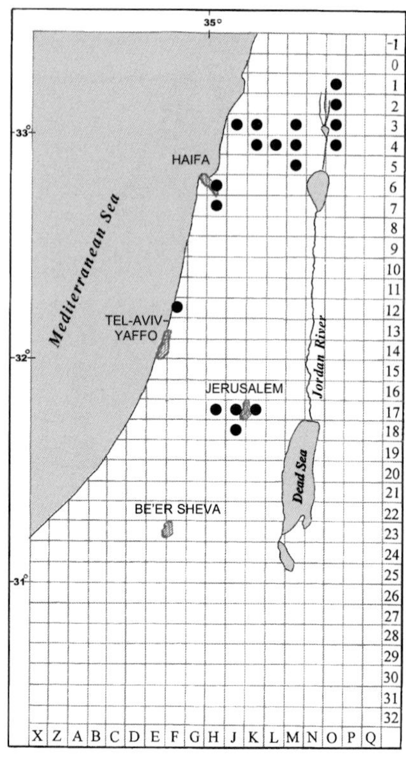

Map 149: *Bryum donianum*

Plants up to 10 mm high, bright to dark green. *Stems* red. *Leaves* shrunken and sometimes twisted when dry, erectopatent when moist, up to 3 mm long, obovate to obovate-lanceolate, acuminate; upper leaves in a distinct comal tuft; margins plane or recurved at base, usually serrate in the upper part (including the acumen), with a very distinct border usually confluent with costa at apex, base red; costa with red base, strong, excurrent into the acumen; cells incrassate, with pitted walls; basal cells rectangular, upper cells rhomboid-hexagonal, in mid-leaf 15–22 μm wide; border two–three-stratose, consisting of two–four rows of yellow, elongate incrassate cells, two–five times the length of adjacent cells. *Rhizoids* papillose; rhizoidal gemmae absent. *Dioicous*. *Setae* 10–20 mm long; capsules pendulous, 3–5 mm long, cylindrical, yellowish to red brown, slightly constricted below mouth, neck up to ⅓ the length of capsule; operculum mammillate; endostome with appendiculate cilia. *Spores* 11–15 μm in diameter.

Habitat and Local Distribution. — Plants growing in tufts, mainly on dark terra rossa and light calcareous soils, on rocks and in rock crevices, often in mountainous areas; sometimes grow with *Bryum caespiticium* and *B. torquescens*. Occasional to common: Sharon Plain, Upper Galilee, Lower Galilee, Mount Carmel, Judean Mountains, and Golan Heights.

General Distribution. — Recorded from Southwest Asia (Cyprus, Turkey, Lebanon, Jordan, and Iraq), scattered around the Mediterranean, Europe (records few, mainly from the south), Macaronesia, North Africa, and eastern Asia.

Bryum donianum is easily recognised by the leaves with two–three-stratose distinct border of elongate, incrassate cells and the strong costa excurrent into the acumen.

Figure 149. *Bryum donianum*: **a** habit, moist with deoperculate capsule (×10); **b** habit, dry with operculate capsule (×10); **c** operculum (×10); **d** leaf (×50); **e** basal cells; **f** marginal upper cells (all leaf cells ×500); **g** cross-section of costa and portion of lamina (×500). (*Crosby & Herrnstadt 79-92-14a*)

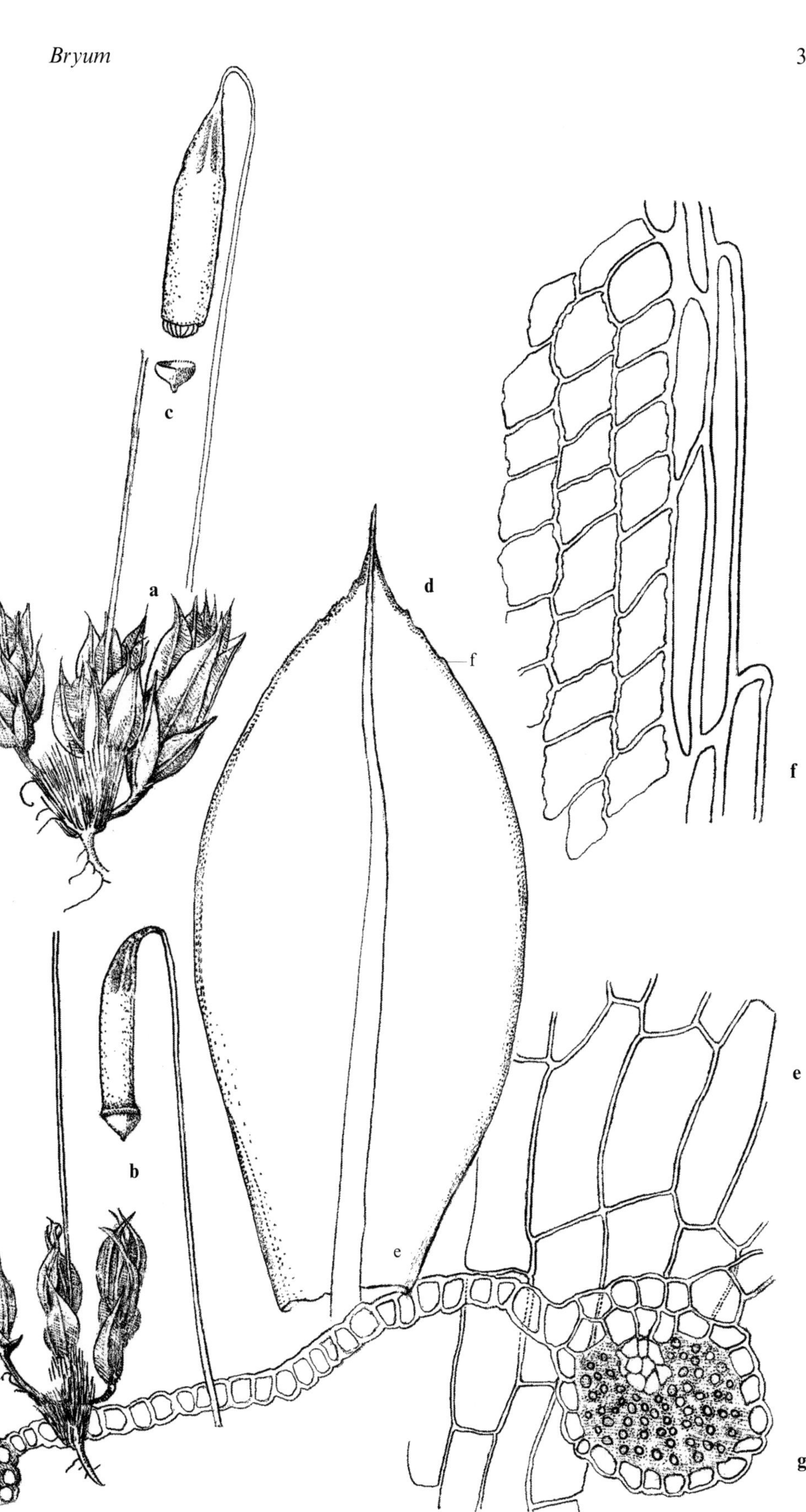
c
a
d
f
f
b
e
e
g

13. Bryum capillare L. ex Hedw., Sp. Musc. Frond. 182 (1801). [Bilewsky (1965): 402; Bizot 1945]

Distribution map 150, Figures 150: a–c (p. 387).

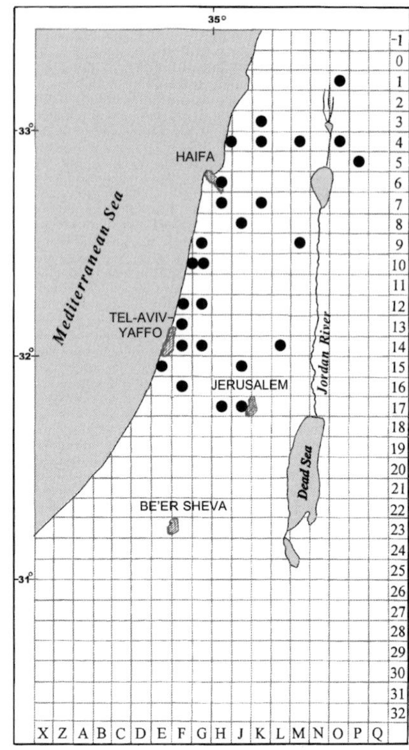

Map 150: *Bryum capillare*

Plants up to 20 mm high, bright to dark green, with reddish tinge towards base. *Leaves* shrunken and twisted spirally around stem when dry, patent when moist, plane to concave, 1.3–3 mm long, obovate-spathulate, widest above middle, acuminate, usually cuspidate or piliferous; margins narrowly recurved, entire or finely serrulate towards apex, with unistratose border; costa excurrent in a long cuspidate point, rarely ending below apex; basal cells short-rectangular, upper cells rhomboid-hexagonal, mid-leaf cells 16–25 µm wide; border of (two–)three–four marginal rows of narrow, elongate and incrassate cells. *Rhizoids* brown to reddish-brown, papillose; rhizoidal gemmae similar in colour to rhizoids, 155–175 µm in diameter, peripheral cells not protuberant. *Dioicous*. *Setae* up to 30 mm long; capsules pendulous or cernuous, up to 5 mm long including operculum, cylindrical, yellowish-brown to dark red-brown (the latter when growing on sand), constricted below mouth, with neck about ¼ the length of capsule, shrunken to various degrees when dry; endostome segments abruptly long pointed. *Spores* 13.5–16 µm in diameter.

Habitat and Local Distribution. — Plants growing in dense or loose tufts on rendzina, terra rossa, and basalt soils, rarely on sand, on and among limestone and basalt rocks (one collection on dead bark). Common: Coastal Galilee, Sharon Plain, Philistean Plain, Upper Galilee, Lower Galilee, Mount Carmel, Mount Gilboa, Samaria, Shefela, Judean Mountains, and Golan Heights.

General Distribution. — A widespread cosmopolitan species; seems to be closely related to *Bryum torquescens* (see the discussion there).

14. Bryum torquescens Bruch ex De Not., Syllab. Musc. no. 163 (1838).
[Bilewsky (1965) : 403; Bizot (1945)]

B. philippianum Müll. Hal., Linnaea 18 : 701 (1845); *B. capillare* L. ex Hedw. var. *torquescens* (De Not) Husn., Muscol. Gall. 240 (1889).

Distribution map 151, Figure 150 : d–i (p. 387).

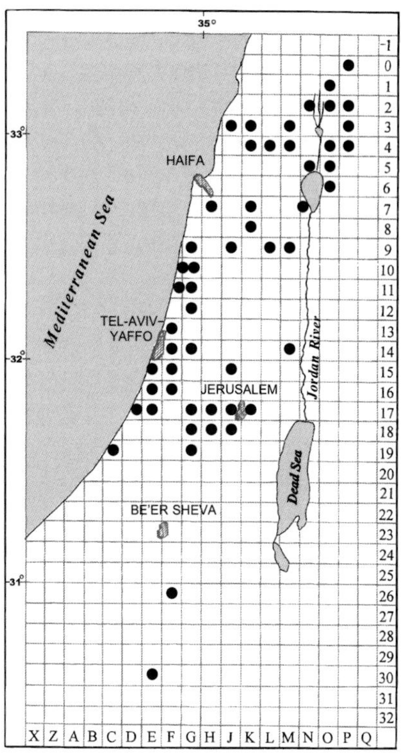

Map 151: *Bryum torquescens*

Plants bright to dark green, very similar to *Bryum capillare*, except in a few characters. *Leaves* twisted around stem when dry, cuspidate but not piliferous; costa excurrent in a cuspidate point, not piliferous; upper leaf cells ± narrower than of *B. capillare*. *Rhizoids* reddish-brown; rhizoidal gemmae red (usually differing in colour from rhizoids), 135–145 μm in diameter. *Synoicous* (in all plants investigated in the local flora); capsules dark red when growing on sand. *Spores* processes widely varying in size, some joined.

Habitat and Local Distribution. — Plants growing in loose or dense tufts, most frequently on sand, sometimes on various other calcareous soils (rendzina, terra rossa, sandy-loam), rarely basalt, in soil pockets, rarely on thin layers of soil on rocks, walls and stones. Very common (more widespread than *Bryum capillare*, perhaps because it is synoicous): Coastal Galilee, Sharon Plain, Philistean Plain, Upper Galilee, Lower Galilee, Mount Carmel, Esdraelon Plain, Mount Gilboa, Samaria, Shefela, Judean Mountains, Central Negev, Dan Valley, Hula Plain, Lower Jordan Valley, Mount Hermon, and Golan Heights.

General Distribution. — Recorded from Southwest Asia (Cyprus, Turkey, Syria, Lebanon, Jordan, Iraq, Iran, and Afghanistan), around the Mediterranean, Europe (throughout, apparently more widespread in the south), Macaronesia, North, tropical, and South Africa, central Asia, Australia, New Zealand, and America (throughout, except for the Caribbean Islands).

Various additional characters, differentiating between *Bryum torquescens* and *B. capillare* and recorded in the literature (e.g., the colour of the whole plant or the costa, size of plants, shape of capsules), are not applied here. In local material, those characters have no diagnostic value.

There does not seem to be a uniform concept of *Bryum torquescens* and *B. capillare* in the literature (as examples, compare Syed 1973, Smith 1978, Magill 1987, and Frahm 1995). Thus, their treatments were not based on the same differential characters. Ochi (*in* Sharp *et al.* 1994) pointed out that the only clear-cut difference between *B. torquescens* and *B. capillare* is that *B. torquescens* is synoicous and *B. capillare* is dioicous. Accordingly, he treated them conspecifically as *B. capillare*. Here, they are treated as two distinct species (nos. 13 and 14 above), although they may represent two extreme forms of a large, cosmopolitan taxon.

Bryum torquescens and *B. capillare* together with *B. donianum* and *B. canariense* (species nos. 12–15) were treated by some authors (e.g., Syed 1973, Demaret 1986b, Frahm 1995) as the "*Bryum capillare* group".

Figure 150. *Bryum capillare*: **a** habit, dry (×10); **b** habit, moist (×10); **c** leaf (×50).
Bryum torquescens: **d** habit, dry (×10); **e** habit, moist (×10); **f** leaf (×50); **g** basal cells; **h** marginal mid-leaf cells; **i** upper cells (all leaf cells ×500).
(**a**–**c**: *Herrnstadt & Crosby 78-21-2*; **d**–**i**: *Danin*, 24 Mar. 1984)

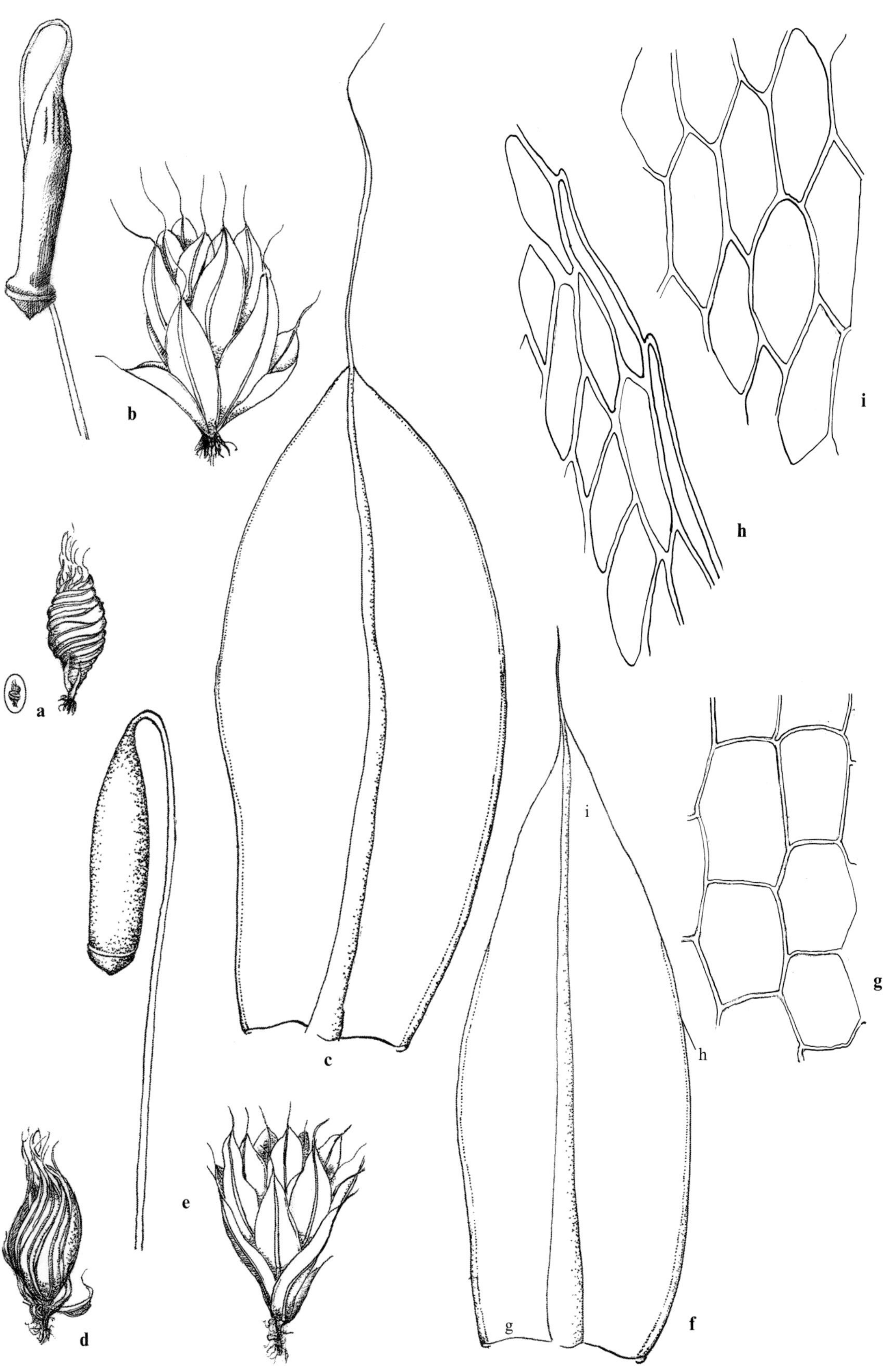
a
b
c
d
e
f
g
h
i
g
h
i

15. Bryum canariense Brid., Sp. Musc. Frond. 3:29 (1817). [Bilewsky (1965):403]
B. provinciale H. Philib. *in* Schimp., Syn. Musc. Eur., Edition 2:432 (1876); *B. canariense* Brid. var. *provinciale* (H. Philib.) Husn., Muscol. Gall. 239 (1889).

Distribution map 152. Figure 151.

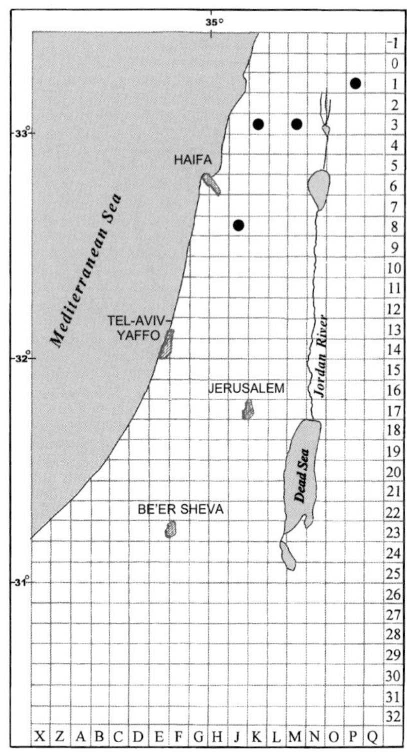

Map 152: *Bryum canariense*

Plants up to 25 mm high, dull green. *Stems* tomentose. *Leaves* often arranged in two or more successive comal tufts (corresponding to annual growth) with scattered smaller leaves between them; imbricate, not twisted when dry, spreading when moist, up to 2.5(–3) mm long, broadly ovate to ovate-oblong or obovate, widest at middle or above, usually acute, cuspidate; margins usually recurved below, distinctly serrulate to serrate above, base red; costa strong, excurrent; cells in mid-leaf 13–15 µm wide, marginal cells narrower, sometimes slightly more incrassate than adjacent cells but not forming a distinct border. *Rhizoids* papillose, deep red; rhizoidal gemmae 200–300 µm in diameter, spherical, with peripheral cells protuberant or not. *Dioicous* or synoicous. *Setae* 12–23 mm long; capsules pendulous or nearly horizontal, widely varying in size, up to 4 mm long including operculum, narrowly pyriform or oblong-cylindrical, yellowish to reddish-brown; neck about ¼ length of capsule; operculum mammillate; endostome with nodulose to appendiculate cilia. *Spores* 13–17 µm in diameter.

Habitat and Local Distribution. — Plants growing on terra rossa and among soft and hard limestone rocks. Rare to occasional: Upper Galilee, Samaria, and Golan Heights.

Bilewsky (1965) recorded *Bryum canariense* from Mount Carmel on the basis of a single collection (Herb. Bilewsky no. 108), which was found to belong to *B. caespiticium*. Although those plants have comal tufts from successive years resembling *B. canariense*, they have to be referred to as *B. caespiticium* because of the shape of the leaves and the recurved entire margins.

General Distribution. — Recorded from Southwest Asia only from Turkey, scattered around the Mediterranean, Europe (mainly the south), Macaronesia, North, tropical, and South Africa, and northern, central, and southern South America.

Bryum canariense can be recognised by the successive annual growth of comal tufts, the distinct serrulation of the upper leaf margins, the revolute margins at the leaf base, and the usually deep-red rhizoids. The type of *B. canariense* in the British Museum was found to be congruent with the usually accepted concept of the species (for a different viewpoint, see Frahm 1995).

Figure 151. *Bryum canariense*: **a** habit, moist (×10); **b** habit, dry (×10); **c** part of seta and capsule (×10); **d** leaves (×50); **e** basal cells; **f** marginal upper cells (all leaf cells ×500); **g** rhizoidal gemma (×125). (*Crosby & Herrnstadt 79-92-13*)

16. Bryum caespiticium Hedw., Sp. Musc. Frond. 180 (1801). [Bilewsky (1965): 400; Bizot (1945)]

B. badium (Brid.) Schimp., Syn. Musc. Eur., Edition 2: 444 (1876).

Distribution map 153, Figure 152, Plate XI: d.

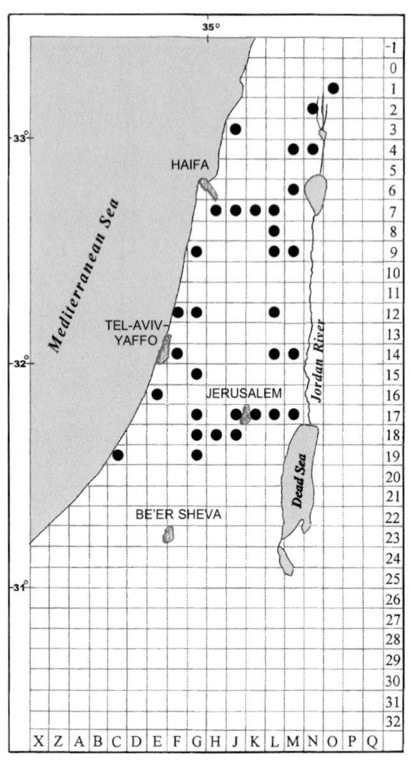

Map 153: *Bryum caespiticium*

Plants 4–10(–15) mm high, short, erect, yellowish-green. *Stems* tomentose below. *Leaves* slightly imbricate, erect, sometimes slightly twisted when dry, erect to erect-spreading when moist, concave, up to 1.2–2 mm long, ovate-oblong to ovate-lanceolate, long acuminate, base reddish; margins ± entire, recurved, obscurely bordered with two–three narrower marginal unistratose rows of cells; costa strong, long-excurrent into an almost smooth acumen, reddish; basal cells rectangular, upper cells narrower, narrowly rhomboid-hexagonal, in mid-leaf 8–11 µm wide, smooth, slightly incrassate (especially in older leaves), gradually becoming narrower and longer towards margins and apex. *Rhizoids* red brown, coarsely papillose, in local plants frequently bearing gemmae, 200–250 µm in diameter. *Dioicous. Setae* 10–15(–20) mm long, slender; capsules yellowish-brown to reddish-brown, darker at mouth, horizontal, cernuous to pendulous, 2–3 mm long including operculum, oblong-pyriform, when dry constricted below wide mouth; neck wrinkled when dry, up to ⅓ the length of capsule; operculum mammillate; exostome orange to pale brown; endostome with appendiculate cilia. *Spores* 9–12 µm in diameter; densely gemmate, processes small with rough surface (seen with SEM).

Habitat and Local Distribution. — Plants forming readily separable tufts; growing on calcareous soil, and rocks (sandstone, limestone), basalt rocks, and waste ground; sometimes growing partially covered by sand; may serve as sand binders. Very common: Sharon Plain, Philistean Plain, Upper Galilee, Lower Galilee, Mount Carmel, Esdraelon Plain, Mount Gilboa, Samaria, Shefela, Judean Mountains Judean Desert, Dan Valley, Lower Jordan Valley, and Golan Heights.

General Distribution. — An almost cosmopolitan species.

Figure 152. *Bryum caespiticium*: **a** habit, moist (×10); **b** habit, dry (×10); **c** deoperculate capsule (×10); **d** leaves (×50); **e** basal cells; **f** marginal mid-leaf cells; **g** upper cells (all leaf cells ×500); **h** part of exostome and endostome teeth (×125); **i** rhizoidal gemma (×125). (*Naftolsky*, 9 Apr. 1927)

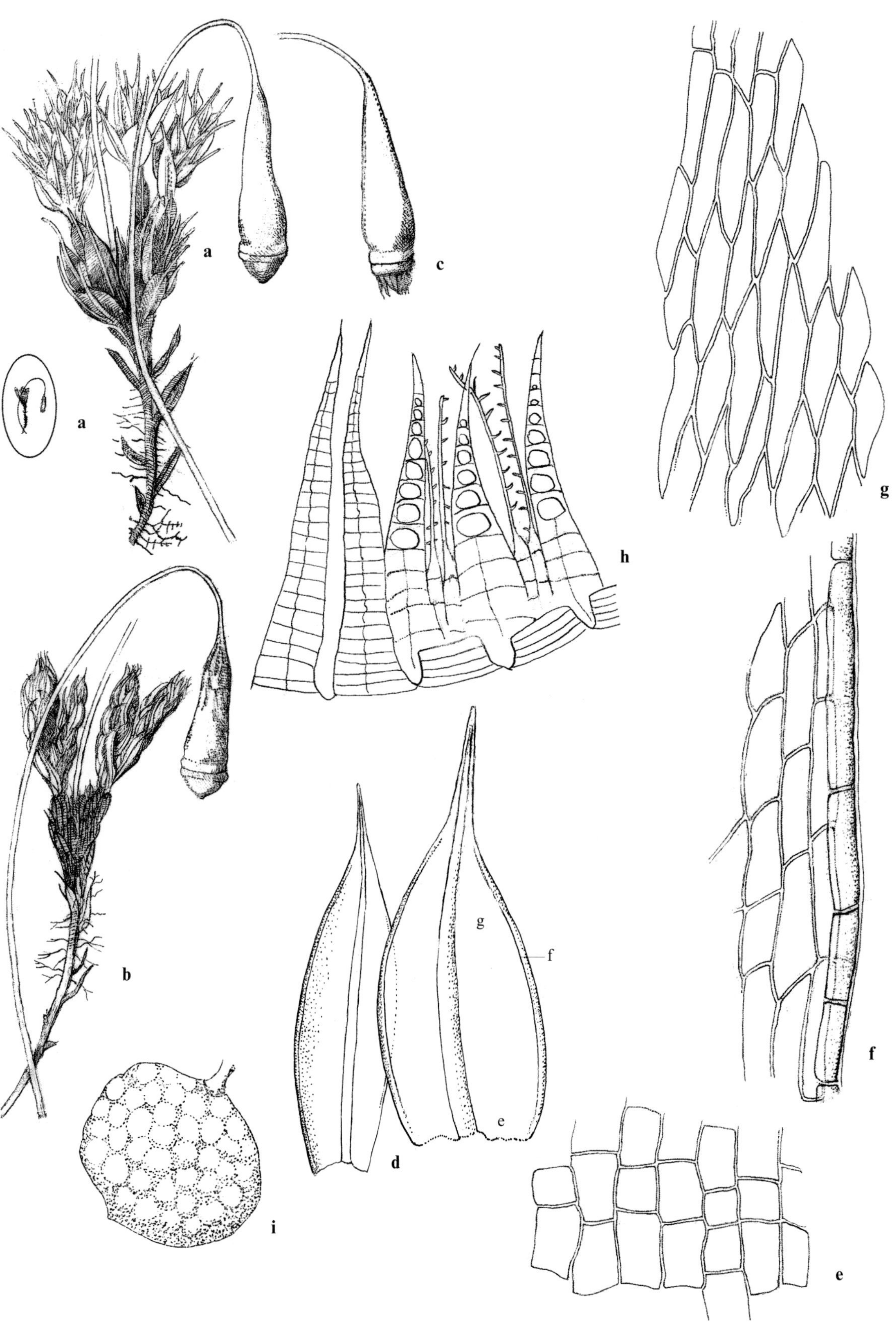
a
a
c
h
g
f
b
g
f
e
d
i
e

Bryum caespiticium can be recognised among other local *Bryum* species by the yellowish-green acuminate leaves with recurved margins and long-excurrent costa. Plants without sporophytes superficially resemble those of *B. dunense*, but the latter has axillary bulbils, usually plane leaf margins (though sometimes revolute at base), and smooth rhizoids. Plants of *B. caespiticium* in Israel frequently bear rhizoidal gemmae contrary to records on the absence of such gemmae in this species (e.g., Crundwell & Nyholm 1964).

Bryum badium, described by some authors as a separate taxon comprising smaller plants with smaller dark reddish-brown capsules, is treated here as a synonym of *B. caespiticium*. Those modified characters are, perhaps, correlated with the dry habitats of the plants (cf. also Crundwell & Nyholm 1964).

17. Bryum kunzei Hornsch., Flora 2 : 90 (1819).

B. caespiticium L. ex Hedw. var. *imbricatum* B.S.G., Bryol. Eur. 4 : 140 (1839); *B. caespiticium* Hedw. var. *kunzei* (Hoppe & Hornsch.) Braithw., Brit. Moss Fl. 2 : 175 (1892). — Bilewsky (1965) : 400 (as "var. *kuntzei* (Hornsch.) Warnst.").

Distribution map 154, Figure 153.

Bryum kunzei seems to be closely related to *B. caespiticium* and mainly differs in the following characters.

Plants smaller, up to 4 mm high. *Leaves* imbricate, 0.5–1 mm long, strongly concave, broadly ovate to ovate-oblong or obovate, abruptly narrowing into an acuminate apex; margins plane, length of acumen varying in plants. *Setae* 8–10(–11) mm long; capsules ovoid, ± 2 mm long including operculum.

Habitat and Local Distribution. — Plants growing on various soils (sand, marl, and rendzina), on calcareous rocks and on walls. Occasional, but may have been overlooked in some localities: Coastal Galilee, Sharon Plain(?), Philistean Plain, Shefela, Judean Mountains, and Golan Heights.

General Distribution. — Recorded from Southwest Asia (Turkey and Afghanistan), around the Mediterranean, Europe (throughout), Macaronesia (few records), North Africa, and central Asia.

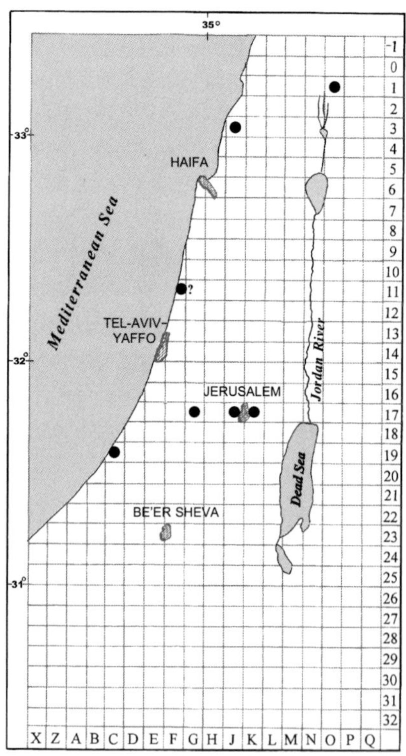

Map 154: *Bryum kunzei*

Bryum kunzei is considered here as a separate species on account of the differential characters listed above. It is often treated in the literature as a variety of *B. caespiticium* (e.g., Dixon 1924, Mönkemeyer 1927, Smith 1978). Nyholm (1954–1969) regards it as a depauperate form of *B. capillare*.

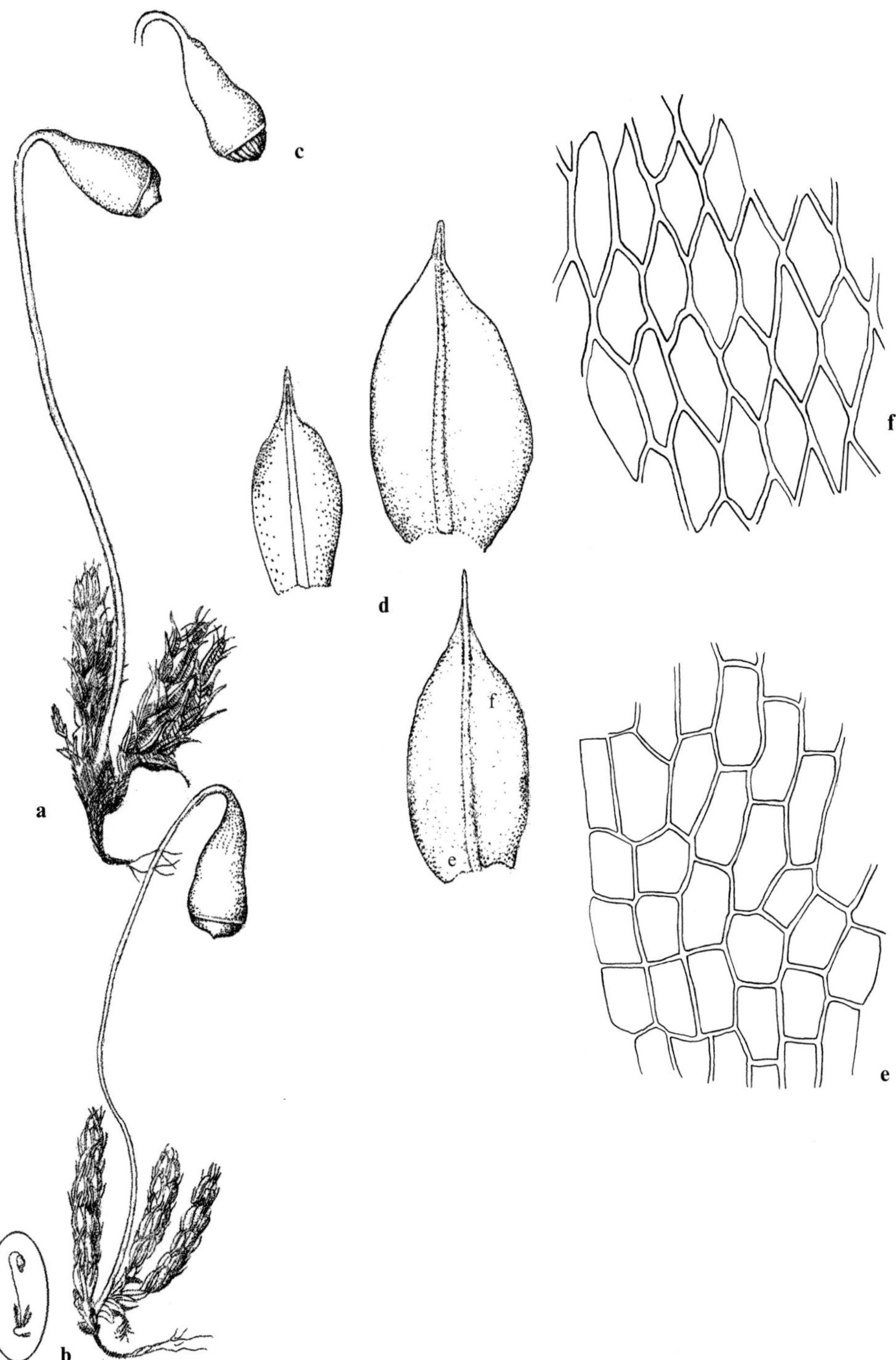

Figure 153. *Bryum kunzei*: **a** habit, moist (×10); **b** habit, dry (×10); **c** deoperculate capsule (×10); **d** leaves (×50); **e** basal cells; **f** upper cells (all leaf cells ×500). (Jerusalem: Bayit WeGan, *Kushnir*, *sine dato*)

18. Bryum pallescens Schleich. ex Schwägr., Sp. Musc. Frond., Suppl. 1 : 107 (1816). [Bilewsky (1965) : 400; Rabinovitz-Sereni (1931)]

Distribution map 155, Figure 154.

Plants up to 7–20(–30) mm high, growing in very dense tomentose tufts, formed by much branched stems of successive years, green above, reddish below. *Stems* radiculose. *Leaves* ± twisted towards apex when dry, erectopatent when moist, up to 2.5(–3) mm long, ovate-lanceolate, acuminate; margins recurved, sometimes plane, entire or finely serrulate towards apex, with distinct unistratose border; upper leaves imbricate, forming comal tufts; costa red, long-excurrent; basal cells short-rectangular, mid- and upper leaf cells rhomboidal to narrow-rhomboidal, in mid-leaf 10–19 μm wide; two–three marginal rows of linear, yellowish cells, longer and more incrassate than adjacent cells. *Autoicous. Setae* up to 23 mm long; capsules pendulous to inclined, 2.5–3.5 mm long including operculum, narrowly and longly pyriform, brown to red-brown; neck ⅓ to nearly ½ length of capsule; endostome segments with elongated perforations, cilia with poorly developed appendages. *Spores* 10–14 μm in diameter.

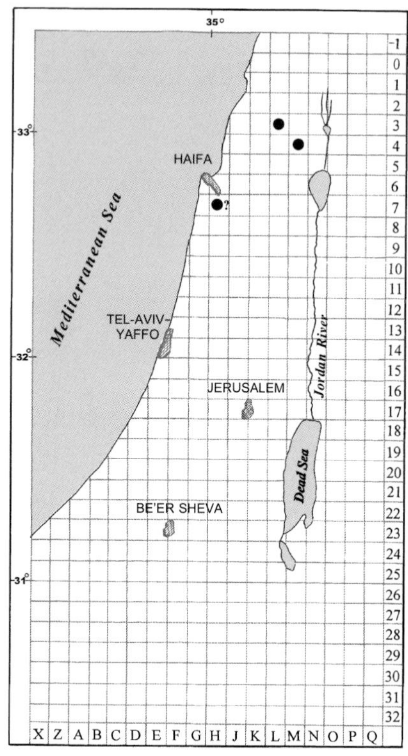

Map 155: *Bryum pallescens*

Habitat and Local Distribution. — Plants growing, in crevices of calcareous rocks. Rare: Upper Galilee and Mount Carmel (?).

General Distribution. — Recorded from Southwest Asia (Cyprus, Turkey, Iraq, Iran, and Afghanistan), around the Mediterranean, Europe (many records from throughout), tropical and North Africa, Asia (throughout, except the southeast), New Zealand, and North, Central, and South America.

Literature. — Demaret (1986a).

Bryum pallescens resembles *B. caespiticium* in general habit but can be distinguished by the distinctly bordered leaves and the absence of rhizoidal gemmae (which can be found in local collections of *B. caespiticium*). Ochi (1980, *in* Sharp *et al.* 1994) considered the larger spores — about 20 μm in diameter — as a character differentiating *B. pallescens* from *B. caespiticium*. In the few local plants seen, the spores hardly differ in size.

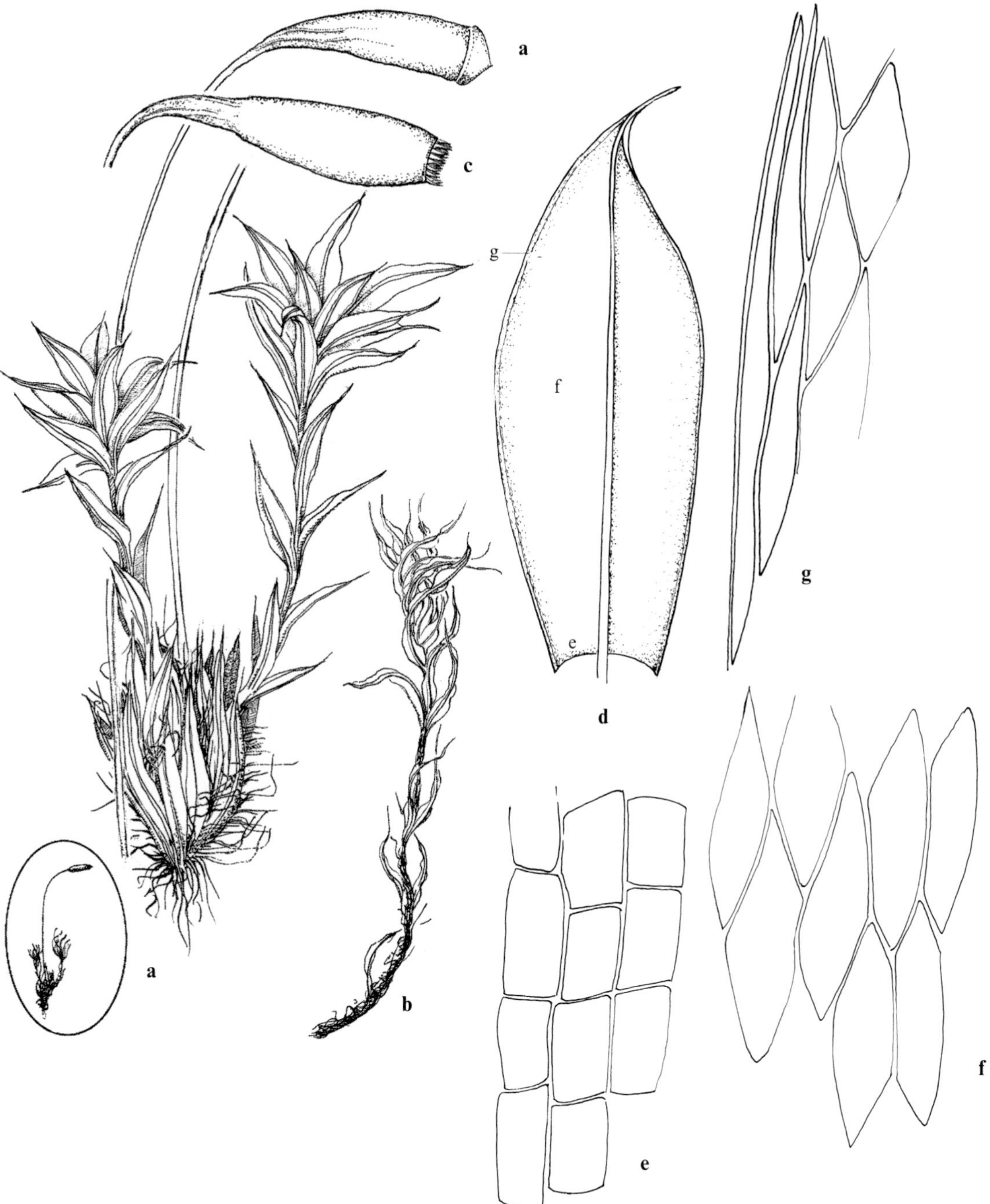

Figure 154. *Bryum pallescens*: **a** habit, moist (×10); **b** habit, dry (×10); **c** deoperculate capsule (×10); **d** leaf (×50); **e** basal cells; **f** mid-leaf cells; **g** marginal upper cells (all leaf cells ×500). (*D. Kaplan*, 6 Feb. 1990)

19. Bryum intermedium (Brid.) Blandow, Ueber. Mecklenb. Moose 6 (1809).
Pohlia intermedia Brid., Muscol. Recent. 2: 144 (1803).

Distribution map 156, Figure 155.

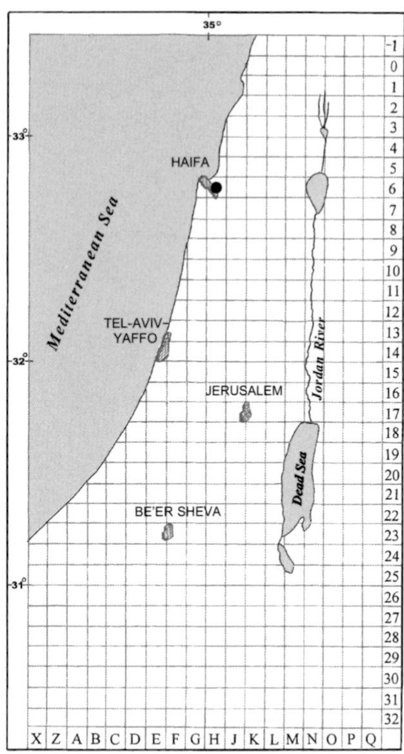

Map 156: *Bryum intermedium*

Plants up to 15 mm high, yellowish-green. *Leaves* appressed and erect when dry, erecto-patent when moist, ± concave, up to 2 mm long, ovate-lanceolate to lanceolate, acuminate; upper leaves imbricate; margins narrowly revolute to near apex, entire, obscurely bordered with one row of narrower cells; costa excurrent into a long point; basal cells rectangular to quadrate, with some enlarged angular cells; upper cells hexagonal, 9–14 µm wide in mid-leaf. *Synoicous*. *Setae* 14–17 mm long; capsules pendulous, ± 3 mm long including operculum, pyriform, slightly asymmetrical with ± oblique mouth when empty, red-brown when mature; neck wrinkled when dry, ⅓ to nearly half the length of capsule; endostome with nodulose cilia. *Spores* about 14 µm in diameter (immature spores seen).

Habitat and Local Distribution. — Plants growing in dense tufts on thin soil layers on rocks. So far identified only in a single locality in Israel, but highly probable to have been confused elsewhere with *Bryum caespiticium*: Mount Carmel.

General Distribution. — Not recorded from Southwest Asia; recorded from the northern and western Mediterranean, also from North Africa, Europe (many records from throughout), northeastern, eastern, and central Asia, southern Australia, Tasmania, North America (doubtful, according to Duell 1992), and Greenland.

Plants of *Bryum intermedium* resemble in gametophytic characters *B. caespiticium* but are synoicous, not dioicous, and the cilia of the endostome are nodulose, not appendiculate.

Figure 155. *Bryum intermedium*: **a** habit, dry (×10); **b** habit, moist with deoperculate capsule (×10); **c** leaf (×50); **d** basal cells; **e** upper cells (all leaf cells ×500); **f** part of endostome teeth (×125). (*Landau*, 2 Apr. 1937)

MNIACEAE

Plants acrocarpous, medium to robust, green to reddish, often growing in tufts. *Stems* usually radiculose or tomentose, at least below; fertile stems erect, sterile stems stoloniform, erect, or sometimes curved or prostrate to plagiotropic; central strand present. *Leaves* of fertile stems often larger and more crowded above; all leaves bordered by two–five rows of long narrow cells, often single- or double-toothed; costa well developed, subpercurrent to shortly excurrent, with one or two stereid bands, or stereid bands absent; cells short, usually smooth, isodiametric or hexagonal, ± incrassate especially at corners; perigonia with clavate paraphyses. *Dioicous* or synoicous. *Setae* elongate, one to many from each perichaetium; capsules inclined to horizontal or pendulous, symmetrical, oblong-cylindrical, with a short neck; annulus frequently differentiated; peristome double, endostome well developed, with distinct basal membrane, usually with nodulose cilia; stomata usually immersed, restricted to the neck; operculum often rostrate. *Calyptrae* cucullate, smooth, naked.

A family comprising ca. 10 genera, mainly distributed in the Northern Hemisphere; represented in the local flora by *Plagiomnium*.

Plagiomnium T. J. Kop.

Plants yellowish-green to green, in most species with plagiotropic stoloniform sterile stems (erect in the local *Plagiomnium undulata*). *Stems* with epidermis of two or more cell layers; rhizoids on stems sometimes up to apex. *Leaves* large, oblong-elliptic to narrow-lingulate, emarginate to obtuse or acute, often apiculate, decurrent; margins unistratose, single-toothed; costa usually percurrent, generally with a single stereid band; cells, except in margins, ± isodiametric throughout leaf. *Sporophytes* single or many from each perichaetium; operculum usually conical, not rostrate.

A genus comprising 20 species (Koponen 1968). One species occurs in the local flora.

Plagiomnium was described by Koponen (1968) as a genus separate from *Mnium*, differing in the green (not reddish) colour, the frequent occurrence of plagiotropic sterile stems, the two or more-layered stem epidermis (vs. one in *Mnium*) and the unistratose, single-toothed leaf margins (vs. double-toothed and often thickened in *Mnium*).

Literature. — Koponen (1968, 1971).

Plagiomnium undulatum (Hedw.) T. J. Kop., Ann. Bot. Fenn. 5: 146 (1968).
Mnium undulatum Hedw., Sp. Musc. Frond. 195 (1801).

Distribution map 157 (p. 400), Figure 156.

Plants up to 5 cm high, pale green to bright green. *Stems* tomentose, mainly in lower part; sterile stems erect, sometimes curved, ± complanate; fertile stems erect. *Leaves* of

---

Figure 156. *Plagiomnium undulatum*: **a** habit of sterile plant, moist (×1); **b** habit of sterile plant, dry (×1); **c** habit of fertile plant, moist (×1); **d** part of habit, dry (×10); **e** upper part of habit, moist (×10); **f** lower part of habit, moist (×10); **g** leaf (×50); **h** marginal basal cells; **i** marginal mid-leaf cells (all leaf cells ×500). (*Kushnir*, 10 Oct. 1943)

i
h
e
i
h
g
f
c
d
a
b

sterile stems crisped and strongly transversely undulate when dry, spreading and weakly undulate when moist, sparsely spaced, up to 5 mm long and 2.5 mm wide, elliptical, apex rounded to obtuse, apiculate; leaves of fertile stems more densely spaced, up to 8 mm long, oblong-lingulate; margins with border of about four rows of long narrow cells, with one-celled prominent teeth from apex to near to base; costa percurrent or slightly excurrent; cells slightly incrassate, not in rows, irregularly hexagonal, in mid leaf 14–21(–27) µm wide; perichaetial leaves forming a terminal rosette at stem apex. *Dioicous*; some plants with clustered archegonia have been observed in local plants but no sporophytes have been found.

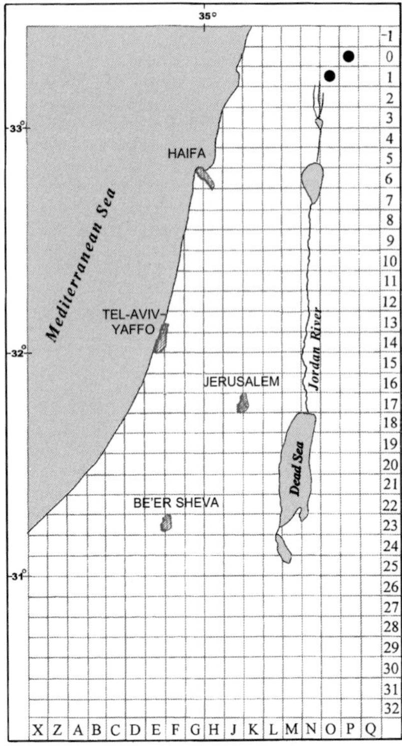

Map 157: *Plagiomnium undulatum*

Habitat and Local Distribution. — Plants growing in loose tufts in damp habitats near running water. Rare, perhaps locally abundant in some localities on Mount Hermon: Dan Valley and Mount Hermon.

General Distribution. — Recorded from Southwest Asia (Turkey, Syria, Lebanon, and Iran), around the Mediterranean, Europe (throughout), Macaronesia, Africa (except tropical Africa), and Asia (except the southeast).

Mnium seligeri (Jur.) Limpr. (Bilewsky 1965 : 404) is based on misidentified specimens of *Plagiomnium undulata.*

BARTRAMIACEAE

Plants acrocarpous, growing in tufts, often glaucous, slowly absorbing water. *Stems* simple or forked, in whorls or fastigiately branched, tomentose, with a central strand. *Leaves* linear to ovate-lanceolate; margins usually serrulate to serrate; cells prorate or papillose near one or both ends, or at cell centre; axillary hairs present; perigonia with filiform or clavate paraphyses. *Setae* terminal; capsules erect to inclined, rarely horizontal, often asymmetrical, subglobose, ovoid or oblong; operculum convex to conical; annulus usually absent; peristome usually double, with exostome of 16 teeth and endostome with a basal membrane, sometimes peristome single or absent. *Calyptrae* small, cucullate, naked.

A family comprising 11 genera (Griffin & Buck 1989); two genera occur in the local flora.

Literature. — Griffin & Buck (1989).

1 Plants growing on soil on rocks, in rock crevices, and on banks of running water; stems often forked, or with few subgametangial innovations; leaves linear to linear-lanceolate, often forming a sheathing base, at least partly bistratose **1. Bartramia**

1 Plants growing in wet, muddy, or marshy habitats; stems with branches often from subgametangial innovations; leaves ovate-lanceolate, never forming a sheathing base, unistratose **2. Philonotis**

1. *Bartramia* Hedw.

Plants small to ± robust. *Stems* erect, forked or with only two subgametangial innovations. *Leaves* linear to linear-lanceolate, often from an erect, sheathing base, at least partly bistratose; margins usually serrulate or serrate above, teeth sometimes double; costa percurrent to excurrent, prominent at back, often toothed on dorsal surface; basal cells rectangular to linear, subquadrate at basal angles, not incrassate, smooth only at extreme base; mid- and upper leaf cells subquadrate to narrow-rectangular, incrassate, prorate or papillose at one end. *Dioicous*, autoicous, or synoicous. *Setae* elongate, usually straight; capsules erect or inclined, symmetrical or asymmetrical, subglobose to cylindrical, with oblique mouth, sulcate when dry; peristome usually double; endostome without cilia or cilia poorly developed. *Spores* spherical to reniform.

A large worldwide genus of ca. 90 species for which records of the number of species widely vary. One species occurs in the local flora.

1. Bartramia stricta Brid., Muscol. Recent. 2:132 (1803).

Distribution map 158, Figure 157, Plate XI:e, f.

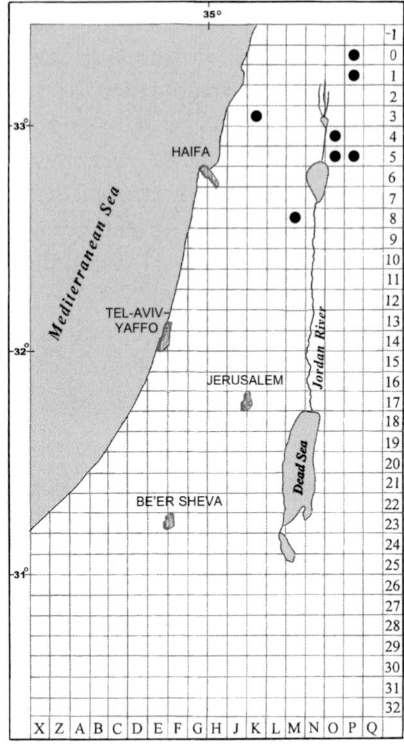

Map 158: *Bartramia stricta*

Plants moderately sized, up to 20(–30) mm high, glaucous, brownish-green. *Leaves* 3–4 mm long, densely spaced, erect when dry, ± erect when moist, linear-lanceolate, without sheathing base, long-subulate, at least partly bistratose; margins bistratose, plane or narrowly recurved below, serrulate in upper half, teeth usually single: costa excurrent; cells narrow-rectangular to rectangular throughout, with small and subquadrate cells at basal angles; basal cells ± translucent, longer than upper cells; mid-leaf cells 5–8 μm wide, opaque. *Synoicous*. *Capsules* erect, symmetrical, subglobose to cylindrical, nearly smooth when moist, deeply sulcate when dry; exostome teeth finely papillose below, smooth above. *Spores* ca. 25 μm in diameter, smooth-papillose, surface greatly varying within capsules; surface foveolate, covered by gemmae and/or granulae, (Plate XI:e), or nearly smooth (Plate XI:f) (seen with SEM).

Habitat and Local Distribution. — Plants growing in dense rigid tufts on soil covering basalt and limestone rocks and in rock crevices, often on banks of running water. Occasional: Upper Galilee, Lower Galilee, Mount Hermon, and Golan Heights.

General Distribution. — Recorded from Southwest Asia (Cyprus, Turkey, Syria, Lebanon, Saudi Arabia, and Afghanistan), around the Mediterranean, Europe (the south in particular), Macaronesia, tropical and North Africa, Australia, and North America.

Figure 157. *Bartramia stricta*: **a** habit, dry (×10); **b** habit, moist (×10); **c** mouth of capsule and detached operculum (×20); **d** leaf (×50); **e** basal cells; **f** mid-leaf cells; **g** apical cells (all leaf cells ×500); **h** cross-section of leaf (×500); **i** peristome teeth (×250). (*Crosby & Herrnstadt 79-77-1*)

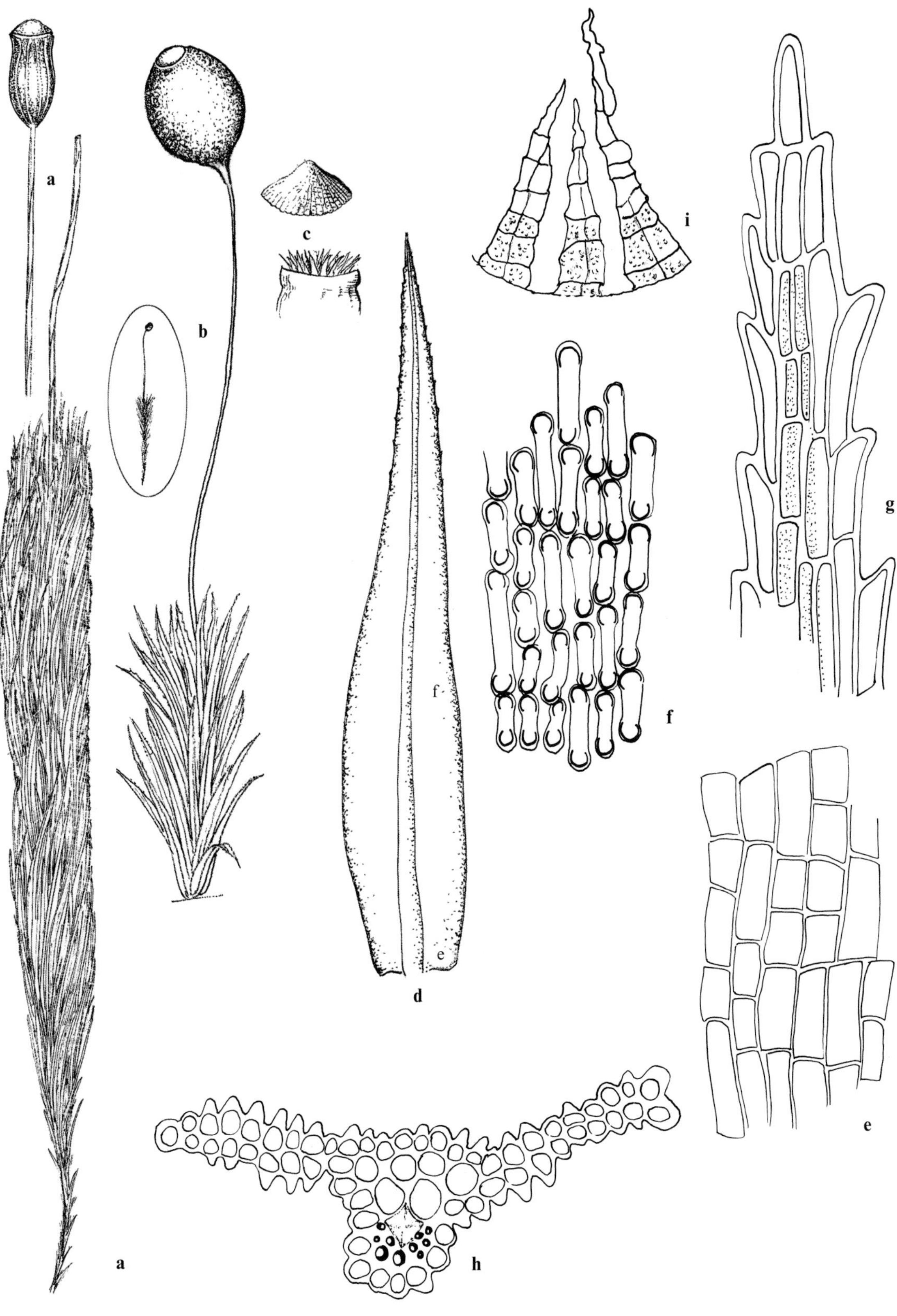
a
a
b
c
d
e
f
f
e
g
h
i

2. *Philonotis* Brid.

Plants small to moderately sized. *Stems* erect, or loosely ascending, sometimes prostrate; often branched in whorls from subperichaetial innovations. *Leaves* ovate-lanceolate, obtuse or acute to long-acuminate; margins crenulate to serrulate nearly throughout, teeth single or double; costa ending below apex to excurrent, often scabrous or toothed on dorsal surface; basal cells rectangular, thin-walled, smooth; upper cells rectangular to oblong hexagonal or rectangular to linear (sometimes narrow, elongate cells forming distinct border), incrassate, smooth, prolate, or papillate. *Axillary propagules* often present. *Dioicous* usually. *Capsules* strongly inclined to horizontal, usually asymmetrical, subglobose, generally sulcate when dry; peristome usually present; endostome with or without cilia. *Spores* reniform.

A large world-wide genus comprising ca. 200 species. Three species occur in the local flora; all restricted to the Syrian-African Rift Valley and growing on shallow soil and rocks in very damp habitats, e.g., behind waterfalls or in marshy habitats. Phenotypic variation within species seems very common and consequently their identification and delimitation may be rather difficult, in particular in sterile plants.

Literature. — Field (1963), Petit (1976), Iwatsuki (1977).

1 Plants fairly robust; leaves secund; costa long-excurrent; cells with proximal papillae **3. P. calcarea**

1 Plants small and slender; leaves not secund; costa ending below apex to short-excurrent; cells smooth or with distal papillae.

  2 Leaves densely spaced; apex acute to long-acuminate; costa usually percurrent to excurrent; cells with distal papillae, in mid-leaf 7–10 μm wide **1. P. marchica**

  2 Leaves laxly spaced; apex obtuse to acute; costa usually ending below apex or percurrent; cells smooth or with weak distal papillae only in upper part of leaf, in mid-leaf ± 15 μm wide **2. P. hastata**

1. Philonotis marchica (Hedw.) Brid., Bryol. Univ. 2:23 (1827).
[Bilewsky (1965):407]

Distribution map 159 (p. 405), Figure 158:e–i (p. 407).

Plants slender, up to 20 mm high, yellowish-green to light green. *Leaves* densely spaced, erectopatent, hardly changing when moist, usually ca. 0.85 mm long, ovate-lanceolate to lanceolate, acute to long-acuminate; margins usually plane, rarely narrowly recurved, serrulate with single teeth or sometimes coarsely crenulate; costa percurrent to short-excurrent, sometimes ending below apex, ± dentate on dorsal surface; cells ± incrassate, with prominent distal papillae in upper and lower part of leaf; basal cells rectangular; upper cells smaller, rectangular to narrow rectangular, in mid leaf 7–10 μm wide. *Propagules* resembling short shoots, varying in size and shape, sometimes formed in leaf axils and may be abundant in some populations. *Dioicous*. *Sporophytes* not seen in local plants.

Habitat and Local Distribution. — Plants growing in tufts or flat patches, sometimes nearly horizontally, on muddy and marshy soil and on rocks exposed to frequent water flushes and spray, sometimes partly submerged: mainly in districts with tropical climate along the Syrian-African Rift Valley. Occasional to common: Judean Desert, Dan Valley, Hula Plain, Upper and Lower Jordan Valley, Dead Sea area, and Golan Heights.

General Distribution. — Recorded from Southwest Asia (Turkey, Iran, and Afghanistan), around the Mediterranean, Europe (widespread throughout), Macaronesia (in part), tropical and North Africa, northeastern and eastern Asia, and North and Central America. According to Griffin (*in* Sharp *et al.* 1994) it is mainly a species of the temperate Northern Hemisphere, but occurs also in the tropics and mountains of Africa.

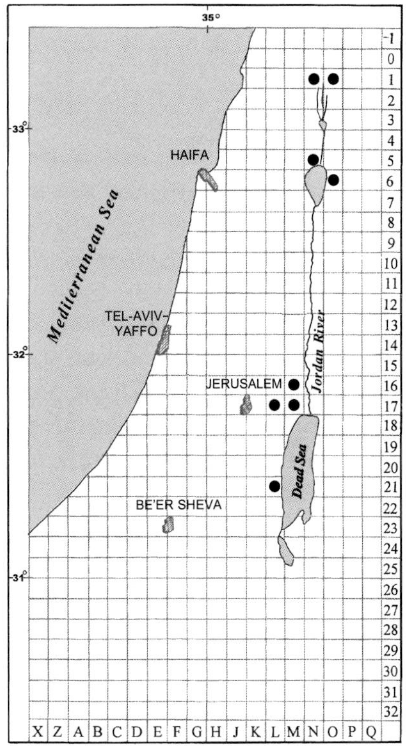

Map 159: *Philonotis marchica*

Petit (1976) used the shape of the propagules as a specific diagnostic character in the genus *Philonotis*. The plants identified here as *P. marchica*, using other characters, do not agree in the morphology of their propagules with Petit's description (they are smaller and usually stalked). Leaves of some plants of *P. marchica*, in which only a single row of marginal cells is recurved, may have the second marginal row protruding and resemble at first glance the paired teeth of the leaves of *P. calcarea* or *P. hastata*.

The previous record of the possible occurrence of *Philonotis arnelli* Husn. in Israel (Herrnstadt *et al.* 1982, 1991) was based on misidentification of a population of *P. marchica*.

2. Philonotis hastata (Duby) Wijk & Margad., Taxon: 8:74 (1959).

Hypnum hastatum Duby *in* Moritzi, Syst. Verz. 132 (1846); *Philonotis laxissima* Mitt., J. Linn. Soc. Bot., Suppl. 1:61 (1859). — Bilewsky (1965):406 (as "*P. laxissima* (Müll. Hal.) Br. Janav. var. *gemmiclada* Bizot"); *P. obtusata* Müll. Hal. ex Renauld & Cardot, Bull. Soc. Roy. Bot. Belgique. 34:61 (1896). — Bilewsky (1970).

Distribution map 160, Figure 158: a–d.

Philonotis hastata resembles *P. marchica*, but differs in the more laxly spaced leaves with obtuse to acute apex, larger and less incrassate cells up to ± 15 µm in mid-leaf, smooth or with weak distal papillae; margins narrowly revolute, usually with paired teeth to below middle of leaf; costa ending below apex to percurrent, never excurrent.

Habitat and Local Distribution. — *Philonotis hastata* was found always growing together with *P. marchica*. Occasional: Lower Jordan Valley, Hula Plain (?), Dead Sea area, and Golan Heights.

General Distribution. — Recorded from Southwest Asia (Turkey and Lebanon), the Mediterranean (few records), Europe (only from Spain), Macaronesia (in part), throughout Africa, Asia (throughout, except the northeast), Australia, Oceania, and America (throughout, except northeastern South America); often considered as a pantropical species.

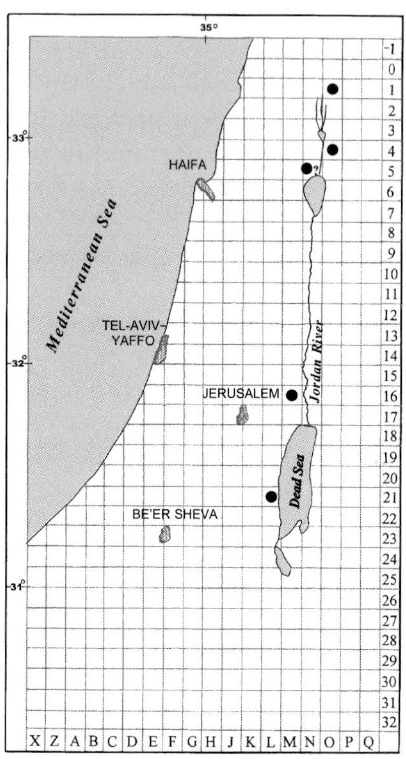

Map 160: *Philonotis hastata*

The plants identified as *Philonotis hastata* and *P. marchica* may, perhaps, represent two extreme forms of a gradient of quantitative morphological characters. Slender plants, resembling *P. hastata* in several characters, were described by Mönkemeyer (1927) as *P. marchica* forma *rivularis* (= *P. laxa* Limpr.) and may perhaps be considered as links between *P. hastata* and *P. marchica*.

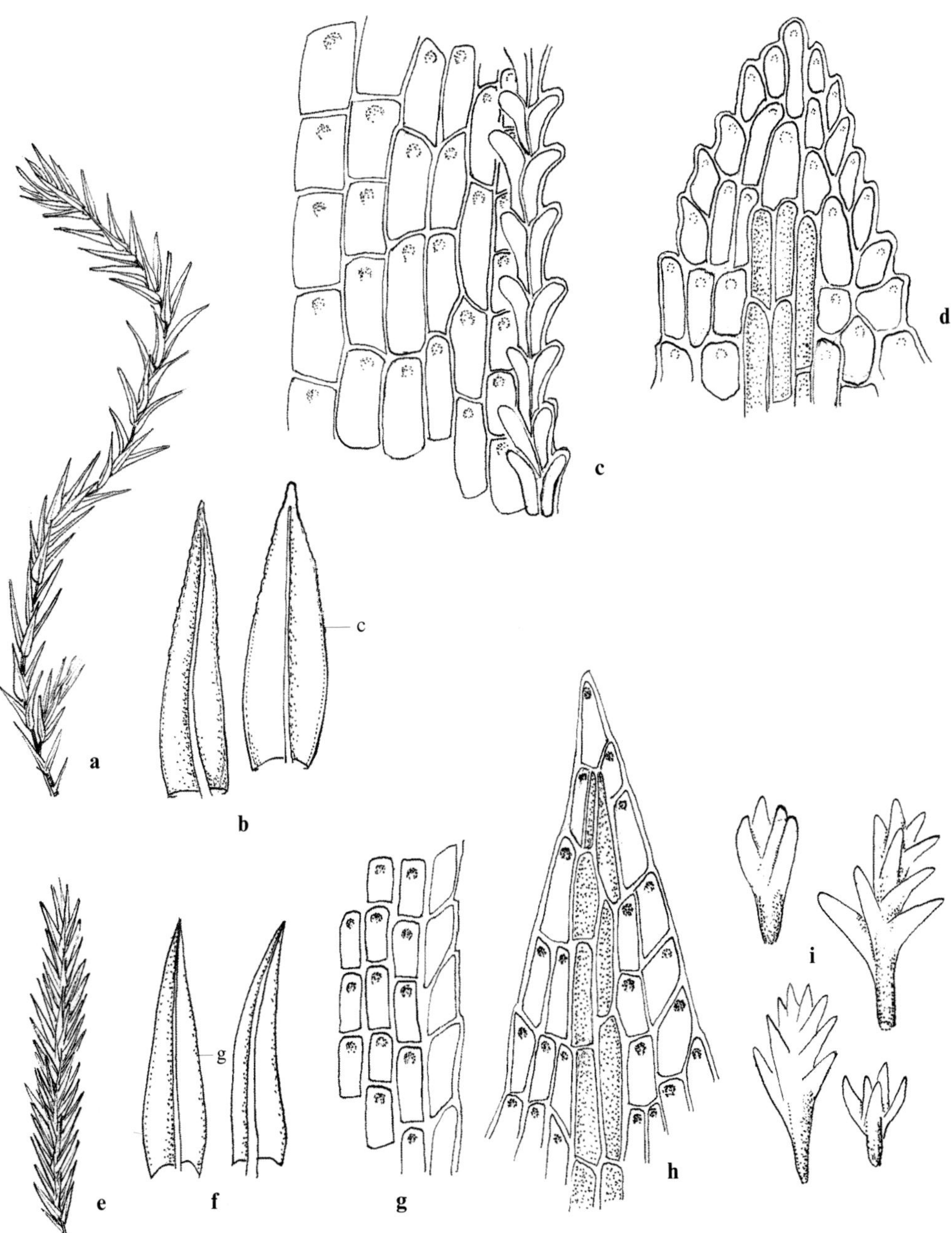

Figure 158. *Philonotis hastata*: **a** habit, moist (×10); **b** leaves (×50); **c** mid-leaf cells with paired teeth at margin; **d** leaf apex (all leaf cells ×500).
Philonotis marchica: **e** habit, moist (×10); **f** leaves (×50); **g** mid-leaf cells with serrulate margin of single teeth; **h** leaf apex (all leaf cells ×500); **i** propagula (×50).
(**a**–**d**: *Herrnstadt 80-11-12*; **e**–**i**: *Herrnstadt* & *Crosby 78-11-6*)

3. Philonotis calcarea (B.S.G.) Schimp., Coroll. Bryol. Eur. 86 (1856).
Bartramia calcarea B.S.G., Bryol. Eur. 4:49 (1842).

Distribution map 161, Figure 159.

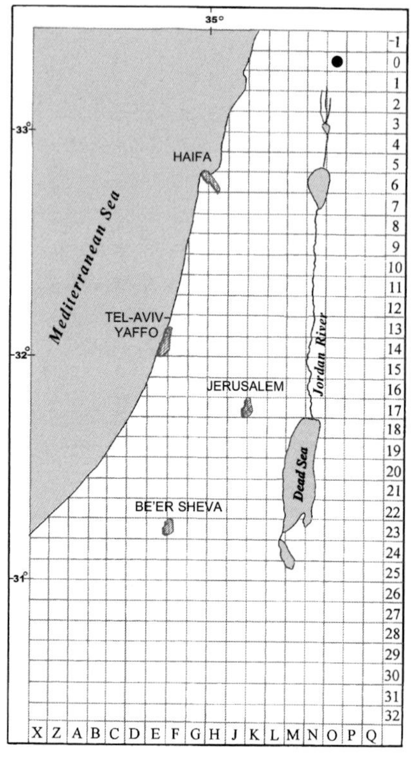

Map 161: *Philonotis calcarea*

Plants ± robust, up to 70 mm high, tomentose below, yellowish-green. *Leaves* falcate-secund to secund when dry, erect-patent to secund when moist, ca. 2.5 mm long, weakly carinate, slightly plicate, lanceolate or ovate-lanceolate, gradually tapering to a long, narrow, acuminate, slightly twisted apex; margins plane or narrowly recurved, serrulate, often with double teeth, except in upper ⅓; costa stout, long-excurrent, toothed on dorsal surface; cells ± incrassate; basal cells rectangular, smooth; mid- and upper leaf cells narrow rectangular to linear in mid-leaf 9–15 μm wide with papillae at proximal end. *Sessile propagules* sometimes present. *Androecium* discoid, perigonial leaves from a wide erect base, ¼–⅓ length of leaf, gradually narrowing to long-acute, reflexed point, cells weakly prorate. *Dioicous*, in local plants some gametangia but no sporophytes were found.

Habitat and Local Distribution. — Plants growing in dense tufts on muddy soil. Found in a single locality on Mount Hermon (Wadi Shibea; this collection was recorded as *Philonotis seriata* Mitt. by Herrnstadt *et al.* 1982).

General Distribution. — Recorded from Southwest Asia (Turkey, Syria, Lebanon, Iraq, Iran, and Afghanistan), around the Mediterranean, Europe (throughout), Macaronesia (in part), North Africa, and Asia (throughout, except the southeast).

Philonotis calcarea is easily distinguished by the falcate-secund leaves, the prominent proximal papillae on upper lamina cells, and the often double teeth at the leaf margins.

Figure 159. *Philonotis calcarea*: **a** part of habit, moist (×10); **b** part of habit, dry (×10); **c** habit, male plant (×10); **d** leaf (×50); **e** basal cells; **f** cells on ventral surface of costa; **g** mid-leaf cells; **h** marginal upper cells (all leaf cells ×500); **i** perigonial leaves (×50); **k** propagula (×10). (*Naftolsky*, 20 Jul. 1924)

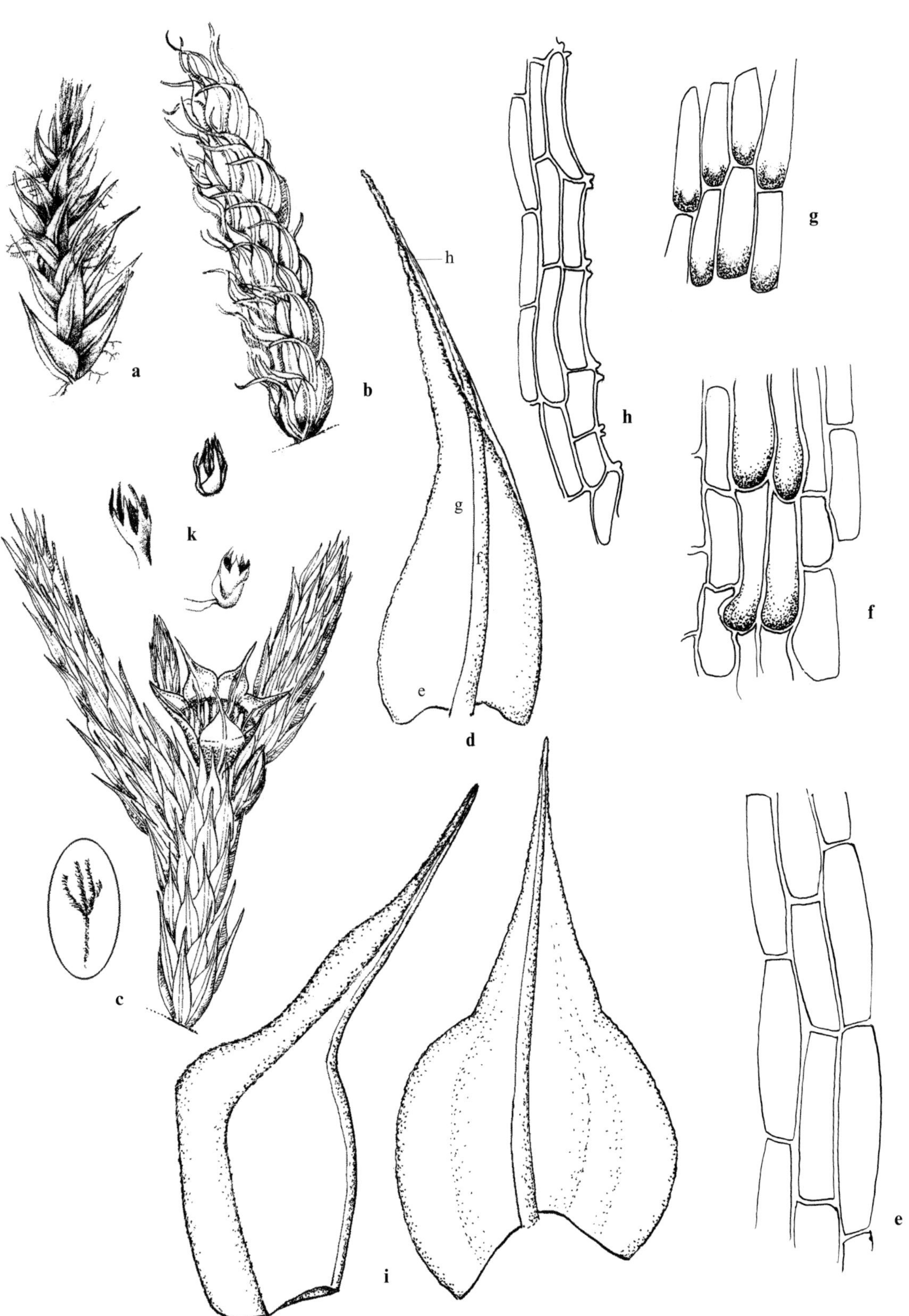
a
b
c
d
e
f
g
h
i
k

Orthotrichales

ORTHOTRICHACEAE

Plants acrocarpous, sometimes apparently pleurocarpous due to pattern of branching, small to moderately robust, forming tufts or cushions. *Stems* erect, ascending or creeping, simple or branched, central strand not distinct. *Leaves* crowded, lingulate, ovate-lanceolate or linear-lanceolate; costa usually single, strong, ending near apex to excurrent; basal cells subquadrate, rectangular, or linear, usually smooth; upper cells incrassate, mostly rounded-hexagonal, smooth or more often papillose with one to many papillae. *Perichaetia* terminal, with perichaetial leaves sometimes larger than other leaves. *Setae* erect, very short to elongate; capsules erect, immersed to exserted, symmetrical, often striate or ribbed; stomata superficial or immersed; operculum convex to conical, rostrate; peristome double with thin endostome segments, alternating with exostome teeth and without cilia, single, or rarely absent; occasionally preperistome present. *Calyptrae* mitrate or cucullate, naked or hairy, usually covering most of the capsule.

A family comprising 13 genera (Vitt 1973), with plants growing mainly on bark of trees or rocks. In the local flora it is represented by two genera: *Zygodon* and *Orthotrichum*.

1 Capsules long-exserted, calyptrae cucullate, not plicate, stomata superficial; gemmae frequent in leaf axils and on rhizoids; upper leaf cells usually with several (three–six), round, usually simple papillae **1. Zygodon**

1 Capsules immersed or short-exserted, calyptrae mitrate, at least ± plicate, stomata superficial or immersed; gemmae rare, if present, only on leaves; upper leaf cells usually with few (one–three) conical and forked papillae, rarely smooth **2. Orthotrichum**

1. *Zygodon* Hook. & Taylor

Plants small to moderately robust, usually up to 30 mm high, rarely higher, light to olive green, growing in tufts on rocks or on bark. *Stems* erect or ascending, simple or forked, tomentose below. *Leaves* imbricate, often twisted when dry, patent or spreading to squarrose-recurved when moist, usually ± keeled, ovate-lanceolate to linear-lanceolate, acute or acuminate, often apiculate; margins usually plane, entire or ± denticulate towards apex; costa usually strong, ending below apex to excurrent; cells ± uniform, except for few, less incrassate, smooth, hyaline cells at leaf insertion, round-quadrate to round hexagonal, usually with many simple conical papillae. *Gemmae* frequently produced in leaf axils and on rhizoids. *Dioicous* or autoicous. *Setae* elongate, erect; capsules long-exserted, erect, ovoid, ellipsoid, or pyriform, often with eight ribs when dry; neck fairly long; annulus persistent, poorly developed; operculum usually long-rostrate; stomata superficial, restricted to the neck; peristome single, double, or absent; exostome teeth usually 16, often joined in pairs below or all along, papillose; endostome segments eight or 16, slender, weakly papillose. *Calyptrae* cucullate, usually smooth, not plicate, naked, rarely hairy. *Spores* small, smooth to papillose.

A genus comprising about 90 species, mainly in North Africa and America. One species occurs in the local flora.

Literature. — Malta (1926).

Zygodon viridissimus (Dicks.) Brid., Bryol. Univ. 1 : 592 (1826). [Bilewsky (1965) : 407 (as "*Z. viridissimus* (Dicks.) R. Brown")]

Distribution map 162, Figure 160 (p. 413), Plate XI : g.

Plants up to 15 mm high, green to yellowish brown-green, male and female plants together within tufts. *Stems* usually forked. *Leaves* in upper part of stem slightly falcate-secund when dry, 1–2 mm long, narrow-lanceolate to linear-lanceolate, acute or acuminate, with apical cell 1(–3) μm long, hyaline; margins plane, entire, papillose; costa strong, ending below apex; cells, except cells at lower base, strongly incrassate, 6–10 μm wide, with numerous papillae (also on cell walls); perichaetial leaves ± longer than vegetative leaves. *Gemmae* ellipsoid, three–five(–eight) cells long, usually without longitudinal walls, sometimes few cells divided into two. *Dioicous*. *Setae* 2–4 mm long; capsules exserted, 1–2 mm long, yellowish brown, ovoid to ellipsoid, with long neck, smooth when moist, ribbed when dry. *Spores* 13–17 μm, papillose; sparsely covered by large, smooth, often fused gemmae (seen with SEM). *Sporophytes* abundant.

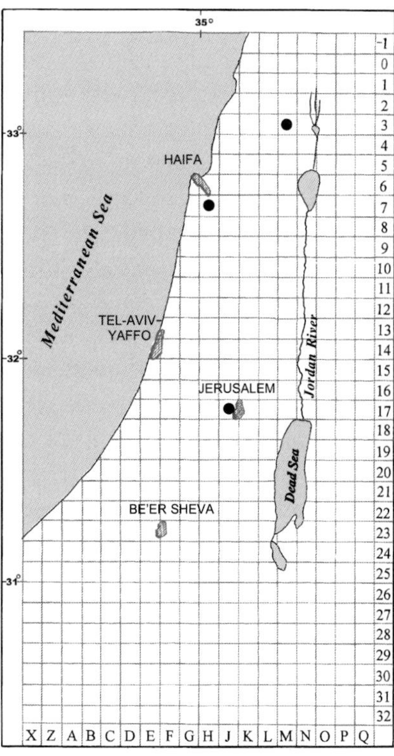

Map 162: *Zygodon viridissimus*

Habitat and Local Distribution. Plants growing in usually dense tufts on *Quercus*, at tree base or above, if shaded by neighbouring rocks. Rare, locally abundant: Upper Galilee, Mount Carmel, and Judean Mountains.

General Distribution. Recorded from Southwest Asia from Turkey, around the Mediterranean (few and scattered records), Europe (mainly northern and southern Europe), Macaronesia, North Africa, northern and eastern Asia, and northern, central and northwestern South America.

Herrnstadt *et al.* (1991) assumed that both typical *Zygodon viridissimus* and *Z. baumgartneri* Malta (= *Z. viridissimus* var. *rupestris* Hartm.) grow in Israel. *Zygodon viridissimus* is usually considered as having septate gemmae, ± secund leaves with a long, single apical cell, whereas *Z. baumgartneri* is described as not or rarely having septate gemmae, secund leaves with one–three apical cells. Malta (1926) pointed out, after describing *Z. baumgartneri* (Malta 1924), that gemmae typical of one of the two species may occur together with leaves typical of the other species. In all local populations, the upper part of leaves tend to be ± secund, the gemmae are without longitudinal walls, though some partial septa can usually be found, and the number and size of the apical leaf cells vary. For these reasons it is preferred here to consider all the local plants of *Zygodon* as a single species: *Z. viridissimus.*

2. *Orthotrichum* Hedw.

Plants small, usually up to 50 mm high, growing on bark of trees or rocks. *Stems* erect or ascending, branched, usually one–four times forked. *Leaves* imbricate, erect, rarely contorted when dry, spreading when moist, ovate-lanceolate or oblong-lanceolate, usually acute or obtuse, rarely rounded, acuminate or piliferous, ± keeled; margins usually reflexed to recurved, entire, notched, or irregularly toothed near apex; costa strong, usually ending near apex, with rectangular-linear cells on ventral surface; basal cells smooth, slightly to distinctly incrassate, sometimes nodulose, hyaline or pellucid, often shorter towards margins; upper cells incrassate, usually with one–three conical and forked papillae, rarely smooth. *Autoicous* (all local species gonioautoicous) or rarely dioicous. *Setae* very short to fairly long; capsules immersed to exserted, ovoid to cylindrical, smooth or eight–sixteen ribbed, sometimes slightly constricted below mouth; neck short; annulus absent or poorly differentiated; stomata usually up to around middle of capsule, immersed or superficial; peristome double (sometimes single, rarely absent), exostome teeth usually 16, often joined in pairs into eight teeth, papillose or striate, reflexed, recurved or erect; endostome segments eight or rarely 16, erect to incurved. *Calyptrae* mitrate, plicate to various degrees, naked or hairy, short to oblong-conical, little lobed at base. *Spores* finely to coarsely papillose; in local species, verrucate, verrucae covered by granules to various extent (seen with SEM) (exception: *Orthotrichum striatum* — see the description there).

A large worldwide genus of ca. 200 species, mainly xerophytes with temperate and boreal distribution. Six species occur in the local flora.

Literature. — Vitt (1973), Lewinsky (1977, 1983).

1 Leaves ending in hyaline hair-point **6. O. diaphanum**
1 Leaves not ending in hyaline hair-point.

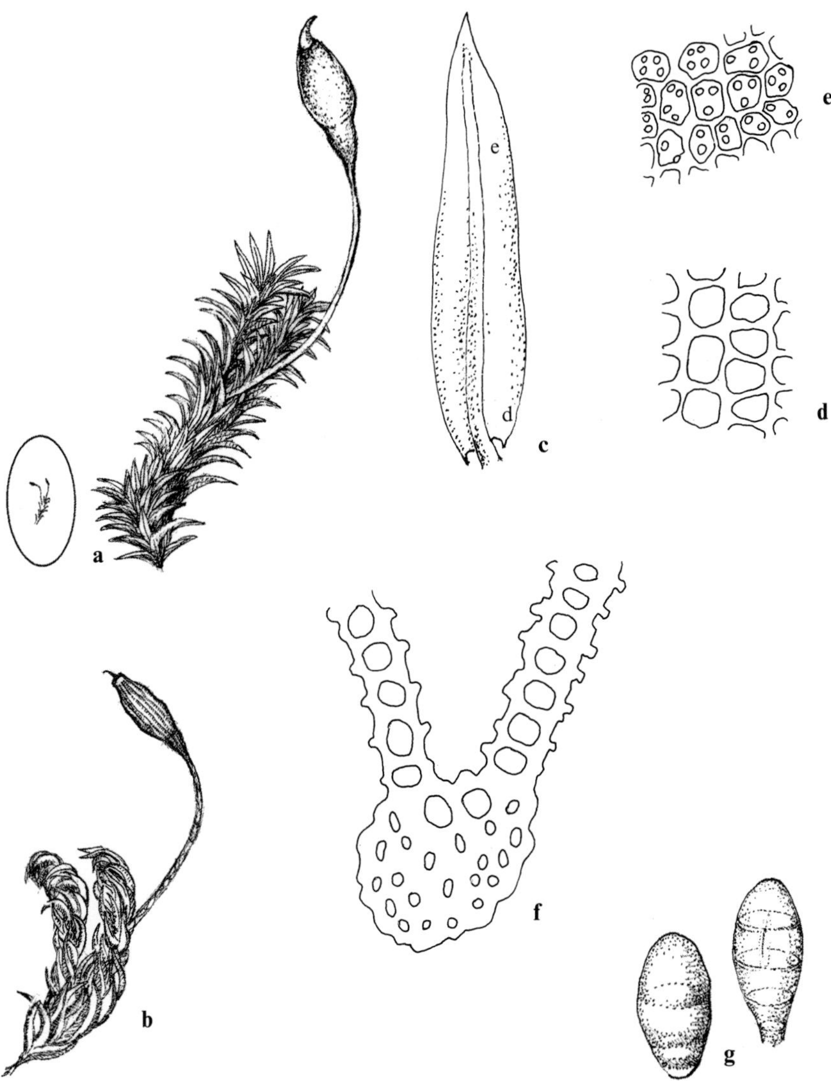

Figure 160. *Zygodon viridissimus*: **a** habit, moist (×10); **b** habit, dry (×10); **c** leaf (×50); **d** basal cells; **e** upper cells (all leaf cells ×500); **f** cross-section of costa and portions of lamina (×500); **g** gemmae (×500). (*Herrnstadt 81-185-3*)

2 Stomata on capsules superficial; basal cells, except near margins, elongate-linear, incrassate, at least ± nodulose.
   3 Mature capsules smooth when dry and empty **1. O. striatum**
   3 Mature capsules ribbed when dry and empty.
      4 Exostome teeth 16, erect or spreading when mature and dry; plants usually growing on rocks **3. O. rupestre**
      4 Exostome teeth eight, recurved-reflexed when mature and dry; plants usually growing on trees **2. O. affine**
2 Stomata on capsules immersed, basal cells rectangular, not incrassate or moderately incrassate, not nodulose.
   5 Setae short, ca. 0.5 mm high; exostome teeth eight, papillose, reflexed when dry; lamina unistratose; plants growing on trees **5. O. pumilum**
   5 Setae usually 1–1.6 mm high; exostome teeth 16, longitudinally striate-papillose, erect or spreading when dry; lamina usually partly bistratose in upper part; plants growing on rocks, rarely on trees **4. O. cupulatum**

1. Orthotrichum striatum Hedw., Spec. Musc. Frond. 163 (1801).

Distribution map 163, Figure 161, Plate XII: a, b.

Plants up to 30 mm high, dark green to brownish. *Stems* branched. *Leaves* erect when dry, spreading when moist, 3–4 mm long, ovate-lanceolate to lanceolate, acute to acuminate; margins recurved nearly up to apex, often crenulate-denticulate near apex; costa ending just below apex; cells incrassate; basal cells elongate, short-rectangular towards margins, nodulose, smooth; upper cells irregularly quadrate-elliptic, often with simple papillae; mid-leaf cells 7–9 µm wide. *Sporophytes* usually abundant. *Setae* 1–1.5(–2) mm long; capsules immersed or slightly emergent, up to 2 mm long without operculum (no mature capsules with opercula could be found in local plants), pale brown, wide oblong to oblong-ovate, smooth (unribbed) when dry and empty, not constricted below mouth; stomata superficial; exostome teeth 16, densely papillose, recurved to spreading when mature and dry; endostome segments 16, densely papillose, irregularly shaped, each formed by two rows of cells.

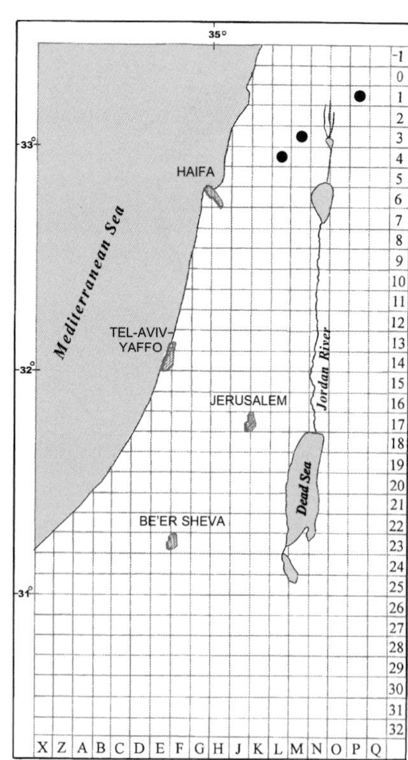

Map 163: *Orthotrichum striatum*

Figure 161. *Orthotrichum striatum*: **a** habit, moist (×10); **b** habit, dry (×10); **c** sporophyte with deoperculate capsule (×10); **d** leaf (×50); **e** basal cells; **f** upper cells, (all leaf cells ×500); **g** leaf apex (×125); **h** cross-section of costa and portion of lamina (×500); **i** portion of peristome teeth (×100); **k** stoma (×500). (*Raven*, 24 Jan. 1975)

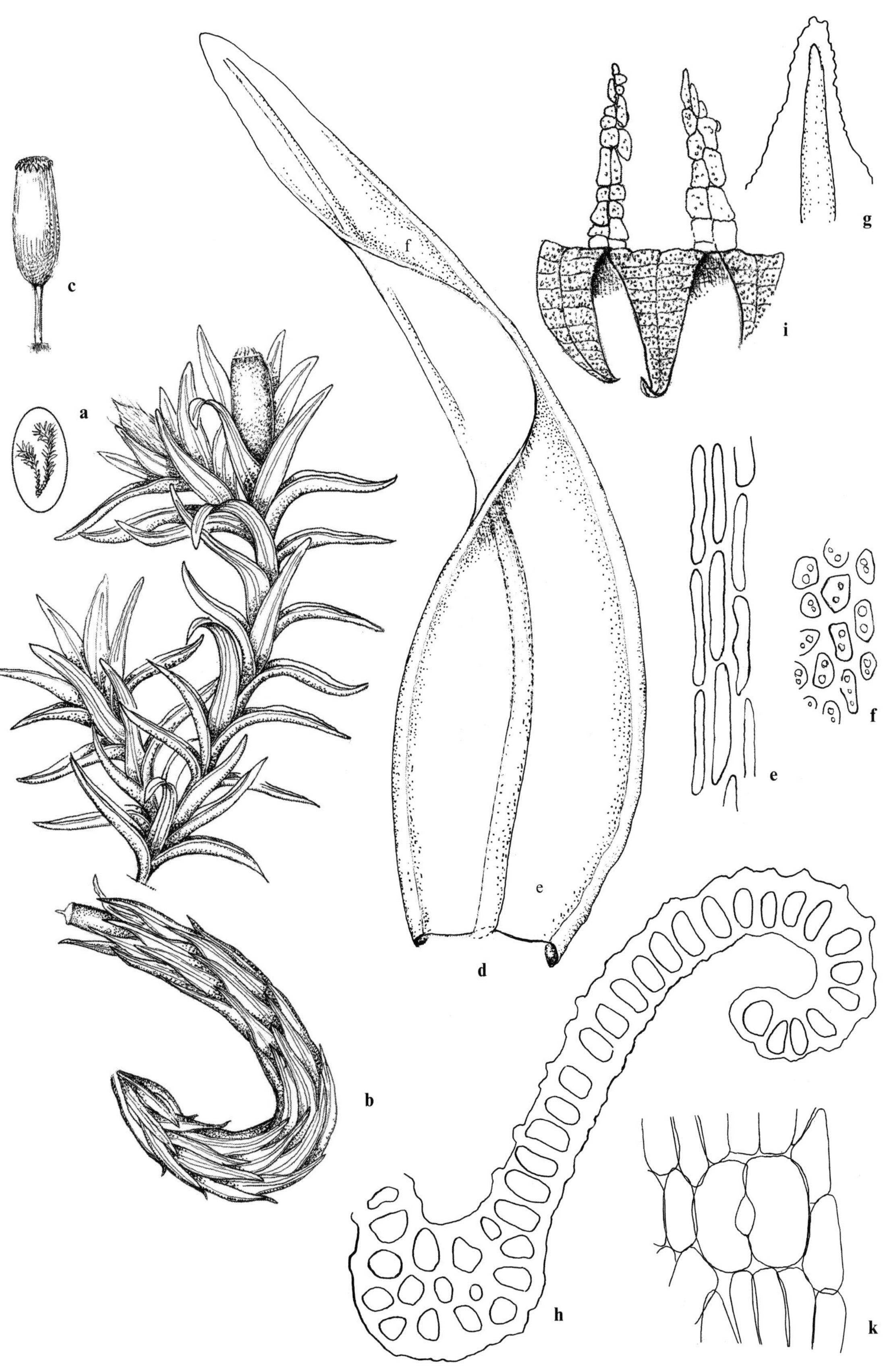
c
a
b
f
e
d
i
g
e
f
h
k

Spores 27–30 µm in diameter; gemmate, processes with rough surface, variously sized and spaced (seen with SEM). *Calyptrae* plicate, usually sparsely hairy.

Habitat and Local Distribution. — Plants growing in small tufts on the bark of trees of *Pistacia* spp. and *Quercus* spp., and on terra rossa in high elevations in the North. Occasional: Upper Galilee and Golan Heights.

General Distribution. — Recorded from Southwest Asia (Cyprus and Turkey), around the Mediterranean, Europe (throughout), Macaronesia, North Africa, north-eastern, eastern, and central Asia, and northern and northwestern South America.

Orthotrichum striatum was found in mixed populations with *O. diaphanum*, but can be easily distinguished by the darker colour of plants and the lack of hyaline hair-points. The unribbed dry capsules can serve as a good distinguishing character of the species in general.

2. Orthotrichum affine Schrad. ex. Brid., Muscol. Recent. 2: 22 (1801).
O. octoblephare Brid., Muscol. Recent. 2: 24 (1801).

Distribution map 164, Figure 162, Plate XII: c.

Plants up to 20(–30) cm high, brown to brownish-green. *Stems* profusely branched. *Leaves* nearly erect, appressed when dry, patent to spreading when moist, 2–3(–4) mm long, lanceolate to ovate-lanceolate; apex variously shaped, obtuse, broadly acute, sharply acute, or acuminate; margins recurved nearly up to apex, entire; costa ending in or near apex; basal and upper cells incrassate and nodulose; basal cells smooth, pellucid, linear, subquadrate towards margins; upper cells irregularly rounded or round-elliptic, with low, forked papillae; mid-leaf cells 8–12(–13) mm wide. *Setae* 1–1.5 mm; capsules emergent, up to 2.5 mm long including operculum, yellowish-brown, cylindrical, narrow cylindrical when dry, with eight prominent ribs down ⅔ length of capsule, constricted below mouth; stomata superficial; exostome teeth eight, papillose, erect when young, recurved-reflexed when mature and dry; endostome segments eight, each formed by one–two rows of papillose cells. *Spores* up to 20 µm in diameter,

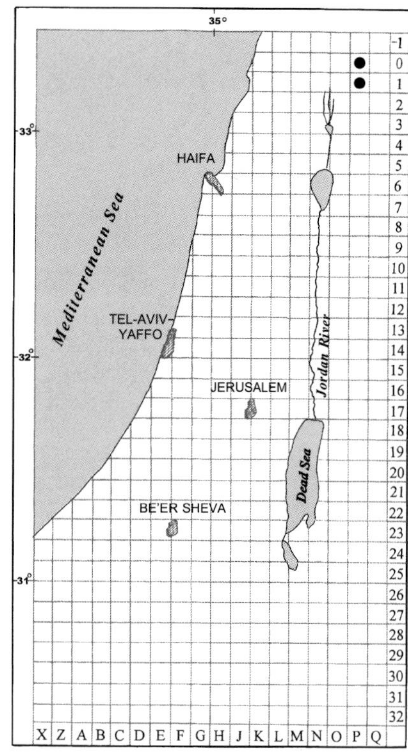

Map 164: *Orthotrichum affine*

Figure 162. *Orthotrichum affine*: **a** habit, moist (×10); **b** habit, dry (×10); **c** calyptra (×10); **d** deoperculate capsule; view from above (×10); **e** leaf (×50); **f** basal cells; **g** upper cells (all leaf cells ×500); **h** cross-section of costa and portion of lamina (×500); **i** portion of peristome teeth (×100); **k** stoma (×500). (*Kushnir*, 10 Oct. 1943)

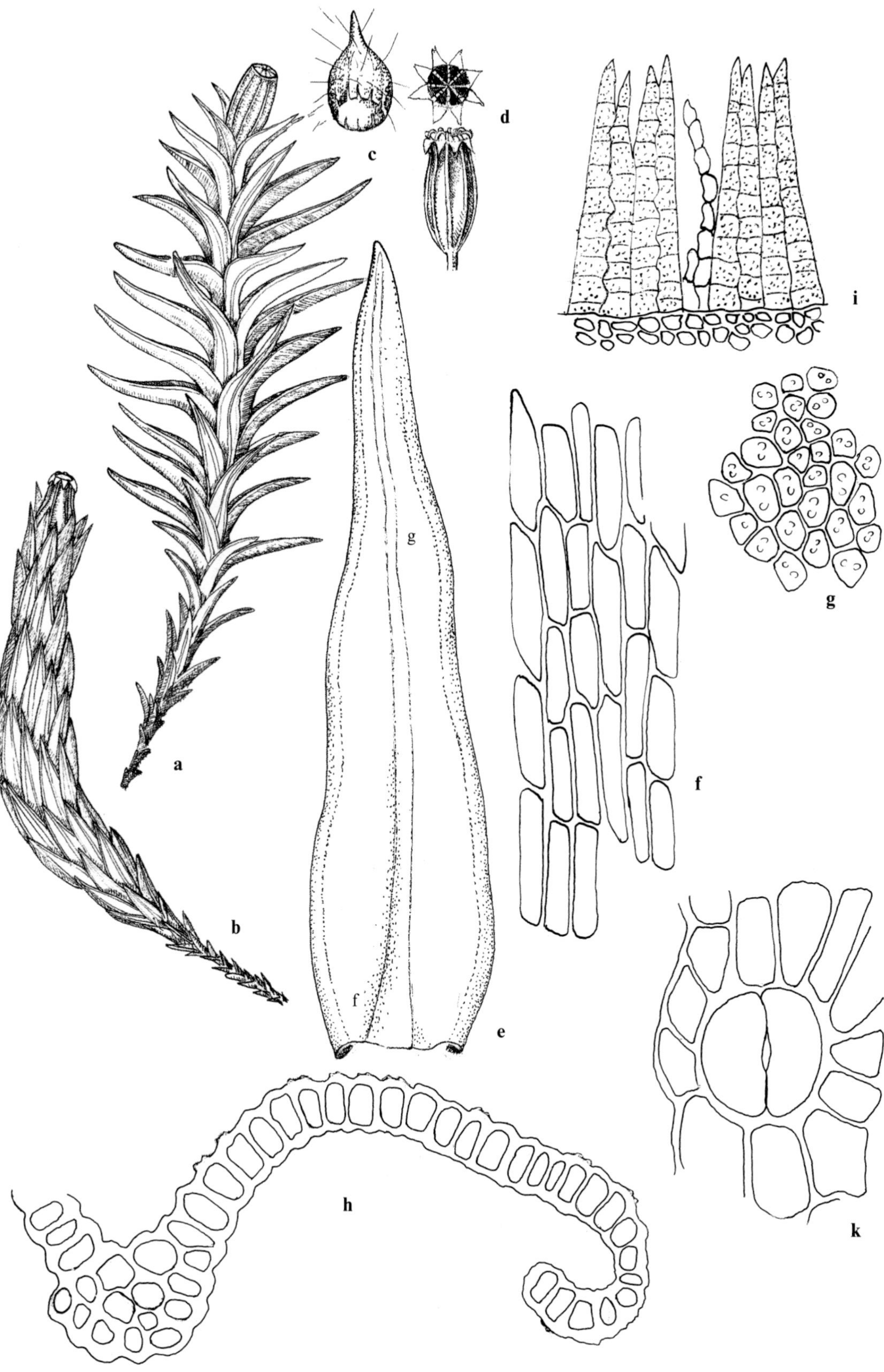
c
d
i
g
f
g
a
f
b
e
h
k

papillose; verrucate, verrucae smooth, greatly varying in size, some fused (seen with SEM). *Calyptrae* (in local plants) sparsely covered with short hairs.

Habitat and Local Distribution. — Plants growing in tufts on the bark of trees of *Quercus* spp.; so far found only in three localities of the high elevations of the North. Rare: Mount Hermon and Golan Heights.

General Distribution. — Recorded from Southwest Asia (Cyprus, Turkey, Syria, Lebanon, and Iraq), around the Mediterranean, Europe (throughout), Macaronesia, tropical and North Africa, Asia (throughout), and North America.

Orthotrichum affine can be distinguished from other local *Orthotrichum* species by a combination of characters of the dry capsules: ribbed, narrowly constricted below mouth down to two-thirds, with superficial stomata and reflexed exostome teeth.

3. Orthotrichum rupestre Schleich. ex Schwägr., Sp. Musc. Frond., Suppl. 1 : 27 (1816).

Distribution map 165, Figure 163.

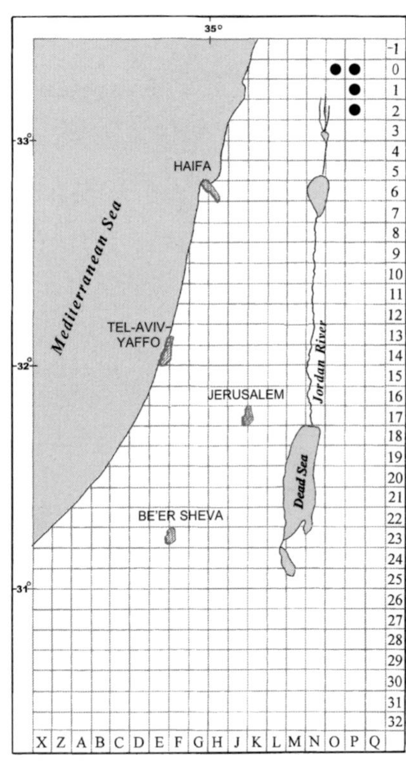

Map 165: *Orthotrichum rupestre*

Plants up to 30 mm high, green to olive above, dark green to brownish-red below. *Stems* often much branched. *Leaves* keeled, erect and appressed when dry, patent to spreading when moist, 3–4 mm long, lanceolate to ovate-lanceolate, narrow-acute; margins strongly recurved to near apex; costa strong, ending just below apex, turning reddish with age; basal cells elongate-linear, shorter towards margins, walls strongly incrassate, ± nodulose, smooth; upper cells irregularly rounded to elongate, with low papillae; mid-leaf cells 7–10 µm wide. *Setae* about 1 mm high; capsules slightly emergent or immersed, 1.5–3 mm long including operculum, light brown, ellipsoid or oblong-ovoid when moist, with eight prominent ribs in upper half and hardly constricted below mouth when dry and empty; neck long, gradually tapering into seta; stomata superficial; operculum convex-rostellate; exostome teeth 16, joined in eight pairs when young, papillose, erect to slightly spreading when mature and dry; endostome segments eight, slender and usually broken or missing in mature capsules; fragments of preperistome sometimes present. *Spores* ± 15 µm in diameter, papillose. *Calyptrae* densely hairy. *Sporophytes* always present.

Habitat and Local Distribution. — Plants growing in tufts usually on basaltic rocks, rarely on trees in shade. Rare, though locally abundant: Mount Hermon, and northern Golan Heights (850–1200 m).

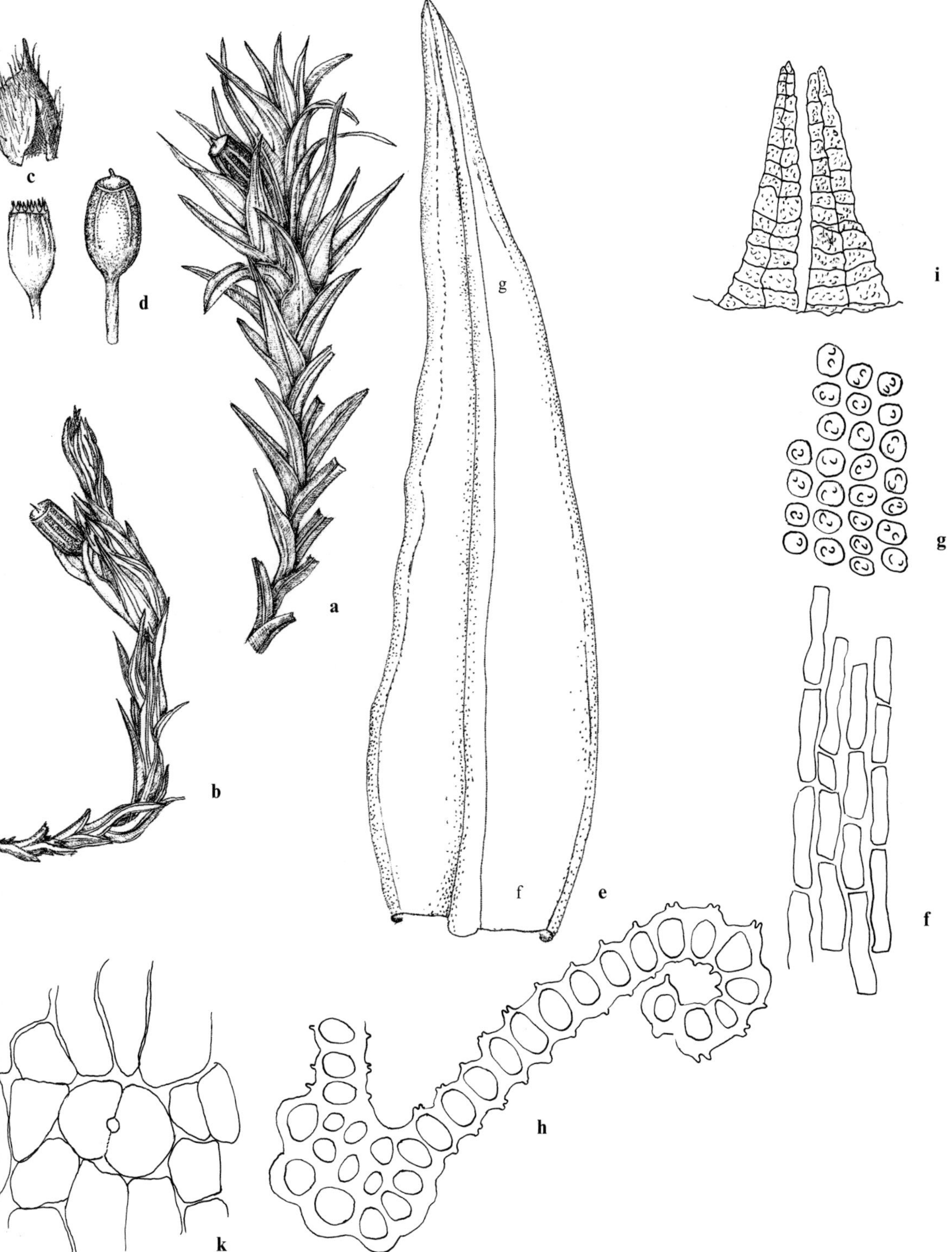

Figure 163. *Orthotrichum rupestre*: **a** habit, moist (×10); **b** habit, dry (×10); **c** deoperculate moist capsule and calyptra (×10); **d** sporophyte (×10); **e** leaf (×50); **f** basal cells; **g** upper cells (all leaf cells ×500); **h** cross-section of costa and portion of lamina (×500); **i** peristome teeth (×125); **k** stoma (×500). (*Markus & Kutiel 77-601-1a*)

General Distribution. — Recorded from Southwest Asia (Cyprus, Turkey, Lebanon, Iraq, and Afghanistan), around the Mediterranean, Europe (throughout), Macaronesia, North, tropical, and South Africa, northeastern, eastern, and central Asia, Australia, New Zealand, and northern, central, and southern South America.

Orthotrichum rupestre can be distinguished from the other local species of the genus by a combination of characters: strongly incrassate, elongate-linear, ± nodulose basal cells of leaves, capsules with usually erect exostome teeth when mature and dry, superficial stomata, and densely hairy calyptrae.

4. Orthotrichum cupulatum Hoffm. Ex Brid., Muscol. Recent. 2:25 (1801).
O. cupulatum var. *papillosum* Gronvall, Öfvers. Förh. Kongl. Svenska Vetensk.-Akad. 46:174 (1889).

Distribution map 166, Figure 164, Plate XII:d.

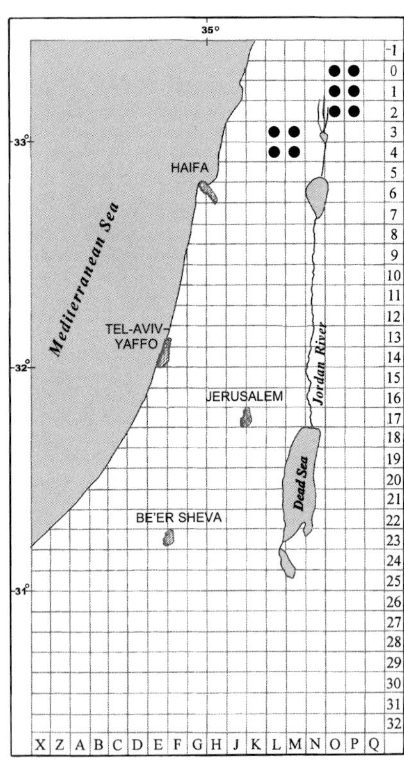

Map 166: *Orthotrichum cupulatum*

Plants small to medium-sized, up to 15 mm high, light green to olive, brownish-green below. *Stems* sparsely branched. *Leaves* erect and appressed when dry, patent to spreading when moist, up to 2.5 mm long, ovate-lanceolate to oblong-lanceolate, acute; margins recurved except near apex, entire, often bistratose in upper half; costa ending just below apex, turning reddish brown with age; cells moderately incrassate, upper usually partly bistratose; basal cells rectangular, smooth; mid- and upper leaf cells irregularly rounded with variously sized simple or forked papillae; mid-leaf cells 7–11 μm wide. *Setae* 1–1.6 mm long; capsules immersed when moist, emergent when dry, up to 2 mm long including operculum, yellowish-brown, oblong-ovoid to pyriform when dry, with 16 ribs — often eight — shorter, slightly constricted below mouth; stomata immersed; exostome teeth 16, spreading and often fracturing when mature and dry, papillose and longitudinally striate; endostome absent or only of few fragmentary segments. *Spores* 12–16 μm in diameter, coarsely papillose; surface rough, covered by very large verrucae, (seen with SEM). *Calyptrae* hairy.

Figure 164. *Orthotrichum cupulatum*: **a** habit (×10); **b** capsule, dry (×10); **c** calyptra (×10); **d** deoperculate capsule, moist (×10); **e** mouth of capsule, dry (×10); **f** leaf (×50); **g** basal cells; **h** upper cells (all leaf cells ×500); **i** peristome teeth (×125); **k** cross-section of costa and portion of lamina near leaf apex (×500); **l** cross-section of costa and portion of lamina at mid-leaf (×500); **m** stoma (×500). (*Kushnir*, 9 Oct. 1943)

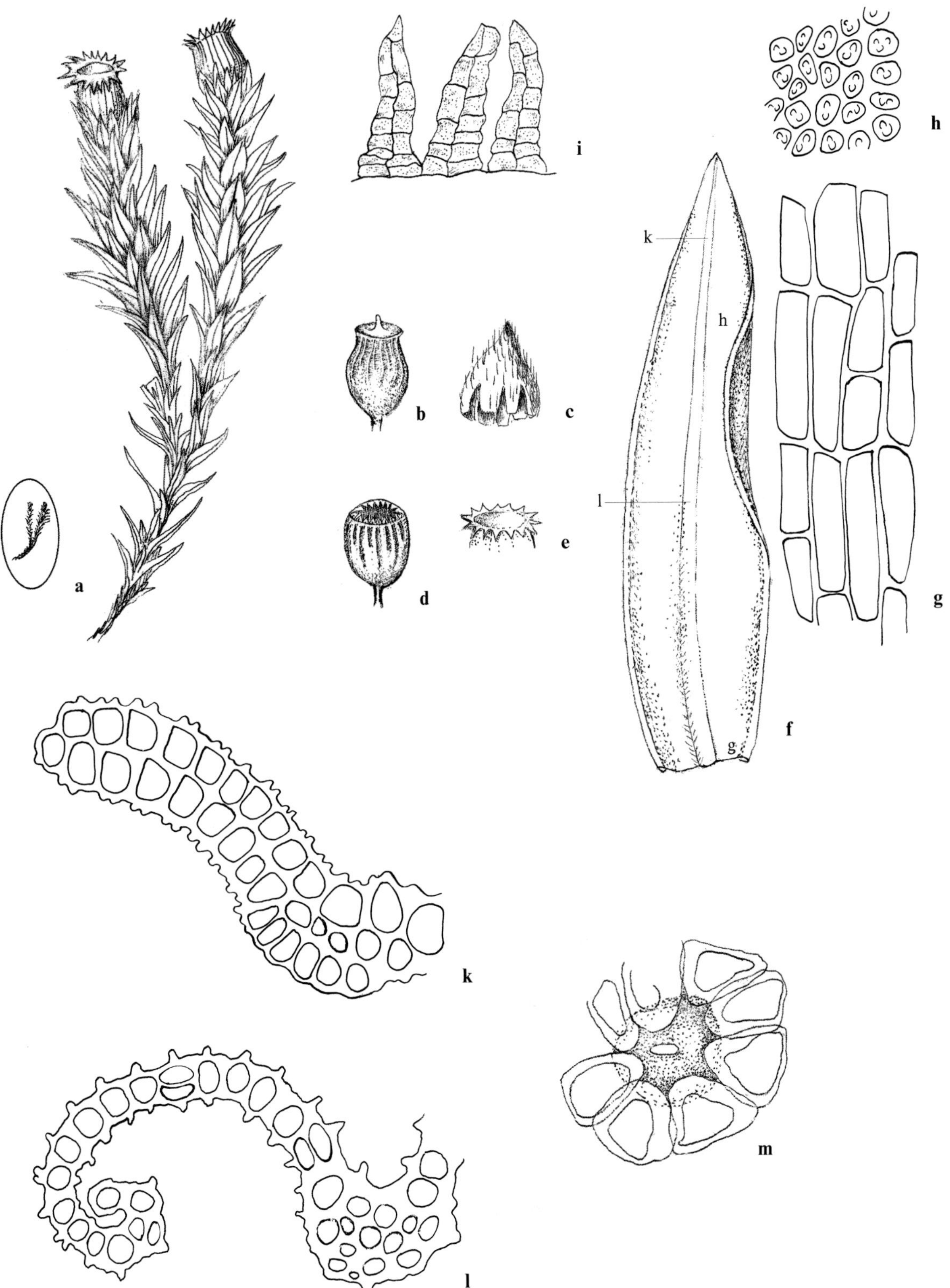
i
h
k
h
b
c
l
e
a
d
g
g
f
k
m
l

Habitat and Local Distribution. — Plants growing in cushions on rocks in the mountainous northern parts. In one locality on Mount Hermon *Orthotrichum cupulatum* was found growing together with *Orthotrichum affine* on the bark of a tree, which is an unusual substrate for *O. cupulatum*. Occasional: Upper Galilee, Mount Hermon, and Golan Heights.

General Distribution. — Recorded from Southwest Asia (Turkey, Syria, Lebanon, Jordan, Iraq, Iran, and Afghanistan), around the Mediterranean, Europe (throughout), Macaronesia, North Africa, northeastern and central Asia, and North and Central America.

Orthotrichum cupulatum var. *papillosum* Gronvall was recorded from Iraq by Agnew & Vondráček (1975). In addition to the larger, often two–three lobed papillae of the mid- and upper leaf cells, this variety has, according to them, also some typical peristome characters, among which the most constant is the occurrence of partly papillose and partly striate exostome teeth. Although large papillae and the peristome characters mentioned by Agnew & Vondráček have been found in some of the local collections, they did not always appear correlated. Therefore, it is preferred here to follow Vitt (1973), who does not accept var. *papillosum* as a separate variety.

5. Orthotrichum pumilum Sw., Monthly Rev. 34 : 538 (1801).
O. schimperi Hammar, Mon. Orthotrivh. Ulot. Suec. 9 (1852). — Bilewsky (1977).

Distribution map 167, Figure 165.

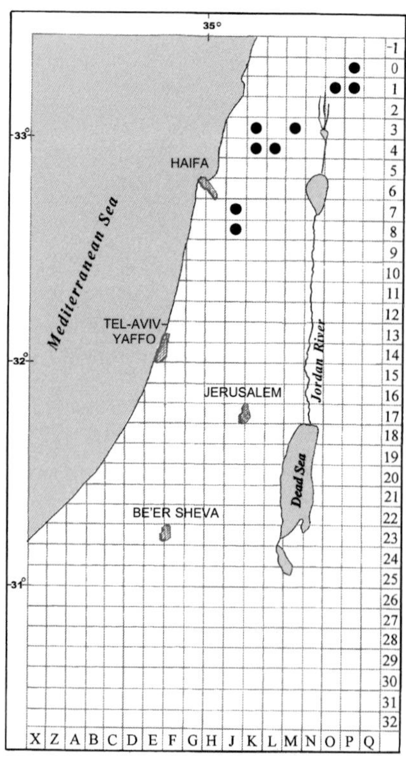

Map 167: *Orthotrichum pumilum*

Plants small, up to 5 mm high, pale to dark green. *Stems* simple or forked. *Leaves* nearly erect when dry, spreading when moist, 2–3 mm long, ovate-lanceolate, acute to obtuse and apiculate; margins recurved nearly to apex, entire below, sometimes slightly denticulate at apex; costa ending just below apex; basal cells rectangular, smooth; mid- and upper leaf cells irregularly rounded, pellucid, not incrassate, usually with one–two (three) small conical papillae; mid-leaf cells 15–21 µm wide. *Gemmae* mainly simple, greenish-brown, filamentous, up to eight cells long, sometimes present on leaves. *Setae* very short, up to 0.5 mm; capsules abruptly differentiated into seta, immersed, up to 1.5 mm long including operculum, yellowish-brown, wide cylindrical to ovoid when moist, narrow cylindrical, strongly and deeply constricted below mouth, with eight prominent dark brown ribs when mature and dry; stomata immersed; exostome teeth eight, papillose, reflexed when dry; endostome segments eight, smooth. *Spores* 12–18 µm in diameter. *Calyptrae* sparsely hairy.

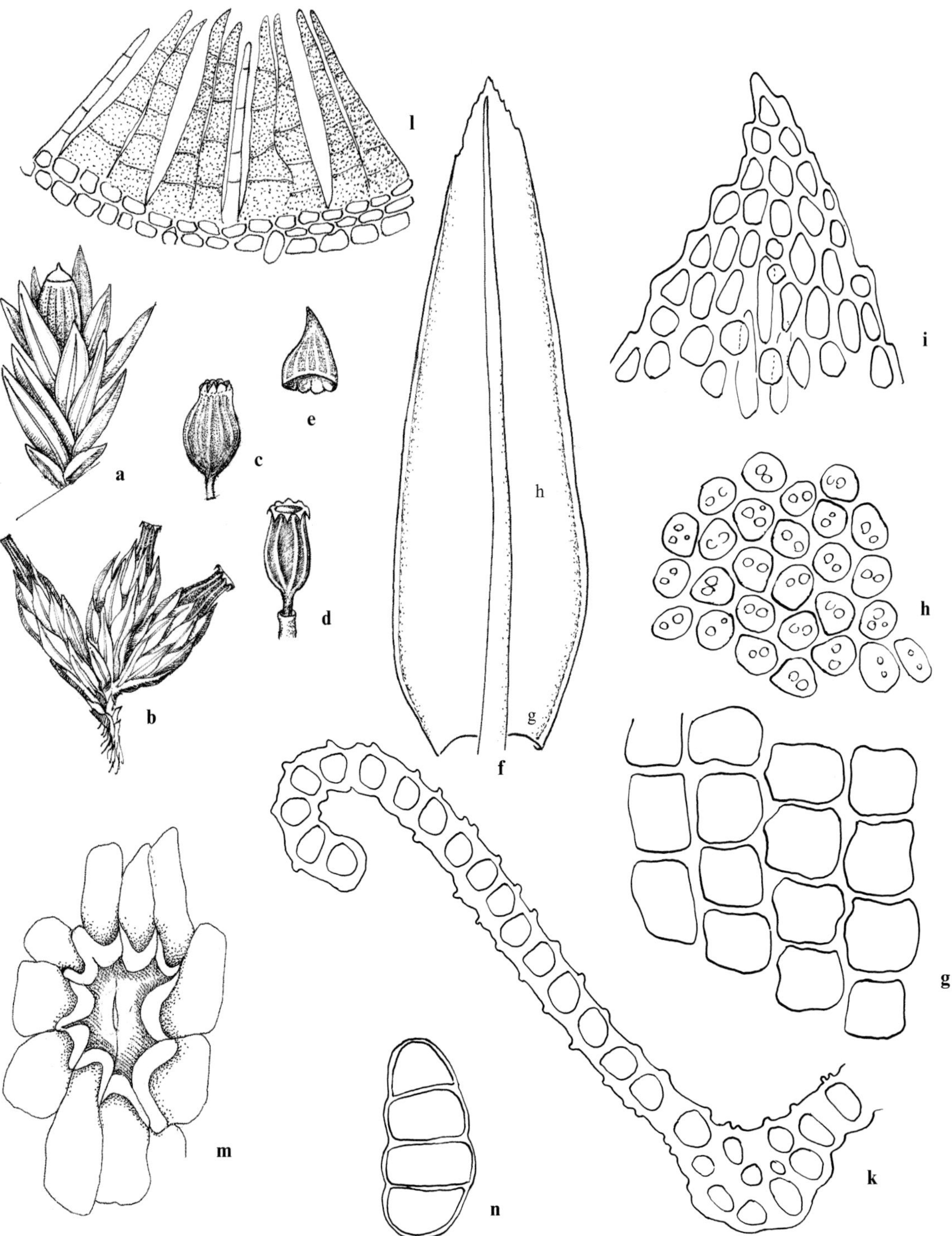

Figure 165. *Orthotrichum pumilum*: **a** habit, moist (×10); **b** habit, dry (×10); **c** sporophyte, moist (×10); **d** sporophyte, dry (×10); **e** calyptra (×10); **f** leaf (×50); **g** basal cells; **h** mid-leaf cells; **i** leaf apex (all leaf cells ×500); **k** cross-section of costa and portion of lamina (×500); **l** portion of peristome teeth (×150); **m** stoma (×500); **n** gemma (×500). (*Bilewsky*, 19 Apr. 1960, No. 120a in Herb. Bilewsky)

Habitat and Local Distribution. — Plants growing in tufts on tree trunks, very often on *Quercus* spp., mainly in the north. Occasional: Upper Galilee, Lower Galilee, Samaria, Mount Hermon, and Golan Heights.

General Distribution. — Recorded from Southwest Asia (Turkey, Iraq, and Iran), around the Mediterranean, Europe (throughout), Macaronesia, North and South Africa, central Asia, and North America.

Plants of *Orthotrichum pumilum* (*Herrnstadt et al. 83-291-2*) were found on the Golan Heights growing together with *O. diaphanum* on some trees of *Quercus calliprinos* Webb. They can be readily distinguished by the shorter capsules, which are strongly constricted at the mouth when dry, and by the reflexed exostome teeth. In addition, the leaves do not have a hyaline hair-point as of *O. diaphanum*. Rarely, the much larger plants of *O. rupestre* may also grow together with *O. pumilum* on *Quercus*.

Orthotrichum schimperi (Pierrot 1978, Duell 1992) is sometimes considered as the valid name of this taxon. Here the treatment of Vitt (1973), who reduced *O. schimperi* to a synonym of *O. pumilum*, is accepted.

6. Orthotrichum diaphanum Schrad. ex Brid., Muscol. Recent. 2 : 29 (1801). [Bilewsky (1965) : 408]

Distribution map 168, Figure 166, Plate XII : e, f.

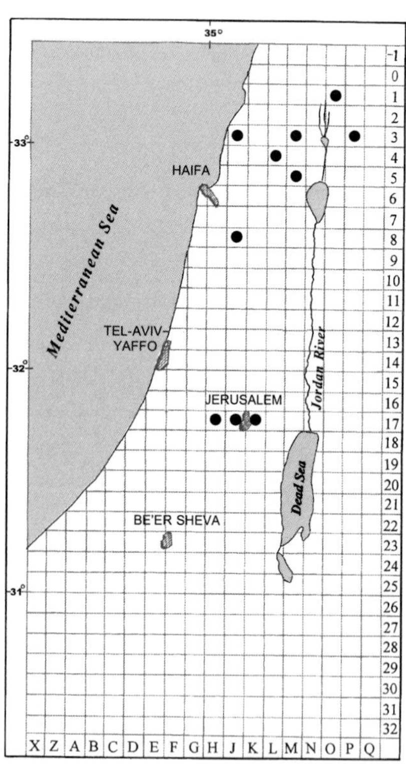

Map 168: *Orthotrichum diaphanum*

Plants small, 3–10(–15) mm high, light to brownish-green. *Stems* sometimes dichotomously branched. *Leaves* erect, rarely slightly contorted when dry, erectopatent to spreading when moist, up to 2.5(–3.5) mm long with hair-point, ovate-lanceolate to oblong, acuminate, with a long, hyaline, usually irregularly toothed hair-point, reaching in upper leaves up to ⅓ the length of leaf; margins recurved, entire; costa ending in hair-point; basal cells rectangular, smooth; upper cells round-hexagonal, pellucid, smooth or weakly papillose, moderately incrassate at corners; mid-leaf cells 12–18 μm wide. *Gemmae* simple or branched, greenish, filamentous, 2–20 cells long often present on leaves. *Setae* up to 1 mm long; capsules slightly emergent or immersed, up to 2 mm long including operculum, yellowish-green, oblong-cylindrical, smooth or wrinkled, more often eight-ribbed when dry; stomata immersed; exostome teeth 16, papillose, erect-spreading, sometimes sharply reflexed when mature and dry; endostome segments 16, very slender, each made of one row of papillose cells. *Spores* 15–18 μm in diameter, papillose; verrucate, gemmate, processes greatly varying in size (seen with SEM). *Calyptrae* naked or with few hairs.

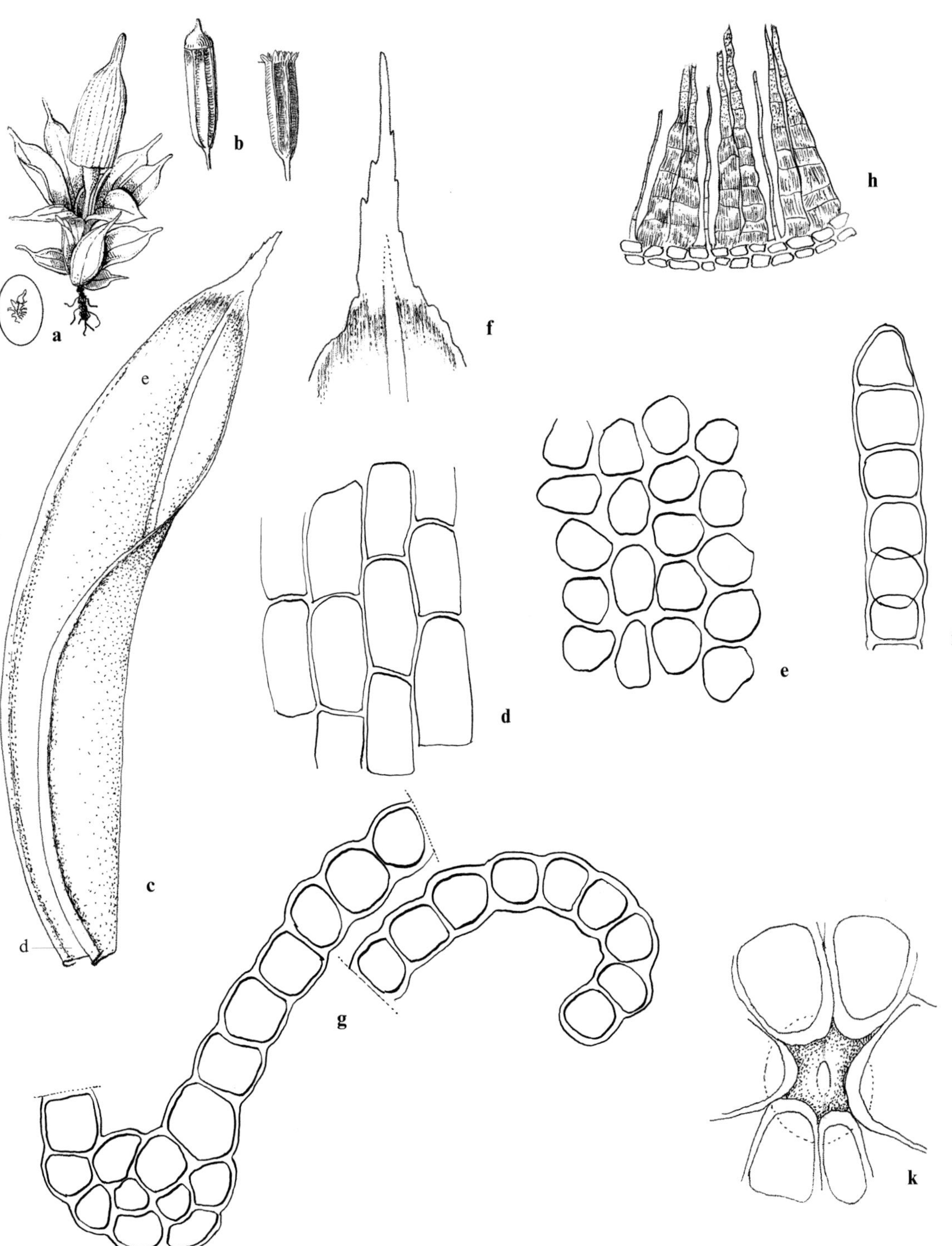

Figure 166. *Orthotrichum diaphanum*: **a** habit, moist (×10); **b** operculate and deoperculate capsules, dry (×10); **c** leaf (×50); **d** basal cells; **e** upper cells (all leaf cells ×500); **f** leaf apex (×125); **g** cross-section of costa and portion of lamina (×500); **h** portion of peristome teeth (×150); **i** gemma (×500); **k** stoma (×500). (*Herrnstadt & Crosby 78-54-1*)

Habitat and Local Distribution. — Plants growing in tufts on the trunk of various trees (including *Olea europaea* L., *Pistacia lentiscus* L., *Quercus* spp., and *Pinus halepensis* Mill.), on detached tree bark or rarely on rocks. Occasional: Upper Galilee, Lower Galilee, Samaria, Judean Mountains, Dan Valley, and Golan Heights.

General Distribution. — Recorded from Southwest Asia (Turkey, Lebanon, Jordan, Afghanistan, and Saudi Arabia), around the Mediterranean. Available records are from Europe (throughout), Macaronesia, tropical and North Africa, Oceania, and North and Central America.

Orthotrichum diaphanum is the most widespread local *Orthotrichum* species and may grow under dryer local conditions than the other species. It may occur together with either *O. striatum* or *O. pumilum*, but can be easily distinguished from all local species by the leaves with hyaline hair-point.

LEUCODONTALES

LEUCODONTACEAE

Plants pleurocarpous, rigid, rather robust, growing in mats or patches. *Stems* without or with rudimentary central strand; primary stems creeping, stoloniform, bearing sparsely spaced rhizoids; secondary stems erect, ascending, rarely pendant, simple or branched, not bearing rhizoids; branch tips sometimes flagelliform. *Paraphyllia* absent. *Leaves* crowded in many rows, ± imbricate, appressed when dry, erectopatent to spreading when moist, often concave, sometimes longitudinally plicate, unbordered, ovate or ovate-lanceolate, apex acute or acuminate, ± decurrent; costa single, double or absent; supplementary costae sometimes present; cells incrassate, usually smooth, rhomboidal-elliptical above, elongate below; irregularly round-quadrate, ± oblate cells, often in oblique rows, at basal margins; perichaetial leaves elongate, erect, sheathing. *Dioicous* usually. *Setae* very short to elongate; capsules exserted or immersed, erect and symmetrical, ovoid to subcylindrical; annulus usually differentiated; exostome with 16 teeth, endostome absent or variously developed, cilia absent; operculum conical or obliquely rostrate. *Calyptrae* cucullate, naked or hairy, smooth.

The Leucodontaceae is represented in the local flora by three species, each belonging to a different genus.

Literature. — Manuel (1974), Crosby (1980), Akiyama (1994).

The concepts adopted for the family differ. Consequently, the Leucodontaceae are variously delimited, dealt with in a wide sense (e.g., Manuel 1974, Crosby 1980), or in a much narrower sense (e.g., Buck 1980). Recently, a new classification of the Leucodontaceae was suggested by Akiyama (1994). *Cryphaea* D. Mohr & F. Weber, a genus sometimes included in Leucodontaceae, is usually considered as belonging to the Cryphaeaceae (e.g., Bowers & Crum *in* Sharp *et al.* 1994). That family differs from the Leucodontaceae *sensu stricto* by being autoicous, and by having immersed capsules and mitrate calyptrae.

Previous records of *Cryphaea heteromalla* (Hedw.) D. Mohr from Israel (Herrnstadt *et al.* 1982, 1991) — considered there as belonging to Leucodontaceae — were based on plants from two localities on Mount Hermon. Upon re-examination of those collections of sterile plants, some doubts arose whether it was indeed possible to distinguish between plants of *Cryphaea* and *Antitrichia californica* without the examination of sporophytes. Therefore, *C. heteromalla* is not included in this Flora.

1 Plants coarsely robust; leaves without costa **1. Leucodon**
1 Plants moderately robust; leaves with costa.
  2 Costa single; sometimes supplementary costae present **2. Antitrichia**
  2 Costa double or forked; supplementary costae absent **3. Pterogonium**

1. *Leucodon* Schwägr.

Plants robust with irregularly branched secondary stems. *Stem* with central strand. *Leaves* sometimes slightly secund, usually longitudinally plicate; costa absent; cells smooth. *Capsules* usually exserted; annulus not differentiated; peristome whitish; exostome teeth coarsely papillose, perforated; endostome rudimentary. *Calyptrae* naked.

A genus comprising ca. 35 species. One species occurs in the local flora.

Leucodon sciuroides (Hedw.) Schwägr., Sp. Musc. Frond., Suppl. 1 : 1 (1816). [Bilewsky (1965) : 409]

Fissidens sciruoides Hedw., Sp. Musc. Frond. 161 (1801).

Distribution map 169, Figure 167, Plate XII : g.

Plants coarsely robust, brownish green. *Secondary stems* erect to ascending often curved at tips, up to 30(–40) mm long. *Leaves* longitudinally plicate; erect, appressed when dry, patent when moist, wide ovate-lanceolate, up to 2.5(–3) mm long, narrow-acuminate; margins entire, plane; cells incrassate, smooth oblong-rhomboidal to linear-rhomboidal from mid-leaf to acumen, longer towards mid-base of leaf, rounded-quadrate to transversely oblong towards margins at lower half of leaf, at mid-leaf 6–8 µm wide; perichaetial leaves without costa, sheathing the setae at least half of its length. *Dioicous. Setae* straight, yellowish brown, 4–8 mm long; capsules emergent to exserted, 3–4 mm long, cylindrical, straight or slightly curved, brown; operculum short-rostrate, slightly oblique. *Spores* 21–27 µm, densely papillose; verrucate-gemmate, processes varying in size, often in small groups (seen with SEM).

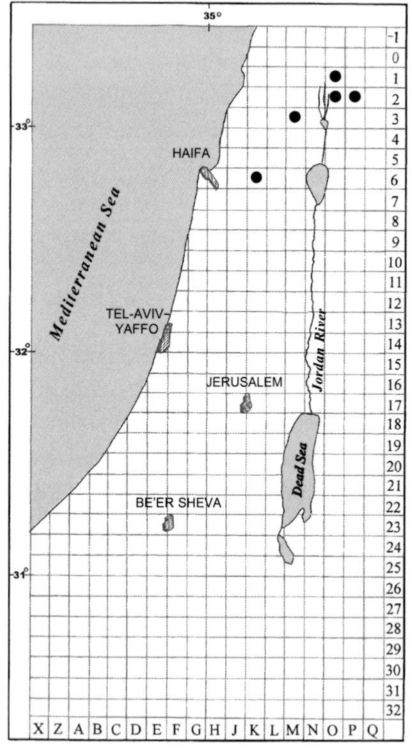

Map 169: *Leucodon sciuroides*

Habitat and Local Distribution. — Plants growing on exposed to partially shaded basaltic and calcareous rocks, and on bark (found on *Olea europaea*). Occasional: Upper Galilee, Lower Galilee, and Golan Heights.

General Distribution. — The typical *Leucodon sciuroides* is recorded from Southwest Asia (Cyprus, Turkey, Lebanon, Jordan, Iran, and Afghanistan), around the Mediterranean, Europe (widespread), Macaronesia, North and tropical Africa, and throughout Asia (except the southeast).

Leucodon sciuroides var. *morensis* (Schwägr.) De Not. (Syllab. Musc. 79. 1838) was described as including larger and more robust plants. Plants in different local populations seem indeed to differ in size, but because of the small number of samples seen, we prefer not to recognise var. *morensis* as a separate taxon.

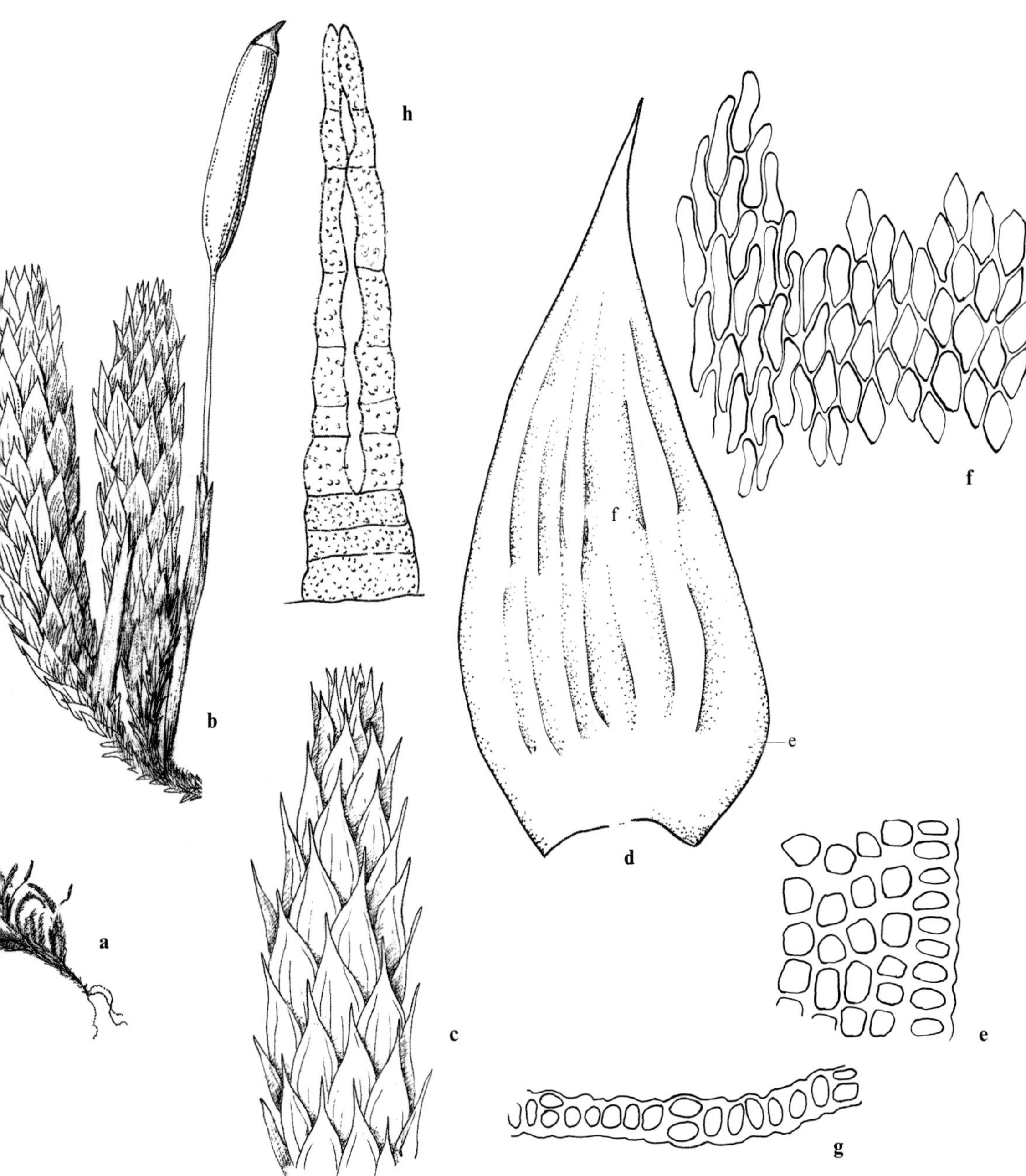

Figure 167. *Leucodon sciuroides*: **a** habit, dry (×1); **b** portion of habit, dry (×10); **c** upper part of stem, moist (×10); **d** leaf (×50); **e** marginal cells at lower half of leaf; **f** mid-leaf cells from centre towards margin (all leaf cells ×500); **g** portion of cross-section of lamina (×500); **h** peristome tooth (×250). (**a**–**c**, **h**: *Raviv*, 3 Feb. 1970; **d**–**g**: *Kushnir*, 5 Oct. 1943)

2. *Antitrichia* Brid.

Plants with irregularly branched stems. *Stem* with small central strand. *Leaves* subsecund and slightly plicate, or plane when dry, sometimes also when moist; costa strong, single, with or without additional costae at base of leaves; cells smooth, strongly incrassate. *Capsules* exserted, ovoid-cylindrical to cylindrical, with stomata at extreme base; annulus narrow; peristome whitish or pale yellow; peristome double, endostome segments ± opposite exostome teeth; cilia and basal membrane absent. *Calyptrae* naked.

A genus comprising four species. One species occurs in the local flora.

Antitrichia californica Sull. *in* Lesq., Trans. Amer. Philos. Soc. 13 : 11 (1865).
A. breidleriana Schiffn., Oesterr. Bot. Z. 58 : 344 (1908).

Distribution map 170, Figure 168.

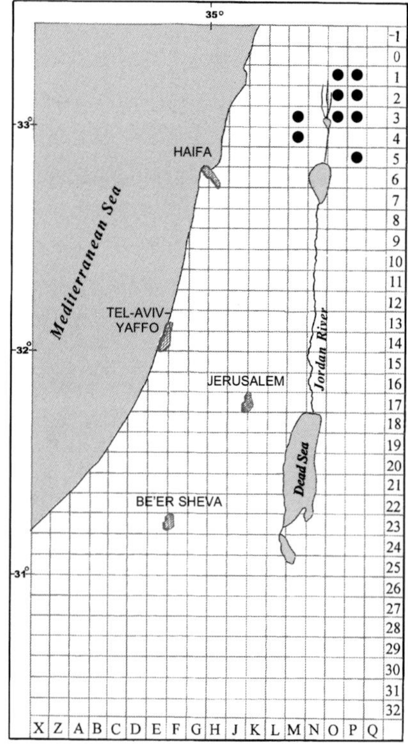

Map 170: *Antitrichia californica*

Plants moderately robust, yellowish-green, darker below. *Secondary stems* usually erect, sometimes ascending, regularly or sometimes rather irregularly branched, 20(–30) mm long. *Leaves* concave, ca. 1.5 mm long, wide-ovate, acuminate; margins strongly recurved from base, up to ¾ length of leaf, distinctly dentate near apex; costa extending up to ¾ length of leaf, without supplementary costa or with weak and short ones at base; cells in mid-leaf oblong-ellipsoid, 6–8 µm wide. *Sporophytes* have not been found in local plants.

Habitat and Local Distribution. — Plants growing in at least partial shade on basaltic and calcareous rocks, and tree trunks. Occasional: Upper Galilee, Mount Hermon, and Golan Heights.

General Distribution. — Recorded from Southwest Asia (Cyprus, Turkey, Syria, Lebanon, Jordan, and Iraq), around the Mediterranean, Europe (mainly the south), Macaronesia, North Africa, and North America.

Literature. — Schiffner (1908).

The type of *Antitrichia breideliana*, described by Schiffner (1908) was found by Townsend (1964–1965) to agree with *A. californica.*

Figure 168. *Antitrichia californica*: **a** habit, dry (×1); **b** part of habit, moist (×10); **c** part of habit, dry (×10); **d** leaf (×50); **e** basal cells; **f** upper cells; **g** leaf apex (all leaf cells ×500); **h** cross-section of costa and portion of lamina (×500). (*Herrnstadt & Crosby 79-77-10a*)

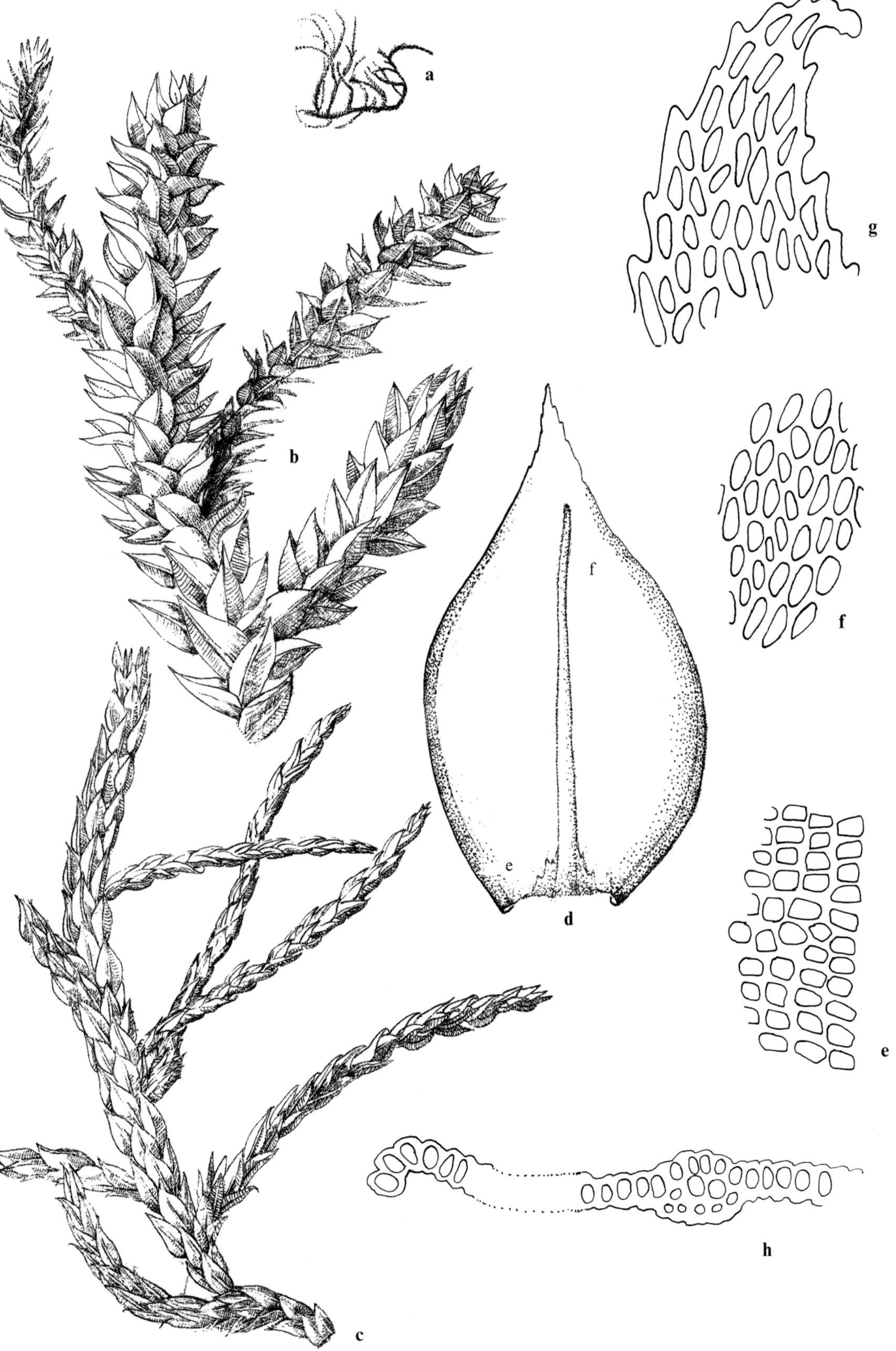
a
b
c
d
e
f
g
h
e
f

3. *Pterogonium* Sw.

Plants with irregularly to regularly branched secondary stems. *Stem* with central strand. *Leaves* not plicate; costa double or forked; cells papillose, elliptical. *Capsule* exserted; annulus present; peristome double, pale; endostome with narrow basal membrane and segments shorter than exostome teeth. *Calyptrae* with few hairs.

A genus comprising five species. One species occurs in the local flora.

Pterogonium gracile (Hedw.) Sm., Engl. Bot. 16 : 1085 (1802). [Bilewsky (1965) : 409]
Pterigynandrum gracile Hedw., Sp. Musc. Frond. 80 (1801).

Distribution map 171, Figure 169.

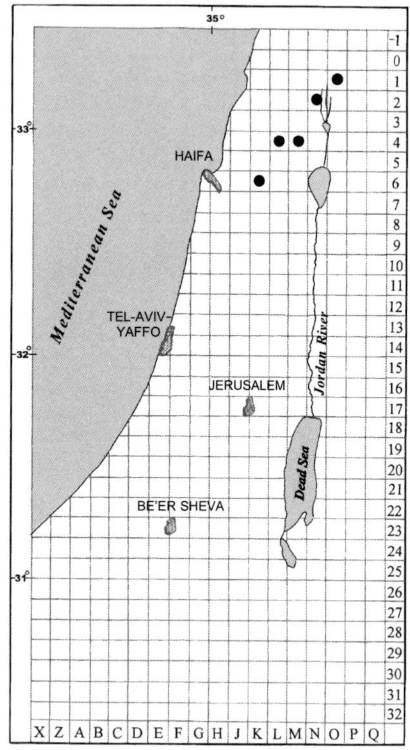

Map 171: *Pterogonium gracile*

Plants moderately robust, yellowish to olive-green turning reddish, dull. *Secondary stems* up to 30 mm long, simple below, ± irregularly branched or clustered above, curved, flagelliform, often attached to substrate; central strand present. *Branch leaves* densely imbricate, concave, not plicate, spreading when moist, up to 1.5 mm long, ovate to wide ovate, acute; margins plane, entire below, serrulate above; costa short, weak, usually double or forked, up to $\frac{1}{4}$–$\frac{1}{3}$ length of leaf; cells incrassate, smooth except for papilla-like projections at upper part of dorsal side of leaf; mid-leaf cells ca. 8 μm wide. *Sporophytes* have not been found in local plants.

Habitat and Local Distribution. — Plants growing at least partially shaded on calcareous or basaltic rocks. Rare, occasional in some areas: Upper Galilee, Lower Galilee, and Golan Heights.

General Distribution. — Recorded from Southwest Asia (Cyprus, Turkey, and Lebanon), around the Mediterranean, Europe (throughout), Macaronesia, Africa (throughout), and North America.

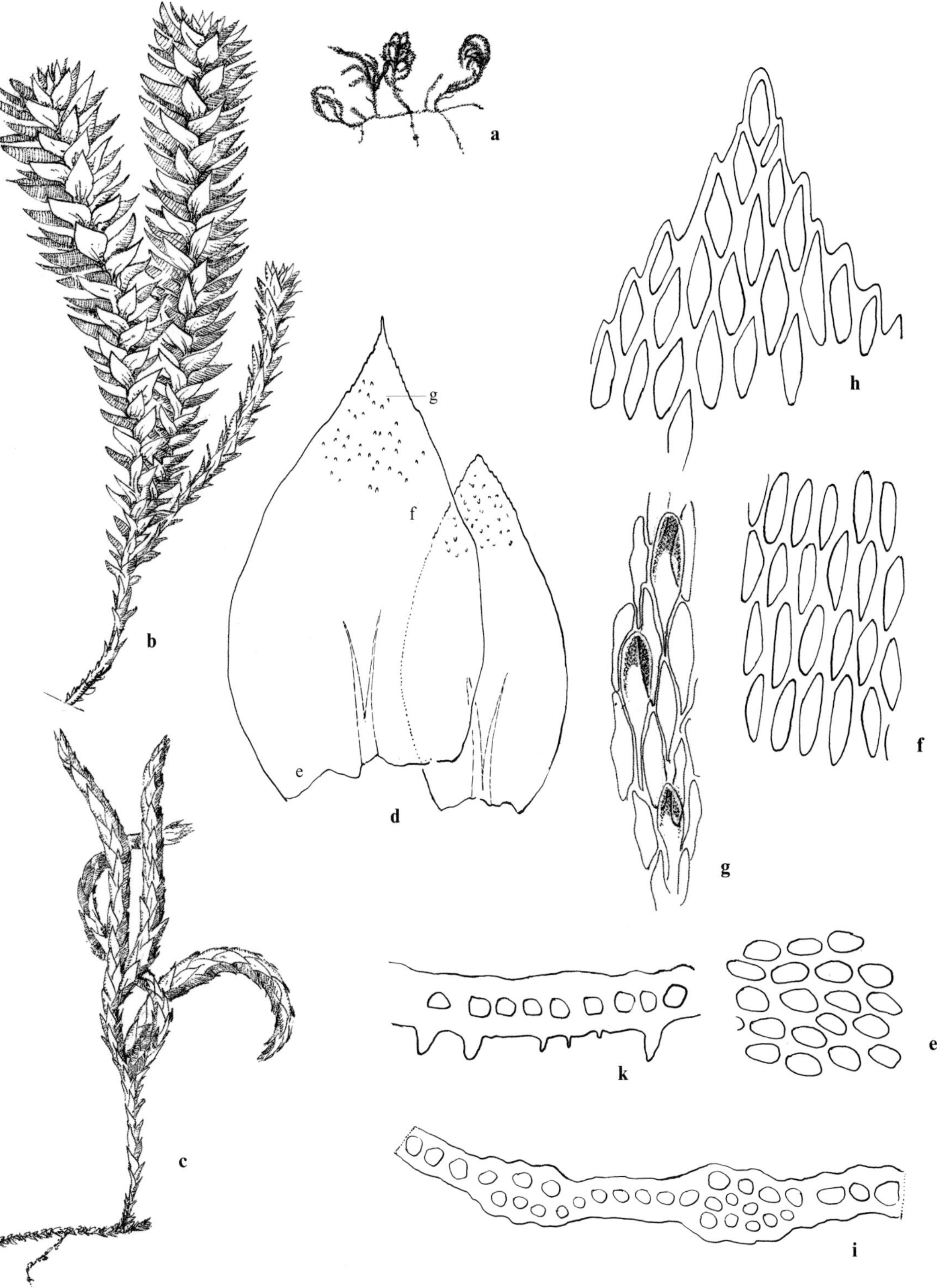

Figure 169. *Pterogonium gracile*: **a** habit, dry (×1); **b** part of habit, moist (×10); **c** part of habit, dry (×10); **d** leaves, dorsal side (×50); **e** basal cells; **f** upper cells; **g** apical cells; **h** leaf apex (all leaf cells ×500); **i** cross-section of costae and portion of lamina in lower part of leaf (×500); **k** cross-section of portion of lamina in upper part of leaf (×500). (*Shmida*, 26 Apr. 1982)

NECKERACEAE

Plants pleurocarpous, often quite robust and ± glossy, growing in mats or patches. *Stems* most often without central strand; primary stems creeping, usually stoloniform; secondary stems prostrate, decumbent, or erect, mostly pinnately or bipinnately branched, flattened, at least when moist. *Paraphyllia* often present. *Leaves* usually complanate, often asymmetrical, plane or undulate, ovate-oblong, elliptical to lingulate, apex truncate to rounded or acute; costa single, double, or absent, when present not exceeding ⅔ the length of leaf; cells usually smooth; basal cells elongate or linear; upper cells round-quadrate to rhomboidal. *Capsules* immersed to exserted, usually erect and symmetrical, oblong-cylindrical or ovoid-cylindrical; annulus usually not differentiated; peristome double or endostome with rudimentary segments, cilia sometimes present; operculum conical, usually obliquely rostrate. *Calyptrae* mitrate or cucullate, naked or hairy, smooth.

The family is variously delimited. *Leptodon* and *Thamnobryum* — two genera which are represented in the local flora — are often included in Neckeraceae or are considered as part of Leptodontaceae or Thamnobryaceae respectively (see also the discussion of that family below, p. 440). *Thamnobryum* is not considered here as part of the Neckeraceae, whereas *Leptodon* is retained in the family. The family Leptodontaceae was originally established by Schimper (1856). Buck (1980) reinstated this family and also accommodated in it the large genus *Forsstroemia* Lindb., which does not occur in the region covered in this Flora. We are not recognising the Leptodontaceae following Smith (1978) and Frahm (1995).

In the local flora, the Neckeraceae are represented by two species, one from each of the two genera *Leptodon* and *Neckera.*

Various concepts on the delimitation of the Neckeraceae were discussed in detail by Buck (1980), Stark (1987), and Enroth (1994); see also the treatment by Smith (*in* Sharp *et al.* 1994).

1 Stems with side-branches rolled inwards to coiled, when dry; leaves not distinctly complanate **1. Leptodon**
1 Stems with side-branches flat; leaves distinctly complanate **2. Neckera**

1. *Leptodon* Mohr

Leptodon, as originally described, included species which were later transferred to *Forsstroemia*. The various treatments of the *Forsstroemia* as a genus separate from *Leptodon* were summarised by Stark (1987). *Leptodon sensu stricto* comprises four or five species, of which only the local *L. smithii* — the type species of the genus — has a wide distribution.

For generic characters, see the following description of *Leptodon smithii.*

Leptodon smithii (Hedw.) F. Weber & D. Mohr, Index Mus. Pl. Crypt. 2 (1803). [Bilewsky (1965): 412 (as "*Leptodon smithii* (Dicks.) Mohr")]
Hypnum smithii Dicks. ex Hedw., Sp. Musc. Frond. 264 (1801).

Distribution map 172, Figure 170 (p. 437), Plate XIII: a.

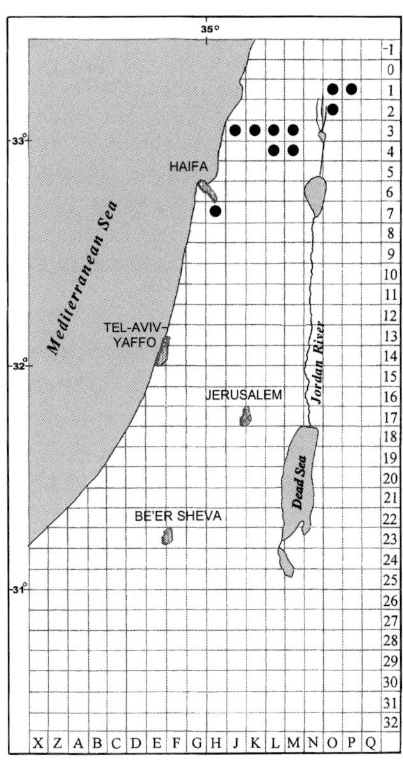

Map 172: *Leptodon smithii*

Plants slender to moderately robust, dark green, with primary and secondary stems stoloniform, pinnately or bipinnately branched; branches often rolled inwards or coiled when dry, incurved when moist; thin, branched innovations with small leaves often present. *Paraphyllia* abundant, linear, simple, or irregularly branched. *Leaves* in eight rows, imbricate when dry, patent to spreading when moist, small, up to 1 mm long, often asymmetric, ovate to shortly wide-lingulate, usually obtuse to rounded; margins recurved on one side, entire; costa single, vanishing above mid-leaf, in local plants usually up to ⅔ the length of leaf; cells ± incrassate, irregularly round-quadrate to hexagonal, 8–12 µm wide, near base more elongate; perichaetial leaves longer than stem and branch leaves, sheathing, lanceolate, long-acuminate, without costa. *Dioicous. Setae* straight, very short, up to 2 mm long, green; capsules straight, up to 3.5 mm long including operculum, slightly exceeding perichaetial leaves in length, ellipsoid, brown; peristome with straight exostome teeth; endostome rudimentary; operculum rostrate. *Spores* 16–26 µm in diameter, irregularly granulate; covered by simple or compound granules (seen with SEM). *Calyptrae* cucullate, hairy. *Vaginula* with hyaline hairs on upper part. *Sporophytes* formed only along upper side of stem, hidden between coiled branches when dry, exposed in moist plants.

Habitat and Local Distribution. — Plants growing in patches on bark of trees, on hard limestone and basaltic rocks, mainly in northern exposure of forests; usually confined to the northern part of Israel. Occasional, locally abundant: Upper Galilee, Mount Carmel, Dan Valley, and Golan Heights.

General Distribution. — A widespread species. Recorded from Southwest Asia (Cyprus, Turkey, Lebanon, and Saudi Arabia), around the Mediterranean, Europe (throughout), Macaronesia, tropical and North Africa, Australia, New Zealand, and North and South America.

2. *Neckera* Hedw.

Primary stems often stoloniform. Secondary stems pinnately or bipinnately branched, complanate when dry and moist, decumbent to erect. *Leaves* inserted in eight rows, strongly complanate, asymmetrical, mostly undulate; costa single, double, or absent. *Dioicous* or autoicous. *Capsules* immersed to exserted, erect and symmetrical; peristome usually double, endostome without cilia. *Calyptrae* usually cucullate, mostly naked.

A worldwide genus of ca. 50 species (Enroth 1994). One species occurs in Israel.

Neckera complanata (Hedw.) Hübener, Muscol. Germ. 576 (1833).
[Bilewsky (1965): 411]
Leskea complanata Hedw, Sp. Musc. Frond. 231 (1801).

Distribution map 173, Figure 171 (p. 439), Plate XIII: b.

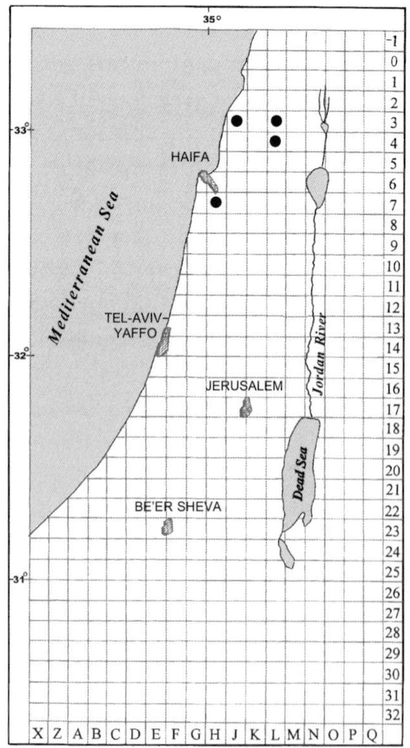

Map 173: *Neckera complanata*

Plants moderately robust, yellowish-green to pale green. *Secondary stems* pinnately branched; stoloniform branches or branches with stoloniform tips sometimes present. *Pseudoparaphyllia* present at base of branches. *Leaves* spreading when dry and moist, slightly undulate when dry, up to 1.5 mm long, asymmetrical, oblong-elliptical to lingulate, obtuse, acute, often abruptly apiculate; margins recurved on one side below, crenulate to finely toothed in upper ⅓ of leaf; costa faint, usually single, short, simple, forked, or absent; cells ± incrassate; basal cells ± vermicular; alar cells short, irregularly subquadrate; mid- and upper leaf cells hexagonal to rhomboidal, 8–10 µm wide in mid-leaf, near base of leaf elongate-rhomboidal; perichaetial leaves longer than stem and branch leaves, sheathing, acuminate, without costa. *Dioicous*. *Setae* yellowish, 4–7 mm long, ± flexuose; capsules exserted, brown, ovate-ellipsoid, 1–1.25 mm (without operculum). *Spores* varying in size, up to 26 µm in diameter; gemmate on granulate surface (seen with SEM). *Sporophytes* only rarely borne.

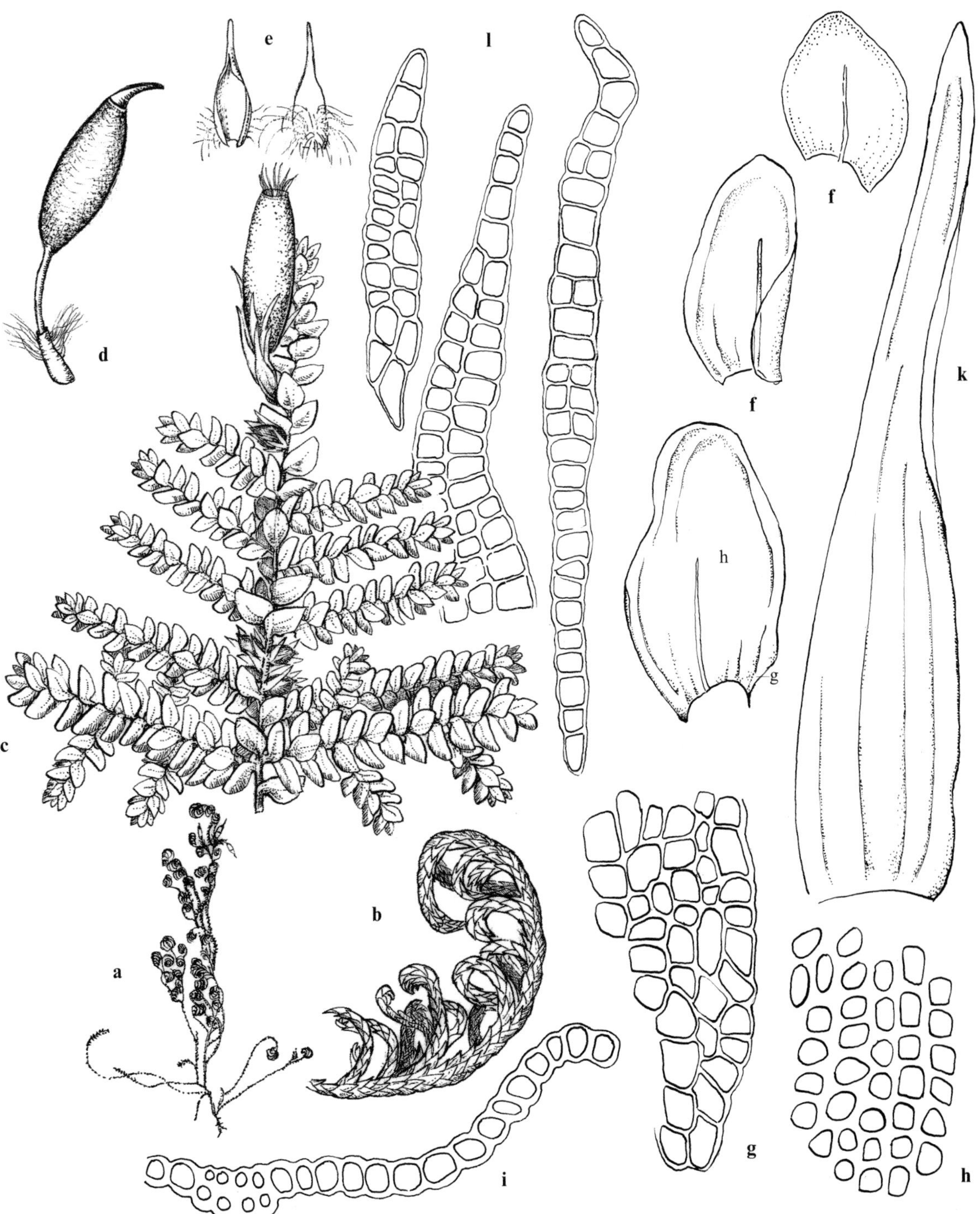

Figure 170. *Leptodon smithii*: **a** habit, dry (×1); **b** part of habit, dry (×10); **c** part of habit, moist (×10); **d** sporophyte (×10); **e** calyptrae (×10); **f** leaves (×50); **g** marginal basal cells; **h** mid-leaf cells (all leaf cells ×500); **i** cross-section of costa and portion of lamina (×500); **k** perichaetial leaf (×50); **l** paraphyllia (×500). (*Herrnstadt* & *Crosby 78-21-5*)

Habitat and Local Distribution. — Plants growing in flat mats on soil and rocks in shaded habitats. Rare, locally abundant: Mount Carmel and Upper Galilee (on Mount Carmel found being used for building nests by the local bird *Troglodytes troglodytes*).

General Distribution. — Recorded from Southwest Asia (Turkey and Iran), around the Mediterranean, Europe (throughout), Macaronesia, North Africa, Madagascar, northeastern and central Asia (Turkey and Iran), and North America.

No specimen of *Neckera complanata* var. *tenella* Schimp., which was mentioned by Bilewsky (1965), could be found in his herbarium.

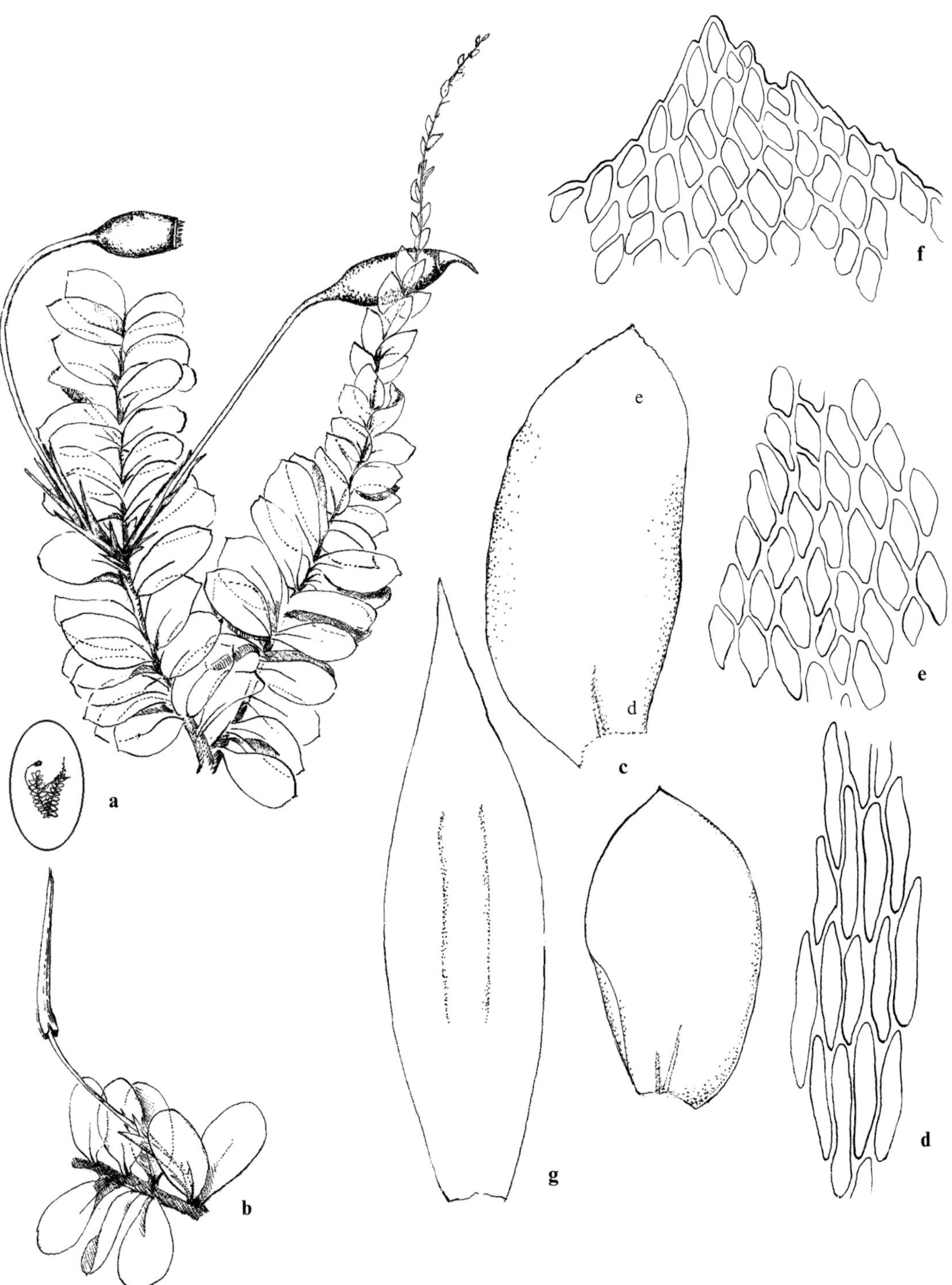

Figure 171. *Neckera complanata*: **a** part of habit, moist (×10); **b** branch with young sporophyte (×10); **c** leaves (×50); **d** basal cells; **e** upper cells; **f** apical cells (all leaf cells ×500); **g** perichaetial leaf (×50). (*Shahar*, 4 Jan. 1981)

THAMNOBRYACEAE (THAMNIACEAE)

Plants pleurocarpous, moderate to robust, growing usually in ± dull mats or patches. *Stems* with central strand; primary stems creeping, usually stoloniform; secondary stems erect, unbranched below, covered by remotely spaced scarious leaves on its lower part, branched above, often dendroid, sometimes flagelliform at branch tips. *Paraphyllia* absent. *Leaves* with acute to obtuse apex; costa ± strong, single, ending below apex; cells short, smooth or papillose. *Capsule* exserted, ± inclined, symmetrical or asymmetrical; annulus usually differentiated; peristome double, well developed, cilia present, often appendiculate; operculum rostrate. *Calyptrae* cucullate, naked, smooth.

A family comprising ca. five genera in tropical and temperate zones, often in wet habitats. One species occurs in the local flora.

The Thamnobryaceae are considered either as part of the Neckeraceae or as a separate family. In a recent classification of Buck & Vitt (1986), the two families are included in two separate orders (Leucodontales and Hypnales, respectively). On the other hand, Enroth (1994) proposed to retain the "thamnoid" plants in the Neckeraceae because of the absence of a clear demarcation between the two families (see also Smith *in* Sharp *et al.* 1994). The "Thamnobryaceae" are considered by Enroth as the primitive component of the family with a combination of the primitive sporophytic characters found in the Neckeraceae, which are linked with some primitive gametophytic characters, e.g., short leaf cells.

Thamnobryum Nieuwl.

Plants moderately robust to robust, rarely slender, dendroid. *Paraphyllia* absent. *Leaves* ovate-lanceolate to ovate-oblong, acute; margins plane, entire below, dentate above. *Dioicous*. *Capsules* usually inclined to horizontal, subcylindrical, slightly asymmetrical; annulus differentiated.

A genus comprising 35 species (Enroth 1994). One species occurs in the local flora.

Literature. — Buck & Vitt (1986), Enroth (1994), Smith (*in* Sharp *et al.* 1994).

Thamnobryum alopecurum (Hedw.) Nieuwl., Amer. Midl. Naturalist 5 : 50 (1917).
Hypnum alopecurum Hedw., Sp. Musc. Frond. 267 (1801); *Thamnium alopecurum* (Hedw.) B.S.G., Bryol. Eur. 5 : 214 (1852).

Distribution map 174 (p. 442), Figure 172.

Plants slender to moderately robust, often dendroid, dark green to yellowish-green. *Secondary stems* sometimes flagelliform. *Leaves* on stem distinctly secund when dry, on lower unbranched part of secondary stem scarious, triangular, without costa; branch leaves densely spaced above, imbricate and secund when dry, patent when moist, ± concave, up to 1.5 mm long, often asymmetric, ovate, obtuse to acute; margins plane, denticulate to dentate from below mid-leaf to apex; costa sparsely dentate on upper dorsal end, abruptly ending below apex, often ending in a distinct dorsal projection; cells ± incrassate; basal cells rectangular to quadrate; mid- and upper leaf cells oblong-hexagonal, 6–9 µm wide in mid-leaf, shorter towards apex; one–two rows of

Figure 172. *Thamnobryum alopecurum*: **a** habit, moist (×1); **b** part of habit, moist (×10); **c** branches, moist and dry (×10); **d** leaves (×50); **e** basal cells; **f** mid-leaf cells; **g** apical cells (all leaf cells ×500); **h** scarious lower leaves (×50). (*Herrnstadt*, 13 May 1988)

small irregularly quadrate cells at leaf margins. *Sporophytes* not found in local material.

Habitat and Local Distribution. — Plants growing in dense mats or patches in wet habitats, submerged or growing under spray, on moist shaded rocks, boulders or tree roots, on banks of running water. Rare: southern Golan Heights.

General Distribution. — Recorded from Southwest Asia (Turkey and Iran), around the Mediterranean, Europe (widespread throughout), Macaronesia, North Africa, Asia (throughout, except the southeast).

Thamnobryum alopecurum is usually considered a most variable species, a characteristic often observed in mosses growing in wet habitats and undergoing temporary dry periods.

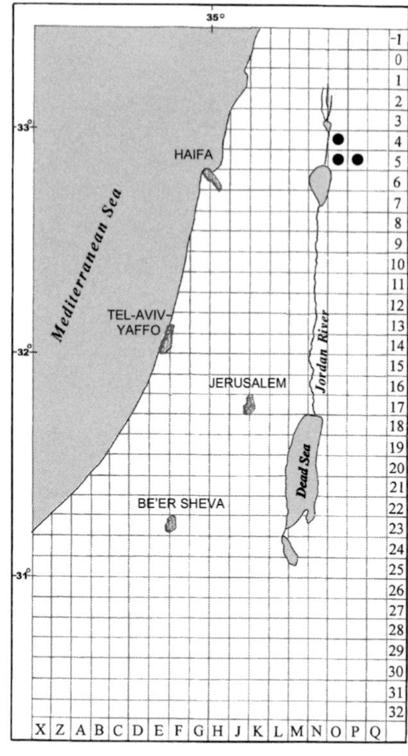

Map 174: *Thamnobryum alopecurum*

HYPNALES

FABRONIACEAE

Plants pleurocarpous, small and slender, growing in green to yellowish mats. *Stems* creeping, sparsely branched; central strand usually absent. *Leaves* of stems and branches usually similar; costa single, short, usually not reaching apex; mid- and upper leaf cells smooth; upper cells hexagonal to rhomboidal; alar cells usually quadrate. *Setae* elongate, smooth; capsules erect and symmetric; annulus absent or poorly differentiated; peristome double (basal membrane sometimes present, cilia absent), single, or absent; operculum usually convex-conical, mammillate, sometimes rostrate. *Calyptrae* cucullate, smooth and naked.

A family widespread mainly in the warmer regions of America and Africa; represented in the local flora by a single genus: *Fabronia*.

The Fabroniaceae are variously delimited from related families. According to Buck & Crum (1978) the family includes eight genera, which are described as small pleurocarps with "non-hypnoid" peristome (see also Buck *in* Sharp *et al.* 1994).

Literature. — Buck & Crum (1978), Buck (*in* Sharp *et al.* 1994).

Fabronia Raddi

Plants very small and slender, growing in soft silky mats. *Stems* with clustered rhizoids (in local species). *Leaves* appressed when dry, erect-spreading, sometimes subsecund when moist, lanceolate, oblong to ovate, gradually or abruptly acuminate, often with long and thin acumen; margins plane, entire, distinctly serrate, or with short to long cilia-like teeth; costa slender, not exceeding ⅔ the length of lamina; quadrate alar cells always present, sometimes extending towards costa; mid- and upper leaf cells rhomboidal to elongate-hexagonal, mostly ± thin walled; perichaetial leaves wider than vegetative leaves. *Paraphyllia* usually absent. *Autoicous. Setae* erect, smooth; capsules ovoid to short-pyriform; neck inconspicuous; annulus absent; operculum apiculate to mammillate; peristome single, rarely absent; teeth 16, fused in pairs in young capsules, incurved when moist, spreading when dry.

A genus comprising ca.100 species. Two species occur in the local flora.

The two local species — *Fabronia pusilla* and *F. ciliaris* — belong to a group of extremely difficult species. They are mainly delimited by the different dentation of their leaf margins. The validity of this character is often doubted (e.g., Agnew & Vondráček 1975, Buck *in* Sharp *et al.* 1994). Grout (1934) suggested to consider *F. pusilla* as a subspecies of *F. ciliaris*. The two taxa, as described below, may represent extreme forms of one complex since some plants with intermediate characters have been found in local populations.

1 Leaf margins with long, cilia-like teeth; teeth often consisting of more than one cell at base **1. F. pusilla**
1 Leaf margins with short, not cilia-like teeth; teeth usually consisting of only one cell at base **2. F. ciliaris**

1. Fabronia pusilla Raddi, Atti Accad. Sci. Siena 9 : 231 (1808). [Bilewsky (1965) : 413]
F. pusilla var. *ciliata* Lesq. & James, Man. Mosses N. America 294 (1884). — Bilewsky (1974); *F. pusilla* var. *schimperi* Venturi ex Limpr., Laubm. Deutschl. 2 : 728 (1895). — Bilewsky (1965) : 413.

Distribution map 175, Figure 173.

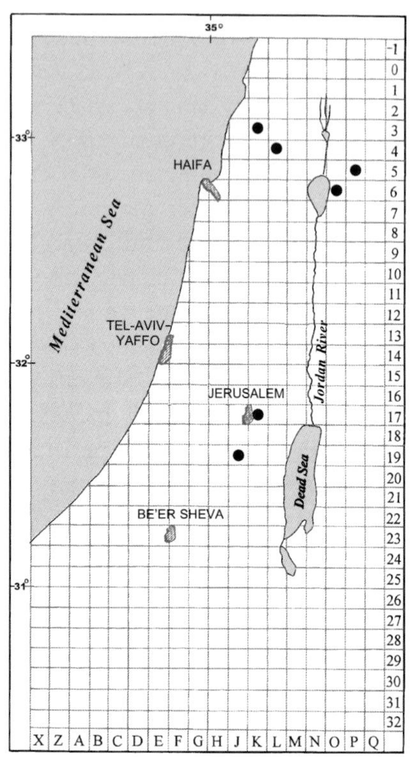

Map 175: *Fabronia pusilla*

Plants yellowish to greyish green. *Leaves*: stem leaves in some plants larger than branch leaves; branch leaves lanceolate to ovate-lanceolate, varying among populations in size, shape, and length of leaf-point, 0.5–1.0 mm long (including leaf-point), abruptly to gradually long-acuminate; margins with long cilia-like teeth, often consisting of more than one cell at base, up to 80 µm long; costa usually weak, from very short to up to half the length of lamina, gradually vanishing; mid-leaf cells 8–12 µm wide.

Habitat and Local Distribution. — Plants growing in shade on rocks and tree trunks (found on *Olea europaea* and *Quercus* spp.), frequently bearing sporophytes. Occasional, perhaps locally abundant in some localities: Upper Galilee, Judean Mountains, and Golan Heights.

General Distribution. — Recorded from Southwest Asia (Cyprus, Turkey, Iraq, and Afghanistan), around the Mediterranean, Europe (especially the south), Macaronesia, North Africa, northeastern and central Asia, and North, Central, and Caribbean America.

In two localities (Judean Mountains: env. of Jerusalem, and Upper Galilee: Adamit) plants of *Fabronia pusilla* have leaves with exceptionally long acumen and marginal teeth, sometimes with a weak and short costa. Such plants were mentioned by Bilewsky (1965) — without specifying localities — as "var. *schimperi* (De Not.) Vent.", which is a synonym of *F. pusilla* var. *ciliata*, and agree with the description of var. *ciliata*. It seems, however, that some of those diagnostic characters are most variable and often may occur independently of others. Therefore, it is rather difficult to delimit var. *ciliata* from the typical *F. pusilla*, and it is preferred here not to consider *F. pusilla* as comprising two separate taxa.

Figure 173. *Fabronia pusilla*: **a** habit, moist (×10); **b** habit, dry (×10); **c** sporophyte with calyptra (×10); **d** stem leaf (×50); **e** alar cells; **f** mid-leaf cells; **g** marginal upper cells (all leaf cells ×500); **h** branch leaf (×50); **i** peristome (×50); **k** peristome teeth (×500); **l** leaves ("var. *ciliata*" type) (×50). (**a**–**k**: *Bilewsky* Apr. 1952, No. 126 in Herb. Bilewsky; **l**: *Bilewsky* 16 Apr. 1960, No. 127 in Herb. Bilewsky)

2. Fabronia ciliaris (Brid.) Brid., Bryol. Univ. 2:171 (1827). [Bilewsky (1974)]
Hypnum ciliaris Brid., Spec. Musc. 2:155 (1812); *F. octoblepharis* Schwägr., Spec. Musc., Suppl. 1:338 (1816). — Bilewsky (1965):414.

Distribution map 176, Figure 174, Plate XIII:c.

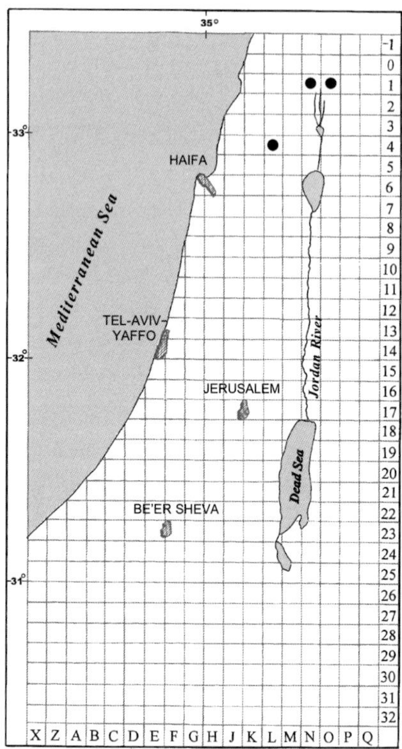

Map 176: *Fabronia ciliaris*

Fabronia ciliaris closely resembles the locally more common *Fabronia pusilla.* Plants differ mainly in the teeth of the leaf margins. *Leaf margins* with teeth, teeth usually consisting of one cell at base, usually shorter than in *F. pusilla*, only sometimes reaching up to 27 µm in length. *Spores* 11–15 µm in diameter, papillose; densely gemmate, processes greatly varying in size (seen with SEM).

Habitat and Local Distribution. — To date, only three local collections of Bilewsky taken off tree trunks (*Olea europaea* and *Quercus* spp.) were confirmed as *Fabronia ciliaris*: one from Upper Galilee and two from Dan Valley; in one of those collections (Upper Galilee: near Meron), plants of *F. ciliaris* grow together with *F. pusilla.*

General Distribution. — Recorded from Southwest Asia (Cyprus, Turkey, and Jordan), around the Mediterranean, Europe (scattered), North Africa (doubtful; cf. Duell 1992), Asia (throughout, except the southeast), Australia, New Zealand, Oceania, and North, Central, and South America (perhaps additional records of the species are included in *Fabronia pusilla*, and vice versa).

Our plants generally agree with the specimen collected in Helvetia in 1807, which is cited by de Bridel (1812) as *Hypnum ciliaris*, and deposited in BM. However, in that specimen the teeth are shorter than in the local plants, and all consist of only one cell.

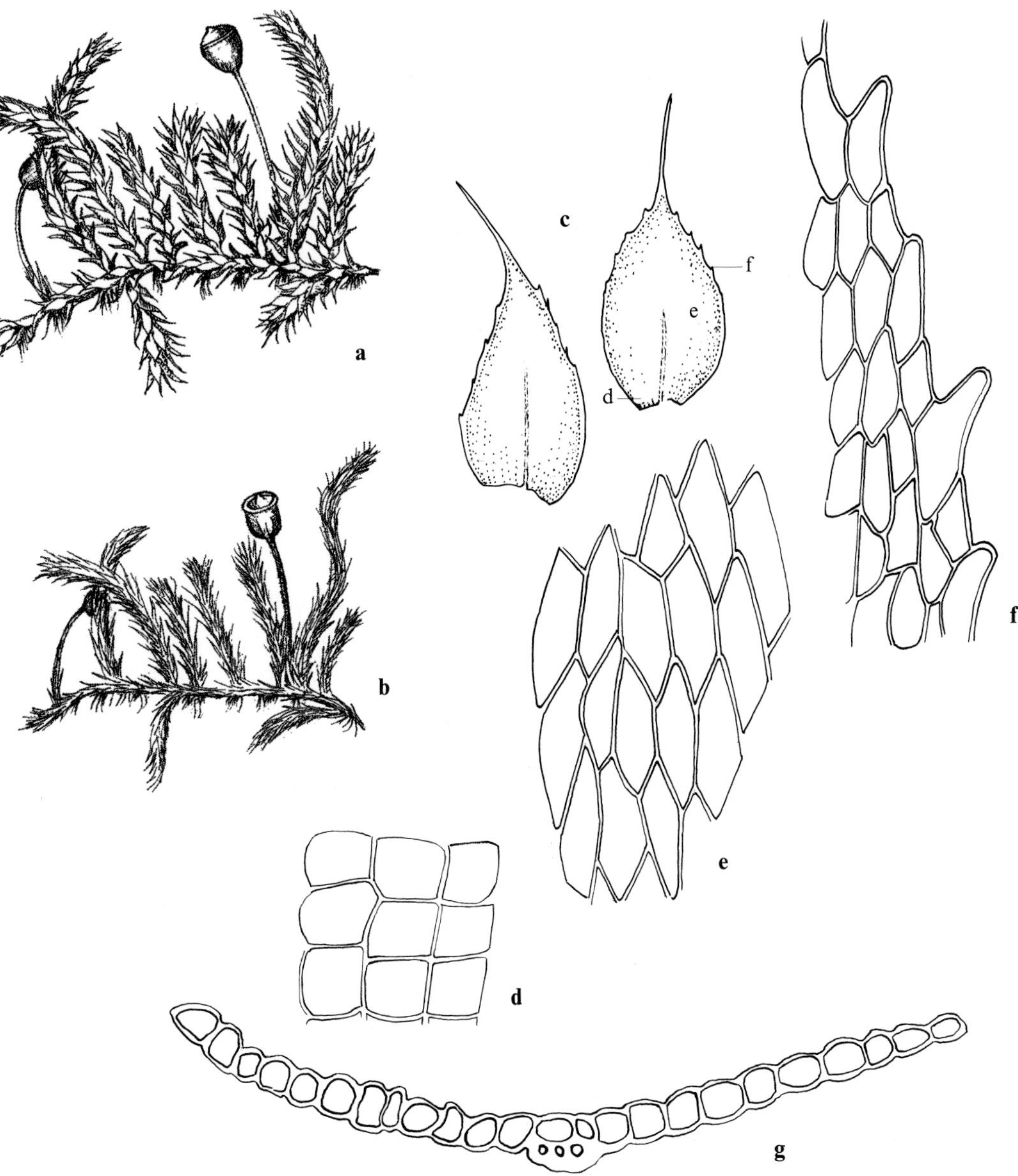

Figure 174. *Fabronia ciliaris*: **a** habit, moist (×10); **b** habit, dry (×10); **c** leaves (×50); **d** alar cells; **e** mid-leaf cells; **f** marginal upper cells (all leaf cells ×500); **g** cross-section of leaf (×500). ('En Meron, *Bilewsky*, *sine dato*)

AMBLYSTEGIACEAE

Plants pleurocarpous, slender to robust, in dense or lax mats, sometimes in tufts. *Stems* creeping to erect, irregularly branched to pinnate; central strand often present. *Leaves*: stem and branch leaves similar in shape; branch leaves usually smaller; costa often strong, single, rarely double or absent; leaf cells mostly smooth, oblong-hexagonal to linear, alar cells often ± distinctly differentiated at basal angles; perichaetial leaves different from stem leaves. *Setae* elongate, usually reddish, smooth; capsules inclined to horizontal, asymmetrical, often distinctly curved, oblong-ellipsoid to cylindrical, with well developed neck, constricted below mouth; annulus usually differentiated; peristome double; exostome teeth 16, usually yellowish-brown; endostome with a high basal membrane, segments 16, broad, sometimes perforated, usually with well developed cilia; operculum shortly conical, usually mammillate or apiculate, rarely short-rostrate. *Calyptrae* cucullate, smooth.

A family that is variously circumscribed and whose members occupy wet habitats; represented in the local flora by three genera: *Cratoneuron*, *Amblystegium*, and *Leptodictyum*.

Literature. — Kanda (1975), Hedenäs (1989), Ochyra (1989).

1 Paraphyllia present; leaves with distinct, inflated alar cells forming auricles **1. Cratoneuron**

1 Paraphyllia usually absent; leaves with alar cells not or hardly distinct, not forming auricles.

2 Plants small and slender; leaf cells two–six times longer than wide **2. Amblystegium**

2 Plants moderately sized; leaf cells six–fifteen times longer than wide **3. Leptodictyum**

1. *Cratoneuron* (Sull.) Spruce

Plants usually moderately robust, growing in lax mats or tufts. *Stems* regularly or irregularly pinnately branched, procumbent to erect, with central strand. *Paraphyllia* present, often abundant. *Leaves* often falcate-secund, usually plicate, broadly ovate to ovate-lanceolate, acuminate, mostly with decurrent base; margins usually plane, entire or serrulate; costa single, sometimes very strong, reaching up to the base of acumen or above; cells shortly oblong-rhomboidal to long and linear, ± incrassate, smooth; alar cells distinct, inflated, thin- or thick-walled forming auricles; perichaetial leaves erect, elongate, lanceolate, acuminate. *Dioicous*. *Capsules* recorded as inclined or horizontal, curved, asymmetrical, often contracted below mouth when dry and empty, annulus present, operculum conical.

The genus is often considered to belong to a separate family: Cratoneuraceae Mönk. (e.g., by Bilewsky 1965, Ochyra 1989, Frahm 1995). Ochyra (1989) proposed to retain only *Cratoneuron filicinum* in *Cratoneuron* as opposed to van der Wijk *et al.* (1959–1969), who recorded 18 species in this genus.

1. Cratoneuron filicinum (Hedw.) Spruce, Cat. Musc. Amaz. And. 21 (1867).
[Bilewsky (1970)]
Hypnum filicinum Hedw., Sp. Musc. Frond. 285 (1801).

Distribution map 177, Figure 175 (p. 451).

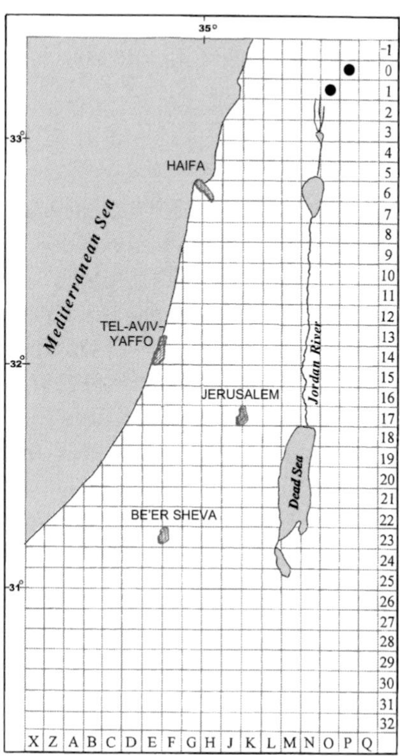

Map 177: *Cratoneuron filicinum*

Plants moderately robust, green to yellowish-green. *Paraphyllia* few to numerous, leaf-like, without costa, denticulate at margin. *Leaves* falcate-secund; stem leaves 1(–1.2) mm long, branch leaves similar in shape but smaller, 0.8–1.0 mm long, ovate-lanceolate, gradually tapering into straight or curved acumen, strongly decurrent; margins serrulate all around; costa strong, reaching base of acumen to percurrent, in cross section biconvex and composed of uniform, strongly incrassate, small cells; lamina cells short, in mid-leaf irregularly-rhomboidal, 7–10 μm wide, two–four times longer than wide; alar cells incrassate, inflated, forming decurrent auricles. *Sporophytes* very rare in local plants.

Habitat and Local Distribution. — Plants of wet habitats, growing in dense mats on damp soil of river banks and on rocks occasionally submerged in running water; found growing together with *Amblystegium tenax*. Rare: Dan Valley (locally abundant), Mount Hermon, and Golan Heights.

General Distribution. — Recorded as a widespread and nearly cosmopolitan species, in particular of temperate regions.

Plants are easily recognised by the leaves with distinct alar cells and by the curved shape of stem and branch apices, which are formed by the falcate-secund leaves.

2. *Amblystegium* B.S.G.

Plants small and slender, growing in mats. *Stems* creeping, irregularly branched, with central strand; branches decumbent to erect. *Paraphyllia* absent (sometimes pseudoparaphyllia present at branch base). *Leaves* erectopatent to spreading, rarely subsecund or ± complanate, ovate to lanceolate, gradually or ± abruptly acuminate; margins plane, entire or ± serrulate; costa slender to strong, reaching mid-leaf or above; cells oblong to rhomboidal or hexagonal, rarely linear, two–six times longer than wide, smooth; alar cells sometimes slightly differentiated. *Autoicous*. *Spores* nearly smooth to minutely papillose.

A genus of worldwide distribution in which up to 95 species, occupying wet habitats, have been recorded. The genus is treated in a wider or narrower sense, mainly by including or separating *Hygroamblystegium* Loeske or *Leptodictyum* (Schimp.) Warnst. (e.g., compare Smith 1978, Sharp & Crum *in* Sharp *et al.* 1994, and Frahm 1995).

1 Costa less than ¾ the length of leaf **3. A. serpens**
1 Costa longer, from ¾ the length of leaf to excurrent.
  2 Stem leaves loosely appressed and slightly incurved when dry; costa usually strong, over 35 µm wide **1. A. tenax**
  2 Stem leaves patent to spreading when dry; costa less than 35 µm wide **2. A. varium**

Five local species were included in *Amblystegium* in Herrnstadt *et al.* (1991). Among those, *A. riparium* is considered here as *Leptodictyum riparium*. *Amblystegium compactum* (Müll. Hal.) Austin — syn. *Conardia compacta* (Müll. Hal.) H. Rob. — was recorded there from the Upper Galilee and the Golan Heights. Re-examination of the collections has shown that the collection from the Upper Galilee (Nahal Keziv, *Nachmony*, 19 Mar. 1955) is a misidentified *Rhynchostegiella letourneuxii* and the plants of the sterile collection from the Golan Heights (Tell el Qadi, *Kushnir*, 8 Apr. 1942) somewhat resemble *Amblystegium compactum* in the serrulate leaves and the shape of pseudoparaphyllia. Propagules resembling those described as borne in clusters on the dorsal side of the costa were found among the plants but detached from the leaves. However, after evaluating the data, this single collection seems rather scanty and consequently the species is not included here.

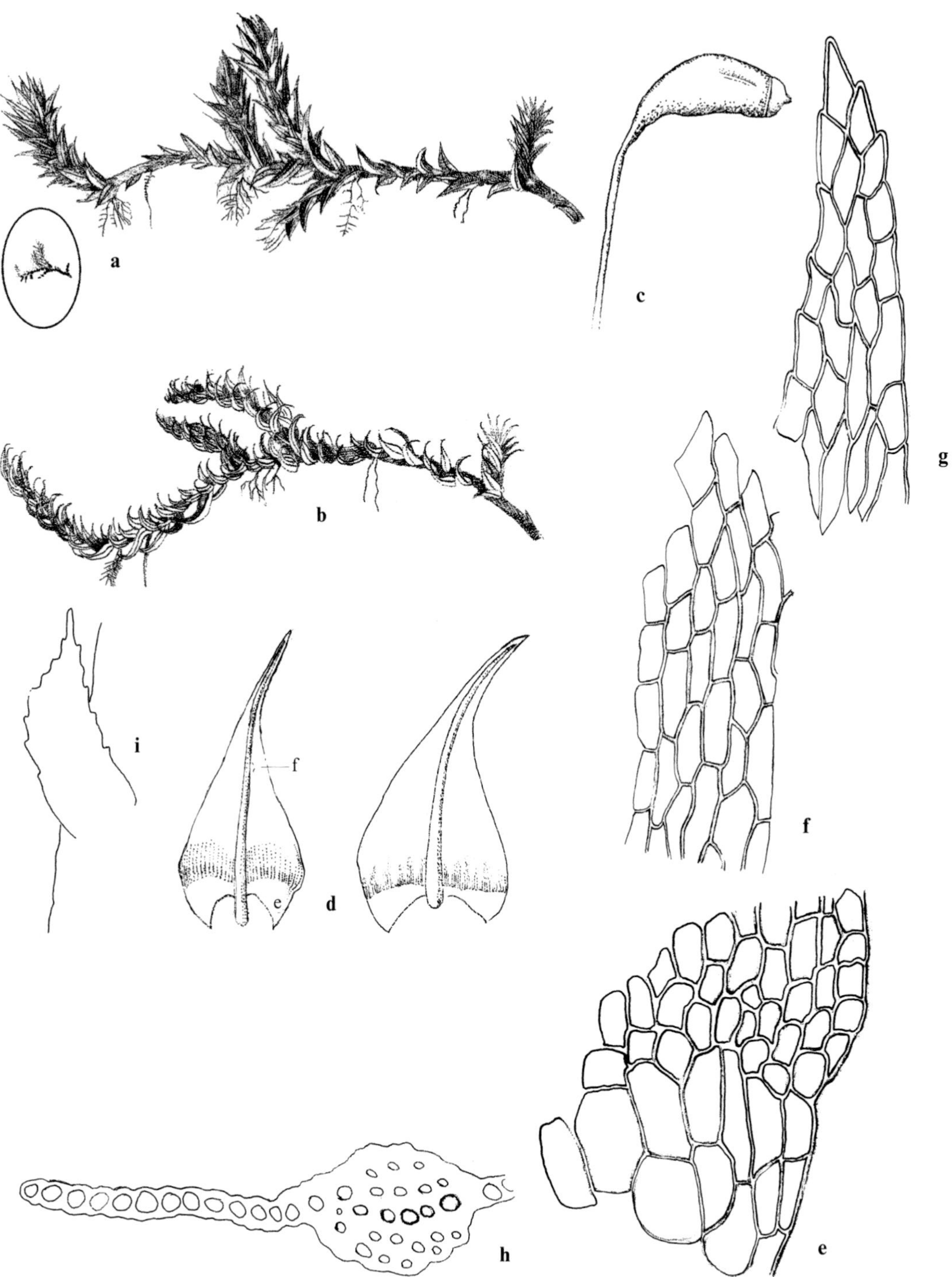

Figure 175. *Cratoneuron filicinum*: **a** habit, moist (×10); **b** habit, dry (×10); **c** capsule and part of seta (×10); **d** leaves (×50); **e** cells of leaf base with alar cells ; **f** upper cells; **g** apical cells (all leaf cells ×500); **h** cross-section of costa and portion of lamina (×500); **i** paraphyllium (×125). (**a**, **b**, **d**–**i**: *Ben-Shaul*, 5 Aug. 1956; **c**: *Herrnstadt* & *Crosby 78-29-27*)

1. Amblystegium tenax (Hedw.) C. E. O. Jensen, Skand. Bladmossfl. 483 (1939).
Hypnum tenax Hedw., Sp. Musc. Frond. 277 (1801); *Hygroamblystegium irriguum* (Hook. & Wilson) Loeske, Moosfl. Harz. 299 (1903). — Bilewsky (1965):417; *Hygroamblystegium tenax* (Hedw.) Jenn., Man. Mosses W. Pennsylvania 227 (1913).

Distribution map 178, Figure 176.

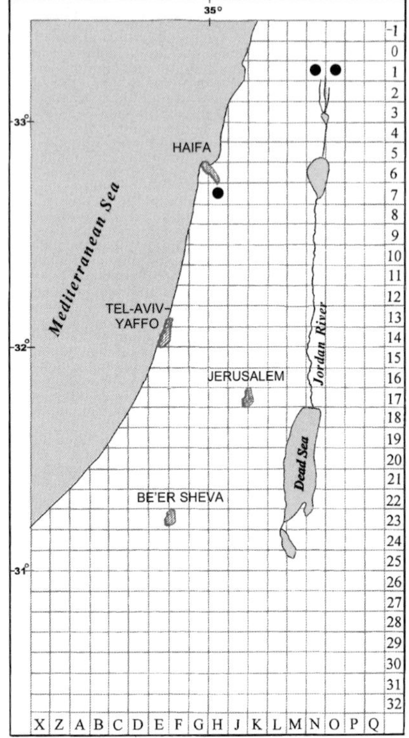

Map 178: *Amblystegium tenax*

Plants small and slender, irregularly pinnately branched. *Leaves* loosely appressed and slightly incurved when dry, erectopatent to spreading, sometimes subsecund when moist, 0.75–1.25 mm long, ovate to ovate-lanceolate, acute to acuminate; margins plane, entire or serrulate above; costa disappearing near the apex, usually strong, over 35 µm wide at leaf base, sometimes flexuose above; basal cells irregularly quadrate to rectangular with two–three rows of enlarged cells extending to basal angles; mid-leaf cells rhomboidal, 8–10 µm wide, two–four times longer wide; cells in acumen sometimes larger, rhomboidal. *Sporophytes* not seen in local plants.

Habitat and Local Distribution. — Plants growing on rocks in running water (sources of the Dan River), or on the bank of a seasonal stream (Mount Carmel). Rare: Mount Carmel and Dan Valley (locally abundant).

General Distribution. — Recorded from Southwest Asia (Turkey, Syria, Lebanon, and Iran), around the Mediterranean, Europe (throughout), Macaronesia, tropical and North Africa, northeastern, eastern, and central Asia, and North and Central America.

Amblystegium tenax is sometimes included in the genus *Hygroamblystegium* Loeske as *H. tenax* or *H. irriguum* (e.g., by Sharp & Crum *in* Sharp *et al.* 1994).

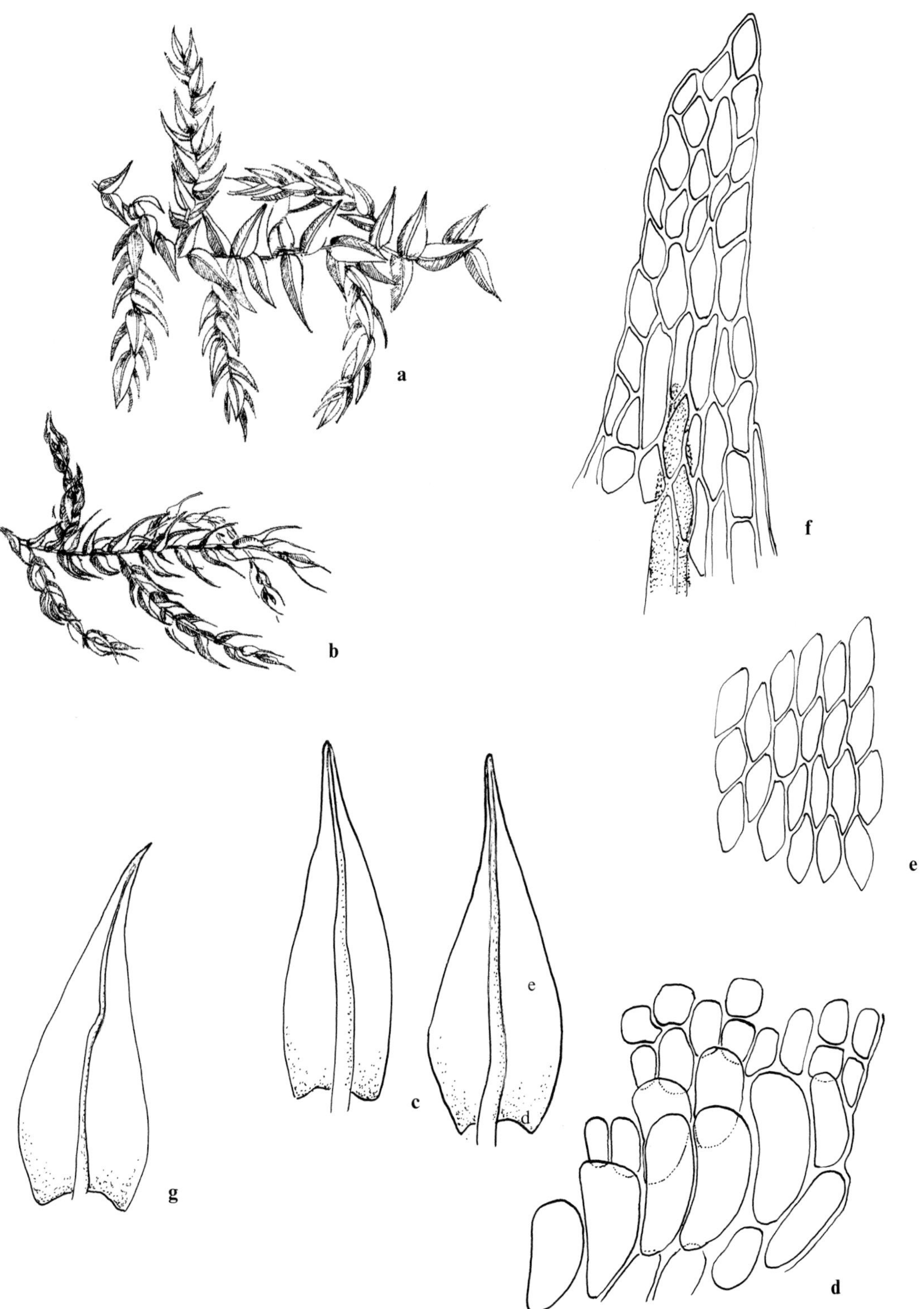

Figure 176. *Amblystegium tenax*: **a** habit, moist (×10); **b** habit, dry (×10); **c** branch leaves (×50); **d** basal cells; **e** mid-leaf cells; **f** apical cells (all leaf cells ×500); **g** branch leaf (×50). (**a**–**f**: 'En Dan, *Hefez*, *sine dato*; **g**: *Nachmony*, May 1954, No. 130 in Herb. Bilewsky)

2. Amblystegium varium (Hedw.) Lindb., Musci Scand. 32 (1879).
Leskea varia Hedw., Sp. Musc. Frond. 216 (1801).

Distribution map 179, Figure 177.

Plants small, irregularly branched, dull yellowish-green. *Leaves* patent to spreading when dry, hardly changing when moist; stem leaves 1.0–1.25 mm long, ovate-lanceolate, gradually tapering into a long acumen; margins plane, ± entire; costa strong, disappearing in the acumen, 27–35 μm wide at leaf base, often flexuous above; basal cells irregularly quadrate to ± rectangular, with one basal row of enlarged cells, not extending to leaf angles; mid-leaf cells hexagonal-rhomboidal, 8–10 μm wide, two–four times longer than wide, apical cells sometimes larger; branch leaves smaller than stem leaves, ca. 0.5 mm long, more loosely arranged, slightly incurved when dry, with costa usually narrower and shorter, ending below apex or excurrent. *Setae* 10–20 mm long; capsules inclined, curved, ca. 3 mm long including operculum, ellipsoid to cylindrical; operculum conical, acute.

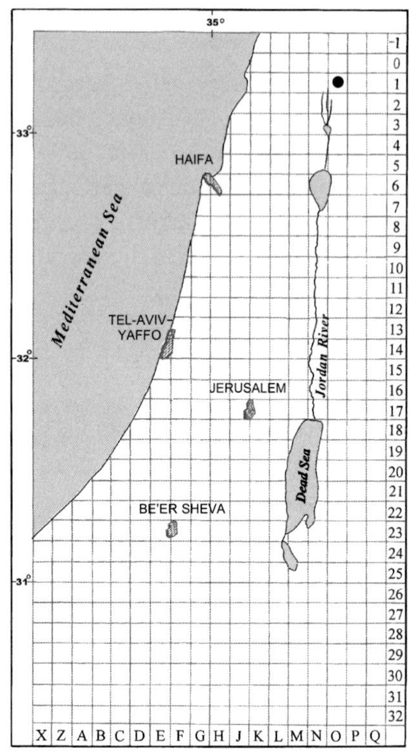

Map 179: *Amblystegium varium*

Habitat and Local Distribution. — Plants forming irregularly branched tufts, greatly varying in size and shape of leaves, on the bank of a seasonal stream. To date, collected only twice in the Dan Valley.

General Distribution. — Recorded from Southwest Asia (Turkey, Lebanon, Iraq, Iran, and Afghanistan), around the Mediterranean, Europe (throughout), Macaronesia, North Africa, northeastern, eastern, and central Asia, and North, Central, and Caribbean America.

Amblystegium varium closely resembles *A. tenax* and the discrimination between them may often be very difficult. The most reliable differential characters seem to be the shorter and narrower costa of branch leaves and the single row of enlarged cells at leaf base in *A. varium*. Nevertheless, these characters also seem to have a wide range of variation.

The previous record of *Amblystegium varium* from Mount Hermon (Herrnstadt *et al.* 1991), following Bilewsky & Nachmony (1955), was found to be based on a misidentified specimen of *Cratoneuron filicinum*.

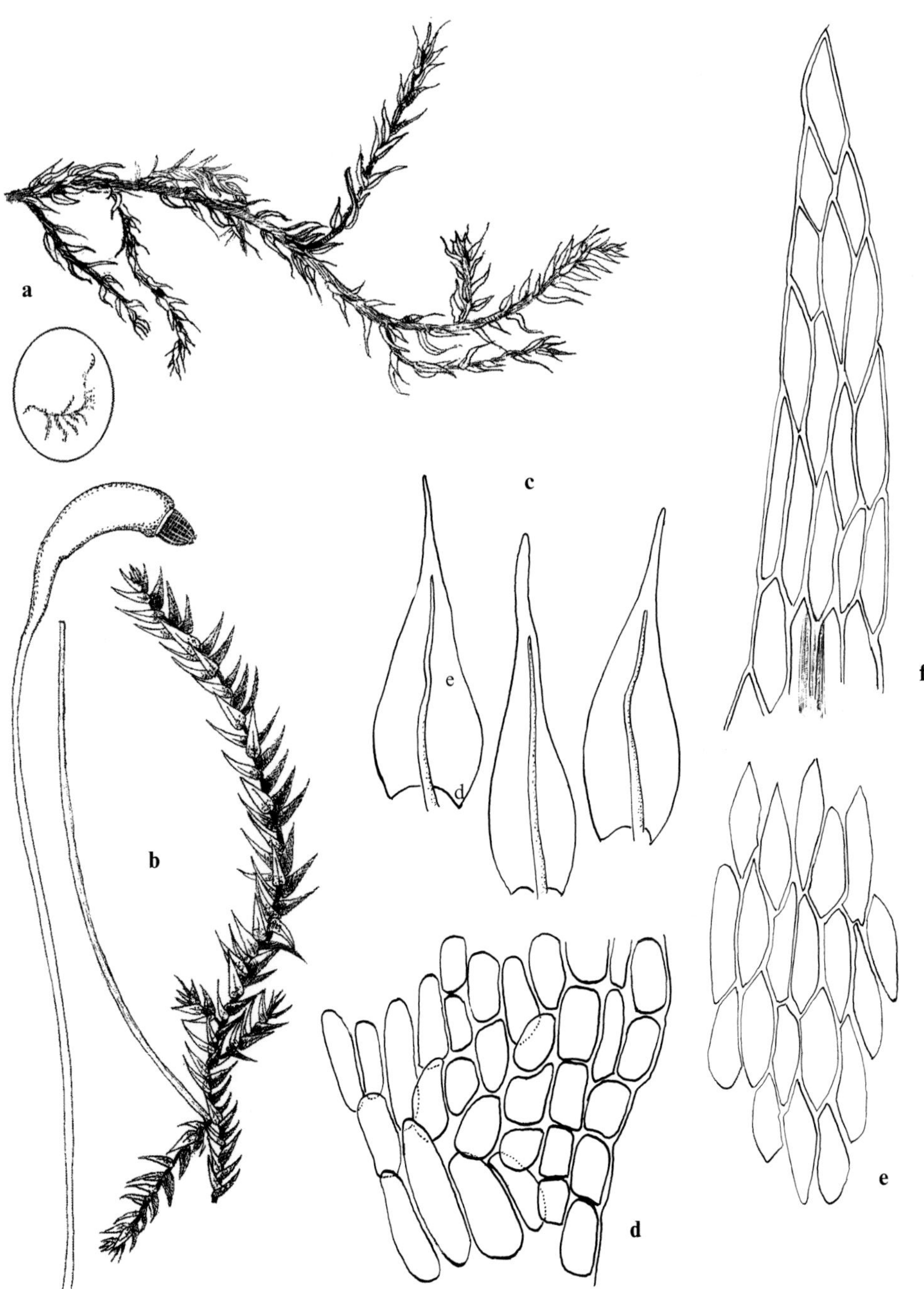

Figure 177. *Amblystegium varium*: **a** habit, dry (×10); **b** habit, moist, sporophyte (×10); **c** leaves (×50); **d** basal cells; **e** mid-leaf cells; **f** apical cells (all leaf cells ×500). ('En Leshem, *Hefez*, 1986)

3. Amblystegium serpens (Hedw.) B.S.G., Bryol. Eur. 6 : 53 (1853).
[Bilewsky (1965) : 417]
Hypnum serpens Hedw., Sp. Musc. Frond. 268 (1801); *A. juratzkanum* Schimp., Syn. Musc. Eur. 693 (1860).

Distribution map 180, Figure 178, Plate XIII : d.

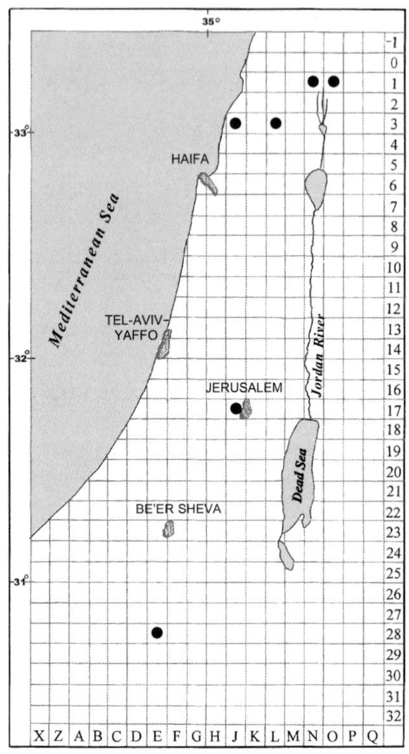

Map 180: *Amblystegium serpens*

Plants small and slender, irregularly pinnately branched, green to yellowish-green. *Leaves* erect to patent when dry, erectopatent to spreading when moist, ovate-lanceolate to lanceolate, tapering into a long acuminate apex; stem leaves similar to branch leaves in shape but ± larger, 1.0–1.2 mm long; branch leaves ca. 0.75 mm long; margins plane, subentire to serrulate all around; costa less than ¾ the length of leaf; basal cells subquadrate to rectangular; mid- and upper leaf cells rhomboidal to oblong-rhomboidal, three–six times longer than wide, in mid-leaf 8–11 μm wide, cells in acumen sometimes longer. *Setae* 14–20 mm long; capsules inclined, curved, 1.5–2.5 mm long including operculum, cylindrical; operculum conical, apiculate. *Spores* up to 17 μm in diameter; sparsely covered by irregular, small granulate clusters (seen with SEM).

Habitat and Local Distribution. — Plants growing in interwoven soft mats on rocks, stones and logs, in damp or shaded habitats, often near running water. Occasional: Upper Galilee, Judean Mountains, Central Negev, Dan Valley, and Golan Heights.

General Distribution. — Recorded as a nearly cosmopolitan species.

Amblystegium juratzkanum, which seems to differ from *A. serpens* only in the wide spreading leaves of dry plants (cf. Sharp & Crum *in* Sharp *et al.* 1994), is sometimes considered as a separate species or a variety of *A. serpens*.

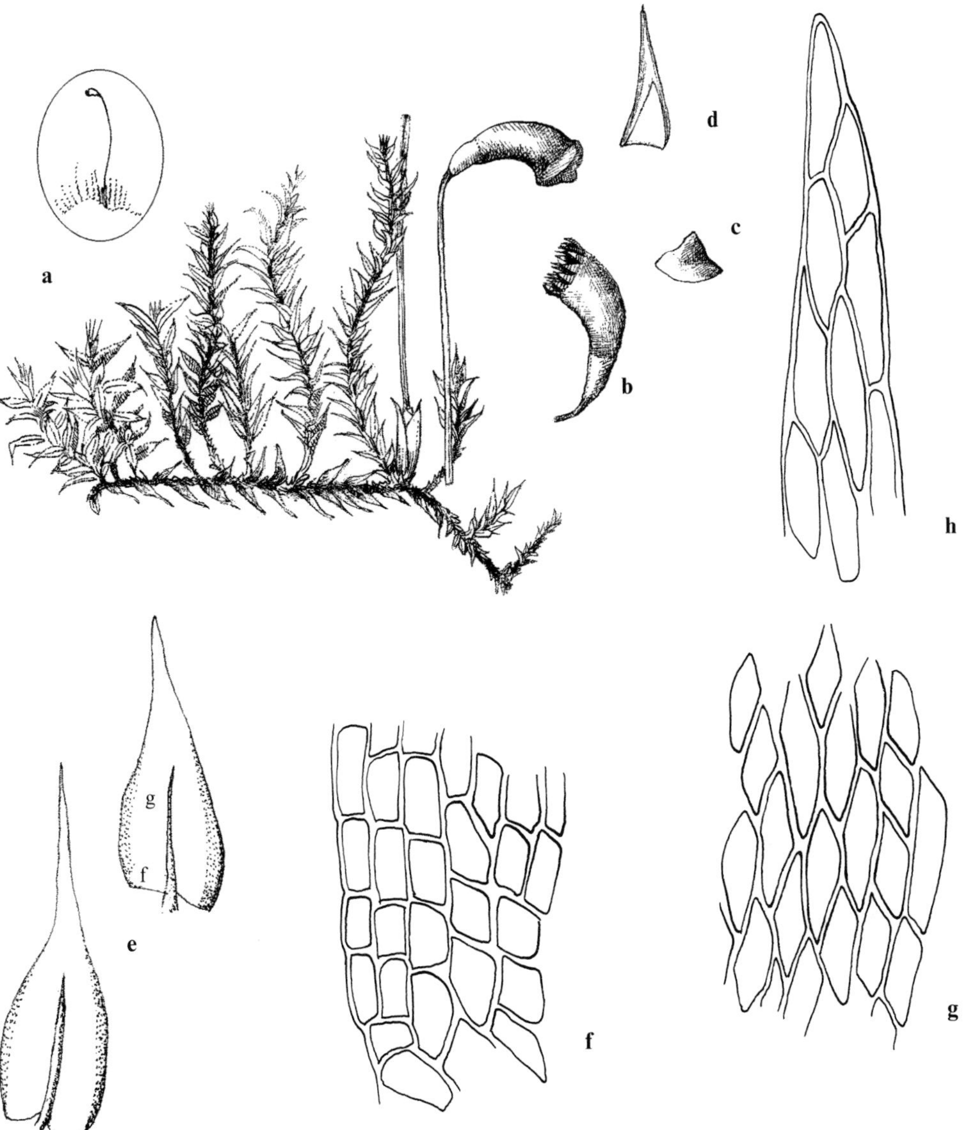

Figure 178. *Amblystegium serpens*: **a** habit, moist (×10); **b** deoperculate capsule (×10); **c** operculum (×10); **d** calyptra (×10); **e** branch leaves (×50); **f** basal cells; **g** mid-leaf cells; **h** apical cells (all leaf cells ×500). (*Herrnstadt* & *Crosby 78-29-19*)

3. *Leptodictyum* (Schimp.) Warnst.

Plants moderately sized, resembling *Amblystegium* in general habit. *Leaves*: cells above base oblong to narrow rhomboidal, six–fifteen times longer than wide.

A genus comprising ca. 12 species, often included in *Amblystegium* (also in Herrnstadt *et al.* 1991). One species occurs in the local flora.

Leptodictyum riparium (Hedw.) Warnst., Krypt.-Fl. Brandenburg 2 : 878 (1906). [Bilewsky (1965) : 417]

Amblystegium riparium B.S.G., Bryol. Eur. 6 : 58 (1853). — Herrnstadt *et al.* (1991); *Hypnum riparium* Hedw., Sp. Musc. Frond. 241 (1801).

Distribution map 181, Figure 179, Plate XIII : e.

Plants usually moderately robust, loosely branched, green or yellowish-green. *Leaves* about 2 mm long, ± laxly arranged, spreading, usually subcomplanate, rarely secund towards stem and branch tips, slightly twisted when dry, narrow-lanceolate to ovate-lanceolate, usually gradually tapering into a long acuminate apex; margins plane, entire; costa ½–¾ the length of leaf, weak or strong; basal cells wide-rectangular, at margins subquadrate to rectangular, lax; mid-leaf cells oblong to narrow-rhomboidal, 6–11 µm wide, six–fifteen times longer than wide; apical cells rhomboidal towards apex. *Setae* 10–20 mm long; capsules inclined, curved, ca. 2.5 mm long including operculum, ellipsoid to short-cylindrical; operculum conical-obtuse. *Spores* ca. 16 µm in diameter, slightly papillose; densely granulate to gemmate (seen with SEM).

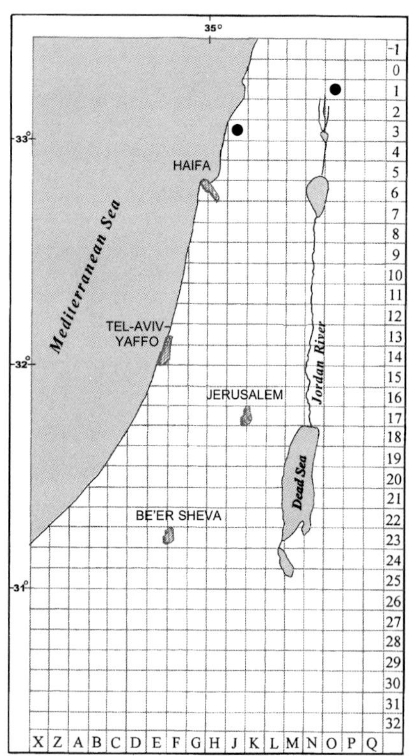

Map 181: *Leptodictyum riparium*

Habitat and Local Distribution. — Plants of wet habitats, growing in loose mats on rocks of stream banks, often temporarily submerged in running water. Rare, but locally abundant: Upper Galilee and Dan Valley.

General Distribution. — Recorded as a widespread, nearly cosmopolitan species.

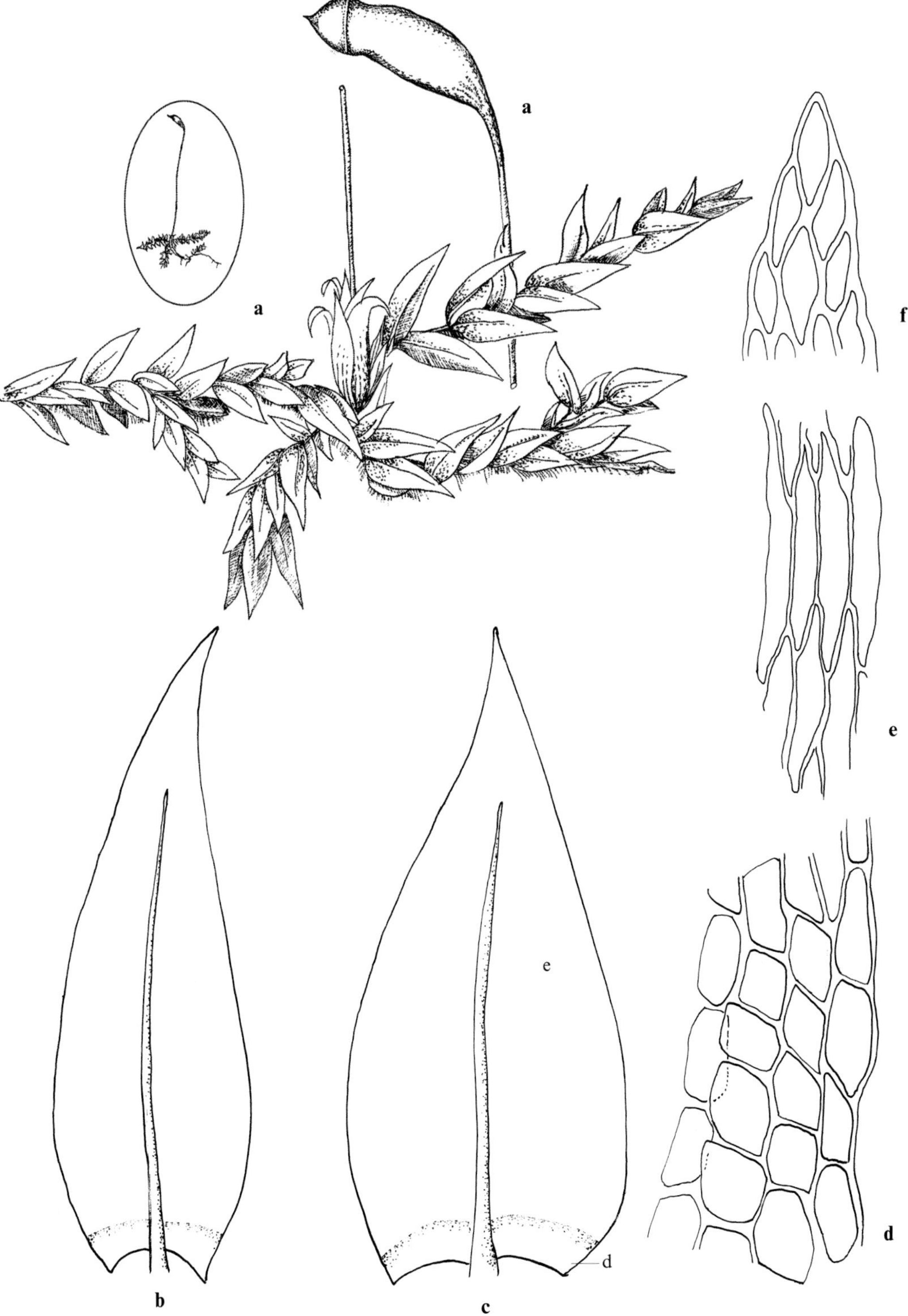

Figure 179. *Leptodictyum riparium*: **a** habit, moist (×10); **b** branch leaf (×50); **c** stem leaf (×50); **d** marginal basal cells; **e** mid-leaf cells; **f** apical cells (all leaf cells ×500). (*Ginzburg & Kushnir*, 21 Nov. 1943)

BRACHYTHECIACEAE

Plants pleurocarpous, small to large, slender to robust, growing in mats or occasionally in tufts. *Stems* mostly with central strand, creeping to ascending, rarely erect, irregularly to pinnately branched, branches straight or curved. *Paraphyllia* usually absent. *Leaves* imbricate to spreading, rarely complanate or ± secund, plane or concave, often plicate, usually broadly ovate to lanceolate, sometimes triangular or cordate, ± decurrent; apex obtuse to long-acuminate; margins plane or recurved at base, sometimes also above, entire to serrulate or serrate; costa single, rarely forked, composed of homogenous cells, often ending in a dorsal projection (rarely more than one projection present); cells rhomboidal to linear-flexuose, usually smooth; basal cells shorter, laxly arranged, sometimes porose, alar cells mostly subquadrate, not or little differentiated; stem and branch leaves similar or different in size or shape or both; perichaetial leaves differentiated. *Setae* elongate, smooth, or rough; capsules inclined, horizontal to cernuous, usually asymmetrical, ± curved, rarely symmetrical, erect, ovoid to oblong-cylindrical, neck short and inconspicuous; annulus differentiated or not; peristome double, exostome teeth 16, endostome with a high basal membrane, segments keeled, cilia usually well developed, nodulose or appendiculate; operculum conical, often rostrate. *Spores* 10–20 µm in diameter, ± smooth to finely papillose. *Calyptrae* cucullate, smooth, usually naked.

The circumscription of the family is rather difficult and the delimitation of several genera is not clear-cut. Some widely used diagnostic characters, e.g., the dorsal projection of the costa, may have different degrees of expression, sometimes linked geographically and are variously weighted in the literature.

The Brachytheciaceae are represented in the local flora by seven genera with 29 species. They are mainly plants occupying comparatively mesic habitats.

Literature. — Robinson (1962), Wigh (1974).

1 Costa ending in a distinct dorsal projection (sometimes two–three projections present).
  2 All branches curved when dry; stem leaves usually shorter than branch leaves; apical cells usually as long as mid-leaf cells or longer, not differing in shape; basal and alar cells ascending along margins **2. Scorpiurium**
  2 All branches straight or some curved when dry; stem leaves as long as or longer than branch leaves; apical cells shorter than mid-leaf cells, differing in shape; basal and alar cells not ascending along margins.
    3 Leaves deeply concave; margins entire or finely serrulate above; operculum conical to short-rostrate **4. Scleropodium**
    3 Leaves plane, sometimes moderately concave; margins usually serrulate, all around; operculum obliquely long-rostrate **6. Eurhynchium**
1 Costa not ending in a dorsal projection.*

* Records of weak projections are available for some of the genera below, at least in parts of their range of distribution.

4 Plants very slender; leaves linear-lanceolate or oblong-lanceolate **7. Rhynchostegiella**

4 Plants slender to robust; leaves broadly ovate to ovate-lanceolate or lanceolate-triangular.

5 Leaves strongly plicate; all cells ± uniformly linear to linear-flexuose **1. Homalothecium**

5 Leaves smooth to slightly plicate; cells at base shorter and wider than cells above.

6 Setae smooth; operculum obliquely long-rostrate **5. Rhynchostegium**

6 Setae smooth or rough; operculum conical to short-rostrate **3. Brachythecium**

1. *Homalothecium* B.S.G.

Plants moderately robust to robust. *Stems* creeping to erect, usually densely and ± pinnately branched, sometimes subsecund when dry; branches straight or often some curved. *Pseudoparaphyllia* generally present. *Leaves*: stem leaves usually similar to branch leaves or larger, imbricate, appressed when dry, erect to erectopatent when moist, strongly two–four plicate, ovate-lanceolate or lanceolate-triangular, acuminate to long-acuminate, slightly decurrent at base; margins entire or serrulate, plane or slightly recurved; costa ⅔ the length of leaf or longer, sometimes extending into the acumen, without a dorsal projection (in local plants); leaf cells all ± uniformly linear to linear-flexuose, becoming shorter towards base; alar cells ± differentiated, ± incrassate, irregularly quadrate; perichaetial leaves sheathing, acuminate or long-subulate, entire or serrulate, with a weak costa or without (in local plants), inner perichaetial leaves long, ± erect, outer perichaetial leaves shorter, ± recurved. *Dioicous*. *Setae* rough or smooth; capsules erect, inclined or horizontal, straight and symmetrical to curved and asymmetrical, oblong-ovoid to cylindrical; annulus differentiated; endostome with short or long segments, cilia well developed and nodulose or absent; operculum conical to rostrate. *Spores* (in local species) finely papillose.

A genus comprising ca. 11 species restricted to the Northern Hemisphere. Four species occur in the local flora.

Homalothecium, in its current circumscription (cf. Dixon 1924, Robinson 1962) is considered by some authors (Husnot 1884–1894, Brotherus 1904, Nyholm 1965, Koponen *et al.* 1977), especially in earlier literature, as comprising two genera separated mainly by sporophytic characters: *Camptothecium* B.S.G. with inclined to horizontal, curved capsules and well-developed endostome, and *Homalothecium sensu stricto* with erect and straight capsules and reduced endostome.

Homalothecium is a well-defined genus, but the delimitation of some of its species may pose considerable difficulties, especially in local material in which sporophytes are rare. The most reliable diagnostic character of the local species seems to be the shape of the perichaetial leaves, which can be often found on short, archegonia-bearing branches. Otherwise, one has to rely on somewhat variable characters, e.g., the leaf shape, the location and degree of leaf serration, and the relative length of the costa. Some of the habit characters, which are often used for diagnostic purposes (e.g., the colour and extent of branch curvature) seem to depend in herbarium material, at least to some degree, on the drying conditions of the investigated samples.

1 Plants often with ± erect branches when dry; costa extending high up into the acumen; setae smooth **2. H. philippeanum**
1 Plants with branches curved to various degrees when dry; costa up to ¾ the length of leaf; setae rough.
  2 Plants often glossy; leaf margins serrulate near base, sometimes also near apex; perichaetial leaves long-acuminate; capsules erect or slightly curved, ± straight **1. H. sericeum**
  2 Plants usually not glossy; leaf margins serrulate in upper part only; perichaetial leaves acuminate or subulate; capsules inclined, ± curved.
    3 Plants green to yellowish-green; stems ± creeping; branches 4–7 mm long; leaves ovate-lanceolate, acuminate, margins sometimes narrowly recurved in upper part; perichaetial leaves entire; operculum conical-obtuse **4. H. aureum**
    3 Plants yellowish-green to golden-brown; stems ascending; branches 10–20 mm long; leaves lanceolate-triangular, long-acuminate, margins narrowly recurved near base; perichaetial leaves subulate, serrulate at base of subula; operculum conical-rostrate **3. H. lutescens**

1. Homalothecium sericeum (Hedw.) B.S.G., Bryol. Eur. 5:93 (1851).

Leskea sericea Hedw., Sp. Musc. Frond. 228 (1801); *Camptothecium sericeum* (Hedw.) Kindb., Canad. Rec. Sci. 6:73 (1894). — Bilewsky (1965):420; "*Camptothecium sericeum* (L.) Kindb. var. *robustum* Moenkem." — Bilewsky (1965):420.

Distribution map 182, Figure 180, Plate XIII:f.

Plants ± robust, yellowish to brownish-yellow, often glossy. *Stems* creeping, tomentose, attached to substrate by groups of rhizoids along the stem, irregularly and densely branched; branches ± straight to curved to various degrees when dry, 8–15 mm long. *Leaves* appressed and densely imbricate when dry, erectopatent when moist, strongly two–four plicate; stem leaves as long as branch leaves, sometimes more abruptly narrowing at apex; ca. 2 mm long, lanceolate-triangular, narrowing into a long filiform acumen; margins often recurved near base, plane above, serrulate near base, sometimes also near apex; costa up to ¾ the length of leaf; mid-leaf cells seven–ten times longer than wide; perichaetial leaves entire, long-acuminate. *Setae* rough; capsules ± straight, 2–3 mm long without operculum, subcylindrical, erect or slightly curved;

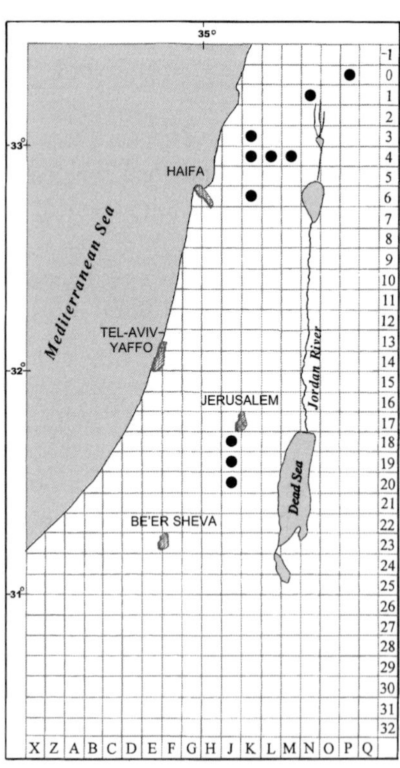

Map 182: *Homalothecium sericeum*

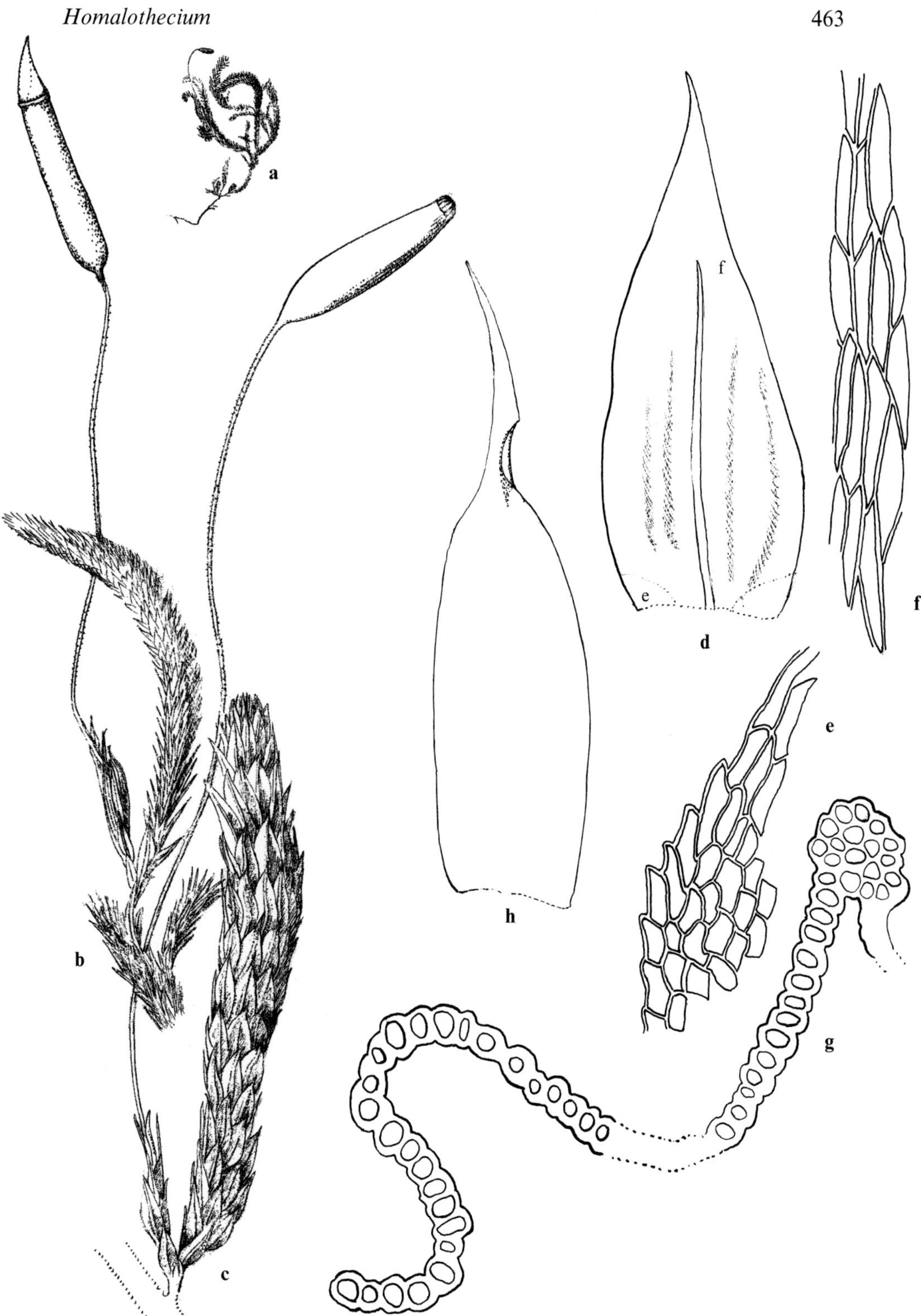

Figure 180. *Homalothecium sericeum*: **a** habit (×1); **b** part of habit, dry (×10); **c** part of habit, moist with deoperculate capsule (×10); **d** leaf (×50); **e** marginal alar cells; **f** upper cells (all leaf cells ×500); **g** cross-section of costa and portion of lamina (×500); **h** perichaetial leaf (×50).
(**a**–**g**: *Danin 80-160-1*; **h**: *Bilewsky*, Apr. 1952, No. 141 in Herb. Bilewsky)

operculum conical to conical-rostrate. *Spores* 13–20 µm in diameter; surface granulate, sparsely covered by irregular, large gemmae (seen with SEM).

Habitat and Local Distribution. — Plants growing in dense mats, on rocks, stones and walls, preferring shade. Occasional: Upper Galilee, Lower Galilee, Judean Mountains, Dan Valley, and Mount Hermon.

General Distribution. — Recorded from throughout the temperate regions of the Northern Hemisphere: Southwest Asia (Cyprus, Turkey, Lebanon, Jordan, Iraq, Iran, and Afghanistan) around the Mediterranean (throughout), Europe (widespread), Macaronesia, tropical and North Africa, eastern and central Asia, and North America.

"*Camptothecium sericeum* var. *robustum* Mönk." mentioned by Bilewsky (1965) refers to some specimens within the range of variation of the species as considered here.

2. Homalothecium philippeanum (Spruce) B.S.G., Bryol. Eur. 5 : 93 (1851).
Isothecium philippeanum Spruce, Musci Pyren. no. 77 (1847); *Camptothecium philippeanum* (Spruce) Kindb., Canad. Rec. Sci. 6 : 3 (1894); *H. philippeanum* var. *maroccanum* Thér. & Trab., Bull. Soc. Hist. Nat. Afrique N. 21 : 31 (1930). — Bilewsky (1970).

Distribution map 183, Figure 181.

Plants robust, green to yellowish-green, sometimes dark green at their base. *Stems* creeping, densely branched; branches ± erect when dry and moist, 8–15 mm long. *Leaves* appressed and imbricate, 1.5–3 mm long, lanceolate, gradually tapering into a very long, narrow acumen; stem leaves similar to branch leaves or sometimes slightly larger; margins plane, sometimes recurved on one side near base, weakly serrulate, in particular near apex; costa extending far up into acumen; mid-leaf cells 10–15 times longer than wide; outer perichaetial leaves abruptly tapering to a long thin acumen with a serrulate base. *Setae* recorded as smooth. *Sporophytes* not seen in local material.

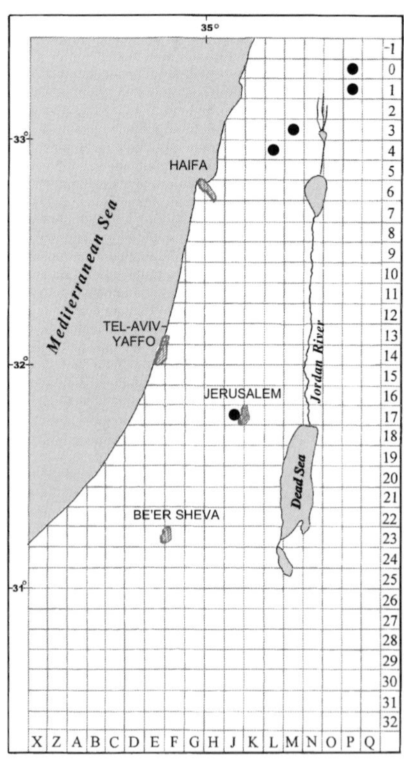

Map 183: *Homalothecium philippeanum*

Habitat and Local Distribution. — Plants growing in mats in mountainous areas, on limestone and basalt rocks, and in rock crevices. Occasional: Upper Galilee, Judean Mountains, and Mount Hermon.

General Distribution. — Recorded from Southwest Asia (Turkey, Syria, Lebanon, Iraq, and Iran), around the Mediterranean (scattered records), Europe (mainly the south), and northeastern, eastern, and central Asia.

Homalothecium philippeanum can be easily distinguished from all other local species of *Homalothecium* by the nearly erect branches, the long costa, and the smooth setae.

Figure 181. *Homalothecium philippeanum*: **a** habit (×1); **b** part of stem and branches, dry (×10); **c** part of branch, moist (×10); **d** leaves (×50); **e** marginal alar cells; **f** mid-leaf cells (all leaf cells ×500); **g** perichaetial leaf (×50). (*D. Kaplan*, 9 Dec. 1982)

3. Homalothecium lutescens (Hedw.) H. Rob., Bryologist 65:98 (1962).
Hypnum lutescens Hedw., Sp. Musc. Frond. 274 (1801); *Camptothecium lutescens* (Hedw.) B.S.G., Bryol. Eur. 6:36 (1853). — Bilewsky (1965):419.

Distribution map 184, Figure 182.

Plants ± robust, resembling *Homalothecium sericeum* in size or larger. *Stems* ascending, attached to substrate only at base; branches usually erect, 10–20 mm long. *Leaves* imbricate, appressed and erect when dry, erectopatent when moist; stem leaves similar to branch leaves, wider; branch leaves ca. 2 mm long, plane to strongly plicate, lanceolate-triangular, gradually narrowing into a long narrow acumen; margins narrowly recurved near base, weakly serrulate, sometimes more distinctly towards apex; costa up to ¾ as long as leaf; mid-leaf cells 10–12 times longer than wide; perichaetial leaves subulate, serrulate at base of subula. *Setae* rough; capsules with conical-rostrate operculum; mature capsules not seen in local material.

Habitat and Local Distribution. — Plants yellowish-green, growing in tufts or patches on rocks, and on soil on rocks. Occasional: Upper Galilee, Judean Mountains, and Golan Heights (?).

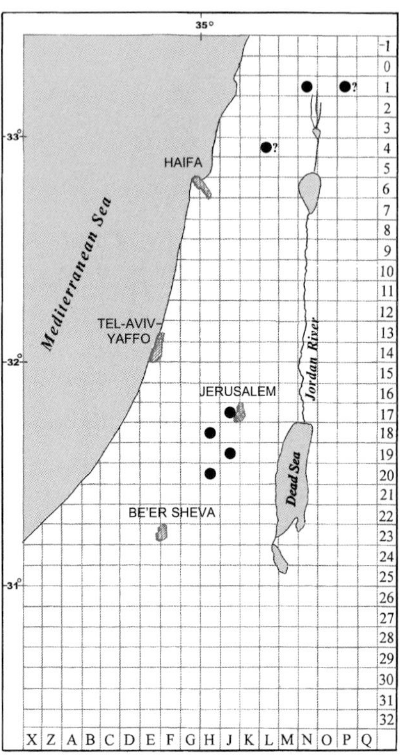

Map 184: *Homalothecium lutescens*

General Distribution. — Recorded from Southwest Asia (Turkey, Syria, Lebanon, Iraq, and Iran), around the Mediterranean (throughout), Europe (throughout), Macaronesia, North Africa, and eastern Asia.

Homalothecium lutescens closely resembles in habit some large forms of *H. sericeum*. The collections identified by Bilewsky as *Camptothecium lutescens* from the Upper Galilee (near Hunin) and as *C. lutescens* var. *fallax* from the Judean Mountains (near Hebron) should perhaps be considered as *H. sericeum*. The identity of *H. lutescens* var. *fallax* (H. Philib. *in* Husn.) Hedenäs & L. Söderstr. (Söderström *et al.* 1992), could not be clarified with local specimens.

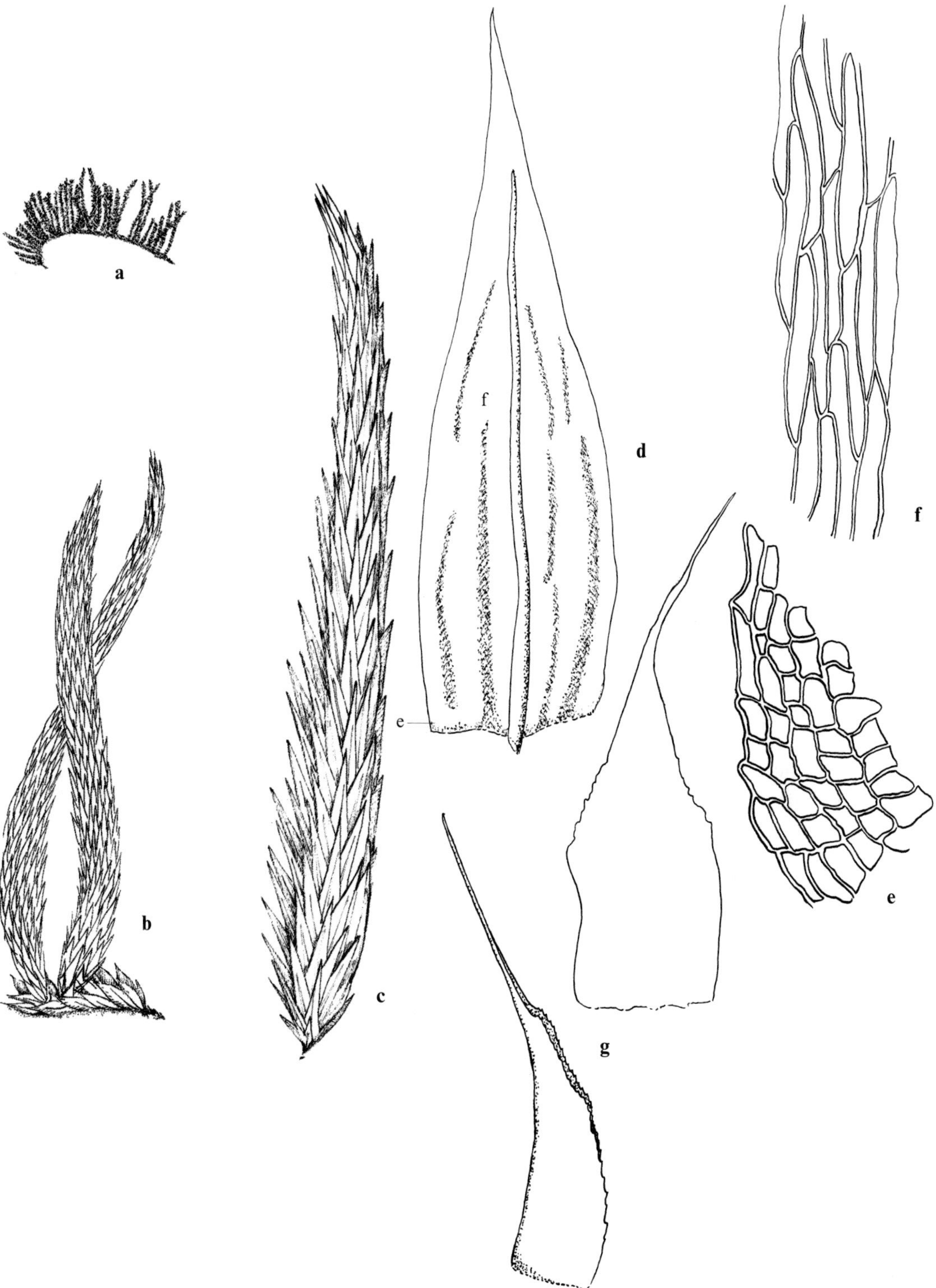

Figure 182. *Homalothecium lutescens*: **a** habit (×1); **b** part of stem and branches, dry (×10); **c** part of branch, moist (×10); **d** leaf (×50); **e** marginal alar cells; **f** mid-leaf cells (all leaf cells ×500); **g** perichaetial leaves (×50). (*Weitz*, 2 Dec. 1970)

4. Homalothecium aureum (Spruce) H. Rob., Bryologist 65:96 (1962).
Isothecium aureum Spruce, Musci Pyren. no. 85 (1847); *Camptothecium aureum* (Lag.) B.S.G., Bryol. Eur. 6:37 (1853). — Bilewsky (1965):419.

Distribution map 185, Figure 183, Plate XIV:a.

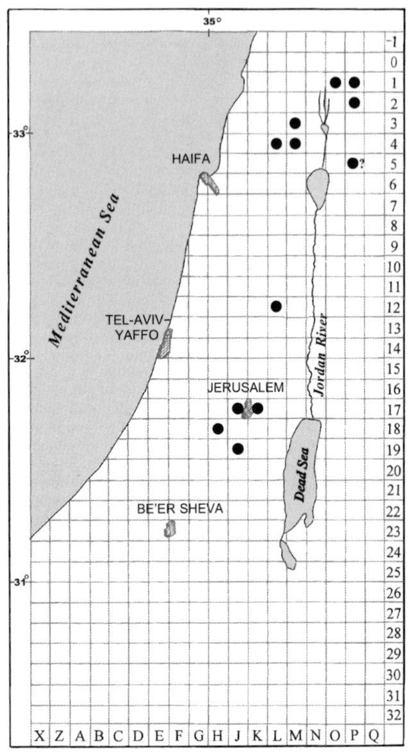

Map 185: *Homalothecium aureum*

Plants medium-sized, green to yellowish-green. *Stems* creeping, ± pinnately branched; branches often curved when dry, short, 4–7 mm long. *Leaves* imbricate, appressed when dry, erectopatent when moist, weakly plicate; stem and branch leaves ± similar, up to 1.5 mm long, ovate-lanceolate, acuminate; margins sometimes narrowly recurved in upper part, serrulate above; costa ending below apex; mid-leaf cells 10–12(–15) times longer than wide; perichaetial leaves acuminate, entire. *Setae* rough; capsules inclined, ca. 2 mm without operculum, oblong to subcylindrical, curved; operculum conical-obtuse. *Spores* 14-20 µm in diameter; surface rough, sparsely covered by gemmae or granulae, often in irregular clusters (seen with SEM).

Habitat and Local Distribution. — Plants growing in mats on rocks and soil in shade of rocks and trees or on north-facing slopes. Occasional, locally common: Upper Galilee, Samaria, Judean Mountains, and Golan Heights.

General Distribution. — Recorded from Southwest Asia (Cyprus, Turkey, Syria, Lebanon, Jordan, and Iraq), around the Mediterranean (throughout), southern Europe, Macaronesia, and North Africa.

2. *Scorpiurium* Schimp.

Plants slender to moderately robust, growing in mats. *Stems* creeping with tufts of rhizoids; secondary stems ascending; branches numerous, all distinctly curved when dry, straightening when moist. *Leaves* appressed and imbricate when dry, patent to spreading when moist, weakly plicate; stem leaves usually shorter and less crowded than branch leaves, ovate-triangular, abruptly acuminate; branch leaves ovate-lanceolate, obtuse, acute to long-acuminate, base decurrent; margins plane or recurved at base, serrulate to serrate above; costa from fairly strong to very strong, extending to acumen or shorter, ending in a dorsal projection, sometimes several dorsal projections at its upper half; mid- and upper leaf cells rhomboidal to elongate-rhomboidal, apical cells as long as mid-leaf cells or longer, not differing in shape; basal cells and alar cells ± quadrate, opaque, ascending along margins; perichaetial leaves distinctly differentiated. *Dioicous*. *Setae* smooth; capsules inclined to cernuous, curved and asymmetrical, or ± straight and almost symmetrical (*S. sendtneri*), ellipsoid or cylindrical,

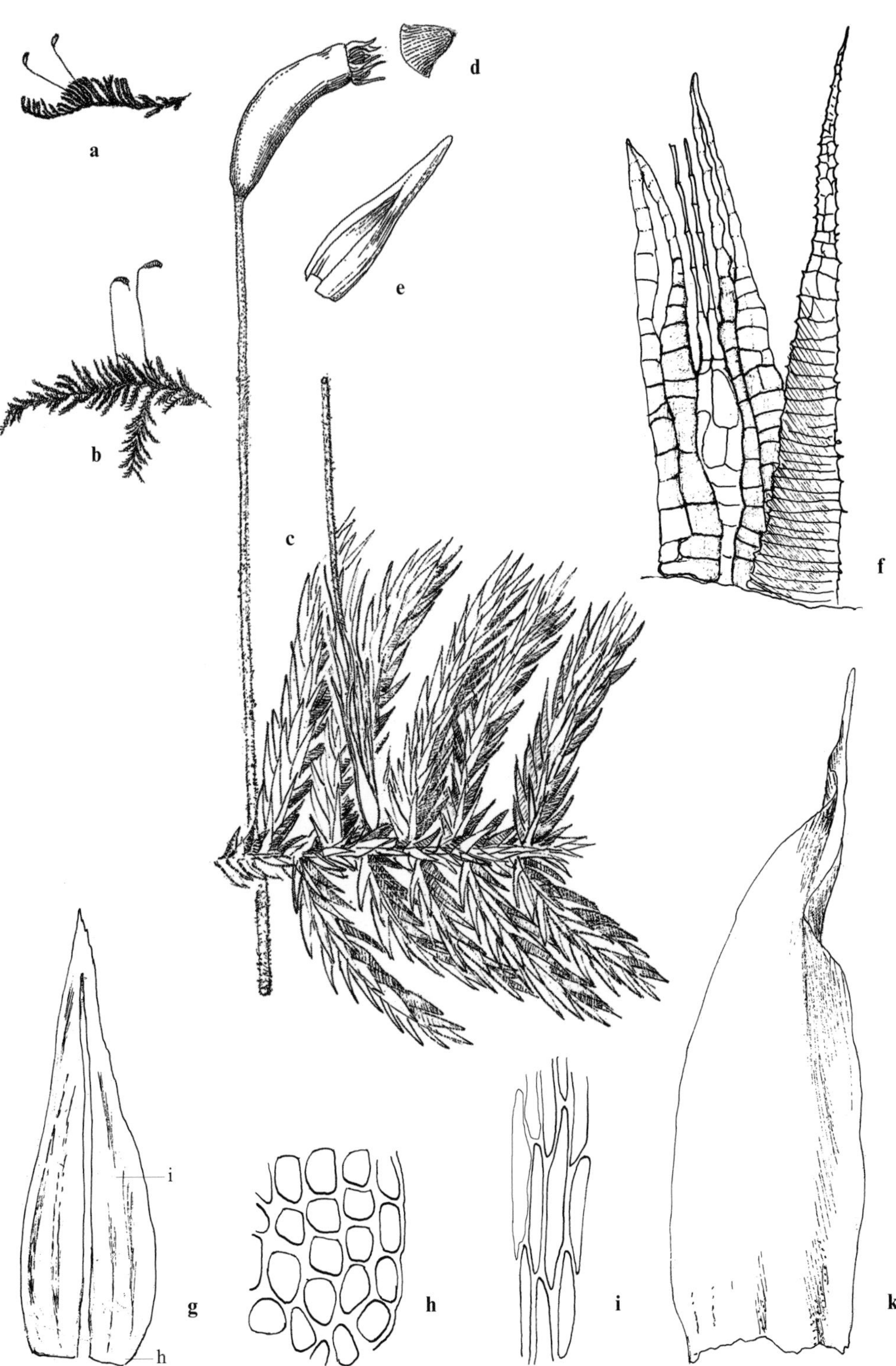

Figure 183. *Homalothecium aureum*: **a** habit, dry (×1); **b** habit, moist (×1); **c** part of habit, moist (×10); **d** operculum (×10); **e** calyptra (×10); **f** portion of peristome teeth (×125); **g** leaf (×50); **h** marginal basal cells; **i** mid-leaf cells (all leaf cells ×500); **k** perichaetial leaf (×50). (*Herrnstadt & Crosby 79-88-4*)

often constricted below mouth when dry; operculum obliquely long-rostrate. *Spores* nearly smooth to ± papillose.

A small, mainly Mediterranean genus comprising three species, all growing in the local flora.

1 Plants small, slender; branches slightly curved; branch leaves long-acuminate; costa ± strong, ending below acumen; margins ± serrulate above; capsules erect, ± straight, almost symmetrical **3. S. sendtneri**

1 Plants usually medium-sized, ± robust; branches often strongly curved; branch leaves acute or acuminate; costa strong, extending to acumen; margins serrulate to serrate above; capsules inclined, cernuous, or horizontal, curved, distinctly asymmetrical.

2 Branch leaves with margins ± recurved at base, short-acute to ± obtuse; mid-leaf cells two–three times longer than wide, apical cells as long as mid-leaf cells **2. S. deflexifolium**

2 Branch leaves with margins plane throughout, acuminate; mid-leaf cells four–six (seven) times as long as wide, apical cells longer than mid-leaf cells **1. S. circinatum**

1. Scorpiurium circinatum (Brid.) M. Fleisch. & Loeske, Allg. Bot. Z. Syst. 13:22 (1907). [Bilewsky (1965):423]

Hypnum circinatum Brid., Sp. Musc. Frond. 2:148 (1812); *Eurhynchium circinatum* (Brid.) B.S.G., Bryol. Eur. 5:221 (1854).

Distribution map 186, Figure 184.

Plants usually medium-sized, ± robust, yellowish-green. *Branches* often strongly curved when dry. *Leaves* densely imbricate, with spreading acumen; branch leaves up to 1.25 mm long, ovate-lanceolate, acuminate; margins plane, serrulate from below mid-leaf; costa strong, extending to acumen, sometimes to apex; mid-leaf cells, four–six (seven) times longer than wide, apical cells usually longer than mid-leaf cells. *Capsules* inclined to cernuous, ellipsoid or sometimes cylindrical, curved and distinctly asymmetrical, varying in size within populations, 1.5–3(–3.5) mm long including operculum.

Habitat and Local Distribution. — Plants growing in dense mats, in moist and shaded habitats, often near running water, on rocks, in rock crevices, on walls, tree trunks and bark. Common: Coastal Galilee, Coast of Carmel, Sharon Plain, Upper Galilee, Lower Galilee, Mount Carmel, Judean Mountains, Dan Valley, and Golan Heights.

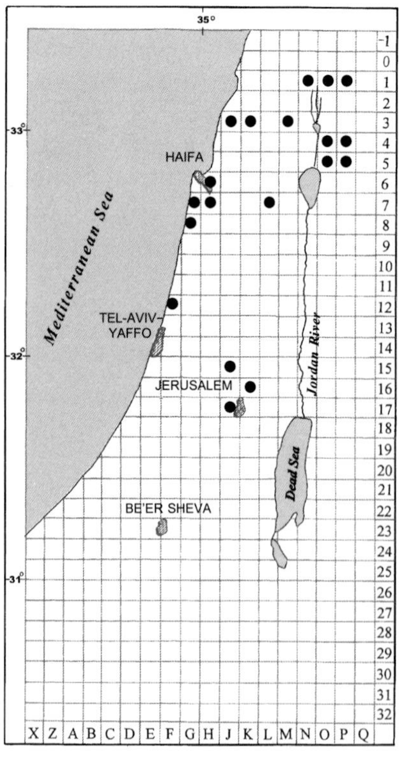

Map 186: *Scorpiurium circinatum*

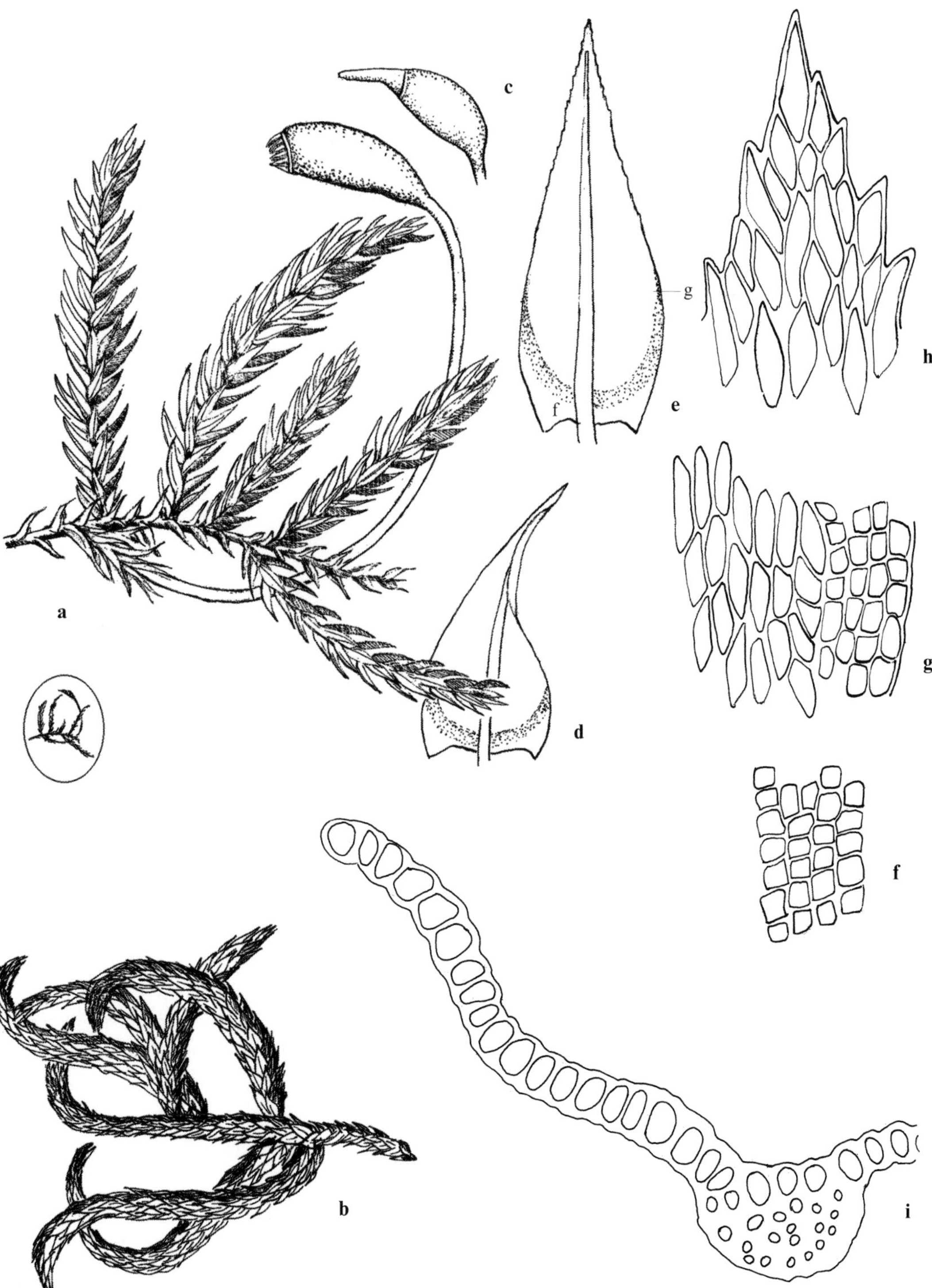

Figure 184. *Scorpiurium circinatum*: **a** habit, moist (×10); **b** habit, dry (×10); **c** immature capsule (×10); **d** stem leaf (×50); **e** branch leaf (×50); **f** basal cells; **g** cells below mid-leaf from margin towards centre; **h** apical cells (all leaf cells ×500); **i** cross-section of costa and portion of lamina (×500). (*D. Zohary*, 15 Mar. 1945)

General Distribution. — Recorded from Southwest Asia (Cyprus, Turkey, Lebanon, Jordan, and Saudi Arabia), around the Mediterranean (throughout), Europe (mainly the south), Macaronesia, North Africa, and eastern Asia.

Plants of *Scorpiurium circinatum* can be easily distinguished by the densely imbricate leaves with a spreading acumen.

2. Scorpiurium deflexifolium (Solms) M. Fleisch. & Loeske, Allg. Bot. Z. Syst. 13 : 22 (1907). [Bilewsky (1965) : 424]

Hypnum deflexifolium Solms, Tent. Bryo-Geogr. Algarv. 40 (1868); *S. circinatum* subsp. *deflexifolium* (Solms) Giacom., Atti Ist. Bot. "Giovanni Briosi" 5, 4 : 266 (1947).

Distribution map 187, Figure 185, Plate XIV : b.

Plants usually medium-sized, ± robust, olive-green. *Branches* often strongly curved when dry. *Leaves*: branch leaves ca. 1 mm long, ovate to ovate-lanceolate, short-acute to ± obtuse; margins ± recurved at base, strongly serrulate or serrate above, weakly below; costa stout, reaching acumen, often bearing two–three dorsal projections at its upper half; mid-leaf cells rhomboidal, two–three times longer than wide, apical cells usually as long as mid-leaf cells. *Capsules* inclined to horizontal, curved, distinctly asymmetrical, 2.0–2.5(–3.0) mm long including operculum, ellipsoid to cylindrical. *Spores* 13–16 μm in diameter, ± smooth; densely granulate, some in small clusters (seen with SEM).

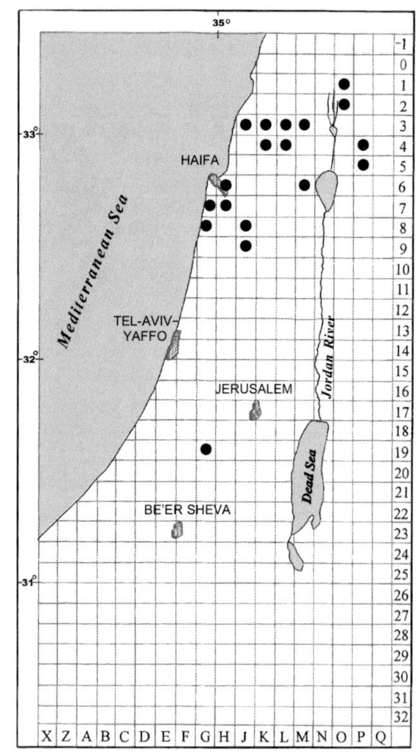

Map 187: *Scorpiurium deflexifolium*

Habitat and Local Distribution. — Plants growing in dense mats on stones, limestone, and basaltic rocks, in shade of trees, on north-facing slopes, and under periodical water spray of small pools. Common: Upper Galilee, Lower Galilee, Mount Carmel, Samaria, Shefela, and Golan Heights.

General Distribution. — Recorded from Southwest Asia (Turkey, Afghanistan, and Lebanon), around the Mediterranean, Europe (only the south), North Africa, and eastern Asia.

Scorpiurium deflexifolium can be distinguished from the other two local *Scorpiurium* species by the olive-green coloured branch leaves, which are somewhat less densely spaced and more strongly serrulate than in the branch leaves of the other species, and have short acute to obtuse — not acuminate — apex. Sporophytes are produced in some local populations, whereas all records of *S. deflexifolium* from Europe refer to sterile plants only. The sporophytes resemble those of *S. circinatum* in the shape and size.

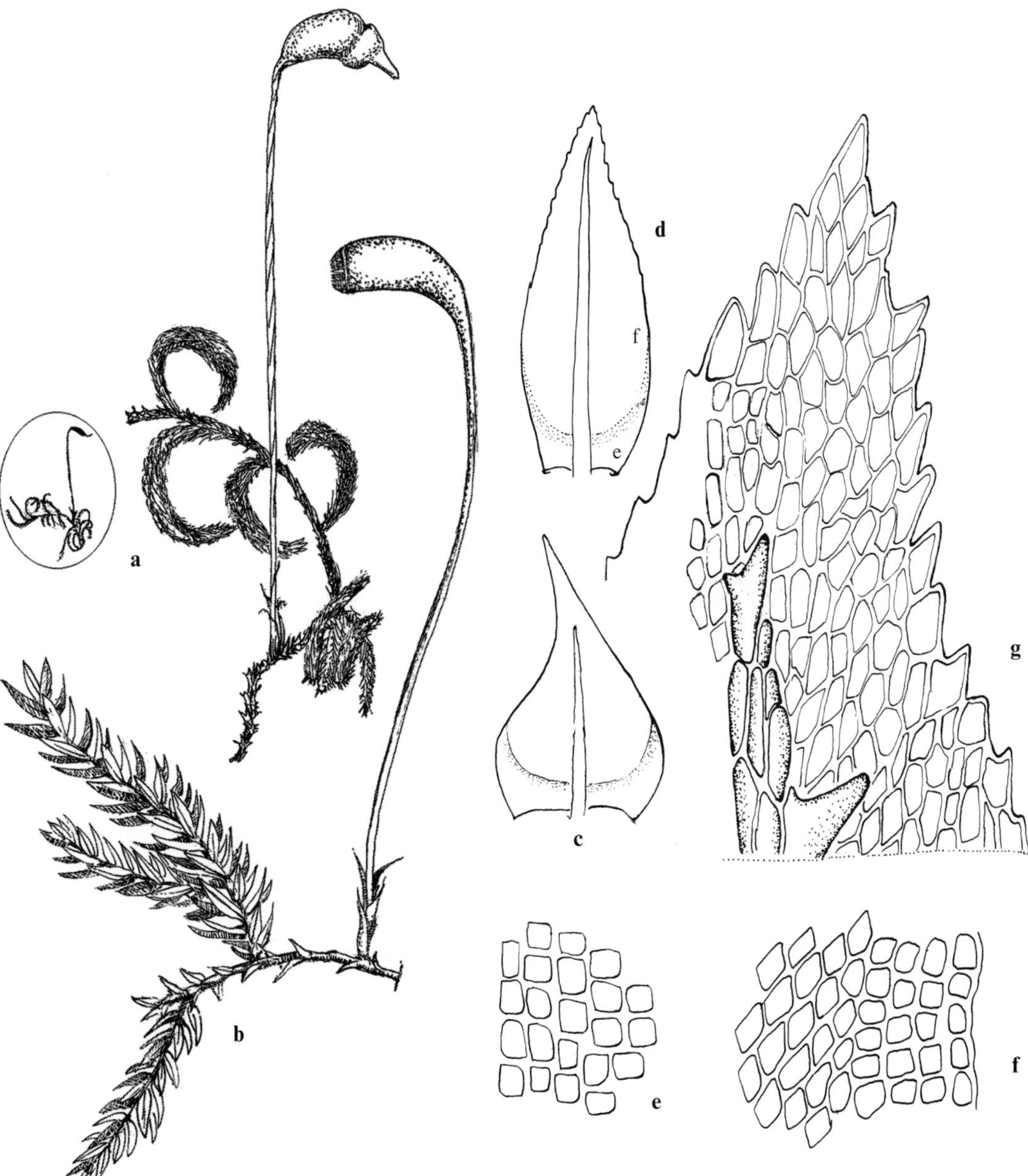

Figure 185. *Scorpiurium deflexifolium*: **a** habit, dry (×10); **b** habit, moist with deoperculate capsule (×10); **c** stem leaf (×50); **d** branch leaf (×50); **e** basal cells; **f** cells at about mid-leaf from margin towards centre; **g** apical cells (all leaf cells ×500). (**a**–**e**: *Herrnstadt* & *Crosby 79-84-20*; **f**, **g**: *Dimentman*, 10 May 1995)

3. Scorpiurium sendtneri (Schimp.) M. Fleisch., Hedwigia 61 : 405 (1920).
[Bilewsky (1965) : 424]

Fabronia sendtneri Schimp., Syn. Musc. Eur., Edition 2 : 585 (1876); *S. leskeoides* Suse, Deutsche Bot. Monatsschr. 28 : 2 (1910).

Distribution ma, 188, Figure 186, Plate XIV : c.

Plants resembling *Scorpiurium circinatum* in habit, but smaller and more slender, yellowish-green to bright green. *Branches* usually less slightly curved. *Leaves*: stem and branch leaves long-acuminate, acumen sometimes ± reflexed; branch leaves usually 0.75–1.0 mm long; margins ± serrulate above; costa ± strong, ending below acumen; mid-leaf cells three–four times longer than wide. *Capsules* erect, ± straight, almost symmetrical, ca. 3 mm long including operculum, narrow-cylindrical. *Spores* 13–19 µm in diameter; densely granulate-gemmate (seen with SEM).

Habitat and Local Distribution. — Plants growing in dense mats on soil on rocks, in rock crevices, and on bark of *Quercus* spp. Occasional: Upper Galilee, Lower Galilee, Mount Carmel, Judean Mountains, Dan Valley, Mount Hermon, and Golan Heights.

General Distribution. — Recorded from Southwest Asia (Cyprus and Turkey), around the Mediterranean (scattered records, mainly from the northern Mediterranean), Madeira, and Europe (few records, from the south only).

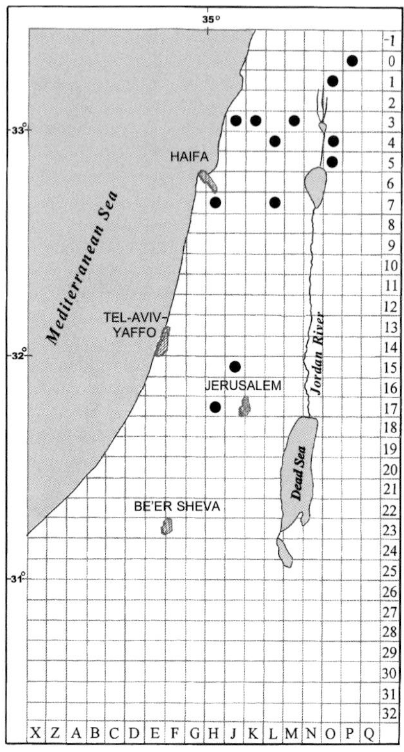

Map 188: *Scorpiurium sendtneri*

Scorpiurium sendtneri can be distinguished from the other species of the genus by the slender habit, the shorter costa, and the erect, ± straight capsules. Records in the literature refer only to sterile plants. In the local flora, sporophytes were observed in one population on the Golan Heights.

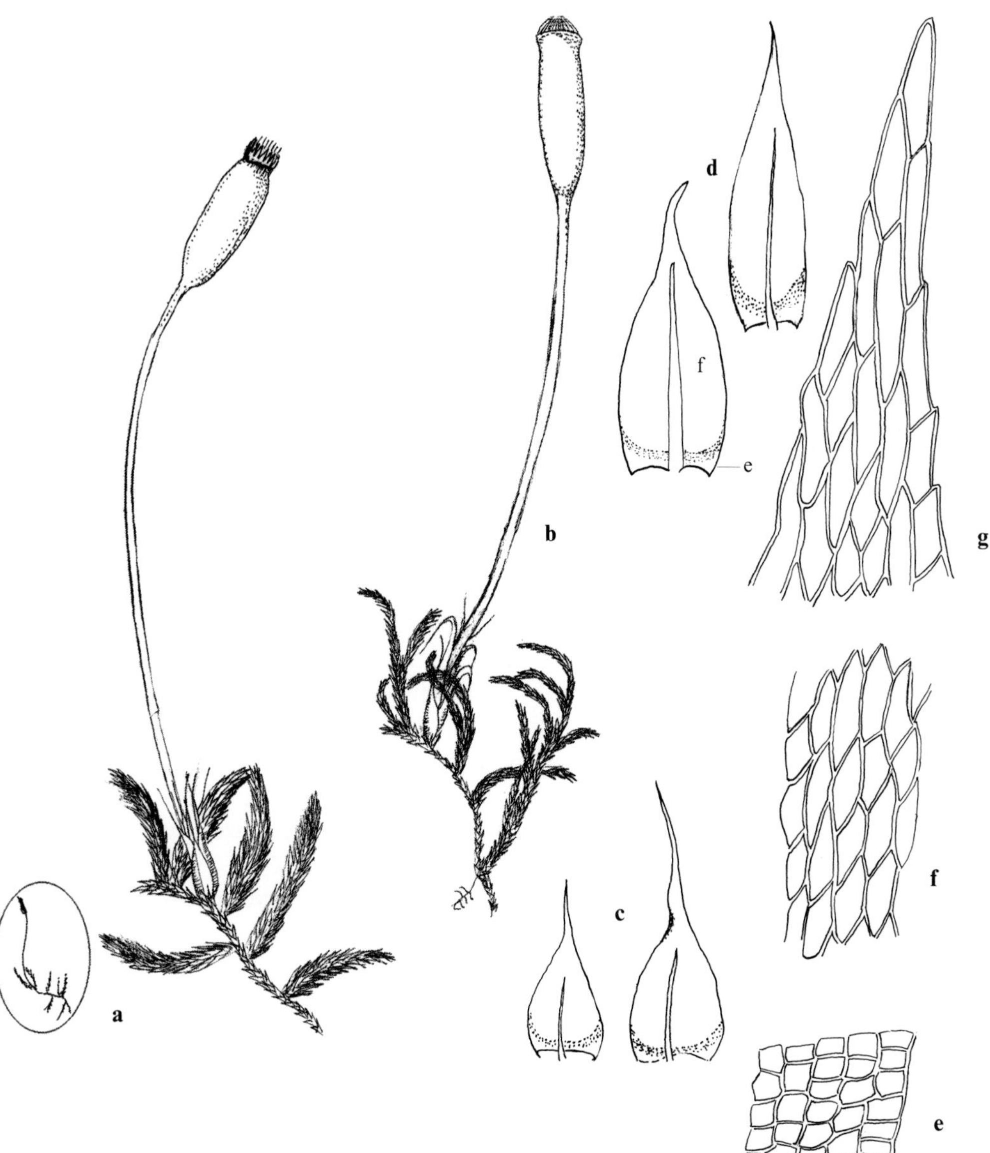

Figure 186. *Scorpiurium sendtneri*: **a** habit, moist with deoperculate capsule (×10); **b** habit, dry (×10); **c** stem leaves (×50); **d** branch leaves (×50); **e** marginal basal cells; **f** mid-leaf cells; **g** apical cells (all leaf cells ×500). (*Kushnir*, 8 Apr. 1945)

3. *Brachythecium* B.S.G.

Plants slender to moderately robust. *Leaves*: stem and branch leaves similar or ± different; leaves imbricate to spreading when dry, spreading when moist, ± concave, usually ± plicate, cordate-triangular to ovate-lanceolate, acute to long-acuminate, ± decurrent at base; margins usually plane, entire to serrulate, often only in upper part; costa from up to half the length of leaf to percurrent; basal cells shorter and wider than cells above, quadrate, rectangular to oblong-rhomboidal; mid- and upper leaf cells linear-fusiform to rhomboidal, alar cells ± conspicuous. *Dioicous* or autoicous, rarely synoicous. *Setae* smooth or rough; capsules inclined to horizontal, rarely erect, usually curved, asymmetric, usually short and broad, oblong-ovoid to cylindrical; dry capsules strongly curved, distinctly shorter than moist capsules, and constricted below mouth; annulus differentiated; operculum conical to short-rostrate; cilia of endostome usually well developed, rarely rudimentary or absent. *Spores* smooth or papillose.

A large cosmopolitan genus. Four species occur in the local flora.

1 Leaves plicate; margins entire or slightly sinuose; setae smooth **1. B. mildeanum**
1 Leaves only slightly or not plicate; margins sinuose to serrulate; setae at least partly rough.
  2 Costa extending into acumen or ± percurrent; mid-leaf cells five–nine times longer than wide **4. B. populeum**
  2 Costa not extending over ¾ the length of leaf; mid-leaf cells eight–eighteen times longer than wide.
    3 Plants slender; stem and branch leaves ± similar, branch leaves 1–1.25 mm long, narrowly to broadly lanceolate **3. B. velutinum**
    3 Plants medium-sized to robust; stem and branch leaves not similar, branch leaves ca. 2 mm long, ovate to ovate-lanceolate **2. B. rutabulum**

1. Brachythecium mildeanum (Schimp.) Schimp. *in* Milde, Bot. Zeitung 20 : 452 (1862).
Hypnum mildeanum Schimp., Syn. Musc. Eur. 694 (1860).

Distribution map 189 (p. 478), Figure 187.

Plants medium-sized, green to yellowish-green. *Stems* creeping, irregularly branched; branches (5–)7–10 mm long. *Leaves* ± spreading and ± contorted when dry, spreading to ascending when moist; stem leaves longer than branch leaves, long-acuminate, branch leaves up to 1.5(–2) mm long, plicate, ovate-lanceolate, acuminate; margins plane, entire, or slightly sinuose; costa ½–⅔ the length of leaf; basal cells rhomboidal, slightly incrassate, alar cells quadrate to rectangular, mid-leaf cells seven–eleven times longer than wide. *Autoicous*. *Setae* ca. 15 mm long, smooth, orange to brown; capsules up to 3 mm long including operculum, curved, oblong-cylindrical, orange to brown.

Habitat and Local Distribution. — Plants growing on a temporarily submerged riverbank. Very rare: Dan Valley (so far collected only in two neighbouring localities).

General Distribution. — Recorded from Southwest Asia (Turkey), around the Mediterranean (records few and scattered), Europe (mainly central and northern Europe, few records from the south), Macaronesia, northeastern and central Asia, and North America.

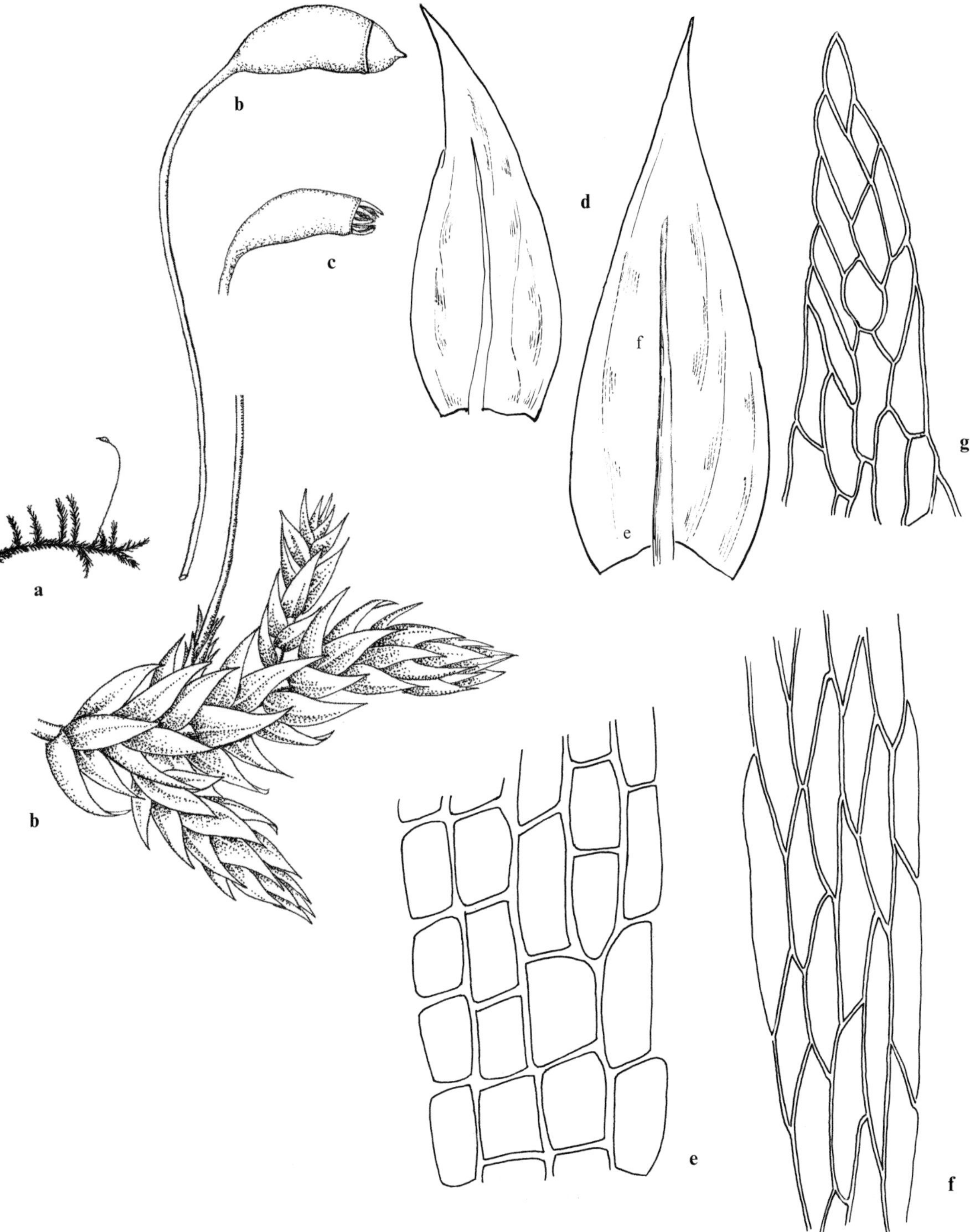

Figure 187. *Brachythecium mildeanum*: **a** habit (×1); **b** part of habit, moist (×10); **c** deoperculate capsule (×10); **d** stem leaves (×50); **e** basal cells; **f** mid-leaf cells; **g** apical cells (all leaf cells ×500). (Ein-el-Bard, *Hefez*, May 1986)

2. Brachythecium rutabulum (Hedw.) B.S.G., Bryol. Eur. 6:15 (1853). [Bilewsky (1965):422]

Hypnum rutabulum Hedw., Sp. Musc. Frond. 276 (1801).

Distribution map 189, Figure 188.

Plants medium-sized to robust, yellowish-green. *Stems* creeping, irregularly to subpinnately branched; branches ascending, sometimes curved, 8–14(–15) mm long. *Leaves* erectopatent, concave, smooth, or slightly plicate when dry and moist; stem leaves longer and wider than branch leaves, decurrent; branch leaves ca. 2 mm long, ovate to ovate-lanceolate, ± decurrent at base, acute to acuminate, apex sometimes twisted; margins plane, serrulate above; costa ⅔–¾ the length of leaf; basal cells rhomboidal, alar cells rectangular, only slightly differing from basal cells, mid-leaf cells linear, ± flexuous, eight–fifteen times longer than wide. Recorded as *autoicous*. *Sporophytes* not found in local material.

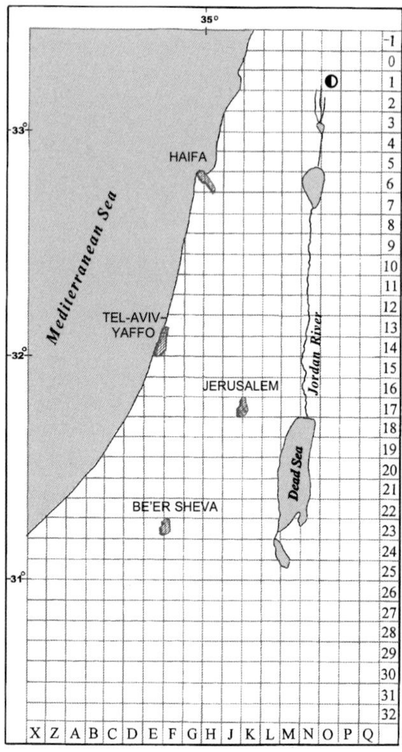

Map 189: ◐ *Brachythecium mildeanum* + *B. rutabulum*

Habitat and Local Distribution. — Plants, growing in or by a stream on soil or on the base of a *Quercus* trees, submerged parts encrusted; sources of the Dan River. Very rare: Dan Valley (so far collected only in two localities).

General Distribution. — Recorded from Southwest Asia (Turkey, Syria, Lebanon, Iran, and Afghanistan), around the Mediterranean, Europe (throughout), Macaronesia, tropical and North Africa, Madagascar, northeastern, eastern, and central Asia, Australia, New Zealand, Oceania, and north, central, and northwestern South America.

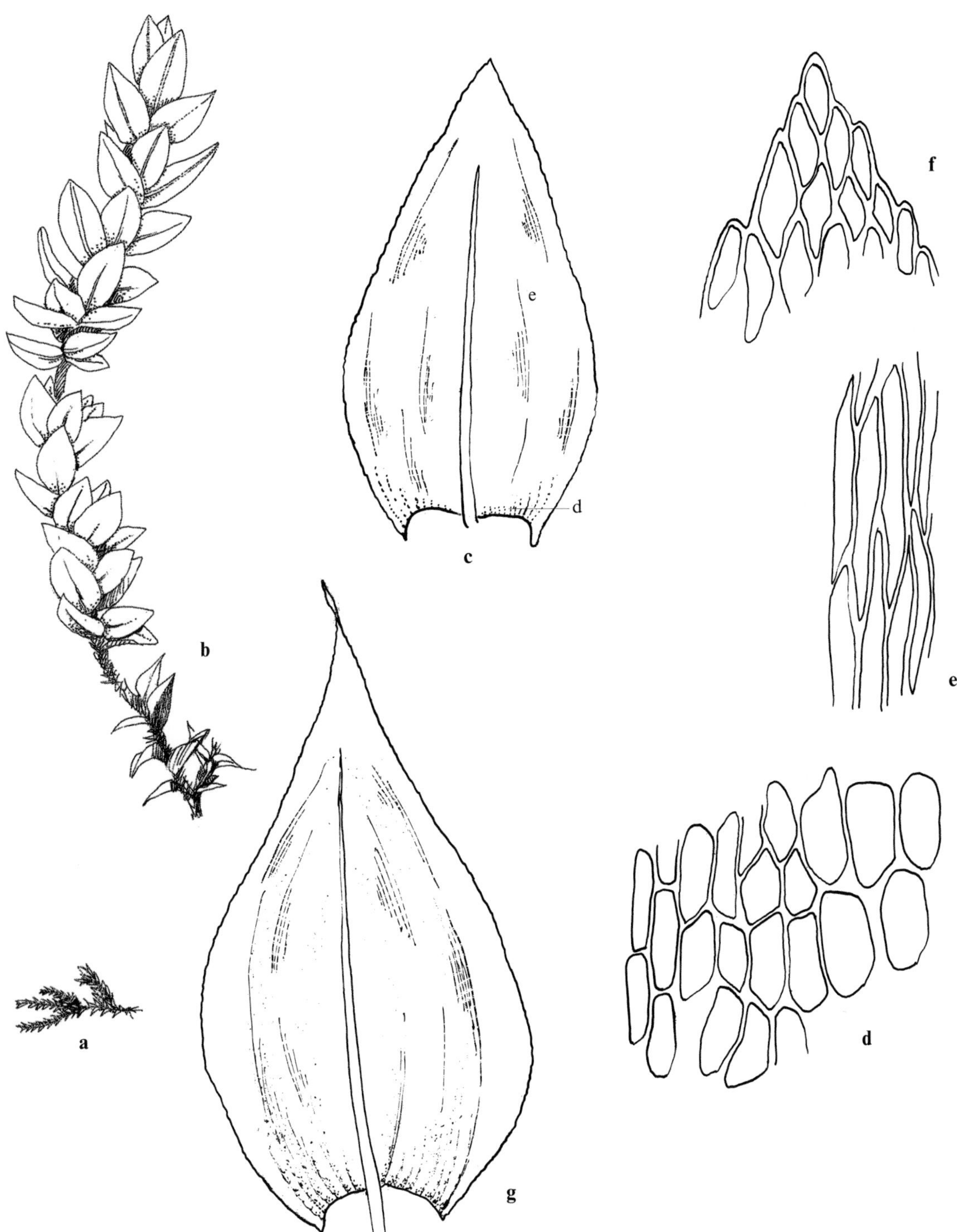

Figure 188. *Brachythecium rutabulum*: **a** habit (×1); **b** branch, moist (×10); **c** branch leaf (×50); **d** basal cells; **e** mid-leaf cells; **f** apical cells (all leaf cells ×500); **g** stem leaf (×50). (*Bilewsky*, 24 Apr. 1954, No. 143 in Herb. Bilewsky).

3. Brachythecium velutinum (Hedw.) B.S.G., Bryol. Eur. 6:9 (1853).
[Bilewsky & Nachmony (1955)]
Hypnum velutinum Hedw., Sp. Musc. Frond. 272 (1801); *B. salicinum* B.S.G., Bryol. Eur. 6:19 (1853).

Distribution map 190, Figure 189.

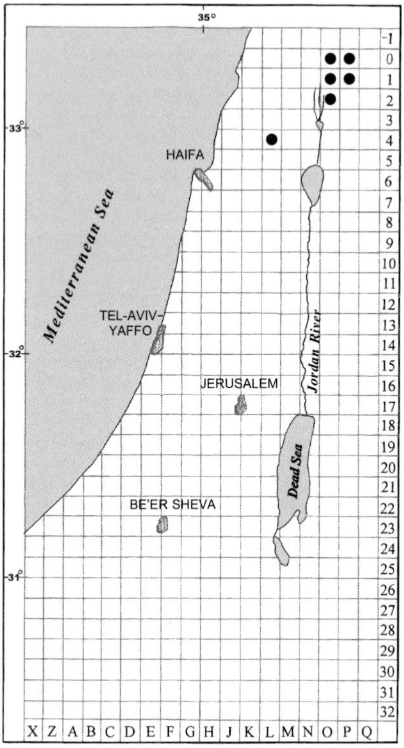

Map 190: *Brachythecium velutinum*

Plants small and slender, yellowish-green to green. *Stems* creeping, irregularly to subpinnately branched. *Branches* 3–6 mm long, ± erect. *Leaves*: stem and branch leaves ± similar; branch leaves usually laxly spaced, erectopatent to spreading, sometimes ± falcate-secund, 1–1.25 mm long, narrowly to broadly lanceolate, shortly decurrent at base, gradually tapering into a long acumen, apex usually twisted; margins plane, serrulate mainly above; costa ½–¾ the length of leaf, dorsal projection sometimes present; basal cells rhomboidal to irregularly rectangular, alar cells few, quadrate, varying in size; mid-leaf cells linear-fusiform, 10–18 times longer than wide, cells at apex rhomboidal. *Autoicous*. *Setae* 6–7 mm long, rough to various degrees, rarely nearly smooth; capsules horizontal to inclined, curved, up to 2 mm long including operculum, oblong-cylindrical.

Habitat and Local Distribution. — Plants growing in dense mats in shaded habitats, on soil, rocks, tuff, tree trunks, and on decaying material; often bearing sporophytes. Rare to occasional, locally abundant: Upper Galilee, Mount Hermon, and Golan Heights.

General Distribution. — Recorded from Southwest Asia (Cyprus, Turkey, Lebanon, Sinai, Iraq, Iran, and Afghanistan), around the Mediterranean (throughout), Europe (throughout), Macaronesia, North Africa (var. *salicinum* only?), and North America.

Brachythecium velutinum is a most variable species for which different diagnostic characters have been used, e.g., the occurrence of branches narrowing towards tips or short capsules (e.g., Frahm 1995). Plants with smooth setae have been described as *B. velutinum* var. *salicinum* (B.S.G.) Mönk., or var. *venustum* (De Not.) De Not., and as *B. olympicum* Jur. Bilewsky & Nachmony (1955) recorded plants from Mount Hermon ("Lebanon") with smooth setae as *B. velutinum* var. *venustum*. In local material (even within populations), setae have been found to vary in the degree of roughness and in the location of the rough section on the setae.

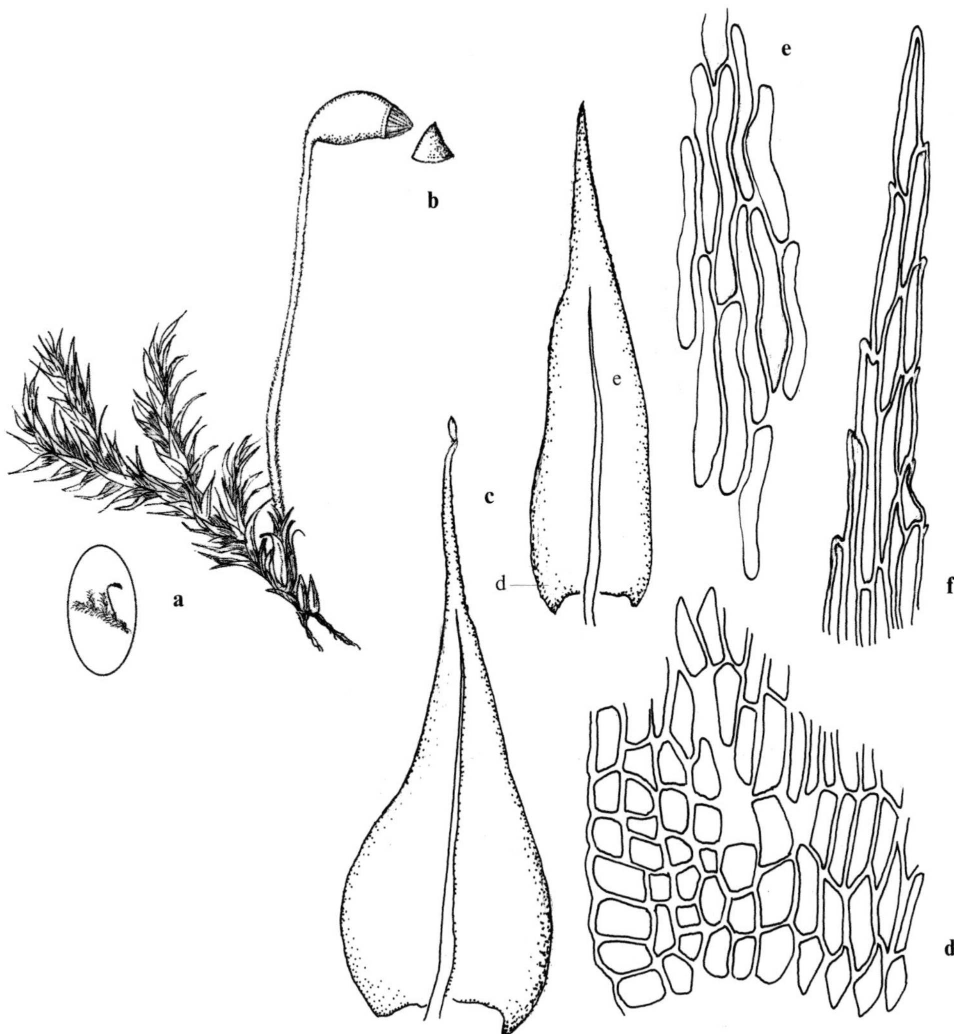

Figure 189. *Brachythecium velutinum*: **a** part of habit, moist (×10); **b** operculum (×10); **c** leaves (×50); **d** basal cells from margin towards costa; **e** mid-leaf cells; **f** apical cells (all leaf cells ×500). (*Kushnir*, Oct. 1943)

4. Brachythecium populeum (Hedw.) B.S.G., Bryol. Eur. 6:7 (1853).
Hypnum populeum Hedw. Sp. Musc. Frond. 270 (1801).

Distribution map 191, Figure 190.

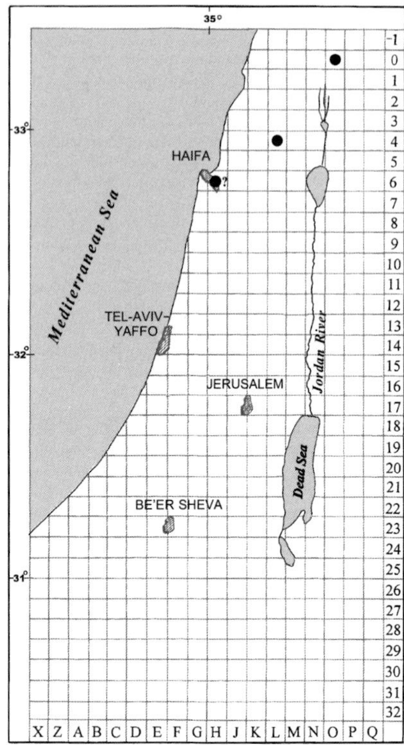

Map 191: *Brachythecium populeum*

Plants slender, light green to yellowish-green. *Stems* creeping, irregularly to subpinnately branched; branches 3–6(–10) mm long, ascending to erect. *Leaves* appressed at their base when dry, erect-spreading when moist; stem leaves larger than branch leaves; branch leaves 1–1.5 mm long, narrow ovate-lanceolate, narrowly long-acuminate; margins plane or slightly recurved below, sinuous to serrulate, often serrate near apex; costa extending into acumen or ± percurrent; basal cells rhomboidal, alar cells rectangular to irregularly quadrate, mid- and upper leaf cells linear-rhomboidal to linear-fusiform, five–nine times longer than wide at mid-leaf; stem leaves 1.5–1.8 mm long, broadly ovate-lanceolate, ending in a long, subulate acumen. *Autoicous*. *Setae* ca. 6 mm long, rough in upper part; capsules short, up to 1.5 mm long including operculum, oblong-cylindrical (only one mature capsule seen in local material). *Spores* nearly smooth; densely granulate (seen with SEM).

Habitat and Local Distribution. — Plants growing in shaded habitats. Rare: Upper Galilee, Mount Carmel (?), and Mount Hermon.

General Distribution. — Recorded from Southwest Asia (Cyprus, Turkey, Lebanon, Iran, and Afghanistan), around the Mediterranean (throughout; scattered records), Europe (mainly central and northern Europe), Macaronesia, North, tropical, and South Africa, northeastern, eastern, and central Asia, and North America.

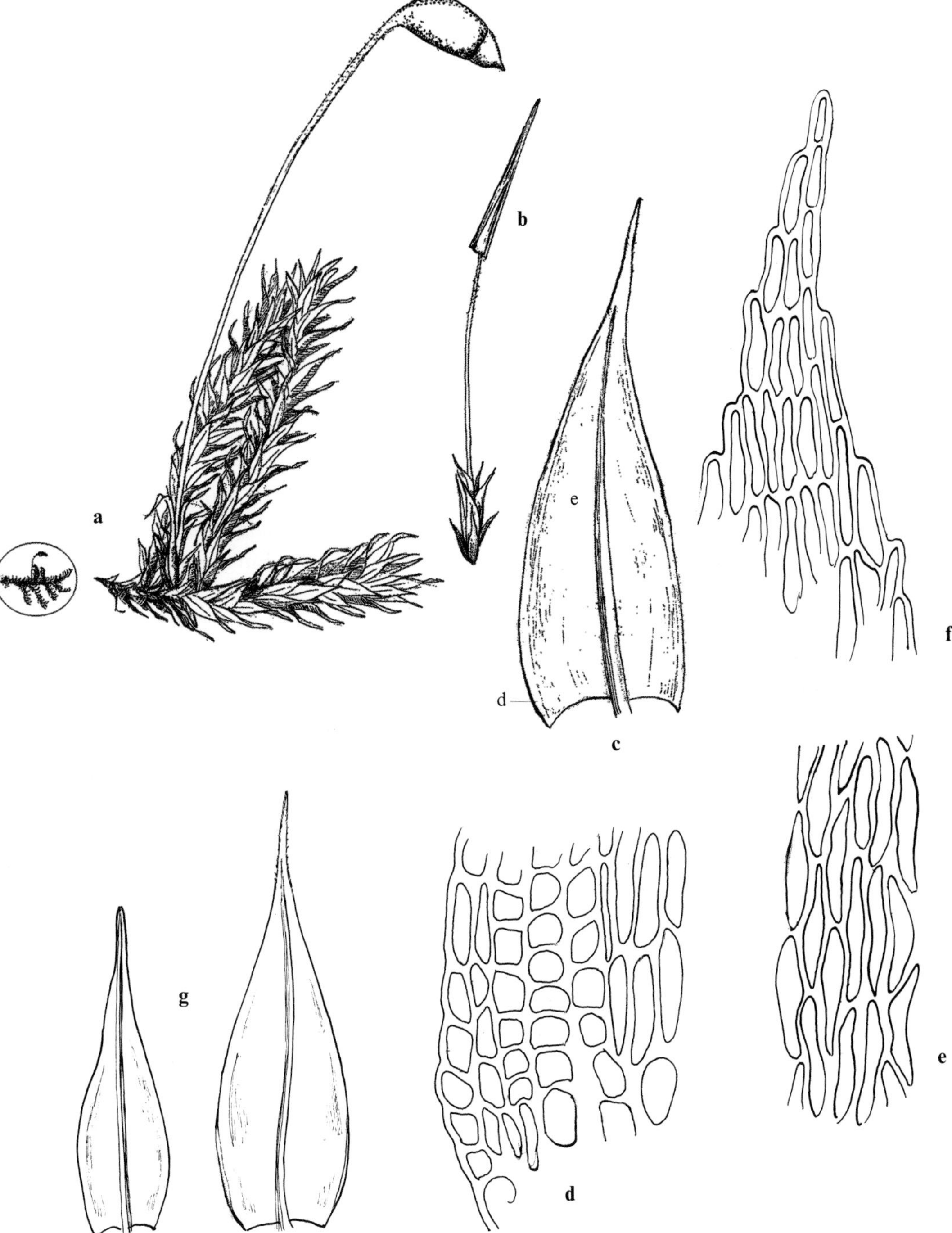

Figure 190. *Brachythecium populeum*: **a** part of habit, moist (×10); **b** young sporophyte with calyptra and perichaetial leaves (×10); **c** stem leaf (×50); **d** basal cells from margin towards costa; **e** mid-leaf cells; **f** apical cells (all leaf cells ×500); **g** branch leaves (×50). (*Kushnir*, 9 Oct. 1943)

4. *Scleropodium* B.S.G.

Plants glossy, slender to robust, growing in mats. *Branches* often distinctly julaceous, sometimes complanate. *Leaves*: stem and branch leaves similar or different in shape and size, usually imbricate when dry, deeply concave and at least ± plicate, broadly ovate to ovate-lanceolate, obtuse and apiculate to acuminate, decurrent at base; margins plane, or recurved at base, entire or finely serrulate above; costa single or forked, when single up to ¾ the length of leaf, often ending in a very short dorsal projection; mid- and upper cells linear-rhomboidal to linear-flexuose, shorter towards base, alar cells quadrate to rectangular, apical cells considerably shorter than upper cells and different in shape, hexagonal-rhomboidal (at least in local plants). *Dioicous*. *Setae* smooth or rough; capsules slightly inclined to horizontal, ± curved, oblong-cylindrical; annulus differentiated; operculum conical to short-rostrate.

A small genus of ca. 10 species, mainly in the Northern Hemisphere. Three species occur in the local flora.

The concept of the genus *Scleropodium* differs in the literature. Often, the widespread *S. purum* is placed together with two other species in *Pseudoscleropodium* (Limpr.) M. Fleisch. *in* Broth. (included by some authors in the Entodontaceae). In local plants, the costa often ends in a short dorsal projection, a character usually not associated with *Scleropodium*. The discrimination of sterile plants of *Scleropodium* from *Rhynchostegium murale* (Hedw.) B.S.G. may be sometimes very difficult (see the discussion of *R. murale* below, p. 492).

1 Stems subpinnately branched, complanate; branch leaves ± laxly spaced, with reflexed apiculus; setae smooth **1. S. purum**

1 Stems irregularly branched, not complanate; branch leaves densely spaced, apiculus not reflexed; setae rough.

  2 Branches ± julaceous; leaves concave, branch leaves gradually tapering to a rather long, acuminate apex, stem leaves abruptly narrowed into apiculus; capsules slightly inclined (sporophytes not seen so far in local plants) **2. S. cespitans**

  2 Branches distinctly julaceous; leaves strongly concave, branch and stem leaves abruptly narrowed into acute apex; capsules inclined to horizontal **3. S. touretii**

1. Scleropodium purum (Hedw.) Limpr., Laubm. Deutschl. 3 : 147 (1896).
[Bilewsky (1965) : 423]

Hypnum purum Hedw., Sp. Musc. Frond.: 253 (1801); *Pseudoscleropodium purum* (Hedw.) M. Fleisch. *in* Broth., Nat. Pflanzenfam., Edition 2, 11 : 395 (1925).

Distribution map 192 (p. 486), Figure 191.

Plants ± robust, pale green to yellowish-green. *Stems* creeping or ascending, subpinnately branched, complanate; branches ascending to nearly erect. *Leaves*: stem and branch leaves ± similar, branch leaves imbricate, laxly spaced, strongly concave, plicate when dry, up to 2 mm long, broad-ovate or rounded, abruptly narrowing into reflexed apiculus; margins plane, sometimes recurved below, entire to weakly serrulate

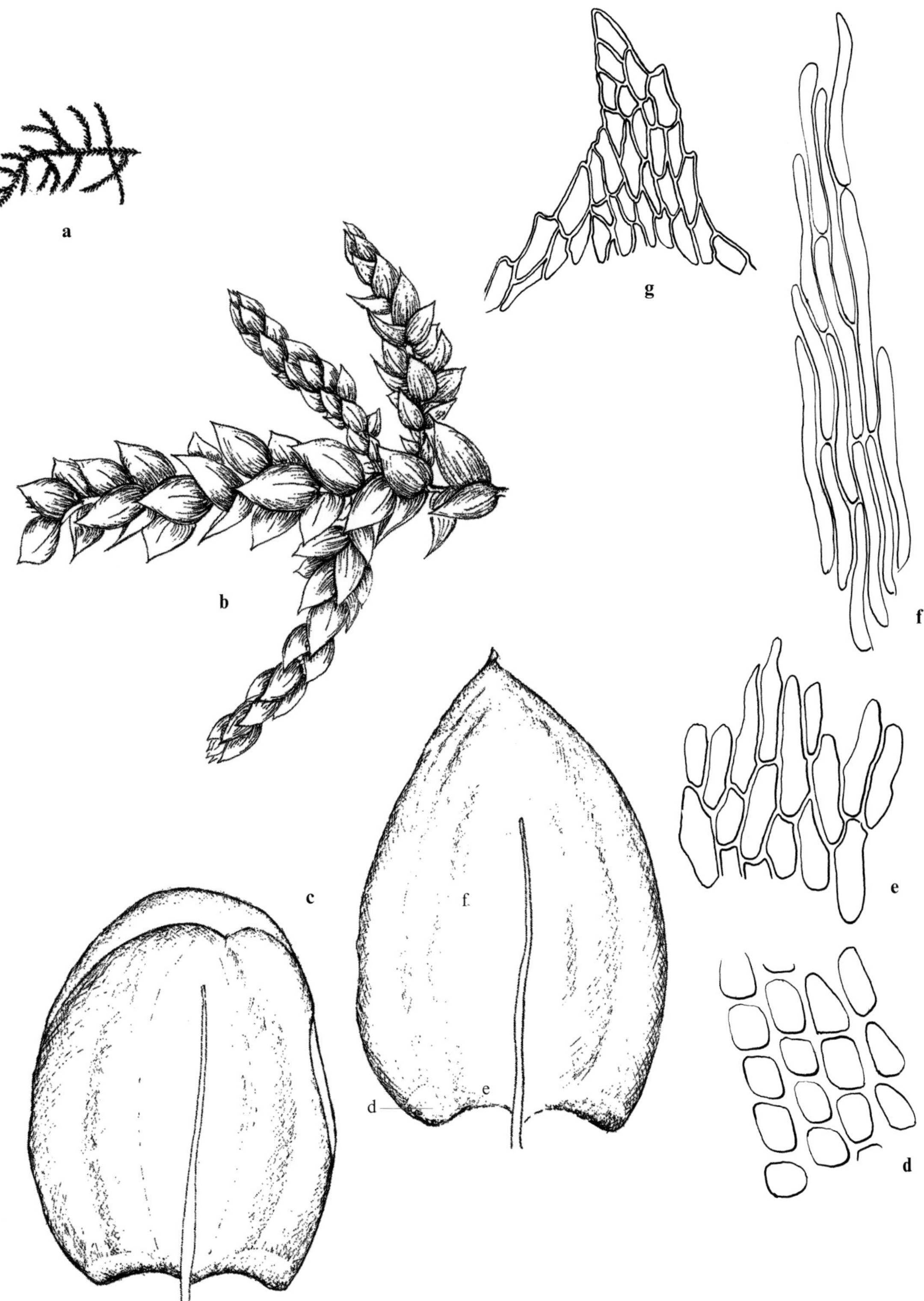

Figure 191. *Scleropodium purum*: **a** habit (×1); **b** part of habit, moist (×10); **c** leaves (×50); **d** alar cells; **e** basal cells; **f** mid-leaf cells; **g** apical cells (all leaf cells ×500). (*Mattatia*, 10 Feb. 1970)

above; costa single, very slender, ½–⅔ the length of leaf, rarely forked; basal cells irregularly rhomboidal, alar cells ± differentiated, subquadrate, mid-leaf cells linear-flexuose, 15–25 times longer than wide. *Sporophytes* not seen in local plants. *Setae* recorded as smooth, with horizontal capsules.

Habitat and Local Distribution. — Plants growing in loosely interwoven mats, on soil, rocks, and tree bark, in at least partly shaded habitats, on northern slopes, near running water, also on road banks, mainly in mountainous areas. Occasional: Upper Galilee, Lower Galilee, Samaria, and Judean Mountains.

General Distribution. — Recorded from Southwest Asia (Cyprus, Turkey, and Iran), around the Mediterranean (throughout), Europe (throughout), Macaronesia, tropical and North Africa, eastern Asia, New Zealand (introduced; Lewinsky & Barlett 1982), and Oceania.

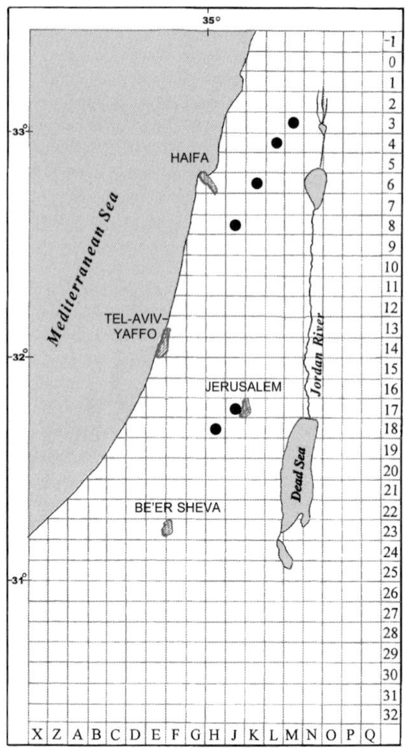

Map 192: *Scleropodium purum*

Scleropodium purum can be identified by the julaceous stems, the ± nearly erect and complanate branches, and the concave, broad-ovate or rounded leaves with reflexed apiculus.

2. Scleropodium cespitans (Müll. Hal.) L. F. Koch, Leafl. W. Bot. 6:31 (1950).
Hypnum cespitans Müll. Hal., Syn. Musc. Frond. 2:354 (1851); *Scleropodium caespitosum* B.S.G., Bryol. Eur. 6:28 (1853).

Distribution map 193 (p. 488), Figure 192.

Plants small to medium-sized, green to yellowish-green. *Stems* irregularly branched; main stems creeping, secondary stems ascending; branches ± erect to slightly curved, ± julaceous, not complanate. *Leaves* densely spaced, imbricate when dry and moist, concave, weakly plicate or not plicate; stem leaves ovate-oblong, abruptly narrowed into apiculus; branch leaves ca. 1.5 mm long, ovate-oblong, gradually tapering into a rather long, acuminate apex; margins plane, entire to weakly serrulate above; costa single, ca. ¾ the length of leaf, sometimes shorter and forked; basal cells short-rhomboidal to rectangular, alar cells quadrate to rectangular, wider, mid-leaf cells very narrow, linear-rhomboidal to linear-flexuose, 15–20(–25) times longer than wide. *Sporophytes* not seen in local plants. *Setae* recorded as rough, with slightly inclined capsules.

Habitat and Local Distribution. — Plants growing in ± loose mats, on calcareous and basaltic soils. Occasional: Upper Galilee, Lower Galilee, Mount Carmel, Judean Mountains, and Mount Hermon.

Figure 192. *Scleropodium cespitans*: **a** habit (×1); **b** part of habit, moist (×10); **c** part of habit, dry (×10); **d** stem leaves (×50); **e** basal cells including alar cells; **f** marginal upper cells; **g** apical cells (all leaf cells ×500). (*Heyn, Herrnstadt & Ben-Sasson 81-172-14*)

General Distribution. — Records few and scattered; recorded from Southwest Asia (Turkey), around the Mediterranean (few records from the south and the east), Europe (few records), Macaronesia, and North America.

3. Scleropodium touretii (Brid.) L. F. Koch, Rev. Bryol. Lichénol. 18 : 177 (1949). *Hypnum touretti* Brid., Sp. Musc. Frond. 2 : 185 (1812); *S. illecebrum* auct., non B.S.G. (1853). — Bilewsky (1965) : 423.

Distribution map 194, Figure 193, Plate XIV : d.

Plants medium-sized, pale green to yellowish-green. *Stems* creeping, irregularly branched; branches distinctly julaceous, shorter than in *Scleropodium purum*, ascending, curved when dry. *Leaves*: stem and branch leaves ± similar; branch leaves imbricate, densely spaced, strongly concave, weakly plicate; ca. 1.5 mm long, ovate, abruptly narrowed into acute apex, apiculate; margins plane, entire below, serrate near apex; costa single ca. ¾ the length of leaf, sometimes shorter and forked; basal cells rectangular to rhomboidal, alar cells quadrate to rectangular, wider, mid-leaf cells linear-flexuose, 10–15(–20) times longer than wide, cells at apex rhomboidal. *Setae* rough; capsules inclined to horizontal, asymmetrical, up to 2.5 mm long including operculum; operculum long-conical. *Spores* 11–16 µm in diameter, finely papillose; densely covered by compound gemmae of various sizes (seen with SEM).

Habitat and Local Distribution. — Plants growing in dense mats, usually in shaded habitats, often in maquis, on soil, rocks, rock crevices, and on tree trunks. Common;

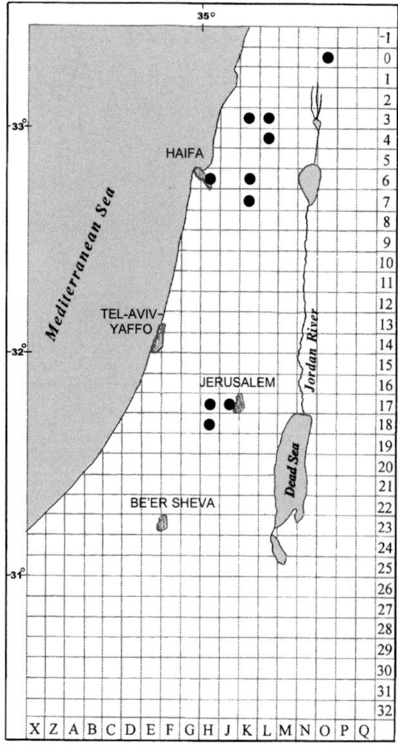

Map 193: *Scleropodium cespitans*

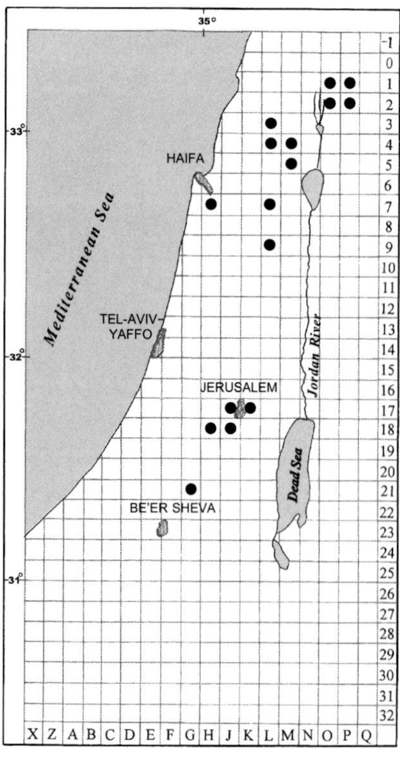

Map 194: *Scleropodium touretii*

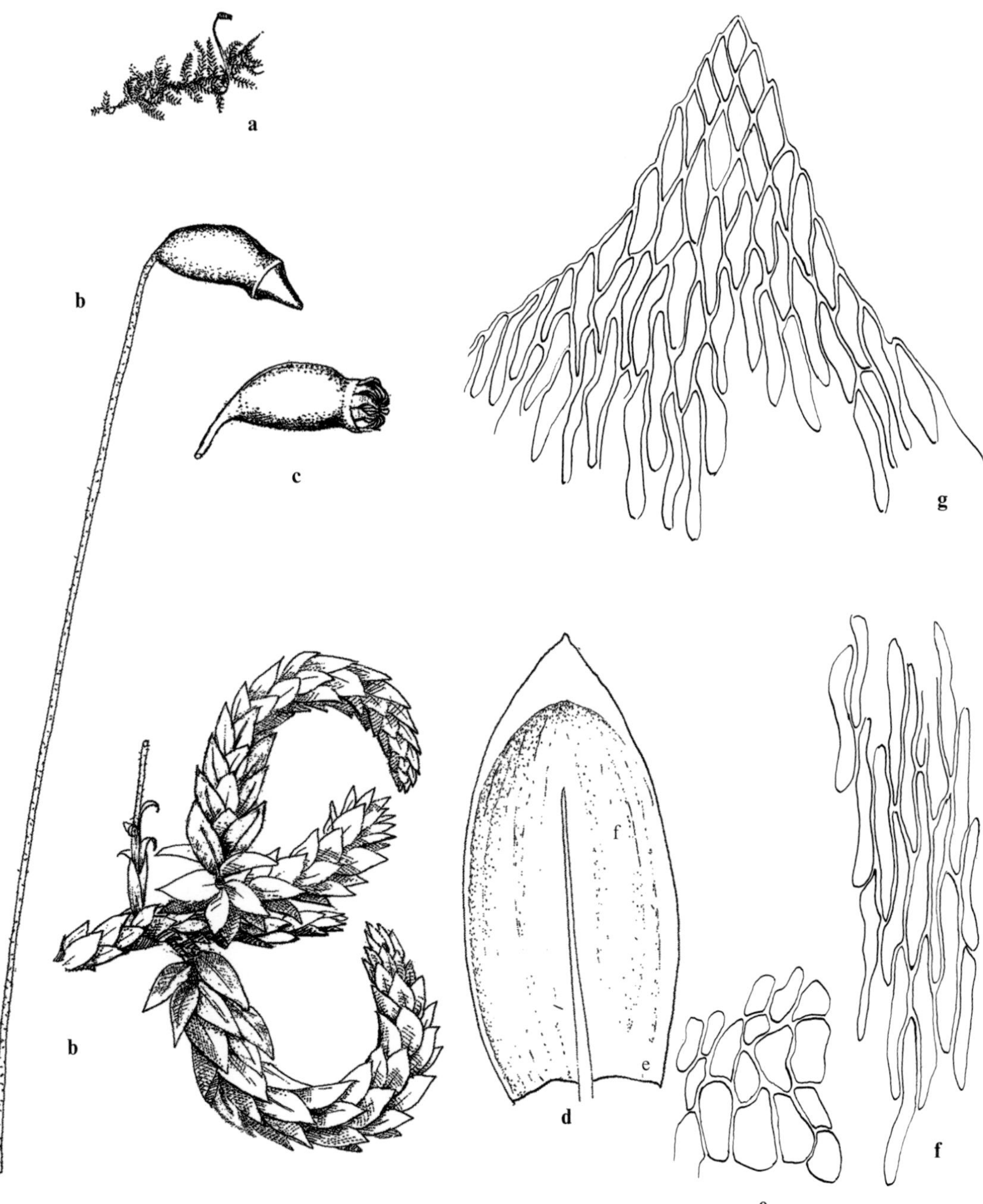

Figure 193. *Scleropodium touretii*: **a** habit (×1); **b** part of habit, moist (×10); **c** deoperculate capsule (×10); **d** stem leaf (×50); **e** alar cells; **f** mid-leaf cells, (all leaf cells ×500); **g** apical cells (×500). (**a**–**c**: *Kushnir*, 5 Jan. 1943; **d**–**g**: *Wilczek*, 28 Aug. 1944)

locally the most widespread *Scleropodium* species: Upper Galilee, Lower Galilee, Mount Carmel, Mount Gilboa, Judean Mountains, Mount Hermon, and Golan Heights.

General Distribution. — Recorded from Southwest Asia (Cyprus, Turkey, Syria, Lebanon, and Jordan), around the Mediterranean (throughout), Europe (throughout), Macaronesia, North Africa, and North and Central America.

Scleropodium touretii differs from *S. cespitans* mainly in the slightly larger size of the plants and the strongly concave leaves, all abruptly narrowed into acute apex. The differentiating character of the length versus width of the mid-leaf cells, used by some authors (e.g., Grout 1928, Smith 1978), was not found by us to be applicable. The cell proportions were found variable within each species, and values are often intermediate.

5. *Rhynchostegium* B.S.G.

Plants slender to moderately robust. *Stems* creeping or ascending, ± irregularly branched. *Leaves*: stem and branch leaves similar or branch leaves smaller; leaves imbricate, erectopatent to spreading, sometimes complanate, ± concave, smooth to plicate, broadly ovate to ovate-lanceolate, obtuse and apiculate, or acute or acuminate; margins plane or ± recurved below, entire to serrulate; costa ½ – ¾ the length of leaf, not ending in a dorsal projection (in local plants); basal cells shorter and wider than cells above; mid- and upper leaf cells narrow-rhomboidal to linear, becoming shorter and wider at base; alar cells sometimes differentiated; apical cells not differentiated. *Autoicous*. *Setae* smooth; capsules inclined to horizontal or pendulous, ± curved, asymmetrical, oblong to cylindrical; annulus differentiated; cilia nodulose or appendiculate; operculum obliquely long-rostrate.

A large genus distributed throughout the Northern Hemisphere and comprising ca. 200 species. Three species occur in the local flora.

1 Branches elongate; rhizoids usually at stem base only; leaves long-acuminate, apex distinctly twisted **3. R. megapolitanum**
1 Branches short; rhizoids along stem; leaves obtuse, acute or short-acuminate, apex plane, sometimes slightly twisted.
  2 Plants medium-sized; leaves not complanate, strongly concave, apex obtuse or short-acute, margins entire below, ± serrulate above **1. R. murale**
  2 Plants small and slender; leaves ± complanate and concave, apex acute to short-acuminate, margins serrulate all around **2. R. confertum**

1. Rhynchostegium murale (Hedw.) B.S.G., Bryol. Eur. 5:207 (1852).
Hypnum murale Hedw., Sp. Musc. Frond. 240 (1801); *Eurhynchium murale* (Hedw.) Milde, Bryol. Siles. 310 (1869).

Distribution map 195 (p. 492), Figure 194.

Plants medium-sized, green to yellowish-green. *Stems* creeping; rhizoids along stem; branches short, sometimes slightly curved. *Leaves* imbricate, not complanate, strongly concave, sometimes plicate when dry, up to 1.5 mm long, ovate, obtuse, or short-acute,

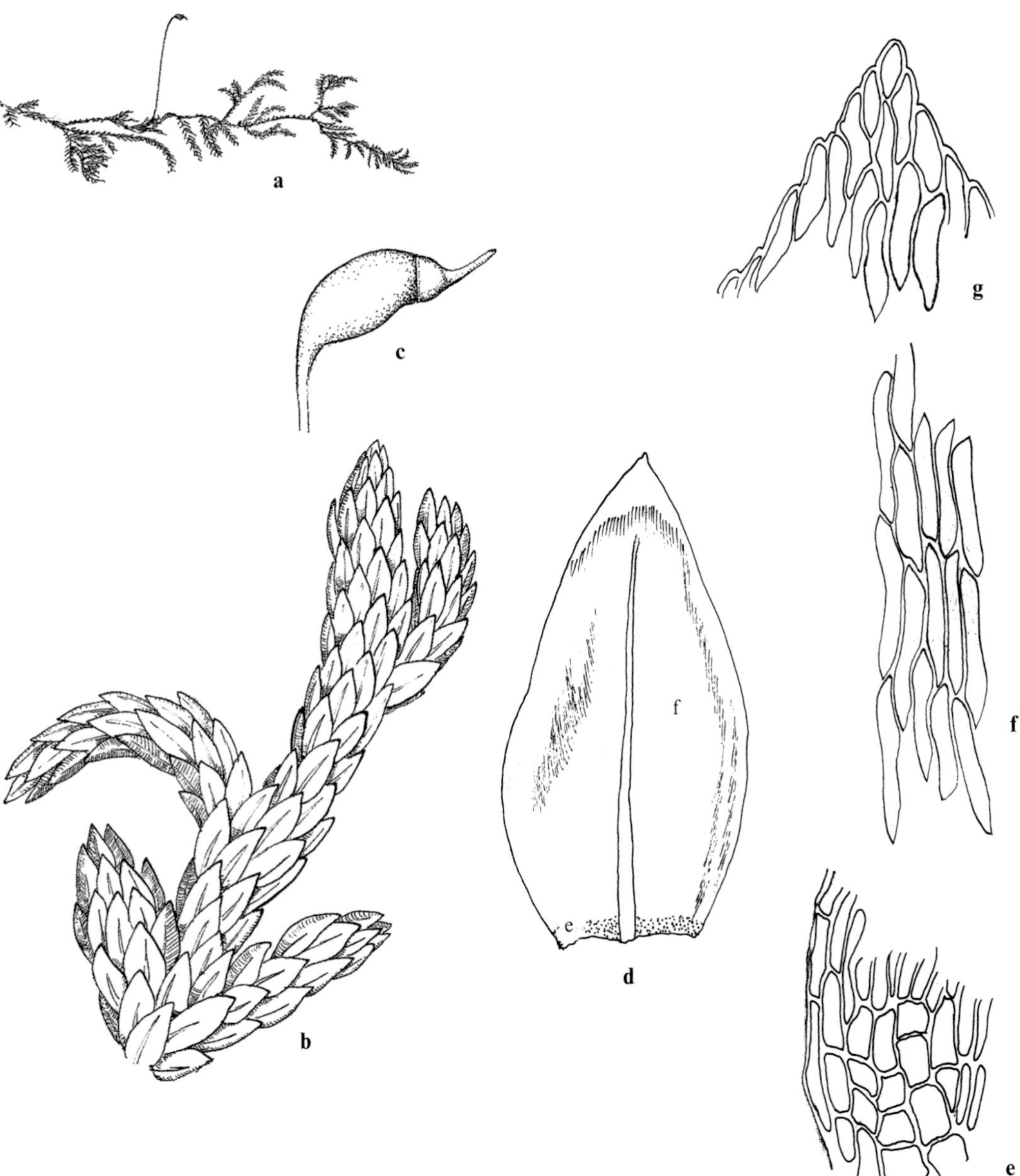

Figure 194. *Rhynchostegium murale*: **a** habit (×1); **b** part of habit, moist (×10); **c** capsule (×10); **d** branch leaf (×50); **e** alar cells; **f** mid-leaf cells; **g** apical cells (all leaf cells ×500).
(*Crosby & Herrnstadt 79-99-9*)

apex plane; margins plane, entire below, ± serrulate above; costa extending to ½ – ⅓ the length of leaf; basal cells rhomboidal, alar cells short-rectangular; mid-leaf cells linear-rhomboidal, eight–fifteen times longer than wide. *Sporophytes* only rarely seen in local plants.

Habitat and Local Distribution. — Plants growing on soil layers on rocks in shaded habitats; hidden among species of *Scleropodium touretii*. Rare to occasional, perhaps more widespread: Lower Galilee, Judean Mountains, Mount Hermon, and Golan Heights (?).

General Distribution. — Recorded from Southwest Asia (Turkey, Syria, and Afghanistan), around the Mediterranean (rare; cf. Duell 1992), Europe (various parts), Macaronesia, North Africa, and eastern Asia.

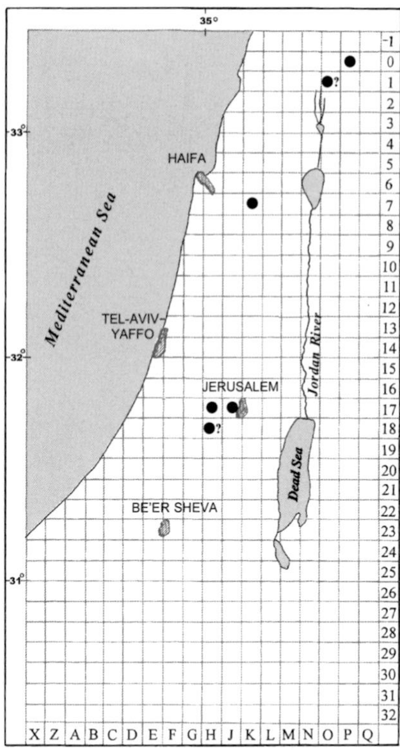

Map 195: *Rhynchostegium murale*

Sterile plants of *Rhynchostegium murale* are easily confused with those of local species of *Scleropodium*, in particular *S. touretii*. They can be distinguished at closer study by several leaf characters: the leaves of *R. murale* have apical cells that are not differentiated from the mid-leaf cells and the costa is devoid of a dorsal projection. Conversely, the leaves of *Scleropodium* have smaller apical cells than mid-leaf cells and often a dorsal projection. However, we have found in some plants, which have generally the suite of characters of *R. murale*, that a few leaves on a stem may bear a short dorsal projection. The long-rostrate operculum of *Rhynchostegium* clearly differs from the conical to short-rostrate operculum of *Scleropodium*, but in local populations of both *S. touretii* and *R. murale* sporophytes are usually not formed.

In one population (Judean Mountains: "Ein Kerem"), plants with young sporophytes, closely resembling *Rhynchostegium murale* in habit, were found to have a long-rostrate operculum typical of the genus but leaves with a dorsal projection of the costa and a rough setae (a character sometimes found in *Scleropodium* but not in *Rhynchostegium*).

2. Rhynchostegium confertum (Dicks.) B.S.G., Bryol. Eur. 5 : 203 (1852). [Bilewsky & Nachmony (1955)]

Hypnum confertum Dicks., Pl. Crypt. Brit. 4 : 17 (1801); *Eurhynchium confertum* (Dicks.) Milde, Bryol. Siles. 309 (1869). — Bilewsky (1965) : 428.

Distribution map 196 (p. 494), Figure 195.

Plants small and slender, dull green. *Stems* creeping; rhizoids along stem; branches short. *Leaves* ± spreading, sometimes complanate, ca. 1 mm long, narrow-ovate to ovate-lanceolate, acute to short-acuminate, with apex sometimes slightly twisted;

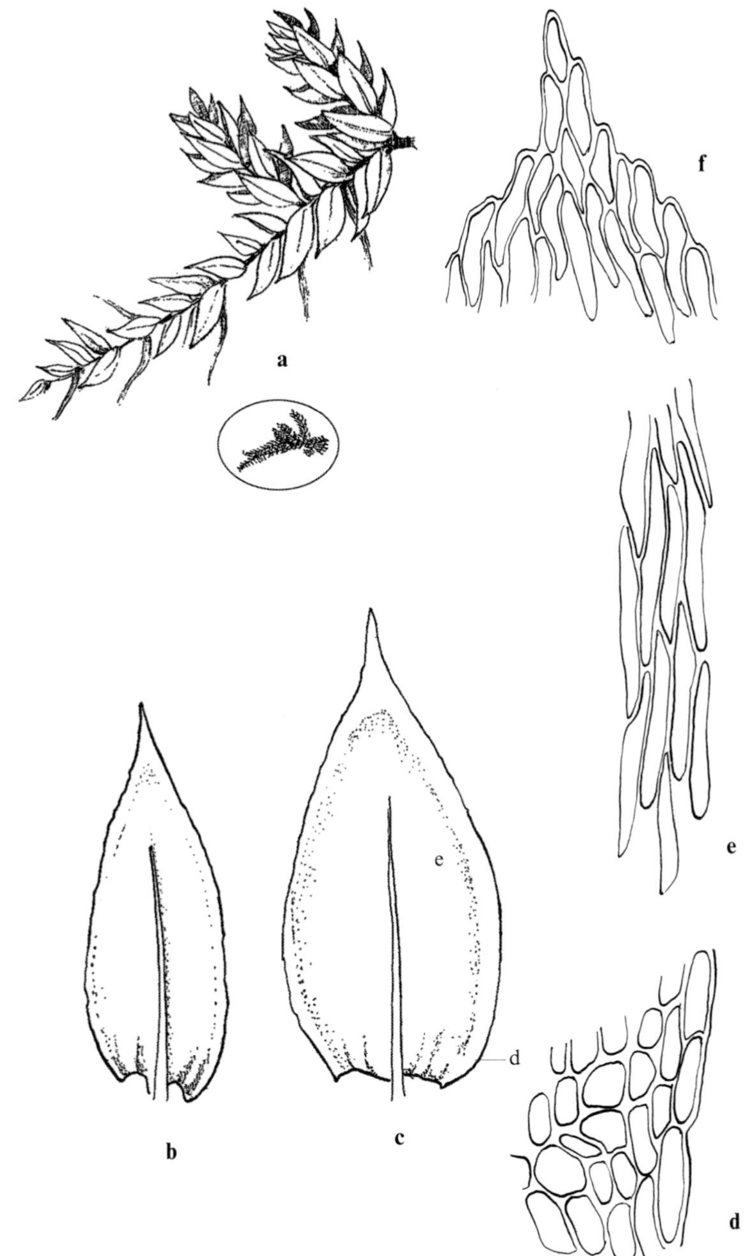

Figure 195. *Rhynchostegium confertum*: **a** part of habit, moist (×10); **b** branch leaf (×50); **c** stem leaf (×50); **d** alar cells; **e** mid-leaf cells; **f** apical cells (all leaf cells ×500). (*Bilewsky*, 10 Apr. 1954, No. 151 in Herb. Bilewsky)

margins plane, serrulate all around; costa extending to ⅔–¾ the length of leaf; basal cells rhomboidal, alar cells irregularly quadrate-rectangular; mid-leaf cells linear-rhomboidal, seven–ten times longer than wide. *Sporophytes* not seen in local plants.

Habitat and Local Distribution. — Plants on rocks in shaded and moist habitats. Rare to occasional: Upper Galilee, Mount Carmel, Shefela, and Judean Mountains.

General Distribution. — Recorded from Southwest Asia (Cyprus and Turkey), Europe (widespread throughout), Macaronesia, North Africa, and northeastern and eastern Asia.

Plants collected in Israel are somewhat smaller in size than plants described from Europe.

3. Rhynchostegium megapolitanum (F. Weber & D. Mohr) B.S.G., Bryol. Eur. 5:204 (1852).

Hypnum megapolitanum Blandow ex F. Weber & D. Mohr, Bot. Taschenb. 325 (1807); *Eurhynchium megapolitanum* (F. Weber & D. Mohr) Milde, Bryol. Siles. 311 (1869). — Bilewsky (1965):427.

Distribution map 197, Figure 196.

Plants medium-sized, vivid green to yellowish-green. *Stems* ascending; rhizoids usually at stem base only; branches elongate. *Leaves* ± shrunken, imbricate with appressed bases and spreading tips when dry, patent to spreading when moist, branch leaves up to 1.5 mm long, stem leaves up to 2 mm long, ovate to broadly ovate with narrowing

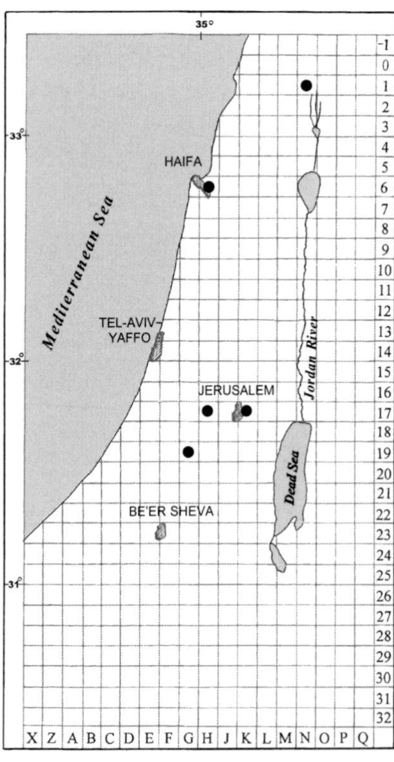

Map 196: *Rhynchostegium confertum*

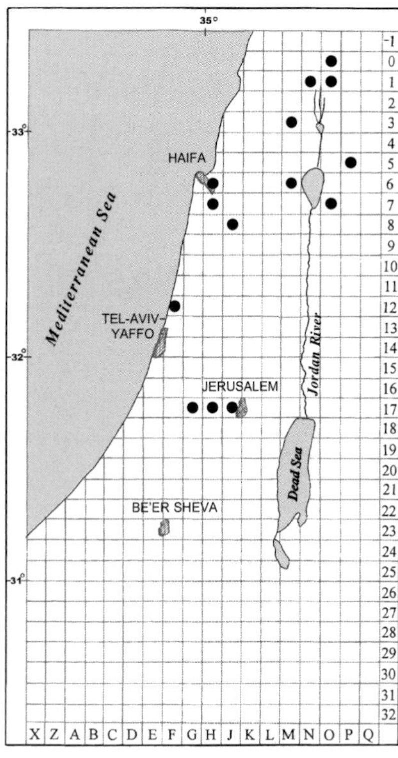

Map 197: *Rhynchostegium megapolitanum*

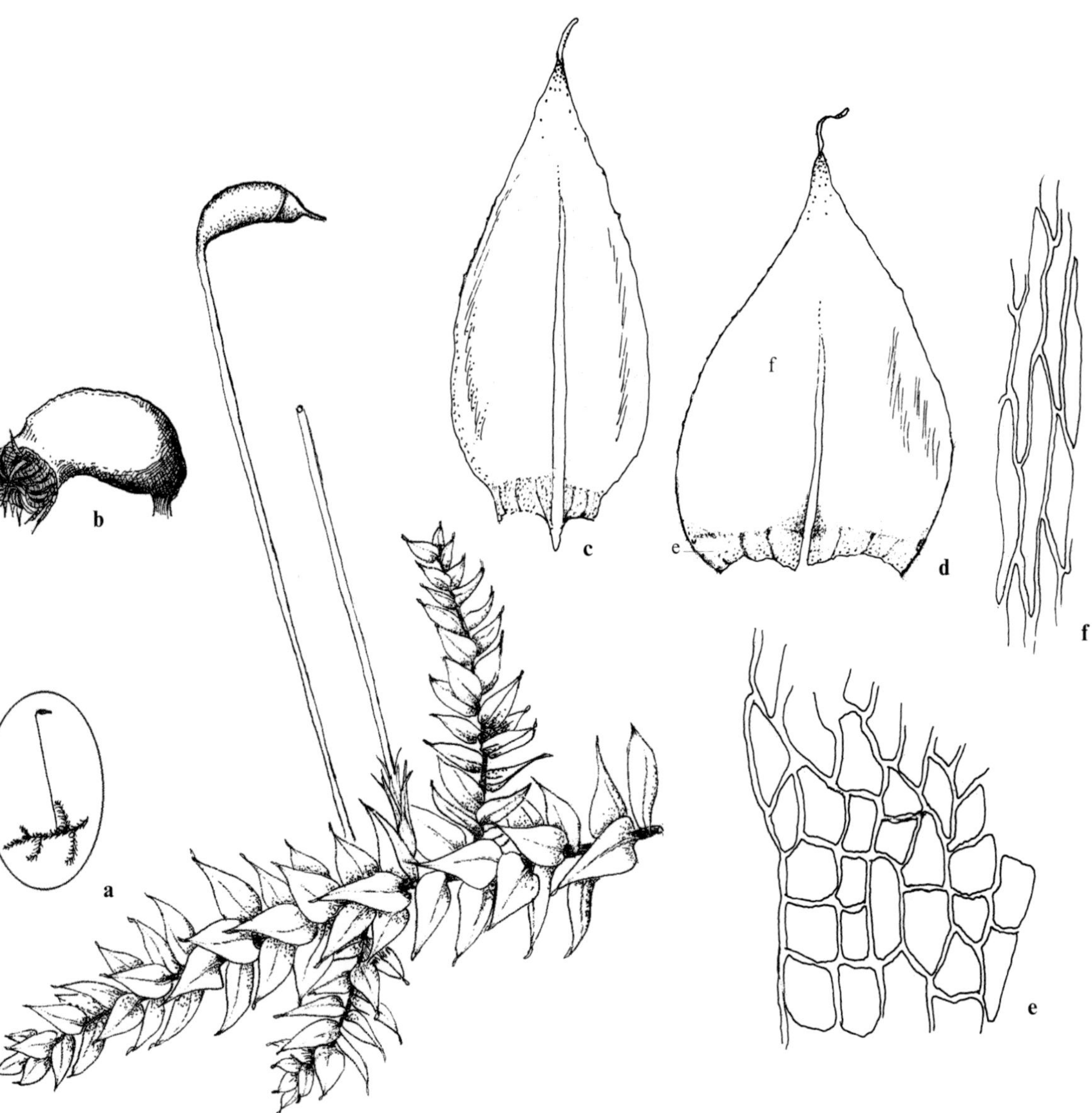

Figure 196. *Rhynchostegium megapolitanum*: **a** part of habit, moist (×10); **b** deoperculate capsule (×20); **c** branch leaf (×50); **d** stem leaf (×50); **e** alar cells; **f** mid-leaf cells (all leaf cells ×500). (*Herrnstadt & Crosby 78-23-10*)

base, long-acuminate, apex distinctly twisted; margins plane, serrate around; costa extending to ca. ⅔ the length of leaf; basal cells numerous, rhomboidal to short-rectangular, much wider than mid-leaf cells; mid-leaf cells linear-rhomboidal, ca. ten times longer than wide. *Sporophytes* common. Spores ± smooth, 11–16 μm in diameter.

Habitat and Local Distribution. — Plants growing in shaded habitats, in soil pockets on rocks, in rock crevices, on walls, and near running water. Occasional to common, the most widespread local *Rhynchostegium* species: Sharon Plain, Upper Galilee, Lower Galilee, Mount Carmel, Samaria, Shefela, Judean Mountains, Mount Hermon, and Golan Heights.

General Distribution. — Recorded from Southwest Asia (Cyprus, Turkey, and Lebanon) the Mediterranean (throughout), Europe (throughout), Macaronesia, North Africa, and eastern Asia.

Rhynchostegium megapolitanum can be easily distinguished by the long branches and the leaves with appressed bases and the distinctly twisted apex. *Rhynchostegium megapolitanum* var. *meridionale* Schimp., recorded by some authors as occurring throughout the Mediterranean, is not recognised here as a separate taxon.

6. *Eurhynchium* B.S.G.

Plants slender to moderately robust, often glossy. *Stems* usually creeping, ± irregularly to pinnately branched; branches straight or some curved when dry. *Leaves*: stem and branch leaves similar or branch leaves ± larger, sometimes different in shape; leaves erect to spreading, plane or sometimes moderately concave, often plicate, ovate-lanceolate to cordate-triangular, acute to long-acuminate, usually ± decurrent; margins plane or recurved below, often serrulate all around; costa half the length of leaf or longer, often ending in a distinct dorsal projection (rarely more than one); mid- and upper leaf cells narrow-rhomboidal to linear, becoming shorter and wider at base, sometimes alar cells differentiated, apical cells much shorter than mid-leaf cells and different in shape. *Dioicous* usually (exceptions among the local species: *Eurhynchium speciosum* and *E. riparioides*). *Setae* elongate, rough or smooth; capsules inclined to horizontal or pendulous, ± curved, asymmetrical, oblong to cylindrical; annulus differentiated; peristome with cross-striolae in lower part, endostome pale, keeled, and perforated, ± equal in length to exostome teeth, cilia nodulose or appendiculate; operculum obliquely long-rostrate. *Spores* ± smooth (in local species).

A genus comprising ca. 85 species. Seven species occur in the local flora.

Bilewsky (1965) recognised the genus *Eurhynchium* and cited *Rhynchostegium* as its synonym. He recorded from Israel six species of *Eurhynchium*. Bilewsky (1974) recorded the same species as belonging to three genera: *Rhynchostegium*, *Oxyrrhynchium*, and *Platyhypnidium*. McFarland (*in* Sharp *et al.* 1994) considered *Rhynchostegium* and *Eurhynchium* from Mexico as part of the single genus *Rhynchostegium*.

As treated here, *Eurhynchium* and *Rhynchostegium* are differentiated by two main characters: in *Eurhynchium* the costa of the leaf ends in a dorsal projection and the leaves have apical cells, which are shorter than the mid-leaf cells, whereas in *Rhynchostegium* the costa lacks a projection and the apical and mid-leaf cells are ± equal. Accordingly, in this treatment *Eurhynchium* includes *E. riparioides*, which is considered as *Rhynchostegium riparioides* in many European accounts of these genera (e.g., Nyholm 1954, Smith 1978, Corley *et al.* 1981, Frahm & Frey 1983, and Frahm 1995) and also in Herrnstadt *et al.* (1991). The record of *E. praelongum* (Hedw.) B.S.G. from Israel in Herrnstadt *et al.* (1991) is based on misidentified herbarium specimens.

1 Plants submerged, robust, forming large mats or wefts, autoicous; setae smooth **1. E. riparioides**
1 Plants of wet habitats but not submerged* usually smaller, not forming large mats or wefts, dioicous**; setae smooth or rough.
  2 Stem leaves cordate-triangular, strongly plicate.
    3 Plants medium-sized; leaves with irregularly quadrate alar cells, about equal in size to basal cells; leaf margins entire or finely serrate **3. E. meridionale**
    3 Plants large; leaves with rectangular alar cells, ± larger than basal cells; leaf margins serrulate to serrate all around **2. E. striatum**
  2 Stem leaves ovate-oblong to ovate-lanceolate, not or weakly plicate.
    4 Stem and branch leaves different, branch leaves distinctly concave; setae smooth **4. E. pulchellum**
    4 Stem and branch leaves ± similar, branch leaves not concave; setae rough.
      5 Autoicous or synoicous; leaves 1.5–2 mm long **7. E. speciosum**
      5 Dioicous; leaves 0.75–1.3 mm long.
        6 Leaf apex not twisted; mid-leaf cells six–eleven times longer than wide **5. E. hians**
        6 Leaf apex often twisted; mid-leaf cells ten–sixteen times longer than wide **6. E. schleicheri**

* In *E. striatum*, periodically submerged.
** Except for *E. speciosum*.

1. Eurhynchium riparioides (Hedw.) P. W. Richards, Ann. Bryol. 9 : 135 (1937).
Hypnum riparioides Hedw., Sp. Musc. Frond. 242 (1801); *E. rusciforme* (B.S.G.) Milde, Bryol. Siles. 312 (1869). — Bilewsky (1965) : 426; *Rhynchostegium riparioides* (Hedw.) Cardot *in* Tourret, Bull. Soc. Bot. France 60 : 231 (1913); *Platyhypnidium riparioides* (Hedw.) Dixon, Rev. Bryol. Lichénol. 6 : 111 (1934). — Bilewsky (1974).

Distribution map 198, Figure 197, Plate XIV : e.

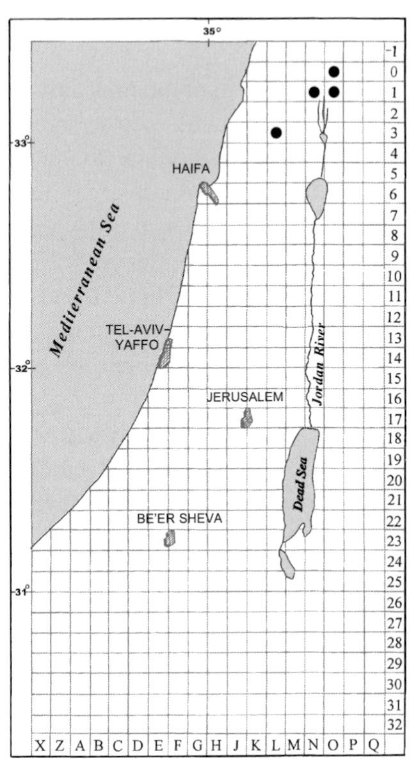

Map 198: *Eurhynchium riparioides*

Plants submerged, robust, forming very large mats or wefts, bright green or yellowish to brown, dark green and denuded below, often encrusted by lime and covered by cyanophytes. *Stems* creeping or ascending; branches elongate, considerably varying in size, up to 90 mm long, straight or sometimes with curved tips. *Leaves*: stem and branch leaves similar in shape, stem leaves 2–2.5 mm, branch leaves smaller, 1.5 mm long, spreading and ± imbricate when dry or moist, slightly contorted when dry, sometimes subsecund, concave, broadly ovate to ovate, acute or obtuse; margins plane, serrulate all around, nearly to base; costa $\frac{2}{3}$–$\frac{3}{4}$ the length of leaf; basal cells rhomboidal to irregularly rectangular, alar cells rectangular, mid-leaf cells seven–ten times longer than wide, narrow rhomboidal to linear, apical cells short rhomboidal. *Autoicous*. *Sporophytes* frequently present. *Setae* 15–23 mm long, smooth, orange, turning reddish; capsules inclined to horizontal, curved, ovate-ellipsoid, 1.5–2 mm long not including operculum. *Spores* 15–21 μm in diameter, nearly smooth; sparsely covered by small, irregularly shaped granules (seen with SEM).

Habitat and Local Distribution. — Plants growing in rigid, loose or dense, large mats or wefts, sometimes pendulous, submerged in standing or running water, attached at base to stones and rocks. Rare to occasional, locally very common in the sources of the Jordan River: Upper Galilee, Dan Valley, Golan Heights, and Mount Hermon.

General Distribution. — Widespread as a species of the temperate regions of the world; many records from throughout temperate Southwest Asia.

Some plants of *Eurhynchium riparioides*, when growing (in rare cases) unsubmerged may resemble *E. pulchellum*, but can be distinguished by the more robust habit and the more laxly spaced leaves along the stem.

Eurhynchium riparioides is often considered as a species of *Rhynchostegium*.

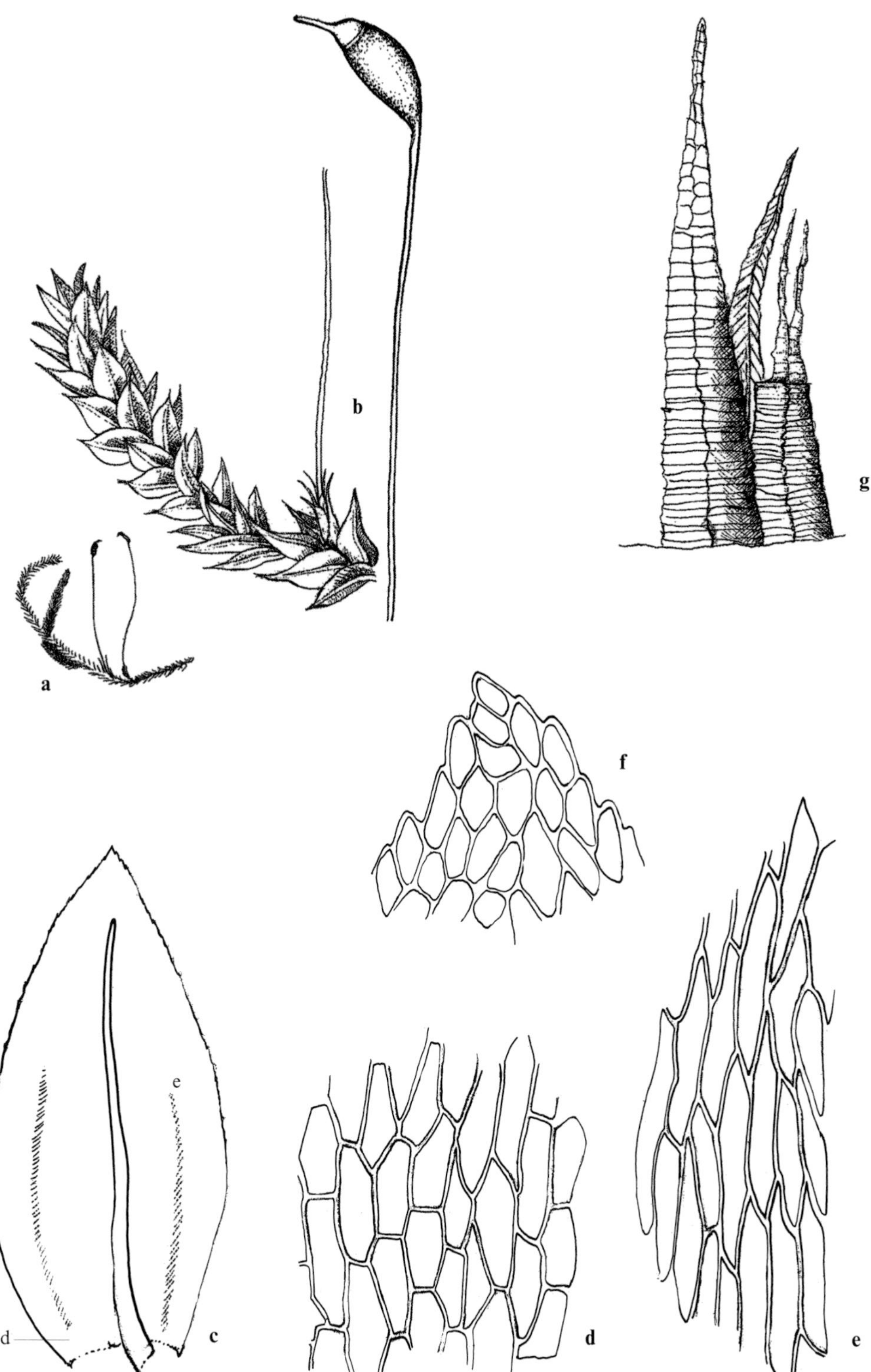

Figure 197. *Eurhynchium riparioides*: **a** habit (×1); **b** part of habit, moist (×10); **c** branch leaf (×50); **d** basal cells; **e** mid-leaf cells; **f** apical cells (all leaf cells ×500); **g** portion of peristome teeth (×150). (*Kushnir*, 10 Sep. 1942, No. 136 in Herb. Bilewsky)

2. Eurhynchium striatum (Hedw.) Schimp., Coroll. Bryol. Eur. 119:(1856).
Hypnum striatum Hedw., Sp. Musc. Frond. 275 (1801).

Distribution map 199, Figure 198.

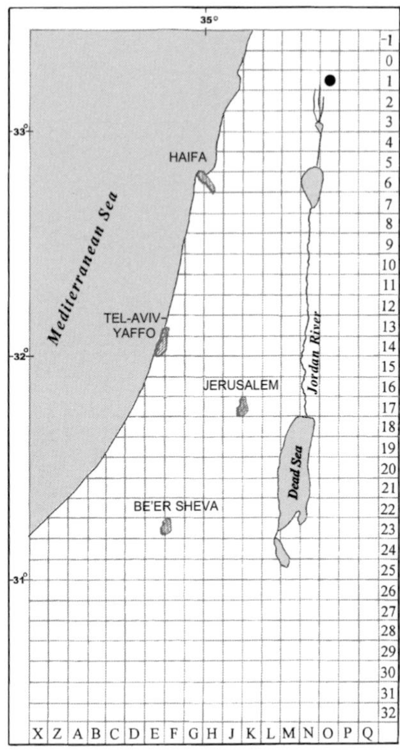

Map 199: *Eurhynchium striatum*

Plants medium-sized to robust, rigid, creeping, green to yellowish green, glossy. *Branches* with densely spaced, not complanate leaves. *Leaves*: stem leaves and branch leaves different; stem leaves plicate, ovate to cordate-triangular, acuminate; branch leaves strongly plicate, slightly contorted when dry, patent to spreading when moist, ca. 2 mm long, ovate-lanceolate, ± acute; margins plane, serrulate to serrate all around; costa extending to $\frac{4}{5}$ the length of leaf; basal cells rhomboidal, alar cells larger, rectangular, incrassate and yellowish, conspicuous; mid-leaf cells linear, 10–15 times longer than wide. *Sporophytes* dark red when mature. *Dioicous*. *Setae* smooth, ca. 10 mm long; capsules inclined, ± curved, ca. 2 mm long (not including operculum), ellipsoid to subcylindrical.

Habitat and Local Distribution. — Plants growing in interwoven mats or wefts on rocks near streams, periodically submerged. Very rare: Dan Valley.

General Distribution. — Recorded from Southwest Asia (few records; Turkey and Iran), around the Mediterranean (throughout), Europe (throughout), Macaronesia, North Africa, and northeastern, eastern, and central Asia.

Eurhynchium striatum is found in the Dan Valley growing together with *E. speciosum*, but can be distinguished by its coarser habit, the shorter, smooth seta, and by being dioicous (not synoicous).

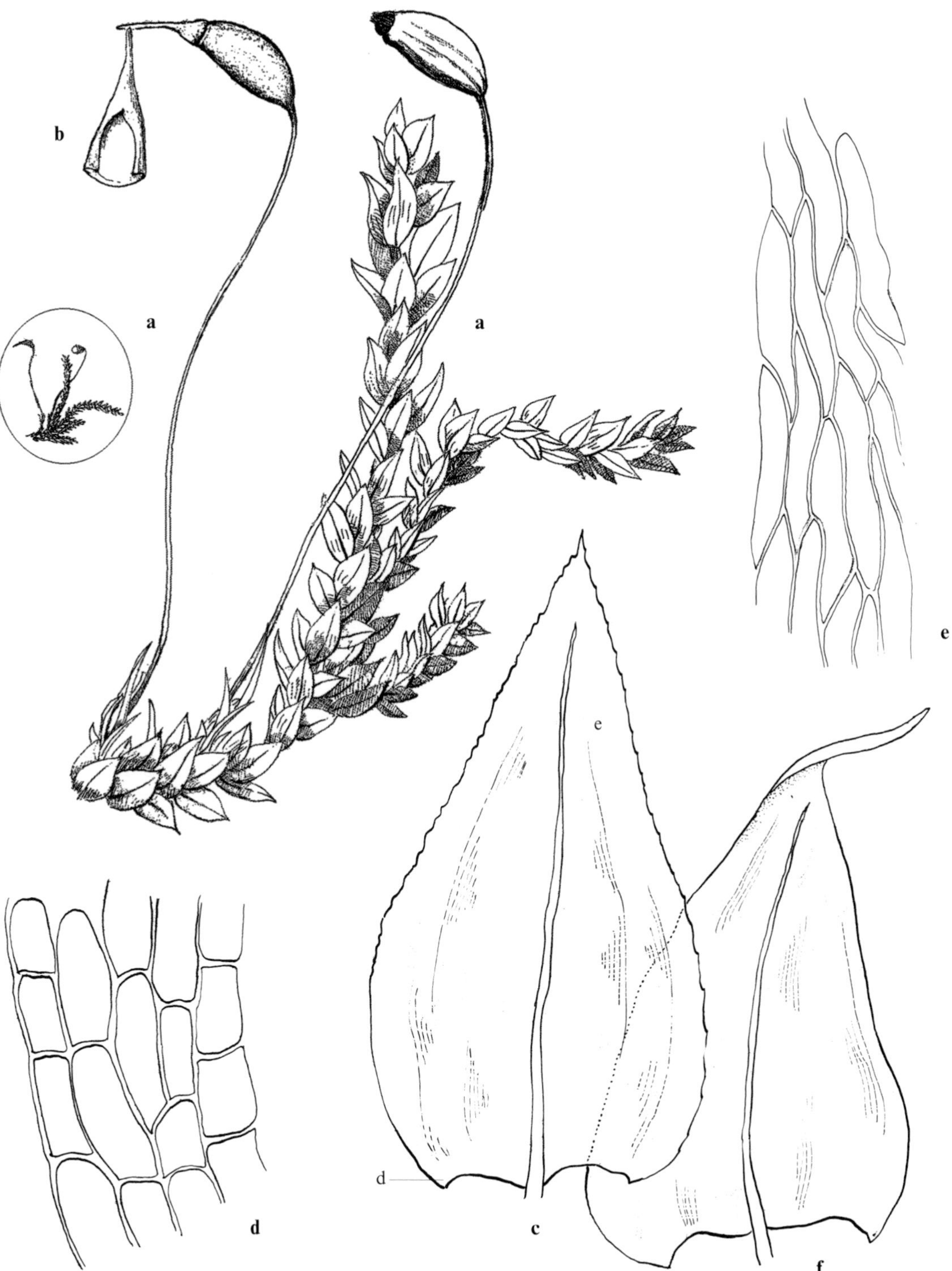

Figure 198. *Eurhynchium striatum*: **a** part of habit, moist (×10); **b** calyptra (×10); **c** branch leaf (×50); **d** alar cells; **e** upper cells (all leaf cells ×500); **f** stem leaf (×50). (*Heyn & Herrnstadt 82-228-2*)

3. Eurhynchium meridionale (B.S.G.) De Not. *in* Picc., Comment. Soc. Critt. Ital. 1 : 249 (1863). [Bilewsky (1965) : 426]

E. longirostre B.S.G. var. *meridionale* B.S.G., Bryol. Eur. 5 : 223 (1854); *Plasteurhynchium meridionale* (B.S.G.) M. Fleisch. *in* Broth., Nat. Pflanzenfam., Edition 2, 11 : 212 (1925). — Bilewsky 1974.

Distribution map 200, Figure 199.

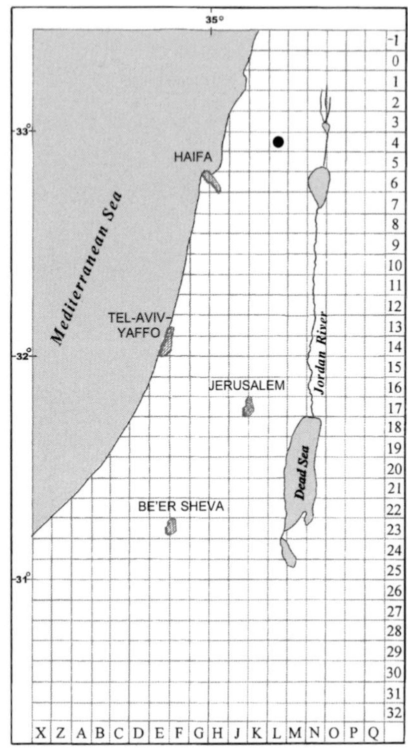

Map 200: *Eurhynchium meridionale*

Plants medium-sized; stems creeping, densely branched. *Leaves*: stem and branch leaves ± similar, some branch leaves shorter and narrower than stem leaves; leaves crowded, spreading, undulate when dry and moist, not complanate, strongly plicate, 1–2 mm long, cordate-triangular, acuminate; margins plane, entire or finely serrate; costa ca. ¾ the length of leaf; basal cells few, irregularly rectangular, alar cells, irregularly quadrate, incrassate; mid-leaf cells seven–ten times longer than wide; perigonial leaves ovate-lanceolate, gradually tapering into a strongly serrate acumen. Recorded as *dioicous*. *Sporophytes* not seen in local plants.

Habitat and Local Distribution. — Plant growing on rocks in shade of trees. Found only in one locality: Upper Galilee.

General Distribution. — Recorded from Southwest Asia (Turkey), around the Mediterranean (scattered data throughout), Europe (mainly the south), Macaronesia, and North Africa.

Eurhynchium meridionale is distinguished from other local *Eurhynchium* species by the spreading, undulate, and crowded leaves.

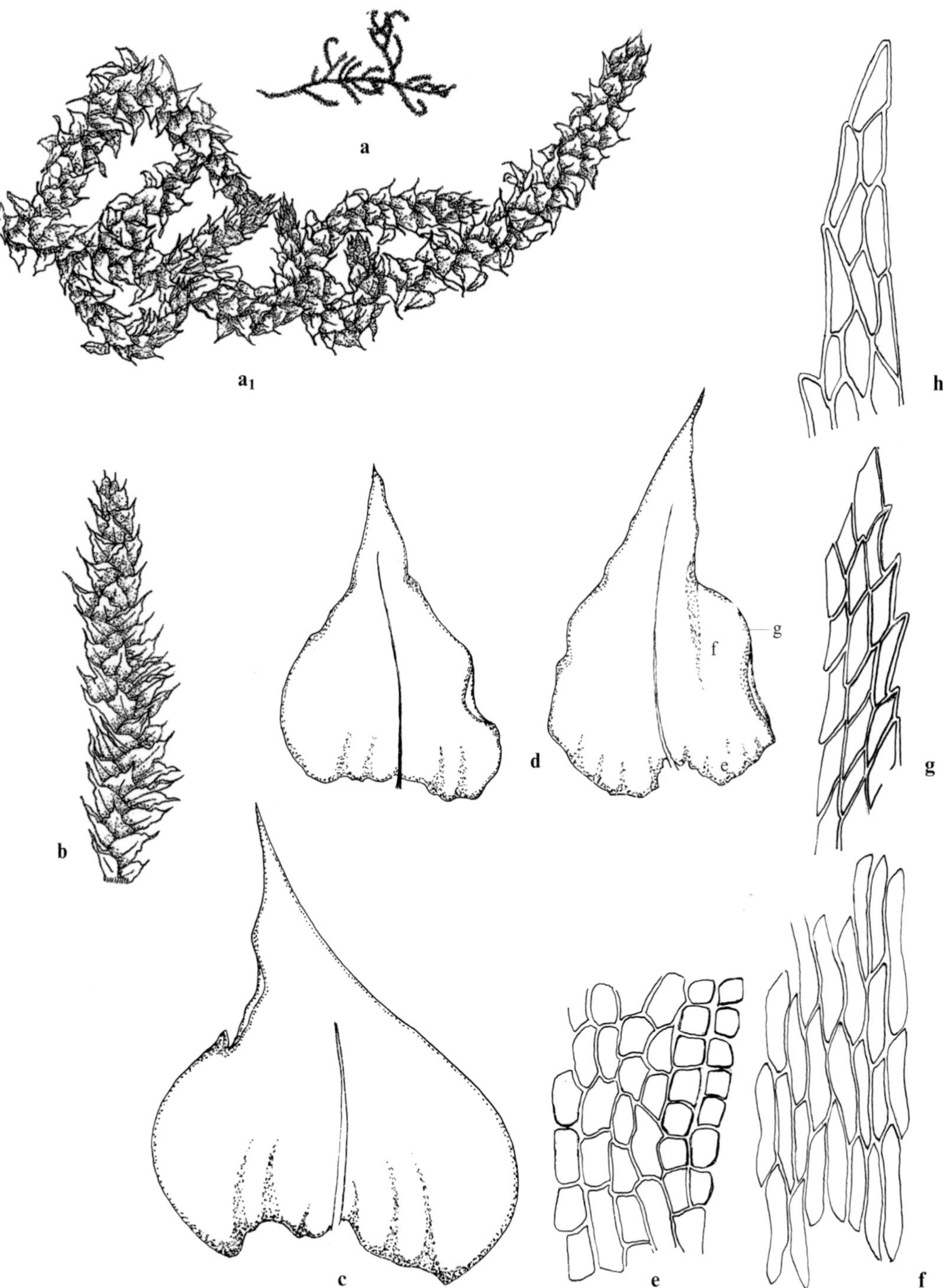

Figure 199. *Eurhynchium meridionale*: **a** habit (×1); **a₁** habit, dry (×10); **b** part of branch, moist (×10); **c** stem leaf (×50); **d** branch leaves (×50); **e** alar and basal cells; **f** mid-leaf cells; **g** marginal mid-leaf cells; **h** apical cells (all leaf cells ×500). (*Danin*, 8 Apr. 1979)

4. Eurhynchium pulchellum (Hedw.) Jenn., Man. Mosses W. Pennsylvania 350 (1913).
Hypnum pulchellum Brid. ex Hedw., Sp. Musc. Frond. 265 (1801); *E. strigosum* (F. Weber & D. Mohr) B.S.G., Bryol. Eur. 5:218 (1854).

Distribution map 201, Figure 200.

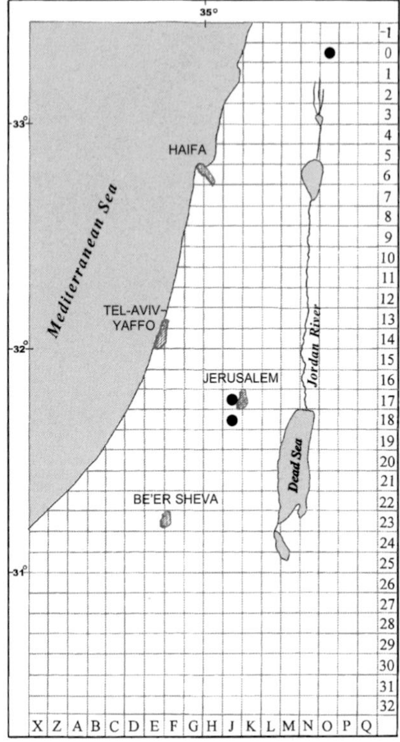

Map 201: *Eurhynchium pulchellum*

Plants small, yellowish-green, turning brown. *Stems* creeping; branches erect, straight. *Leaves*: stem and branch leaves different; branch leaves ± patent when dry and moist, ± complanate, distinctly concave, 1.5–2 mm long, ovate, obtuse to acute; stem leaves usually larger, ovate-lanceolate, acuminate; margins plane, usually serrulate in upper part; costa up to ¾ the length of leaf; basal cells rhomboidal, slightly incrassate; alar cells oval to irregularly rectangular, incrassate; mid-leaf cells linear-rhomboidal, 10–13 times longer than wide. *Setae* 10–20 mm long, smooth; capsules inclined to horizontal, ca. 2 mm long not including operculum, constricted below mouth, neck short.

Habitat and Local Distribution. — Plants growing among rocks in shaded habitats. Rare: Judean Mountains and Mount Hermon.

General Distribution. — Recorded from Southwest Asia (Turkey and Afghanistan), around the Mediterranean (scattered records), Europe (scattered records throughout, often from the south), North Africa, northeastern, eastern, and central Asia, North America (very widespread), and central and northwestern South America.

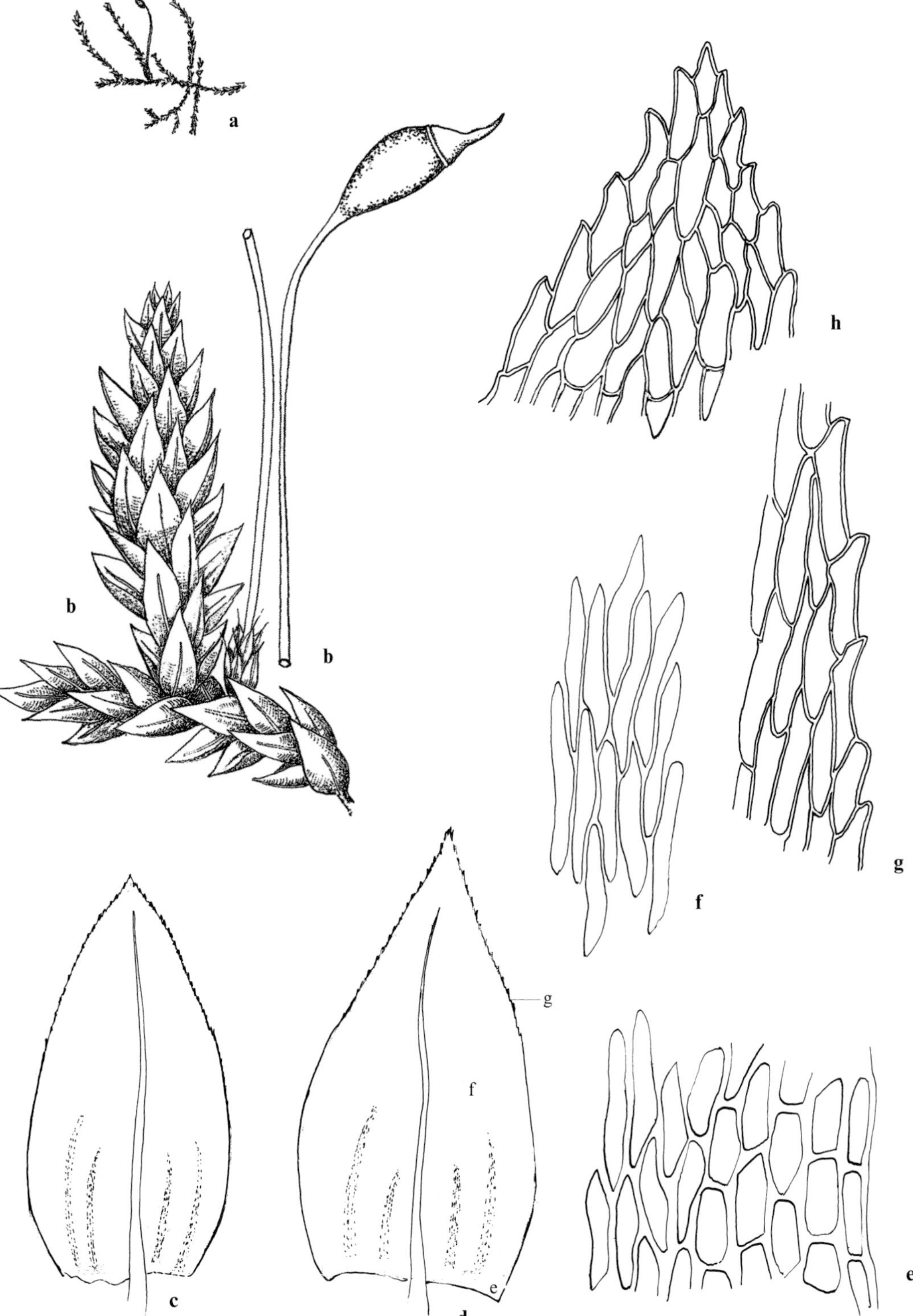

Figure 200. *Eurhynchium pulchellum*: **a** habit (×1); **b** part of habit, moist (×10); **c** branch leaf (×50); **d** stem leaf (×50); **e** alar and basal cells; **f** mid-leaf cells; **g** marginal upper cells; **h** apical cells (all leaf cells ×500). (*Kushnir*, 9 Oct. 1943)

5. Eurhynchium hians Sande Lac., Ann. Mus. Bot. Lugduno-Batavum 2:299 (1866).
Hypnum hians Hedw., Sp. Musc. Frond. 272, Tab. 70 (1801); *E. swartzii* (Turner) Curn. *in* Rabenh., Bryoth. Eur. 12:593 (1862). — Bilewsky (1965):426; *Oxyrrhynchium swartzii* (Turner) Warnst., Krypt.-Fl. Brandenburg, Laubm. 2:784 (1905). — Bilewsky (1974).

Distribution map 202, Figure 201.

Plants small, yellowish-green. *Stems* creeping. *Leaves*: stem and branch leaves ± similar; leaves ± complanate, 0.75–1 mm long, ovate-lanceolate, acuminate; margins plane, serrulate; costa ¾–⅘ the length of leaf; basal cells few, rhomboidal, alar cells rectangular, incrassate, mid-leaf cells linear to linear-rhomboidal, 6–11 times longer than wide, apical cells shorter, rhomboidal. Recorded as *dioicous*. *Sporophytes* not seen in local plants.

Habitat and Local Distribution. — Plants growing in moist and shaded habitats, on rocks. Rare: Upper Galilee and Mount Hermon.

General Distribution. — Recorded from Southwest Asia (Cyprus, Turkey, Syria, Lebanon, and Iran), around the Mediterranean (throughout), Europe (throughout), Macaronesia, tropical and North Africa, northeastern, eastern, and central Asia, and North, Central, and Caribbean America.

Literature. — Touw & Knol (1978).

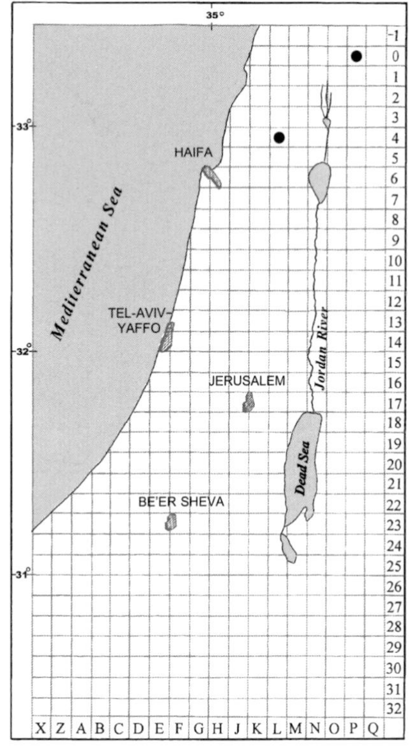

Map 202: *Eurhynchium hians*

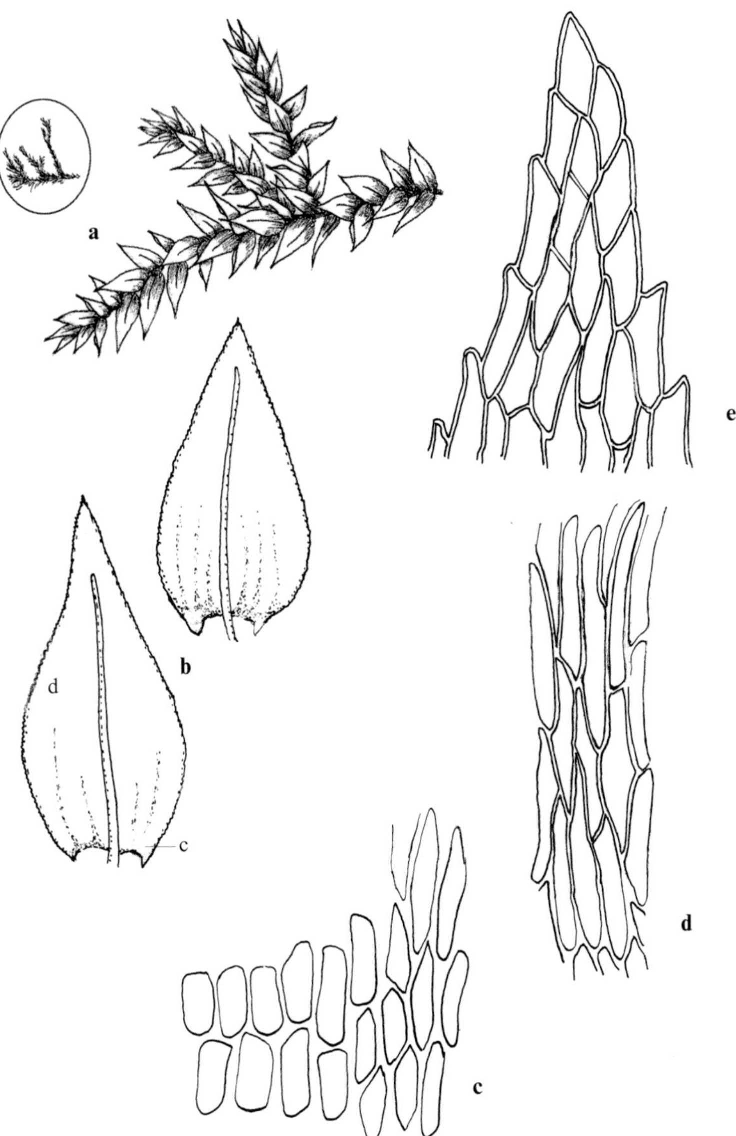

Figure 201. *Eurhynchium hians*: **a** part of habit, moist (×10); **b** leaves (×50); **c** alar and basal cells; **d** mid-leaf cells; **e** apical cells (all leaf cells ×500). (Shibea, *Kushnir*, No. 153 in Herb. Bilewsky)

6. Eurhynchium schleicheri (R. Hedw.) Jur. *in* C. Röm., Verh. Zool. Bot. Ges. Wien 16 : 942 (1866).

Hypnum schleicheri R. Hedw. *in* F. Weber & D. Mohr, Beitr. Naturk. 1 : 128 (1806); *Oxyrrhynchium schleicheri* (R. Hedw.) Roell, Hedwigia 56 : 249 (1915).

Distribution map 203, Figure 202.

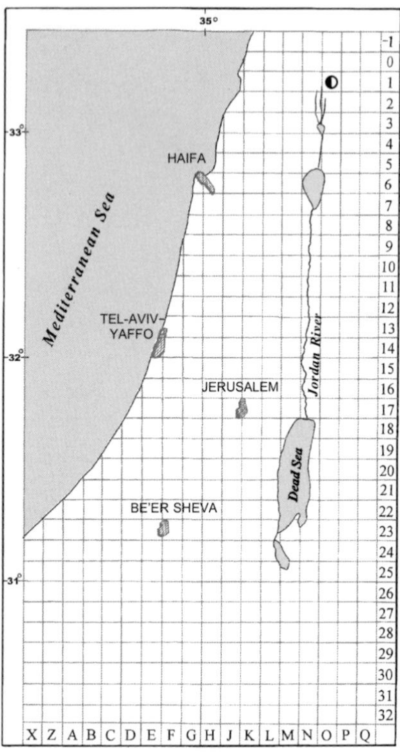

Map 203: ◐ *Eurhynchium schleicheri* + *E. speciosum*

Plants medium-sized, yellowish-green, with subterranean stolons and stems. *Branches* erect, short. *Leaves*: branch and stem leaves ± similar; branch leaves not complanate, 1–1.3 mm long, ovate-oblong, acuminate, apex often twisted; margins plane, serrulate all around; costa (in single local sample) ca. ⅘ the length of leaf; basal cells few, irregularly rhomboidal; alar cells few, little differentiated, rectangular; mid-leaf cells linear-rhomboidal, 10–16 times longer than wide. *Sporophytes* dark red to red-brown when mature. *Dioicous*. *Setae* 20–25 mm long, rough; capsules inclined to horizontal, curved, ovate to ellipsoid, ca. 2 mm long not including operculum.

Habitat and Local Distribution. — Plants growing on soil on a stream bank. Found only in one locality: Dan Valley (together with *Eurhynchium speciosum*).

General Distribution. — Recorded from Southwest Asia (Turkey and Iran), around the Mediterranean (records few, none from North Africa), Europe (throughout), Macaronesia, and northeastern Asia.

The description of *Eurhynchium schleicheri* above is based on the only local collection from Israel. The plants considered here as *E. schleicheri* clearly differ from the other *Eurhynchium* species but do not fully agree with available descriptions of *E. schleicheri* in the literature. The length of the costa in the majority of leaves reaches ca. ⅘ of the length of leaf (not ⅔ as described), and it is difficult to discern between denuded primary stems (as the result of the dry habitat conditions) and subterranean stolons, considered as characteristic of *E. schleicheri*.

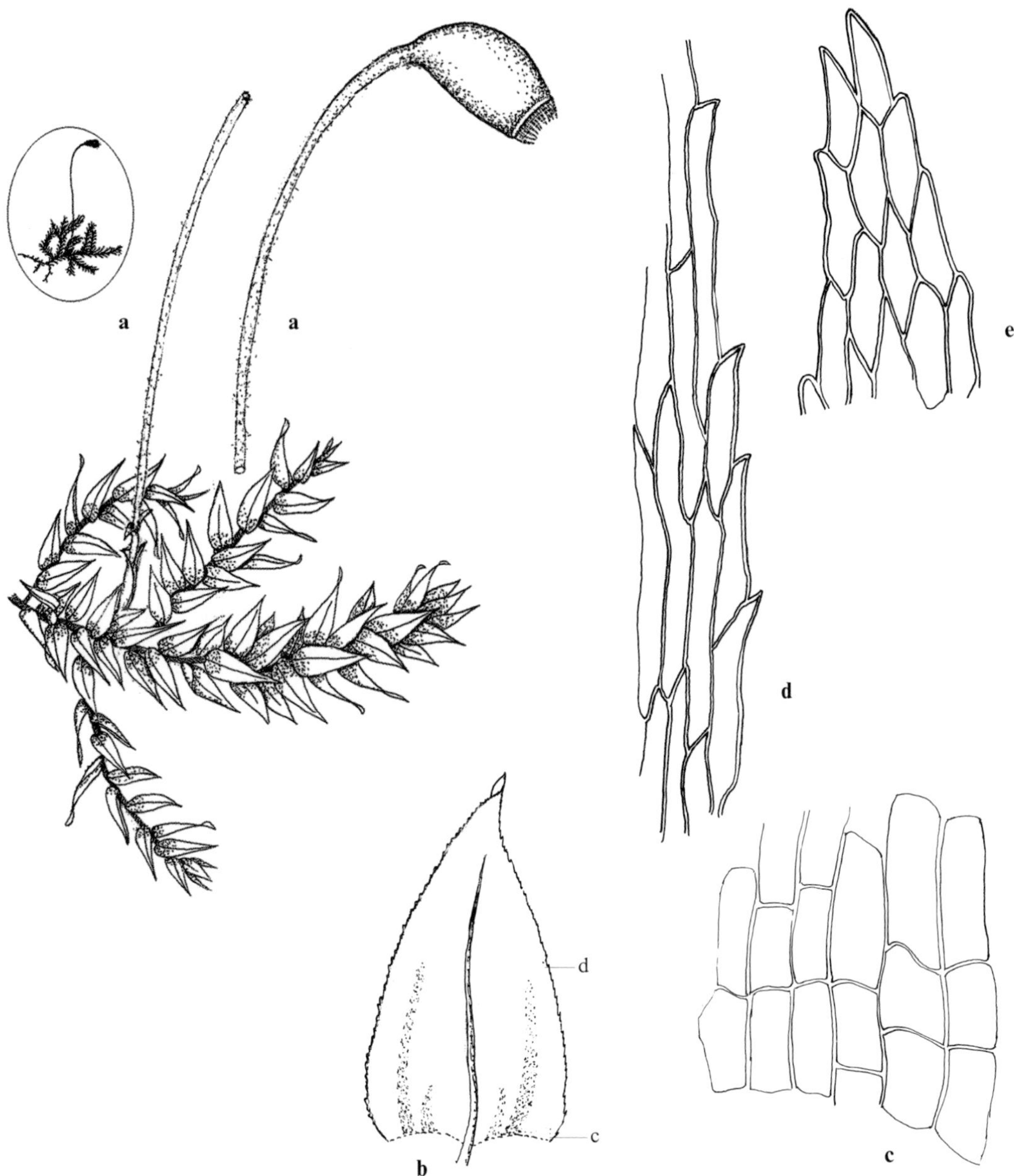

Figure 202. *Eurhynchium schleicheri*: **a** part of habit, moist with deoperculate capsule (×10); **b** leaf (×50); **c** alar cells; **d** marginal mid-leaf cells; **e** apical cells (all leaf cells ×500). (*Heyn & Herrnstadt*, 3 Mar. 1982)

7. Eurhynchium speciosum (Brid.) Jur., Verh. Zool. Bot. Ges. Wien 13 : 500 (1863). [Bilewsky (1965) : 428 (as *Eurhynchium speciosum* "(Brid.) Milde")]
Hypnum speciosum Brid., Spec. Musc. 2 : 105 (1812); *Oxyrrhynchium speciosum* (Brid.) Warnst., Krypt.-Fl. Brandenburg, Laubm. 2 : 786 (1905).

Distribution map 203 (p. 508), Figure 203.

Plants medium-sized, dull green to yellowish-green. *Stems* creeping; branches spreading, with laxly spaced leaves. *Leaves*: stem and branch leaves ± similar, but only branch leaves ± complanate; leaves weakly plicate, with twisted apex when dry, patent when moist, 1.5–2 mm long, ovate and acuminate or ovate-oblong and acute; margins plane, serrulate to serrate all around; costa extending high into acumen, $\frac{4}{5}$ the length of leaf; basal cells few, irregularly rhomboidal, alar cells irregularly rectangular, differentiated, mid-leaf cells linear-rhomboidal, seven–fifteen times longer than wide, apical cells shorter, rhomboidal. *Autoicous* or synoicous. *Sporophytes* as in *Eurhynchium schleicheri.*

Habitat and Local Distribution. — Plants growing on exposed roots and rocks, near streams, often encrusted with lime and covered by cyanophytes, attached at base to stones. Rare though locally abundant in a restricted area: Dan Valley.

General Distribution. — Recorded from Southwest Asia (Cyprus, Turkey, Iran, Iraq, and Saudi Arabia), around the Mediterranean (throughout), Europe (throughout the temperate parts), Macaronesia, and North Africa; recorded with some doubts from North America (Duell 1992).

Eurhynchium speciosum is the only synoicous local *Eurhynchium* species. It can be readily identified by the laxly spaced, ± complanate branch leaves, toothed all around. Some denuded primary stems may resemble stolons.

Figure 203. *Eurhynchium speciosum*: **a** part of habit, moist (×10); **b** calyptra (×10); **c** branch leaf (×50); **d** alar cells; **e** mid-leaf cells; **f** marginal mid-leaf cells; **g** apical cells (all leaf cells ×500). (*Kushnir*, 21 Nov. 1943, No. 152 in Herb. Bilewsky)

7. *Rhynchostegiella* (B.S.G.) Limpr.

Plants very slender. *Stems* creeping, irregularly to pinnately branched. *Leaves*: stem and branch leaves ± similar or branch leaves ± larger; leaves ± patent to spreading, small, linear-lanceolate or oblong-lanceolate, acute to long-acuminate; margins plane, entire or serrulate in upper part; costa from half as long as leaf to exceeding to base of acumen or above, end often gradually vanishing; mid-leaf cells narrow-rhomboidal to linear, apical cells equal to mid-leaf cells or shorter and rhomboidal, alar cells little differentiated. *Autoicous* (in local species). *Setae* smooth or rough, ± curved; capsules inclined to horizontal, ovate to ellipsoid; annulus differentiated, operculum long-rostrate. *Spores* finely papillose (in local species).

A cosmopolitan genus comprising ca. 50 species. Five species occur in the local flora.

Rhynchostegiella is generally considered as a difficult genus in which delimitation between species poses many problems. This is due to some local species that seem to be rather similar in habit and can be identified only when examined under high magnification. Often, the length of the costa cannot be estimated because of its gradual disappearance. Although sporophytes are quite frequently formed, it is not always easy to use the texture of the setae for identification of species because of the various expressions of seta roughness.

1 Leaves linear-lanceolate, costa extending into acumen.
  2 Plants usually silky; leaves six–ten times longer than wide, margins entire to sinuate; setae smooth **1. R. tenella**
  2 Plants not silky, leaves three–five times longer than wide, margins serrulate in upper part; setae rough **4. R. jacquinii**
1 Leaves oblong-lanceolate to ovate-lanceolate, costa not extending into acumen.
  3 Mid-leaf cells ca. 10–12(–15) times as longer than wide.
    4 Setae rough **2. R. curviseta**
    4 Setae smooth **3. R. letourneuxii**
  3 Mid-leaf cells ca. five–eight times longer than wide **5. R. teesdalei**

1. Rhynchostegiella tenella (Dicks.) Limpr., Laubm. Deutschl. 3 : 209 (1890). [Bilewsky (1965) : 429]

Hypnum tenellum Dicks., Pl. Crypt. Brit. 4 : 16 (1801); *Eurhynchium tenellum* (Dicks.) Milde, Bryol. Siles. 308 (1869); *R. algiriana* (P. Beauv.) Warnst., Krypt.-Fl. Brandenburg, Laubm. 2 : 800 (1906).

Distribution map 204 (p. 514), Figure 204, Plate XIV : f.

Plants silky, bright green to yellowish-green. *Leaves*: stem leaves with wider base than branch leaves; branch leaves ± patent, ca. 1.5 mm long, six–ten times longer than wide, linear-lanceolate, long-acuminate; margins almost entire to sinuate; costa extending at least into base of acumen, often terminal part gradually vanishing; basal cells few, rhomboidal to irregularly rectangular; mid-leaf cells linear, 10–20 times longer than wide; apical cells equal to mid-leaf cells. *Sporophytes* abundant. *Setae* smooth. *Spores* 10–16 μm in diameter; covered by gemmae of various sizes (seen with SEM).

Habitat and Local Distribution. — Plants growing in dense mats on limestone and chalk rocks and stones, exposed or shaded; near running water on soil on rocks and on

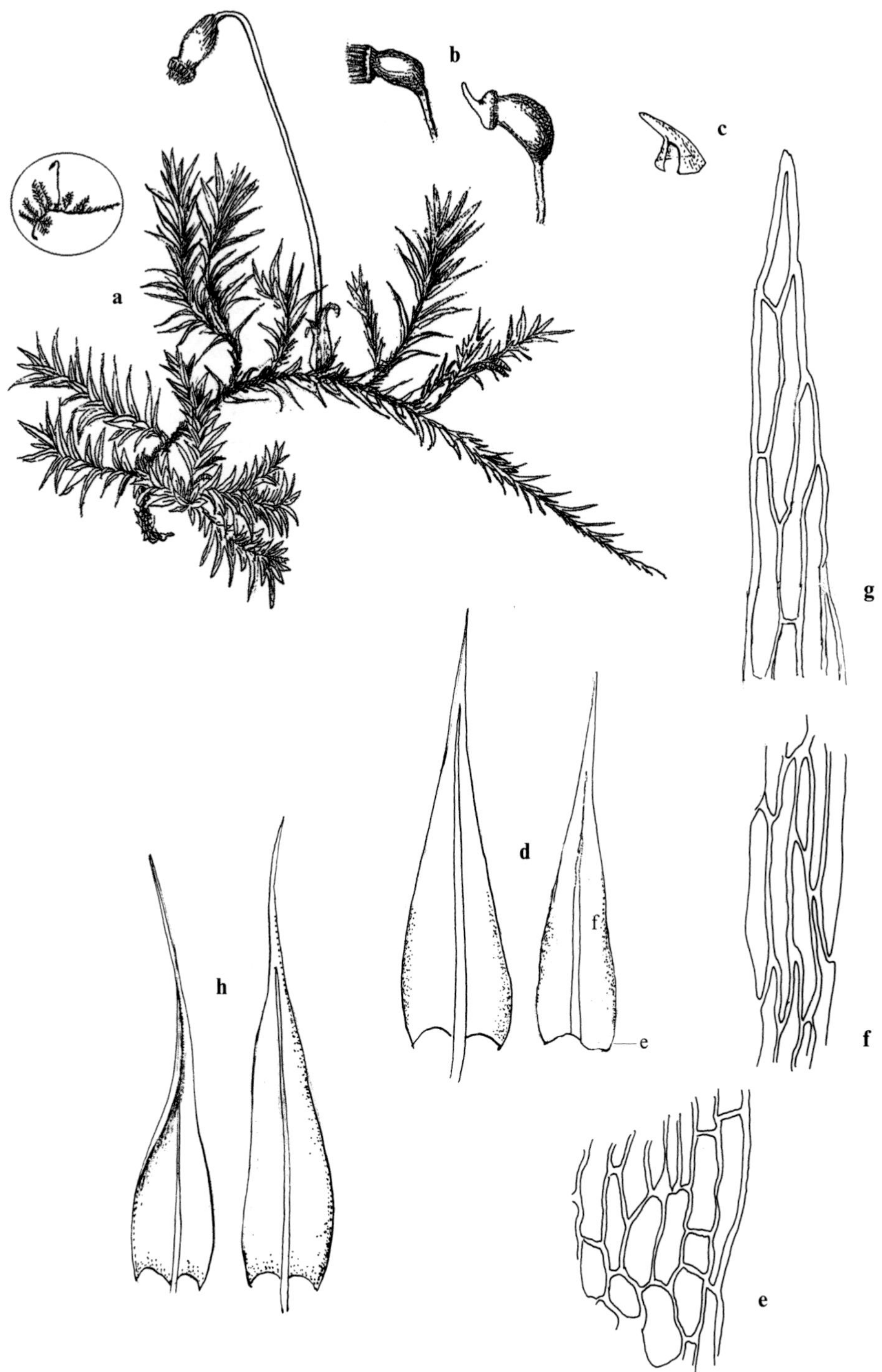

Figure 204. *Rhynchostegiella tenella*: **a** habit, moist (×10); **b** deoperculate and operculate capsules (×10); **c** calyptra (×10); **d** leaves (×50); **e** marginal basal cells; **f** mid-leaf cells; **g** apical cells (all leaf cells ×500); **h** leaves (×50). (**a**, **d**–**g**: *Herrnstadt & Crosby 79-83-7*; **b**, **c**, **h**: *Nachmony*, 6 Feb. 1955)

rotting logs. Common: Coastal Carmel, Upper Galilee, Lower Galilee, Mount Carmel, Mount Gilboa, Samaria, Shefela, Judean Mountains, Dan Valley, and Mount Hermon.

General Distribution. — Recorded from Southwest Asia (Cyprus, Turkey, Lebanon, Sinai, Jordan, and Iran), around the Mediterranean (throughout), Europe (throughout), Macaronesia, North Africa, and eastern Asia.

Rhynchostegiella tenella is the most widespread species of *Rhynchostegiella* in the local flora, and occurs in a wide variety of habitats. It can be distinguished by the long-acuminate leaves, the long costa that reaches the acumen, and the very elongate mid-leaf cells.

It is assumed here that *Rhynchostegiella tenella* var. *litorea* (De Not.) P. W. Richards & E. C. Wallace does not grow in Israel. In its shorter leaves and the rough setae it bears some resemblance to *R. jaquinii*, but differs in some other characters.

2. Rhynchostegiella curviseta (Brid.) Limpr., Laubm. Deutschl. 3 : 207 (1896). [Bilewsky (1965) : 429]

Hypnum curvisetum Brid., Spec. Musc. Frond. 2 : 111 (1812); *Eurhynchium curvisetum* (Brid.) Delogne, Ann. Soc. Belge Microscop. 9 : 126 (1885).

Distribution map 205, Figure 205 : a–e.

Plants bright green to yellowish-green. *Leaves*: stem and branch leaves ± similar; branch leaves ± patent, 0.75–1 mm long, ovate-lanceolate, acute to short acuminate; margins serrulate mainly above; costa short, extending to ½–⅔ the length of leaf,

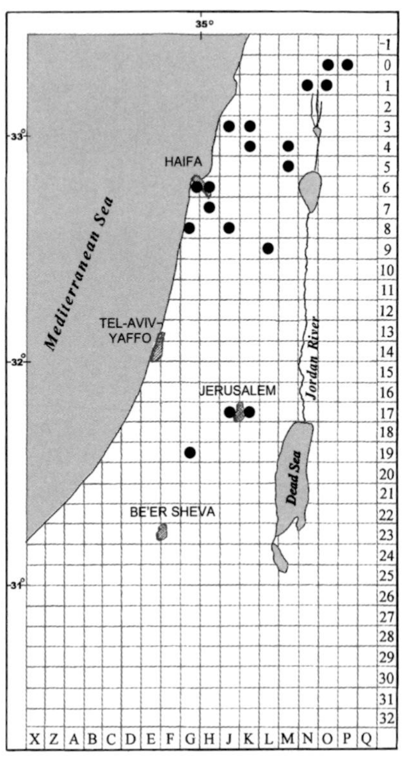

Map 204: *Rhynchostegiella tenella*

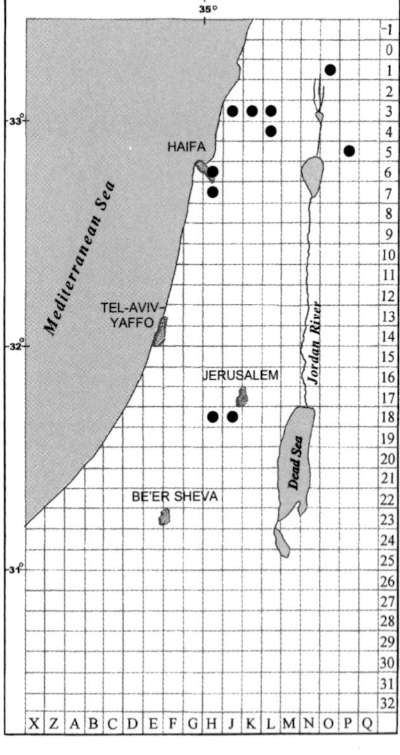

Map 205: *Rhynchostegiella curviseta*

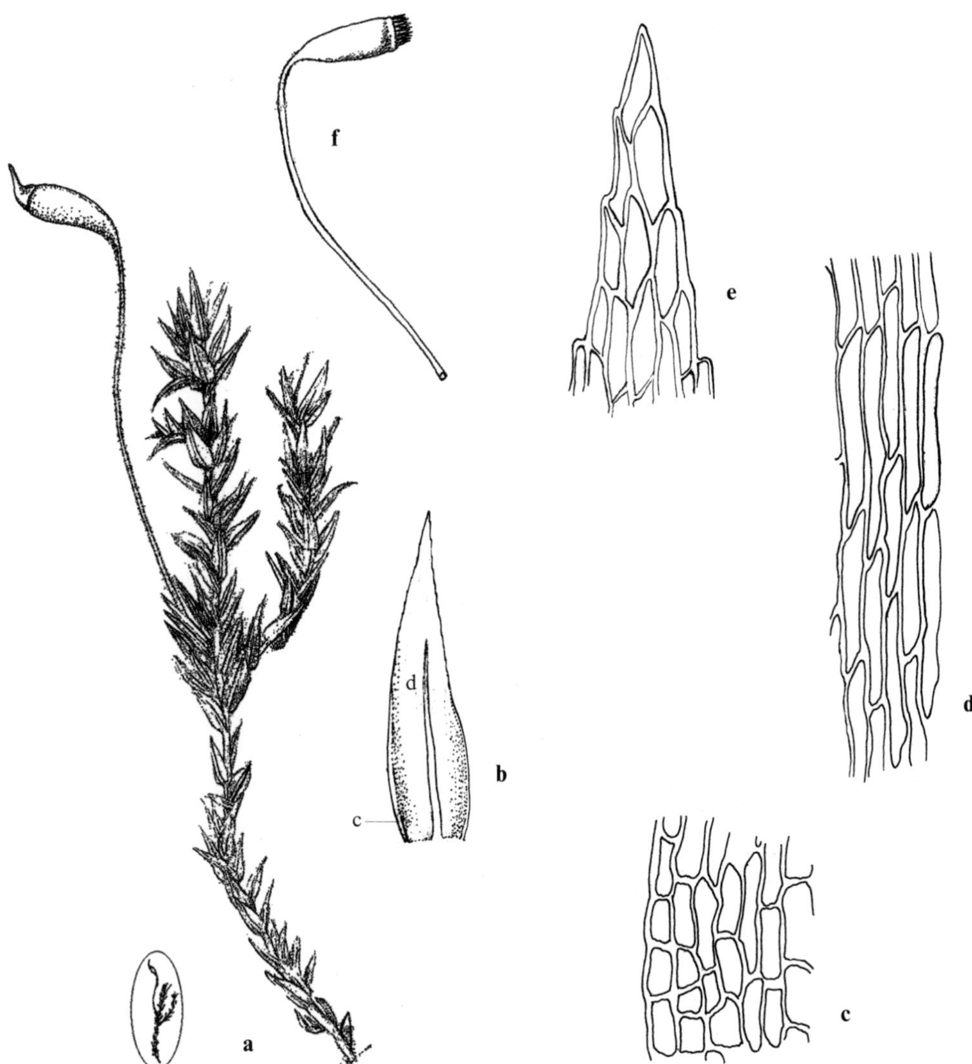

Figure 205. *Rhynchostegiella curviseta*: **a** part of habit, moist (×10); **b** leaf (×50); **c** marginal basal cells; **d** mid-leaf cells; **e** apical cells (all leaf cells ×500);
Rhynchostegiella letourneuxii: **f** deoperculate capsule with part of seta (×10).
(**a**–**e**: *D. Zohary*, 17 Mar. 1951; **f**: Mount Carmel, *Galili*, No. 156 in Herb. Bilewsky)

often terminal part gradually vanishing; basal cells few, rhomboidal to irregularly rectangular; mid-leaf cells linear, 10–12(–15) times longer than wide; apical cells rhomboidal, shorter than mid-leaf cells. *Setae* rough.

Habitat and Local Distribution. — Plants growing on rocks and soil in shaded habitats or near water. Occasional: Upper Galilee, Mount Carmel, Judean Mountains, Dan Valley, and Golan Heights.

General Distribution. — Recorded from Southwest Asia (Cyprus, Turkey, Lebanon, Jordan, and Iraq), around the Mediterranean (scattered throughout), Europe (scattered), Macaronesia, and North Africa.

3. Rhynchostegiella letourneuxii (Besch.) Broth., Nat. Pflanzenfam. 1:1162 (1909). [Bilewsky (1974)]

Rhynchostegium letourneuxii Besch., Cat. Mous. Algérie 38 (1882); *Rhynchostegiella curviseta* (Brid.) Limpr. var. *laeviseta* (W. E. Nicholson & Dixon) Podp., Consp. Musc. Eur. 642 (1954).

Distribution map 206, Figure 205:f (p. 515).

Plants resembling *Rhynchostegiella curviseta* in habit and leaf shape. The main differential character is the smooth setae of *R. letourneuxii.*

Habitat and Local Distribution. — Plants growing on shaded rocks and walls. Rare to occasional: Upper Galilee, Mount Carmel, and Judean Mountains.

General Distribution. — Very few existing records: Southwest Asia (Turkey and Lebanon), southern Europe (Portugal, Spain, and France), and North Africa.

Rhynchostegiella letourneuxii is perhaps only a variety of *R. curviseta* (as often considered, e.g., Frahm 1995). More samples are needed in order to decide whether the smooth seta is correlated with some additional constant differential characters.

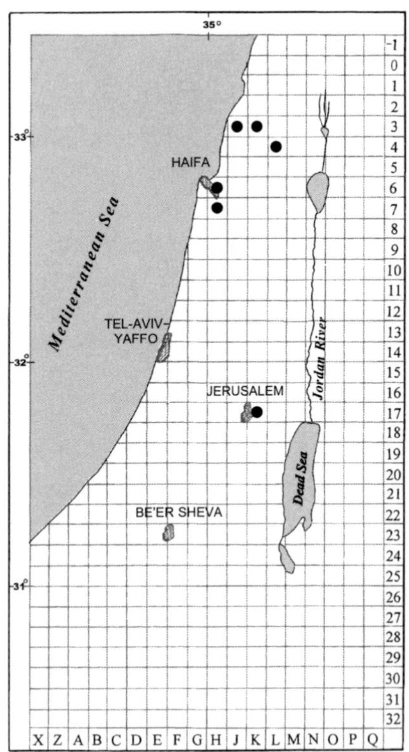

Map 206: *Rhynchostegiella letourneuxii*

4. Rhynchostegiella jacquinii (Garov.) Limpr., Laubm. Deutschl. 3:215 (1896). [Bilewsky (1970)]

Hypnum jacquinii Garov., Bryol. Austr. Excurs. 82 (1840); *Eurhynchium jacquinii* (Garov.) Velen., Rozpr. Ceské Akad. Ved. Tr. 2, 7:16 (1898).

Distribution map 207 (p. 518), Figure 206.

Plants green to yellowish-green, with interwoven branches. *Leaves*: stem leaves ± larger than branch leaves; branch leaves ± patent, ca. 0.8 mm long, three–five times longer than wide, linear-lanceolate, short acuminate; margins serrulate in upper part, sinuate

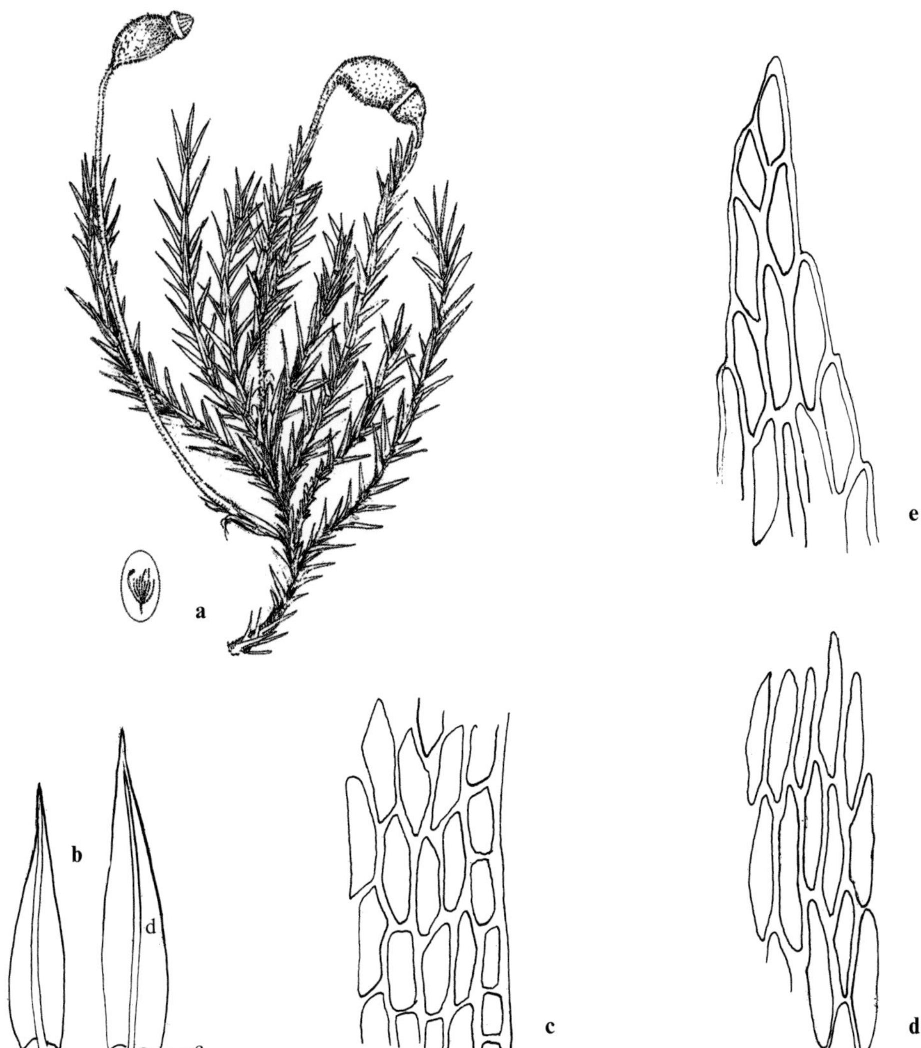

Figure 206. *Rhynchostegiella jacquinii*: **a** habit, moist (×10); **b** leaves (×50); **c** marginal basal cells; **d** mid-leaf cells; **e** apical cells (all leaf cells ×500). (Tel-el-Kadi, *Bilewsky*, No. 157 in Herb. Bilewsky)

below; costa usually extending into acumen; basal cells few, rhomboidal to irregularly rectangular; mid-leaf cells linear, 6–10(–12) times longer than wide; apical cells rhomboidal, shorter than mid-leaf cells. *Sporophytes* abundant. *Setae* distinctly rough.

Habitat and Local Distribution. — Plants growing in dense mats on tree barks, rotting logs, and rocks, often near water. Rare to occasional: Upper Galilee, Mount Carmel, and Dan Valley.

General Distribution. — Records few: Southwest Asia (Turkey and Iran), Europe (few localities), and the Canary Islands.

Plants of *Rhynchostegiella jacquinii* resemble those of *R. tenella* — the most widespread local species — but differ in the smaller, serrulate leaves and the rough setae.

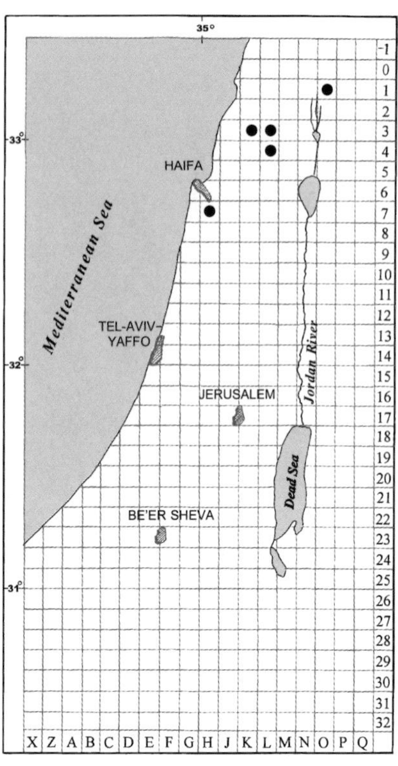

Map 207: *Rhynchostegiella jacquinii*

5. Rhynchostegiella teesdalei (Schimp. *in* B.S.G.) Limpr., Laubm. Deutschl. 3:217 (1896).

Rhynchostegium teesdalei Schimp. *in* B.S.G., Bryol. Eur. 5:202 (1857); *Eurhynchium teesdalei* (Schimp. *in* B.S.G.) Milde, Bryol. Siles. 313 (1869).

Distribution map 208, Figure 207.

Plants green to yellowish-green. *Leaves*: stem and branch leaves ± similar; branch leaves laxly spaced, patent to spreading, ca. 0.8 mm long, oblong-lanceolate to ovate-lanceolate, acute to short acuminate; margins serrulate in upper part; costa ca. ¾ the length of leaf, terminal part often gradually vanishing; basal cells few, rhomboidal to irregularly rectangular; mid-leaf cells linear-rhomboidal, five–eight times longer than wide; apical cells rhomboidal, shorter than mid-leaf cells. *Setae* rough.

Habitat and Local Distribution. — Plants growing on rocks in shaded or moist habitats. Rare to occasional: Upper Galilee, Mount Carmel, Dan Valley, and Mount Hermon.

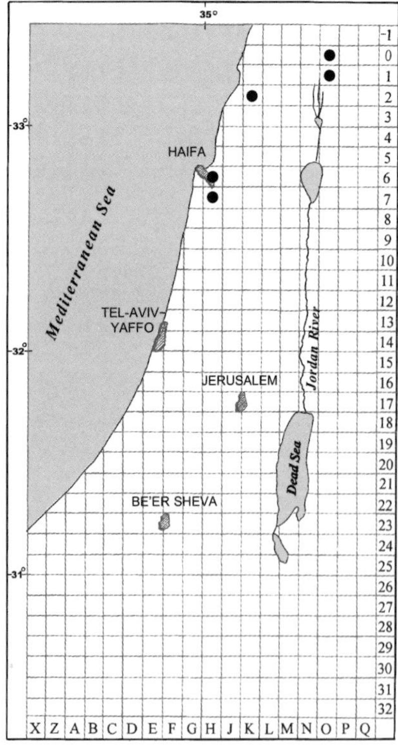

Map 208: *Rhynchostegiella teesdalei*

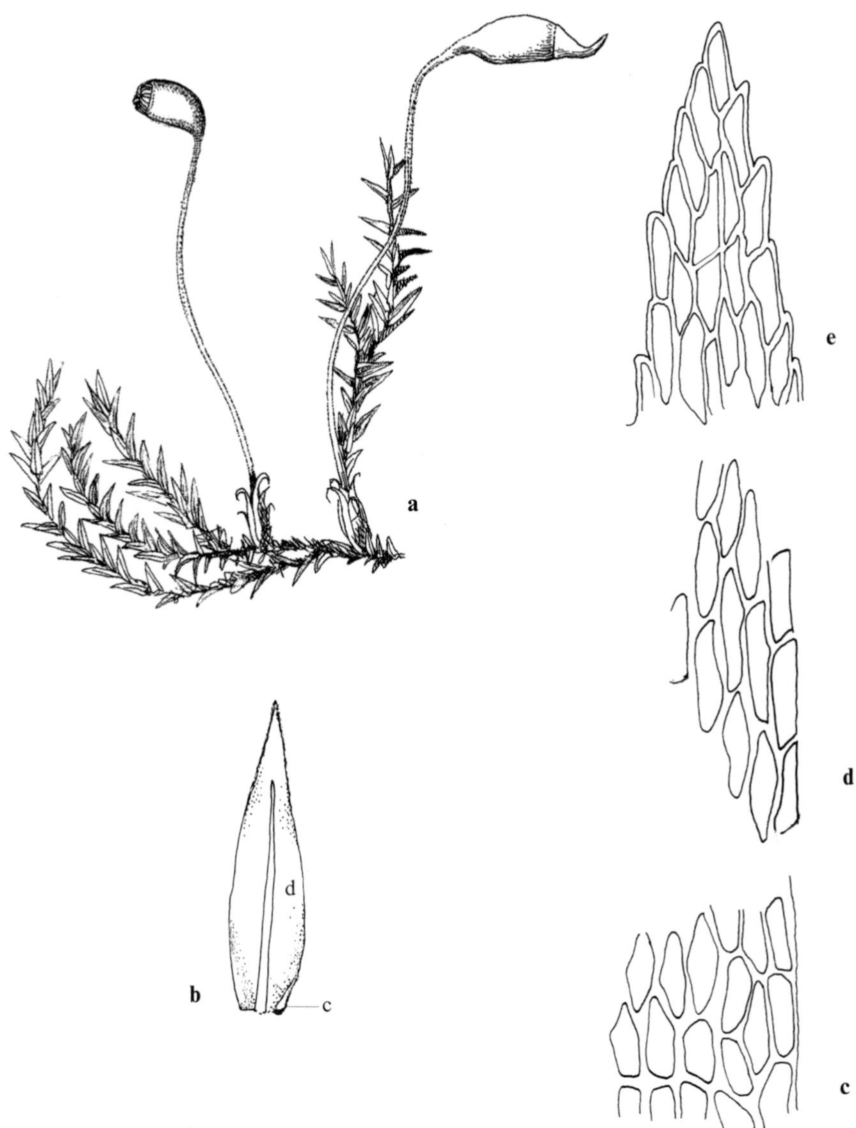

Figure 207. *Rhynchostegiella teesdalei*: **a** habit, moist (×10); **b** leaf (×50); **c** marginal basal cells; **d** mid-leaf cells; **e** apical cells (all leaf cells ×500). (*Herrnstadt 79-75-2b*)

General Distribution. — Records few: Southwest Asia (Turkey), around the Mediterranean (scattered throughout), Europe (not widespread), Macaronesia, and North Africa.

Rhynchostegiella teesdalei can be distinguished from the other local species of *Rhynchostegiella* by the more laxly spaced and spreading leaves and the shorter mid-leaf cells.

Athalamia spathysii (p. 590)

Lunularia cruciata (p. 580)

Sphaerocarpos michelii (p. 570)

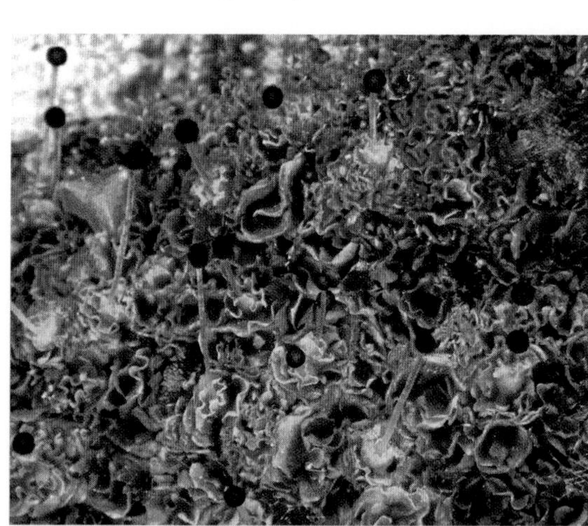

Fossombronia caespitiformis (p. 556)

Riccia crystallina (p. 630)

Riccia crustata (p. 604)

Part II

ANTHOCEROTOPSIDA (HORNWORTS)
and
MARCHANTIOPSIDA (LIVERWORTS)

by

Helene Bischler and Suzanne Jovet-Ast

INTRODUCTION

This part of the Flora summarises the up-to-date knowledge on the local hornworts and liverworts. Those plants, lacking roots and water-storage capabilities, are able to survive and spread under extreme climatic conditions. Although relatively few species and no endemics exist locally, the flora is of special interest to bryologists and ecologists investigating drought tolerance and colonisation of harsh biotopes.

Many new records can be expected because a substantial number of species vanish during several years of unfavourable conditions. Those species re-appear only during a period of a few weeks following heavy rains when the conditions for germination and growth become favourable. We hope that this treatment of the local hornworts and liverworts would encourage additional studies of those plants, and promote exploration of new biotopes and areas in which collections have rarely or never been made.

THE HISTORY OF COLLECTIONS AND STUDIES OF HORNWORTS AND LIVERWORTS IN ISRAEL

The first four liverworts collected in Judea and the Northern Negev were recorded by Hart (1891): *Fossombronia angulosa* (Dicks.) Raddi (in fact, *F. caespitiformis* De Not. ex Rabenh.), *Lunularia cruciata* (L.) Lindb., *Plagiochasma rupestre* (J. R. Forst. & G. Forst.) Steph., and *Riccia lamellosa* Raddi. Forty years later, Rabinovitz-Sereni (1931) published six additional species from the coast and Judea: *Athalamia spathysii* (Lindenb.) S. Hatt., *Fossombronia caespitiformis* De Not. ex Rabenh., *Mannia fragrans* (Balb.) Frye & L. Clark (in fact, *Reboulia hemisphaerica* (L.) Raddi), *Riccia fluitans* L. (not recorded since), *Ricciocarpos natans* (L.) Corda, and *Targionia hypophylla* L. Washburn & Jones (1937) gathered the rare *Riella helicophylla* (Bory & Mont.) Mont. Proskauer (1953) published 15 species on the basis of a collection made by D. Zohary between 1949–1951 on the coast, in Judea, and in the Galilee. Ten of those species were new for Israel: one *Cephalozia* (probably identical with *Leiocolea turbinata* (Raddi) Steph.), *Corsinia coriandrina* (Spreng.) Lindb., *Oxymitra incrassata* (Brot.) Sérgio & Sim-Sim, *Phaeoceros bulbiculosus* (Brot.) Prosk., *Southbya nigrella* (De Not.) Henriq., *S. tophacea* (Spruce) Spruce, *Riccia crozalsii* Levier, *R. gougetiana* Durieu & Mont., *R. sorocarpa* Bisch., and *R. trabutiana* Steph. Soon after, Arnell (1957) described a new species from the Negev, *Riccia palaestina* S. W. Arnell, and recorded the African *Riccia terracianoi* Gola from the Shefela and Upper Galilee. Both were shown to be species widely distributed in the Mediterranean Region, the first *R. frostii* Austin and the latter *R. crystallina* L. emend. Raddi (Jovet-Ast & Bischler 1966). Rungby (1959) added *Sphaerocarpos michelii* Bellardi, Jovet-Ast (1961) added *Targionia lorbeeriana* Müll. Frib. (now considered as conspecific with *Targionia hypophylla*), and Baum & Jovet-Ast (1962) added four *Riccia* species: *R. atromarginata* Levier, *R. macrocarpa* Levier, *R. nigrella* DC., and *R. subbifurca* Warnst. ex Croz. to the local liverwort flora. The interest in bryophytes increased, and in 1963 Bilewsky published his "Introduction to Bryophytes of Israel" with three new records: *Marchantia polymorpha* L., *Phaeoceros laevis* (L.) Prosk., and *Riccia glauca* L. (a species not recorded since). Finally, Lipkin & Proctor (1975) published their discovery of the rare *Riella cossoniana* Trab. and *R. affinis* M. Howe & Underw. We have had the opportunity to examine most of these collections.

Between 1966 and 1982, we collected 830 specimens from 274 sampling plots, adding 12 species to the local liverwort list: *Cephaloziella baumgartneri* Schiffn., *Frullania dilatata* (L.) Dumort., *Leiocolea turbinata* (as *Cephalozia bicuspidata* (L.) Dumort.), *Mannia androgyna* (L. emend. Lehm. & Lindenb.) A. Evans, *Metzgeria furcata* (L.) Dumort., *Pellia endiviifolia* (Dicks.) Dumort., *Porella cordaeana* (Huebener) Moore, *Riccia bicarinata* Lindb., *R. crystallina*, *R. crustata* Trab., *R. frostii*, and *R. michelii* Raddi. However, we were unable to collect several previously recorded species during the course of our fieldwork, which was limited to only a few weeks.

The exploration of the flora of Israel was initially part of our research program on the hornworts and liverworts of the Mediterranean Region. From 1969 to 1982 we set up sampling plots, collected plants in many Mediterranean countries, and established a data bank (Baudoin *et al.* 1984). Taxonomic and distribution data, which accumulated during that period, permitted the completion of the descriptions and the distribution maps of the local species.

THE COMPOSITION OF THE HORNWORT AND LIVERWORT FLORA OF ISRAEL

The 39 species of the local flora belong to five orders, 19 families, and 22 genera. Most families (17 out of 19) are represented by a single genus, and most genera (18 out of 22), are represented by a single species. The genus *Riccia*, with 14 species, is well represented. The tally listed above is based on collections deposited in BM, HUJ, TELA (now in HUJ), and UC, and on our collections deposited in HUJ and PC. Some specimens gathered recently, along with bibliographic records, completed the data.

The local flora comprises species distributed over several continents (80%) and species with wide ranges in the Mediterranean Region (20%). The phytogeographical subdivisions defined for phanerogams are meaningless in this context (see the section Phytogeography in the Preface). To date, not a single endemic taxon has been recorded. The flora appears to be a mixture of elements from various sources that found suitable habitats in Israel, but subsequent diversification has not occurred.

The orders Anthocerotales, Sphaerocarpales, and Marchantiales, which are characterised by drought tolerant species or species with large, resistant spores, comprise together 77% of the total number of species in the local flora compared to only 34% of the total in the Mediterranean Region, and 18% of the total in Europe. The Jungermanniales (the largest order in Europe and in the Mediterranean Region) and the Metzgeriales, which are characterised mainly by sciophilous species, comprise only 23% of the total number of species in the local flora compared to 66% of the total in the Mediterranean Region and 82% of the total in Europe (Table 2).

In addition, the local flora appears depleted as compared to the Mediterranean and European floras (Table 2). Nine percent of European species and only 17.5% of the species known from the Mediterranean Region are recorded from Israel.

The low species diversity seems to be mainly due to the paucity of acidic soils. Acidophilous species (e.g., *Phaeoceros bulbiculosus*, *Corsinia coriandrina*, and *Riccia gougetiana*), which are frequent among Mediterranean hornworts and liverworts, are rare. The absence of extended maquis and forests and of high altitudes in most of the area precludes the establishment of species with higher moisture requirements. Human activity, especially in the Mediterranean zone, may also have lowered species diversity. The taxa inhabiting temporary pools, e.g., *Riella* spp. and *Ricciocarpos natans*, have become extremely rare or might be even extinct because of extended habitat destruction.

Table 2

A summary of the orders and species treated in this Flora, and a comparison of the local hornwort and liverwort flora with the floras of Europe and the Mediterranean Region. Percent of the total number of species in each order is provided in parentheses.

Order	Number of species in the local flora		Number of species in the Mediterranean Region		Number of species in Europe	
Anthocerotales	2	(5.1%)	5	(2.2%)	8	(1.9%)
Jungermanniales	6	(15.4%)	124	(55.8%)	320	(74.4%)
Metzgeriales	3	(7.7%)	23	(10.4%)	32	(7.4%)
Sphaerocarpales	4	(10.3%)	11	(5.0%)	5	(1.2%)
Marchantiales	24	(61.5%)	59	(26.6%)	65	(15.1%)
Total number of species	39		222		430	

ECOLOGY

Hornworts and liverworts show high sensitivity to microhabitat conditions: temperature, moisture, topography, pH of the substrate, and shelter. Each species has specific requirements, but none can survive on saline or unstabilised substrates. However, many can stay in open and exposed sites with significant temperature fluctuations.

Two main life-history traits encountered in liverworts from areas with a Mediterranean climate (i.e., a long dry season) enable us to divide the liverworts of the local flora into two groups. The first group comprises species characterised by a long life cycle and inhabiting soil pockets in rock and wall crevices. The thalli coil up, and their usually dark-coloured scales and ventral face protect the assimilatory layer and the growing point from desiccation. Recovery of active growth takes place in a few hours as soon as water becomes available, sometimes after weeks or even years of drought. The following genera of the Marchantiales, which possess large thalli, belong to this group: *Athalamia*, *Lunularia*, *Mannia*, *Plagiochasma*, *Reboulia*, and *Targionia*. The second group comprises species characterised by a life cycle shortened to a few weeks and inhabiting exposed ledges, soil depressions, run-off slopes, temporary ponds or their margins, and wadis. Germination, gametophyte development, gamete diversification, fertilisation, and ripening of the sporophytes hardly take more than two or three weeks. The gametophyte vanishes when the spores are ripe, and the spores alone persist. In addition, tubers, which are produced during vegetative growth and penetrate into the soil, guarantee survival. The following genera belong to this group: *Corsinia*, *Fossombronia*, *Oxymitra*, *Phaeoceros*, *Riccia*, *Ricciocarpos*, *Riella*, and

Sphaerocarpos. Both groups are terrestrial, in both, sexual reproduction is frequent, sporophytes are often produced in sequence on the dorsal side of thallus, and their spores are large, mechanically resistant, withstanding desiccation, and have long-term germination ability. Those spores do not require special conditions for germination (such as dormancy or seasonality), and can produce a new gametophyte whenever the moisture conditions are favourable.

The flora is predominantly made up of calcicolous, xerophytic species. In areas with less than 100 mm of annual precipitation, *Athalamia*, *Riccia atromarginata*, *R. lamellosa*, *R. michelii*, *R. nigrella*, *R. sorocarpa*, *R. trabutiana*, *Sphaerocarpos*, and *Targionia* still survive. In areas with slightly higher amounts of annual precipitation (ca. 200 mm), *Cephaloziella baumgartneri*, *Fossombronia caespitiformis*, and *Southbya nigrella* appear. Sciophilous species that require more and almost permanent moisture, e.g., *Frullania*, *Leiocolea*, *Marchantia*, *Metzgeria*, *Pellia*, *Porella*, and *Southbya tophacea* are rare. They grow on ground, rocks, soil, or as epiphytes, and have been found only in dense maquis or forests, always in sterile condition. Their local habitats seem to correspond to the limits of their ecological tolerance.

DISTRIBUTION

Most of the species exist in nearly all districts, including the Northern Negev, e.g., *Athalamia spathysii*, *Lunularia cruciata*, *Sphaerocarpos michelii*, *Targionia hypophylla*, and most *Riccia* species. The remainder of the species seem not to extend to the Negev, and appear to be limited to the Mediterranean Region (e.g., *Cephaloziella baumgartneri*, *Reboulia hemisphaerica*, and *Southbya nigrella*). *Mannia androgyna* has not been recorded from the coast, but there is no reason why this species should not exist there. Other species (e.g., *Plagiochasma rupestre*, *Riccia crustata*, and *R. frostii*) have been found only in dry areas in the Judean and Negev deserts. Yet these species have wide geographical ranges elsewhere in the Mediterranean Region, and are not specific to dry zones. Thus, additional collections are needed to assess horizontal and altitudinal distribution of the local species.

DELIMITATION OF ORDERS, FAMILIES AND SPECIES

In contrast to the mosses, the taxonomic boundaries of orders, families and genera of hornworts and liverworts are mostly well circumscribed.

Homogeneity of species has also been repeatedly questioned (Boisselier-Dubayle & Bischler 1998, 1999; Boisselier-Dubayle *et al.* 1995b; Zielinski 1987). Some species have been shown to include morphologically indistinct but genetically different entities, especially among haploid-polyploid siblings. Therefore, a broad taxonomic species concept has been adopted here.

GUIDE TO PART II

Descriptions. The descriptions of families and genera account for the characters of all known members of those taxa worldwide. The descriptions of species are based on local material. The descriptions of sporophytes, if not observed in local material, have been added on the basis of other Mediterranean collections. The chromosome counts listed have been made by us on those Mediterranean specimens that could be collected in an appropriate condition.

Classification. The classification of Grolle (1983a, b), in current use, is adopted. Recent publications are cited to account for the total number of species within each genus. The species within genera are listed in alphabetical order, except in *Riccia*, where the arrangement is based on the author's concepts of the affinities among the species.

Nomenclature. Nomenclature follows Grolle & Long (2000). Many nomenclatural types and other original materials have been examined in the course of formerly published monographic studies, i.e., *Plagiochasma* (Bischler 1978), *Riccia* (Jovet-Ast 1986), and *Marchantia* (Bischler 1993). No new species or new combinations are proposed.

Dubia and Excludenda. References to formerly published erroneous records are included in the sections Dubia and Excludenda.

Keys. If not otherwise indicated, the keys to orders, families and species apply to the entire geographical range of the taxa.

Distributional and ecological data. Distributional and ecological data for Israel are based on the herbarium collections and on our sampling plots. The general distributions of the species were retrieved from our data base on Mediterranean hornworts and liverworts, and from the floristic literature. For *Riccia*, Jovet-Ast (1986) was used.

Drawings and SEM photographs. Each of the drawings and SEM micrographs has been made from a single local collection. However, in species collected sterile, only the characters of the sporophytes are illustrated on the basis of other Mediterranean collections (cited in the legends). If not otherwise indicated, voucher specimens of the drawings are deposited in PC. Scale bars were used to indicate magnifications. The figures of the species of *Riccia* were made by S. Jovet-Ast, whereas the figures of the other species are based on sketches made by Helene Bischler and inked by Michal Boaz-Yuval.

CONSPECTUS

The orders, families, and genera of Anthocerotopsida and Marchantiopsida included in the Flora are listed below. The number of species in each genus follows in parentheses.

Division BRYOPHYTA

Subdivision HEPATICAE

Class ANTHOCEROTOPSIDA

Order ANTHOCEROTALES
ANTHOCEROTACEAE: *Phaeoceros* (2)

Class MARCHANTIOPSIDA

Order JUNGERMANNIALES
CEPHALOZIELLACEAE: *Cephaloziella* (1)
LOPHOZIACEAE: *Leiocolea* (1)
ARNELLIACEAE: *Southbya* (2)
PORELLACEAE: *Porella* (1)
FRULLANIACEAE: *Frullania* (1)

Order METZGERIALES
CODONIACEAE: *Fossombronia* (1)
PELLIACEAE: *Pellia* (1)
METZGERIACEAE: *Metzgeria* (1)

Order SPHAEROCARPALES
RIELLACEAE: *Riella* (3)
SPHAEROCARPACEAE: *Sphaerocarpos* (1)

Order MARCHANTIALES
CORSINIACEAE: *Corsinia* (1)
TARGIONIACEAE: *Targionia* (1)
LUNULARIACEAE: *Lunularia* (1)
AYTONIACEAE: *Plagiochasma* (1), *Mannia* (1), *Reboulia* (1)
CLEVEACEAE: *Athalamia* (1)
MARCHANTIACEAE: *Marchantia* (1)
OXYMITRACEAE: *Oxymitra* (1)
RICCIACEAE: *Ricciocarpos* (1), *Riccia* (14)

KEY TO CLASSES, ORDERS, AND FAMILIES

1 Plants thallose; chloroplast single per cell; capsule slenderly cylindrical, green when mature, with two-celled stomata and central columella, splitting at maturity from apex downwards into two valves; pseudoelaters multicellular, without helical bands Class ANTHOCEROTOPSIDA, Order ANTHOCEROTALES, **1. Anthocerotaceae**

1 Plants thallose or leafy; chloroplasts several per cell; capsule not cylindrical, not green when mature, without stomata and columella, opening irregularly or splitting into four valves; elaters absent or unicellular, with helical bands Class MARCHANTIOPSIDA

  2 Plants leafy, leaves one-layered throughout Order JUNGERMANNIALES

    3 Leaves succubous or transversely inserted, entire or divided up to half of their length into lobes of the same size and shape; underleaves absent or very small; rhizoids scattered over the ventral side of stem.

      4 Plants very small, 0.25–0.35 mm wide; leaves bilobed; mid-leaf cells 12–14 μm wide in local representatives **2. Cephaloziellaceae**

      4 Plants 0.6–1.2 mm wide; leaves lobed or entire; mid-leaf cells 25–50 μm wide in local representatives.

        5 Plants 0.6–1 mm wide in local representatives; stem not swollen underneath gynoecium; leaves alternate, two–four-lobed; leaf cells with two–five oil bodies; female bracts free basally; perianth longly exserted from the bracts, with short beak apically **3. Lophoziaceae**

        5 Plants 1–1.2 mm wide in local representatives; stem swollen underneath gynoecium; leaves opposite, entire; leaf cells with five–ten oil bodies in local representatives; female bracts connate basally; perianth included in the bracts, wide open apically **4. Arnelliaceae**

    3 Leaves incubous, divided to near base into two lobes of different size and shape, the dorsal leaf lobe larger; underleaves large; rhizoids only at base of underleaves.

      6 Ventral leaf lobe ligulate; stylus absent; underleaves entire or toothed **5. Porellaceae**

      6 Ventral leaf lobe galeate or lanceolate; stylus subulate; underleaves bilobed in local representatives **6. Frullaniaceae**

  2 Plants thallose or with leaf-like outgrowth, several-layered near axis.

    7 Plants thallose or with leaf-like outgrowth, without epidermal pores and air chambers on dorsal side; rhizoids all smooth; capsule-wall several-layered.

      8 Plants thallose or with leaf-like outgrowth; oil cells absent, but often each cell with several oil bodies; archegonia not in pyriform or bottle-shaped involucres; elaters with helical band Order METZGERIALES

        9 Plants a creeping axis with erect, crisped, leaf-like outgrowths; rhizoids usually purplish; elaters free **7. Codoniaceae**

        9 Plants thallose; rhizoids hyaline or brownish; elaters fixed on elaterophores.

          10 Thallus throughout several-layered, with wide median band of small cortical and large inner cells; mucilage hairs two–five-celled, present only at thallus apex; antheridia and archegonia dorsal on main thallus; elaterophores at bottom of capsule **8. Pelliaceae**

10 Thallus delicate, with sharply defined, narrow median band of large cortical and small inner cells, and one-layered wings; mucilage hairs one-celled, present everywhere on the ventral side of thallus; antheridia and archegonia on small, modified ventral branches; elaterophores at top of capsule valves **9. Metzgeriaceae**

8 Plants with leaf-like outgrowth; oil cells with single oil-body present or absent, but no cells with several oil bodies; archegonia in pyriform or bottle-shaped involucres; elaters absent, but spores mixed with minute, green sterile cells without helical bands Order SPHAEROCARPALES

11 Plants with an axis, a large, wavy, one-layered dorsal wing and two irregular rows of small, leaf-like, green scales on ventral side of axis; oil cells present; antheridia immersed into dorsal side of apical wing margin; archegonia in pyriform or bottle-shaped involucres, arising from axis on both sides of wing; spore tetrads separating at maturity **10. Riellaceae**

11 Plants a thick axis with leaf-like, suborbicular outhgrowth, several-layered in the centre; scales and oil cells absent; antheridia and archegonia bounded by large, bottle-shaped involucres, closely grouped on dorsal side of thallus, hiding it nearly completely; spores remaining in tetrads often permanently **11. Sphaerocarpaceae**

7 Plants thallose, with epidermal pores and air chambers on dorsal side; some rhizoids smooth, some tuberculate; capsule-wall one-layered Order MARCHANTIALES

12 Thallus branching dichotomous and ventral and/or apical; archegonia grouped in receptacles; elaters with helical bands.

13 Archegonia and sporophyte in sessile, bivalved, mussel-shaped, black-purplish involucre underneath thallus apex **13. Targioniaceae**

13 Archegonia in several involucres.

14 Epidermal pores bounded by one or several concentric rings of cells; gemma cups crescent-shaped or absent; scales in two or in irregular rows on ventral side of thallus; archegonial receptacle stalked, antheridial receptacle sessile, or antheridia scattered on dorsal side of thallus.

15 Oil cells present in thallus and scales; capsule wall without ring-like thickenings.

16 Air chambers in single layer, with assimilatory filaments; scales with large, reniform appendage; gemma cups crescent-shaped; epidermal pores in female receptacle absent **14. Lunulariaceae**

16 Air chambers in several layers, without assimilatory filaments; scales with triangular, lanceolate or filiform appendages; gemma cups absent; female receptacle with compound epidermal pores **15. Aytoniaceae**

15 Oil cells present in thallus, absent in scales; capsule wall with ring-like thickenings **16. Cleveaceae**

14 Epidermal pores compound; gemma cups cup-shaped; scales in

six rows on ventral side of thallus in local representatives; archegonial and antheridial receptacles stalked **17. Marchantiaceae**

12 Thallus branching only dichotomous; archegonia in irregular groups on dorsal side of thallus, or embedded in dorsal thallus tissue along median groove, or scattered; elaters absent but sometimes spores mixed with sterile cells.

17 Archegonia irregularly grouped on dorsal side of thallus; sporophyte bounded by a thick, warty calyptra, or by a pyriform involucre.

18 Air chambers with assimilatory filaments; oil cells present in thallus and scales; involucre a posterior, several-layered scale; capsule bounded by calyptra; calyptra thickened, smooth, warty in local representatives **12. Corsiniaceae**

18 Air chambers without assimilatory filaments; oil cells absent in thallus and scales; involucre large, pyriform; capsule bounded by involucre; calyptra thin **18. Oxymitraceae**

17 Archegonia embedded along median groove or scattered on dorsal side of thallus; sporophyte embedded in thallus; calyptra and involucre absent **19. Ricciaceae**

TAXONOMIC TREATMENT

ANTHOCEROTOPSIDA

ANTHOCEROTALES

1. ANTHOCEROTACEAE

Gametophyte thallose, dichotomously branched; air chambers absent; scales absent; chloroplast single per cell (rarely a few in tropical species); oil bodies absent; oil cells absent. *Rhizoids* smooth. *Antheridia* dorsal, single in antheridial cavities. *Archegonia* dorsal, embedded in thallus. *Sporophyte* dorsal, erect, with conspicuous foot and apical meristem allowing indeterminate growth, bounded by a cylindrical involucre. *Capsule* slenderly cylindrical, green, wall four–six-layered, with stomata bounded by two sausage-shaped guard cells and central columella, splitting at maturity from apex downwards into two valves. *Sterile cells* beside spores, undergoing repeated divisions before maturing into multicellular pseudoelaters without helical bands. *Chromosome number* $x=5$ or 6.

A family comprising five–six genera with worldwide distribution, but growing mainly in warmer regions on soil or rocky soil. A single genus is represented in the local flora.

Phaeoceros Prosk.

Thallus dark green, prostrate, variously lobed and often forming rosettes, without distinct median band, thickened in the central part. *Perennating tubers* stalked, frequent on ventral side of thallus. *Capsule* green, long-exserted; wall chlorophyllose. *Spores* yellow. *Pseudoelaters* one–six-celled, geniculate, without helical bands.

A genus characterised by yellow spores. It is sometimes included in the genus *Anthoceros* L. (e.g., Schuster 1992b) and comprises 10–15 species. Two species, *Phaeoceros bulbiculosus* and *P. laevis*, are frequent in the Mediterranean Region and are also present in Israel.

Literature. — Hasegawa (1988), Haessel de Menéndez (1988), Schuster (1992b).

1 Spores smooth, with distinct, central ornamentation on distal face, faintly vermiculate on proximal face; capsule 5–25 mm long, usually rather thick; stalked tubers on ventral side of thallus frequent **1. P. bulbiculosus**

1 Spores tuberculate on distal face, irregularly papillose on proximal face; capsule 20–50 mm long, usually slender; stalked tubers on ventral side of thallus occasional **2. P. laevis**

1. Phaeoceros bulbiculosus (Brot.) Prosk., Rapp. Comment. 8° Congr. Int. Bot. Paris 14–16:69(1954).

Anthoceros bulbiculosus Brot., Fl. Lusit. 2:430 (1804); *A. dichotomus* Raddi, Atti Accad. Sci. Siena 9:239 (1808). — Jovet-Ast & Bischler (1966):106.

Distribution map 209, Figure 208, Plate XV:a.

Thallus 5–20 mm in diameter, usually divided into irregular segments with entire, undulate margins. *Stalked tubers* frequent on ventral side of thallus. *Dioicous. Involucres* 2–3.5 mm long. *Capsule* 5–25 mm long, usually rather thick. *Spores* 42–63 μm in diameter; distal face smooth with distinct, central ornamentation; proximal face faintly vermiculate (seen with SEM).

Habitat and Local Distribution. — Plants growing in dense colonies or scattered on compact, moist sandy clay or basaltic soil with low pH (5.5), overlying basaltic rock, exposed or at base of cliffs, in grassland and open *Quercus* forest, associated with *Riccia*. Altitude: 50–450 m. Rare; so far collected only in two localities: Sharon Plain and Hula Plain.

General Distribution. — Recorded from Southwest Asia, around the Mediterranean, Europe, Macaronesia, and North and Central America.

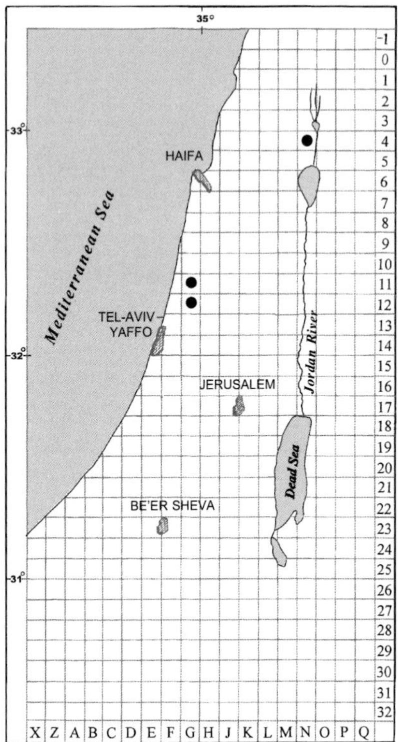

Map 209: *Phaeoceros bulbiculosus*

Tubers are more frequent on the ventral side of the thallus in *Phaeoceros bulbiculosus* than in *P. laevis*. The capsules of *P. bulbiculosus* are usually thicker. However, these two characters do not permit to keep the two taxa apart with certainty. Only spore coat ornamentation clearly separates them.

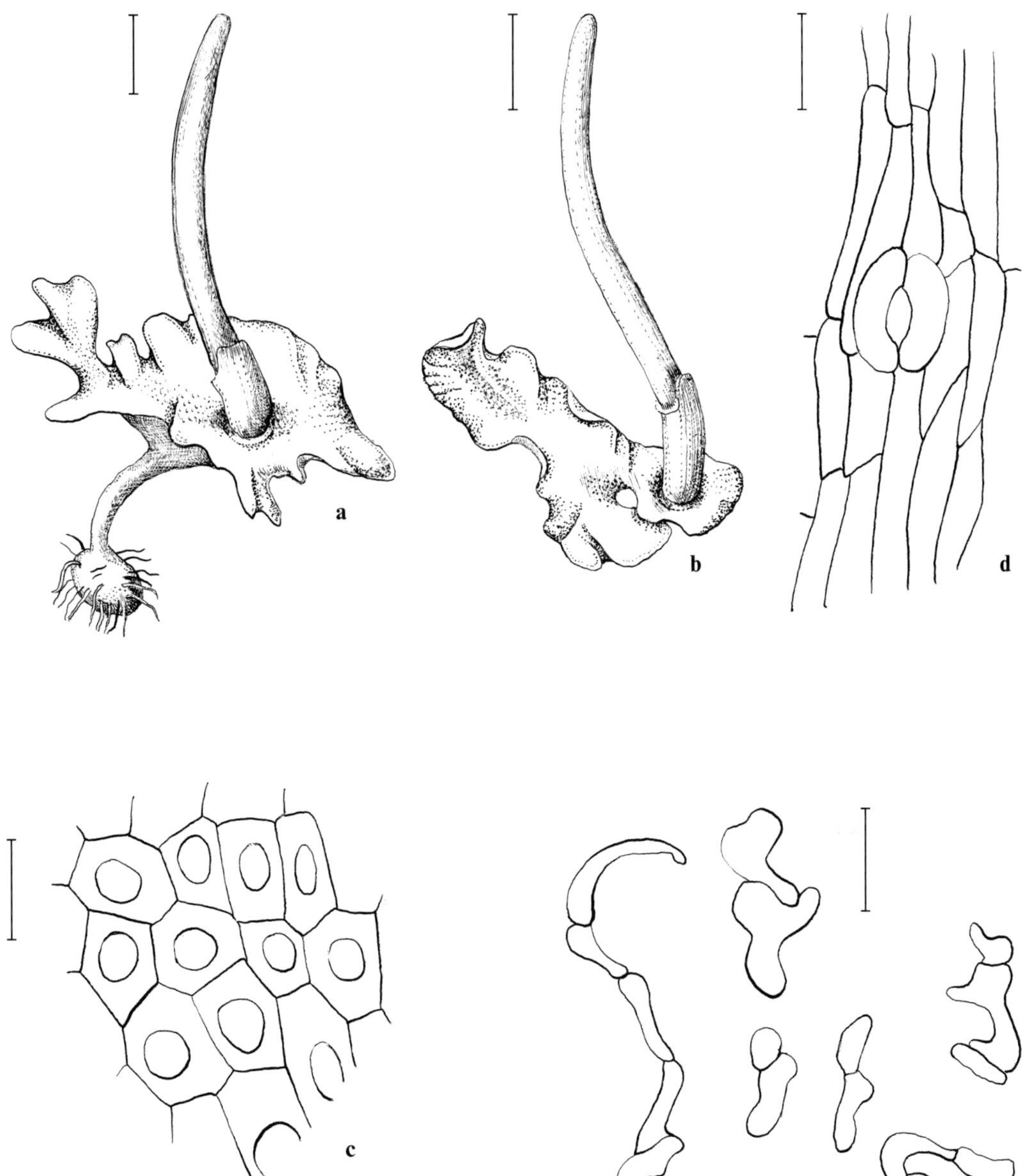

Figure 208. *Phaeoceros bulbiculosus*: **a** thallus with sporophyte and ventral tuber; **b** thallus with sporophyte; **c** thallus cells with single chloroplast; **d** cells and stoma of capsule wall; **e** pseudoelaters. Scale bars: **a**, **b** = 2 mm; **c**–**e** = 50 µm. (*Jovet-Ast* & *Bischler 64003*)

2. Phaeoceros laevis (L.) Prosk., Bull. Torrey Bot. Club 78 : 347 (1951).
Anthoceros laevis L., Sp. Pl., Edition 1 : 1139 (1753). — Jovet-Ast & Bischler (1966) : 106.

Distribution map 210, Figure 209, Plate XV : b.

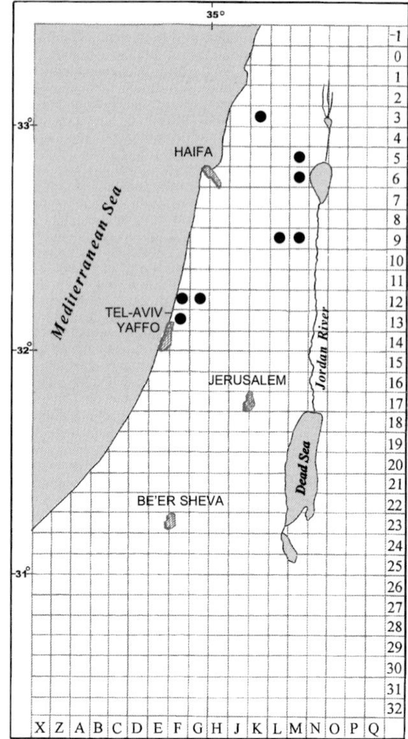

Map 210: *Phaeoceros laevis*

Thallus 20–30 mm in diameter, divided into broad segments with entire, sinuose margins. *Stalked tubers* occasionally present. *Monoicous* or dioicous. *Involucres* 2–3.5 mm long. *Capsule* 20–50 mm long, usually slender. *Spores* 30–60 µm in diameter; distal face with acute tubercles, proximal face with irregular, rounded papillae (seen with SEM).

Habitat and Local Distribution. — Plants growing in colonies or scattered on compact, moist or wet clay, sandy clay, or basaltic soil (pH 5–8), overlying sandstone, limestone or basalt, on exposed rocks or wadi banks, in grassland, open *Quercus* and *Eucalyptus* groves, associated with *Riccia* and *Sphaerocarpos*. Altitude: -200–250 m. Occasional: Sharon Plain, Upper Galilee, Lower Galilee, Samaria, and Bet She'an Valley.

General Distribution. — Nearly cosmopolitan in tropical and temperate parts of the world. Recorded from Southwest Asia, around the Mediterranean, Europe, Macaronesia, throughout Africa and the Mascarene Islands, eastern Asia, India, Oceania, and North, Central, and South America.

In sterile condition, *Phaeoceros laevis* is easily distinguished from sterile *Pellia* by its cells containing a single chloroplast.

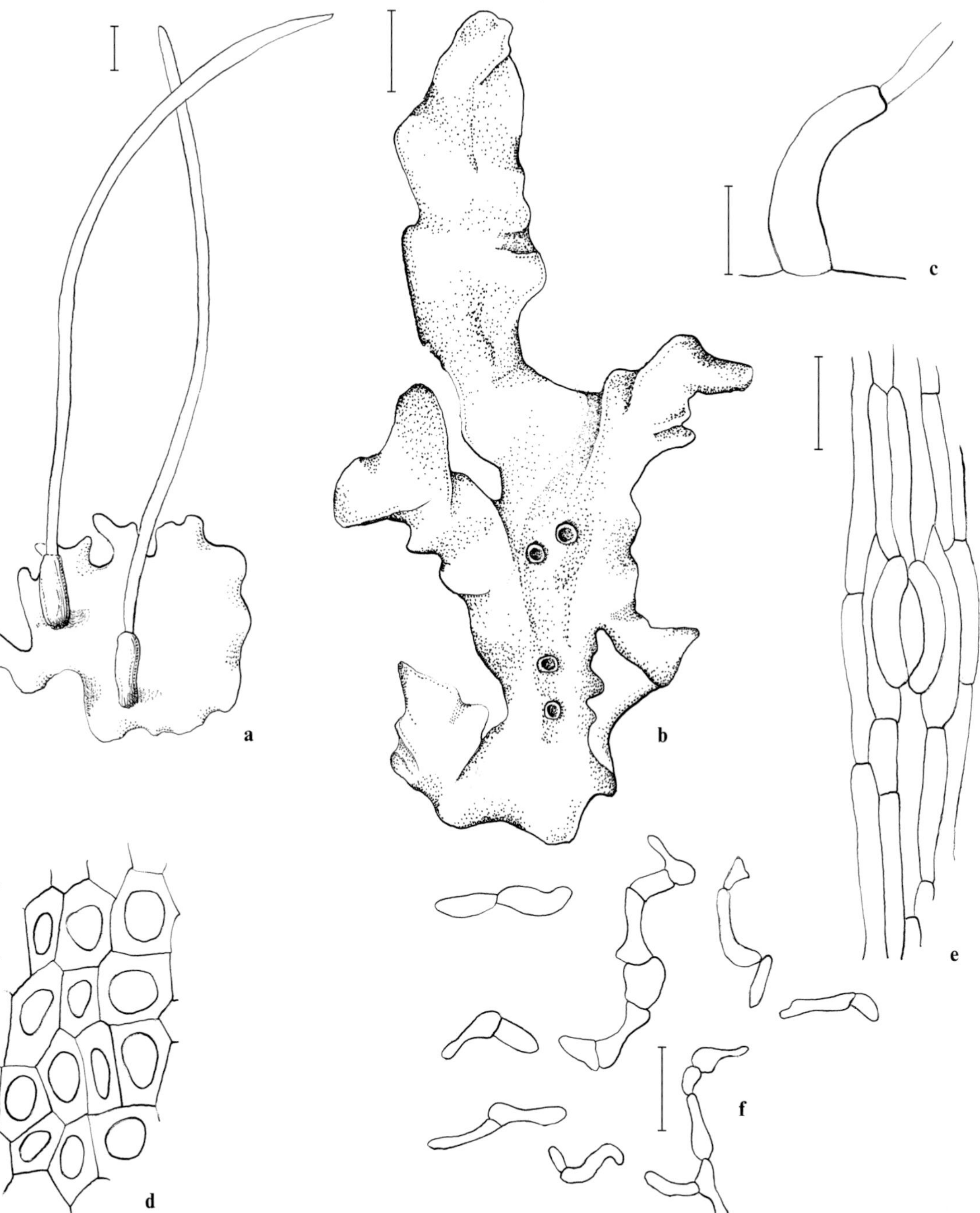

Figure 209. *Phaeoceros laevis*: **a** thallus with sporophytes; **b** thallus with antheridial cavities; **c** involucre; **d** thallus cells with single chloroplast; **e** cells and stoma of capsule wall; **f** pseudoelaters. Scale bars: **a**–**c** = 1 mm; **d**–**f** = 50 µm. (*Jovet-Ast* & *Bischler 64002*)

MARCHANTIOPSIDA

JUNGERMANNIALES

2. CEPHALOZIELLACEAE

Plants leafy, minute. *Stems* prostrate, irregularly branched; cortical stem cells of similar size than inner cells. *Leaves* alternate, usually distant, transversely or obliquely inserted, succubous, shallowly to deeply bilobed, margins entire to spinose; cell walls thin or thickened; oil bodies small, spherical; underleaves small or missing. *Rhizoids* scattered over the ventral side of stem, smooth. *Gemmae* often present, at apex of stem or at margins of upper leaves and underleaves, one–two-celled. *Male bracts* bilobed, with solitary, axillary antheridia. *Female bracts* larger than leaves, connate or not for part of their length to the bracteole. *Perianth* four–five-plicate, crenulate apically. *Sporophyte* seta of four cortical cells, enclosing a single inner cell; capsule ovoid to ellipsoidal, wall two-layered. *Elaters* with helical bands. *Chromosome number* $x = 9$.

A family comprising seven genera with a worldwide distribution. A single genus is represented in the local flora.

Cephaloziella (Spruce) Schiffn.

Plants less than 1 mm wide, creeping. *Leaves* nearly transversely inserted, remote, divided up to ½–¾ of leaf length into two lobes of similar size and shape; leaf cells 10–14 µm wide; oil bodies three–nine per cell, nearly homogeneous; underleaves absent or present, very small. *Gemmae* two-celled. *Monoicous* or dioicous. *Male bracts* in three–six pairs, resembling in shape the stem leaves but larger, imbricate and ± inflated. *Female bracts* larger than leaves but similar in shape, more or less highly connate to bracteole, entire or toothed on margins. *Perianth* exserted, narrowed at apex. *Capsule* exserted, elliptical, opening at maturity into four valves; wall two-layered, both layers with nodulose thickenings. *Spores* small, 6–12 µm in diameter. *Elaters* with two helical bands.

A genus comprising about 30 species, phenotypically very variable and difficult to discern. Eleven species occur in the Mediterranean Region. A single species occurs in the local flora.

Literature. — Müller (1951–1958), Schuster (1980), Paton (1999).

Cephaloziella baumgartneri Schiffn., Verh. K. K. Zool.-Bot. Ges. Wien 56 : 273 (1906). [Jovet-Ast & Bischler (1966) : 105; Bischler & Jovet-Ast (1975) : 19]

Distribution map 211 (p. 540), Figure 210.

Plants yellowish-green when moist, black when dry. *Stems* 5–10 mm long, 0.25–0.35 mm wide, more or less branched. *Leaves* divided up to half of their length into two triangular, acute lobes, four–six cells wide basally, often ending in a row of two cells apically, sinus acute, ca. 90°; cells 12–14 µm wide in mid-leaf, with thickened walls; oil bodies spherical, three–eight per cell, small; cuticle smooth; underleaves

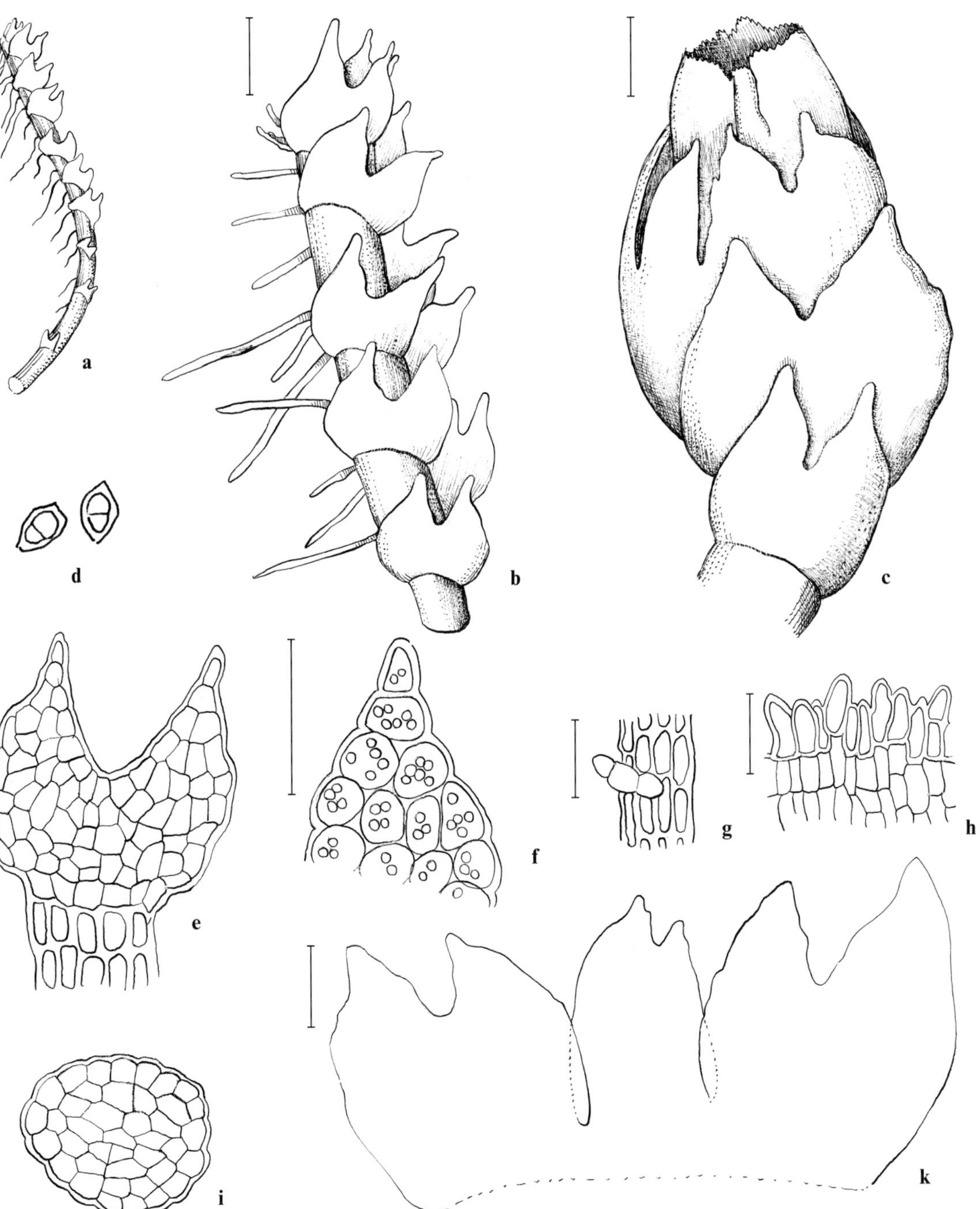

Figure 210. *Cephaloziella baumgartneri*: **a**, **b** stems, dorsal side; **c** bracts and perianth; **d** gemmae; **e** leaf; **f** leaf lobe; **g** underleaf; **h** cells at apex of perianth; **i** cross-section of stem; **k** bracts and bracteole. Scale bars: **a**–**c**, **k** = 100 μm; **e**–**i** = 50 μm; **d** = 25 μm. (*Jovet-Ast* & *Bischler 82365*)

absent on sterile stems, very small, filiform (two–three cells in a row) on fertile and gemmiferous stems. *Gemmae* in clusters at stem apex or on leaf margins, elliptical, yellowish-green, smooth, 10×20 μm long, with slightly thickened walls. *Autoicous. Male branches* intercalary, bracts in four–ten pairs, imbricate, with a single antheridium per bract. *Female bracts* terminal, in two–three pairs, larger than leaves, connate to bracteole on ventral side, free on dorsal, bilobed up to ⅓–½ of length, lobes acute, sinus acute, margins entire or with a few teeth. *Bracteole* large, shortly bilobed. *Perianth* exserted up to ⅓–½ of length, deeply five-plicate, with hyaline and crenulate upper margin built up of cells with slightly thickened walls, twice as long as wide. *Sporophyte* ephemeral (not seen in local specimens). *Capsule* ovate, inner wall layer with ring-like thickenings. *Spores* red-brown, 8–12 μm in diameter, finely granulose in other Mediterranean material.

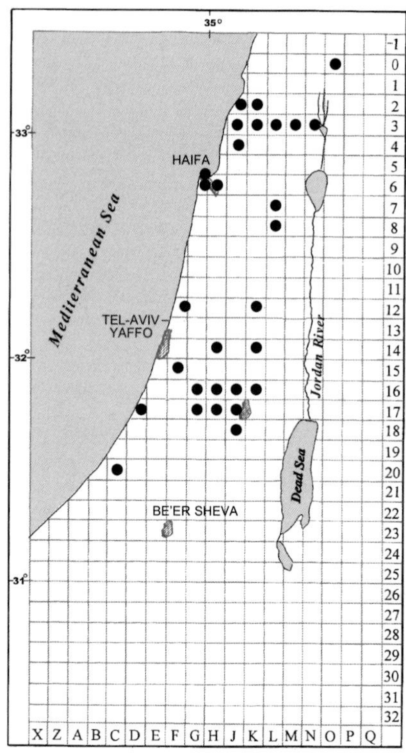

Map 211: *Cephaloziella baumgartneri*

Habitat and Local Distribution. — Plants growing as scattered stems among mosses or other liverworts, on moist clay, sandy clay, or terra rossa (pH 7–8), overlying limestone, dolomite, or other calcareous rocks, in rock crevices or on stone walls, on exposed wadi banks or cliffs, in grassland, open shrubland, and *Pinus* forests; associated with *Southbya* and *Fossombronia.* Altitude: 10–1100 m. Common, except along most of the part of the Syrian-African Rift Valley in Israel and in the Negev: Coast of Galilee, Coast of Carmel, Sharon Plain, Philistean Plain, Upper Galilee, Lower Galilee, Mount Carmel, Esdraelon Plain, Samaria, Shefela, Judean Mountains, Judean Desert, Hula Plain, and Golan Heights.

General Distribution. — Recorded from Southwest Asia, around the Mediterranean, the Atlantic coast of Europe, and Macaronesia.

A plant often overlooked due to its small size. It resembles *Leiocolea turbinata* (Raddi) Steph., but in that species the leaves are narrowed towards base, the leaf lobes are obtuse, and the leaf cells are distinctly larger than the leaf cells in *Cephaloziella baumgartneri.*

3. LOPHOZIACEAE

Plants leafy, minute to large. *Stems* prostrate or ascending, not swollen underneath gynoecium, branching mainly axillary, from the lower half of the axil of normal leaves; cortical stem cells of similar size to inner cells but often thickened. *Leaves* alternate, succubous, plane to very concave, more or less transversely to very obliquely inserted, mostly two–four-lobed; cell walls thin or thickened; oil bodies two–five, finely papillose or smooth; underleaves deeply bifid, sometimes reduced and present only on fertile female branches, or absent. *Rhizoids* scattered over the ventral side of stem, smooth. *Gemmae* often present, one–two-celled and angular, from leaf margins. *Male bracts* spicate, terminal, sometimes intercalary, or below female bracts, of the same size as normal leaves but more or less inflated ventrally, each with one–four antheridia. *Female bracts* terminal, free basally, usually larger than leaves. *Perianth* exserted, ovoid to cylindrical, inflated, narrowed into a short beak and smooth or plicate at apex, not compressed. *Sporophyte* seta with many cells of similar size in cross section; capsule ovoid, opening into four valves, wall two–five-layered, outer layer with nodulose, inner layer usually with ring-like thickenings. *Spores* small. *Elaters* with helical bands. *Chromosome number* $x=9$.

A family comprising about 19 genera, sometimes treated as a subfamily of the Jungermanniaceae (Grolle 1983a). Its range comprises cooler and cold portions of the Northern Hemisphere. A single genus occurs in Israel.

Leiocolea (Müll. Frib.) H. Buch

Stems usually prostrate, irregularly branched. *Leaves* obliquely inserted, two-lobed, with lobes of similar size; cells 30–50 μm wide in mid-leaf, walls thin or thickened; cuticle smooth or striate-papillose; oil bodies two–five per cell, large; underleaves small or absent. *Rhizoids* hyaline. *Gemmae* present or absent. *Monoicous* or dioicous. *Male bracts* terminal or intercalary, inflated, often ventrally with an irregular additional tooth or lobe. *Female bracts* bilobed. *Perianth* exserted, smooth, plicate in apical part and abruptly contracted at apex into a short, tubular beak, crenulate or ciliate at top. *Seta* long; capsule opening at maturity to near bottom into four valves; wall two–three-layered, innermost layer with ring-like thickenings. *Spores* tuberculate-papillose. *Elaters* with two helical bands.

A genus comprising about nine species. One species, which occurs in Israel, is widespread in the Mediterranean Region.

Literature. — Müller (1951–1958), Schuster (1969), Paton (1999).

Leiocolea turbinata (Raddi) H. Buch *in* Buch *et al.*, Ann. Bryol. 10 : 4 (1938).
Lophozia turbinata (Raddi) Steph., Spec. Hep. 2: 128 (1901); *Jungermannia turbinata* Raddi, Jungermanniogr. Etrusca 18 (1818). — Jovet-Ast & Bischler (1966) : 105 (as "*Cephalozia bicuspidata* (L.) Dum.").

Distribution map 212, Figure 211.

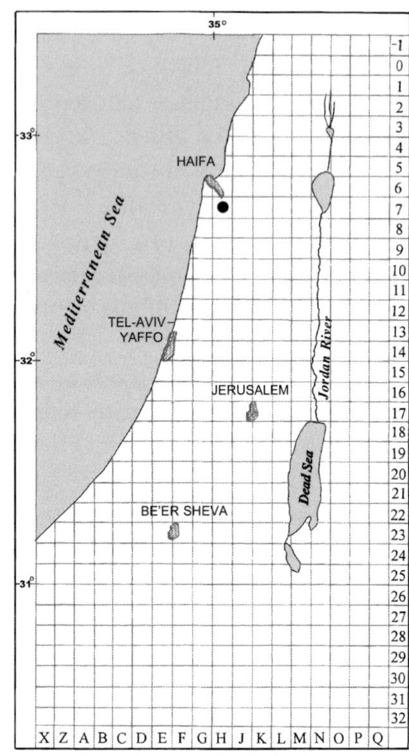

Map 212: *Leiocolea turbinata*

Plants pale green or blackish. *Stems* usually unbranched, 4–12 mm long, 0.6–1 mm wide, translucent. *Leaves* remote, imbricate only in upper part of fertile stems, flat, narrowed towards base, not decurrent, divided up to ⅓ of length into two rounded or obtuse lobes; cells 35–45 μm wide, thin-walled, without trigones; oil bodies four–five per cell, ovoid, hyaline; cuticle smooth; underleaves absent. *Gemmae* absent. *Dioicous*. Sex organs not seen in local specimens; in other Mediterranean material — *Female bracts* loosely bounding perianth, divided up to ¼ of length into two rounded, obtuse or acute lobes; an additional, small lobe or tooth often borne on the ventral side of one or both bracts. *Bracteole* usually absent. *Perianth* long-exserted, pyriform, beak ciliolate. *Capsule* reddish-brown; cells of outer layer with nodulose thickenings, of inner layer with ring-like thickenings. *Spores* 15–18 μm in diameter, brown; finely tuberculate-papillose on both faces.

Habitat and Local Distribution. — Plants growing in loose colonies or scattered stems among mosses on moist, shaded clay overlying calcareous rocks (pH 7), in dense garrigue, associated with *Metzgeria*. Altitude: 300 m. Collected only once: Mount Carmel.

General Distribution. — Recorded from Southwest Asia, around the Mediterranean, and the Atlantic coast of Europe.

Leiocolea turbinata resembles superficially *Cephaloziella baumgartneri* (see the discussion of that species above, p. 540).

In sterile condition, *Cephalozia bicuspidata* (L.) Dumort. is easily confused with *Leiocolea turbinata*. The two species have the same colour and translucence. However, in *Cephalozia* the leaves are not narrowed basally, the leaf lobes are acute, ending often in a row of two cells, and oil bodies are absent.

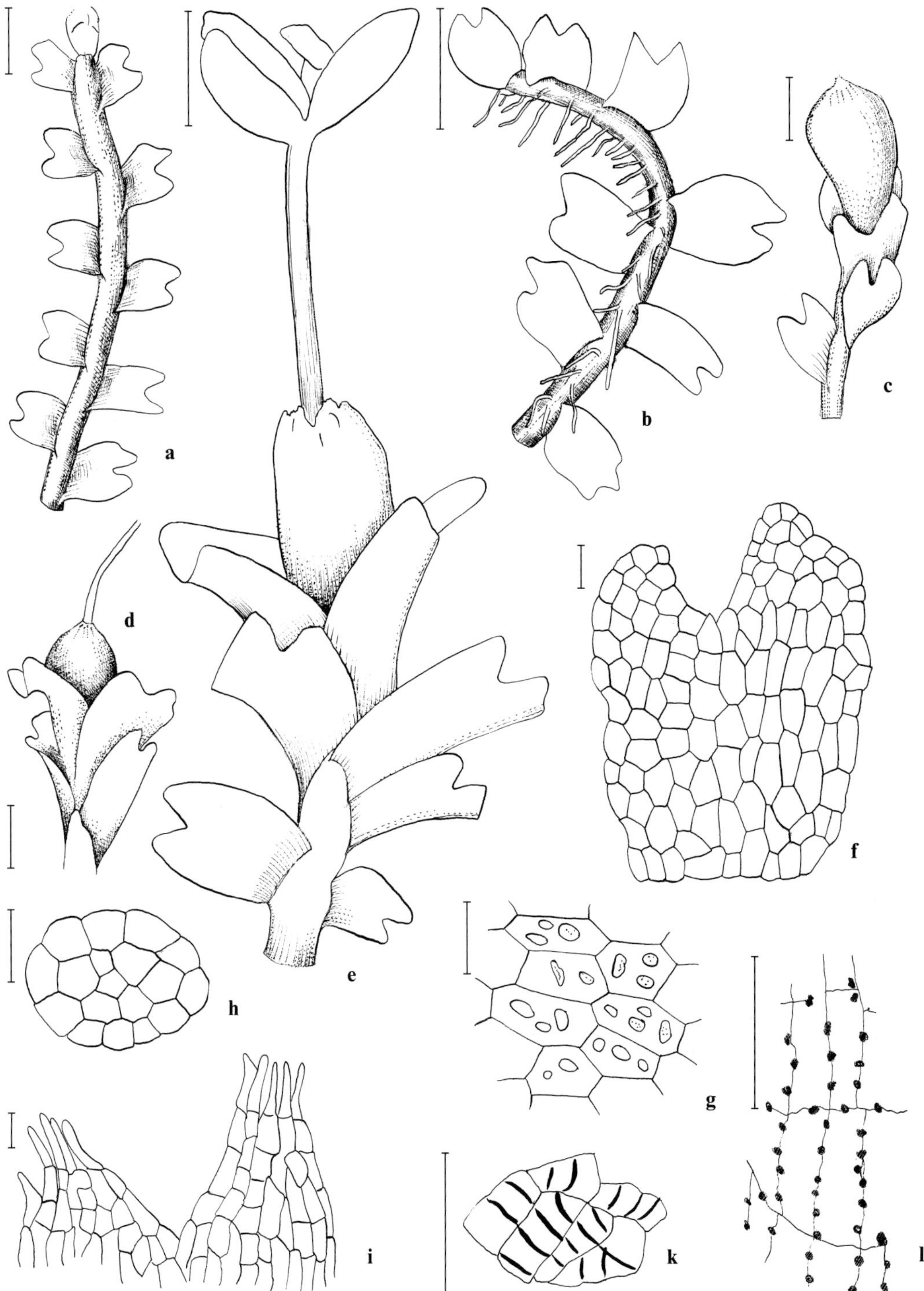

Figure 211. *Leiocolea turbinata*: **a** stem, dorsal side; **b** stem, ventral side; **c**, **d** stems with bracts and perianth; **e** stem with bracts, perianth and sporophyte; **f** leaf; **g** leaf cells; **h** cross-section of seta; **i** cells at apex of perianth; **k** cells of inner layer of capsule wall; **l** cells of outer layer of capsule wall. Scale bars: **a**–**e** = 500 µm; **f**–**i**, **k**, **l** = 50 µm. (**a**, **b**, **f**, **g**: *Jovet-Ast* & *Bischler 77436*; **c**–**e**, **h**, **i**, **k**, **l**: Crete, *Jovet-Ast* & *Bischler 7743*)

4. ARNELLIACEAE

Plants leafy, small to robust, prostrate, simple or sparingly branched. *Leaves* opposite, succubous, imbricate, more or less transversely inserted, orbicular or ovate, with entire margins, adjacent or connate on dorsal side; cells thin-walled, with or without trigones; cuticle often verrucose; underleaves small, often restricted to female stems, or absent. *Rhizoids* scattered over ventral side of stem. *Male bracts* intercalary. *Female bracts* terminal, connate basally, on top of a swollen part of stem (perigynium) sunken into the substratum, nearly at right angles to the prostrate main axis. *Perianth* included in the bracts, wide open and crenulate apically. *Capsule* spherical to cylindrical, wall two-layered, outer wall with nodulose trigones, inner layer with ring-like thickenings. *Chromosome number* $x = 9$.

A family of three genera with restricted ranges in the Holarctic, the Mediterranean Region, and tropical Asia.

Southbya Spruce

Stems usually unbranched or with few ramifications. *Leaves* opposite, strongly imbricate, orbicular to elliptical, entire, contiguous basally; leaf cell walls with small to conspicuous trigones; oil bodies five–ten per cell, hyaline, spherical to ovoid; cuticle smooth or papillose; underleaves absent on sterile stems, present below female bracts. *Rhizoids* hyaline, smooth. *Antheridia* single in bract axils. *Female bracts* erect, connate at base to each other, with erose-toothed margins. *Perianth* shorter than bracts, apical opening wide, bilobed and irregularly toothed. *Capsule* spherical, opening at maturity into four valves. *Spores* indistinctly trilete, with small, rounded tubercles on both faces. *Elaters* with two helical bands.

A genus comprising three–four exclusively calcicolous species. Two species are widely distributed in the Mediterranean Region, and are also present in the local flora.

Literature. — Müller (1951–1958), Paton (1999).

1 Plants dark green, blackish when dry; leaf cells elongated at ventral leaf margin; cuticle smooth; spores 20–25 µm in diameter **1. S. nigrella**

1 Plants yellowish-green, sometimes bright green, brownish when dry; leaf cells hardly elongated at ventral leaf margin; cuticle densely papillose, with hemispherical papillae; spores 18–20 µm in diameter **2. S. tophacea**

1. Southbya nigrella (De Not.) Henriq., Bol. Soc. Brot. 4:244 (1887).

Jungermannia nigrella De Not., Mem. Reale Accad. Sci. Torino, ser. 2, 1:315 (1839). — Jovet-Ast & Bischler (1966):105, Bischler & Jovet-Ast (1975):19.

Distribution map 213 (p. 546), Figure 212, Plate XV:g.

Plants dark green, blackish when dry, 3–5 mm long, 1–1.2 mm wide. *Stem* in section flat on dorsal, convex on ventral side. *Leaves* orbicular, frequently recurved, ventral margin with two–five rows of elongated cells, 50–70 µm long (seen with SEM); mid-leaf cells 25–35 µm in diameter, thin-walled, trigones small; oil bodies five–eight per cell; cuticle smooth. *Monoicous* (paroicous). *Female bracts* toothed or eroded on

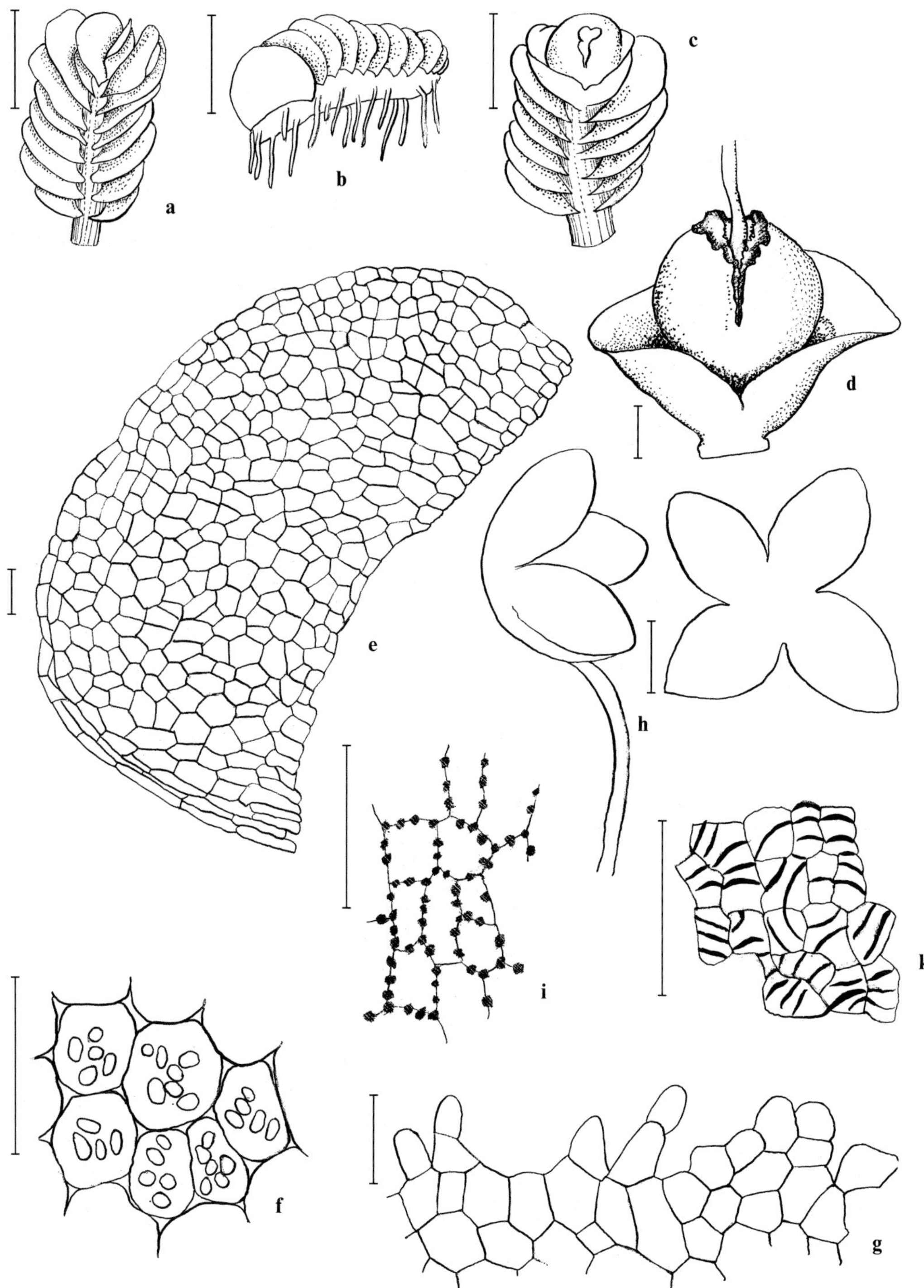

Figure 212. *Southbya nigrella*: **a** stem, dorsal side; **b** stem, lateral view; **c** stem with bracts, perianth, and sporophyte; **d** bracts and perianth; **e** leaf; **f** leaf cells; **g** cells at margin of bracts; **h** open capsules, side and top view; **i** cells of outer layer of capsule wall with nodular thickenings; **k** cells of inner layer of capsule wall with ring-like thickenings. Scale bars: **a**–**c** = 1 mm; **e**–**g**, **i**, **k** = 50 μm; **d**, **h** = 250 μm. (*Jovet-Ast* & *Bischler 82299*)

margins. *Perianth* bilabiate with toothed margins apically. *Capsule* cells of outer layer of wall with nodulose thickenings, cells of inner layer with ring-like thickenings. *Spores* yellow-brown, 20–25 μm in diameter; with numerous, irregular tubercles (seen with SEM).

Habitat and Local Distribution. — Plants growing in loose colonies or scattered stems among mosses, on moist clay or terra rossa (pH 7–8), overlying dolomite or other calcareous rocks, in rock crevices or directly on rock, sometimes in stone walls, on exposed wadi banks and cliffs, in grassland, open shrubland, *Pinus* and *Quercus* forests, *Olea europaea* and *Eucalyptus* groves, associated with *Cephaloziella* and *Fossombronia*. Altitude: -100–900 m. Common, except in the south: Coastal Galilee, Coast of Carmel, Sharon Plain, Philistean Plain, Upper Galilee, Lower Galilee, Mount Carmel, Esdraelon Plain, Samaria, Shefela, Judean Mountains, Judean Desert, Hula Plain, Bet She'an Valley, and Golan Heights.

General Distribution. — A Mediterranean-Atlantic species, recorded from Southwest Asia, around the Mediterranean, the Atlantic coast of Europe, and Macaronesia.

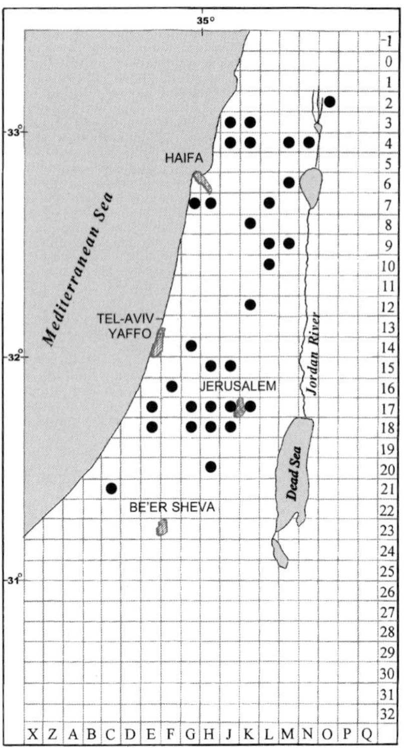

Map 213: *Southbya nigrella*

Southbya nigrella is easily distinguished from the other local leafy liverworts by its opposite leaves. Only *Southbya tophacea* may be confused with it, but in that species the cuticle is densely papillose, the ventral leaf margin is built up of short cells, and its colour is yellowish-green to bright green, becoming brown when dry.

2. Southbya tophacea (Spruce) Spruce, Ann. Mag. Nat. Hist., ser. 2, 3 : 501 (1849).
Jungermannia tophacea Spruce, Hep. Pyren. exs. no. 23 (1847); *Southbya stillicidiorum* (De Not.) Lindb. *in* C. Massal., Ann. Ist. Bot. (Rome) 2 : 12 (1886). — Jovet-Ast & Bischler (1966) : 105.

Distribution map 214, Figure 213, Plate XV : h.

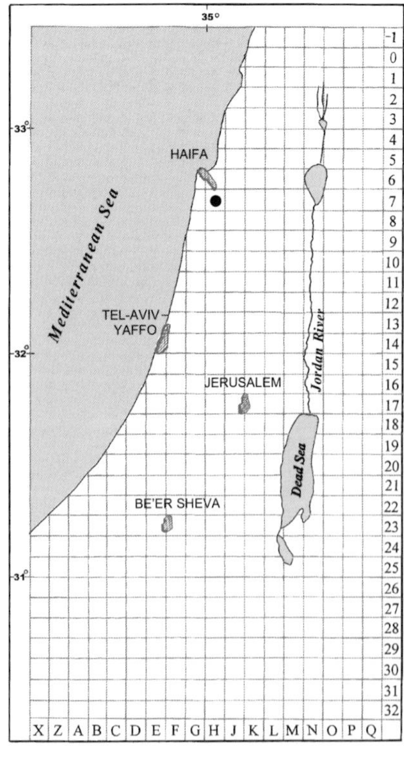

Map 214: *Southbya tophacea*

Plants yellowish-green or bright green, brownish when dry, 3–4 mm long, 1–1.2 mm wide.

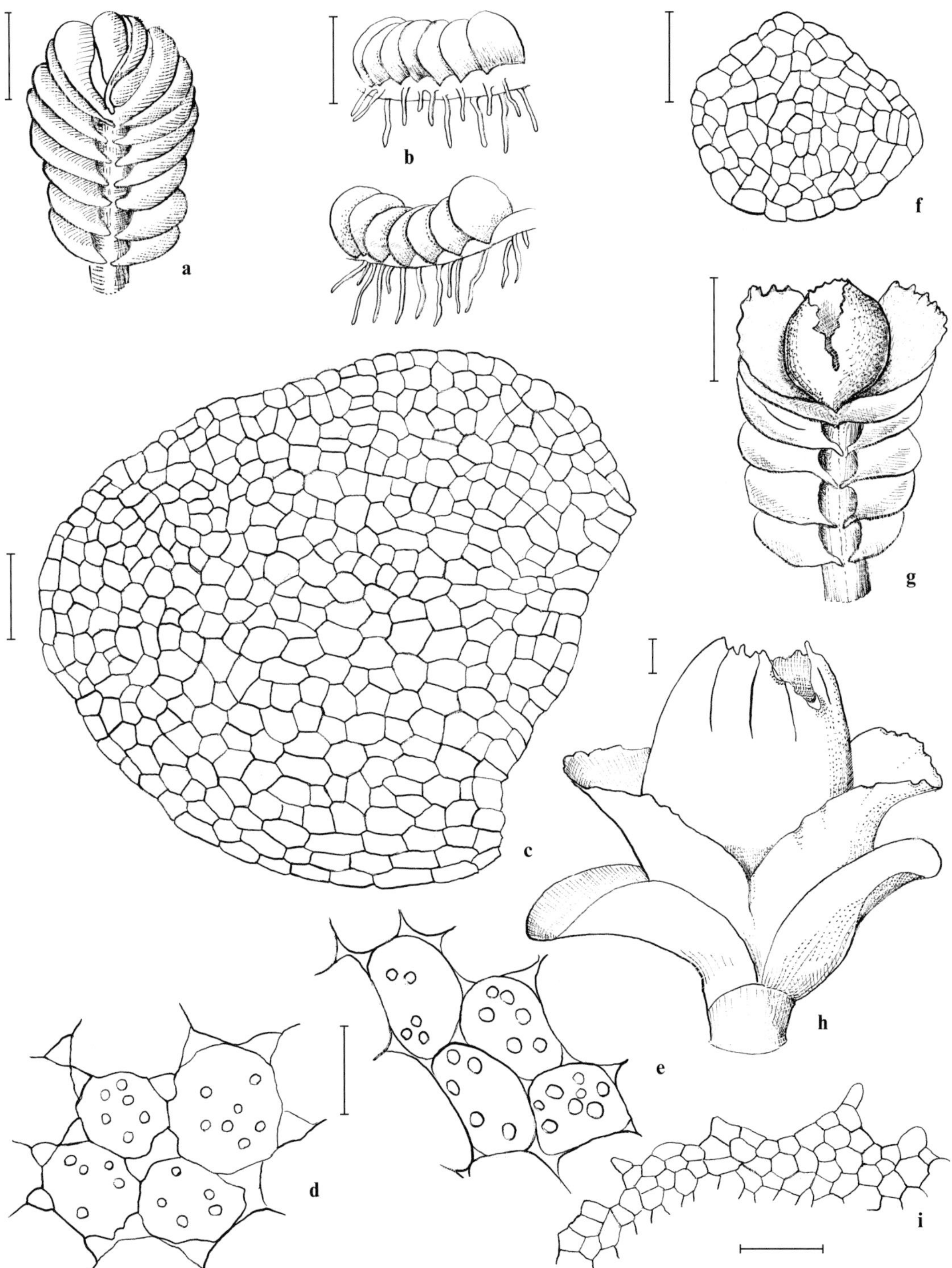

Figure 213. *Southbya tophacea*: **a** stem, dorsal side; **b** stems, lateral view; **c** leaf; **d** mid-leaf cells; **e** cells near ventral leaf margin; **f** cross-section of stem; **g** female stem; **h** bracts and perianth; **i** margin of bract. Scale bars: **a**, **b**, **g** = 1 mm; **c**, **f**, **h**, **i** = 100 µm; **d**, **e** = 50 µm. (Crete, *Jovet-Ast* & *Bischler 77436*)

Stem in section orbicular. *Leaves* usually not recurved, elliptical; mid-leaf cells isodiametric, 30–35 µm in diameter, hardly elongated on ventral leaf margin (seen with SEM), walls thin, trigones small; oil bodies five–ten per cell; cuticle densely papillose, with hemispherical papillae. *Dioicous*. Sex organs not seen in local specimens; in other Mediterranean material — *Female bracts* with eroded or toothed margins. *Perianth* crenulate apically. *Spores* reddish-brown, 18–20 µm in diameter, with numerous, irregular tubercles (seen with SEM).

Habitat and Local Distribution. — Plants growing in loose colonies or scattered stems among mosses, over moist calcareous rocks, in deep shade of dense shrubland, associated with *Fossombronia*. Altitude: 450 m. Collected only once: Mount Carmel.

General Distribution. — An Atlantic-circum-mediterranean species, recorded from Southwest Asia, around the Mediterranean, the Atlantic coast of Europe, and Macaronesia.

5. PORELLACEAE

Plants leafy, robust. *Stems* prostrate, one–three-pinnate, the branches arising in replacement of a ventral leaf lobe; cortical stem cells thickened. *Leaves* alternate, imbricate, divided up to near base into two lobes of different size and shape with very short, vestigial keel; dorsal lobe incubous, ventral lobe small, ligulate, parallel to stem, stylus absent; underleaves large, entire. *Rhizoids* restricted to the base of underleaves, hyaline, smooth. *Dioicous*; androecia and gynoecia terminal on very short lateral branches. *Male bracts* spicate, imbricate, each with a single antheridium. *Female bracts* smaller or of same size than leaves. *Perianth* usually dorsiventrally compressed. *Capsule* wall two–six-layered. *Elaters* with two–three helical bands. *Chromosome number* $x = 8$ or 9.

A family comprising three genera of saxicolous or epiphytic species, mainly distributed in eastern Asia.

Porella L. (*Madotheca* Dumort.)

Plants yellowish-green, dark green or pale brown, pinnately branched. *Leaves* conduplicate; dorsal leaf lobe large, obliquely ovate, entire or toothed on margins; ventral leaf lobe smaller, ligulate, entire or toothed, frequently decurrent at the external base, without stylus at its junction with stem; oil bodies small, spherical to ovoid, 15–40 per cell; cuticle smooth; underleaves large, entire or toothed. *Dioicous. Male bracts* spicate, in three–six pairs, imbricate, inflated, bilobed, connate to bracteole on ventral side. *Female bracts* with entire or toothed margins. *Perianth* long-exserted, apical part compressed, bilabiate. *Seta* short, hardly longer than perianth; capsule short-exserted, ovoid or spherical, opening at maturity into 4–16 valves; capsule wall two–four-layered, without ring-like thickenings. *Spores* verrucose or echinulate.

A genus comprising ca. 100 species, mostly temperate-east Asiatic in distribution. Five species are present in the Mediterranean Region. A single species occurs in Israel.

Literature. — Müller (1951–1958), Schuster (1980), Boisselier-Dubayle & Bischler (1994), Paton (1999).

Porella cordaeana (Huebener) Moore, Proc. Roy. Irish Acad. Sci., ser. 2, 2 : 618 (1876).

Jungermannia cordaeana Huebener, Hepaticol. Germ. 291 (1834); *Madotheca cordaeana* (Huebener) Dumort., Bull. Soc. Roy. Bot. Belgique 13 : 25 (1874).

Distribution map 215, Figure 214.

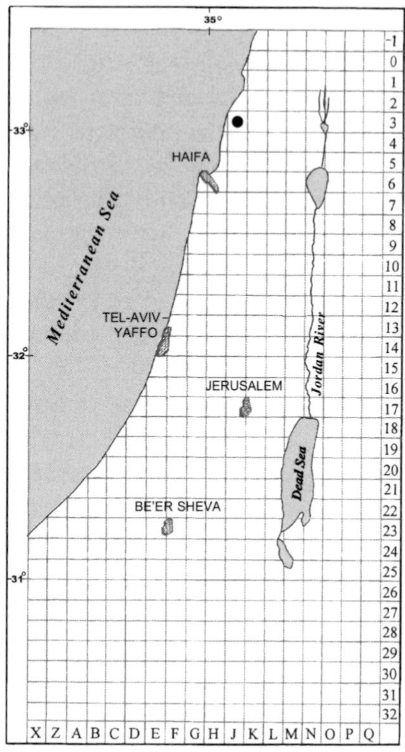

Map 215: *Porella cordaeana*

Plants nearly regularly pinnate, dark green. *Stems* 3–7 cm long, 2 mm wide. *Leaves* conduplicate; dorsal leaf lobe entire, often with one–two isolated, pluricellular teeth on dorsal margin, apex obtuse, often decurved; cells 25–33 µm in diameter in mid-lobe, walls thin, trigones small or absent; oil bodies 30–40 per cell; ventral leaf lobe small-ligulate, acute, narrower than stem, margins entire, undulate-plicate to coiled, strongly decurrent with toothed-undulate margin on the external base along stem; underleaves 1.5–2 times as wide as stem, strongly decurrent, margins entire, apex rounded and decurved. Sex organs not seen in local specimens; in other Mediterranean material — *Male bracts* shortly connate to bracteole. *Female bracts* and bracteole smaller than leaves, with entire margins. *Perianth* upper margin crenulate. *Capsule* ovoid, opening at maturity into four valves. *Spores* yellowish-green, 30–40 µm in diameter.

Habitat and Local Distribution. — Plants growing in loose colonies over moist calcareous rock in dense forest, on a wadi bank, probably restricted to dense shade. Altitude: 200 m. Collected only once: Upper Galilee.

General Distribution. — Recorded from Southwest Asia, reaching Afghanistan in the east, around the Mediterranean (rare), Europe, Macaronesia, and northwestern America.

Figure 214. *Porella cordaeana*: **a** stem, dorsal side; **b** two pairs of leaves, ventral side; **c** ventral lobe; **d** mid-leaf cells of dorsal lobe; **e** male bracts and bracteole; **f** bracts and perianth, ventral side; **g** bracts and perianth, dorsal side; **h** bracteole; **i** cross-section of perianth. Scale bars: **a** = 1 mm; **b**, **e**–**i** = 500 µm; **c** = 100 µm; **d** = 25 µm. (**a**, **e**: France, *Culman*; **b**–**d**: *Danin 79-110-9* (HUJ); **f**–**i**: Corsica, *Camus*)

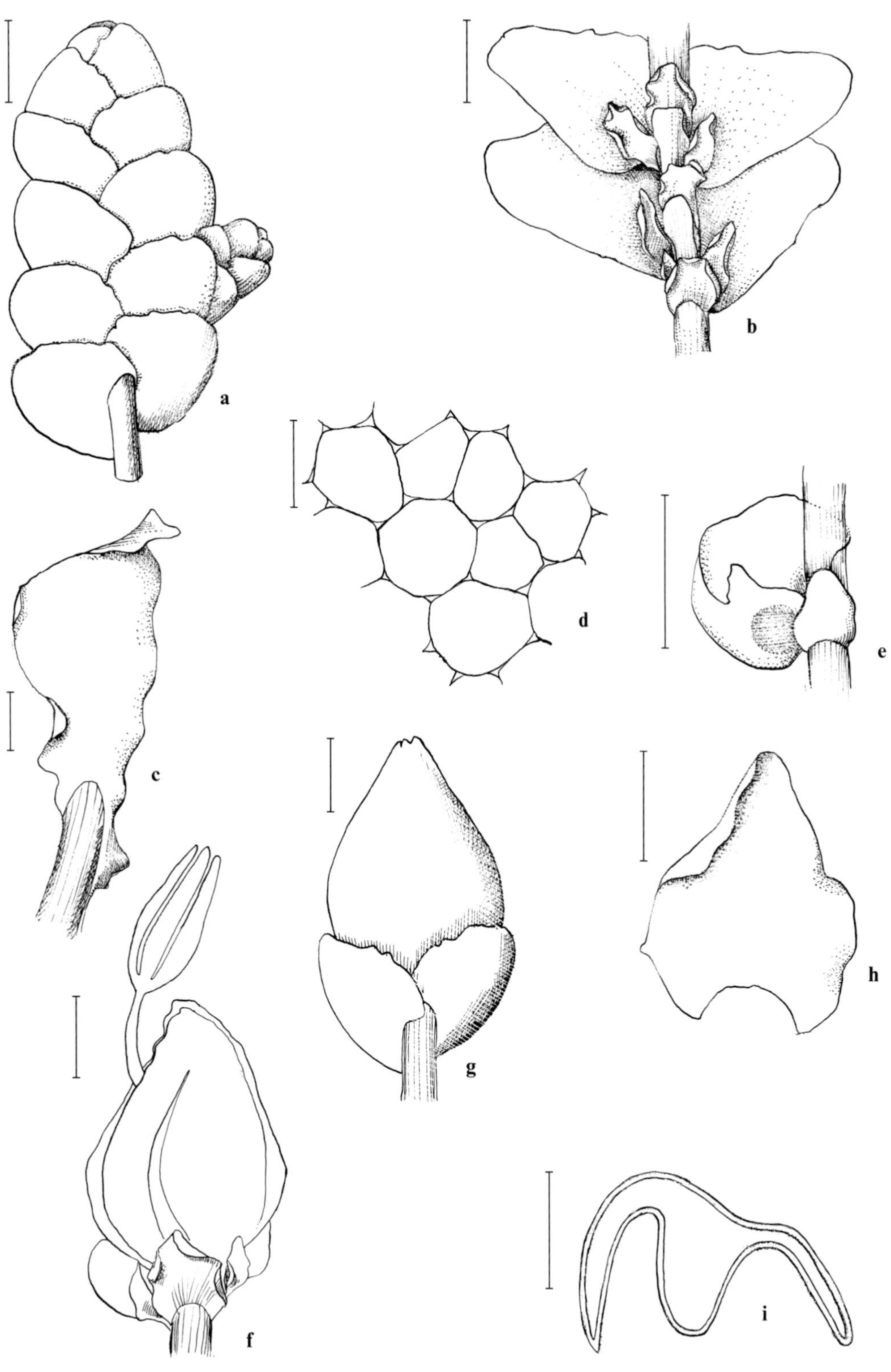
a
b
c
d
e
f
g
h
i

6. FRULLANIACEAE

Plants leafy, dark green to reddish-brown. *Stem* with or without distinct cortex, irregularly one–three-pinnate; branches arising terminally, each replacing the ventral lobe of the associated leaf. *Leaves* conduplicate, alternate, divided up to near base into two lobes of different size and shape; dorsal lobe incubous, large, imbricate, overlapping the stem dorsally, margins usually entire; ventral lobe transversely inserted, galeate or lanceolate, keel very short, with minute, subulate process (stylus) at its junction with stem; leaf cell walls often with trigones; underleaves large, entire or bilobed. *Rhizoids* restricted to underleaf base, smooth. *Gemmae* frequent. *Male branches* spicate, bracts bilobed with lobes nearly equal in size, each with one–two antheridia; bracteoles usually absent. *Female bracts* and bracteoles in two–five pairs, larger than leaves, entire or toothed, ventral lobe not galeate. *Archegonia* two–four per gynoecium. *Perianth* ovoid to obovoid, three–five-gonous with one–three keels on ventral side, abruptly narrowed to a beak at top. *Capsule* wall two-layered, inner layer thin-walled. *Spores* with irregular rosette-like ornamentations. *Chromosome number* $x=8$ or 9.

A family with worldwide distribution comprising three genera, common in tropical and subtropical areas. A single genus is represented in Israel.

Frullania Raddi

Plants pinnately branched. *Leaves* conduplicate; dorsal lobe large, obliquely ovate or orbicular, entire; ventral lobe smaller, galeate or ligulate, attached to dorsal lobe with narrow base, with a minute stylus at its junction with stem; leaf cells usually with thickened walls, well marked trigones and intermediate wall thickenings; oil bodies two–eight per cell, finely papillose; ocelli often present in dorsal lobe; underleaves usually bilobed. *Asexual reproduction* by caducous leaves or discoid gemmae from leaf margins or perianth surface. *Dioicous*. *Male bracts* in 4–20 pairs, on lateral branches, imbricate, subglobose, divided up to ⅓–½ of their length into two lobes. *Gynoecia* terminal. *Female bracts* shortly adnate to bracteole, divided up to ⅓–⅔ of length into two lobes, the ventral lobe ligulate, margins usually with some teeth. *Perianth* exserted, compressed, trigonous. *Sporophyte* with short seta; capsule slightly exserted, spherical, opening at maturity into four valves. *Spores* 40–55 µm in diameter. *Elaters* with helical bands.

A genus comprising ca. 400 mainly tropical species. Six species are found in the Mediterranean Region. A single species occurs in Israel.

Literature. — Müller (1951–1958), Schuster (1980), Paton (1999).

Frullania dilatata (L.) Dumort., Recueil Observ. Jungerm. 13 (1835).
Jungermannia dilatata L., Sp. Pl., Edition 1 : 1133 (1753). — Jovet-Ast & Bischler (1966) : 106.

Distribution map 216 (p. 554), Figure 215.

Plants dark green or blackish, 2–4 mm long, 1–1.3 mm wide. *Stems* irregularly once or twice pinnate. *Leaves* conduplicate; dorsal leaf lobe orbicular, rounded apically; ventral leaf lobe galeate, rarely ligulate, as long as wide, reaching ¼–⅓ of size of dorsal lobe, as wide or wider than underleaf, stylus filiform; mid-lobe cells nearly isodiametric, 20×25 µm

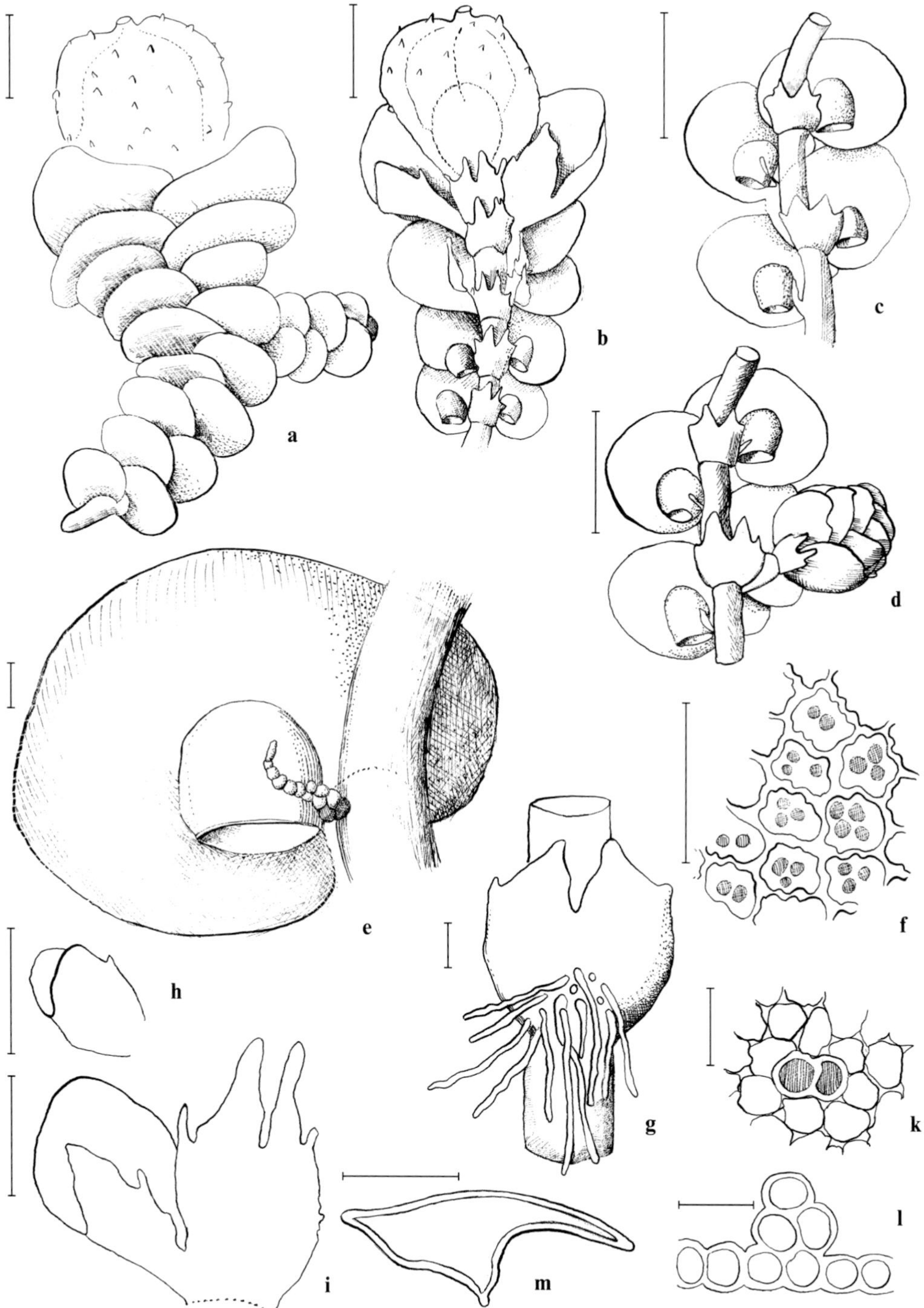

Figure 215. *Frullania dilatata*: **a** stem, dorsal side; **b** stem, ventral side; **c** two pairs of leaves, ventral side; **d** male plant, ventral side; **e** leaf; **f** mid-leaf cells of dorsal lobe; **g** underleaf; **h** male bract; **i** female bract and bracteole; **k** gemma from perianth, front view; **l** gemma from perianth, cross-section; **m** cross-section of perianth. Scale bars: **a**–**d**, **h**, **i**, **m** = 500 μm; **e**–**g**, **k**, **l** = 50 μm. (*Jovet-Ast & Bischler 82117*)

in dorsal lobe, with trigones; oil bodies three–four per cell, greyish or hyaline; ocelli absent; cuticle smooth; underleaves slightly wider than stem, bilobed up to ⅓ of length, lobes and sinus acute, outer margin flat, with a blunt tooth on each side. *Gemmae* discoid, two–six-celled, from leaf and perianth cells. *Dioicous. Male bracts* on short lateral branches, in 6–20 strongly imbricate pairs. *Female bracts* unequally bilobed, dorsal lobe oblong-ovate, rounded at apex, entire, ventral lobe reaching half of size of dorsal lobe, broad-lanceolate, acute, with one to several teeth on free margin. *Bracteole* divided up to ⅓ of its length, lobes and sinus acute, outer margins on one or both sides with one or several teeth. *Perianth* roughened by gemmae, beak short. *Capsule* outer layer of wall with nodulose thickenings. *Spores* pale-brown, 45–55 µm in diameter; with rosette-like ornamentations on both faces, indistinctly trilete. *Elaters* papillose.

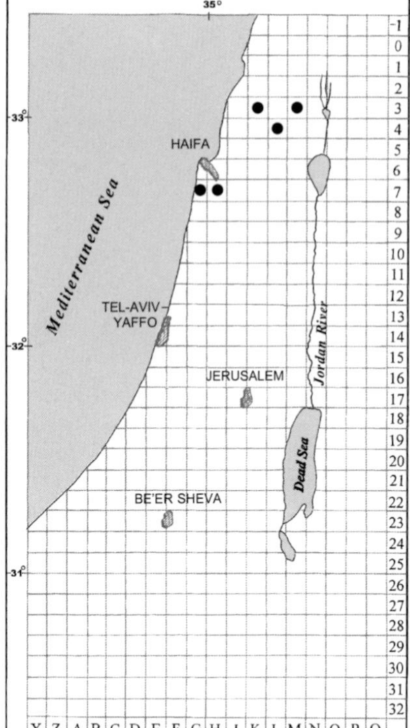

Map 216: *Frullania dilatata*

Habitat and Local Distribution. — Plants growing in loose colonies on bark of *Quercus* spp., in dense forest, in shade; restricted to forests and mountainous areas. Altitude: 250–1200 m. Occasional and only in northern Israel: Upper Galilee and Mount Carmel.

General Distribution. — Recorded from Southwest Asia, around the Mediterranean, Europe, Macaronesia, and northern and eastern Asia.

METZGERIALES

7. CODONIACEAE

Plants with axis and leaf-like, erect, succubous, obliquely inserted and crisped outgrowths, sometimes connate with each other, several-layered near axis; branching dichotomous; mucilage papillae two–three-celled, only near apex; cells thin-walled, each with numerous oil bodies. *Rhizoids* scattered along ventral side of axis, often purplish. *Asexual reproduction* by specialised propagules absent. *Perennating tubers* frequent. *Antheridia* and archegonia on dorsal side of axis, scattered, in three–four irregular rows; antheridia yellow to orange, sometimes shielded by a small scale; archegonia bounded by a campanulate involucre after fertilisation. *Sporophyte* with bulbous foot and short or long seta; capsule spherical, wall two–four-layered, breaking at maturity into irregular plates or splitting irregularly. *Spores* trilete. *Elaters* free, with two–four helical bands. *Chromosome number* $x=9$.

A family of worldwide distribution, comprising three–four genera.

Fossombronia Raddi

Axis creeping, flattened dorsally, convex ventrally; leaf-like outgrowth erect, in two rows, succubously and obliquely inserted, alternate, decurrent on dorsal side of axis, more or less square or slightly wider than long, two–four-layered near stem, one-layered at apex, irregularly undulate, with crisped and lobed margins; cells large, thin-walled; oil bodies 10–30 per cell, small, homogeneous or fine-segmented. *Perennating tubers* from ventral side of axis. *Antheridia* and archegonia dorsal near apex of axis; antheridia free or shielded by a scale; archegonia in campanulate, longitudinally plicate involucres with wide, sinuose and lobed opening at top. *Sporophyte* seta 5–18 mm long; capsule exserted at maturity, spherical, opening at maturity into irregular valves; wall two-layered, outer layer thin-walled, inner layer often with incomplete, ring-like thickenings. *Spores* 25–55 μm in diameter, alveolate, or spiny, or with irregular lamellae on distal face; ornamentation on proximal face similar but shallower.

A genus comprising approximately 30 species from which six–seven are found in the Mediterranean Region. A single species is frequent in the local flora.

Literature. — Müller (1951–1958), Schuster (1992a), Paton (1999).

Fossombronia caespitiformis De Not. ex Rabenh., Hep. Eur. no. 123 (1861). [Jovet-Ast & Bischler (1966): 103; Bischler & Jovet-Ast (1975): 19]

Distribution map 217, Figure 216, Plate XV: d.

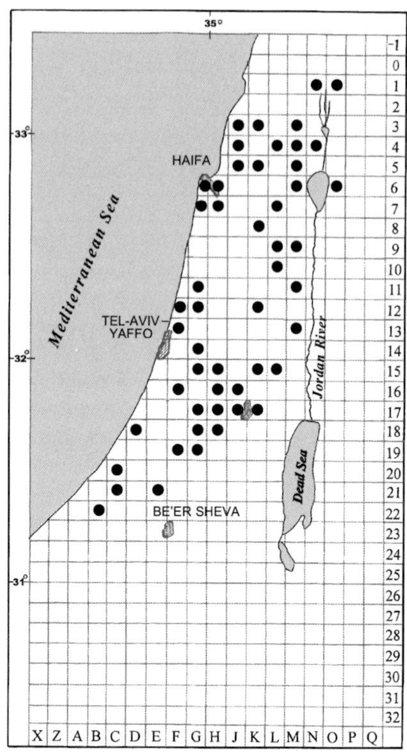

Map 217: *Fossombronia caespitiformis*

Plants pale green. *Axis* up to 10 mm long; leaf-like outgrowth undulate-crisped, closely imbricate, of isodiametric, thin-walled cells. *Rhizoids* dark purplish. *Monoicous*. *Involucre* broadly campanulate. *Capsule* wall inner layer with incomplete, ring-like thickenings. *Spores* brown or reddish-brown, 45–55 μm in diameter; distal face with conspicuous, truncate lamellae appearing on margin as 18–25 rectangular teeth; proximal face with lamellae reduced in size, wing absent (seen with SEM).

Habitat and Local Distribution. — Plants growing in colonies or scattered among mosses, on moist clay, sandy clay, sandy loam, terra rossa, or basaltic soil (pH 6–8) overlying basalt, dolomite, or other calcareous rocks, in rock crevices, at base of cliffs, or in stone walls, on exposed wadi banks, in grassland, open maquis, *Pinus* forests, *Olea europaea* and *Eucalyptus* groves, associated with several other thallose and leafy liverworts; adapted to a large variety of habitats. Altitude: -200–1200 m. Very common: Coastal Galilee, Acco Plain, Coast of Carmel, Sharon Plain, Philistean Plain, Upper Galilee, Lower Galilee, Mount Carmel, Esdraelon Plain, Samaria, Shefela, Judean Mountains, Judean Desert, Northern Negev, Dan Valley, Hula Plain, Bet She'an Valley, and Golan Heights.

General Distribution. — Recorded from Southwest Asia, around the Mediterranean, Europe, Macaronesia, and tropical eastern Africa.

Fossombronia angulosa (Dicks.) Raddi may also be found in the local flora. The morphological characters of the gametophyte, very variable in both species, do not permit a clear distinction between *F. angulosa* and *F. caespitiformis* and consequently, sterile specimens cannot be named with confidence. Only the ornamentation of the spore coat separates them unambiguously. The spores of *F. angulosa* are smaller — 30–45 μm in diameter — with three–four alveolae across the diameter of the distal face, and have a broad, translucent wing. Their margin appears crenulate. *Fossombronia angulosa* is usually larger in size and grows on acidic substrates.

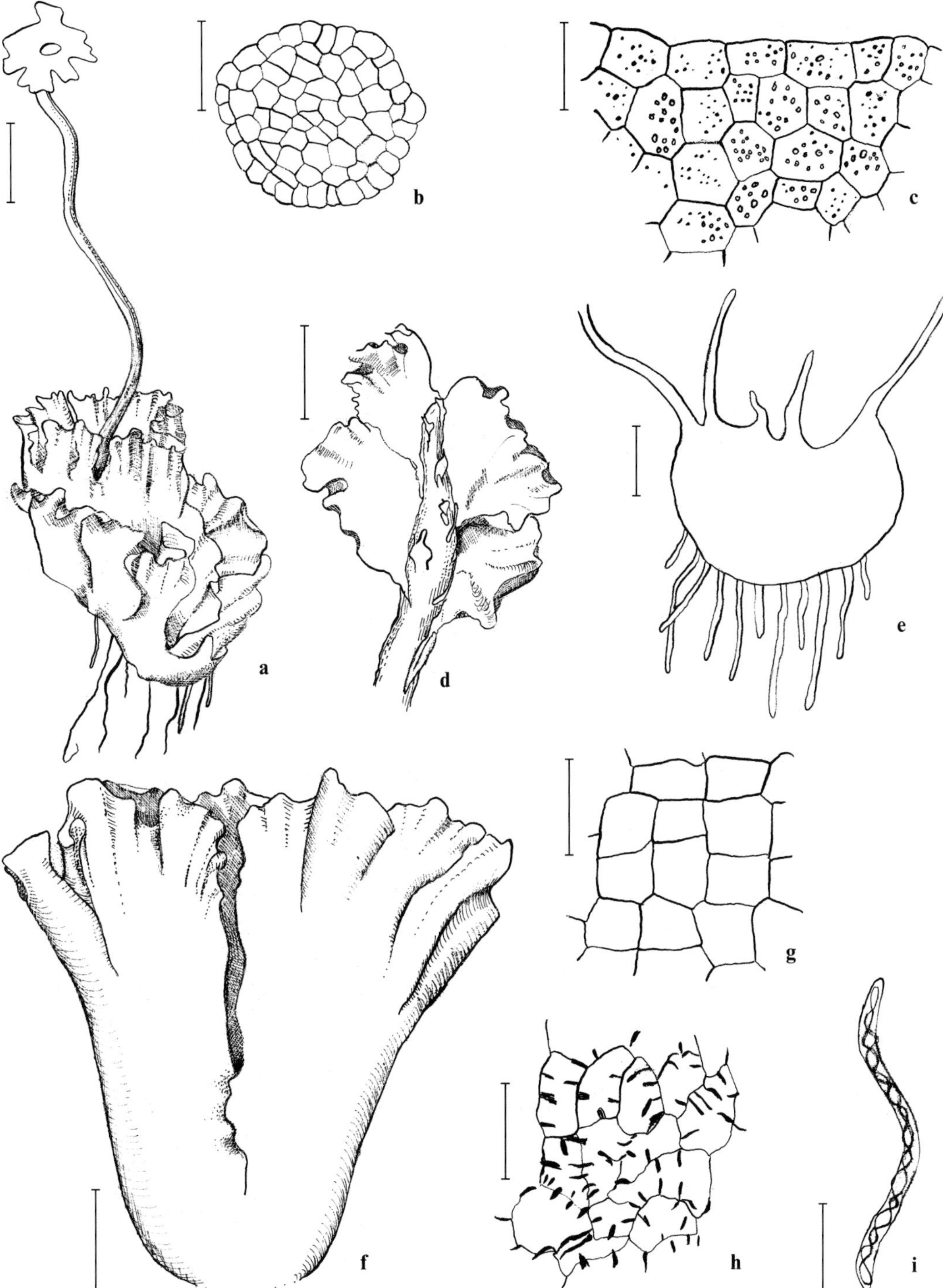

Figure 216. *Fossombronia caespitiformis*: **a** female plant with sporophyte; **b** seta, cross-section; **c** cells of lobe margin; **d** stem, dorsal side; **e** outline of cross-section of stem; **f** involucre; **g** capsule, cells of outer wall layer; **h** capsule, cells of inner wall layer; **i** elater. Scale bars: **a**, **d** = 1 mm; **b**, **c**, **g**–**i** = 50 µm; **e**, **f** = 250 µm. (*Jovet-Ast* & *Bischler 82275*)

8. PELLIACEAE

Plants thallose, vigorous, several-layered with a wide median band of small cortical and large inner cells; branching dichotomous; epidermal pores, air chambers, and scales absent; thallus apex with two–five-celled mucilage hairs. *Rhizoids* smooth, hyaline or brownish. *Antheridia* and archegonia developing on dorsal side of main thallus; antheridia scattered, in two–three irregular rows, each antheridium embedded in small cavity; archegonia 10–12 per involucre. *Involucre* single, tubular. *Sporophyte* with long, massive seta; capsule spherical, splitting at maturity into four valves; wall two–four-layered. *Spores* large, pluricellular, chlorophyllose, not trilete, hardly ornamented. *Elaterophore* at bottom of capsule. *Elaters* with helical bands. *Chromosome number* $x=9$.

A family of two genera, one distributed in the Northern Hemisphere, the other mainly in the Southern Hemisphere.

Pellia Raddi

Thallus irregularly branched, with one-layered, sinuose margin; median band not sharply defined, composed of small cortical and large inner cells, which may have thickened walls; oil bodies small, 15–35 per cell, spherical to ovoid. *Monoicous* or dioicous. *Antheridia* immersed in thallus tissue, ostioles often reddish. *Archegonia* four–twelve in a dorsal pocket, bounded by a short-tubular involucre with fimbriate-laciniate margin. *Calyptra* two-layered, hyaline, included in the involucre or shortly exserted. *Sporophyte* seta often up to 10 cm long; capsule splitting at maturity almost to near bottom into four ovate, acute valves; outermost layer with or without nodulose thickenings, innermost with or without ring-like thickenings. *Spores* ellipsoidal. *Elaters* with two–four helical bands.

A genus comprising approximately six species distributed all over the Northern Hemisphere. Two species are present in wet places in the Mediterranean Region. A single species occurs in Israel.

Literature. — Müller (1951–1958), Zielinski (1987), Schuster (1992a), Paton (1999).

Pellia endiviifolia (Dicks.) Dumort., Recueil Observ. Jungerm. 27 (1835).
Jungermannia endiviifolia Dicks., Fasc. Quart. Plant. Crypt. Brit. 19 (1801); *Pellia fabroniana* Raddi, Jungermaniogr. Etrusca 23 (1818).

Distribution map 218 (p. 560), Figure 217, Plate XV : i.

Thallus 4–10 mm wide, dark green, often reddish or blackish on dorsal side, with six–ten layers of isodiametric cells in cross section, without thickened bands, margins translucent, undulate; epidermal cells thin-walled; oil bodies 25–30 per cell; ventral side of apical part of thallus with numerous, two–five-celled mucilage hairs. *Asexual reproduction* by caducous, repeatedly forked thallus apices. *Dioicous.* Sex organs not seen in local material; in Mediterranean specimens — *Involucre* nearly erect, 2–4 mm long, margin ciliate or laciniate. *Calyptra* included in the involucre, cylindrical. *Capsule* wall outer layer with nodulose thickenings, inner layer thin-walled, without

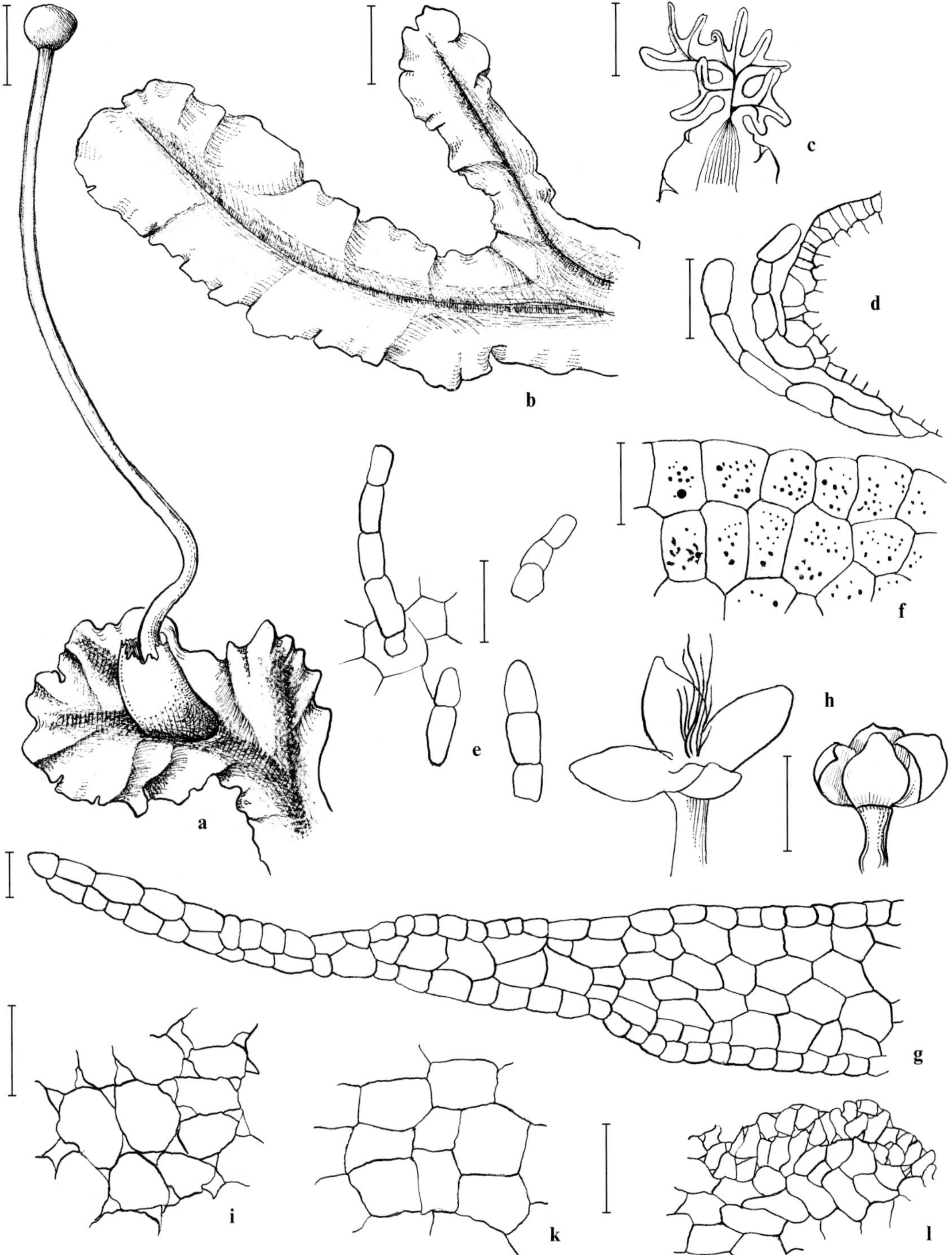

Figure 217. *Pellia endiviifolia*: **a** thallus with sporophyte; **b** thallus, dorsal side; **c** branched thallus apex; **d** thallus apex with mucilage hairs, longitudinal section; **e** mucilage hairs of ventral thallus apex; **f** cells of thallus margin; **g** thallus, cross-section; **h** open and closed capsules; **i** capsule, cells of outer wall layer; **k** capsule, cells of inner wall layer; **l** cross-section of seta. Scale bars: **a**–**c**, **h** = 2 mm; **d**–**g**, **i**, **k**, **l** = 50 µm. (**b**–**g**: *Jovet-Ast* & *Bischler 82179*; **a**, **h**, **i**, **k**, **l**: France, *Delacour*)

ring-like thickenings. *Spores* pluricellular, 40–70 µm in diameter; granulose-tuberculate on both faces (seen with SEM).

Habitat and Local Distribution. — Plants growing in dark green colonies on wet clay or basaltic soil (pH 7), overlying basalt, or directly fixed on basaltic rocks, sometimes in running water, in shade, in dense *Quercus* forest. Altitude: 100 m. Found in a single locality: Dan Valley.

General Distribution. — Recorded from Southwest Asia, around the Mediterranean, Macaronesia, throughout Europe, northern and eastern Asia, India, and North America.

Pellia epiphylla (L.) Corda grows on acidic substrates. It can be distinguished from *P. endiviifolia* by its mucilage hairs, nearly always two-celled and present on the ventral as well as on the dorsal side at thallus apex. The thallus often shows brown, thickened bands in cross section, the calyptra sticks out of the involucre, and the inner layer of the capsule wall has ring-like thickenings.

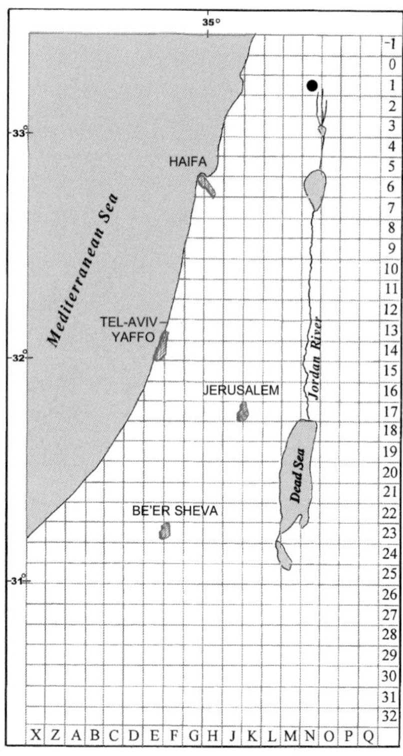

Map 218: *Pellia endiviifolia*

9. METZGERIACEAE

Plants thallose, delicate, branching dichotomous and ventral; median band sharply defined, narrow, of large cortical and small inner cells; wings one-layered; epidermal pores, air chambers, and scales absent; cells thin-walled. *Thallus* apex on ventral side with one-celled mucilage hairs. *Rhizoids* smooth, hyaline or brownish. *Asexual reproduction*, if present, by means of discoid or elliptical propagules from dorsal side or margin of thallus. *Sexual branches* ventral, small, and hidden underneath thallus. *Male branches* usually without hairs, coiled, with incurved lateral margins forming a subglobose pocket. *Antheridia* spherical, in two rows. *Female branches* convex, obcordate to obovate, or bilobed. *Archegonia* in groups at base of branch, without involucres. *Sporophyte* bounded by a large, fleshy, chlorophyllose and hairy calyptra. *Seta* of variable length; capsule ovoid to subspherical. *Spores* small. *Elaters* fixed on an elaterophore, persistent as erect tufts at top of capsule valves, with single helical band. *Chromosome number* $x=9$.

A family of worldwide distribution, comprising a single genus and two–four subgenera, which are sometimes considered as distinct genera.

Metzgeria Raddi

Thallus differentiated into a several-layered, sharply defined costa with larger, thin-walled cortical cells and smaller, often thickened inner cells; lateral wings one-layered; oil bodies absent or structureless; ventral side, margins, and (rarely also) dorsal side of thallus with bristle-like, unicellular, setose trichomes. *Capsule* exserted at maturity, opening into four reddish-brown valves; capsule wall two-layered, outer wall with nodulose, inner wall sometimes with ring-like thickenings. *Spores* 18–28 µm in diameter.

A genus compring ca. 100, mostly tropical, species. A single species occurs in Israel, and is scattered elsewhere in the Mediterranean Region.

Literature. — Müller (1951–1958), Kuwahara (1978), Schuster (1992a), Paton (1999).

Metzgeria furcata (L.) Dumort., Recueil Observ. Jungerm. 26 (1835).
Jungermannia furcata L., Sp. Pl., Edition 1 : 1136 (1753). — Jovet-Ast & Bischler (1966) : 103.

Distribution map 219, Figure 218.

Thallus green or yellowish-green, 2–10 mm long, 0.6–1 mm wide, flat, irregularly branched; laminal cells hexagonal, 30–40 μm wide; dorsal side of costa two cells wide, slightly convex, ventral side two–four cells wide, distinctly convex, inner cells in three layers in section; oil bodies absent; ventral side of costa and lamina with more or less abundant, scattered, straight trichomes, margin and dorsal side without trichomes. *Dioicous.* Sex organs not seen in local specimens; in Mediterranean material — *Female branch* densely hairy. *Seta* 1–2.5 mm long; capsule ovoid, reddish-brown, outer wall layer with nodulose, inner wall layer with ring-like thickenings. *Spores* brownish, 20–25 μm in diameter, indistinctly trilete, short-spinulose on both faces.

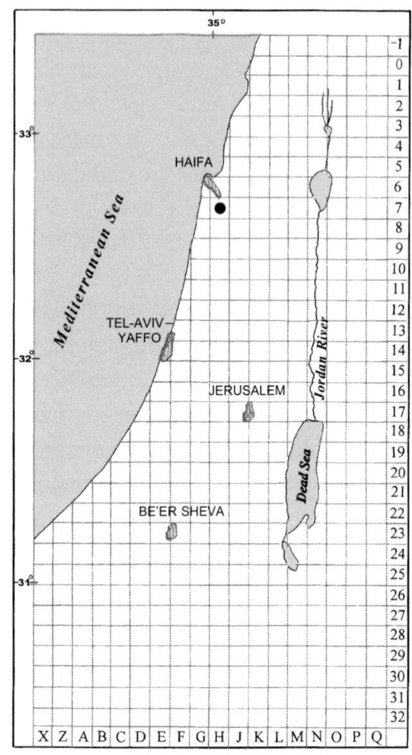

Map 219: *Metzgeria furcata*

Habitat and Local Distribution. — Plants growing in colonies over moist calcareous rocks in dense *Quercus* garrigue, associated with *Leiocolea turbinata.* Altitude: 300 m. Found only once: Mount Carmel.

General Distribution. — Recorded from Southwest Asia, around the Mediterranean, Europe, Macaronesia, eastern, tropical, and South Africa, northern and eastern Asia, India, Australia, New Zealand, and northeastern, Central, and South America.

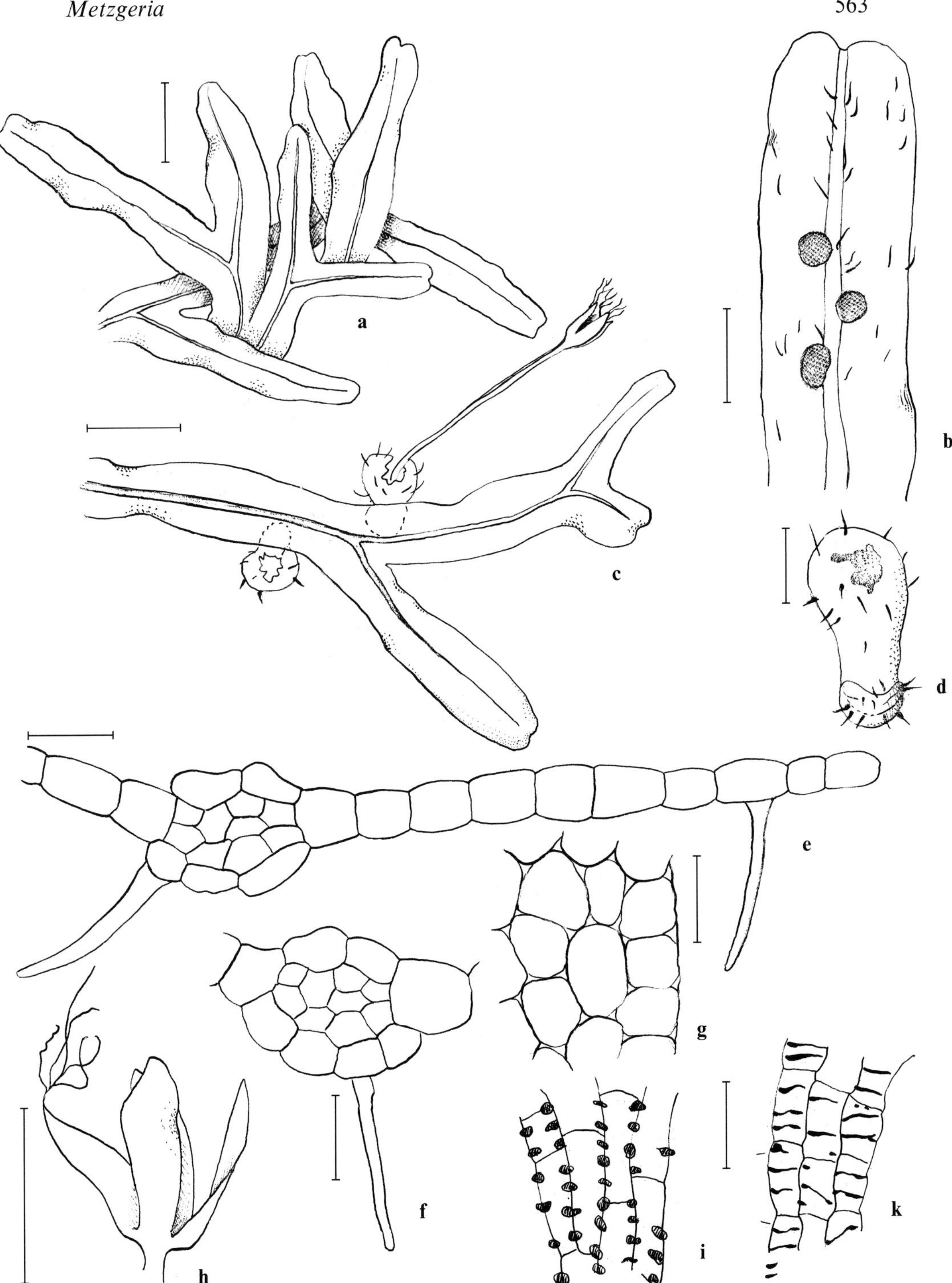

Figure 218. *Metzgeria furcata*: **a** thalli; **b** male thallus with antheridial branches; **c** thallus with female branches and sporophyte; **d** female branch; **e** thallus, cross-section; **f** thallus, cross-section of median band; **g** thallus, laminal cells; **h** open capsule with elaterophore; **i** capsule, outer wall layer with nodulose thickenings; **k** capsule, inner wall layer with ring-like thickenings. Scale bars: **a**, **c** = 1 mm; **b**, **d**, **h** = 500 μm; **e**–**g**, **i**, **k** = 50 μm. (**a**, **e**–**g**: *Jovet-Ast* & *Bischler 64016*; **b**–**d**, **h**, **i**, **k**: Morocco, *Pitard 341*)

SPHAEROCARPALES

10. RIELLACEAE

Plants usually aquatic. *Gametophyte* with large, undulate, one-layered wing along one side of axis, which overarches the apex of the postically coiled axis, or wound around axis in a helical fashion, and with two irregular rows of small, leaf-like, green scales on ventral side; cells thin-walled; epidermal pores absent; air chambers absent; oil cells present. *Rhizoids* smooth. *Asexual reproduction* by propagules from ventral side of axis. *Dioicous* or monoicous; antheridia ovate, individually embedded into the free margin of the thickened, apical part of wing; archegonia single, sessile, in pyriform or bottle-shaped involucres, arising from axis on both sides of wing. *Sporophyte* with spherical foot and short seta; capsule not exserted, cleistocarpous; spore tetrads separating at maturity. *Spores* large, mixed with spherical, green, sterile cells. *Elaters* absent. *Chromosome number* $x=9$.

A family comprising a single genus with a disjunct range in Mediterranean climates. Recorded from the Mediterranean Region, Macaronesia, South Africa, southeastern Australia, and western United States.

Riella Mont.

Axis orbicular or ovate in section, with one-layered, leaf-like, green scales on ventral side. *Archegonia* in sessile involucres open at top, located on axis at right and left of wing. *Capsule* wall one-layered, without ring-like thickenings. *Spores* spiny, 60–130 µm in diameter.

A genus comprising about 18 species, all of scattered occurrence and rare, growing erect in temporary or permanent ponds, or prostrate on sometimes slightly saline silty or muddy soil. Seven species have been recorded from the Mediterranean Region, and three species occur in the local flora. We have not succeeded in collecting the local species.

Literature. — Wigglesworth (1937), Trabut (1941), Lipkin & Proctor (1975), Ros Espin (1987), Schuster (1992a).

1 Involucres with eight (or more) wing-like longitudinal ridges.
  2 Monoicous; spores brown, 85–120 µm in diameter **1. R. affinis**
  2 Dioicous; spores yellow-brown, 70–80 µm in diameter **2. R. cossoniana**
1 Involures without wing-like longitudinal ridges **3. R. helicophylla**

1. Riella affinis M. Howe & Underw., Bull. Torrey Bot. Club 30 : 221 (1903).

Distribution map 220 (p. 566), Figure 219.

Thallus erect, 6–15 mm high, simple or dichotomously branched; axis flattened; wing deeply lobed; wing cells isodiametric, thin-walled; scales conspicuous, linguiform, obtuse or acute. *Monoicous. Involucre* ovate, eight-winged, constricted apically. *Capsule* subglobose. *Spores* brown, 85–120 µm in diameter, distal face with truncate

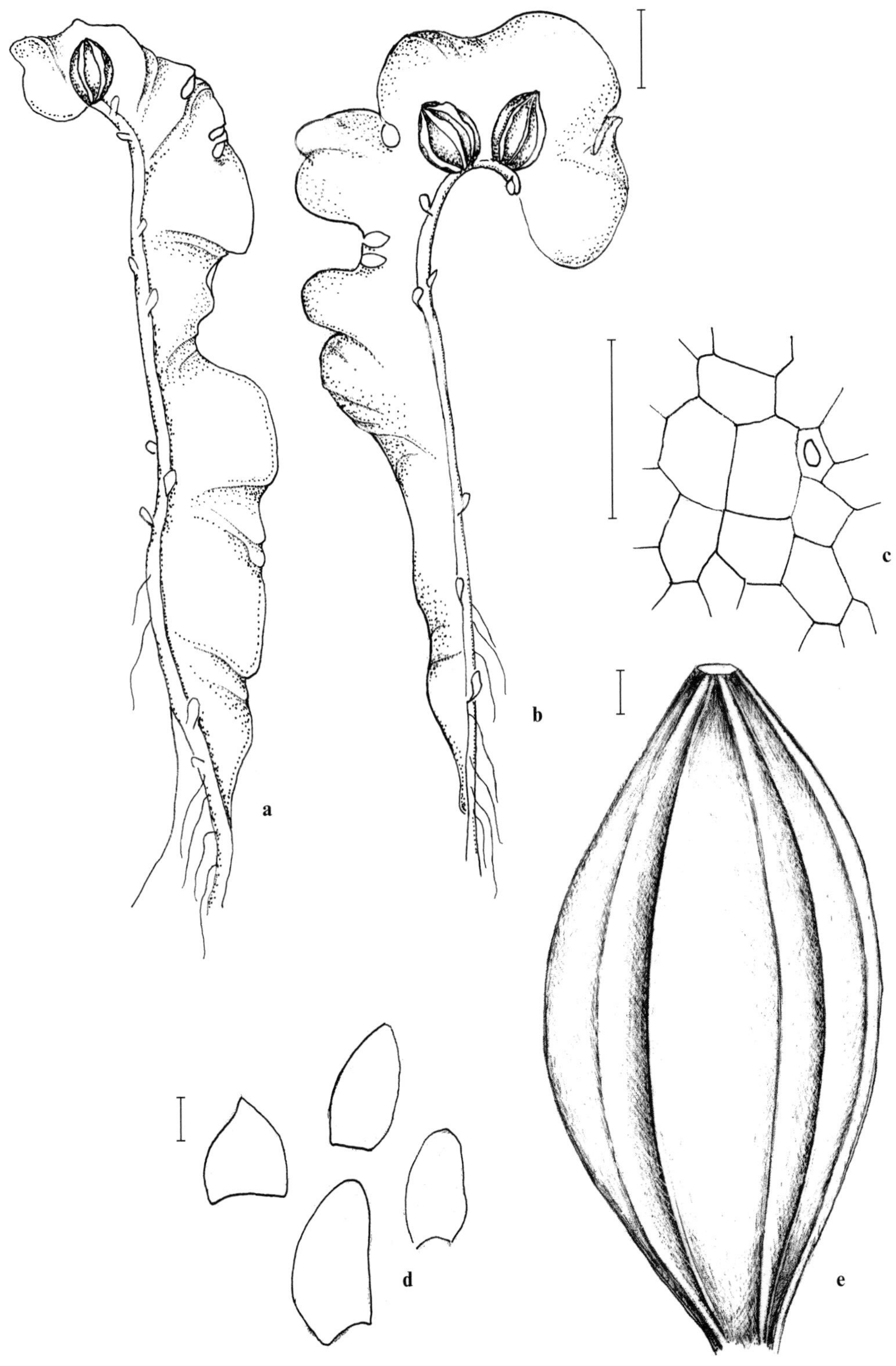

Figure 219. *Riella affinis*: **a, b** thalli with antheridia in wing margin and involucres on axis; **c** cells from wing; **d** scales; **e** involucre. Scale bars: **a**, **b** = 1 mm; **c**–**e** = 100 µm. (Canary Islands, *Cook 729*)

spines, often slightly broadened apically; proximal face with much shorter, truncate or obtuse spines or warts.

Habitat and Local Distribution. — Plants growing scattered in water reservoirs and seasonal pools. Rare: Upper Galilee and Samaria.

General Distribution. — Recorded from Southwest Asia, around the Mediterranean, Macaronesia, South Africa, India, and northwestern North America.

2. Riella cossoniana Trab. *in* Batt. & Trab., Atlas Fl. Algérie 6, Tab. 2 (1886).

Distribution map 221, Figure 220, Plate XV: c.

Thallus erect, 9–13 mm high, unbranched, male thalli often smaller than female; wing undulate, narrowed towards bottom of axis; scales numerous, variable in size and shape. *Dioicous*; antheridia numerous, orange when mature. *Involucre* eight-winged, constricted apically into a small ostiole. *Capsule* subglobose. *Spores* yellow-brown, 70–80 μm in diameter, distal face with long, truncate spines, more or less broadened apically; proximal face with shorter spines or tubercles (seen with SEM).

Habitat and Local Distribution. — Plants growing in old quarries. Rare: Sharon Plain and Mount Carmel.

General Distribution. — Recorded from Southwest Asia, around the Mediterranean, Macaronesia (uncertain), and central Asia.

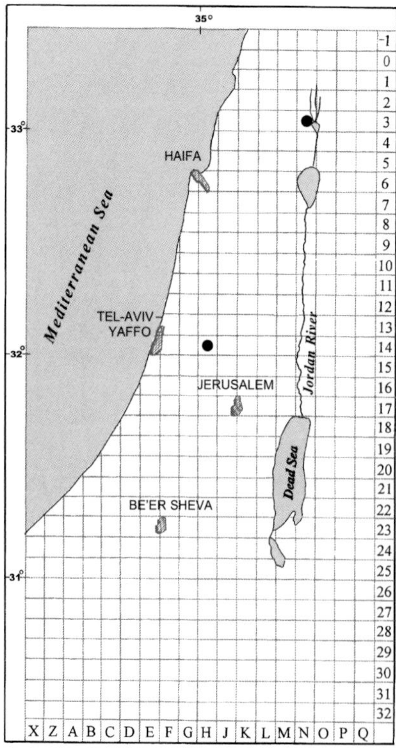

Map 220: *Riella affinis*

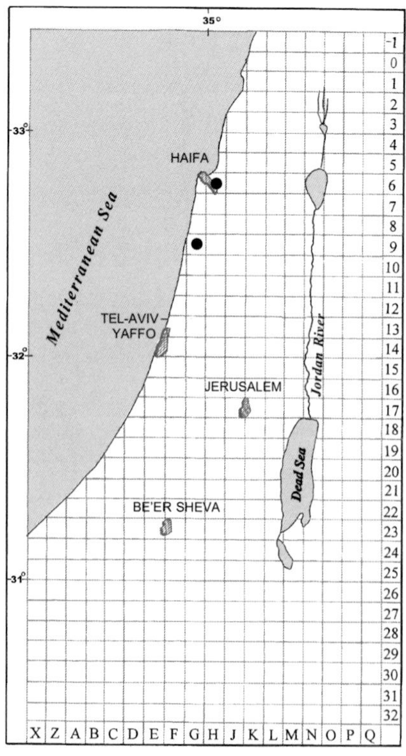

Map 221: *Riella cossoniana*

Figure 220. *Riella cossoniana*: **a** male thallus with antheridia in wing margin; **b** female thallus with involucres; **c** scales; $\mathbf{d_1}$, $\mathbf{d_2}$ involucres. Scale bars: **a**, **b**, $\mathbf{d_1}$ = 1 mm; **c**, $\mathbf{d_2}$ = 100 μm. (Spain, *Llimona*)

3. Riella helicophylla (Bory & Mont.) Mont., Ann. Sci. Nat., Bot., sér. 3, 18 : 12 (1852).

Duriaea helicophylla Bory & Mont., Compt. Rend. Hebd. Séanc. Acad. Sci. (Paris) 16 : 1114 (1843).

Distribution map 222, Figure 221, Plate XV : f.

Thallus erect, light green, 1–10 mm high, with broad wing; cells isodiametric, thin-walled; scales small, lanceolate or ovate, rounded, or obtuse apically. *Dioicous. Involucre* ovate, without longitudinal ridges, constricted apically into a short, tubular beak. *Spores* yellowish-brown, 80–100 µm in diameter; proximal face short-spinulose, distal face with larger spines truncate at apex (seen with SEM).

Habitat and Local Distribution. — Plants growing scattered in old quarries, in seasonal pools, and in water reservoirs. Rare: Sharon Plain, Samaria, Judean Mountains, and Central Negev (?).

General Distribution. — Recorded from Southwest Asia, around the Mediterranean, and Europe.

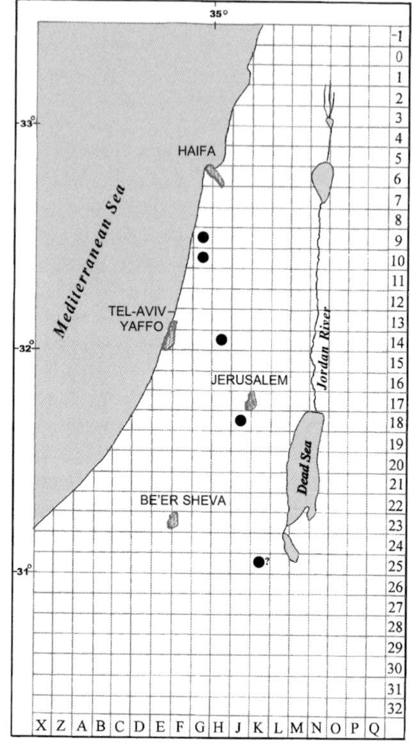

Map 222: *Riella helicophylla*

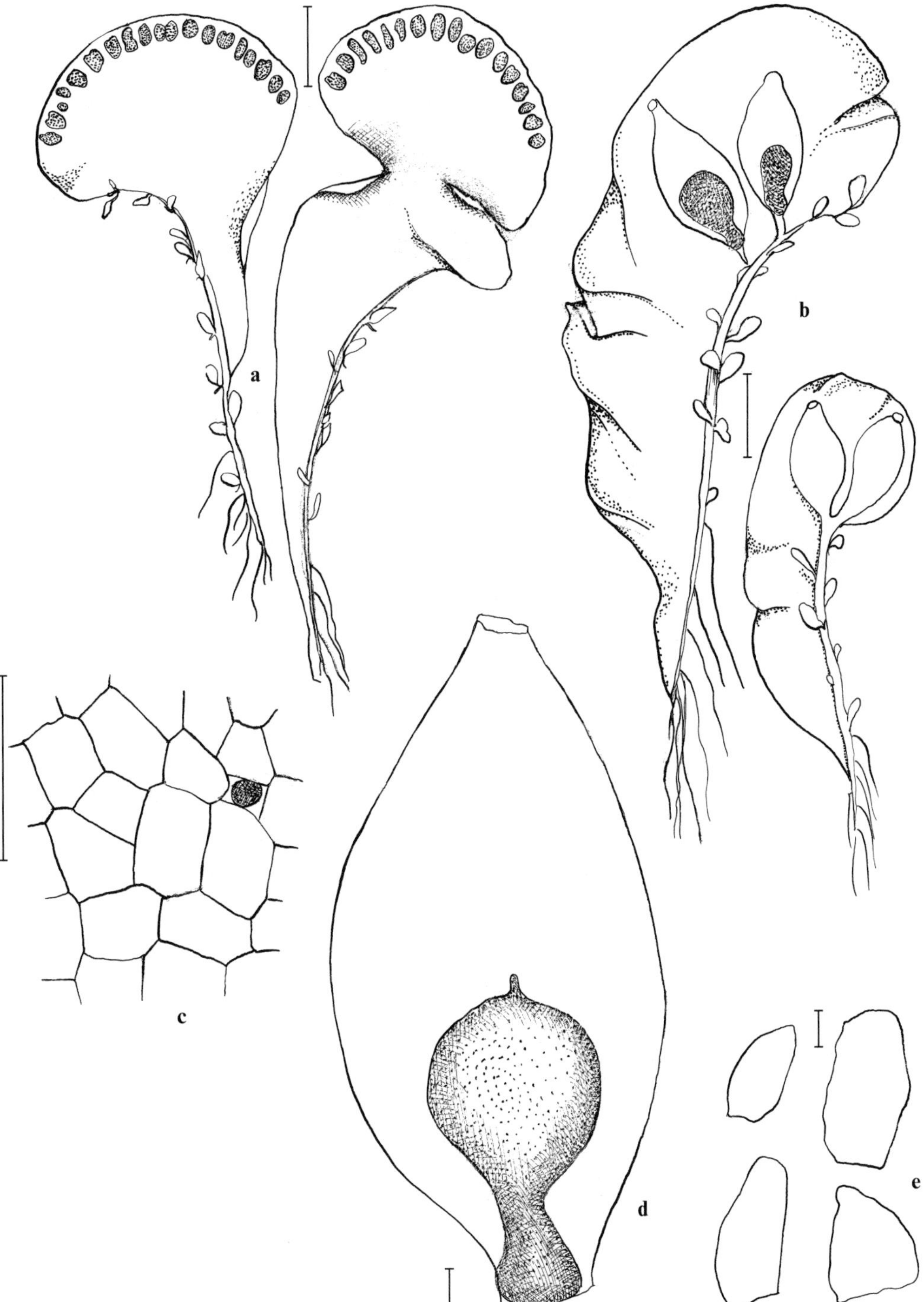

Figure 221. *Riella helicophylla*: **a** male thalli with antheridia in wing margin; **b** female thalli with involucres; **c** cells from wing; **d** outline of involucre and sporophyte; **e** scales. Scale bars: **a**, **b** = 1 mm; **c**–**e** = 100 µm. (Tunisia, *Labbe*)

11. SPHAEROCARPACEAE

Plants terrestrial, prostrate, suborbicular. *Gametophyte* a dichotomously branched axis with suborbicular, leaf-like, and crisped outgrowth, several-layered in the centre; epidermal pores, air chambers, and scales absent; cells thin-walled; oil cells absent. *Rhizoids* smooth. *Asexual reproduction* by specialised propagules absent. *Dioicous*; antheridia and archegonia dorsal, single in large one-layered, bottle-shaped, green involucres, open at top. *Sporophyte* with small foot and short, non-elongating seta, necrotic at spore maturity; capsule cleistocarpous. *Spores* either remaining permanently in tetrads, rarely separating. *Chromosome number* $x = 8$.

A family comprising two genera with worldwide distribution in warmer regions, especially under climates with dry seasons.

Sphaerocarpos Boehm.

Plant suborbicular, rosette-like, with leaf-like, irregular lobes; male plants often smaller and reddish; axis several-layered, flattened; oil cells absent. *Rhizoids* smooth, hyaline. *Gametophyte* almost completely hidden under the closely grouped involucres. *Male involucres* conical or pyriform. *Antheridia* single per involucre, spherical, shortly stalked. *Female involucres* pyriform, each enclosing a single archegonium and subsequently a single sporophyte. *Capsule* spherical, bounded by a one–two-layered calyptra; wall one-layered, without ring-like thickenings, disintegrating at spore maturity. *Spores* remaining often in tetrads, distal face alveolate. *Elaters* absent, but green, spherical, sterile cells intermixed with spores or spore tetrads.

A genus comprising eight–nine species, with disjunct ranges in northern and southern temperate regions. Two of them are widely distributed in the Mediterranean Region, and one species, *Sphaerocarpos michelii*, occurs in Israel.

Literature. — Müller (1951–1958), Proskauer (1954), Schuster (1992a), Paton (1999).

Sphaerocarpos michelii Bellardi, Appendix Fl. Pedemont.: 52 (1792); Mém. Acad. Roy. Sci. Turin 5: 258 (1792).

Sphaerocarpos terrestris Sm. *in* Smith & Sowerby, Engl. Bot., Tab. 299 (1796). — Jovet-Ast & Bischler (1966): 92, Bischler & Jovet-Ast (1975): 18.

Distribution map 223 (p. 572), Figure 222, Plate XV: e.

Plant pale green, delicate, prostrate. *Female thallus* 5–15 mm in diameter, *male thallus* 1–3 mm in diameter. *Axis* four–five cell layers thick in section; cells isodiametric, uniform in size. *Male involucres* conical, often tinged with red. *Female involucres* pyriform. *Spores* remaining in tetrads, dark brown, 90–120 µm in diameter; with eight–ten alveolae across distal face of each spore, wall corners tuberculate, margins appearing shortly spinose; occasionally, an isolated median papilla is seen in each alveola (seen with SEM).

Habitat and Local Distribution. — Plants growing in scattered rosettes on compact, moist sand, sandy clay, clay, terra rossa, or basaltic soil (pH 6.3–8) overlying

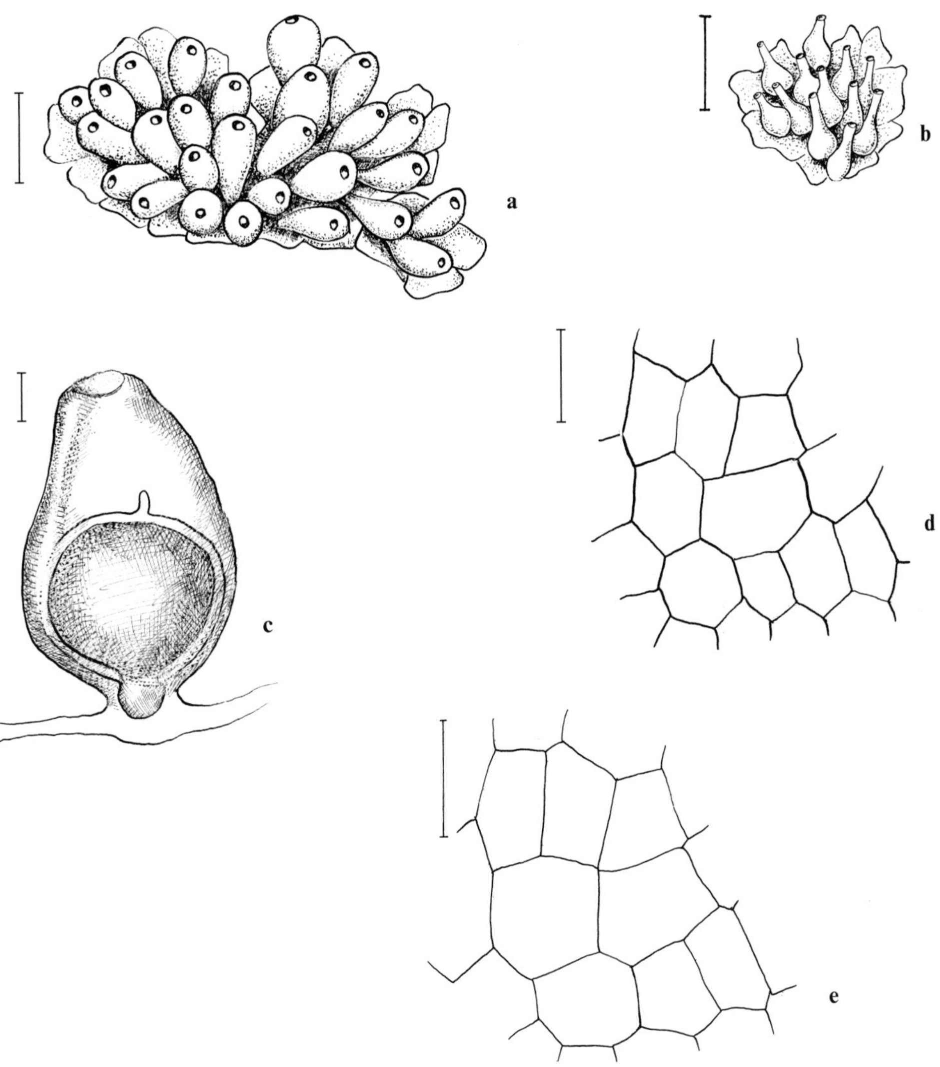

Figure 222. *Sphaerocarpos michelii*: **a** female thallus; **b** male thallus; **c** involucre with capsule; **d** epidermal cells of thallus; **e** cells of capsule wall. Scale bars: **a**, **b** = 1 mm; **c**–**e** = 100 μm. (*Jovet-Ast* & *Bischler 82109*)

limestone or basalt, on exposed soil tracks and wadi banks, or in grassland and open *Quercus* and *Pinus* forests, associated with *Riccia* and *Fossombronia*; likely to survive in very harsh climatic conditions. Altitude: -300–750 m. Common: Sharon Plain, Upper Galilee, Lower Galilee, Shefela, Judean Mountains, Judean Desert, Northern Negev, Dan Valley, Hula Plain, Bet She'an Valley, and Lower Jordan Valley.

General Distribution. — Recorded from Southwest Asia, around the Mediterranean, Europe, Macaronesia, southern North and southern South America.

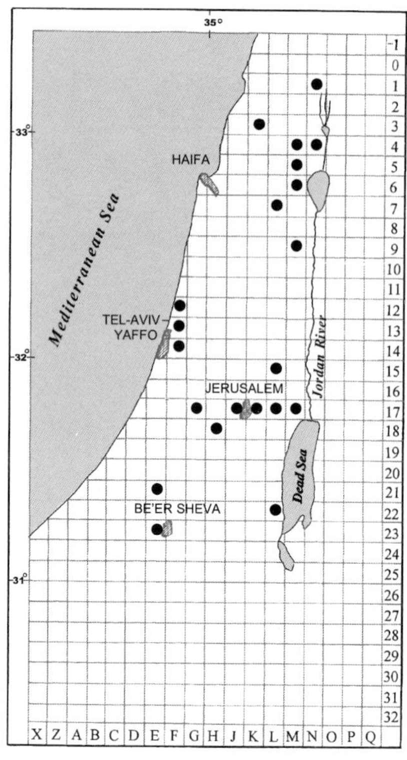

Map 223: *Sphaerocarpos michelii*

Sphaerocarpos texanus Austin may occur but has not been discovered in Israel. It has red-brown spore tetrads, 140–175 µm in diameter, with five–seven alveolae across each spore. The spore margins appear crenulate rather than spinulose. We have collected this species in many localities in the Mediterranean Region.

MARCHANTIALES

12. CORSINIACEAE

Thallus dichotomously branched; epidermal pores bounded by one ring of slightly differentiated cells, radial walls thin; air chambers in single layer, with well developed or rudimentary chlorophyllose assimilatory filaments; oil cells numerous; scales small, lanceolate, in several irregular rows, with oil cells. *Rhizoids* smooth and tuberculate. *Asexual reproduction* by specialised propagules absent. *Monoicous* or dioicous. *Antheridia* in several rows in receptacles along median groove, bounded by a crest. *Archegonia* in dorsal cavities in median groove, with paraphyses. *Involucre* a posterior, several-layered scale. *Calyptra* becoming thickened after fertilisation, smooth or warty, with pluricellular processes. *Capsule* spherical, not exserted, cleistocarpous, wall with or without thickenings. *Spores* on distal face with shallow protuberances or plates. *Elaters* short, with helical bands, or elaters absent but sterile cells among spores. *Chromosome number* $x=8$ or 9.

A family comprising two genera. One genus is known only from the Neotropics.

Corsinia Raddi

Thallus blue-green; epidermal pores bounded by a single ring of cells; air chambers with well developed assimilatory filaments; scales hyaline. *Dioicous* or monoicous. *Calyptra* becoming thickened and warty after fertilisation, wall with pluricellular processes. *Capsule* wall without thickenings; sterile cells among spores. *Chromosome number* $x=8$.

A genus distributed in warm-temperate areas in the Northern Hemisphere and disjunct in southern South America.

Literature. — Müller (1951–1958), Haessel de Menéndez (1963), Schuster (1992b), Boisselier-Dubayle & Bischler (1998).

Corsinia coriandrina (Spreng.) Lindb., Hep. Utveckl. Helsingfors: 30 (1877).
Riccia coriandrina Spreng., Anl. Kenntnis Gewächse 3: 320 (1804); *Corsinia marchantioides* Raddi, Opusc. Sci. (Bologna) 2: 354 (1818). — Jovet-Ast & Bischler (1966): 97.

Distribution map 224 (p. 574), Figure 223 (p. 575), Plate XV: k.

Thallus 3–6 mm wide, not tinged with purple; epidermal pores with ring of five–seven cells, radial walls thin; scales with marginal papillae, lanceolate; oil cells in thallus and scales. *Antheridia* embedded in median groove, bordered by a ciliate crest. *Archegonia* one–ten per archegonial cavity; cavities with slightly proliferating edges forming a toothed-ciliate crest. *Calyptra* becoming massive and several-layered after fertilisation, warty. *Sporophytes* one–three per involucre; mature sporophyte with few-celled, bulbous foot and short seta, four–six cells across diameter. *Spores* 90–150 µm in diameter, light to dark red-brown, distal face with seven–nine shallow, tuberculate projections or plates across diameter; proximal face faintly tuberculate, triradiate mark distinct (seen with SEM).

Habitat and Local Distribution. — Plants growing on sand or clay of low pH, often shaded, on rocks or in rock crevices. Altitude: 50–750 m. Rare: Sharon Plain and Judean Mountains.

General Distribution. — Recorded from Southwest Asia, around the Mediterranean, Europe, Macaronesia, North and southern South America.

Corsinia coriandrina resembles *Riccia*, but it differs by its scales in irregular rows, the structure of its epidermal pores, and the characters of the male and female receptacles. Furthermore, the spore coat ornamentation distinguishes it from all *Riccia* species.

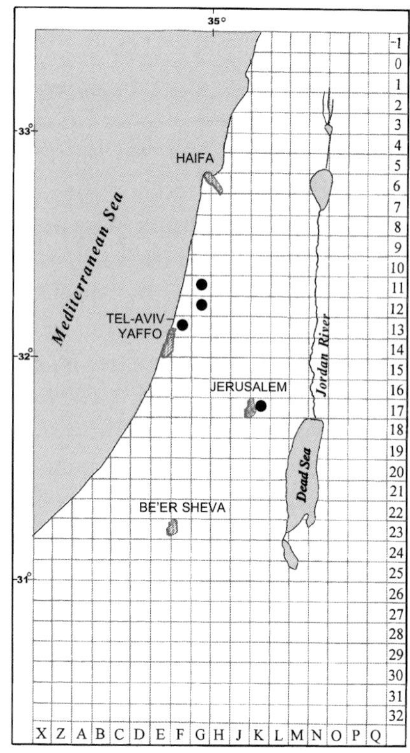

Map 224: *Corsinia coriandrina*

Figure 223. *Corsinia coriandrina*: **a, b** male thalli, dorsal side; **c** female thallus with two calyptrae; **d** calyptra; **e** section of calyptra (cal) and enclosed capsule (cap); **f** four calyptrae on thallus; **g** cross-section of male thallus with three antheridia; **h** cross-section of sterile thallus; **i** cross-section of dorsal tissue (ep = epidermis, ac = air chambers, f = assimilatory filament, oc = oil cell); **k** epidermal pore; **l, m** scales; **n** spore, distal face; **o, p** spores, side view. Scale bars: **a** = 3 mm; **b, c, f** = 1 mm; **d, g, h** = 0.5 mm; **e, i** = 100 μm; **k** = 50 μm; **l** = 100 μm; **m** = 200 μm; **n–p** = 30 μm. (drawn from various sources)

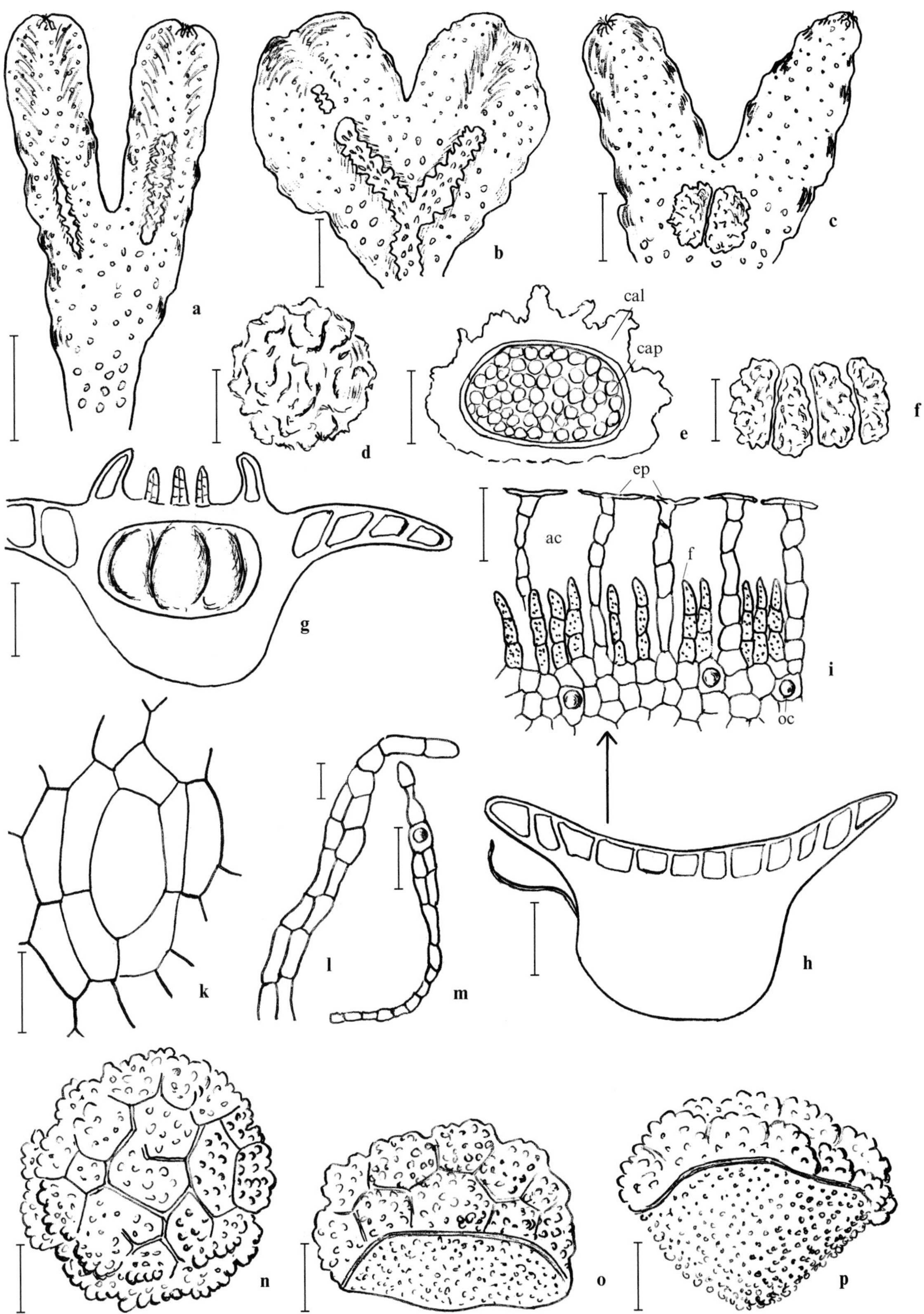
a
b
c
d
cal
cap
e
f
g
ep
ac
f
i
oc
h
k
l
m
n
o
p

13. TARGIONIACEAE

Plants thallose, green to deep green. *Thallus* margins and ventral side purplish, branching dichotomous and ventral; epidermal pores bounded by one–three concentric rings of cells, radial walls thin; air chambers in single layer, with assimilatory filaments; scales in two rows, with oil cells. *Rhizoids* smooth and tuberculate. *Asexual reproduction* by specialised propagules absent. *Monoicous* or dioicous. *Antheridia* loosely aggregated, embedded in thallus, dorsal in main segment or in small ventral branches. *Archegonia* in sessile, bivalved, mussel-shaped, black-purplish involucre on ventral side of thallus apex. *Sporophyte* with foot and short seta; capsule wall one-layered, with ring-like thickenings, opening at maturity into irregular valves. *Elaters* with two–three helical bands. *Chromosome number* $x = 9$.

A family comprising one or two genera, distributed in warm and tropical areas.

Targionia L.

Thallus thick, without distinct median band, margin and ventral side blackish-purple; epidermal cells with nodulose trigones; epidermal pores bounded by two–three concentric rings of six–nine cells; scales purplish-black, with single appendage; oil cells in thallus and scales. *Antheridia* embedded in dorsal side of small, disciform ventral branches, not bounded by scales. *Female involucre* purplish or blackish-purple, laterally compressed and keeled. *Capsule* spherical, not exserted.

A genus comprising three–four species and distributed worldwide in warmer areas. One species is frequent in the Mediterranean Region.

Literature. — Müller (1951–1958), Haessel de Menéndez (1963), Sérgio & Queiroz Lopes (1972), Schuster (1992b), Perold (1993b), Boisselier-Dubayle & Bischler (1999), Paton (1999).

Targionia hypophylla L., Sp. Pl., Edition 1 : 1136 (1753). [Jovet-Ast & Bischler (1966) : 93; Bischler & Jovet-Ast (1975) : 18]

Distribution map 225 (p. 578), Figure 224, Plate XV : 1.

Plants dark or bright green. *Thallus* 2–5 mm wide; epidermal cells with slightly thickened walls and nodulose trigones; epidermal pores conspicuous, radial walls thin; scales purplish-black, with lanceolate-triangular appendage, entire, crenulate or irregularly toothed on margins. *Monoicous. Female involucre* purplish-black. *Capsule* wall with ring-like thickenings. *Spores* brown, 65–75(–100) μm in diameter; alveolate on proximal and distal faces with about five alveolae across diameter, alveolae and ridges reticulate (seen with SEM).

Habitat and Local Distribution. — Plants growing in dense colonies on moist or wet clay, sandy clay, sand, terra rossa, or desert rendzine (pH 6.5–8), overlying limestone, dolomite, tuff or other calcareous rocks, or basalt, in rock crevices, on exposed rocks and wadi banks, in grassland, open maquis, or *Quercus*, *Pinus*, *Eucalyptus*, and *Olea europaea* groves; associated with *Lunularia*, *Reboulia*, *Athalamia*, *Mannia*, and *Fossombronia*. Altitude: -250–1200 m. Very common: Coast of Carmel, Sharon Plain, Upper Galilee, Lower Galilee, Mount Carmel, Esdraelon Plain, Samaria,

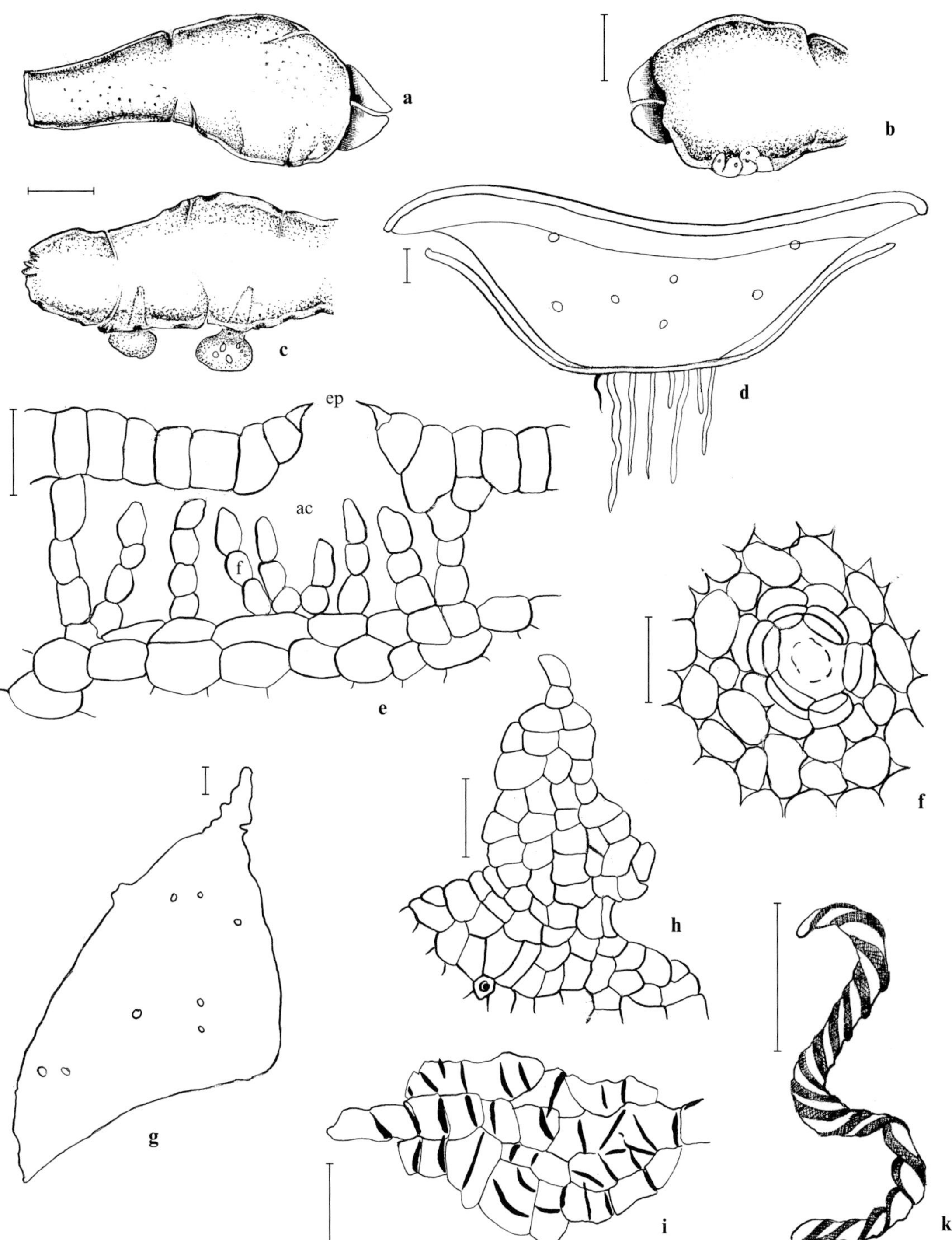

Figure 224. *Targionia hypophylla*: **a, b** female thalli with apical involucres; **c** male thallus with antheridial branches; **d** thallus, cross-section; **e** cross-section of assimilatory layer, with air chamber (ac), filaments (f), and epidermal pore (ep); **f** epidermal pore, front view; **g** scale; **h** appendage of scale; **i** capsule wall cells; **k** elater. Scale bars: **a**–**c** = 2 mm; **d**, **g**, **h** = 100 µm; **e**, **f**, **i**, **k** = 50 µm. (*Jovet-Ast* & *Bischler 82243*)

Shefela, Judean Mountains, Judean Desert, Central Negev, Dan Valley, Hula Plain, Bet She'an Valley, Lower Jordan Valley, and Golan Heights.

General Distribution. — Distributed throughout the warm temperate and tropical parts of the world. Recorded from Southwest Asia, around the Mediterranean, Europe, Macaronesia, tropical and South Africa, Mascarene Islands, eastern Asia, India, tropical Asiatic Islands, Australia, New Zealand, and Oceania, western North America, Central America, tropical Andes, and southern South America.

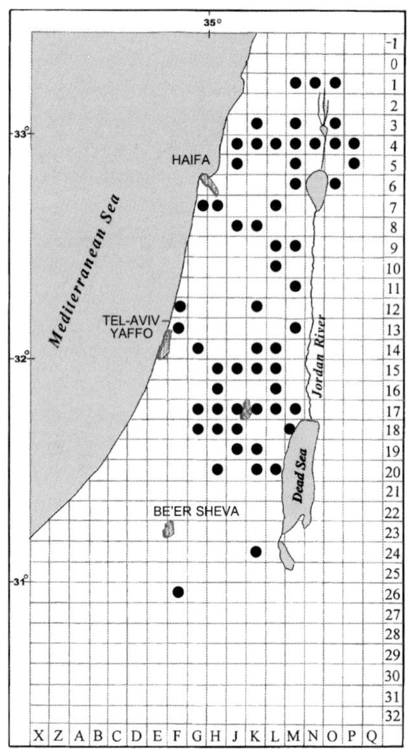

Map 225: *Targionia hypophylla*

Targionia hypophylla in fertile condition is easily distinguished from other liverworts in the local flora by its bivalved, blackish-purple involucres at the apex of the ventral side of the thallus, or by its small, discoid male branches. In sterile condition, it resembles the local *Mannia androgyna* in thallus size and shape, but the scales of this species are light purplish and are often provided with two appendages.

Targionia lorbeeriana Müll. Frib. was separated from *Targionia hypophylla* on the basis of chromosome number. However, the morphological characters (structure of epidermal pores, scales, size of spores) of *T. lorbeeriana* overlap with those of *T. hypophylla*. Despite important genetic differences (Boisselier-Dubayle & Bischler 1999), the two taxa cannot be distinguished on a morphological basis and thus, their taxonomic status remains uncertain.

14. LUNULARIACEAE

Plants thallose, large. *Thallus* branching dichotomous or by apical innovations; epidermal pores bounded by one or several concentric rings of cells; assimilatory tissue one-layered, air chambers with assimilatory filaments; scales in two rows, each with a large, reniform appendage constricted basally; oil cells in thallus and scales. *Rhizoids* smooth and tuberculate. *Asexual reproduction* by discoid, pluricellular gemmae from crescent-shaped gemma cups. *Dioicous*; antheridial receptacle terminal on short lateral branches, discoid, sessile, bounded by a membrane, scales absent; archegonial receptacle four-lobed with four groups of archegonia, terminal on short lateral branches, bounded by scales; archegoniophore, when mature, with very reduced receptacle without epidermal pores, and four tubular involucres spreading obliquely, each with one–two sporophytes. *Stalk* without rhizoid furrow. *Sporophyte* with bulbous foot, massive seta, and spherical capsule, slightly exserted at maturity, opening into four valves; capsule wall one-layered, without ring-like thickenings. *Spores* small. *Elaters* with two helical bands. *Chromosome number* $x = 9$.

A monotypic family of worldwide distribution.

Lunularia Adans.

Thallus without distinct median band, irregularly dichotomously branched or innovating from apex; epidermal pores conspicuous. *Antheridial receptacle* terminal but appearing lateral in position, the branch carrying it very short. *Archegonial cluster* in similar position, bounded by involute, white scales. *Archegoniophore* stalk with filiform, white scales. *Capsule* slightly exserted at maturity, spherical, splitting to near bottom into four valves. *Spores* faintly tuberculate on proximal and distal faces.

A monotypic genus, distributed worldwide in warmer areas, and very common in the Mediterranean Region.

Literature. — Müller (1951–1958), Haessel de Menéndez (1963), Schuster (1992b), Boisselier-Dubayle *et al.* (1995a), Paton (1999).

Lunularia cruciata (L.) Lindb., Not. Sällsk. Fauna Fl. Fenn. Förh. 9 : 298 (1868). *Marchantia cruciata* L., Sp. Pl., Edition 1 : 1137 (1753); *Lunularia vulgaris* Raddi, Opusc. Sci. (Bologna) 2 : 355 (1818). — Jovet-Ast & Bischler (1966) : 95, Bischler & Jovet-Ast (1975) : 18.

Distribution map 226, Figure 225, Plate XV : m.

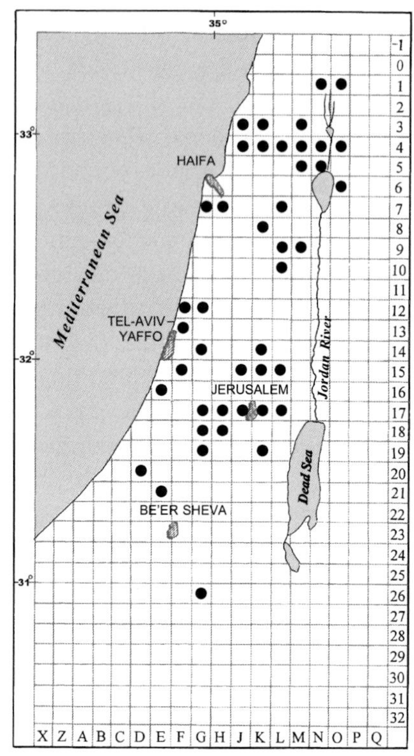

Map 226: *Lunularia cruciata*

Thallus light green, 6–11 mm wide; margins one-layered, hyaline, sinuose; epidermal cells with thin or thickened walls; epidermal pores bounded by four–five concentric rings of six–ten elongated cells, radial walls thin; ventral side of thallus usually green, rarely purplish in plants growing in exposed situations; scales hyaline, appendage with entire margins. *Archegoniophores* very rare. *Stalk* 2–5.5 mm long. *Capsule* dark brown. *Spores* yellowish-green to light brown, 15–20 µm in diameter; indistinctly trilete, tuberculate (seen with SEM).

Habitat and Local Distribution. — Plants growing in dense colonies or scattered among other thallose liverworts, on moist or wet sand, sandy clay, sandy loam (kurkar), terra rossa, or basaltic soil (pH 6.5–8), overlying limestone, dolomite, tuff or other calcareous rocks, or basalt, in rock crevices, at base of cliffs, in stone walls, on exposed wadi banks, in caves, grassland, open or dense maquis, *Pinus* and *Quercus* forests, *Olea europaea*, and *Eucalyptus* groves; associated with *Targionia*, *Reboulia*, *Athalamia*, and *Fossombronia*. Altitude: -200–1200 m. Very common: Coastal Galilee, Coast of Carmel, Sharon Plain, Philistean Plain, Upper Galilee, Lower Galilee, Mount Carmel, Esdraelon Plain, Samaria, Shefela, Judean Mountains, Judean Desert, Northern Negev, Central Negev, Dan Valley, Hula Plain, Bet She'an Valley, and Golan Heights.

General Distribution. — Nearly cosmopolitan in warmer areas, but rare in the tropical belt. Recorded from Southwest Asia, around the Mediterranean, Europe, Macaronesia, eastern Asia, India, tropical and South Africa, Mascarene Islands, Australia, New Zealand, North and Central America, tropical Andes, and southern South America.

Lunularia cruciata is easily distinguished by its crescent-shaped gemma cups, which are usually present.

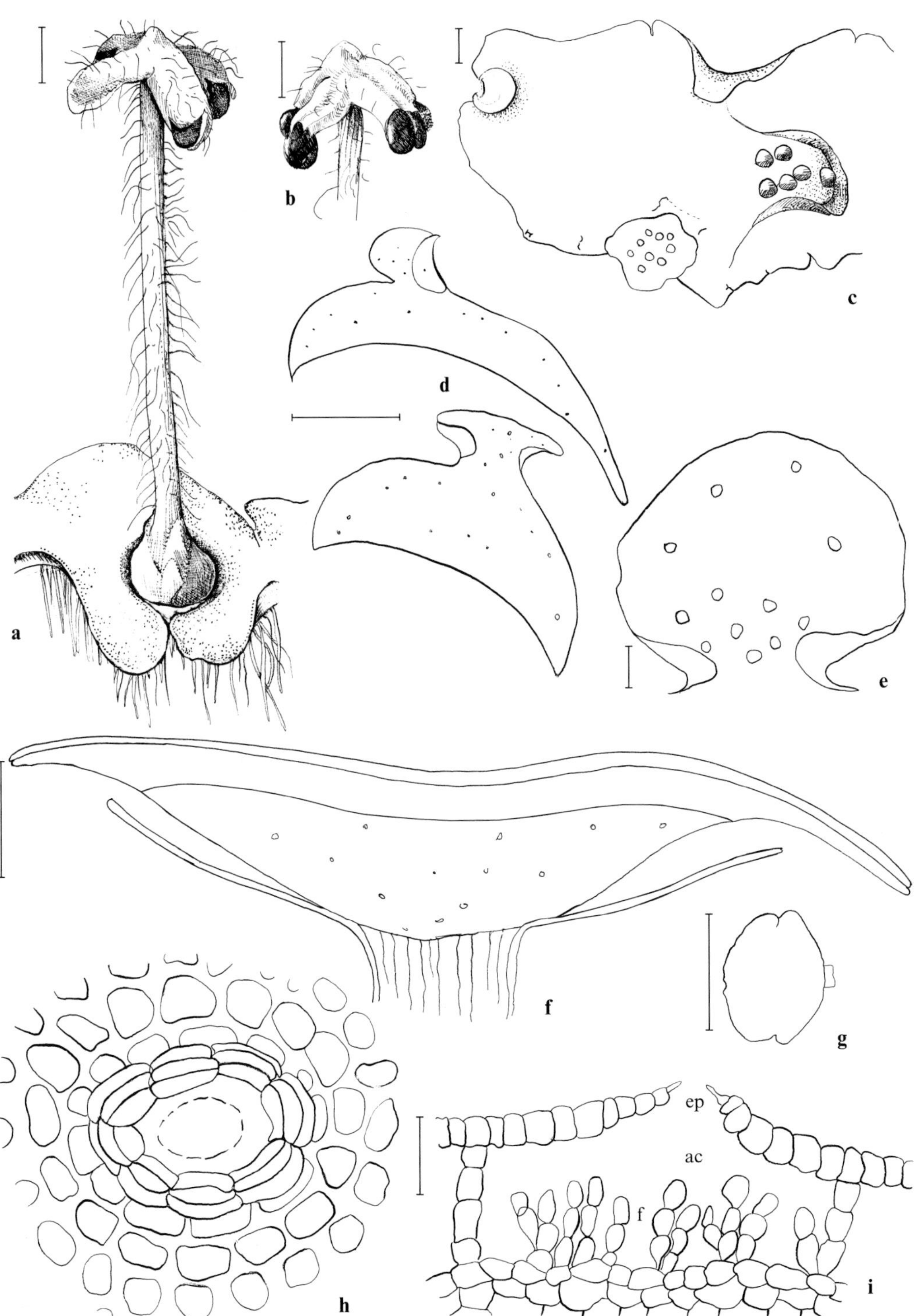

Figure 225. *Lunularia cruciata*: **a** thallus with archegoniophore; **b** female receptacle with sporophytes; **c** male thallus with antheridial receptacle and gemma cup; **d** scales; **e** appendage of scale; **f** thallus, cross-section; **g** gemma; **h** epidermal pore, front view; **i** cross-section of assimilatory layer, with air chamber (ac), filaments (f), and epidermal pore (ep). Scale bars: **a**–**c** = 1 mm; **d**, **f**, **g** = 100 μm; **e**, **h**, **i** = 50 μm. (**a**, **b**: *Herrnstadt*; **c**–**i**: *Jovet-Ast* & *Bischler 82393*)

15. AYTONIACEAE

Plants thallose, medium-sized to small, without differentiated median band. *Thallus* dichotomously and ventrally branched, sometimes with apical innovations; epidermal cells often with thickened walls; epidermal pores bounded by one–six concentric rings of five–eight cells, radial walls thin or thickened; assimilatory tissue in several layers, air chambers empty; oil cells in thallus and scales; ventral scales in two rows, large, with one–three triangular, lanceolate or filiform appendages. *Rhizoids* smooth and tuberculate. *Asexual reproduction* by specialised propagules absent. *Monoicous* mostly. *Antheridia* in cushions, bounded or not by scales, or in irregular dorsal groups. *Archegonia* in cushions bounded by scales. *Archegoniophore*, when mature, dorsal or terminal, stalk without or with single rhizoid furrow. *Female receptacle* two–seven-lobed, with compound epidermal pores on dorsal side and bivalved or cup-shaped involucre containing a single sporophyte under each lobe. *Seta* short; capsule spherical, not exserted, opening at maturity by an apical lid; wall one-layered, without ring-like thickenings. *Spores* usually large. *Elaters* with two–three helical bands. *Chromosome number* $x = 9$.

A family of worldwide distribution in warmer areas, comprising five genera. Three genera are present in the Mediterranean Region and in the local flora.

1 Epidermal pores bounded by a single ring of cells in local species; archegoniophore dorsal; stalk without rhizoid furrow **1. Plagiochasma**

1 Epidermal pores bounded by several concentric rings of cells; archegoniophore terminal; stalk with one rhizoid furrow.

- 2 Epidermal pores bounded by two–three concentric rings of cells, radial walls thin; epidermal cells with hardly distinct trigones in local species; scales with one–three lanceolate appendages; dorsal side of thallus areolate **2. Mannia**
- 2 Epidermal pores bounded by three–six concentric rings of cells, radial walls strongly thickened; epidermal cells with large, often bulging trigones; scales with two–four filiform appendages; dorsal side of thallus smooth **3. Reboulia**

1. *Plagiochasma* Lehm. & Lindenb.

Thallus thick, ventral side nearly black, glossy; epidermis not areolate; scales with one–three appendages. *Monoicous. Antheridia* in dorsal, sessile, rounded or cordate receptacles bounded by filiform scales. *Female receptacle* dorsal, becoming stalked after fertilisation of archegonia. *Stalk* without rhizoid furrow, with filiform scales. *Archegoniophore*, when mature, with one–four large, lateral, bivalved involucres, each with a single sporophyte. *Capsule* urn-shaped, opening at maturity through fragmentation of the apical lid. *Spores* alveolate.

A genus comprising 16 species, with worldwide distribution in warmer areas. Only one species is present in the Mediterranean Region and in the local flora.

Literature. — Müller (1951–1958), Haessel de Menéndez (1963), Bischler (1978).

Plagiochasma rupestre (J. R. Forst. & G. Forst.) Steph., Spec. Hep. 1 : 80 (1898).
Aytonia rupestris J. R. Forst. & G. Forst., Char. Gen. Pl., Edition 1 (folio) : 74 (1775); Edition 2 : 148 (1776).

Distribution map 227, Figure 226 (p. 585), Plate XVI : a.

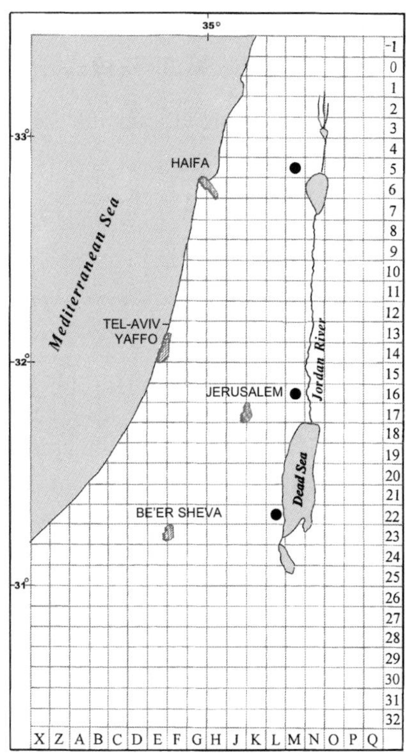

Map 227: *Plagiochasma rupestre*

Thallus bluish-green with black-purplish borders, smooth, 3–7 mm wide; epidermis covered with granulose material; epidermal cells with small trigones; epidermal pores very small, bounded by a single ring of four–six cells, often irregular in shape, radial walls thickened or thin; ventral side of thallus glossy, black; scales large, dark red, with one–two red or hyaline, triangular appendages. *Antheridia* and archegonia in cushions, lined up on midline of thallus. *Female receptacle* stalk 1–3 mm long. *Spores* brown, 70–90 µm in diameter; with five–six alveolae across diameter, ridges and alveolae papillose-tuberculate (seen with SEM).

Habitat and Local Distribution. — Plants growing in dense colonies on moist sandy clay (pH 7–8), in crevices of dolomite or calcareous rocks, on exposed wadi banks, usually facing south, associated with *Targionia, Lunularia, Mannia*, and *Athalamia*. Altitude: -200 to -150 m. Rare; restricted to the part of the Syrian-African Rift Valley in Israel: Lower Galilee, Judean Desert, and Lower Jordan Valley (Hart 1891).

General Distribution. — Widespread in warmer regions of the world, absent from eastern Asia and the tropical Asiatic Islands. Recorded from Southwest Asia, around the Mediterranean, Macaronesia, Europe, India, tropical and South Africa, Mascarene Islands, southern North America, throughout South America, Australia, New Zealand, and Oceania.

Plagiochasma rupestre is easily distinguished by its bluish, smooth dorsal and glossy, black ventral side, and by its male and female receptacles that are lined up on midline of dorsal side of thallus.

2. *Mannia* Opiz (*Grimaldia* Raddi)

Thallus bright or dark green; epidermal cell walls thin or thickened, without or with small trigones; epidermal pores bounded by two–three concentric rings of five–eight narrow cells, radial walls thin; scales with one–three lanceolate appendages. *Monoicous* (mostly). *Antheridia* dorsal on main thallus, in irregular groups, or in cushions, bounded or not by scales. *Archegoniophore* stalked, terminal; stalk with one rhizoid furrow and filiform scales. *Female receptacle* hemispherical, shortly divided into three–five rounded lobes. *Involucres* cup-shaped. *Capsule* urn-shaped, opening at maturity by an apical lid. *Spores* on distal face with large, hemispherical projections, or alveolate.

A genus comprising ca.15 species. Its distribution is nearly worldwide, up to the Arctic regions. However, it is absent from tropical eastern Asia, Australia, and New Zealand. One species is frequent in the Mediterranean Region and is found also in the local flora.

Literature. — Müller (1951–1958), Schuster (1992b).

Mannia androgyna (L. emend. Lehm. & Lindenb.) A. Evans, Chron. Bot. 4:225 (1938).

Marchantia androgyna L., Sp. Pl., Edition 1:1138 (1753); *Grimaldia dichotoma* Raddi, Opusc. Sci. (Bologna) 2:356 (1818). — Bischler & Jovet-Ast (1975):18.

Distribution map 228, Figure 227 (p. 587), Plate XVI:b.

Thallus 2–3 mm wide; dorsal side areolate; epidermal cells thin-walled, with hardly distinct trigones; ventral side of thallus purplish; scales purplish, with one–two short, lanceolate, acute appendages with entire margins. *Monoicous*; antheridia in irregular groups, not bounded by scales; archegoniophore stalk 2–15 mm long, with a few hyaline scales. *Capsule* wall cells with small trigones. *Spores* yellowish-brown, 60–75 μm in diameter; with large, rounded projections on distal face, irregularly reticulate on proximal face, with larger meshes around the pole (seen with SEM).

Habitat and Local Distribution. — Plants growing in dense colonies or scattered among other thallose liverworts, on moist or wet clay, sandy clay, or basaltic soil (pH 6.8–7.5), overlying calcareous rocks or basalt, in rock crevices, on exposed wadi banks, or in grassland and open *Quercus* forests, associated

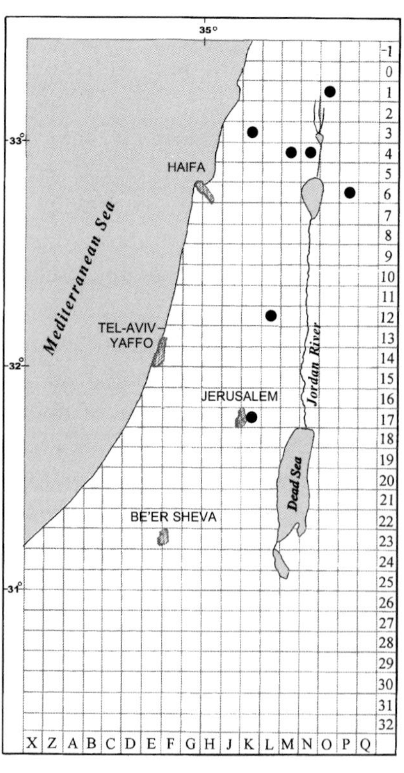

Map 228: *Mannia androgyna*

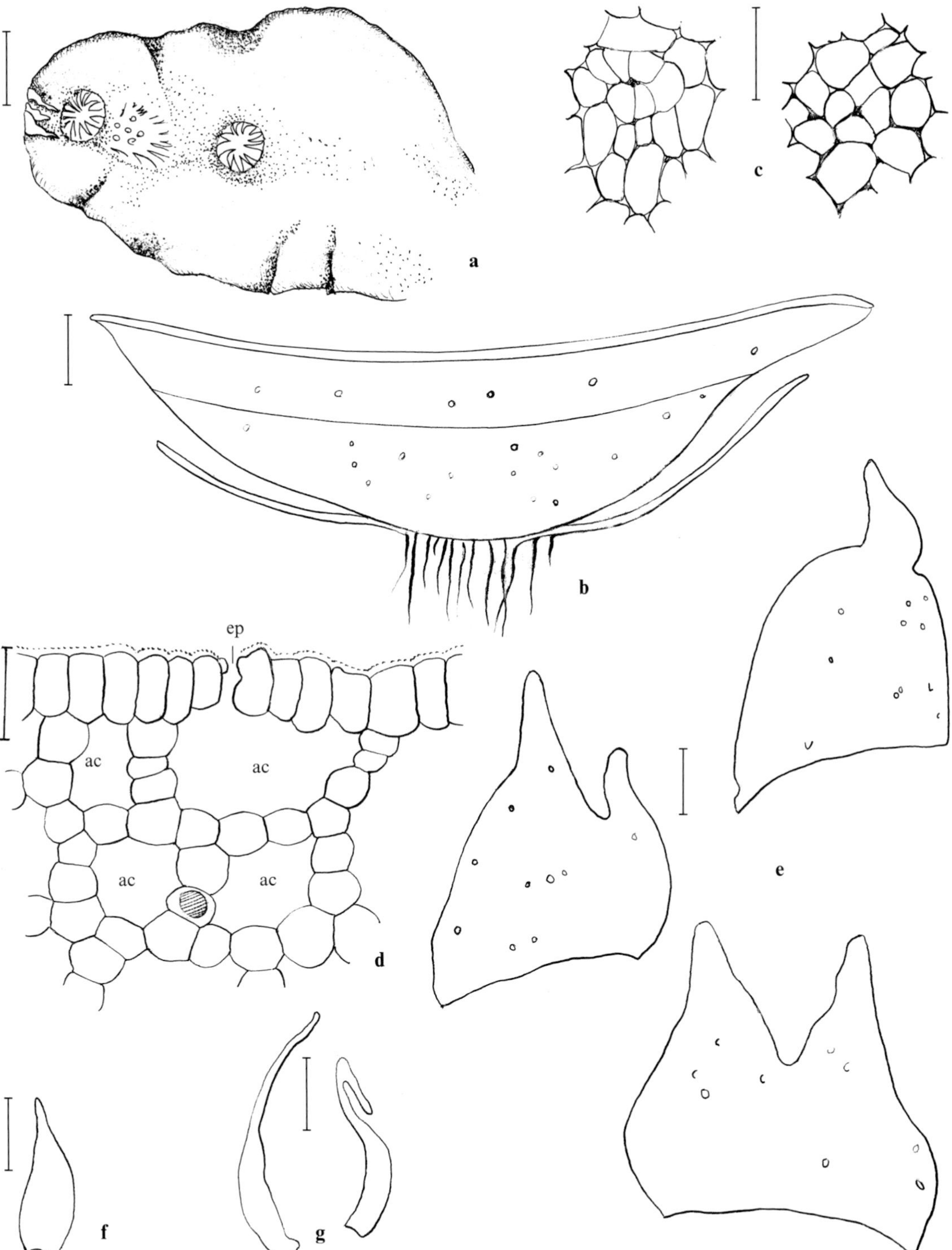

Figure 226. *Plagiochasma rupestre*: **a** thallus with two female and one male receptacle; **b** thallus, cross-section; **c** epidermal pores, front view; **d** air chambers (ac) and epidermal pore (ep), cross-section; **e** scales; **f** scale of male receptacle; **g** scales of female receptacle. Scale bars: **a** = 2 mm; **b**, **e** = 250 μm; **c**, **d**, **f**, **g** = 50 μm. (*Jovet-Ast* & *Bischler 82236*)

with *Targionia*, *Lunularia*, *Reboulia*, and *Athalamia*. Altitude: -100–700 m. Occasional (not yet recorded from the coast): Upper Galilee, Samaria, Judean Desert, Dan Valley, and Golan Heights.

General Distribution. — Recorded from Southwest Asia, around the Mediterranean, Europe, Macaronesia, tropical eastern Africa, eastern Asia and India.

The thallus of *Mannia androgyna* is very similar in size and shape to the thallus of *Targionia hypophylla*. In *Mannia*, the scales are less dark and, at least some scales, have two appendages.

3. *Reboulia* Raddi

Thallus bright green, 3–9 mm wide; epidermis smooth, cell walls thickened and with large, sometimes nodulose trigones; epidermal pores bounded by three–six concentric rings of cells, radial walls strongly thickened; scales large, with two–four filiform appendages. *Monoicous. Antheridia* sessile, in dorsal or terminal receptacles, ovate, orbicular, or reniform, sometimes inconspicuous, bounded or not by filiform scales. *Archegoniophore* terminal, stalked, stalk with a single rhizoid furrow, with filiform scales at top and bottom. *Female receptacle* hemispherical or conical, divided up to half of height into four–seven rounded lobes. *Involucres* bivalve. *Capsule* opening at maturity through fragmentation of the apical lid.

A genus comprising several genetically distinct entities, considered at present to belong to a single, polymorphic species. It is frequent in the Mediterranean Region and widely distributed in warmer regions.

Literature. — Müller (1951–1958), Haessel de Menéndez (1963), Schuster (1992b), Boisselier-Dubayle *et al.* (1998), Paton (1999).

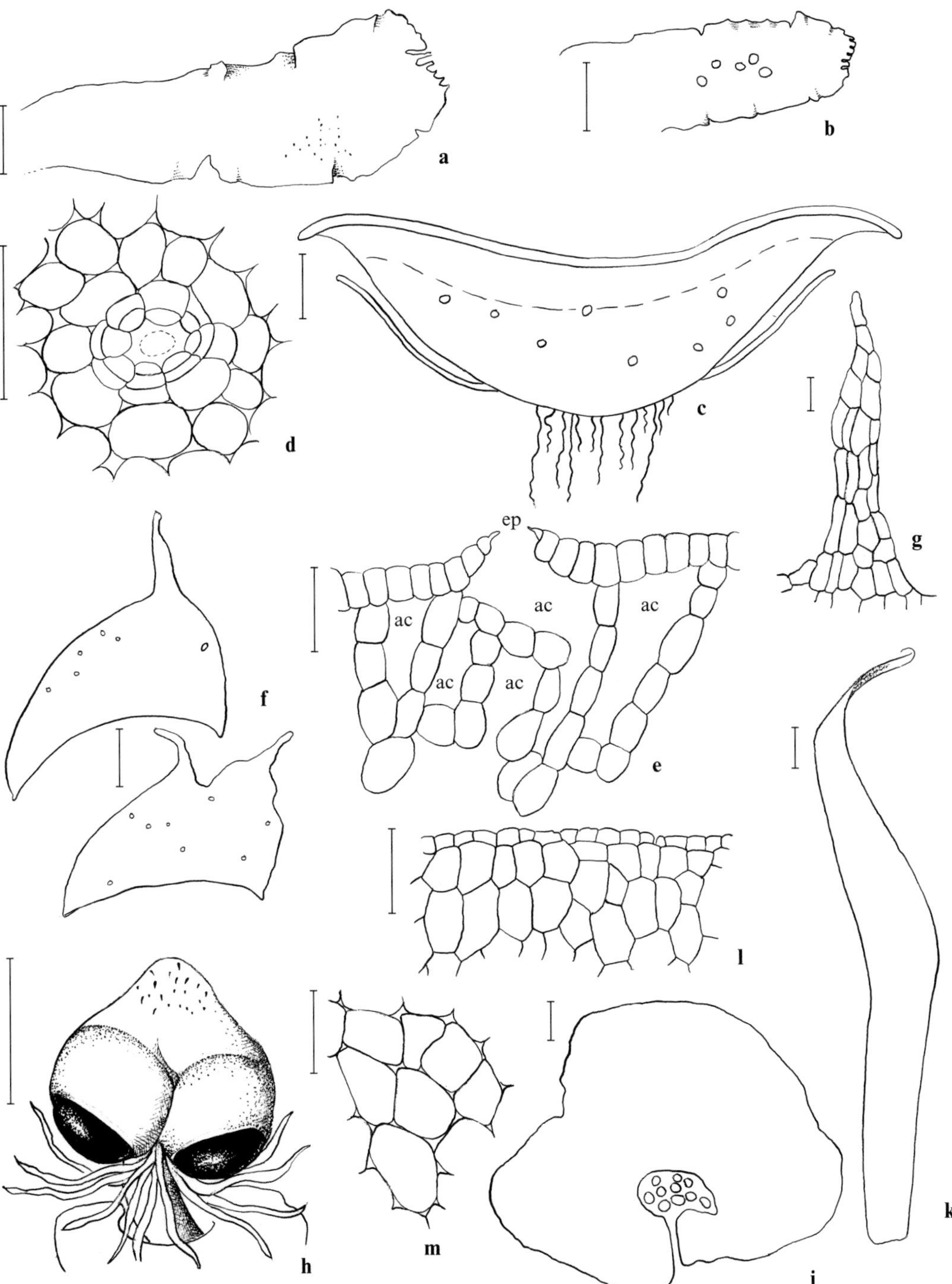

Figure 227. *Mannia androgyna*: **a** sterile thallus; **b** portion of male thallus with antheridia; **c** thallus, cross-section; **d** epidermal pore, front view; **e** air chambers (ac) and epidermal pore (ep), cross-section; **f** scales; **g** appendage of scale; **h** archegoniophore; **i** archegoniophore stalk, cross-section; **k** scale of archegoniophore; **l** capsule, cells of upper margin; **m** capsule wall cells. Scale bars: **a**, **b**, **h** = 2 mm; **c**, **f** = 250 μm; **d**, **e**, **g**, **i**, **k**–**m** = 50 μm. (*Jovet-Ast* & *Bischler 82087*)

Reboulia hemisphaerica (L.) Raddi, Opusc. Sci. (Bologna) 2 : 357 (1818).
Marchantia hemisphaerica L., Sp. Pl., Edition 1 : 1138 (1753). — Jovet-Ast & Bischler (1966) : 95, Bischler & Jovet-Ast (1975) : 18.

Distribution map 229, Figure 228, Plate XVI : c.

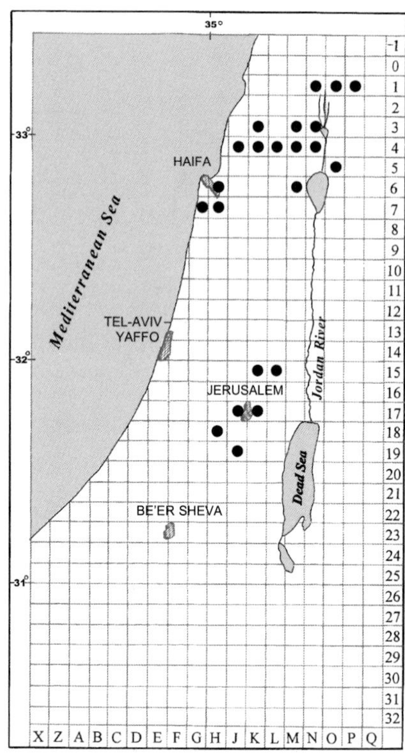

Map 229: *Reboulia hemisphaerica*

Thallus 3–8 mm wide, with narrow, purplish borders which are often lobulate-sinuose and ascending; epidermal cells with thin to slightly thickened walls and conspicuous trigones; epidermal pores bounded by four–six concentric rings of cells, each ring of five–nine cells; ventral side of thallus purplish; scales purplish, imbricate. *Archegoniophore stalk* 6–25 mm long. *Spores* yellowish-brown, 60–90 μm in diameter; alveolate on both faces, with four–five large alveolae across diameter and broad wing, ridges of alveolae and wing finely tuberculate (seen with SEM).

Habitat and Local Distribution. — Plants growing in dense colonies on moist or wet clay, sandy clay, or terra rossa (pH 6.5–8), overlying limestone, dolomite, or other calcareous rocks (but not basalt), in rock crevices or stone walls, on exposed wadi banks, in grassland, open maquis, or *Quercus* and *Olea europaea* groves, associated with *Targionia*, *Lunularia*, *Athalamia*, and sometimes *Fossombronia*. Altitude: 150–1200 m. Common, except in the south; especially frequent at higher elevations: Coastal Galilee, Coast of Carmel, Upper Galilee, Lower Galilee, Mount Carmel, Judean Mountains, Judean Desert, Dan Valley, and Golan Heights.

General Distribution. — Widespread in warm-temperate areas all over the world. Recorded from Southwest Asia, around the Mediterranean, Europe, Macaronesia, eastern, tropical, and South Africa, northern and eastern Asia, India, tropical Asiatic Islands, eastern Australia, northern New Zealand, North and Central America, and tropical and southern South America.

Reboulia hemisphaerica is easily distinguished by its scales with several filiform appendages.

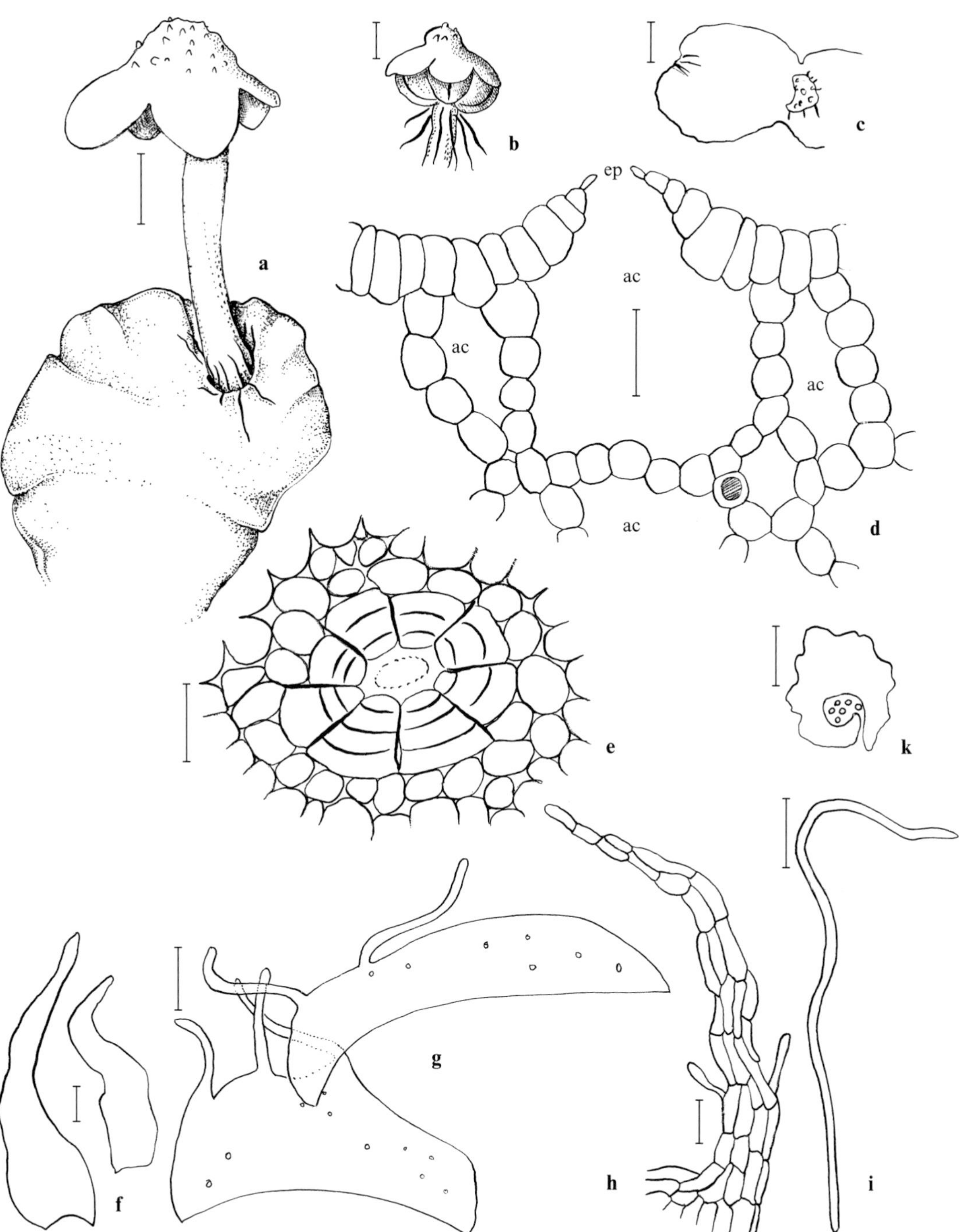

Figure 228. *Reboulia hemisphaerica*: **a** thallus with archegoniophore; **b** female receptacle; **c** male thallus with antheridial receptacle; **d** epidermal pore (ep) and air chambers (ac), cross-section; **e** epidermal pore, front view; **f** scales of male receptacle; **g** scales; **h** appendage of scale; **i** scale of archegoniophore; **k** archegoniophore stalk, cross-section. Scale bars: **a**–**c** = 1 mm; **d**–**f**, **h** = 50 μm; **g**, **i**, **k** = 250 μm. (*Jovet-Ast* & *Bischler 82118*)

16. CLEVEACEAE

Plants thallose, medium-sized, light green with hyaline or purplish borders. *Thallus* branching mainly dichotomous, sometimes with apical innovations; epidermal pores bounded by a single ring of four–seven cells, radial walls thin or thickened; assimilatory tissue in one–four layers of empty air chambers; ventral scales in several, irregular rows, each with one–two acute, lanceolate, often inconspicuous appendages, not constricted basally; oil cells in thallus, absent in scales. *Rhizoids* smooth and tuberculate. *Asexual reproduction* by specialised propagules absent. *Dioicous* or monoicous; antheridia in loose groups on dorsal thallus side, without scales; archegonia in dorsal or terminal cushions, surrounded by scales; archegoniophore stalk, when mature, without or with single rhizoid furrow. *Receptacles* without or with epidermal pores similar to those of thallus on dorsal side, divided into two–eight lobes, each with an obliquely spreading, tubular involucre, bilabiate at top. *Seta* short; capsule spherical, dehiscing at maturity into irregular valves; wall one-layered, with ring-like thickenings. *Spores* 40–80 μm in diameter. *Elaters* with two–four helical bands. *Chromosome number* $x=9$.

A family of worldwide distribution comprising three genera of which one genus is represented in the Mediterranean Region and in Israel.

Athalamia Falc. (*Clevea* Lindb.)

Thallus without distinct median band; branching dichotomous and by apical innovations; epidermal cells thin-walled; scales with single, lanceolate or filiform appendage. *Monoicous* or dioicous; antheridia in dorsal, loose groups; archegonia dorsal, in cushions. *Archegoniophore* stalk, when mature, without rhizoid furrow, with filiform scales at top and bottom. *Female receptacle* without or with epidermal pores of similar structure to those of thallus, with two–eight large, short-tubular involucres, each with a single sporophyte. *Capsule* not or hardly exserted at maturity. *Spores* with large projections on proximal and distal faces.

A genus comprising ca. 12 species, distributed worldwide in cold and temperate areas. Two species are found in the Mediterranean Region. A single species occurs in the local flora.

Literature. — Müller (1951–1958), Schuster (1992b), Perold (1993a).

Athalamia spathysii (Lindenb.) S. Hatt. *in* Shimizu & S. Hatt., J. Hattori Bot. Lab. 12: 54 (1954).

Marchantia spathysii Lindenb., Nova Acta Phys.-Med. Acad. Caes. Leop.-Carol. Nat. Cur. 14 (Suppl.): 104 (1829); *Clevea spathysii* (Lindenb.) Müll. Frib., Hedwigia 79: 75 (1940). — Jovet-Ast & Bischler (1966): 96, Bischler & Jovet-Ast (1975): 18.

Distribution map 230 (p. 592), Figure 229, Plate XVI: d.

Thallus dark green to blackish, 3–5 mm wide, with crisped, dark purplish borders; epidermal cells with small trigones; epidermal pores bounded by four–six cells, radial walls usually thin; ventral side of thallus dark red; scales purplish, with single, purplish or hyaline, lanceolate appendage. *Monoicous. Archegoniophore stalk* 0.5–5 mm long. *Involucres* usually two–three. *Spores* reddish-brown, 50–55 μm in diameter; distal

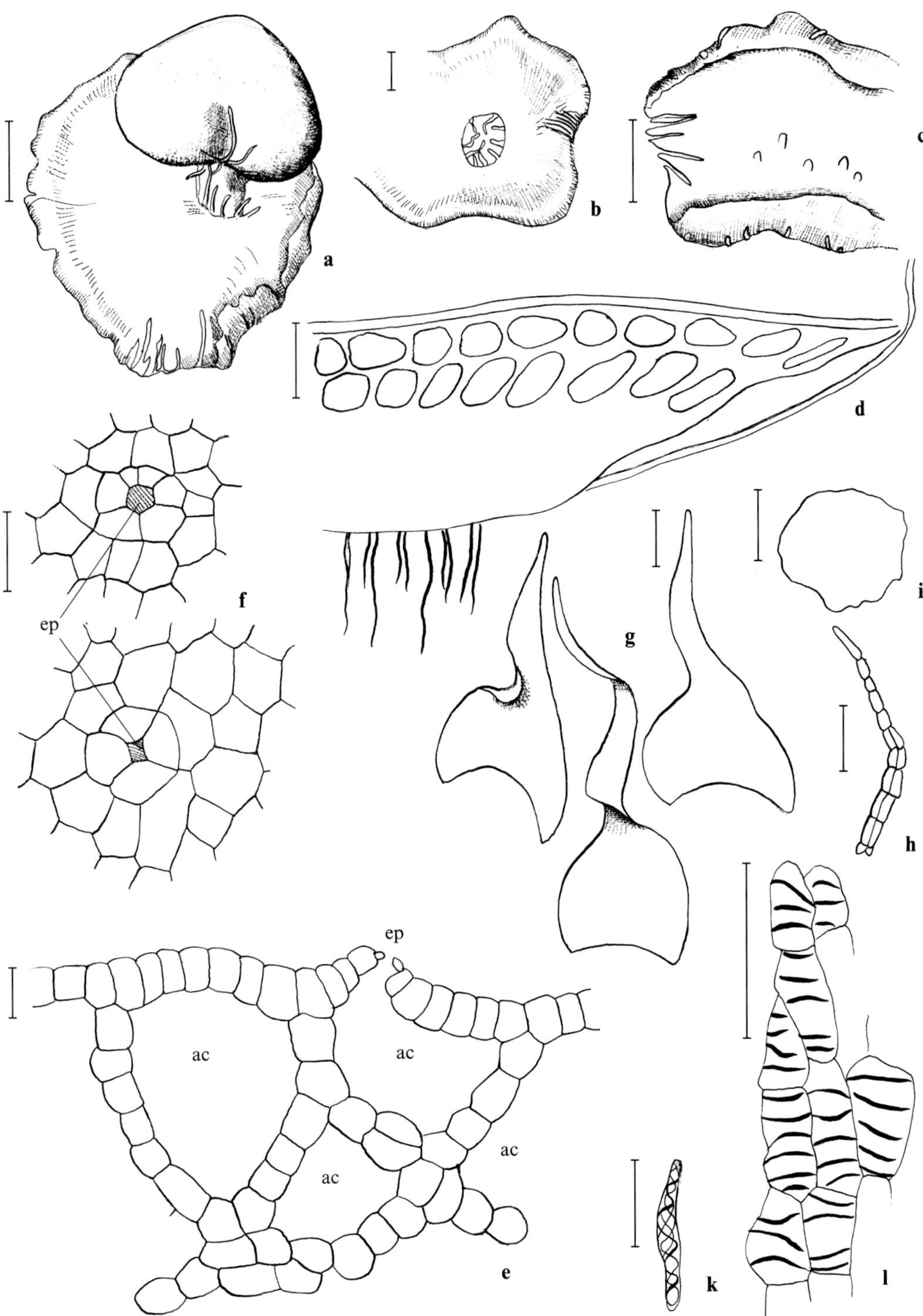

Figure 229. *Athalamia spathysii*: **a** thallus with archegoniophore; **b** thallus with female receptacle; **c** male thallus with antheridia; **d** portion of thallus, cross-section; **e** epidermal pore (ep) and air chambers (ac), cross-section; **f** epidermal pores (ep), front view; **g** scales; **h** scale of archegoniophore; **i** outline of archegoniophore stalk, cross-section; **k** elater; **l** capsule wall cells. Scale bars: **a–c** = 1 mm; **d**, **g–i** = 250 μm; **e**, **f**, **k**, **l** = 50 μm. (*Jovet-Ast* & *Bischler 82230*)

and proximal faces with six–seven rounded projections across diameter, irregularly pitted at top (seen with SEM).

Habitat and Local Distribution. — Plants growing in dense colonies on moist or wet clay, sandy clay, basaltic soil, or terra rossa (pH 7–8), overlying dolomite or other calcareous rocks, or basalt, in rock crevices, or at base of cliffs, on exposed wadi banks, in grassland, open maquis, *Quercus* forests, and *Eucalyptus* groves; associated with *Targionia*, *Reboulia*, *Lunularia*, and *Fossombronia*. The species can stand very harsh climatic conditions. Altitude: -200–970 m. Common; scattered over the country: Sharon Plain, Upper Galilee, Lower Galilee, Mount Carmel, Samaria, Judean Mountains, Judean Desert, Western Negev, Central Negev, and Golan Heights.

General Distribution. — Recorded from Southwest Asia, around the Mediterranean, Europe, Macaronesia, and South Africa.

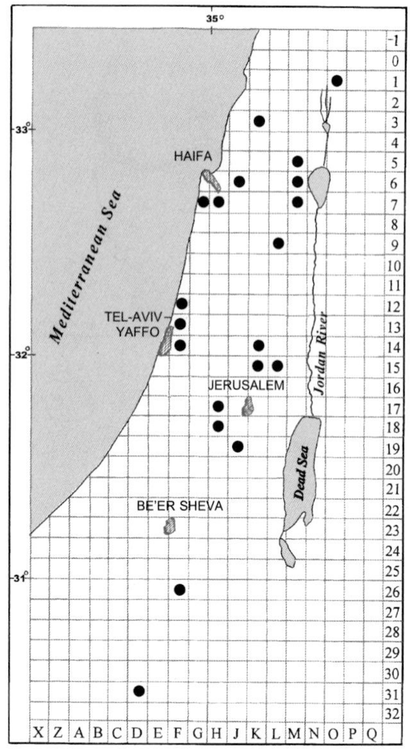

Map 230: *Athalamia spathysii*

Athalamia spathysii is easily distinguished by its thallus with dark, crisped borders, scales lacking oil cells, and, in fertile condition, by dorsal archegonia and archegoniophores.

17. MARCHANTIACEAE

Plants thallose, usually robust. *Thallus* branching dichotomous; epidermal cells thin-walled; epidermal pores compound, of two–eight superimposed concentric rings of cells, each ring of four–eight cells; assimilatory tissue in single layer, air chambers with filaments; ventral scales in two, four, or six rows; oil cells in thallus and scales. *Rhizoids* smooth and tuberculate. *Dioicous* or monoicous; antheridia and archegonia in stalked, terminal receptacles, stalks with one–four rhizoid furrows. *Female receptacles* two–eleven-lobed. *Involucres* bivalve. *Sporophytes* one or several per involucre; each fertilised archegonium usually bounded by a campanulate pseudoperianth. *Seta* short; capsule spherical, dehiscing irregularly; wall one-layered, with ring-like thickenings. *Spores* small or medium sized. *Elaters* with two–three helical bands. *Chromosome number* $x=9$.

A family of worldwide distribution, comprising five genera of which one genus is represented in the local flora.

Marchantia L.

Thallus usually large, with or without distinct median band; epidermal pores bounded by four–seven concentric rings of cells, radial walls thin; scales in four or six rows on ventral side of thallus, the median, laminal, and marginal of different shape and size, the median with single, large appendage. *Gemma cups* on dorsal side of thallus cup-shaped, with lenticular gemmae. *Dioicous*; antheridiophore and archegoniophore stalks with filiform scales, two–four rhizoid furrows and zero–two strips of air chambers. *Antheridial receptacle* peltate or palmate, four–eight-lobed, with antheridia in cavities on dorsal side. *Female receptacle* lobed or stellate, flat or hemispherical, with several involucres, each with several sporophytes. *Capsule* short-exserted at maturity. *Spores* 8–36 µm in diameter.

A genus comprising 36 species, with a worldwide distribution, up to the Arctic and the Antarctic Peninsula. Two species are distributed in the Mediterranean Region. A single species occurs in the local flora.

Literature. — Müller (1951–1958), Schuster (1992b), Bischler (1993), Boisseleir-Dubayle *et al.* (1995b), Paton (1999).

Marchantia polymorpha L., Sp. Pl., Edition 1 : 1137 (1753).
[Jovet-Ast & Bischler (1966) : 97]

Distribution map 231, Figure 230, Plate XVI : e.

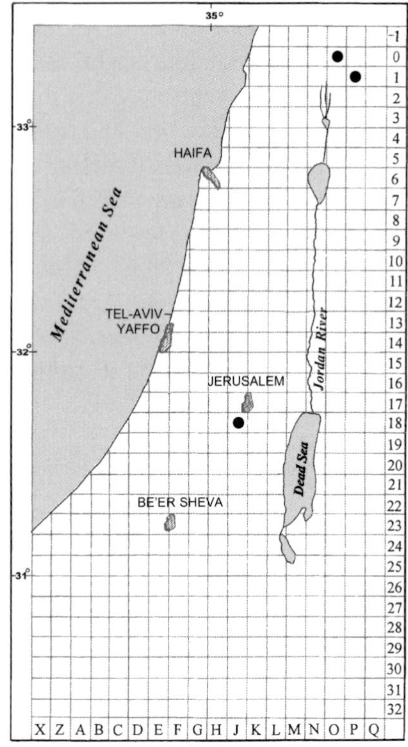

Map 231: *Marchantia polymorpha*

Thallus bright green, 7–20 mm wide; median band hardly distinct or large, loosely delimited and darker; borders hyaline or light red, usually crenulate; epidermal pores bounded by four–six concentric rings of elongated cells, two–three above epidermis, two–three bending into air chambers, uppermost and innermost ring composed usually of four cells, inner walls of innermost ring cells strongly convex, radial walls thin; ventral side of thallus green or brownish, purplish in plants growing in harsh habitat conditions; scales in six rows, the two median hyaline or purplish, with reniform appendage constricted basally, usually with faintly toothed margins; marginal scales passing beyond thallus margin. *Gemma cups* with lobed-ciliate upper margin and papillae on outer side. In Mediterranean specimens (only immature female receptacles were seen in local material) — *Antheridiophore stalk* 2.5–10 mm long, without strips of air chambers. *Antheridial receptacle* peltate, shortly six–eight-lobed. *Archegoniophore stalk* 10–50 mm long, with large strip of air chambers on dorsal side. *Female receptacle* deeply nine–eleven-lobed, with terete, papillose rays. *Involucre* margins with ciliate lobes. *Spores* yellow, 10–12 μm in diameter; hardly trilete, with vermiculate ornamentation on both faces (seen with SEM).

Habitat and Local Distribution. — Plants growing in dense colonies on moist, vertical, shaded tuff in an orchard (Golan Heights), but may be found in other habitats. Altitude: 900 m. Rare; recently recorded as a weed in nurseries, probably imported with peat: Judean Mountains, Mount Hermon (1943, *Zohary*, HUJ), and Golan Heights.

General Distribution. — Nearly cosmopolitan, up to the Arctic and Antarctic Peninsula, absent only from tropical lowland areas. In the Mediterranean Region, mainly occurring on mountains and in colder zones.

A species easily recognised, even in sterile condition, by its cup-shaped gemma cups and its scales in six rows on ventral side of thallus. *Marchantia polymorpha* comprises three genetically distinct subspecies, which are difficult to recognise morphologically. The local specimens have firm thalli without distinct median band. They are purplish on the ventral side and the thallus margin is hardly crenulate. Specimens collected in nurseries have the same characters. They probably belong to subsp. *ruderalis* Bischl. & Boisselier (Boisselier-Dubayle *et al.* 1995b).

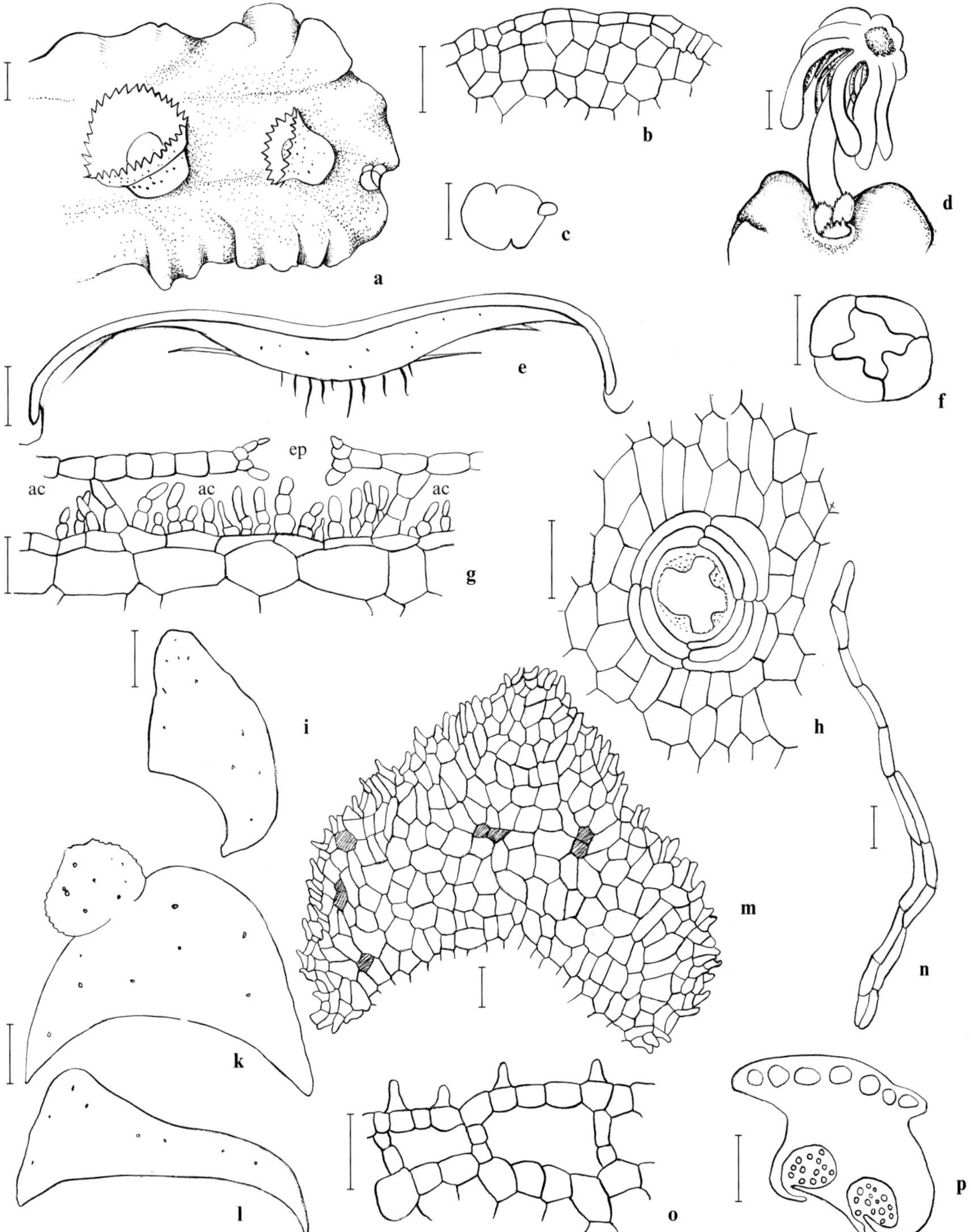

Figure 230. *Marchantia polymorpha*: **a** thallus with gemma cups; **b** portion of thallus margin; **c** gemma; **d** thallus with young archegoniophore; **e** thallus, cross-section; **f** epidermal pore, inner opening; **g** epidermal pore (ep) and air chambers (ac), cross-section; **h** epidermal pore, front view; **i** marginal scale; **k** median scale; **l** laminal scale; **m** appendage of median scale; **n** scale of archegoniophore; **o** ray of female receptacle, cross-section; **p** archegoniophore stalk, cross-section. Scale bars: **a**, **d**, **e** = 1 mm; **b**, **f**–**h**, **m**–**o** = 50 μm; **c**, **i**, **k**, **l**, **p** = 250 μm. (*Kaplan*, HUJ)

18. OXYMITRACEAE

Plants thallose, *Riccia*-like, in partial rosettes. *Thallus* dichotomously branched; epidermal pores bounded by a single ring of isodiametric cells; assimilatory tissue in one layer of narrow, empty air chambers; scales in two rows; oil cells absent. *Rhizoids* smooth and tuberculate. *Asexual reproduction* by specialised propagules absent. *Antheridia* dorsal, in median groove. *Archegonia* in dorsal thallus depression, bounded by large, pyriform involucres (fused to form a crest in a South African species), with epidermal pores and air chambers, open at top; calyptra thin. *Sporophyte* one per involucre, without foot and seta; capsule spherical, cleistocarpous, wall disintegrating at spore maturity. *Spores* large. *Elaters* absent. *Chromosome number* $x=9$.

A monotypic family distributed in warmer regions.

Oxymitra Bisch.

Thallus sharply grooved dorsally; epidermal pores with thickened radial walls; ventral scales large, triangular-lanceolate, acute apically, projecting beyond thallus margins. *Antheridia* in irregular groups bordered by a ciliate ridge. *Spores* alveolate on distal face.

A genus comprising two–three species with limited or disjunct ranges, distributed in the Mediterranean Region, warmer parts of Europe, South Africa, and North and South America.

Literature. — Müller (1951–1958), Haessel de Menéndez (1963), Perold (1993a).

Oxymitra incrassata (Brot.) Sérgio & Sim-Sim, J. Bryol. 15 : 662 (1989).
Riccia incrassata Brot., Fl. Lusit. 2 : 428 (1804); *Oxymitra paleacea* Bisch. *in* Lindenb., Nova Acta Phys.-Med. Acad. Caes. Leop.-Carol. Nat. Cur. 14 (Suppl.) : 124 (1829). — Jovet-Ast & Bischler (1966) : 97.

Distribution map 232 (p. 598), Figure 231, Plate XVI : f.

Thallus greyish-green or dark green, often tinged with purple or pink, light brown on margins and at base, one–three times branched, often forming hemirosettes; lobes up to 10 mm long, 2.5–7 mm wide; epidermal cells thin-walled; epidermal pores bounded by five–seven cells; cross section of lobes ± higher than wide; median groove deep; ventral scales large, white, acute, passing beyond thallus margin. *Dioicous*; archegonia enclosed in pyriform involucres bounded by white, filiform scales. *Spores* black, 100–175 µm in diameter; distal face with four–five shallow, faintly tuberculate alveolae across diameter; proximal face faintly tuberculate, with distinct triradiate mark (seen with SEM).

Habitat and Local Distribution. — Plants growing on compact sandy soil or clay (pH (6–)7–7.5) at base of rocks, on soil between calcareous rock blocks, in maquis and grassland under *Quercus ithaburensis* Decne., associated with *Riccia* species and *Sphaerocarpos michelii*. Altitude: 50–450 m. Occasional: Coast of Galilee, Coast of Carmel, Sharon Plain, Upper and Lower Galilee, Shefela, and Golan Heights.

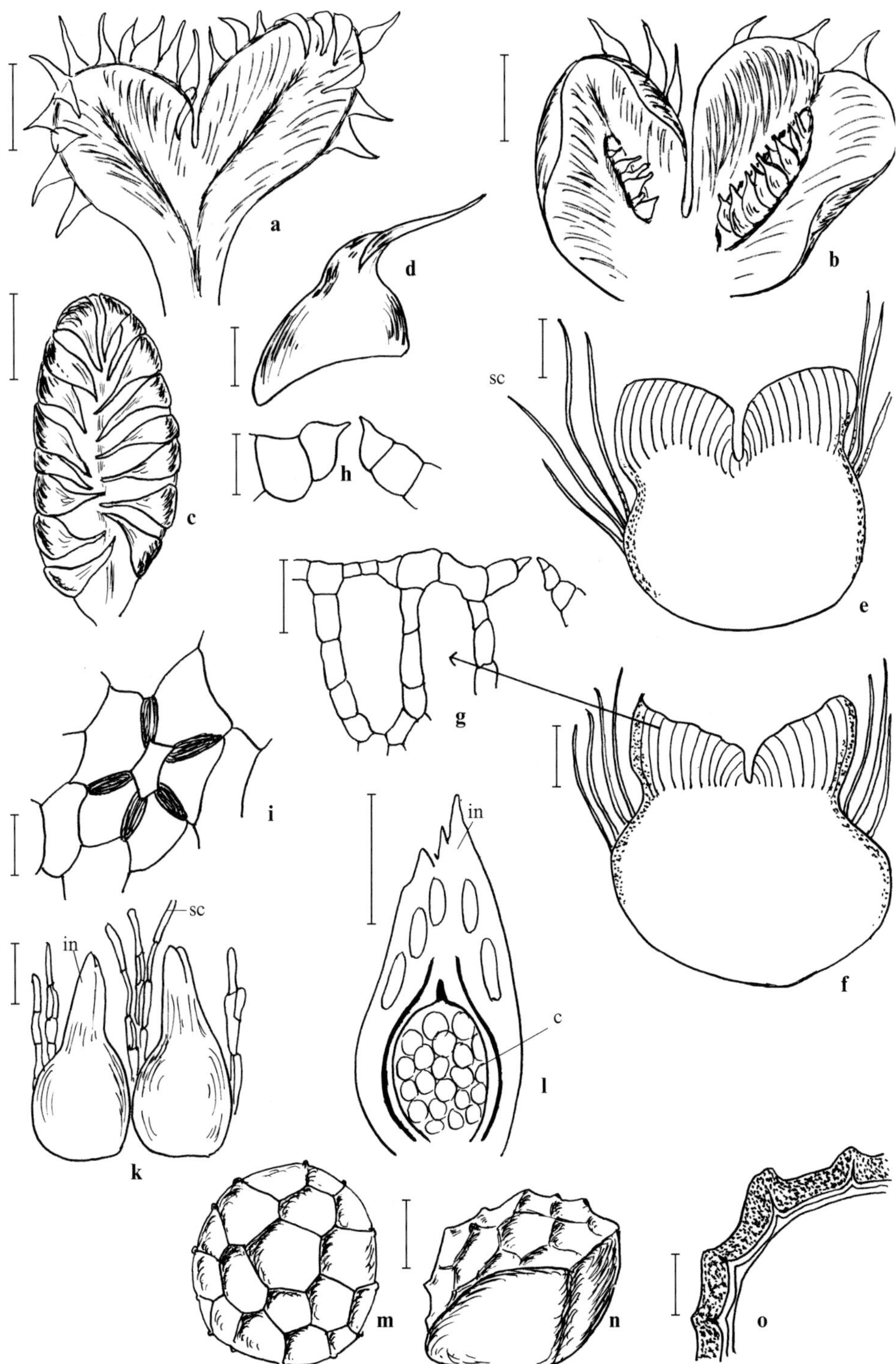

Figure 231. *Oxymitra incrassata*: **a** thallus, moist, sterile; **b** female thallus, moist; **c** thallus, dry; **d** ventral scale; **e**, **f** thallus, cross-sections (sc = scale); **g** assimilatory tissue; **h** epidermal pore, section; **i** epidermal pore, front view; **k** two involucres (in) with scales (sc); **l** involucre (in), longitudinal section (c = capsule); **m** spore, distal face; **n** spore, side view; **o** spore wall, section of distal face. Scale bars: **a**–**c** = 2 mm; **d**–**f**, **k**, **l** = 1 mm; **g** = 100 μm; **h**, **i** = 30 μm; **m**, **n** = 50 μm; **o** = 10 μm. (drawn from various sources)

General Distribution. — A Mediterranean species with major disjunctions in southern North and southern South America. Recorded from Southwest Asia, around the Mediterranean, southern and eastern Europe, Macaronesia, and North and South America.

Oxymitra incrassata resembles a *Riccia* but differs by its large, white, acute ventral scales passing beyond thallus margin, and by its pyriform involucres enclosing each sporophyte.

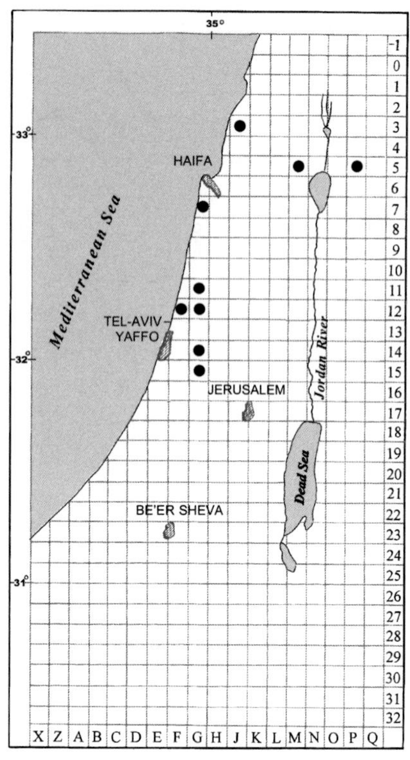

Map 232: *Oxymitra incrassata*

19. RICCIACEAE

Plants thallose, forming more or less complete rosettes. *Thallus* dichotomously branched, often grooved dorsally. *Rhizoids* smooth and tuberculate. *Asexual reproduction* by specialised propagules absent. *Antheridia* dorsal, embedded in thallus groove or scattered, without scales or cilia. *Archegonia* dorsal, singly embedded in median groove or scattered. *Involucre* absent. *Calyptra* absent. *Sporophyte*, when mature, without foot and seta; capsule spherical, cleistocarpous, wall disintegrating at spore maturity. *Chromosome number* $x=8$ or 9.

A family of worldwide distribution comprising two genera; both genera are represented in the local flora.

1 Epidermal pores bounded by a ring of five–six (ten) cells; scales in several irregular, transverse rows, with toothed margins; oil cells in thallus and scales present; chromosome number $x=9$ **1. Ricciocarpos**

1 Epidermal pores mere openings among epidermal cells, or bounded by cells of irregular shape, usually not forming a ring in local species; scales absent or in two rows in local species, scale margins not toothed; oil cells in thallus and scales absent; chromosome number $x=8$ **2. Riccia**

1. *Ricciocarpos* Corda

Thallus epidermis well developed; epidermal pores with single ring of cells; assimilatory tissue in two or several layers of empty air chambers; basal tissue in three–four layers; oil cells present in epidermis, ventral tissue, and scales; scales in several irregular, transverse rows, with toothed margins. *Monoicous*; antheridia and archegonia in dorsal groove, in one–three rows. *Spores* with more or less complete alveolae. *Chromosome number* $x=9$.

A monotypic genus, distributed worldwide, rare everywhere, and rarely bearing sporophytes.

Literature. — Müller (1951–1958), Haessel de Menéndez (1963), Litav & Agami (1976), Paton (1999).

Ricciocarpos natans (L.) Corda *in* Opiz, Beitr. Naturgesch. 12 : 651 (1829).
Riccia natans L., Syst. Nat., Edition 10 : 1339 (1759). — Jovet-Ast & Bischler (1966) : 97.

Distribution map 233, Figure 232, Plate XVI : g.

Thallus yellowish-green, often tinged with purple on margins, one–two times branched; lobes 4–9 mm long, 3–4 mm wide, with narrow median groove, lacunose at base; epidermal pores with single ring of five–six (ten) cells, radial walls thin; scales blackish or hyaline, often very long (5–6 mm) in aquatic forms, shorter in terrestrial, without differentiated appendage, with teeth of one–three cells, with oil cells but without marginal papillae. *Capsules* black, more or less projecting dorsally. *Spores* 42–72 μm in diameter; brown or nearly black, distal face with eight–ten more or less complete, tuberculate-papillose alveolae across diameter, wing narrow, sinuose, ridges tuberculate; proximal face similar but with lower ridges, with distinct triradiate mark (seen with SEM).

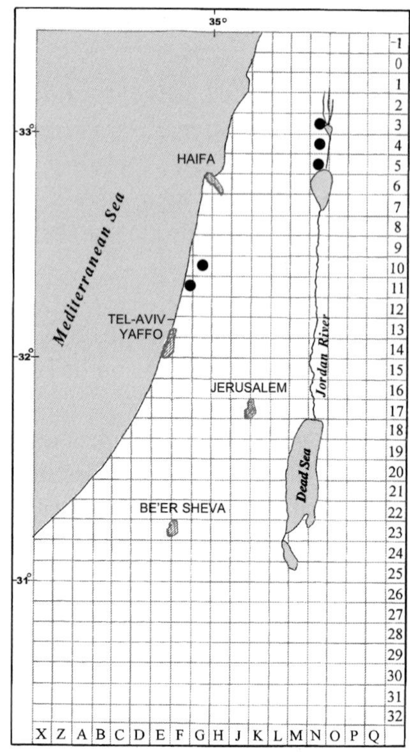

Map 233: *Ricciocarpos natans*

Habitat and Local Distribution. — Plants floating on standing or slowly running water, sometimes on very wet soil, among hydrophilous phanerogams (*Cyperus papyrus* L. and *Phragmites australis* (Cav.) Steud.). Rare: Sharon Plain, Hula Plain, and Upper Jordan Valley.

General Distribution. — Recorded from Southwest Asia, around the Mediterranean, central Europe, tropical and South Africa, northern and eastern Asia, India, Australia, New Zealand, North and Central America, eastern, tropical, and southern South America.

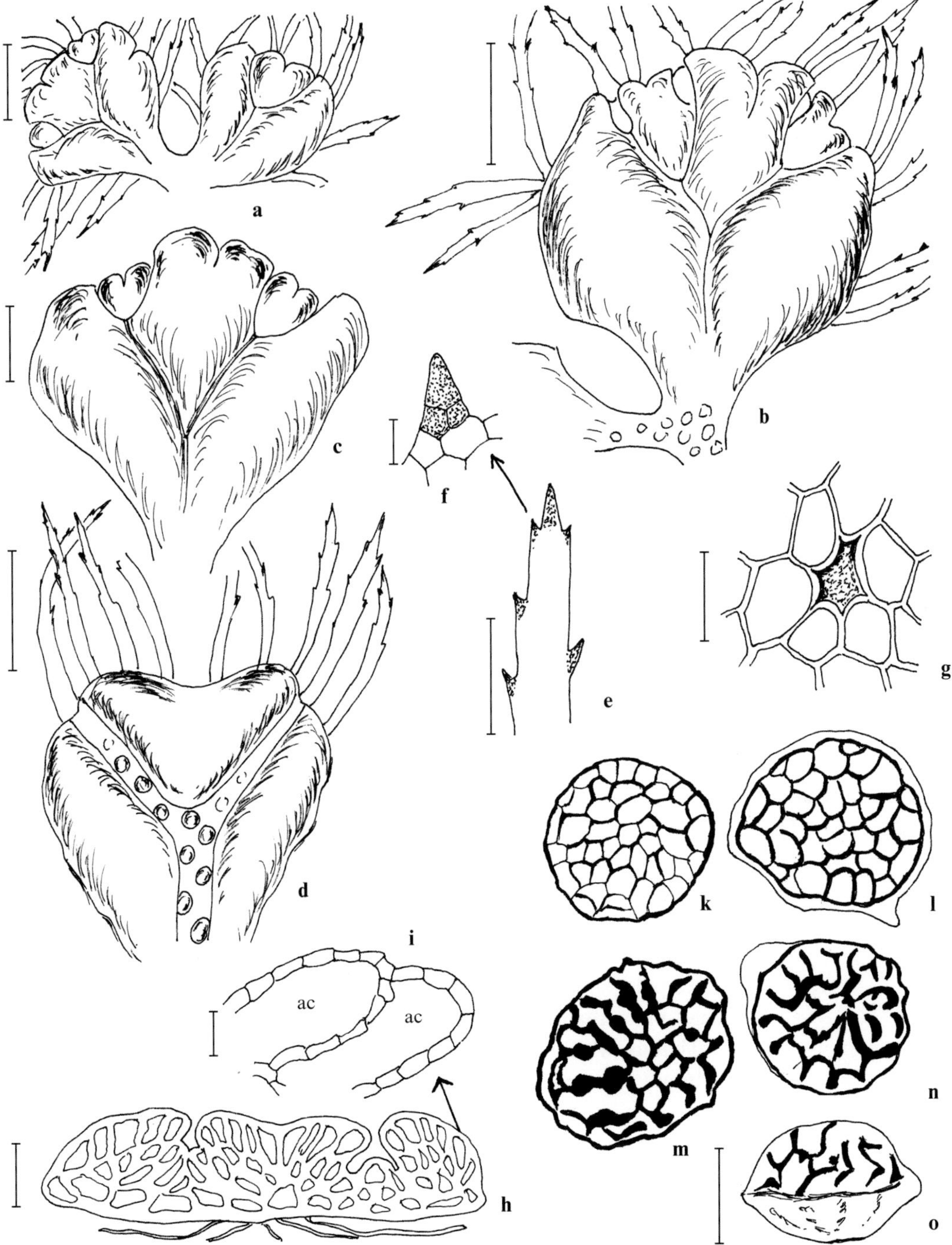

Figure 232. *Ricciocarpos natans*: **a**–**c** thalli; **d** thallus with sporophytes; **e** scale, apical part; **f** scale, tooth of margin; **g** epidermal pore; **h** cross-section of lobe; **i** part of cross-section of lobe (ac = air chamber); **k** immature spore, alveolae hardly distinct; **l** older spore; **m**, **n** mature spores, black, with thick crests; **o** mature spore, side view. Scale bars: **a**–**d** = 3 mm; **e**, **h** = 1 mm; **f**, **i** = 50 µm; **g** = 20 µm; **k**–**o** = 25 µm. (drawn from various sources)

2. *Riccia* L.

Thallus 0.5–4 mm wide, dichotomously branched, often forming rosettes, green or tinged with purple, red or violet; epidermis persistent, or decaying (or completely absent and chamber partitions free-standing, not roofed by epidermal cells in South African species); upper side and margins sometimes with cilia or papillae; epidermal pores a mere opening among epidermal cells, rarely with one–two incomplete rings of four–six cells, radial walls thin; assimilatory tissue compact, with tubular aeriferous channels, open at top, or loose, with one–three layers of polygonal, empty air chambers; oil cells absent; scales usually in two rows on lateral sides, persistent or deciduous, sometimes absent, or in single, median row, without appendage. *Asexual reproduction* by specialised propagules absent, but sometimes thallus with one or several apical or ventral tubers, or with tuberous, perennating apical parts. *Monoicous* or dioicous; gametangia singly embedded in thallus; antheridial and archegonial necks often protruding. *Capsule* dorsally or ventrally protruding. *Spores* 45–215 μm in diameter (remaining in tetrads in some tropical species); distal and proximal faces alveolate, vermiculate, tuberculate, or nearly smooth, with distinct triradiate mark. *Chromosome number* $x = 8$.

A genus comprising 150 or more species with a worldwide distribution up to the Arctic and Antarctic, but more frequent in areas with Mediterranean-type climates. Most species are terrestrial, some floating on water.

Literature. — Jovet-Ast (1986), Schuster (1992b), Paton (1999).

1 Thallus with filiform aeriferous channels open at top; epidermal pores mere openings; capsules protruding on dorsal side of thallus or not protruding; growing on intermittently wet soils — subgen. **Riccia**
- 2 Thallus without cilia or papillae.
  - 3 Thallus white when dry, covered with chalk — **1. R. crustata**
  - 3 Thallus not white, not or not entirely covered with chalk.
    - 4 Lateral side of lobes with large, white scales rounded apically and passing beyond lobe margins — **2. R. lamellosa**
    - 4 Lateral side of lobes without large, white scales.
      - 5 Scales pale; subepidermal cells with thickened walls — **3. R. sorocarpa**
      - 5 Scales pigmented; subepidermal cells with thin walls.
        - 6 Thallus with two short lateral wings; scales with purplish and bright orange spots; spores 95–108(–115) μm in diameter — **4. R. macrocarpa**
        - 6 Thallus without lateral wings; scales purplish-black; spores 60–90 μm in diameter.
          - 7 Dorsal side of thallus dark blue-green with border and base red-brown; scales shiny; spores 60–80 μm in diameter, with 8–9(–10) incomplete alveolae — **6. R. nigrella**
          - 7 Dorsal side of thallus bright blue-green, without differentiated border; scales not shiny; spores (67–)72–85(–90) μm in diameter, with 9–12(–15) complete alveolae — **7. R. trabutiana**
- 2 Thallus with cilia or papillae.
  - 8 Thallus with papillae — **8. R. atromarginata**
  - 8 Thallus ciliate or at least with a few cilia.

9 Thallus with lateral wings; spores (130–)170–180(–215) μm in diameter **9. R. gougetiana**

9 Thallus without lateral wings; spores (65–)75–150 μm in diameter.

10 Some cilia free, some connate at their base in groups of two–three **10. R. bicarinata**

10 Cilia free.

11 Cross section of thallus lobes three–four times wider than high; cilia extending from top to base of lobes; spores 90–125(–140) μm in diameter, with seven–nine alveolae **11. R. michelii**

11 Cross section of thallus lobes 1–2.5 times wider than high; cilia only at upper part of lobes; spores (65–)75–96(–115) μm in diameter, with 8–13 alveolae.

12 Thallus with numerous marginal cilia up to 500 μm in length; spores (65–)75–85(–90) μm in diameter **12. R. crozalsii**

12 Thallus hardly ciliate, only with a few marginal cilia up to 120 or 350 μm in length; spores (75–)84–96(–115) μm in diameter **5. R. subbifurca**

1 Thallus with one–three layers of air chambers; epidermal pores mere openings, or bounded by a ring of cells; capsules protruding on ventral side of thallus, or not protruding; growing on standing water, mud or wet soil subgen. **Ricciella**

13 Dioicous, female thallus two–four times larger than male thallus; thallus blue-green with pink spots; spores vermiculate **13. R. frostii**

13 Monoicous; female thallus not larger than male thallus; thallus glaucous or light blue-green; spores alveolate **14. R. crystallina**

1. Riccia crustata Trab. ex Grolle, Lindbergia 3 : 54 (1975).
Riccia crustata Trab., Bull. Soc. Hist. Nat. Afrique N. 7 : 87 (1916), *nom. inval.* — Bischler & Jovet-Ast (1975) : 19.

Distribution map 234, Figure 233, Plate XVI : h.

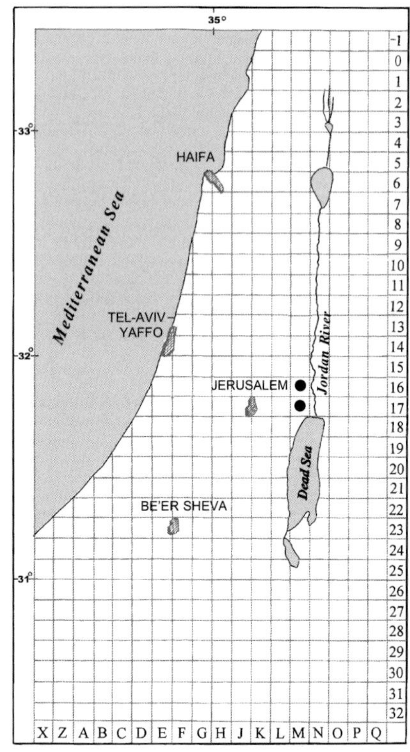

Map 234: *Riccia crustata*

Thallus pale blue-green when wet, becoming white when drying, pure white when dry, light green and less calcified in culture, two–three times branched, forming rosettes; lobes 1.5–4 mm long, 1–1.6 mm wide, rounded apically, margins obtuse and erect or faintly incurved or recurved, cacified on dorsal side and stiff, never upfolded; cross section of lobes as wide as high or 1.5–2 times wider than high; median groove deep; epidermal cells large (63–90×45–70 μm), rounded and convex on dorsal side, dorsal side of lobes appearing mammillate; assimilatory layer often in fan-like rows; scales absent or reduced. *Dioicous*. *Spores* (55–)72–88(–93) μm in diameter, light yellow when young, red-brown when mature with dark brown or black ornamentation, without wing but with thickening at equator, pores absent; distal face slightly convex to nearly flat at pole, in the centre with faintly marked, ramified or radiating lines forming rarely some incomplete alveolae, periphery smooth when wet, wrinkled when dry; proximal face smooth, or covered with small tubercles, or granular, or rarely wrinkled and appearing rugulose, without crests along the triradiate mark (seen with SEM). *Chromosome number* $n = 8$.

Habitat and Local Distribution. — Plants growing on sandy clay or loess over Lisan marl (pH 7), exposed, on slopes and in beds of little wadis, among grasses. Altitude: -150–50 m. Rare: Judean Desert and Lower Jordan Valley.

General Distribution. — Recorded from Southwest Asia, around the Mediterranean, eastern Europe, southern north Asia (Kazakhstan), and Australia. This range includes the probably synonymous species *R. albida* Sull. ex Austin.

Riccia crustata is easily distinguished by the calcareous deposit on the dorsal thallus side, and spores that are only faintly marked with branching or radiating lines in the centre of the distal face and are not alveolate.

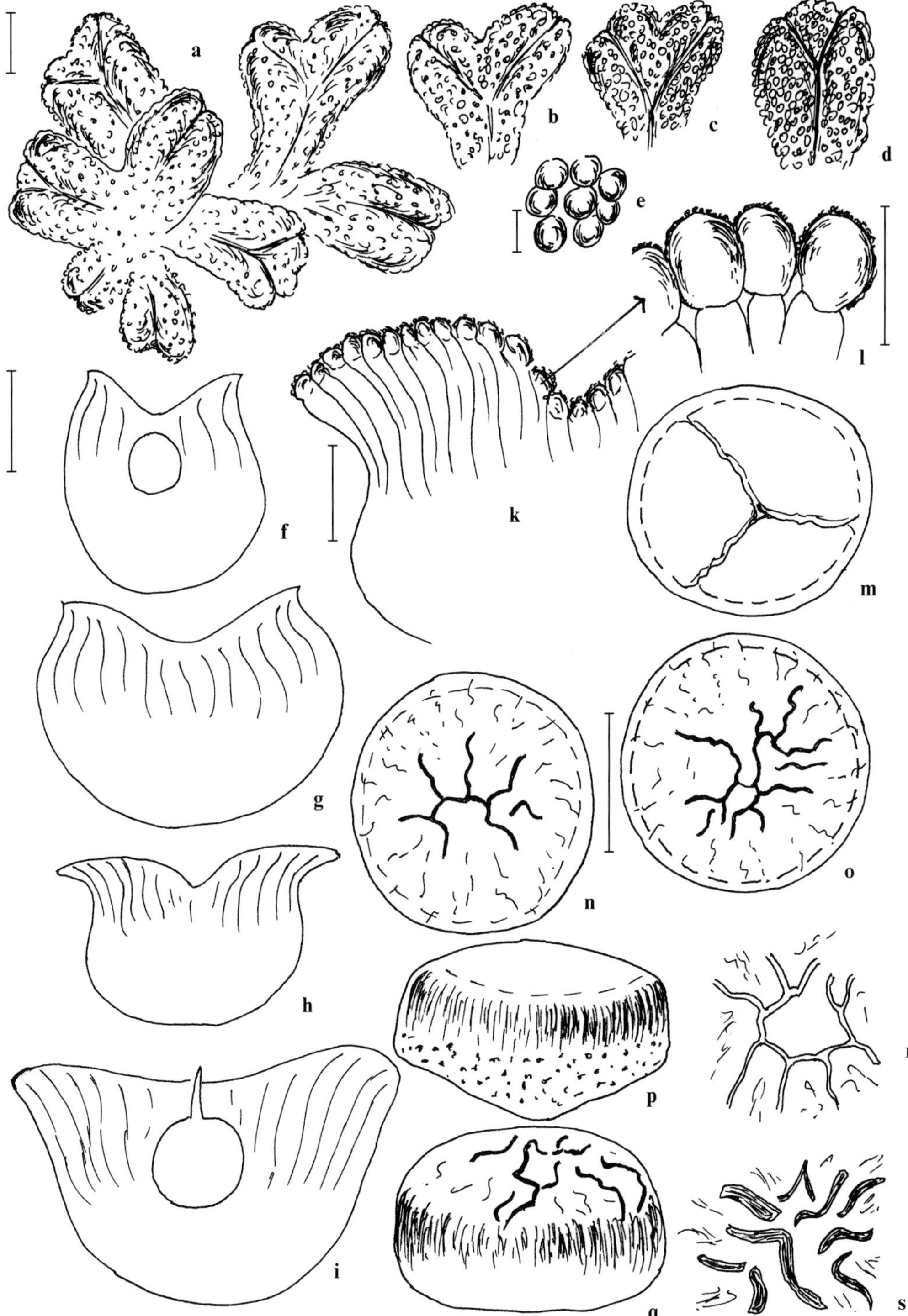

Figure 233. *Riccia crustata*: **a**–**d** thalli; **e** epidermal cells with chalk, front view; **f**–**i** cross-sections of lobe from top to base; **k** part of cross-section; **l** epidermal cells in section; **m** spore, proximal face; **n**, **o** spores, distal face; **p**, **q** spores, side view; **r**, **s** spores, central ornamentation of distal face. Scale bars: **a**–**d** = 1 mm; **e**–**i** = 0.5 mm; **k** = 0.2 mm; **l**–**s** = 40 µm. (drawn from various sources)

2. Riccia lamellosa Raddi, Opusc. Sci. (Bologna) 2 : 351 (1818).
[Jovet-Ast & Bischler (1966) : 101; Bischler & Jovet-Ast (1975) : 19]

Distribution map 235, Figure 234.

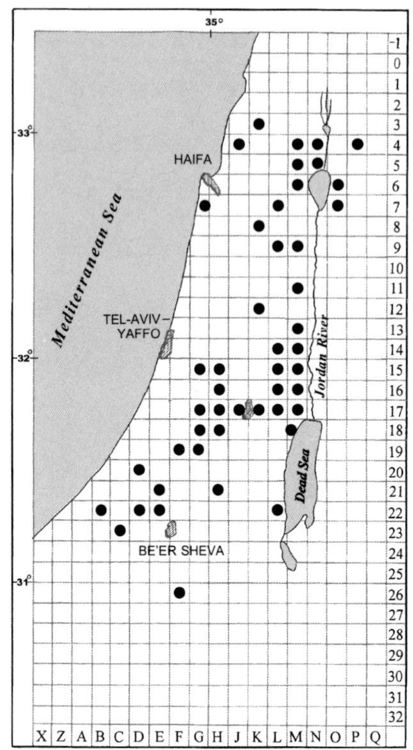

Map 235: *Riccia lamellosa*

Thallus pale green or bluish, three–four times branched, forming complete or incomplete rosettes, sometimes partly covered with chalk; lobes 20 mm long, 2–3 mm wide, rounded or obtuse apically; median groove shallow, distinct all along the lobes; lateral sides of lobes entirely covered with large, pure white scales, rounded apically and passing beyond lobe margins, built up of large cells (145×84 μm); cross section of lobes as wide as high apically, below 1.5–2 times wider than high, upper edge divided into three convex parts, lateral edges erect and in upper part spread out; assimilatory layer made up of short, rectangular, thin-walled cells; epidermal cells convex, persistent in groove. *Monoicous. Spores* (85–) 96–108(–120) μm in diameter, wing often incomplete, rarely entire, ± crenulate and with thickened network; distal face with (9–)10–12(–15) alveolae limited by thick walls with tubercles at wall corners and, sometimes, near wing the alveolae replaced by a rectangular structure limited by straight walls directed towards the equator; proximal face with numerous alveolae incompletely delimited by rather weak walls, with thick triradiate mark. *Chromosome number* $n=8$, 16, 24.

Habitat and Local Distribution. — Plants growing on compact sandy soil or clay (pH (5.5–)7–8), at base of calcareous or basaltic rocks, on soil between rock blocks, on rocky slopes, among grasses in meadows, on sheltered sandy paths, in maquis with *Olea europaea* or under *Quercus calliprinos*, exposed or sheltered by *Sarcopoterium spinosum* (L.) Spach or *Tamarix* spp.; associated with other *Riccia* species (chiefly *R. sorocarpa* and *R. atromarginata*), *Oxymitra*, and *Sphaerocarpos*. Altitude: -20–750 m. Common: Coastal Galilee, Coast of Carmel, Philistean Plain, Upper Galilee, Lower Galilee, Esdraelon Plain, Samaria, Shefela, Judean Mountains, Judean Desert, Northern Negev, Central Negev, Hula Plain, Upper Jordan Valley, Bet She'an Valley, Lower Jordan Valley, and Golan Heights.

General Distribution. — Recorded from Southwest Asia, around the Mediterranean, Europe, Macaronesia, dry tropical Africa, Australia, and North and South America.

Riccia lamellosa is easily recognised by its pale green or bluish colour and by its pure white scales covering the lateral lobe sides and passing beyond the lobe margins.

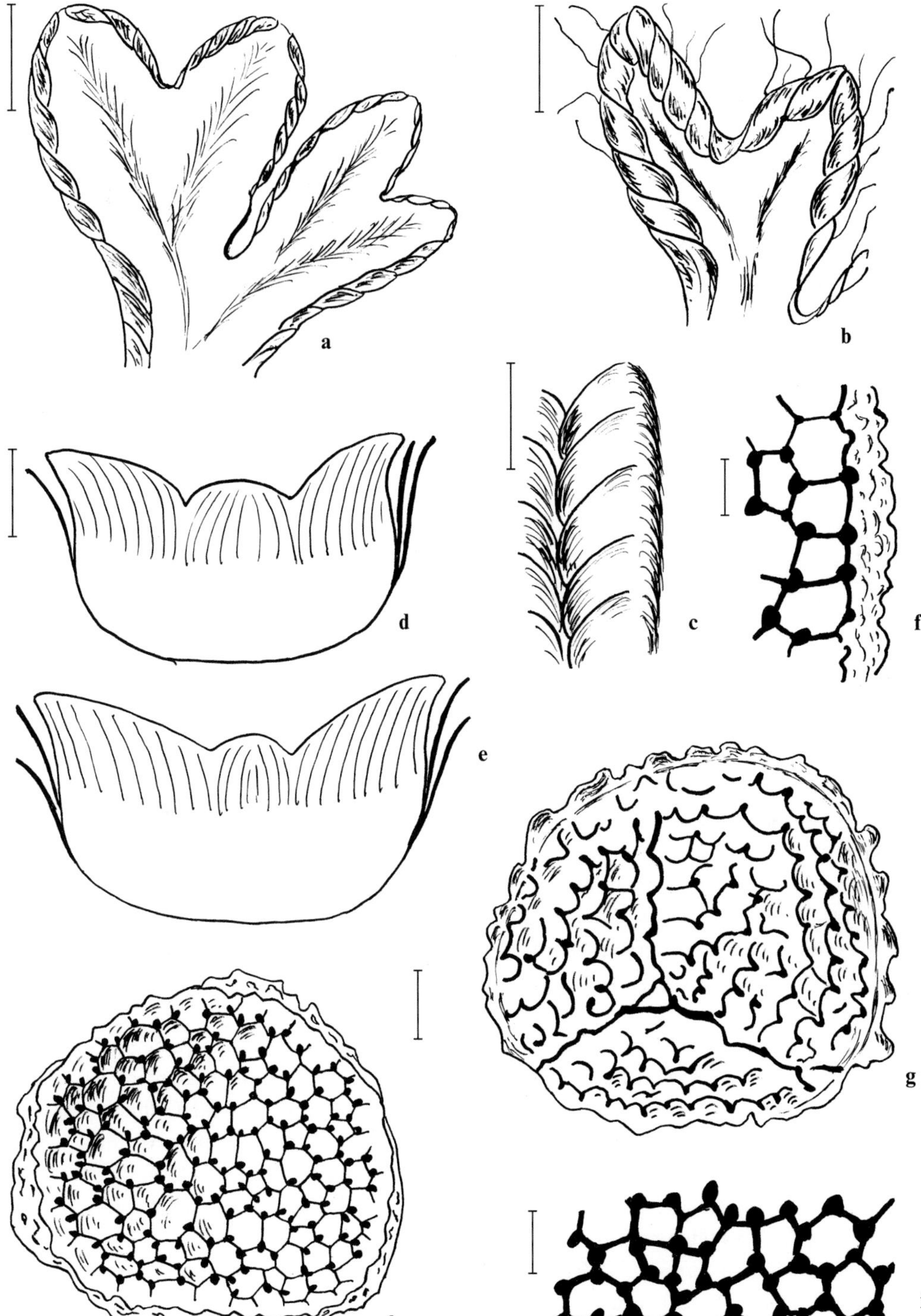

Figure 234. *Riccia lamellosa*: **a** thallus, moist; **b** thallus, dry; **c** portion of thallus, dry; **d**, **e** cross-sections of lobe; **f** portion of spore wing; **g** spore, proximal face; **h** spore, distal face; **i** spore, distal face, alveolae and tubercles. Scale bars: **a**–**c** = 2 mm; **d**, **e** = 0.5 mm; **f**, **i** = 10 µm; **g**, **h** = 25 µm. (drawn from various sources)

3. Riccia sorocarpa Bisch., Nova Acta Phys.-Med. Acad. Caes. Leop.-Carol. Nat. Cur. 17: 1053 (1835). [Jovet-Ast & Bischler (1966): 102; Bischler & Jovet-Ast (1975): 19]

Distribution map 236, Figure 235, Plate XVI: m.

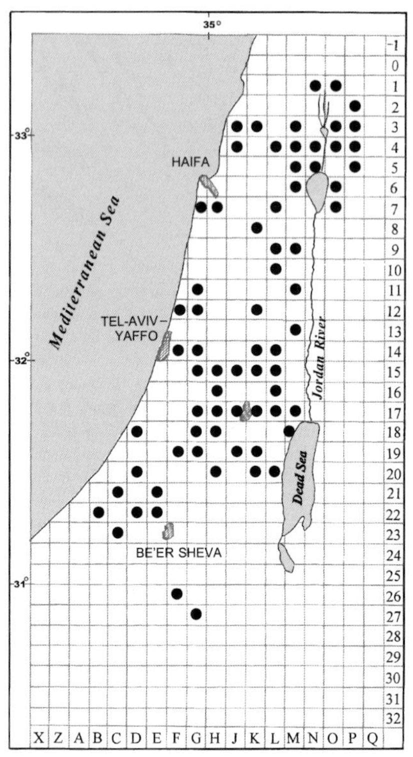

Map 236: *Riccia sorocarpa*

Thallus pale green, sometimes bluish-green, smooth and nearly velvety on dorsal side with narrow, hyaline margin, rarely with lateral sides and margins tinged with purple, two–three times branched, forming rosettes of 2 cm in diameter; lobes 3–10 mm long, 0.5–2.5 mm wide, rounded apically; median groove deep at apex, becoming shallower towards base; cross section of lobes ± wider than high in young parts, two–three times wider than high in older parts, forming there a very open V; free row of cells ending in long or short cells corresponding to the hyaline lobe margin along lateral edges; epidermal cells convex or pyriform, decaying early; lateral walls of broken epidermal cells and of upper subepidermal cells strongly thickened; scales pale, reaching usually the lobe margins but not passing beyond it, upper scale reaching more or less the level of subepidermis. *Monoicous*. *Spores* 70–96 μm in diameter, triangular to spherical, wing 4–5 μm wide, ± irregular, granulose when wet, bordered with minute papillae; distal face with 8–10(–14) alveolae across diameter, limited by rather thin walls with tubercles at wall corners; proximal face covered with granules, or vermiculate, or with short, sinuose crests, with narrow triradiate mark (seen with SEM). *Chromosome number* $n = 8$.

Habitat and Local Distribution. — Plants growing on flat or steep, sandy soil, clay, or silt (pH (5.5–)7–8), at base or in cavities of rocks, exposed or shaded by shrubs, or on bare soil in deserts. Altitude: -100–1200 m. Common: Coastal Galilee, Coast of Carmel, Sharon Plain, Philistean Plain, Upper Galilee, Lower Galilee, Mount Carmel, Esdraelon Plain, Samaria, Shefela, Judean Mountains, Judean Desert, Northern Negev, Central Negev, Dan Valley, Hula Plain, Bet She'an Valley, Lower Jordan Valley, and Golan Heights.

General Distribution. — Recorded from Southwest Asia, around the Mediterranean, Europe, Macaronesia, South Africa, northern and eastern Asia, Australia, New Zealand, North America, and southern South America.

Riccia sorocarpa has a smooth thallus surface with a very narrow, hyaline margin, and epidermal and subepidermal cells with thickened walls. These characters distinguish it from all other *Riccia* species of the local flora.

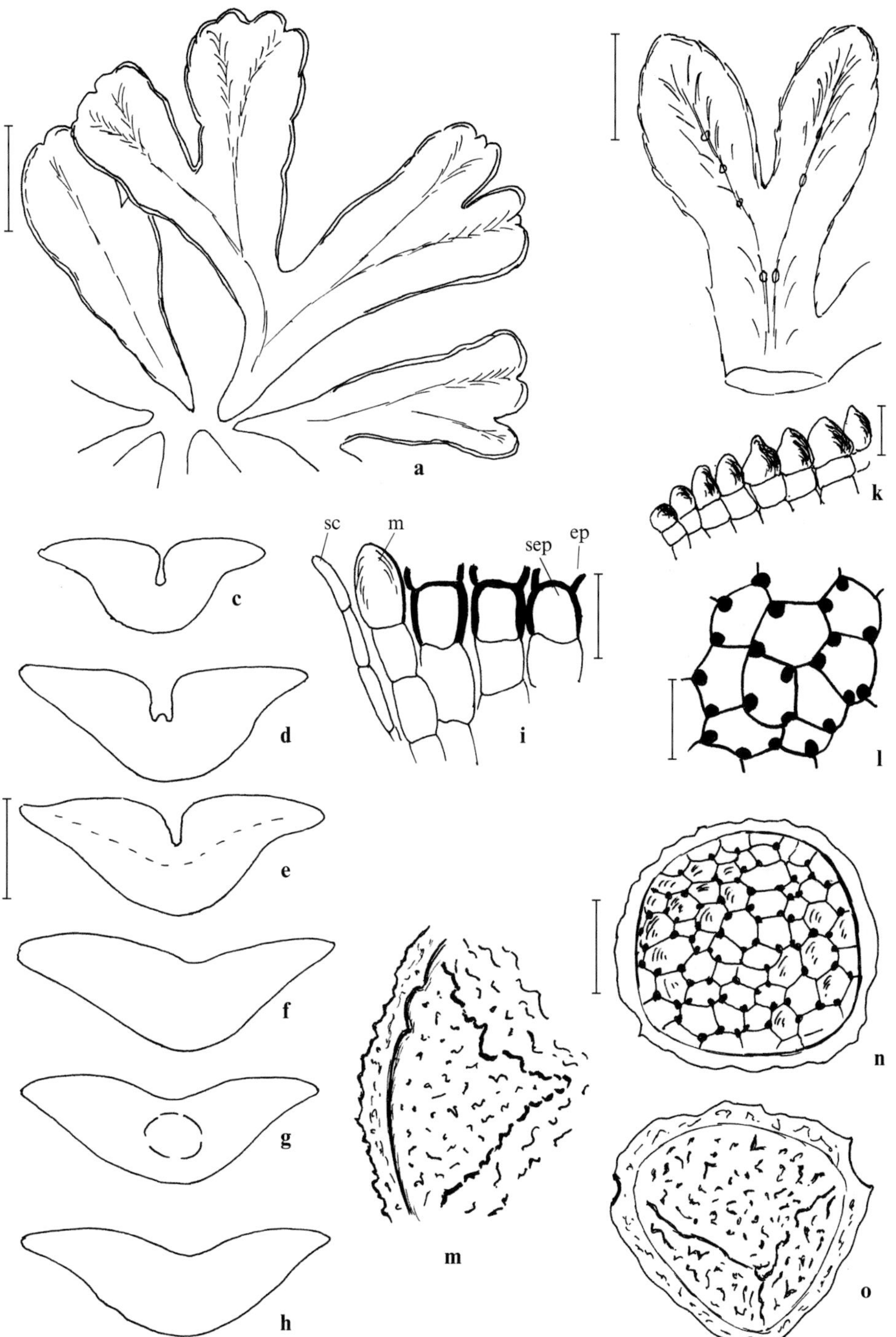

Figure 235. *Riccia sorocarpa*: **a**, **b** thalli; **c**–**h** cross-sections of lobe, from top to base; **i** lobe, cross-section of upper part (ep = dorsal epidermis, sep = subepidermis; m = cell of hyaline margin, sc = scale); **k** epidermis in median groove; **l** spore, central alveolae of distal face; **m** spore, ornamentation of portion of wing and proximal face; **n** spore, distal face; **o** spore, proximal face. Scale bars: **a**, **b** = 2 mm; **c**–**h** = 800 μm; **i** = 50 μm; **k** = 100 μm; **l** = 15 μm; **m**–**o** = 30 μm. (drawn from various sources)

4. Riccia macrocarpa Levier, Bull. Soc. Bot. Ital. 1894 : 114 (1894).
[Jovet-Ast & Bischler (1966) : 101]

Distribution map 237, Figure 236.

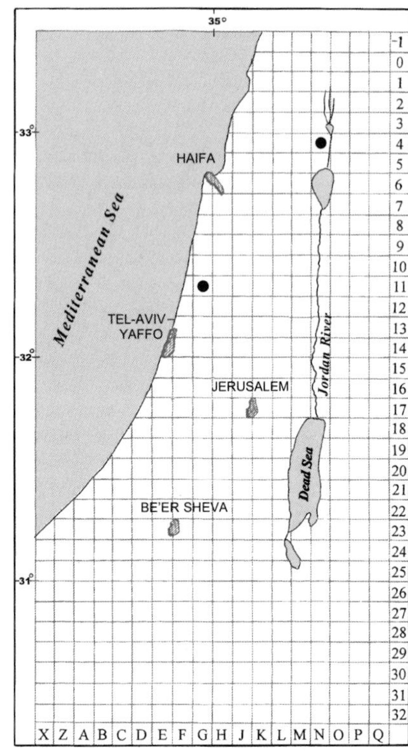

Map 237: *Riccia macrocarpa*

Thallus grey-green, yellow-green in median groove, red-brown in older parts, orange or orange-brown on borders, two–three times branched, gregarious and crowded, usually not forming rosettes; lobes narrow, sublinear, gradually narrowed from apex towards base, sometimes plicate at base, with acute margins, 5–8 mm long, 1(–1.6) mm wide; median groove distinct all along the lobes; epidermal cells in median groove spherical or obtuse; lateral sides tinged with orange-brown, folded up when dry; cross section of lobes higher than wide apically, two–three times wider than high and forming an open V below, in older parts with upper edge nearly flat; lateral edges erect with upper part extending into a short wing (corresponding to the orange border of the lobes when seen from dorsal side); assimilatory layer and tissue near basal epidermis with scattered cells with orange content; scales hardly reaching thallus margins, with groups of colourless, isodiametric cells intermixed with groups of cells with purplish and bright orange content of the same size and shape. *Dioicous* (sometimes monoicous?). *Spores* more or less triangular, 95–108(–115) µm in diameter, red-brown, wing 5 µm wide, finely crenulate at margin, ± granulose; distal face with 8–12 alveolae limited by very thin and sometimes evanescent walls with big tubercles, often coalescent and forming then thick, sinuose ridges; proximal face similarly ornamented, with thick triradiate mark. *Chromosome number* $n = 8$.

Habitat and Local Distribution. — Plants growing on moist, sandy soil (pH 4.2–6.5) at base of basaltic rocks, sheltered by shrubs, and in moist meadows under *Quercus ithaburensis*. Altitude: 50–450 m. Rare: Sharon Plain and Hula Plain.

General Distribution. — Recorded from Southwest Asia, around the Mediterranean, Europe, Macaronesia, southern Africa, and North and South America.

Riccia macrocarpa is easily recognised by its long and narrow lobes with red- or orange-brown borders, and the purplish and bright orange spots in its scales. The lobes are provided with short lateral wings and consequently, this species is often placed near *R. gougetiana*. However, the two species cannot be mistaken as *R. macrocarpa* has much shorter thallus wings, a strikingly different thallus morphology, and different spores.

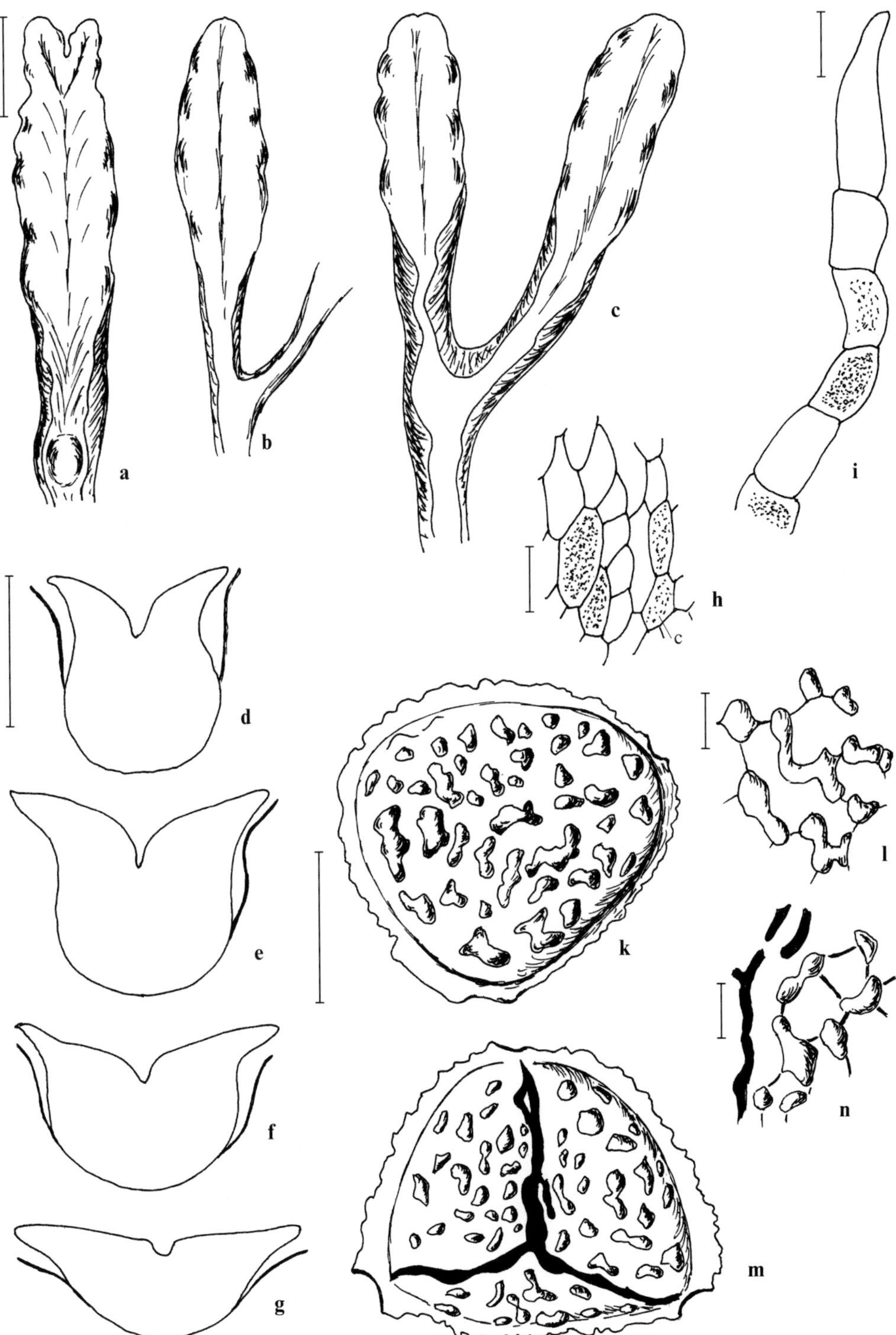

Figure 236. *Riccia macrocarpa*: **a**–**c** thalli; **d**–**g** cross-sections of lobe, from top to base; **h** cells of scale, some tinged with orange or purple (c); **i** scale, section; **k** spore, distal face; **l** spore, ornamentation of distal face; **m** spore, proximal face; **n** spore, ornamentation of proximal face. Scale bars: **a**–**g** = 1 mm; **h** = 100 μm; **i**, **k**, **m** = 50 μm; **l**, **n** = 10 μm. (drawn from various sources)

5. Riccia subbifurca Warnst. ex Croz., Rev. Bryol. 30 : 62 (1903).
[Jovet-Ast & Bischler (1966) : 102]

Distribution map 238, Figure 237.

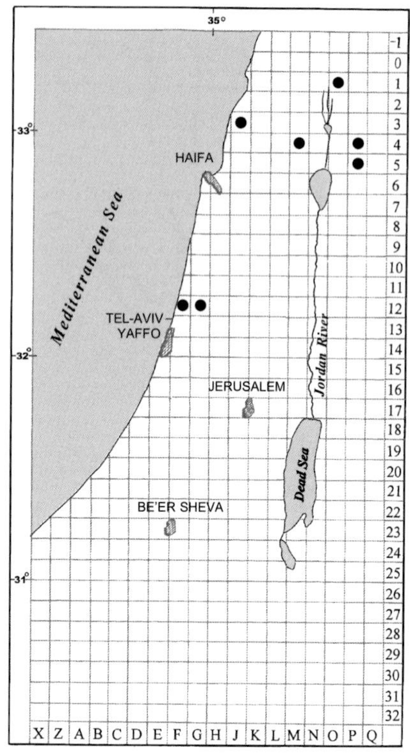

Map 238: *Riccia subbifurca*

Thallus blue-green, more or less purplish near apex, tinged with purple on lateral sides, three–four times branched, forming incomplete rosettes, or gregarious and crowded, usually with a few marginal short (70–120 μm long) or longer (up to 350 μm long) cilia, with tubercles in the upper ½–⅔, or smooth, seldom glabrous; lobes 5–7 mm long, 0.5–1 mm wide, rounded apically; median groove shallow apically; cross section of lobes 1–2.5 times wider than high, with two–three convexities on upper edge near apex, in older parts asymmetrical (wide and rounded on one side, narrower and obtuse on the other); epidermal cells in groove rounded or pyriform; assimilatory layer and ventral tissue loose, of thin-walled, rectangular cells; scales hyaline or with pale purplish spots. *Monoicous*; antheridia and archegonia with red-purplish necks. *Spores* (75–)84–96(–115) μm in diameter, dark brown, wing 5–6 μm wide, irregular, light brown with darker margin; distal face with (7–)8–13 alveolae with rather thin walls and big tubercles at wall corners; proximal face similarly ornamented, often with nine alveolae across largest part of each triangular facet, with distinct triradiate mark.

Habitat and Local Distribution. — Plants growing on clay or compact, sandy soil (pH 6.5–7.5) between calcareous rock blocks or among grasses in meadows, sheltered. Altitude: 0–700 m. Occasional: Coastal Galilee, Sharon Plain, Upper Galilee, and Golan Heights.

General Distribution. — Recorded from Southwest Asia, around the Mediterranean, Europe, Macaronesia, and central Asia.

Riccia subbifurca and *R. crozalsii* share the habit of the thallus and colour, but in *R. subbifurca* the lobes are narrower and the cilia are less numerous and shorter.

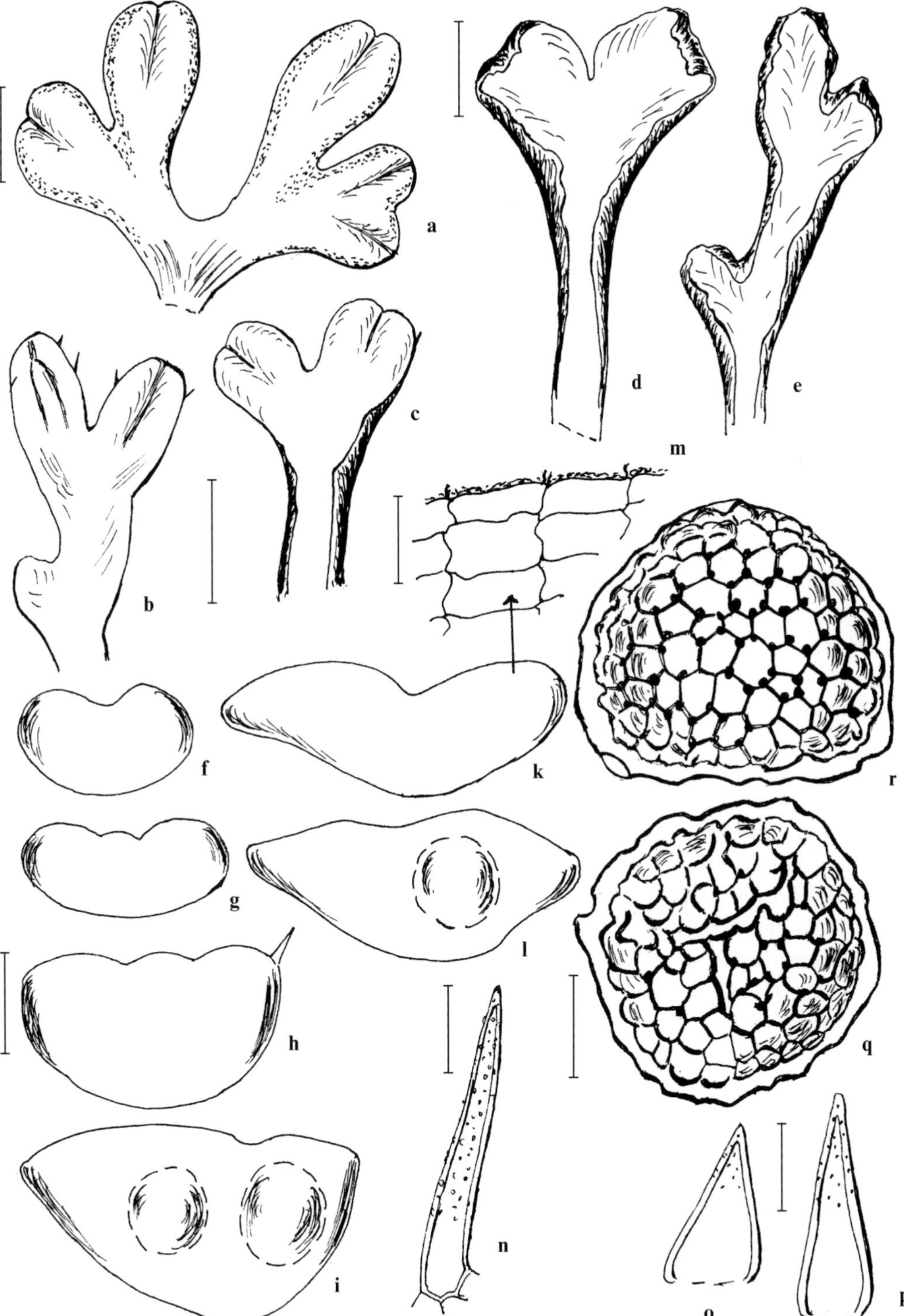

Figure 237. *Riccia subbifurca*: **a–c** thalli, moist; **d**, **e** thalli, dry; **f–l** cross-sections of lobe, from top to base; **m** assimilatory layer; **n–p** marginal cilia of lobe; **q** spore, proximal face; **r** spore, distal face. Scale bars: **a–e** = 1 mm; **f–i**, **k**, **l** = 0.3 mm; **m–p** = 80 μm; **q**, **r** = 30 μm. (drawn from various sources)

6. Riccia nigrella DC. *in* DC. & Lam., Fl. Franç., Edition 3, 5 (Vol. 6): 193 (1815). [Jovet-Ast & Bischler (1966): 101; Bischler & Jovet-Ast (1975): 19]

Distribution map 239, Figure 238.

Thallus dark blue-green on dorsal side, orange-brown or red-brown on borders and basally, two–four times branched, forming rosettes, or gregarious and intermingled, rolled up when dry and showing only the black lateral sides, appearing then as black threads; lobes obtuse or, seldom, rounded apically, 1–2.5 mm long, 1–1.2 mm wide; median groove deep, narrow; lateral lobe sides covered with scales; cross section of lobes wider than high (1–1.2×0.8 mm); lateral edges nearly straight and acute apically; epidermal cells in front view polygonal, with small, triangular pores in the angles, persistent, convex, without chloroplasts, 1.5 times higher than the subepidermal cells; scales large, not passing beyond thallus margins, purplish-black, shiny, imbricate, of lunate shape, cells isodiametric, purplish-black except marginal cells with pink walls. *Monoicous*; dorsal tissue covering the sporophytes usually red-brown; archegonia with long, black necks. *Spores* 60–80 μm in diameter, light brown, wing 5–7 μm wide, faintly crenulate, with a few dark, vermicular ridges; distal face with 8–9(–10) alveolae across diameter, incompletely limited by irregular, thick, sinuose and confluent walls; proximal face often showing thick crests bounding partially or completely the alveolae, with more or less thickened triradiate mark. *Chromosome number* $n=8$.

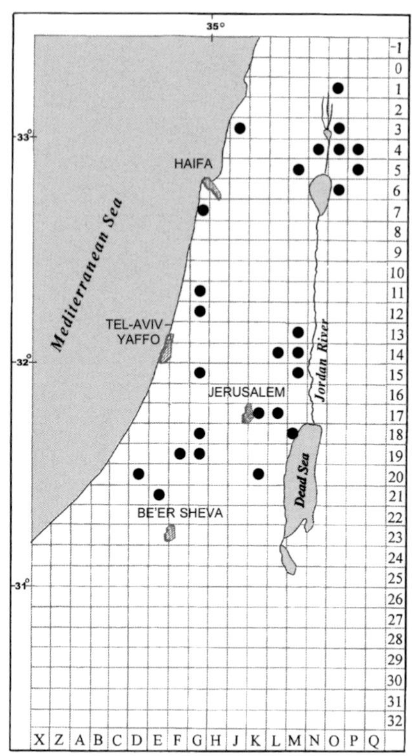

Map 239: *Riccia nigrella*

Habitat and Local Distribution. — Plants growing on compact sand or clay (pH (5–)7–8) at base of calcareous or basaltic rocks, among rock blocks, between grasses in meadows, in maquis of *Olea europaea* or under stands of *Quercus ithaburensis*, associated with other *Riccia* species, especially *R. sorocarpa* and *R. lamellosa*. Altitude: -200–750 m. Common: Coastal Galilee, Coast of Carmel, Sharon Plain, Philistean Plain, Lower Galilee, Samaria, Shefela, Judean Mountains, Judean Desert, Northern Negev, Hula Plain, Lower Jordan Valley, and Golan Heights.

General Distribution. — Recorded from Southwest Asia, around the Mediterranean, Europe, Macaronesia, South Africa, Australia, and North and Central America.

Riccia nigrella is dark blue-green tinged with orange-brown or red-brown on borders and older parts of thalli when moist, and forms nearly black and shiny threads when dry. No other *Riccia* from the Mediterranean Region shows those characters.

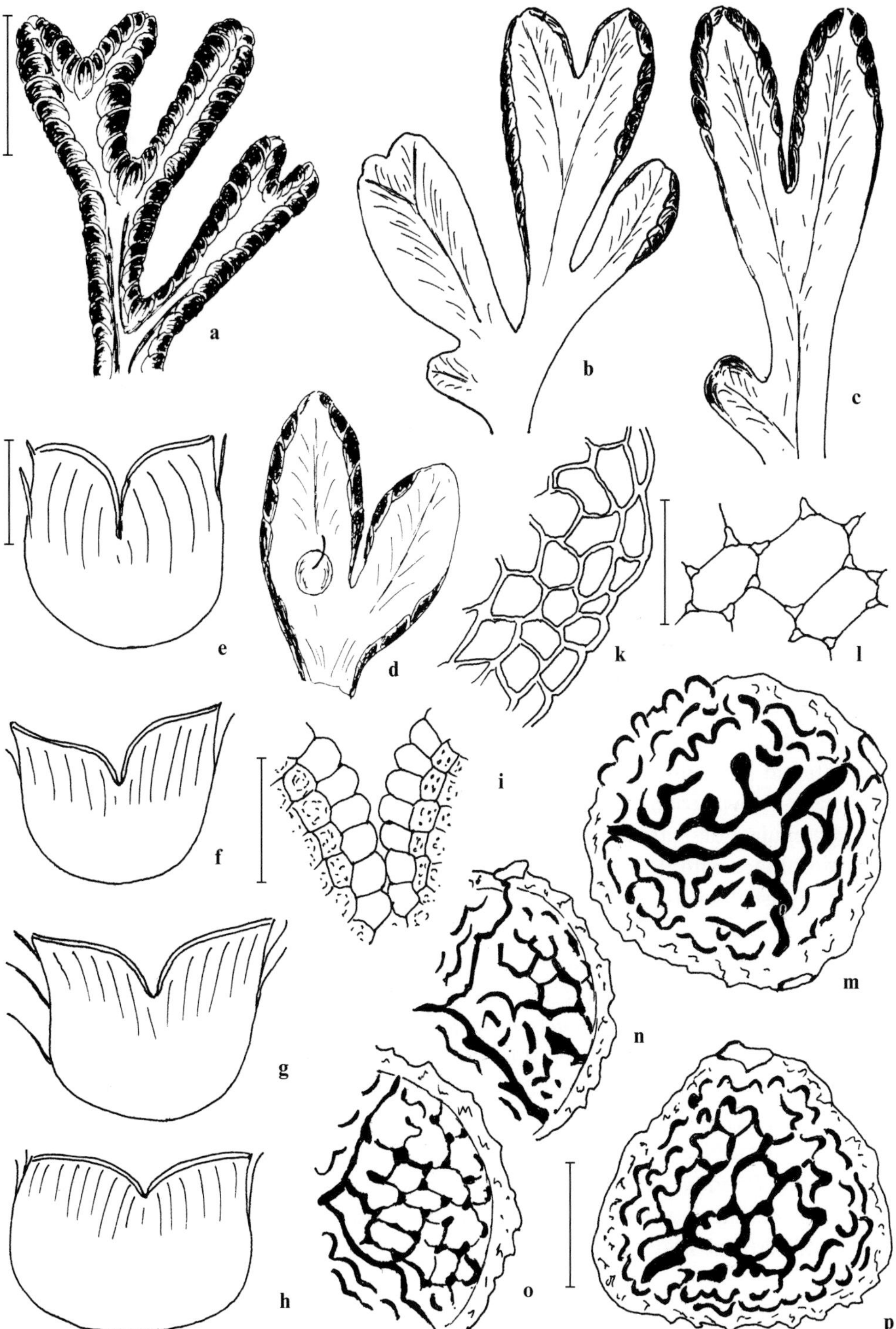

Figure 238. *Riccia nigrella*: **a** dry thallus; **b–d** moist thalli; **e–h** cross-sections of lobe, from top to base; **i** dorsal epidermis in groove; **k** cells of dorsal epidermis, front view; **l** scale cells; **m–o** spores, proximal face; **p** spore, distal face. Scale bars: **a–d** = 2 mm; **e–h** = 0.5 mm; **i**, **k**, **l** = 100 μm; **m–p** = 30 μm. (drawn from various sources)

7. Riccia trabutiana Steph., Rev. Bryol. 16:65 (1889).
Riccia atromarginata Levier var. *glabra* Levier ex Müll. Frib., Rabenhorst's Krypt.-Fl., Edition 2, 6:203 (1907). — Jovet-Ast & Bischler (1966):102, Bischler & Jovet-Ast (1975):19.

Distribution map 240, Figure 239.

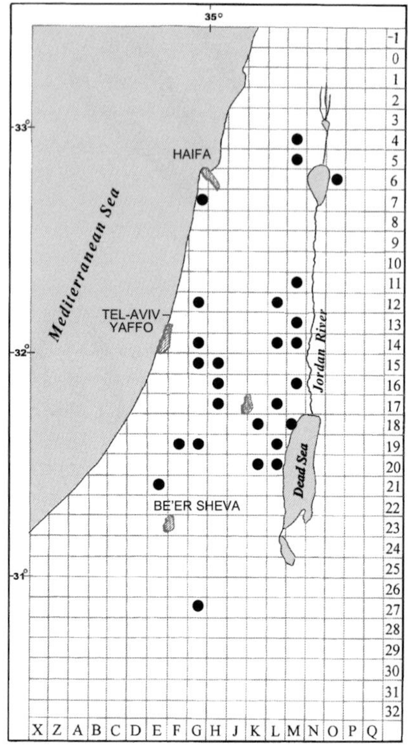

Map 240: *Riccia trabutiana*

Thallus bright blue-green on dorsal side, sometimes tinged with purple, purplish-black on acute margins, red-brown at base, two–three times branched, rarely forming rosettes, usually gregarious and crowded; lobes up to 6 mm long, 0.7–1.2(–1.6) mm wide, rounded-obtuse apically, upfolded when dry and showing the black-purplish lateral sides that are whitish at base; median groove long and narrow or with convex sides; cross section of lobes as wide as high apically, 0.7–1.6 mm wide, and 0.5–1 mm high towards base; lateral edges erect, surpassing the epidermal layer by a purplish tip (corresponding to the purplish, acute lobe margin when seen from above); epidermal cells in groove convex or pyriform, sometimes slightly mucronate, never papilliform; scales purplish, reaching thallus margin or shorter. *Monoicous. Spores* (67–)72–85(–90) μm in diameter, red-brown, becoming black-brown, triangular-rounded, without wing, crenulate on margin by projecting tubercles; distal face with 9–12(–15) alveolae across diameter, usually with thin walls and strongly projecting tubercles; proximal face similarly ornamented, with rather faint triradiate mark. *Chromosome number* $n=8$.

Habitat and Local Distribution. — Plants growing on clay, compact sand, sandy soil, or red soils (pH (6.3–)7–8), at base and in cavities of calcareous rocks or rock blocks, in batha under *Sarcopoterium spinosum*, in southern steppes, exposed or sheltered by shrubs; associated with other *Riccia* species, especially *R. atromarginata*, *R. lamellosa*, and *R. sorocarpa*. Altitude: -400–400 m. Common: Coast of Carmel, Sharon Plain, Lower Galilee, Samaria, Shefela, Judean Mountains, Judean Desert, Northern Negev, Central Negev, Upper Jordan Valley, and Lower Jordan Valley.

General Distribution. — Recorded from Southwest Asia, around the Mediterranean, Europe, and Macaronesia. The range of *Riccia trabutiana* remains uncertain and needs critical study.

Riccia trabutiana differs from *R. atromarginata* by the deeper blue-green colour of thallus, the acute lobe margins, the absence of papillae, and the slightly smaller size of the spores.

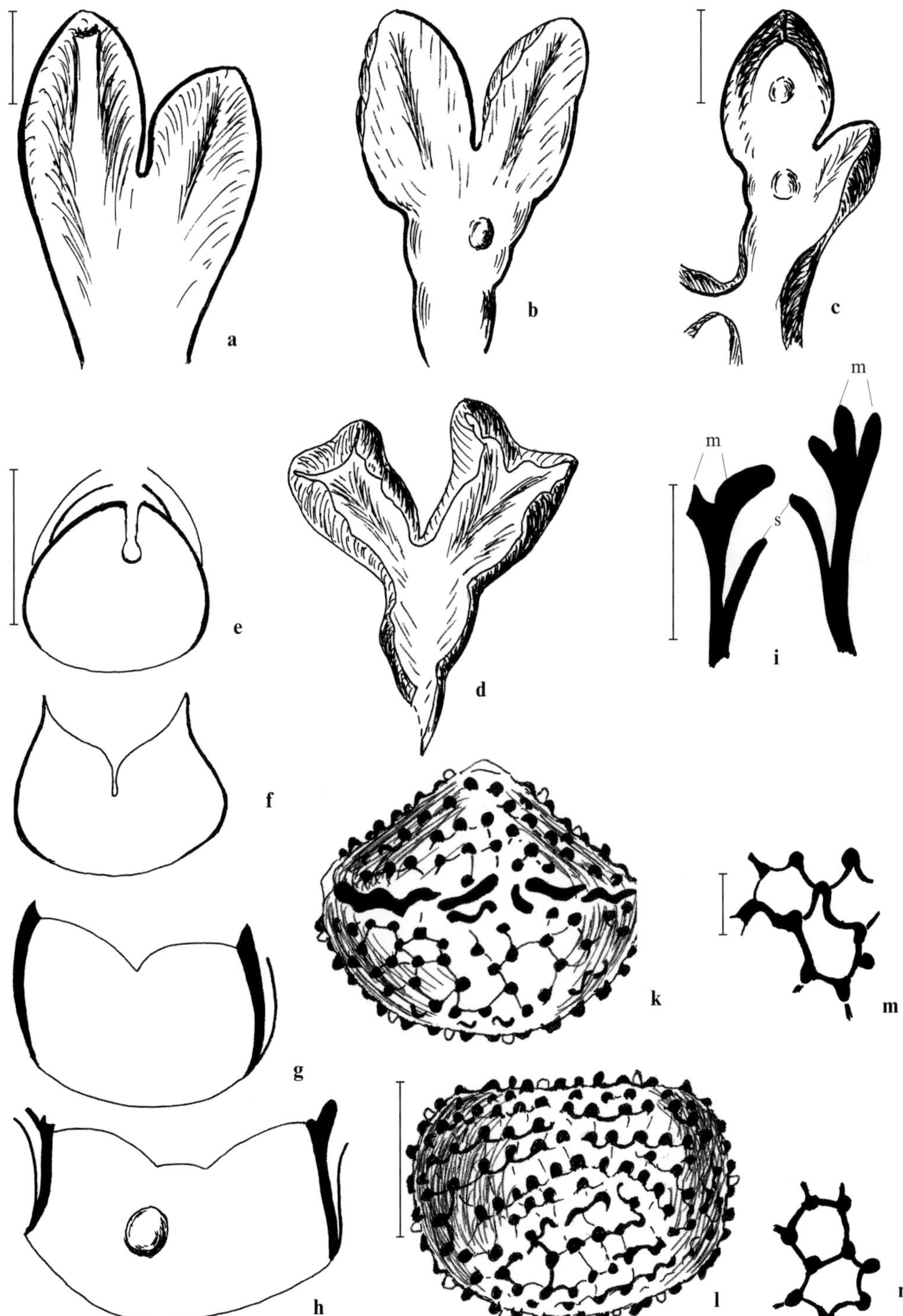

Figure 239. *Riccia trabutiana*: **a**–**c** thalli, moist; **d** thallus, dry; **e**–**h** cross-sections of lobe, from top to base; **i** cross-section of lobe margin (m) and scales (s); **k** spore, side view; **l** spore, distal face; **m**, **n** spores, alveolae of distal face. Scale bars: **a**–**d** = 1 mm; **e**–**i** = 0.5 mm; **k**, **l** = 50 µm; **m**, **n** = 10 µm. (drawn from various sources)

8. Riccia atromarginata Levier *in* Martelli, Nuovo Giorn. Bot. Ital. 21 : 291 (1889). [Jovet-Ast & Bischler (1966) : 98; Bischler & Jovet-Ast (1975) : 18]

Distribution map 241, Figure 240.

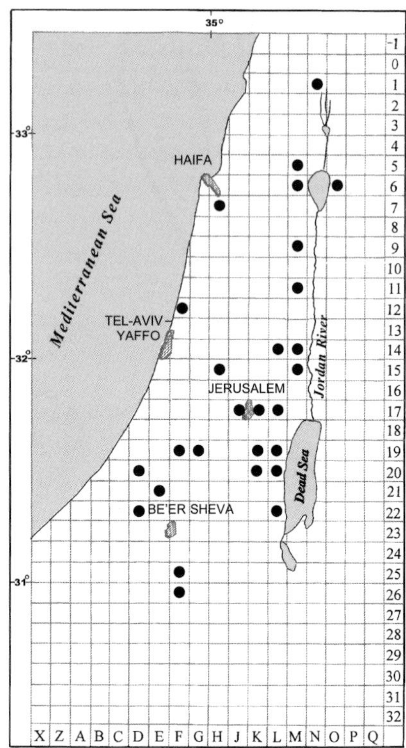

Map 241: *Riccia atromarginata*

Thallus green or bluish-green, dark purplish on the rounded margins, sometimes with purplish borders, two–three times branched, forming incomplete rosettes or crowded mats; lobes 3–6 mm long, 1–1.8 mm wide; dorsal side often slightly convex; cross section of lobes thick, as wide as high near top of lobes, 1.5 times wider than high below; lateral edges black-purplish, rounded apically, upper edge forming a V with convex branches or showing three convex parts; median groove shallow, often narrow; epidermal cells of groove rounded or pyriform; papillae numerous, 60–210 μm long, hyaline, finger-like, rounded or obtuse at top, always present on margins, often also on lateral sides, on top of lobes, or on the entire dorsal side; scales purplish. *Dioicous*; archegonial necks prominent, dark purplish. *Spores* (75–) 80–100(–122) μm in diameter, red-brown or nearly black, without wing, crenulate on margin by projecting tubercles; distal face with 12–18 alveolae across diameter, with thick black walls and big black, sometimes confluent tubercles; proximal face similarly ornamented, with weak triradiate mark. *Chromosome number* $n = (8)\ 16$.

Habitat and Local Distribution. — Plants growing on sand, clay, and red soil, at base of calcareous rocks (pH 6.7–8), on mud of dry wadis, pastures and steppes, exposed or sheltered by shrubs, often together with *Riccia trabutiana* and other *Riccia* species. Altitude: -200–700 m. Common: Sharon Plain, Lower Galilee, Mount Carmel, Samaria, Shefela, Judean Mountains, Judean Desert, Northern Negev, Western Negev, Central Negev, Dan Valley, Bet She'an Valley, Lower Jordan Valley, and Golan Heights.

General Distribution. — A Mediterranean species, recorded also from Southwest Asia, Europe, Macaronesia, southern North America, and Central America. *Riccia atromarginata*'s distribution needs confirmation because it has often been confused with related species.

Riccia atromarginata can be distinguished from all the other local species of *Riccia* by its finger-like papillae on thallus. However, the number, size, and distribution of papillae vary, and in other areas they may be longer. The species is closely allied to *R. trabutiana* from which it differs in colour of the thallus, the presence of papillae, and the rounded margins of lobes.

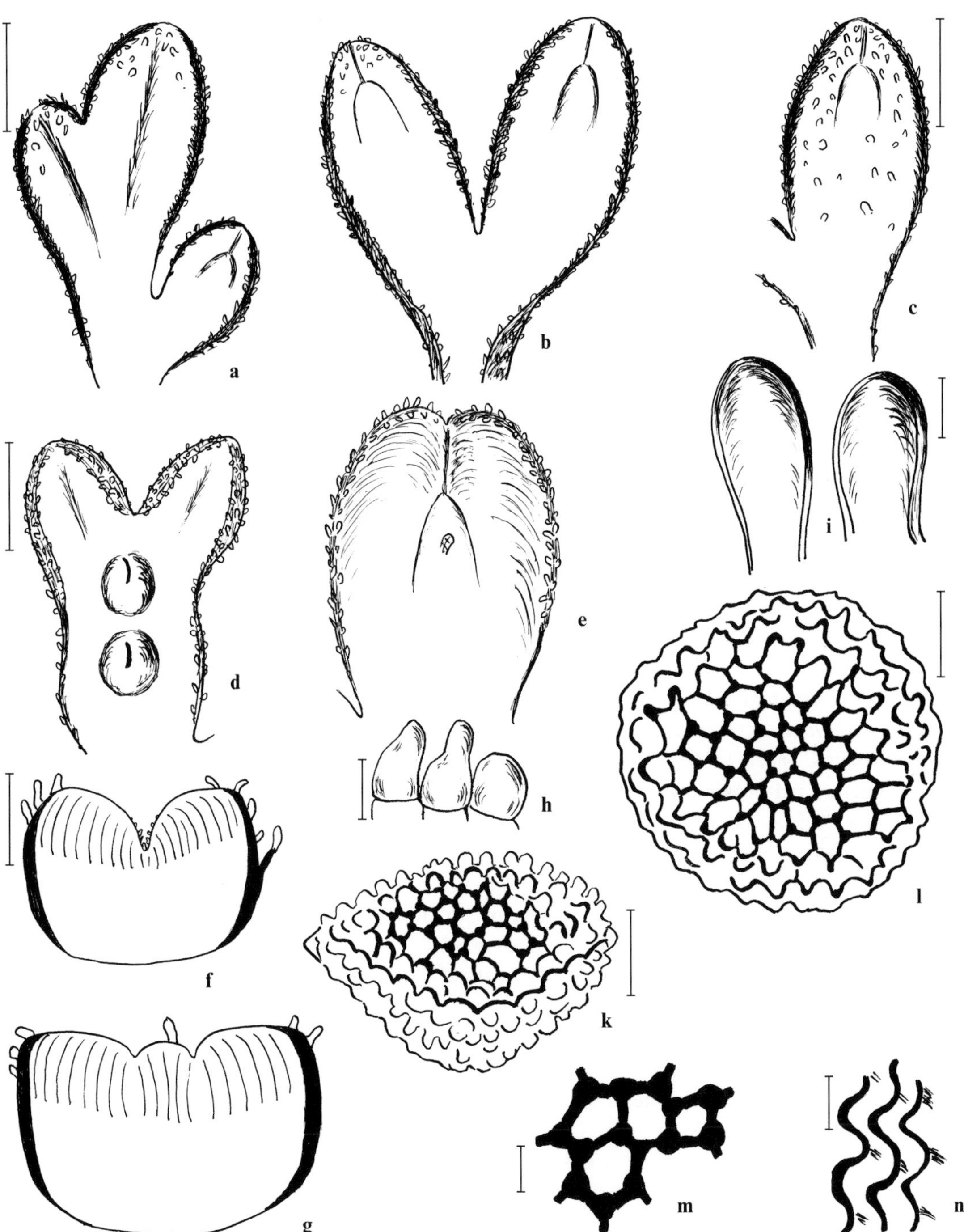

Figure 240. *Riccia atromarginata*: **a–c** thalli, more or less papillose; **d** thallus with two sporophytes; **e** lobe with antheridium; **f** cross-section of lobe near top; **g** cross-section of lobe near base; **h** epidermal cells in groove; **i** two epidermal papillae; **k** spore, side view; **l** spore, distal face; **m** walls and tubercles of alveolae of distal spore face; **n** walls of alveolae, side view. Scale bars: **a–e** = 1 mm; **f**, **g** = 0.3 mm; **h**, **i** = 30 μm: **k**, **l** = 25 μm; **m**, **n** = 10 μm. (drawn from various sources)

9. Riccia gougetiana Durieu & Mont. *in* Mont., Ann. Sci. Nat. Bot., sér. 3, 11 : 35 (1849). [Jovet-Ast & Bischler (1966) : 100]

Distribution map 242, Figure 241, Plate XVI : 1.

Thallus light green or glaucous, brown-yellowish on borders and in older parts, rarely slightly tinged with pink-purple near bottom and on ventral side, one–two times branched, gregarious, rarely forming rosettes; lobes 7–30 mm long, 2.5–6 mm wide, thick in median part, expanded into two thin lateral wings, rounded and sometimes strongly thickened apically; median groove narrow at top, broad at base; margins either naked or, more often, with white, ± numerous, short or long, acute or obtuse cilia, single or connate per pairs; cross section of lobes wider than high, very thick in the median part and abruptly passing into the thin, acute wings; epidermal cells spherical or pyriform, often long in median groove; scales large, scarcely reaching thallus margins, white or occasionally spotted with pink-purple, with rectangular cells of 25–70×95–160(–230) µm. *Dioicous*; female thallus larger than male thallus. *Spores* (130–)170–180(–215) µm in diameter, red-brown, wing 9–10(–15) µm wide; distal face with (8–)10–16(–20) obscurely defined alveolae limited by incomplete walls, which are more or less reduced to form thickened islets, or tubercles, or granules disappearing near wing; proximal face either with incomplete alveolae or with big or small tubercles, with distinct triradiate mark (seen with SEM). *Chromosome number* $n=8$.

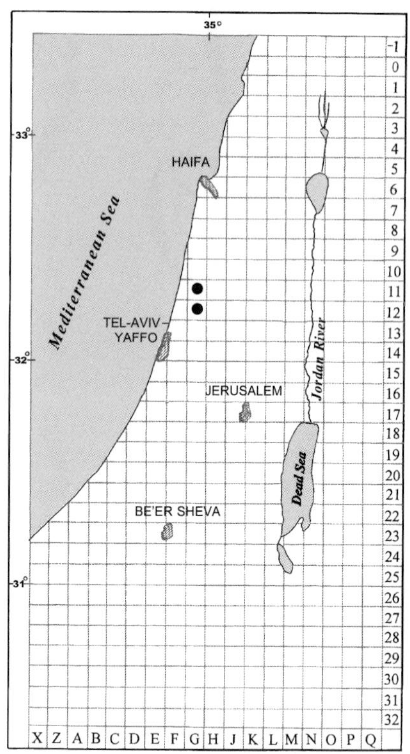

Map 242: *Riccia gougetiana*

Habitat and Local Distribution. — Plants growing on compact, moist, sandy clay (pH (5.3–)7) in a meadow with *Quercus ithaburensis*; associated with other *Riccia* species. Rare: Sharon Plain, found twice at 20–100 m.

General Distribution. — Recorded from Southwest Asia, around the Mediterranean, Europe, Macaronesia, and central Asia.

Riccia gougetiana has the widest thallus among the Mediterranean species of *Riccia*, and well developed lateral wings. It is closely allied to *R. ciliifera* Link, which has not been recorded yet in the local flora. The thallus of *R. ciliifera* is bluish, purple in older parts, and the spores have a different ornamentation pattern. *Riccia gougetiana* var. *armatissima* Levier ex Müll. Frib. should also be searched for in the local flora. In this variety, the dorsal side of the lobes is abundantly covered with long, white and shiny cilia.

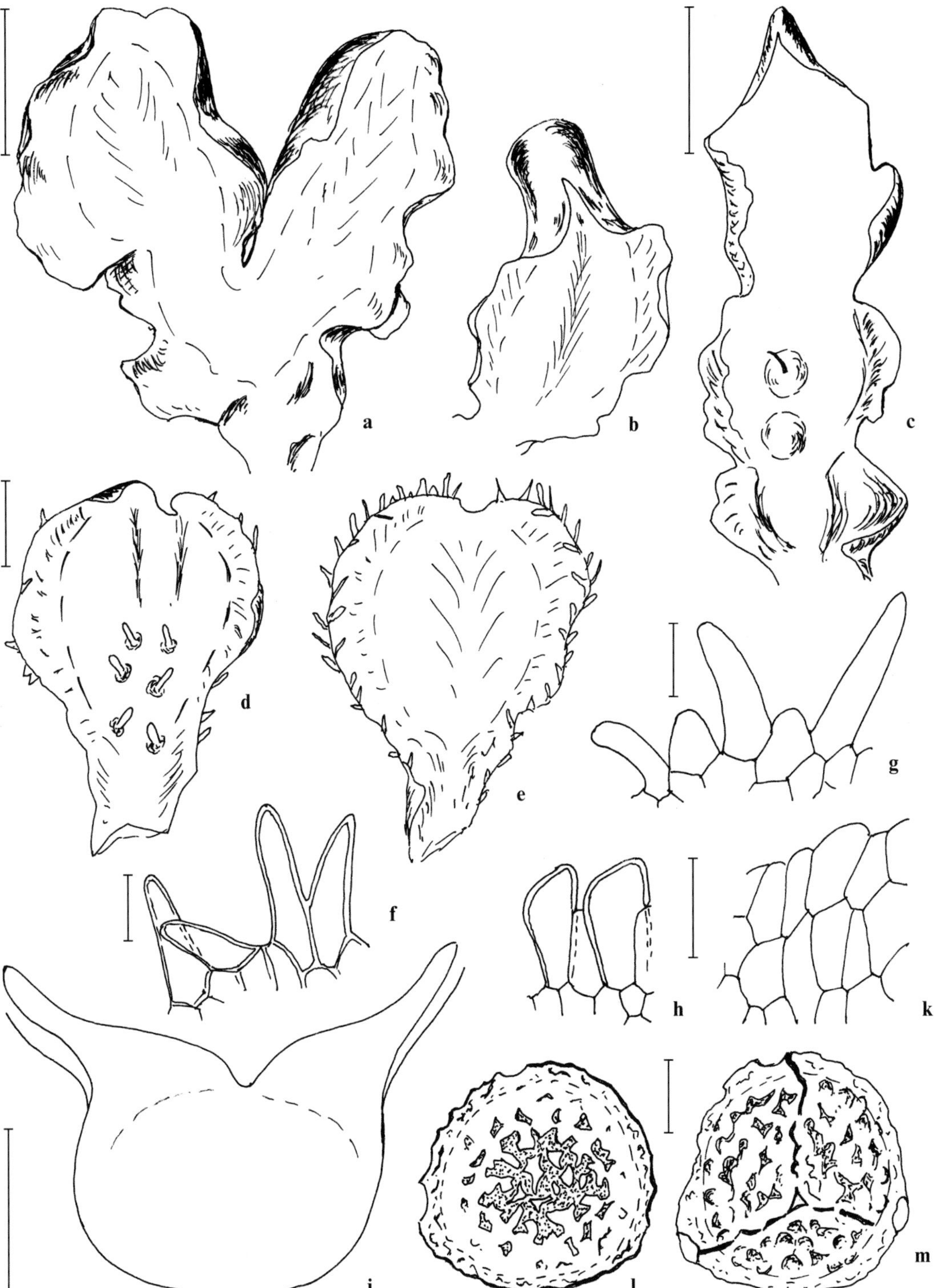

Figure 241. *Riccia gougetiana*: **a–e** different shapes of thalli, with and without cilia; **c** with capsules; **d** with antheridia; **f–h** marginal cilia of lobes; **i** cross-section of lobe; **k** cells of scale; **l** spore, distal face; **m** spore, proximal face. Scale bars: **a–c** = 4 mm; **d**, **e**, **i** = 1 mm; **f–h**, **k** = 200 μm; **l**, **m** = 50 μm. (drawn from various sources)

10. Riccia bicarinata Lindb., Rev. Bryol. 4:41 (1877).
Riccia henriquesii Levier, Bull. Soc. Bot. Ital. 1894:199 (1894). — Jovet-Ast & Bischler (1966):98.

Distribution map 243, Figure 242.

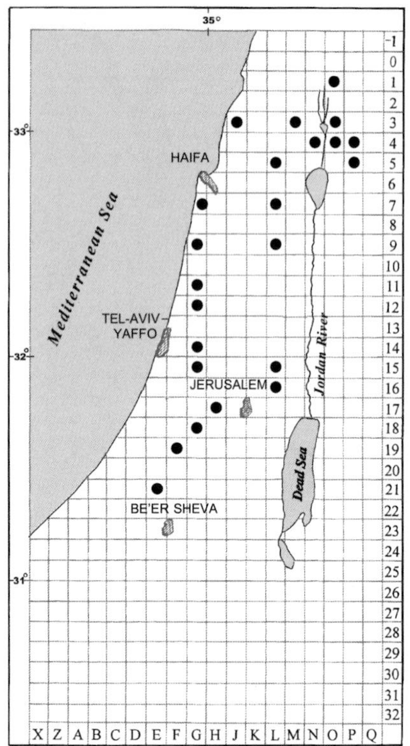

Map 243: *Riccia bicarinata*

Thallus bluish-green or grey-green, grey and tinged with light purple or pink violet in older parts, lateral edges often brown-purplish, two–three times deeply branched, forming more or less complete rosettes; lobes 7–10 mm long, 1–1.6(–1.8) mm wide, obtuse apically and narrowed below, shaped like a spoon; median groove deep, narrow, or flat, and more or less convex; cilia white, on margins of lobes from top to base, either free or connate at base in groups of two–three, either up to 420 μm long, acute, with minute papillae in upper ⅔, or shorter, triangular, 70–300 μm long, smooth or papillose in upper ⅔; cross section of lobes 1.5–2 times wider than high except apically, as high as wide, with a purplish line along the lateral edges; upper edge forming a V, or with median groove wider, or flat; epidermal cells of groove rounded or ovate; scales often purplish, sometimes hyaline, not reaching lobe margin. *Monoicous*. *Spores* 80–150 μm in diameter, dark brown, wing 7 μm wide with thickened, sinuose margin; distal face with 8–9(–10) alveolae across diameter, with conspicuous and sometimes bifid tubercles at wall corners; proximal face similarly ornamented, with thick triradiate mark.

Habitat and Local Distribution. — Plants growing on compact moist sand, clay, and red soils (pH (5.3–)7–8), at base of calcareous or basaltic rocks, on mud of drying wadis, in meadows between grasses, and in maquis of *Olea europaea* and *Quercus calliprinos* Webb. Altitude: 0–750 m. Common: Coastal Galilee, Coast of Carmel, Sharon Plain, Upper Galilee, Lower Galilee, Esdraelon Plain, Shefela, Judean Mountains, Judean Desert, Northern Negev, Hula Plain, and Golan Heights.

General Distribution. — A Mediterranean species, recorded also from Southwest Asia, Europe, and Macaronesia.

Riccia bicarinata is recognised by its colour, by its spoon-like thallus lobes, and by its marginal cilia, often connate basally.

Figure 242. *Riccia bicarinata*: **a**, **b** thalli; **c**–**g** cross-sections of lobe from top to base; **h**, **i** cross-section of lobe below top; **k** spore, proximal face; **l** spore distal face; **m**–**o** marginal cilia; **m** simple; **n** bifid; **o** trifid; **p** epidermal cells of groove. Scale bars: **a**, **b** = 1 mm; **c**–**i** = 0.5 mm; **k**, **l**, **n** = 50 µm; **m** = 100 µm; **o** = 70 µm; **p** = 25 µm. (drawn from various sources)

11. Riccia michelii Raddi, Opusc. Sci. (Bologna) 2:352 (1818).
[Bischler & Jovet-Ast (1975): 19]

Distribution map 244, Figure 243.

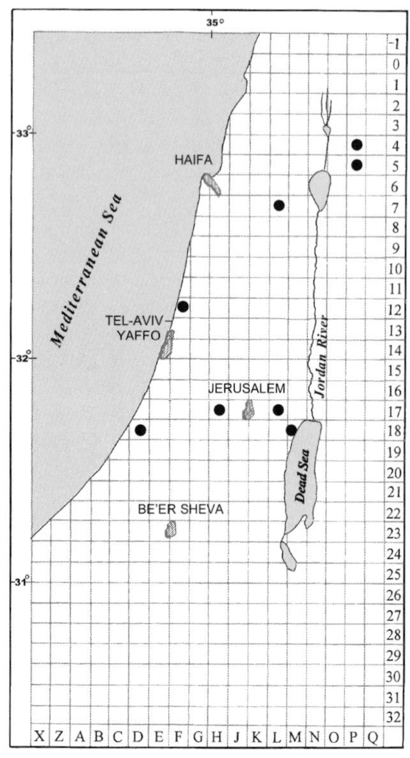

Map 244: *Riccia michelii*

Thallus pale green, rarely tinged with purple on borders and lateral sides, two–three times branched, forming often dense rosettes; lobes 8–15 mm long, 1.4–1.8 mm wide, rarely up to 2–2.5 mm wide, rounded apically, gradually narrowed from apex to base; median groove wide, bordered by two convex ridges; cilia silver-white, on margins of lobes from top to base but more numerous at top, sometimes scarce, usually arranged in two opposite rows, short (250 μm long) and wide at base (then subtriangular), or longer (500–600 μm long), smooth or papillose in upper ⅔; cross section of lobes three–four times wider than high; upper edge with three convexities in the younger part of lobes, nearly flat in older; epidermal cells persistent in groove, convex, ovate, or pyriform; scales small, white. *Dioicous*. *Spores* 90–125(–140) μm in diameter, light brown, obscurely angled or spherical, wing broad, covered with granules, finely papillose on margin; distal face with seven–nine alveolae across diameter, limited by thin walls with tubercles at wall corners, but usually incomplete or nearly absent near wing; proximal face with incomplete alveolae with walls reduced to tubercles or sinuose lines, and with strong triradiate mark. *Chromosome number* $n = 16$.

Habitat and Local Distribution. — Plants growing on moist sand and clay (pH 7–7.5) in meadows, maquis, or north exposed slopes, sheltered by *Sarcopoterium spinosum*. Altitude: -250–550 m. Occasional: Sharon Plain, Philistean Plain, Lower Galilee, Judean Mountains, Judean Desert, and Golan Heights.

General Distribution. — Recorded from Southwest Asia, around the Mediterranean, Europe, and Macaronesia.

Riccia michelii is easily recognised by its pale green thallus, the wide median groove, and the large spores with finely papillose wings. Significant variation in the size of the thallus and in the number of marginal cilia is observed. The margins can be nearly glabrous, with few cilia, or with numerous, long and white cilia in two rows. The cilia may be light golden in older collections.

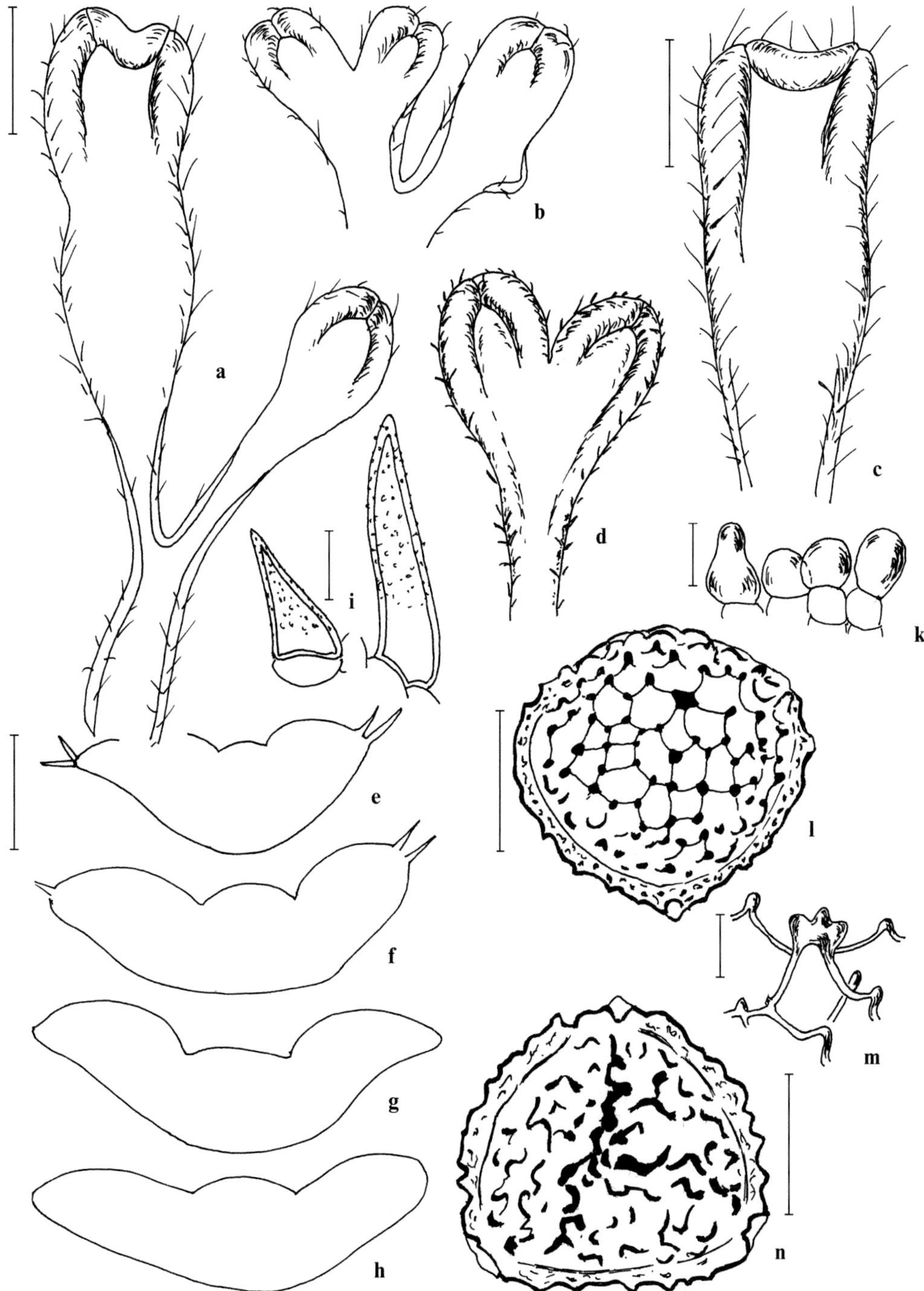

Figure 243. *Riccia michelii*: **a**–**d** thalli; **e**–**h** cross-sections of lobe, from top to base; **i** marginal cilia of lobe; **k** epidermal cells of groove; **l** spore, distal face; **m** spore, ornamentation of distal face; **n** spore, proximal face. Scale bars: **a**–**d** = 2 mm; **e**–**h** = 0.5 mm; **i** = 100 μm; **k**, **m** = 25 μm; **l**, **n** = 50 μm. (drawn from various sources)

12. Riccia crozalsii Levier, Rev. Bryol. 29 : 73 (1902).
[Jovet-Ast & Bischler (1966) : 98]

Distribution map 245, Figure 244.

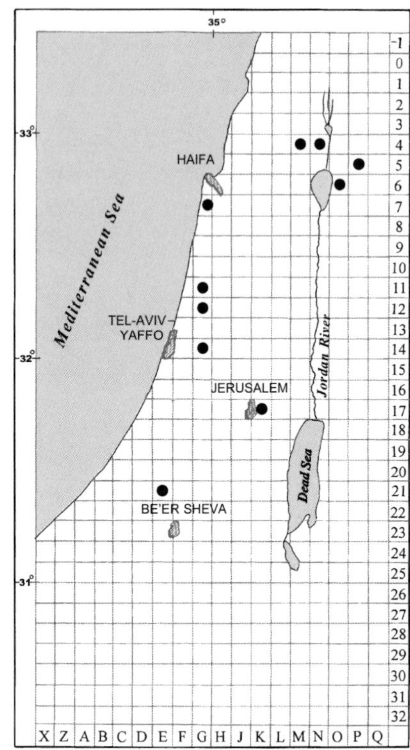

Map 245: *Riccia crozalsii*

Thallus dark green, sometimes blue-green, more or less purplish towards base on dorsal side, purplish or black-red on lateral sides, three times branched, forming rosettes or gregarious; lobes up to 4 mm long, 0.4–1 mm wide, rounded or obtuse apically, sometimes with a very short segment at right angle near base; cross section of lobes thick, as wide as high at top of lobes, 1.5–2 times wider than high below; upper part of lateral edges rounded; epidermal cells large, convex; median groove deep apically, becoming wider and shallower towards base; cilia more or less numerous, in upper part of lobes only, white, inserted on a short cell, 120–300(–500) μm long, covered with fine tubercles in the upper ⅔; scales black-purplish. *Monoicous*; archegonial necks purplish. *Spores* (65–)75–85(–90) μm in diameter, dark brown or black, wing smooth, irregularly toothed, with narrow and thick margin; distal face with 8–10(–12) alveolae across diameter, limited by thin walls with big tubercles; proximal face similarly ornamented, with conspicuous triradiate mark. *Chromosome number* $n=8$.

Habitat and Local Distribution. — Plants growing on compact moist sand, clay, and red soils (pH 6–7), at base of calcareous rocks and between rock blocks, occasionally in maquis. Altitude: -200–750 m. Occasional: Coast of Carmel, Sharon Plain, Upper Galilee, Judean Mountains, Northern Negev, Hula Plain, and Golan Heights.

General Distribution. — A Mediterranean species, recorded also from Southwest Asia, Europe, Macaronesia, tropical and South Africa, India, Australia, and New Zealand.

Riccia crozalsii is characterised by the purplish colour of the ventral thallus side, the frequent presence of a short segment near base of lobes, the thick cross section of lobes, and the aspect and location of cilia.

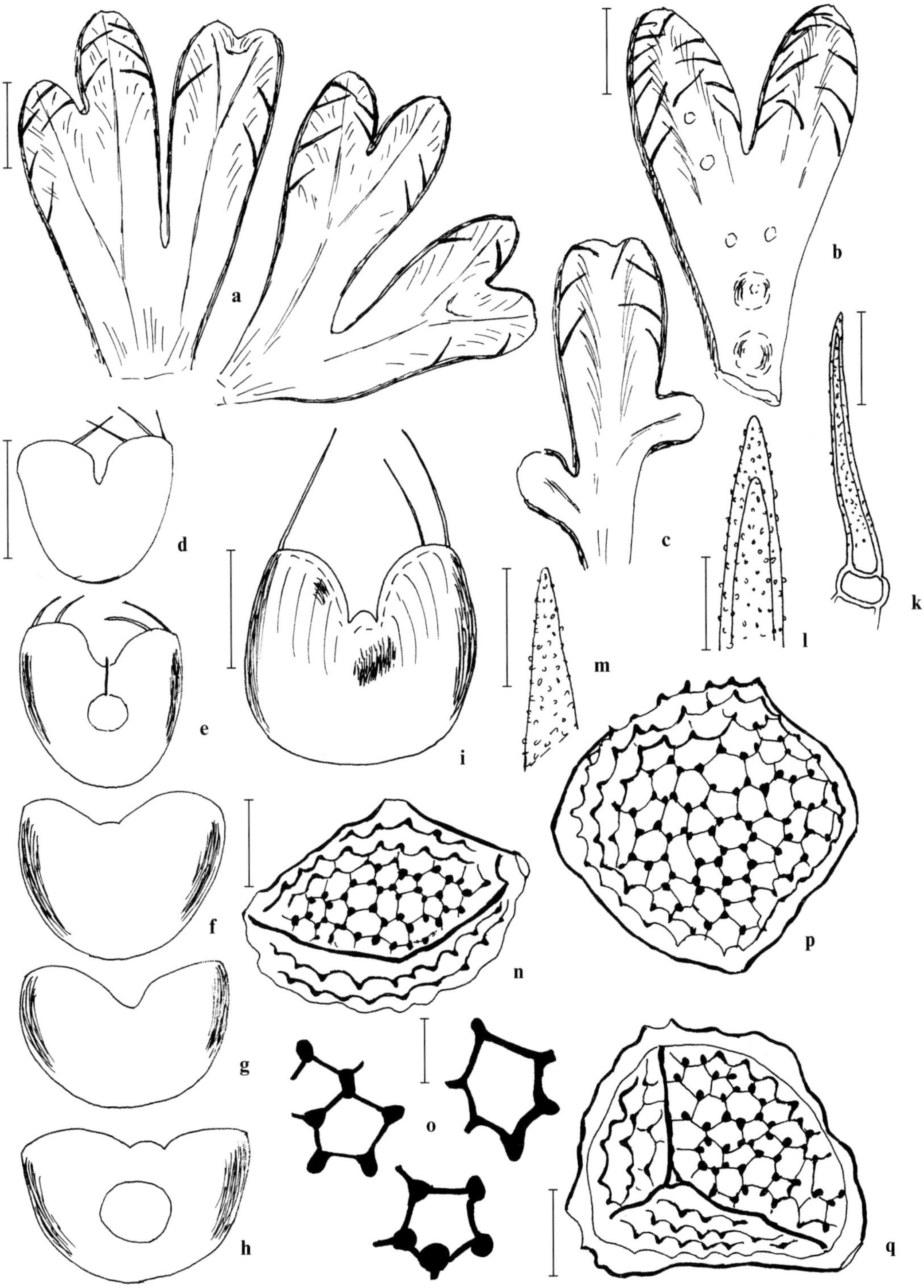

a b c d e f g h i k l m n o p q

Figure 244. *Riccia crozalsii*: **a**–**c** thalli; **d**–**h** cross-sections of lobe, from top to base; **i** cross-section in the ciliate part of lobe; **k** marginal cilium; **l** apex of cilium at lowest range of focus; **m** apex of cilium at highest range of focus; **n** spore, side view; **o** alveolae and tubercles of three different spores; **p** spore, distal face; **q** spore proximal face. Scale bars: **a**–**c** = 1 mm; **d**–**i** = 0.5 mm; **k**–**m** = 50 μm; **n**, **p**, **q** = 30 μm; **o** = 10 μm. (drawn from various sources)

13. Riccia (Ricciella) frostii Austin, Bull. Torrey Bot. Club 6 : 17 (1875).
Riccia palaestina S. W. Arnell, Bull. Res. Council Israel, Sect. D, Bot. 6 : 56 (1957). — Jovet-Ast & Bischler (1966) : 100.

Distribution map 246, Figure 245, Plate XVI : k.

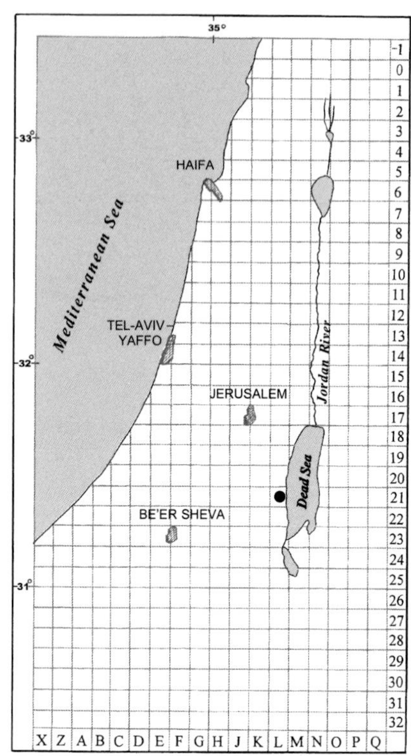

Map 246: *Riccia frostii*

Female thallus blue-green with pink spots on borders, spongy, deeply divided and forming rosettes up to 8 mm in diameter; two–four times larger than male thallus; lobes faintly incised apically, rounded or truncate; perforations small in young parts, wide open in older; median groove very short or absent; dorsal epidermis discontinuous; cross section of lobes usually wider than high, 0.3–2 mm wide, 0.3 mm high, with one–two layers of lacunae in upper ⅔–¾ and compact ventral tissue built up of large cells below; scales absent. *Male thallus* often found near female, smaller, tinged with pink on borders or on the whole of dorsal side, forming a spongy body of 0.3–1.5 mm in diameter, two–four times branched, often cross-shaped. *Dioicous*; antheridia and archegonia scattered; antheridial necks often purplish, archegonial necks white, more or less projecting. *Capsules* numerous, more conspicuous on ventral than on dorsal side of thallus. *Spores* 40–65 μm in diameter, yellow-brown to red-brown, wing 2 μm wide; distal face vermiculate, with branched, low crests radiating sometimes from the distal pole; proximal face less vermiculate and with thin triradiate mark (seen with SEM).

Habitat and Local Distribution. — Plants growing on sandy, intermittently moist, exposed mud. Rare: found only once in Nahal Arugot in the Judean Desert at -300 m.

General Distribution. — Recorded from Southwest Asia, around the Mediterranean, Europe, northern and eastern Asia, dry tropical Africa, North and Central America, and southern South America.

Riccia frostii can be distinguished from *R. crystallina* — the other local species of subgen. *Ricciella* — by the difference in size between male and female thalli and by the ornamentation of the spore coat.

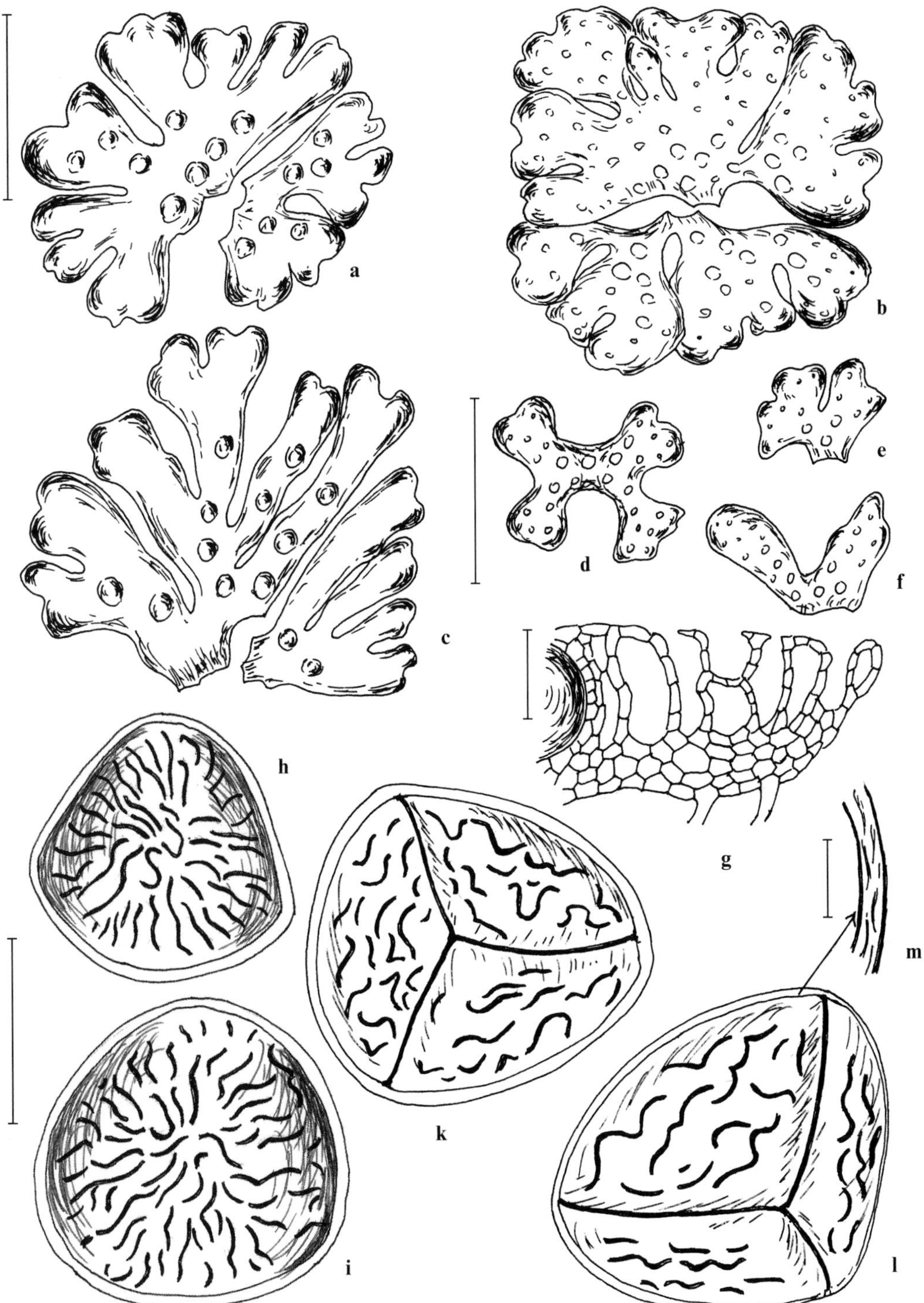

Figure 245. *Riccia frostii*: **a**–**c** female thalli, with or without sporophytes; **d**–**f** male thalli; **g** female thallus, section; **h**, **i** spores, distal face; **k**, **l** spores, proximal face; **m** wing of spore. Scale bars: **a**–**f** = 4 mm; **g** = 0.5 mm; **h**, **i**, **k**, **l** = 20 μm; **m** = 10 μm. (drawn from various sources)

14. Riccia (Ricciella) crystallina L. emend. Raddi, Opusc. Sci. (Bologna) 2 : 351 (1818). *Riccia plana* Taylor, London J. Bot. 5 : 414 (1846). — Jovet-Ast & Bischler (1966) : 99.

Distribution map 247, Figure 246, Plate XVI : i.

Thallus glaucous or light blue-green, crystalline when moist, blue-grey when dry, spongy, forming dense rosettes of 10–15 mm in diameter, or in crowded mats; lobes 2 mm wide, two times branched, broad, truncate or rounded apically, the main lobes deeply incised, with borders raised upwards, the other lobes very short; dorsal epidermis without pores in the young parts, finely perforate in median parts, with wider openings near the centre of the rosettes; cross section of lobes up to six times wider than high, with numerous air chambers in upper three quarters and compact ventral tissue below; scales absent. *Monoicous*; antheridia and archegonia scattered; antheridial necks white, projecting; archegonial necks bright pink, not or little projecting. *Capsules* more conspicuous on ventral than on dorsal side of lobes. *Spores* 60–87 µm in diameter, light yellow-brown, wing 4–5 µm wide, finely crenulate; distal face with 9–10(–11) alveolae across diameter, limited by regular walls provided at wall corners with obtuse and often bifid tubercles; proximal face with similar ornamentation, and with rather thick triradiate mark (seen with SEM).

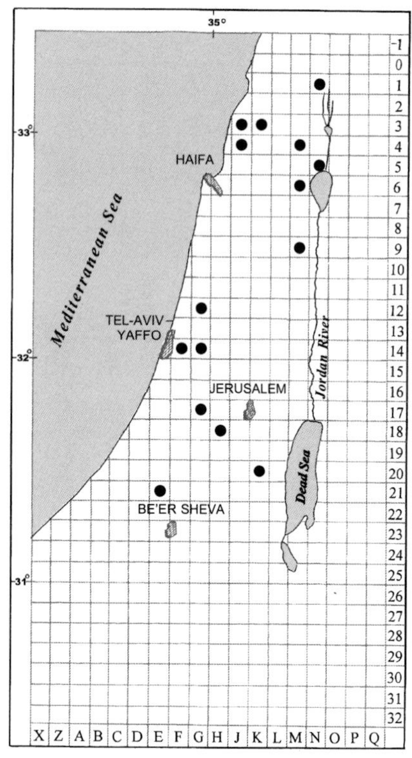

Map 247: *Riccia crystallina*

Habitat and Local Distribution. — Plants growing on sandy clay, moist sand of wadis, compact sandy soil of paths (pH (6–)6.5–7.5), exposed or sheltered by *Sarcopoterium spinosum* and *Tamarix* spp.; associated with *Sphaerocarpos*. Altitude: -200–700 m. Occasional: Coastal Galilee, Sharon Plain, Upper Galilee, Lower Galilee, Shefela, Judean Mountains, Judean Desert, Northern Negev, Dan Valley, and Bet She'an Valley.

General Distribution. — Recorded from Southwest Asia, around the Mediterranean, Europe, India, Africa, Australia, and South America. However, the exact range of the species cannot be provided because it is known under different names and was often confused with *Riccia cavernosa* Hoffm.

Riccia crystallina could be mistaken with *R. cavernosa* Hoffm. emend. Raddi — a Mediterranean species not yet recorded from the local flora. In *R. crystallina*, the thallus is bluish or glaucous, never yellow-green, more compact, with small perforations. Ornamentation of the yellow-brown spores is regular, without specific marking in the centre of distal face. *Riccia terracianoi* Gola, a synonym of *R. cavernosa* (Jovet-Ast 1986), was cited from Israel by Arnell (1957). The specimens *Nachmony 80, 81* (S) belong to *R. crystallina.*

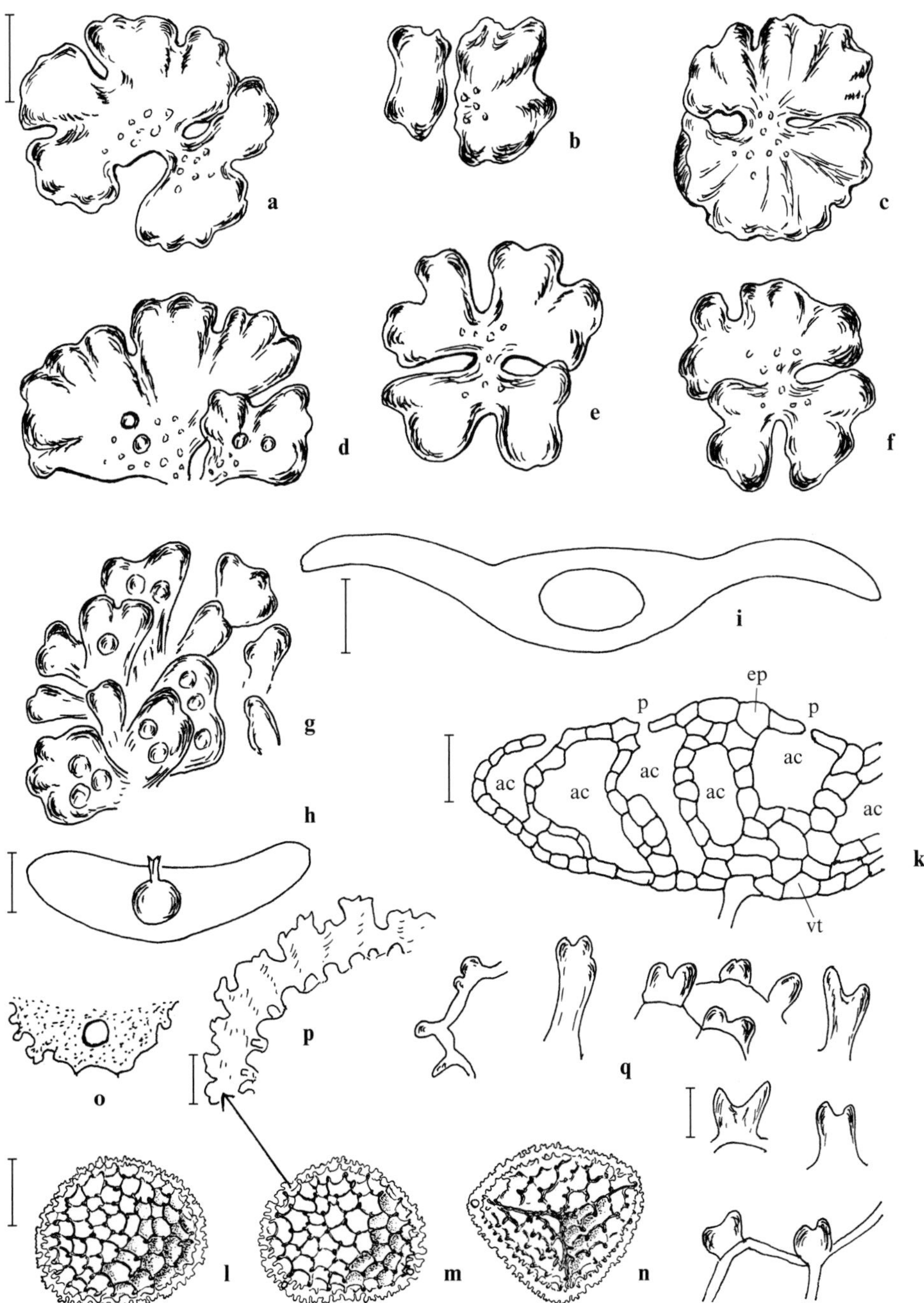

Figure 246. *Riccia crystallina*: **a**–**g** different shapes of thalli; **h** cross-section of narrow lobe; **i** cross-section of wide lobe; **k** lobe, part of cross-section (ep = dorsal epidermis; p = pore; ac = air chambers; vt = ventral tissue); **l**, **m** spores, distal faces; **n** spore, proximal face; **o** spore, wing; **p** spore, part of wing; **q** spore, tubercles of distal face. Scale bars: **a**–**g** = 4 mm; **h**, **i** = 1 mm; **k** = 0.2 mm; **l**–**n** = 30 µm; **o**–**q** = 5 µm. (drawn from various sources)

DUBIA

Several other *Riccia* species could be present in Israel, but have not been recorded recently with certainty or might correspond to misidentifications. Those include:

Riccia canaliculata Hoffm., Deutsch. Fl. 2:96 (1796). A Mediterranean species recorded also from elsewhere in Southwest Asia, Europe, Macaronesia, and North America (disjunct).

Riccia ciliata Hoffm., Deutsch. Fl. 2:95 (1796). This species is recorded from elsewhere in Southwest Asia, around the Mediterranean, Europe, and Macaronesia.

Riccia fluitans L., Sp. Pl., Edition 1:1139 (1753). The species has not been found again since it was recorded by Rabinovitz-Sereni (1931). The two localities where it was collected have been destroyed. Recorded from Southwest Asia, around the Mediterranean, Europe, Macaronesia, tropical and South Africa, eastern Asia, India, the tropical Asiatic Islands, Australia, New Zealand, North and Central America, and eastern, tropical, and southern South America.

Riccia glauca L., Sp. Pl., Edition 1:1139 (1753). Cited by Bilewsky (1963), probably a misidentification (Jovet-Ast & Bischler (1966):92, 100). Recorded from Southwest Asia, around the Mediterranean, Europe, Macaronesia, eastern Asia, New Zealand, and North America.

Riccia ligula Steph., Bull. Herb. Boissier 6:315 (1898). A small, sterile collection from Israel (PC) could correspond to this western Mediterranean species (Jovet-Ast 1986).

EXCLUDENDA

Cephalozia bicuspidata (L.) Dumort. Jovet-Ast *et al.* (1965) and Jovet-Ast & Bischler (1966) recorded this species. The corresponding specimens (PC) contain *Leiocolea turbinata* (Raddi) Steph. The specimen (not seen) cited by Proskauer (1953) as *Cephalozia* spp. could belong to the same species.

Fossombronia angulosa (Dicks.) Raddi. Hart (1891) recorded this species from Gaza. The corresponding specimen (BM) contains *Fossombronia caespitiformis* De Not.

Mannia fragrans (Balb.) Frye & L. Clark. Rabinovitz-Sereni (1931) recorded this species as *Grimaldia fragrans* (Balb.) Corda. The corresponding specimen (HUJ) belongs to *Reboulia hemisphaerica* (L.) Raddi.

PLATES OF SEM MICROGRAPHS

COLOUR PLATES

PLATE I

SEM micrographs of spore surfaces

a–c: *Fissidens bambergeri* (*Kushnir*, Jan. 1943)
- **a** spore (×4,800)
- **b** part of spore surface (×10,000)
- **c** part of spore surface (×20,000)

d: *Pleuridium subulatum* (*Kushnir*, 7 Feb. 1943)
part of spore surface (×10,000)

e: *Dicranella howei* (*Nachmony*, 19 Mar. 1955)
part of spore surface (×20,000)

f: *Weissia controversa* (*Kushnir,* 10 Oct. 1943)
part of spore surface (×10,000)

g: *Weissia condensa* (*Nachmony*, 4 Mar. 1955)
part of spore surface (×10,000)

h: *Weissia breutelii* (*Rigik*, 27 Aug. 1943)
part of spore surface (×10,000)

i: *Weissia rutilans* (*Kushnir*, 21 Mar. 1943)
spore (×2,000)

k, l: *Astomum crispum* (*Heyn & Herrnstadt 80-146-6*)
- **k** spores (×1,300)
- **l** part of spore surface (×10,000)

m: *Aschisma carniolicum* (*Nachmony*, 14 Mar. 1954)
spores (×1,300)

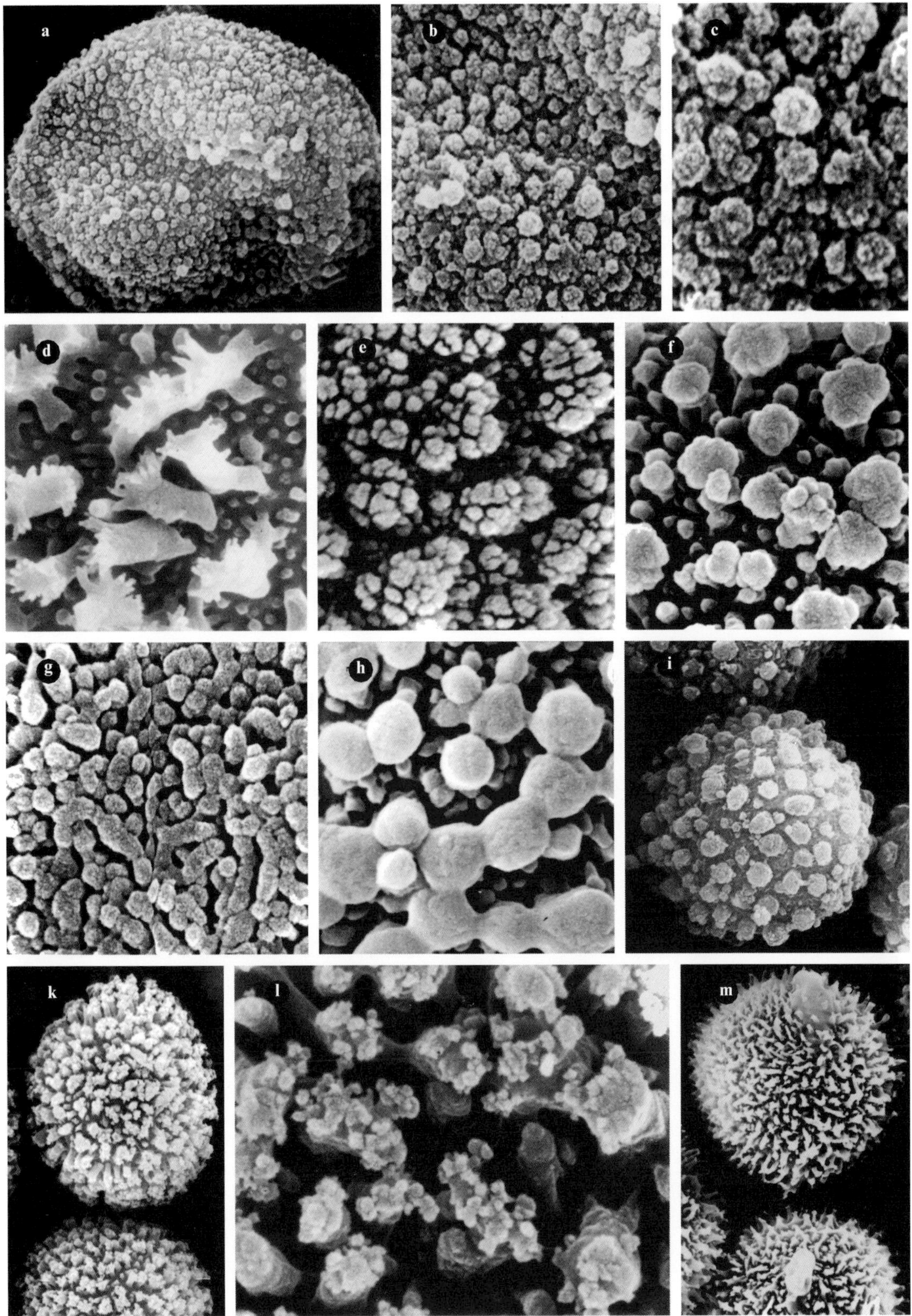
a
b
c
d
e
f
g
h
i
k
l
m

PLATE II

SEM micrographs of spore surfaces

a: *Trichostomum crispulum* (*Nachmony*, 8 Apr. 1945)
spores (×4,800)
b: *Oxystegus tenuirostris* (*Bilewsky*, Apr. 1952, No. 230 in Herb. Bilewsky)
part of spore surface (×10,000)
c: *Timmiella barbuloides* (*Markus* & *Kutiel 77-582-1*)
part of spore with margin (×6,000)
d: *Eucladium verticillatum* (*Crosby* & *Herrnstadt 78-6-6*)
part of spore with margin (×6,000)
e: *Gymnostomum calcareum* (*Herrnstadt* & *Crosby 78-15-4*)
part of spore with margin (×6,000)
f: *Gymnostomum aeruginosum* (*Danin*, 8 Apr. 1979)
part of spore with margin (×6,000)
g: *Gyroweisia reflexa* (*Herrnstadt, 28 May 1984*)
spore (×5,400)
h: *Barbula convoluta* (*D. Zohary*, 28 Mar. 1943)
part of spore surface (×20,000)
i: *Barbula hornschuchiana* (*Pazy* & *Heyn*, 2 Mar. 1976)
part of spore surface (×20,000)
k: *Barbula fallax* (*Landau*, 2 Mar. 1938)
part of spore surface (×20,000)
l: *Leptobarbula berica* (*Bilewsky*, 11 Sep. 1943, No. 9 in Herb. Bilewsky)
part of spore surface (×20,000)
m: *Tortula muralis* var. *israelis* (*Herrnstadt* & *Crosby*, 21 Feb. 1978)
part of spore with margin (×10,000)

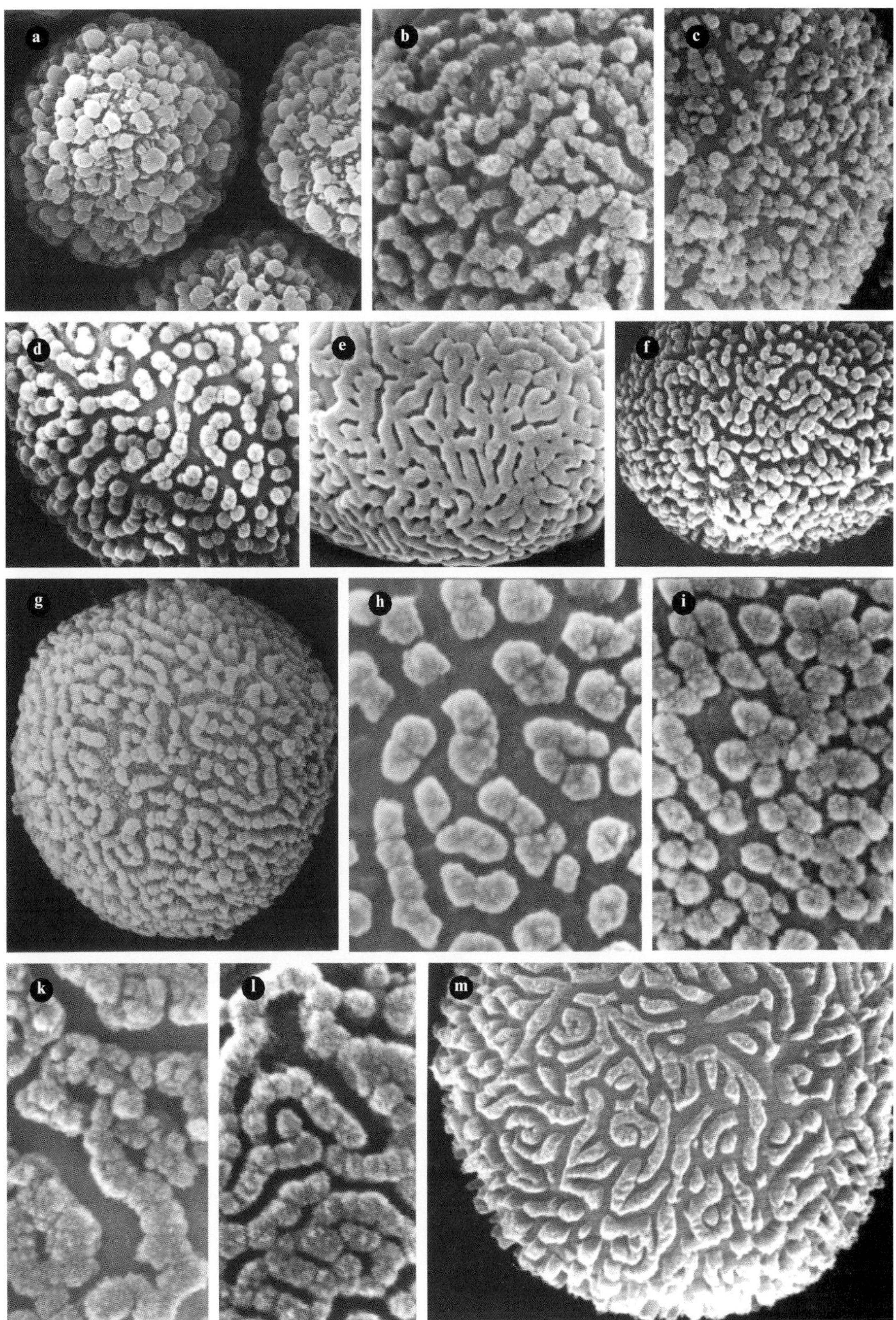
a
b
c
d
e
f
g
h
i
k
l
m

PLATE III

SEM micrographs of spore and leaf surfaces

a: *Tortula muralis* var. *muralis* (*Markus* & *Kutiel 77-599-2*)
part of leaf surface (×2,000)
b: *Tortula muralis* var. *israelis* (*Herrnstadt* & *Crosby*, 21 Feb. 1978)
part of leaf surface (×2,000)
c: *Tortula vahliana* (*Landau* (?), Tiberias, *sine dato*)
part of spore surface (×20,000)
d: *Tortula handelii* (*Herrnstadt* & *Crosby 78-48-12*)
part of spore surface (×20,000)
e: *Tortula princeps* (*Herrnstadt* & *Crosby 79-95-5*)
part of spore surface (×20,000)
f: *Aloina bifrons* (*Oysermann 79-106-1*)
part of spore with margin (×4,400)
g: *Crossidium aberrans* (*Bilewsky*, 3 Jan. 1957, No. 655 in Herb. Bilewsky)
spore (×4,400)
h: *Crossidium crassinervium* var. *crassinervium* (*Herrnstadt* & *Crosby 79-87-13*)
part of spore surface (×20,000)
i: *Crossidium squamiferum* var. *squamiferum* (*Herrnstadt* & *Crosby 78-30-5*)
part of spore with margin (×4,400)
k: *Pterygoneurum ovatum* (*Herrnstadt* & *Crosby 79-97-1*)
part of spore surface (×10,000)
l, **m**: *Pottia lanceolata* (*Kushnir*, 4 Apr. 1943)
l spores (×2,000)
m part of spore surface (×10,000)

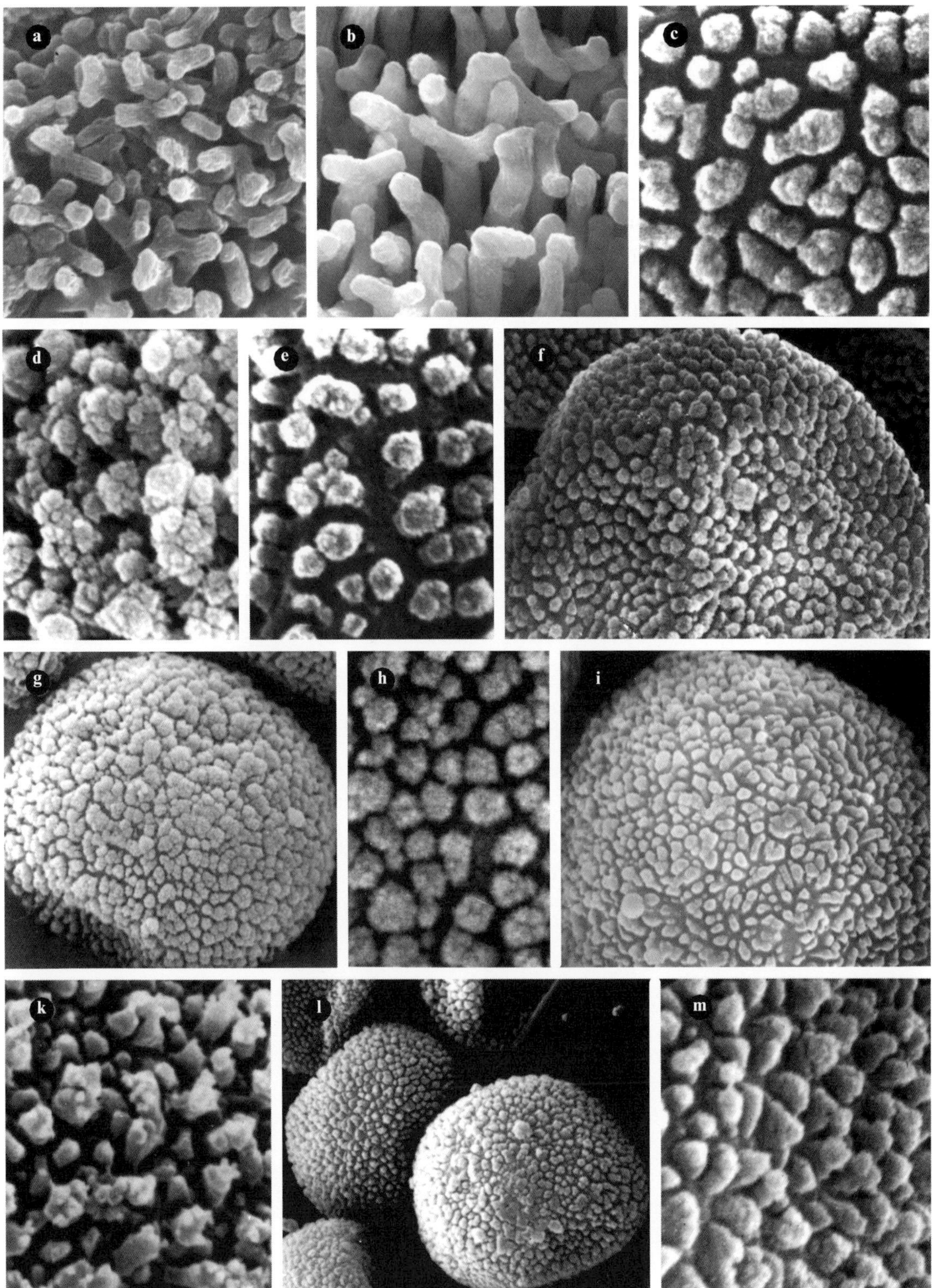
a
b
c
d
e
f
g
h
i
k
l
m

PLATE IV

SEM micrographs of spore surfaces

a, **b**: *Pottia starckeana* (*Herrnstadt* & *Crosby 78-26-3*)
a spores (×2,000)
b part of spore surface (×10,000)
c, **d**: *Pottia commutata* (*Kushnir*, 26 Feb. 1944)
c spores (×2,000)
d part of spore surface (×10,000)
e: *Pottia mutica* (*Kushnir*, 23 Jan. 1943)
spores (×2,000)
f, **g**: *Pottia davalliana* (*Kushnir*, "Jericho", *sine dato*)
f spores (×2,000)
g part of spore surface (×10,000)

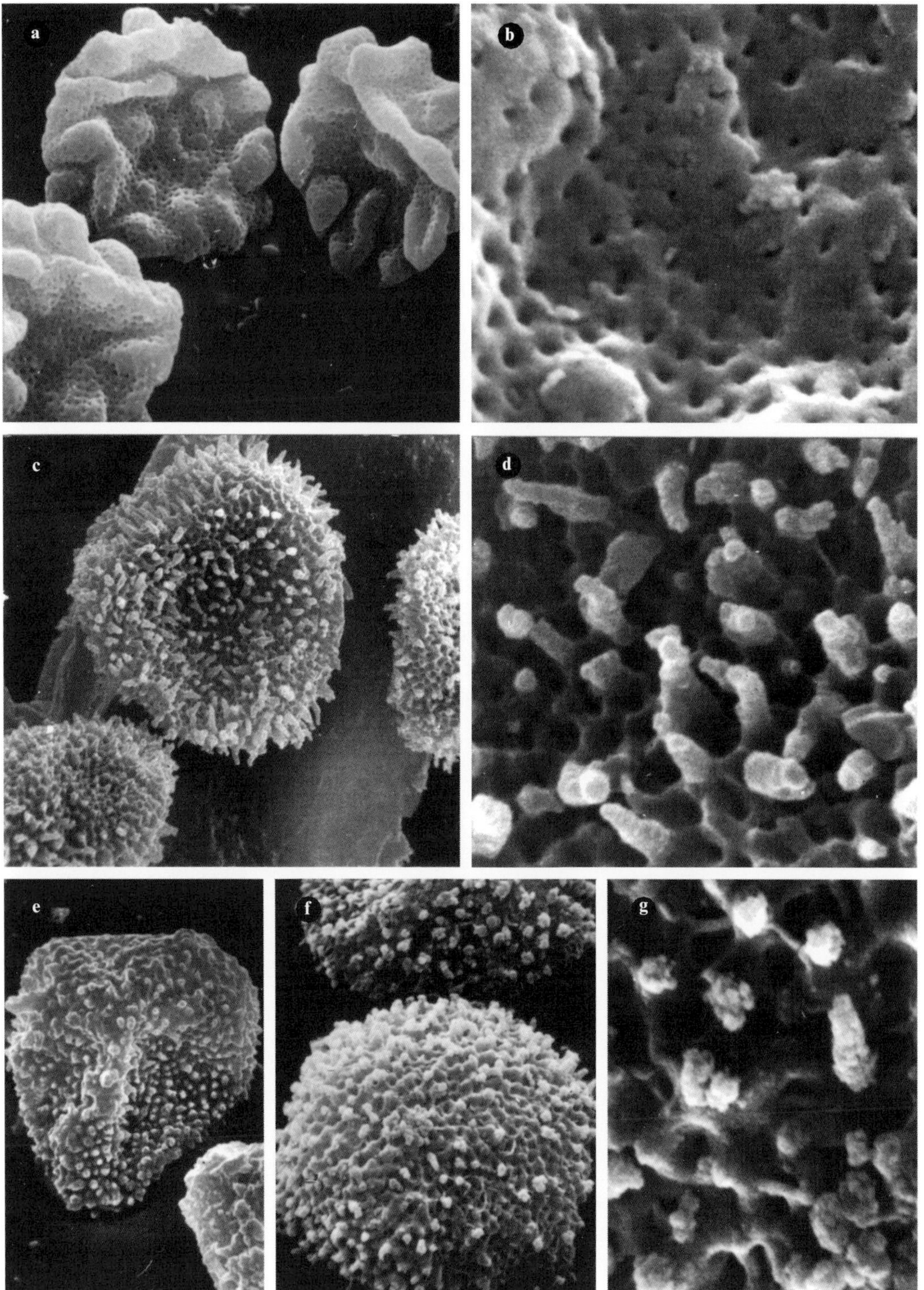
a
b
c
d
e
f
g

PLATE V

SEM micrographs of spore surfaces

a, **b**: *Pottia intermedia* (*Kushnir*, 30 Mar. 1945)
  a spores (×2,000)
  b part of spore surface (×10,000)
c: *Pottia gemmifera* (*Bilewsky*, "Sachne", *sine dato*)
part of spore surface (×10,000)
d: *Pottia recta* (*Markus & Kutiel 77-555-2*)
spore (×2,000)
e: *Phascum cuspidatum* var. *arcuatum* (*Herrnstadt & Crosby 78-48-3*)
spore (×3,000)
f: *Phascum galilaeum* (*Herrnstadt 79-93-1*)
spore (×2,200)

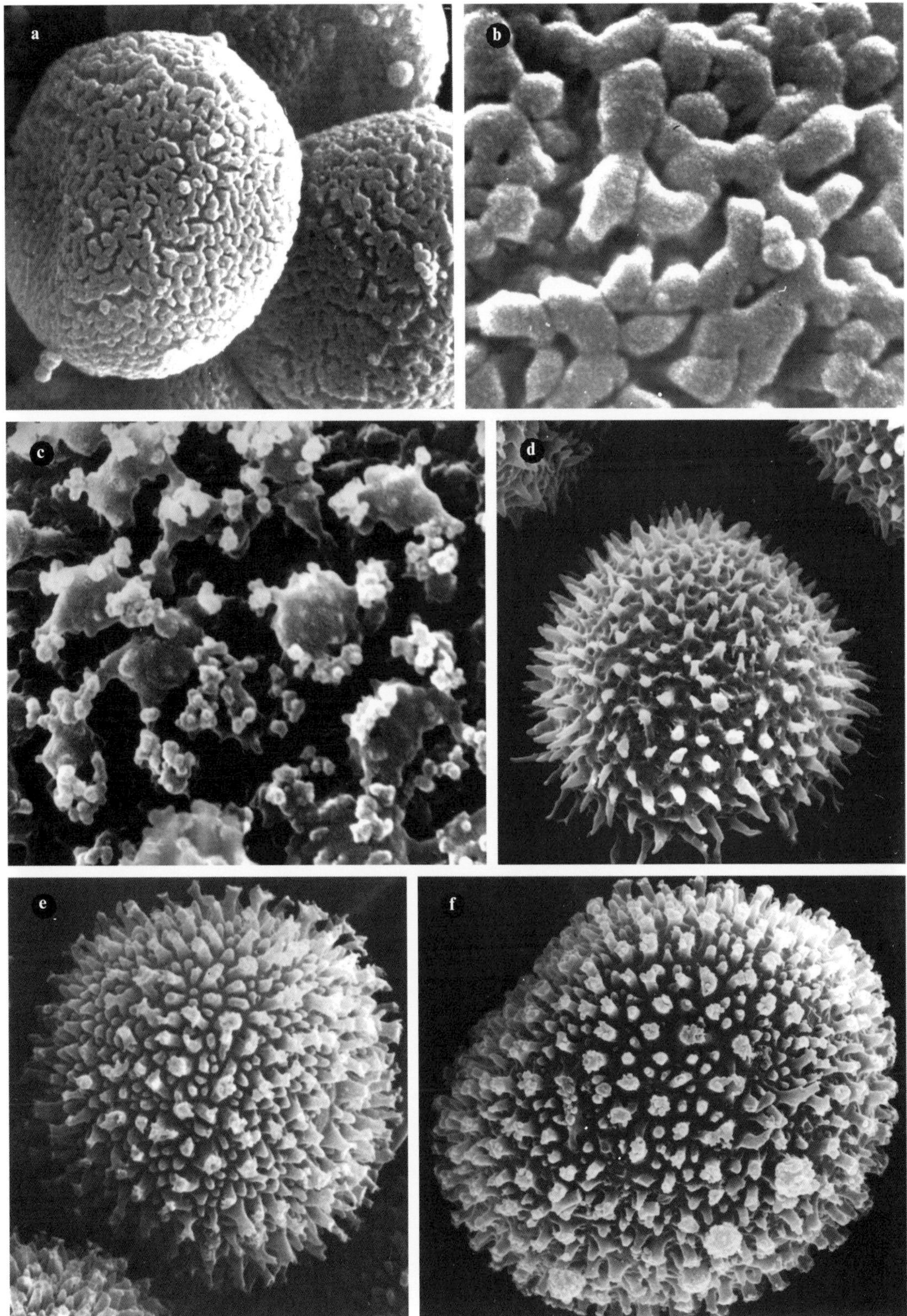
a
b
c
d
e
f

PLATE VI

SEM micrographs of spore surfaces

a: *Phascum curvicolle* (*Bilewsky*, Jerusalem: "Mazleva", No. 47 in Herb. Bilewsky) spore (×2,600)
b: *Phascum floerkeanum* (*Herrnstadt* & *Crosby 78-33-2*) spores (×3,000)
c, **d**: *Acaulon muticum* (*Kushnir*, 21 Mar. 1943)
- **c** spores (×1,300)
- **d** part of spore surface (×10,000)

e, **f**: *Acaulon longifolium* (*D. Zohary*, Mar. 1943, No. 45 in Herb. Bilewsky)
- **e** spores (×1,300)
- **f** part of spore surface (×10,000)

g, **h**: *Acaulon triquetrum* (*Heyn* & *Herrnstadt 80-136-1*)
- **g** spore, not complete (×2,200)
- **h** part of spore surface (×10,000)

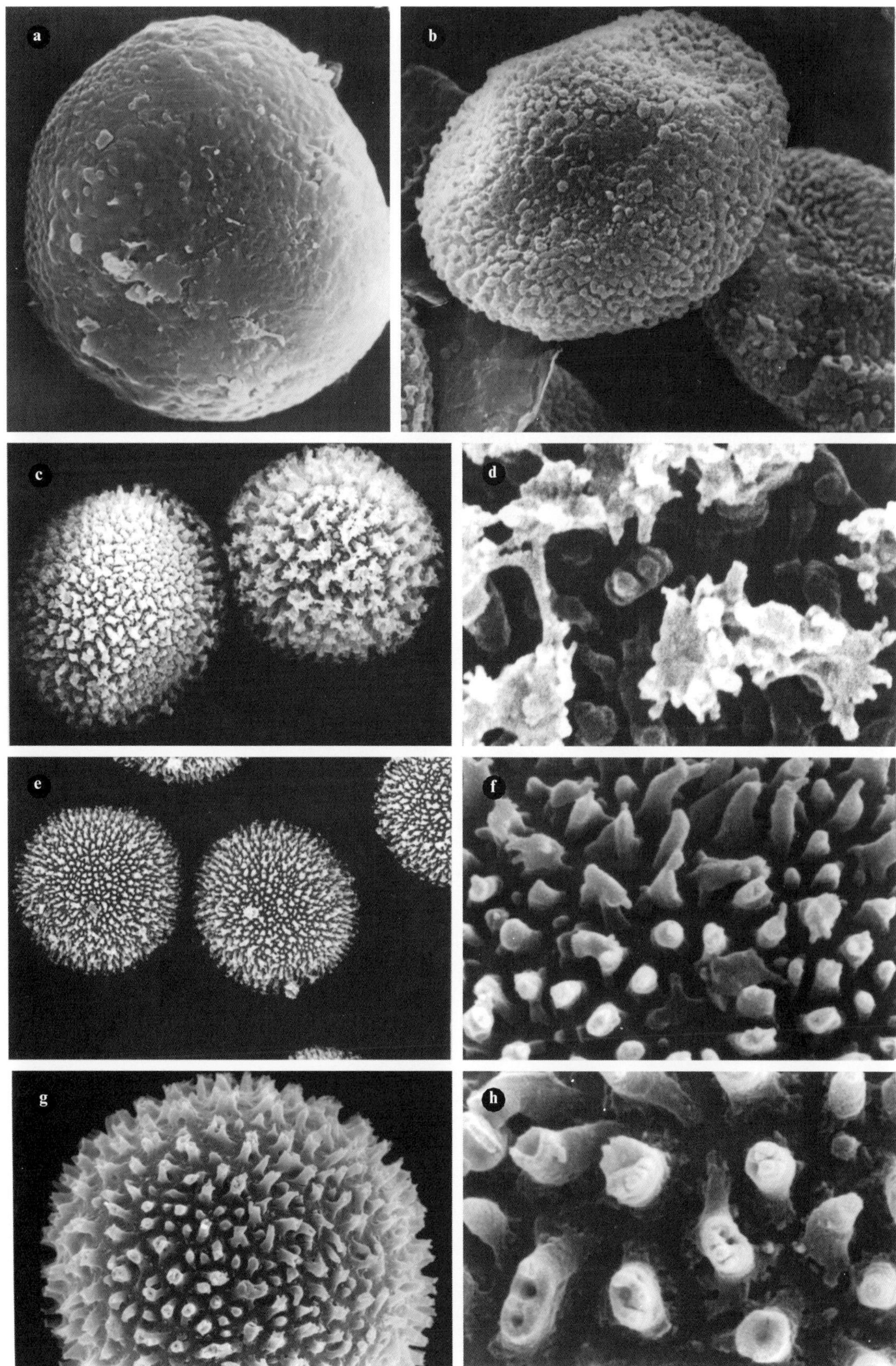
a
b
c
d
e
f
g
h

PLATE VII

SEM micrographs of spore surfaces

a: *Schistidium apocarpum* (*Herrnstadt* & *Crosby 79-98-4*)
part of spore surface (×20,000)

b: *Grimmia anodon* (*Herrnstadt 79-96-3*)
part of spore surface (×20,000)

c: *Grimmia pitardii* (*D. Zohary*, 23 Feb. 1943)
part of spore surface (×20,000)

d: *Grimmia crinita* (*Kushnir*, 15 Dec. 1943)
part of spore surface (×20,000)

e: *Grimmia mesopotamica* (*Herrnstadt 80-158-1*)
part of spore surface (×20,000)

f: *Grimmia laevigata* (*Herrnstadt* & *Crosby 78-28-2*)
part of spore surface (×20,000)

g: *Grimmia pulvinata* (*Herrnstadt* & *Crosby 79-95-10*)
part of spore with margin (×10,000)

h: *Grimmia trichophylla* (*Kushnir*, 9 Oct. 1943)
part of spore surface (×20,000)

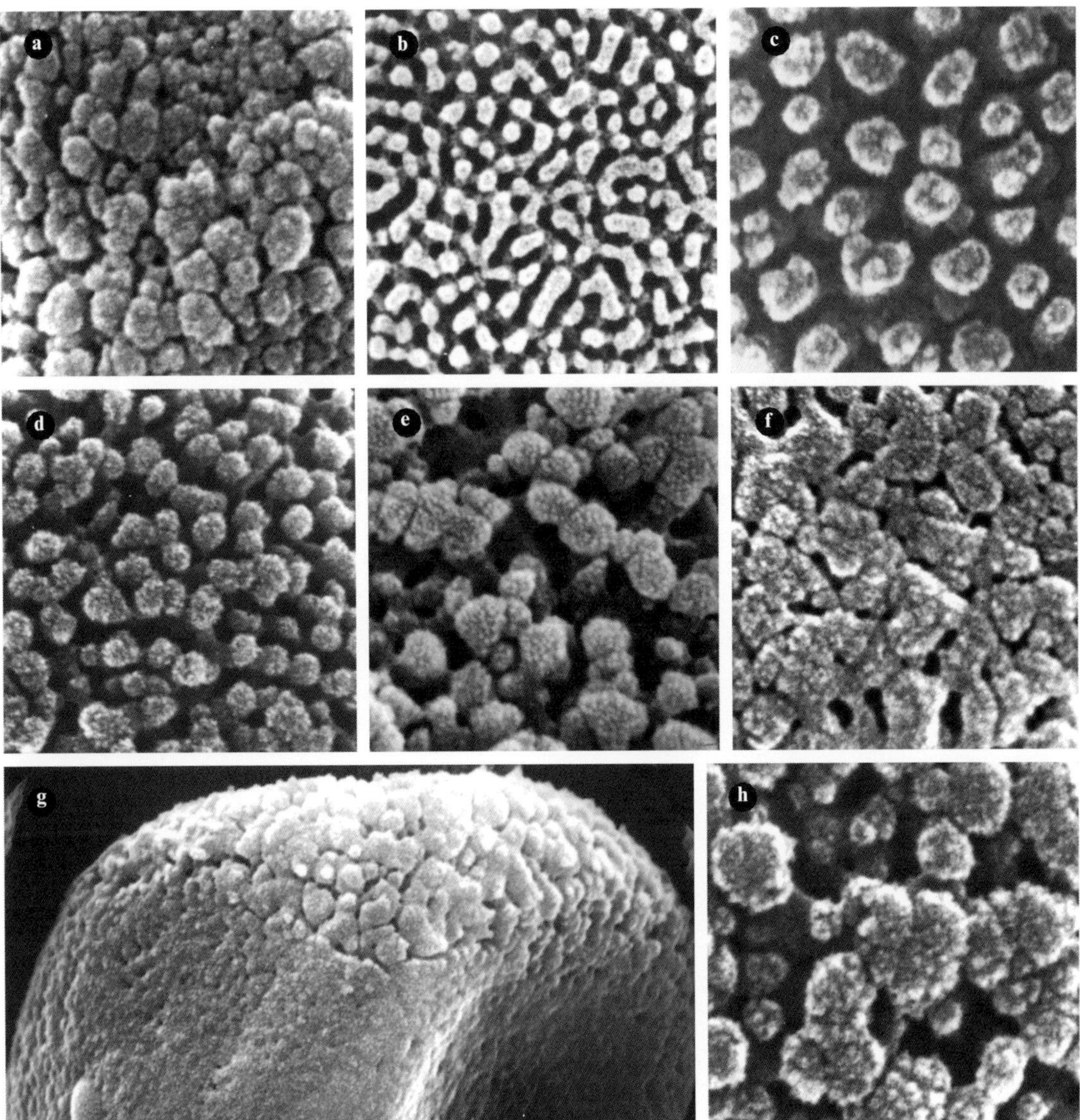
a
b
c
d
e
f
g
h

PLATE VIII

SEM micrographs of spore surfaces and capsule

a: *Encalypta rhaptocarpa* (*Herrnstadt*, 22 Apr. 1983)
spores, proximal and distal view (×1,000)
b: *Encalypta vulgaris* (*Danin*, 2 Oct. 1969)
spores, proximal and distal view (×1,000)
c: *Encalypta rhaptocarpa* (*Danin* & *Liston 83-297-6*)
part of spore surface (×4,000)
d: *Encalypta vulgaris* (*Herrnstadt* & *Crosby 78-45-12*)
part of spore surface (×5,500)
e–g: *Gigaspermum repens* (Australia, *Seppelt 0385*)
e spore (×700)
f part of spore surface (×5000)
g open capsule with spores and operculum (×33)
h, **i**: *Physcomitrium eurystomum* (*Kushnir*, 19 Mar. 1943)
h spores (×1,000)
i part of spore surface (×4,700)

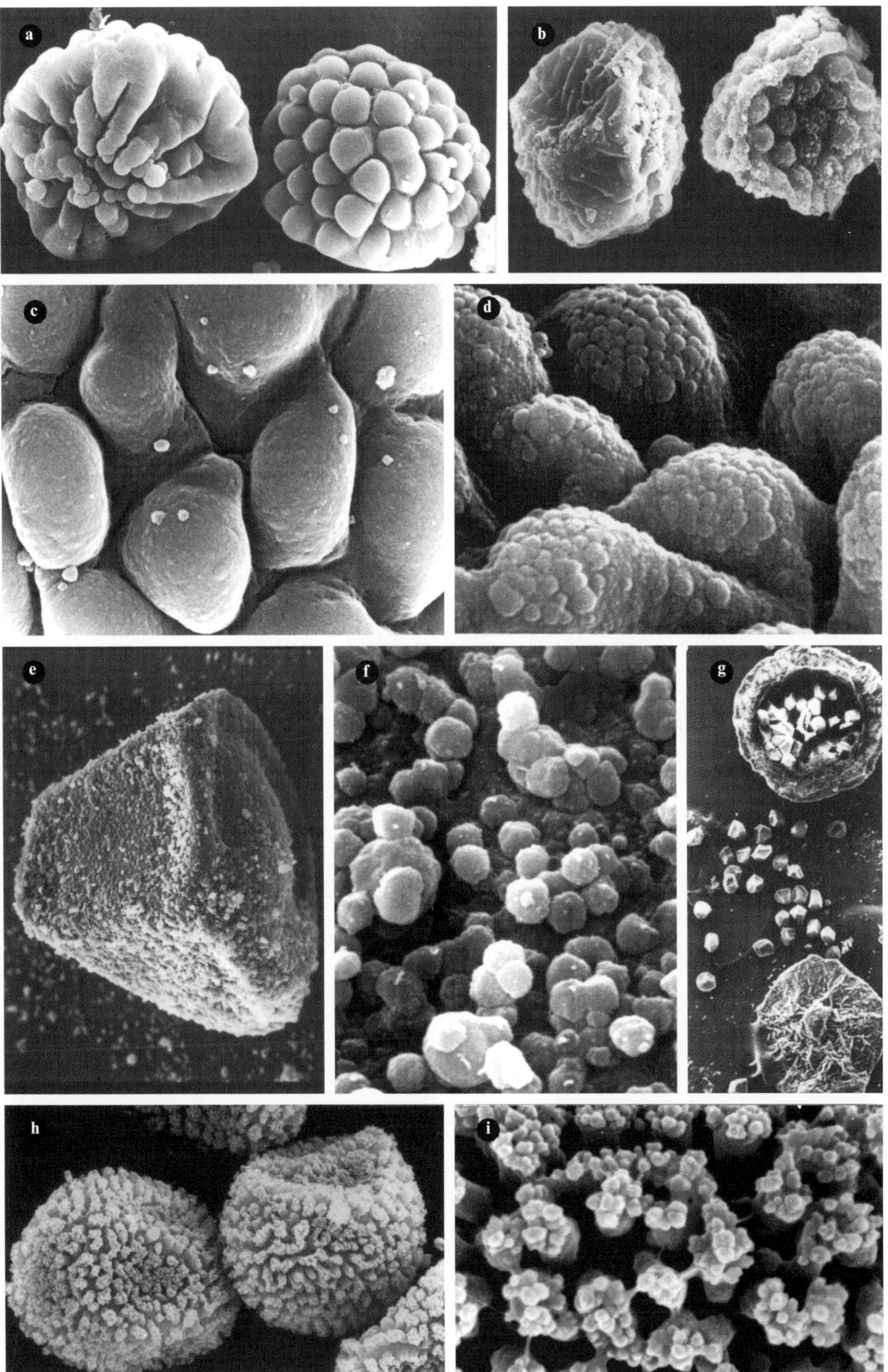
a
b
c
d
e
f
g
h
i

PLATE IX

SEM micrographs of spore surfaces

a, **b**: *Funaria convexa* (a: *Herrnstadt* & *Crosby 79-92-7*; b: *Herrnstadt* & *Crosby 79-99-7*)
 a spores (×1,800)
 b part of spore surface (×10,000)

c: *Funaria pulchella* (*Herrnstadt* & *Crosby 79-92-6*)
spores (×2,000)

d: *Funaria hygrometrica* (*Markus* & *Kutiel 77-594-1*)
part of spore surface (×10,000)

e: *Entosthodon attenuatus* (*Danin* & *Liston 83-296-4*)
part of spore with margin (×2,700)

f: *Entosthodon durieui* (*Bilewsky*, Apr. 1952, No. 89 in Herb. Bilewsky)
part of spore with margin (×4,000)

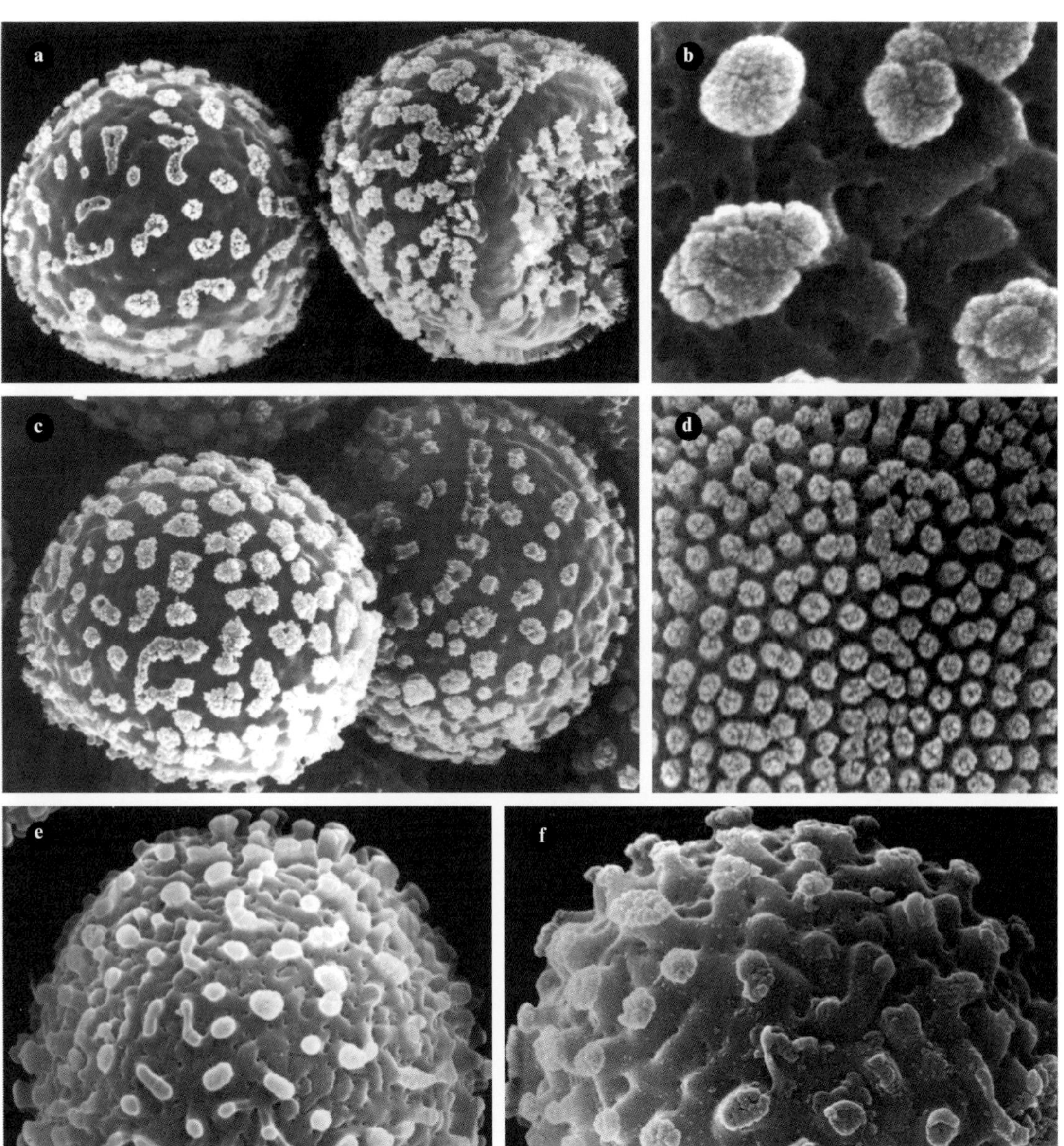
a
b
c
d
e
f

PLATE X

SEM micrographs of spore surfaces

a: *Entosthodon hungaricus* (*Bilewsky*, Mar.–Apr. 1957, No. 721 in Herb. Bilewsky) spores (×720)

b, c: *Entosthodon curvisetus* (b: *Herrnstadt, Liston* & *Boaz 83-282-1*; c: *D. Zohary*, 15 Feb. 1945)

b spores (×2,000)

c part of spore surface (×10,000)

d: *Ephemerum sessile* (*Kushnir*, 21 Mar. 1943, No. 88 in Herb. Bilewsky) part of spore with margin (×940)

e, f: *Ephemerum serratum* var. *minutissimum* (*Kushnir*, 21 Mar. 1943)

e: spore (×940)

f part of spore surface (×10,000)

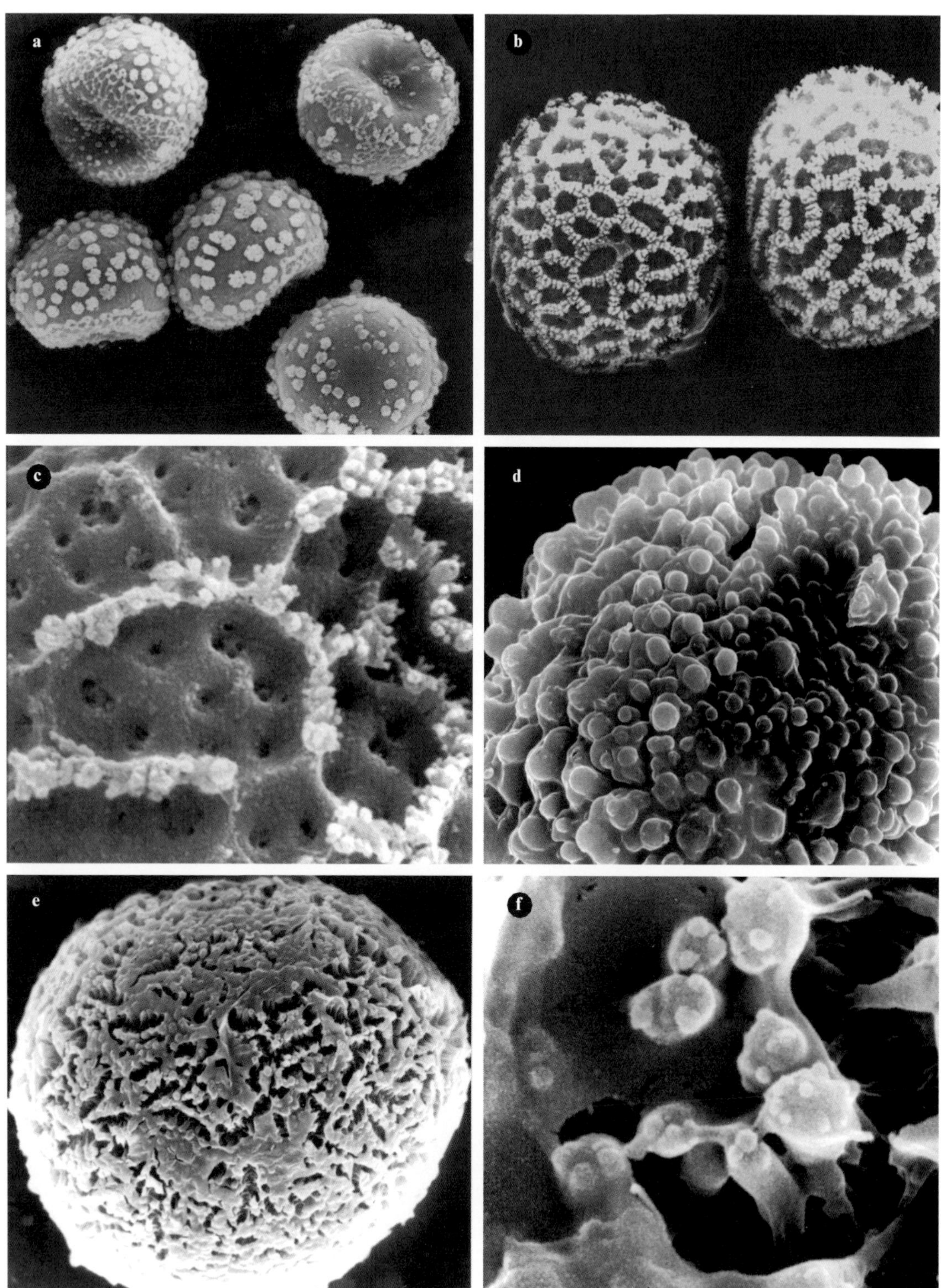
a
b
c
d
e
f

PLATE XI

SEM micrographs of spore surfaces

a: *Pohlia wahlenbergii* (*Kushnir*, 5 Feb. 1943)
part of spore surface (×10,000)
b: *Bryum cellulare* (*Nachmony*, 10 Mar. 1957)
part of spore surface (×10,000)
c: *Bryum subapiculatum* (*Landau*, 2 Mar. 1938)
part of spore surface (×10,000)
d: *Bryum caespiticium* (*Naftolsky*, 9 Apr. 1927)
part of spore surface (×10,000)
e, **f**: *Bartramia stricta* (*Crosby & Herrnstadt 79-70-6*)
e spores from a single capsule (×2,000)
f spores from same capsule as in (e) (×2,000)
g: *Zygodon viridissimus* (*Herrnstadt*, 21 Apr. 1981)
spore (×4,800)

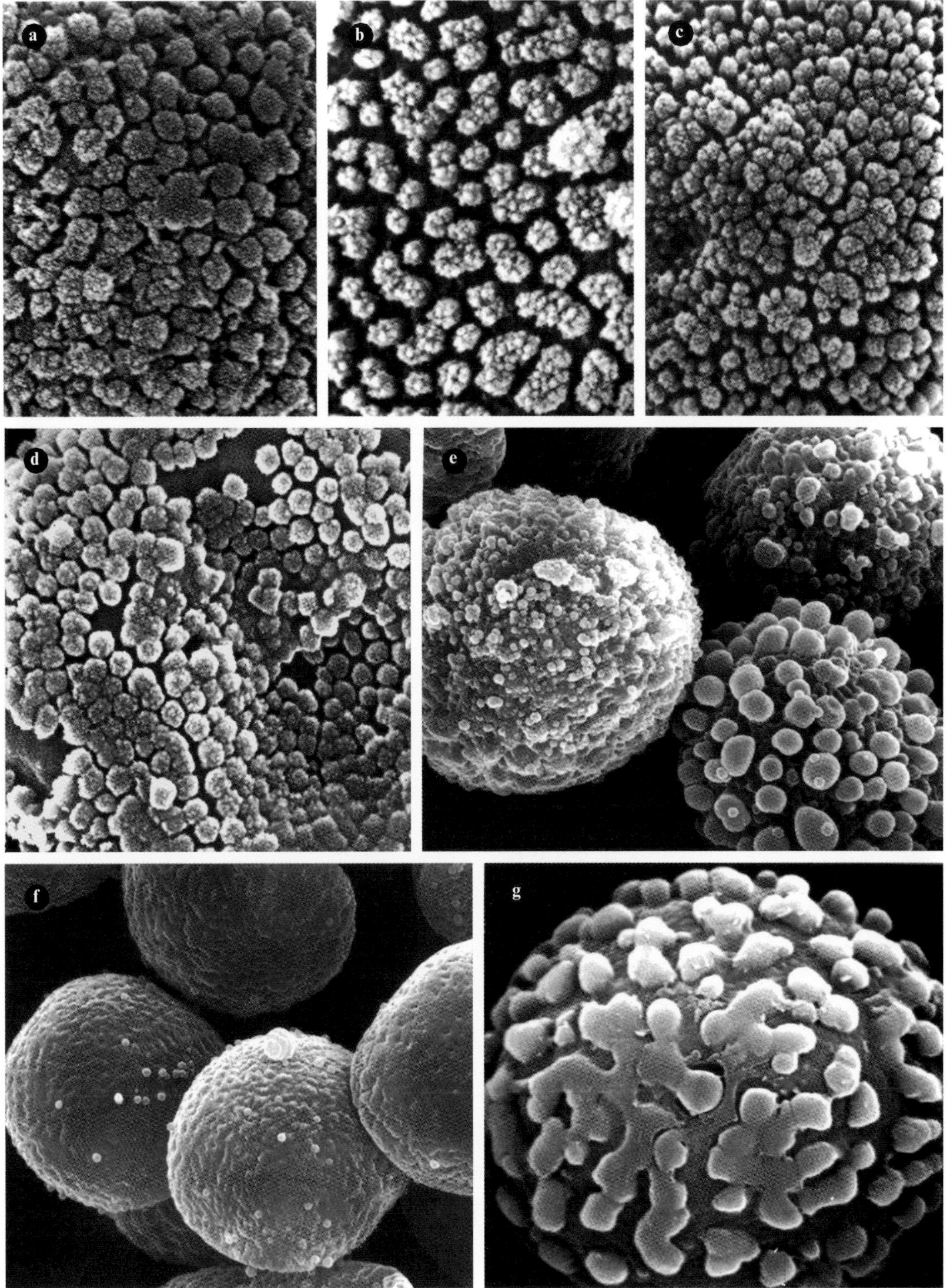
a
b
c
d
e
f
g

PLATE XII

SEM micrographs of spore surfaces

a, b: *Orthotrichum striatum* (*Raven*, 24 Jan. 1975)
  a spore (×3,000)
  b part of spore surface (×10,000)
c: *Orthotrichum affine* (*Kushnir*, 10 Oct. 1943 [mixed sample])
  part of spore surface (×10,000)
d: *Orthotrichum cupulatum* (*Kushnir*, 10 Oct. 1943 [mixed sample])
  part of spore surface (×10,000)
e, f: *Orthotrichum diaphanum* (*Herrnstadt & Crosby 78-54-1*)
  e spore (×4,700)
  f part of spore surface (×10,000)
g: *Leucodon sciuroides* (*Kushnir*, 5 Oct. 1943)
  part of spore surface (×10,000)

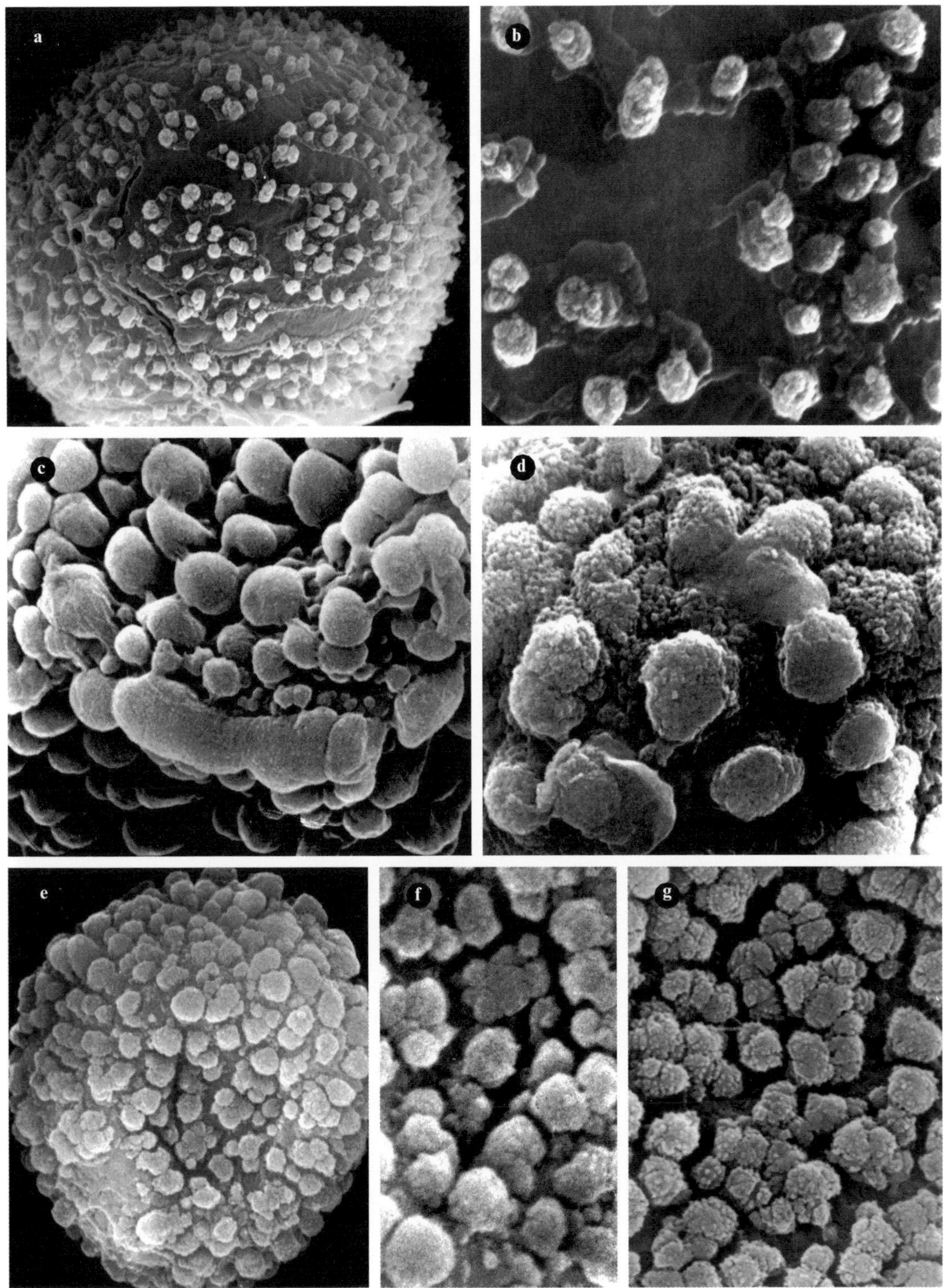
a
b
c
d
e
f
g

PLATE XIII

SEM micrographs of spore surfaces

a: *Leptodon smithii* (*Herrnstadt* & *Crosby 78-21-5*)
part of spore surface (×10,000)
b: *Neckera complanata* (*Shahar*, 4 Jan. 1981)
spores (×1,600)
c: *Fabronia ciliaris* (Meron, *Bilewsky sine dato*)
spore (×4,800)
d: *Amblystegium serpens* (*Herrnstadt* & *Crosby 78-29-19*)
part of spore surface (×10,000)
e: *Leptodictyum riparium* (*Ginzburg* & *Kushnir*, 21 Nov. 1943)
part of spore surface (×10,000)
f: *Homalothecium sericeum* (*Kushnir*, 9 Oct. 1943)
part of spore surface (×10,000)

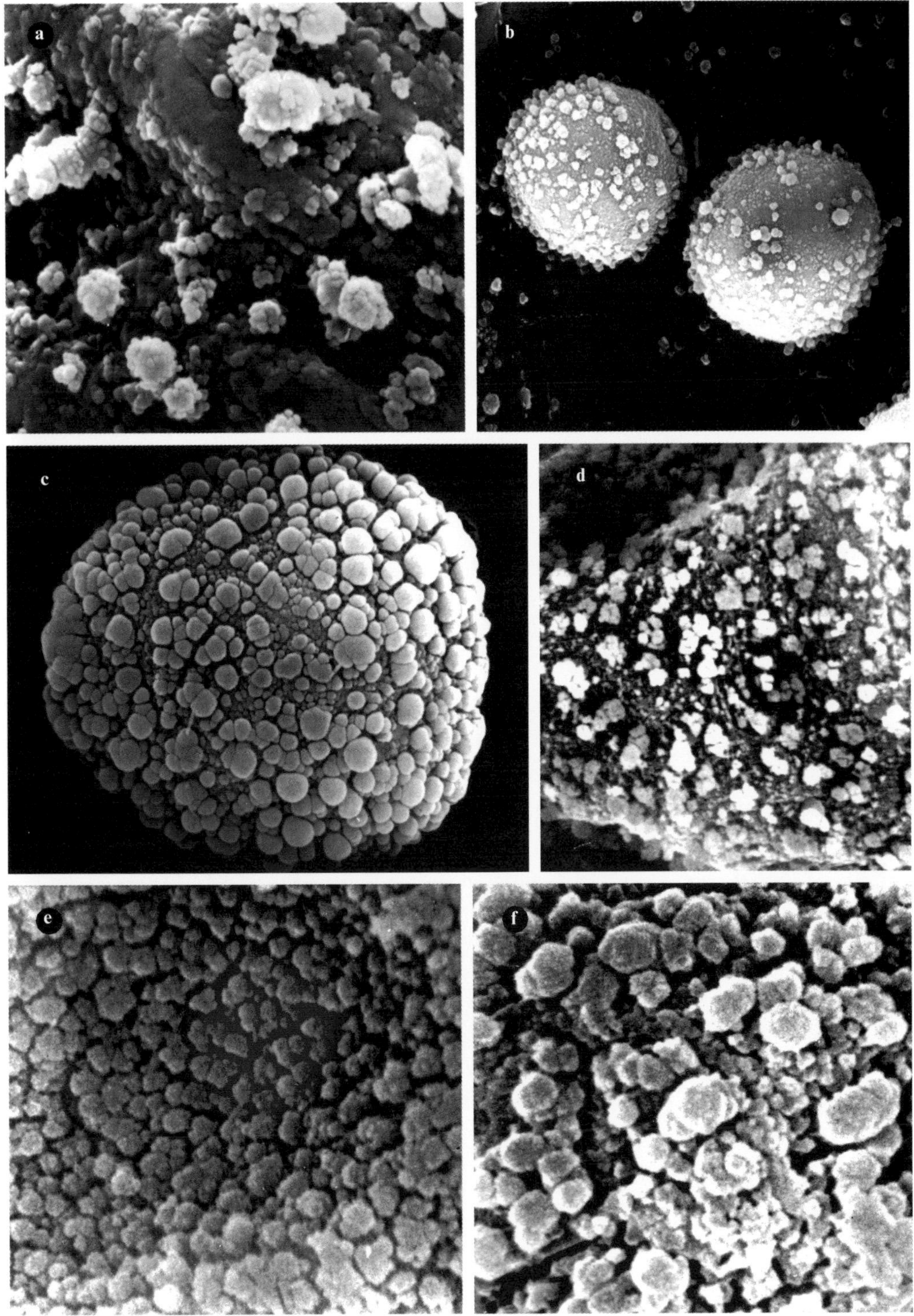
a
b
c
d
e
f

PLATE XIV

SEM micrographs of spore surfaces

a: *Homalothecium aureum* (*Herrnstadt & Crosby 79-88-4*)
part of spore with margin (×5,400)
b: *Scorpiurium deflexifolium* (*Crosby & Herrnstadt 79-84-20*)
part of spore surface (×10,000)
c: *Scorpiurium sendtneri* (*Kushnir*, 8 Apr. 1945)
part of spore surface (×10,000)
d: *Scleropodium touretii* (*Kushnir*, 5 Jan. 1943)
part of spore surface (×10,000)
e: *Eurhynchium riparioides* (*Kushnir sine dato*, No. 136 in Herb. Bilewsky)
part of spore surface (×10,000)
f: *Rhynchostegiella tenella* (*Nachmony*, 6 Feb. 1955)
part of spore surface (×10,000)

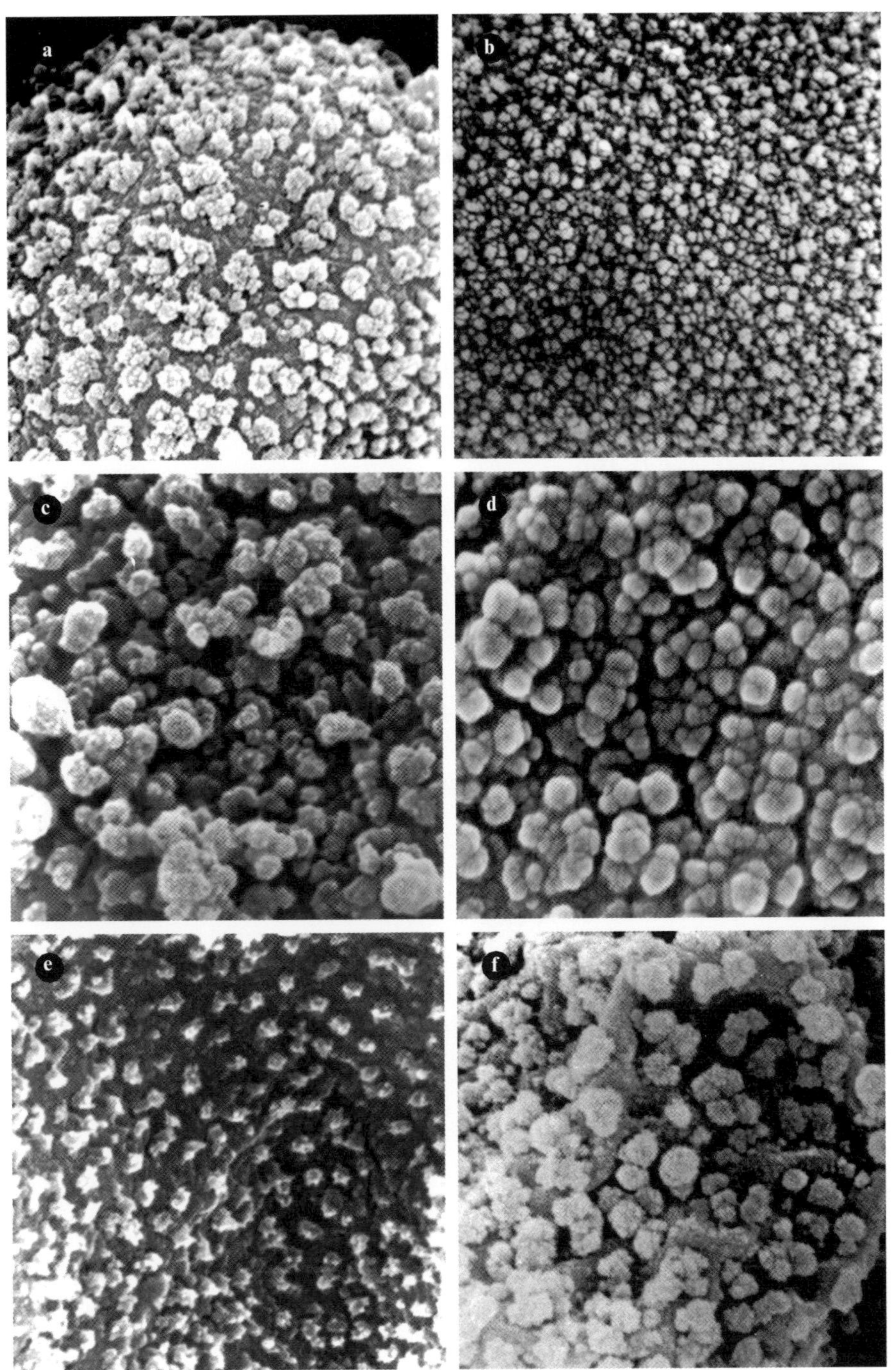
a
b
c
d
e
f

PLATE XV

SEM micrographs of spore surfaces and leaf margins

a: *Phaeoceros bulbiculosus* (*Jovet-Ast & Bischler 64003*)
spore, distal face (×775)

b: *Phaeoceros laevis* (*Jovet-Ast & Bischler 64002*)
spore, distal face (×1150)

c: *Riella cossoniana* (Spain, *Llimona*)
spore, distal face (×600)

d: *Fossombronia caespitiformis* (*Jovet-Ast & Bischler 82275*)
spore, distal face (×725)

e: *Sphaerocarpos michelii* (*Jovet-Ast & Bischler 82109*)
spore tetrad (×400)

f: *Riella helicophylla* (Spain, *Casas*)
spore, lateral view (×450)

g: *Southbya nigrella* (*Jovet-Ast & Bischler 82299*)
leaf margin (×800)

h: *Southbya tophacea* (Crete, *Jovet-Ast & Bischler 77436*)
leaf margin (×725)

i: *Pellia endiviifolia* (France, *Delacour*)
pluricellular spore (×740)

k: *Corsinia coriandrina* (Tunisia, *Jovet-Ast & Bischler 70795*)
spore, distal face (×420)

l: *Targionia hypophylla* (*Jovet-Ast & Bischler 82087*)
spore, distal face (×600)

m: *Lunularia cruciata* (*Herrnstadt*)
spore, distal face (×2350)

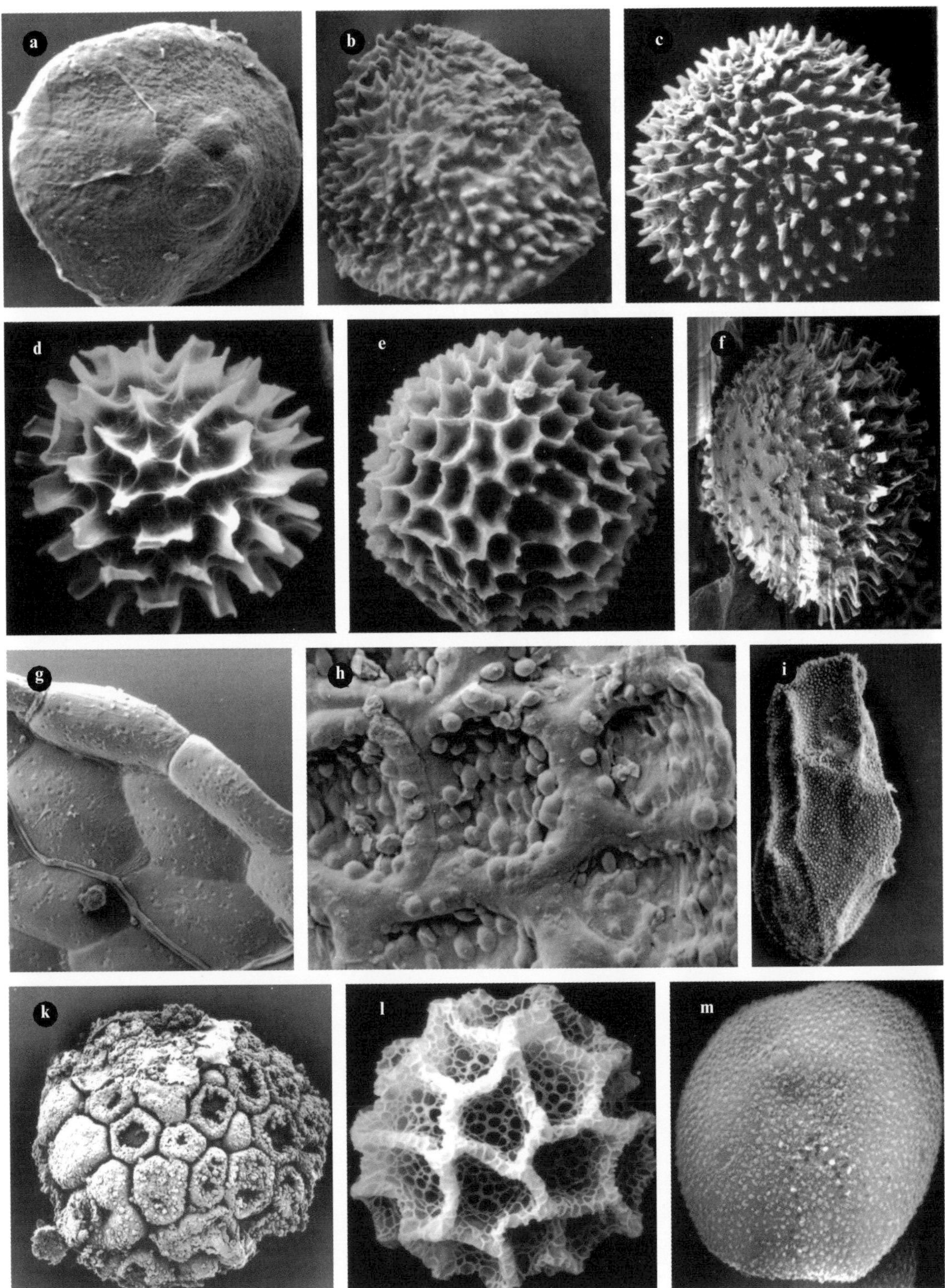
a
b
c
d
e
f
g
h
i
k
l
m

PLATE XVI

SEM micrographs of spore surfaces

a: *Plagiochasma rupestre* (*Zehavi* [HUJ])
spore, distal face (×600)
b: *Mannia androgyna* (*Jovet-Ast* & *Bischler 82087*)
spore, distal face (×685)
c: *Reboulia hemisphaerica* (*Crosby* & *Herrnstadt 78-48-29* [HUJ])
spore, distal face (×600)
d: *Athalamia spathysii* (*Jovet-Ast* & *Bischler 82088*)
spore, distal face (×800)
e: *Marchantia polymorpha* (France, *Bischler*)
spore, distal face (×4500)
f: *Oxymitra incrassata* (Algeria, *Baudoin 4*)
spore, distal face (×350)
g: *Ricciocarpos natans* (Japan, *Mizutani 944*)
spore, distal face (×850)
h: *Riccia crustata*
spore, distal face (×600)
i: *Riccia crystallina*
spore, distal face (×600)
k: *Riccia frostii*
immature spores in tetrad (×850)
l: *Riccia gougetiana*
spore, distal face (×250)
m: *Riccia sorocarpa*
spore, distal face (×600)

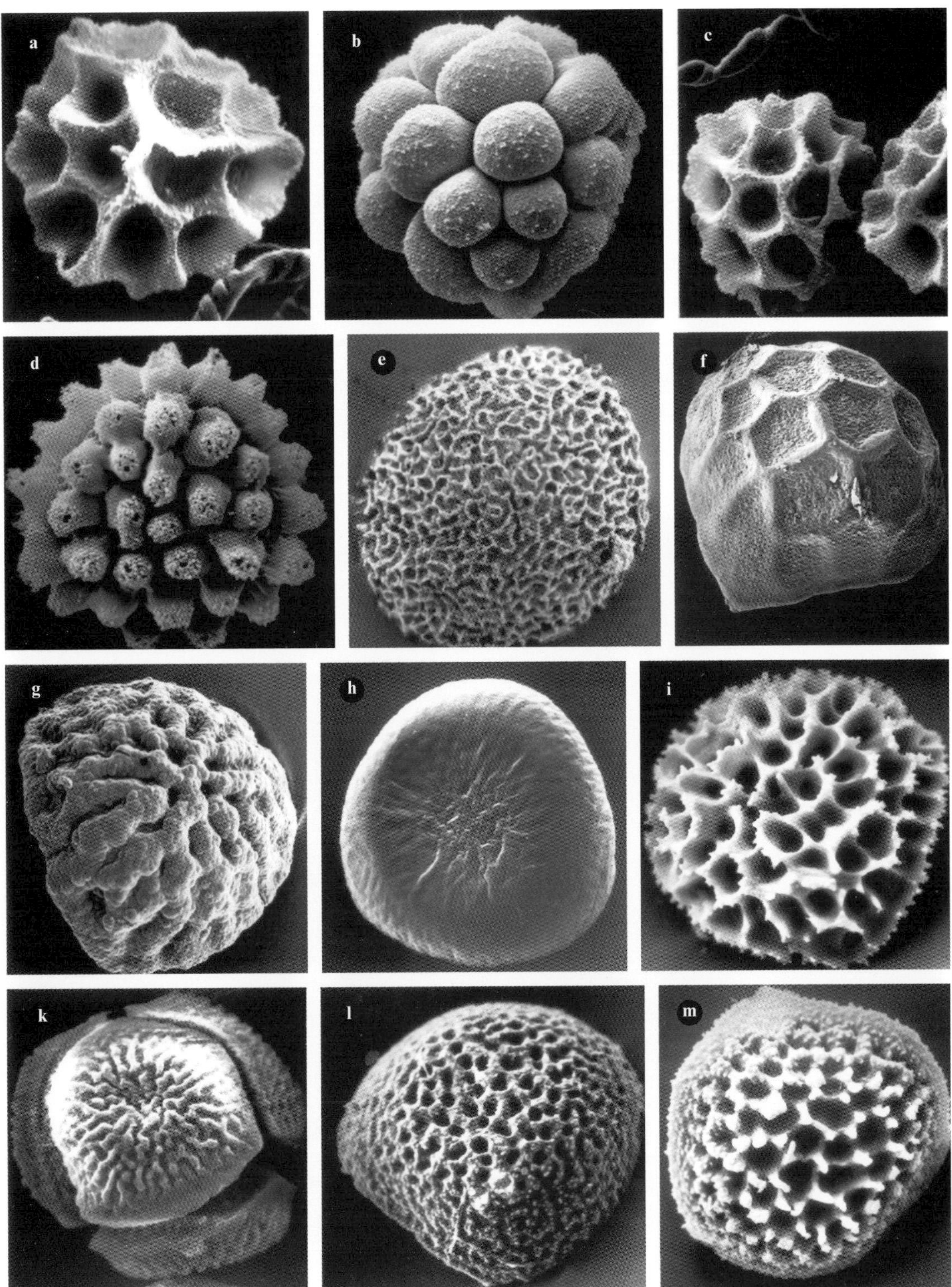
a
b
c
d
e
f
g
h
i
k
l
m

ANNOTATED LIST OF COLOUR PLATES OF SELECTED SPECIMENS

Frontispiece, Part I (following p. 10)
Grimmia pulvinata, Judean Mountains, *Darom*
Bryum caespiticium, Sharon Plain, *Darom*
Gigaspermum mouretii, Dead Sea area, *Herrnstadt & Darom*, 20 Feb. 1987
Tortula inermis, Samaria, *Heyn, Herrnstadt & Darom 81-184-3*
Acaulon longifolium, Philistean Plain, *Herrnstadt*, 15 Feb. 1992
Grimmia mesopotamica, Judean Desert, *Herrnstadt & Danin*, 9 Mar. 1988

Frontispiece, Part II (following p. 520)
Athalamia spathysii, Mount Carmel, *Herrnstadt & Darom*, 4 Mar. 1987
Lunularia cruciata, Judean Mountains, *Heyn & Herrnstadt*, 12 Dec. 1983
Sphaerocarpos michelii, Lower Jordan Valley, *Herrnstadt*, 20 Feb. 1987
Fossombronia caespitiformis, Mount Carmel, *Herrnstadt*
Riccia crystallina, Upper Galilee, *Herrnstadt 85-50-1*
Riccia crustata, Lower Jordan Valley, *Herrnstadt*, 25 Jan. 1992

Plate 1
Fissidens crassipes subsp., *crassipes*, Upper Galilee, *Herrnstadt 81-179-2*
Pleuridium subulatum, Sharon Plain, *Fragman*, 1 Mar. 1992
Dicranella howei, Samaria, *Heyn, Herrnstadt & Darom*, 4 Apr. 1981
Weissia condensa, Mount Carmel, *Herrnstadt & Darom*, 4 Mar. 1987
Gymnostomum calcareum, Mount Carmel, *Herrnstadt & Darom 81-180-4*
Anoectangium handelii, Central Negev, *Danin & Liston*, 21 Mar. 1983

Plate 2
Barbula ehrenbergii var., *ehrenbergii*, Upper Galilee, *Herrnstadt*, 20 Mar. 1981
Barbula revoluta, Judean Mountains, *Herrnstadt & Darom*, 8 Nov. 2001
Barbula vinealis, Golan Heights, *Herrnstadt & Darom*, 25 Mar. 1992
Tortula atrovirens, Judean Desert, *Herrnstadt*, 20 Mar. 1992
Tortula fiorii, Dead Sea area *Herrnstadt & Darom*, 20 Feb. 1987
Tortula muralis var. *muralis*, Mount Carmel, *Herrnstadt*

Plate 3
Tortula intermedia, Judean Mountains, *Herrnstadt*, Apr. 1987
Aloina bifrons, Lower Jordan Valley *Herrnstadt*, 15 Jan. 1992
Tortula princeps, Golan Heights, *Herrnstadt*, 1 Apr. 2003
Crossidium crassinervium var. *laevipilum*, Lower Jordan Valley, *Herrnstadt & Darom*, 20 Feb. 1987
Crossidium squamiferum var. *squamiferum*, Upper Galilee, *Herrnstadt*
Pterygoneurum crossidioides, Lower Jordan Valley, *Herrnstadt*, 25 Jan. 1992

Plate 4

Pterygoneurum subsessile, Northern Negev, *Herrnstadt*
Pottia commutata, Mount Carmel, *Herrnstadt & Darom*, 4 Mar. 1987
Phascum cuspidatum var. *cuspidatum*, Philistean Plain, *Herrnstadt*, 18 Feb. 1992
Encalypta vulgaris, Samaria, *Heyn, Herrnstadt & Darom 81-184-6*
Physcomitrium eurystomum, Golan Heights, *Herrnstadt*, 25 Mar. 1992
Funaria pulchella, Dead Sea area, *Herrnstadt*, 20 Mar. 1992

Plate 5

Funaria hygrometrica, Golan Heights (left), *Herrnstadt*, 1 Apr. 2003; Judean Mountains (right), *Weitz*
Entosthodon curvisetus, Mount Carmel, *Herrnstadt*
Entosthodon attenuatus, Judean Mountains, *Heyn & Herrnstadt*, 8 Mar. 1992
Bryum cellulare, Dead Sea area, *Herrnstadt 85-11-1*
Bryum argenteum, Golan Heights (left), *Herrnstadt*, 25 Mar. 1992; Golan Heights (right), *Herrnstadt*, 1 Apr. 2003
Bryum torquescens, Judean Mountains, *Herrnstadt & Darom*, 4 Mar. 1987

Plate 6

Bartramia stricta, Golan Heights, *Herrnstadt*, 25 Mar. 1992
Philonotis marchica, Upper Jordan Valley, *Herrnstadt*, 2 Apr. 2003
Orthotrichum rupestre, Golan Heights, *Herrnstadt & Darom*, 25 Mar. 1992
Leucodon sciuroides, Golan Heights, *Herrnstadt & Darom*, 25 Mar. 1992
Leucodon sciuroides, Golan Heights Heights, *Herrnstadt & Darom*, 25 Mar. 1992
Leptodon smithii, Upper Galilee, *Herrnstadt*, 2 Apr. 2003

Plate 7

Neckera complanata, Upper Galilee, *Herrnstadt*, 19 Jan. 2002
Amblystegium tenax, Dan Valley, *Herrnstadt*
Homalothecium aureum, Upper Galilee, *Herrnstadt*, 19 Jan. 2002
Scorpiurium deflexifolium, Coast of Carmel, *Herrnstadt & Darom 81-180-5*
Scleropodium touretii, Judean Mountains, *Herrnstadt & Darom*, 18 Mar. 1987
Rhynchostegiella tenella, Mount Carmel, *Herrnstadt & Darom 81-180-6*

Plate 8

Frullania dilatata, Upper Galilee, *Herrnstadt & Darom 81-185-11*
Targionia hypophylla, Mount Carmel *Herrnstadt 81-180-7*
Plagiochasma rupestre, Lower Galilee, *Herrnstadt 83-282-7*
Reboulia hemisphaerica, Upper Galilee, *Herrnstadt 81-179-18*
Oxymitra incrassata, Coastal Galilee, *Herrnstadt 82-203-7*
Ricciocarpos natans, Hula Plain, *Kaplan*, 8 Dec. 1991

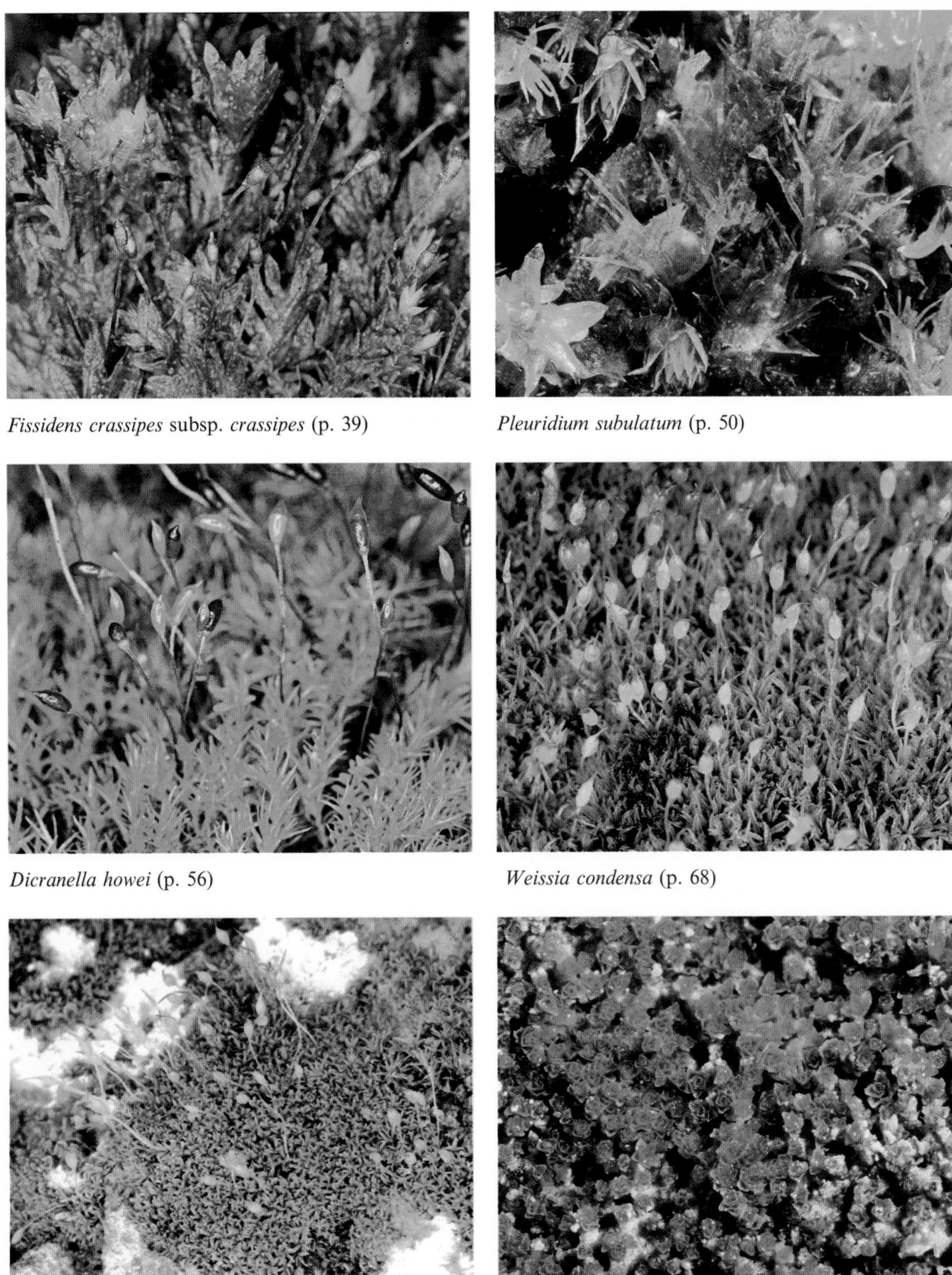

Fissidens crassipes subsp. *crassipes* (p. 39)

Pleuridium subulatum (p. 50)

Dicranella howei (p. 56)

Weissia condensa (p. 68)

Gymnostomum calcareum (p. 112)

Anoectangium handelii (p. 124)

Barbula ehrenbergii var. *ehrenbergii* (p. 133)

Barbula revoluta (p. 138)

Barbula vinealis (p. 152)

Tortula atrovirens (p. 164)

Tortula fiorii (p. 168)

Tortula muralis var. *muralis* (p. 177)

Tortula princeps (p. 194)

Tortula intermedia (p. 190)

Aloina bifrons (p. 206)

Crossidium crassinervium var. *laevipilum* (p. 216)

Crossidium squamiferum var. *squamiferum* (p. 218)

Pterygoneurum crossidioides (p. 222)

PLATE 4

Pterygoneurum subsessile (p. 226)

Pottia commutata (p. 234)

Phascum cuspidatum var. *cuspidatum* (p. 248)

Encalypta vulgaris (p. 302)

Physcomitrium eurystomum (p. 311)

Funaria pulchella (p. 322)

Funaria hygrometrica with young (left) and mature (right) sporophytes (p. 324)

Entosthodon curvisetus (p. 340)

Entosthodon attenuatus (p. 328)

Bryum cellulare (p. 359)

Bryum argenteum with (left) and without (right) sporophytes (p. 362)

Bryum torquescens (p. 385)

Bartramia stricta (p. 402)

Philonotis marchica (p. 404)

Orthotrichum rupestre (p. 418)

Leucodon sciuroides (dry plants) (p. 428)

Leucodon sciuroides (p. 428)

Leptodon smithii (p. 435)

Neckera complanata (p. 436)

Amblystegium tenax (p. 452)

Homalothecium aureum (p. 468)

Scorpiurium deflexifolium (p. 472)

Scleropodium touretii (p. 488)

Rhynchostegiella tenella (p. 512)

PLATE 8

Frullania dilatata (p. 552)

Targionia hypophylla (p. 576)

Plagiochasma rupestre (p. 583)

Reboulia hemisphaerica (p. 588)

Oxymitra incrassata (p. 596)

Ricciocarpos natans (p. 600)

GLOSSARY

Adapted and abridged with permission, and with modifications, from Magill (1990), Glossarium Polyglottum Bryologiae: A multilingual glossary for bryology.

abaxial side away from stem or axis; back, dorsal, or lower surface of leaf or costa (opposed to adaxial).

acidophilous acid-loving.

acrocarpous with gametophyte producing sporophyte at apex of a stem or main branch. Acrocarpous mosses generally grow erect in tufts (rather than mats), and are sparsely or not branched (opposed to pleurocarpous).

acrogynous having archegonia produced at the apex of a stem or branch, accompanied by loss of apical cell and termination of further shoot growth (opposed to anacrogynous).

acuminate slenderly tapered with an angle of less than 45°; longer than acute.

acute sharp pointed, with terminal angle less than 90° but greater than 45°.

adaxial side toward stem or axis; ventral or upper surface of leaf or costa (opposed to abaxial).

adnate fused together; the fusion of unlike parts, i.e., perianth and bract (cf. **coalescent, connate**).

air chamber specialised internal air-containing cavity common in most complex thalloid liverworts, e.g., Marchantiales.

air pore minute opening in the upper epidermis of most complex thalloid liverworts, functioning in gas exchange and water regulation.

alar cells referring to cells at basal margins (angles) of a leaf; these cells are often differentiated in size, shape, or colour from other leaf cells (see **auricle**).

alveolate (alveolus) with depressions on the surface, sometimes applied to spores, e.g., species of *Fossombronia* (cf. **foveolate, areolate**).

alveola (pl. alveolae) small pit or cavity such as a honeycomb cell.

amplexicaulous clasping the stem, as the base of a leaf.

anacrogynous having archegonia producted in a lateral position on a stem, branch, or thallus from superficial initials and without loss of apical cell function (opposed to acrogynous).

androecium (pl. androecia) antheridia and surrounding bracts (perigonium or involucre); the "male inflorescence".

angular cells see **alar cells**.

annulus a zone of variously differentiated cells between the capsule urn and operculum that facilitates the opening of the capsule in stegocarpous mosses (cf. **valve**).

antheridiophore specialised antheridium-bearing branch, e.g., *Marchantia*; the male gametangiophore (cf. **archegoniophore**).

antheridium (pl. antheridia) male gametangium; a multicellular globose to broadly cylindrical, stalked structure containing spermatozoids (syn. androgonium; cf. **archegonium**).

antherozoid see **spermatozoid**.

antical the dorsal surface of a stem; the leaf margin oriented towards the shoot apex of a longitudinal or obliquely inserted leaf (opposed to postical).

anticlinal perpendicular to the surface; radial; at right angles to the free or exposed surface; applied to cell divisions that are perpendicular to the free surface (opposed to periclinal).

apical cell a single meristematic cell at the apex of a shoot, thallus, leaf, or other organ that divides repeatedly to form new cells.

apical lamina in *Fissidens*, the part of

the leaf above or distal to the vaginant and dorsal laminae.

apiculate abruptly short-pointed (mucronate is shorter; cuspidate is longer and stouter).

apiculus (pl. apiculi) a short, abrupt point (see **mucro**).

appendiculate with short, thin, transverse projections formed from horizontal wall-pairs, often borne on endostomial cilia, e.g., *Bryum*.

appressed closely applied, as leaves lying close or flat against the stem.

arboreal growing on trees (cf. **corticolous**).

archegoniophore a specialised archegonia-bearing branch; the female gametangiophore, e.g., *Marchantia* (cf. **antheridiophore**).

archegonium (pl. archegonia) female gametangium or sex organ; a multicellular, flask-shaped structure consisting of a stalk, venter, neck, and containing an ovum (see **gametangium**; cf. **antheridium**).

arcuate curved.

areolate with small angular or polygonal surface areas differing in colour or structure from the surrounding area, forming a pattern or network, e.g., thallus surface in many Marchantiales (cf. **alveolate**).

areolation the cellular network of a leaf or thallus.

arista (pl. aristae) see **awn**.

arthrodontous having a peristome type consisting of one or two rings of triangular or linear appendages (i.e., teeth, segments), consisting essentially of differentially thickened periclinal wall-pairs (lamellae).

articulate jointed; with thickened joints; setae with epidermal cells in distinct, even tiers.

ascending pointing obliquely upward; away from the substrate.

assimilatory filaments photosynthetic filaments of cells in the air chambers.

attenuate slenderly tapering.

auricle a small, ear-like lobe; often present at the basal margins of a leaf in mosses (see **alar cells**); also on thalli and other organs in liverworts.

auriculate with auricles.

autoicous (autoecious*) with archegonia and antheridia in separate clusters (gametoecium) on the same plant;
cladautoicous with the androecium on a separate branch;
gonioautoicous with the androecium bud-like and axillary on the same stem or branch as the gynoecium;
pseudautoicous with dwarf male plants epiphytic on the female;
rhizautoicous with the androecium on a very short branch attached to the female stem by rhizoids and appearing to be a separate plant (cf. **dioicous**, **monoicous**, **paroicous**, **synoicous**).

awn hair-point, usually formed by an excurrent costa, e.g., *Tortula*.

axil the upper angle formed by the axis and any organ that arises from it, e.g., leaf and stem, stem and branch.

axillary in the leaf axils.

axillary hair uniseriate hair found in the leaf axils, generally inconspicuous and well concealed by the leaf bases. The apical and basal cells of the hairs are frequently differentiated in size or colour.

baculate spore surface with pillar-like processes (bacula, sing. baculum), always longer than broad and higher than 0.5(−1) µm.

basal at the base, bottom or proximal end (opposed to terminal, distal).

basal cell cell at the base; in leaves,

* The suffix "-oecious" strictly applies only to sporophytic or diploid sexuality, and thus is inapplicable to bryophytes; however, it is nonetheless frequently employed (autoecious) as an alternative spelling and pronunciation to "-oicous".

frequently differentiated cells of the lower $\frac{1}{4}$–$\frac{1}{3}$ of a leaf; cell at point of attachment of an axillary hair.

basal membrane a short tube or cylinder often supporting segments and cilia of the endostome.

beak (rostrum) elongated apex of an operculum, calyptra, or perianth.

bifid divided into two parts (cf. **bifurcate**).

bifurcate Y-shaped or forked; leaves divided into more or less equal parts.

bilabiate two-lipped.

bipinnate with both primary and secondary stems pinnate; twice-pinnately branched.

bistratose composed of two cell layers, e.g., leaf blades that are two cells thick.

bordered having margins differentiated from the rest of the structure in shape, size, colour, or thickness, e.g., leaves, thalli, and peristome teeth (cf. **limbate**).

bract a modified leaf associated with a gametangium (cf. **perichaetial leaf**, **perigonial leaf**; see **perichaetium**).

bracteole a modified underleaf associated with a gametangium in liverworts (cf. **perichaetium**).

brood body a generalised term used to denote various types of specialised vegetative reproductive structures, e.g., reduced buds, leaves, branches, or plant fragments (propagules); small, globose, ellipsoidal or cylindrical to filamentous septate bodies (gemmae) (cf. **diaspore**, **propagulum**).

bryoid areolation having lax rhomboid-hexagonal cells, as in *Bryum*.

bryoid peristome having a "perfect" diplolepidous peristome, i.e., with a fully formed exostome and endostome having well developed segments and cilia, as in *Bryum, Hypnum*.

bulbiform bulb-like; with imbricate radical leaves forming bulb-like plant.

bulbil vegetative propagule; a small, deciduous, bulb-like axillary propagulum, or rhizoidal gemma (cf. **brood body**, **tuber**).

bulbiform bulb-like; with imbricate leaves forming bulb-like plant.

C-shaped papilla (pl. papillae) papilla (or mammilla) appearing crescent-shaped (occasionally circular) when viewed from above; such papillae may themselves be smooth or rough and variously interlinked, but are typically arranged in groups with the open ends facing inwards.

caducous detaching or falling off very early; usually in reference to leaves, leaf tips, or perianths (cf. **deciduous**, **fugacious**).

caespitose tufted; growing in cushions or sods.

calcicolous growing on chalky, sandstone, or limestone substrates.

calcifuge growing on acidic or poor in base substrates.

calyptra (pl. calyptrae) a membranous covering of haploid tissue over the developing sporophyte, derived largely from the archegonial venter (see also **vaginula**).

campanulate bell-shaped; referring to a calyptra that is elongated and cylindrical; a campanulate-cucullate calyptra is cylindrical and split along one side only; a campanulate-mitrate calyptra is cylindrical and undivided or equally lobed at the base.

canescent grey or hoary; usually caused by long hyaline leaf awns, e.g., *Grimmia*.

capsule the sporangium; terminal spore-producing part of the sporophyte; in most mosses it is differentiated into an apical operculum, central urn (spore-bearing region), and a sterile basal neck or hypophysis; in most liverworts and hornworts it is a uniform structure containing spores and elaters or pseudoelaters.

carinate with a keel.

central strand a small group of elongate cells forming a central axis of some stems and thalli, usually coloured and thin-walled in transverse section; also called axial strand.

cernuous nodding or drooping.

channelled hollowed out like a gutter and semicircular in cross-section (cf. **keeled**).

cilium (pl. cilia) delicate, hair-like or thread-like structure mostly one cell wide and unbranched (opposed to lacinia); in peristomes, the structures frequently found singly or in groups alternating with the segments of the inner peristome.

chlorophyllose containing chlorophyll; generally green unless masked by some other pigments.

circinate curved in a circle, e.g., leaves of *Hypnum circinale*.

cladautoicous see **autoicous**.

cladocarpous form of pleurocarp (see **pleurocarpous**) in which sporophytes are borne terminally on short lateral branches.

clavate, claviform club-shaped; thickened towards the apex; spore surface with processes (clave, sing. clava) higher than broad, with swollen heads that are slightly tapering towards the base, higher than 0.5(−1) μm.

cleistocarpous indehiscent; capsule without a regular mechanism for opening (opposed to stegocarpous).

coalescent fusing together; union of like parts (cf. **adnate**, **connate**).

columella the central, sterile tissue in the sporogenous region of a capsule in most mosses and hornworts.

comal tuft a tuft of leaves at tip of a stem or branch, e.g., *Bryum*.

comose with larger and more crowded leaves forming tufts or comae (sing. coma) at the stem tips.

complanate flattened or compressed, such as leaves flattened into more or less one plane.

compound pore air pore bordered by both superficial cells arranged in concentric rings and an internal, cylindrical, or barrel-shaped structure of epidermal origin; e.g., in *Marchantia*.

conduplicate strongly folded longitudinally along the middle.

conical cone-shaped, e.g., the operculum of *Bryum*.

connate growing together; the fusion of like parts (cf. **adnate**, **coalescent**).

constricted abruptly narrowed; tightened, e.g., urn below capsule mouth.

contorted irregularly curved or twisted (cf. **tortuose**).

contracted abruptly narrowed or shortened.

convolute rolled together and forming a sheath, e.g., the perichaetial leaves of *Barbula convoluta*.

cordate heart-shaped (in the traditional playing-card sense); leaf shape with the large, rounded ends at the base or point of attachment (cf. **auriculate**).

corticolous growing on bark (cf. **arboreal**).

costa (pl. costae) nerve or midrib of a leaf, always more than one cell thick.

crenate with rounded teeth (cf. **crenulate**).

crenulate with minute, rounded teeth (cf. **crenate**).

crisped (crispate) wavy; often used more loosely to mean variously curled, twisted, and contorted.

cucullate hooded or hood-shaped; a calyptra split along one side only; also used to describe leaves strongly concave and erect or inflexed at the tips, like a monk's cowl.

cushions in mosses, clusters formed by plants with stems ± erect, tightly clustered, and somewhat radiating at edges, e.g., *Grimmia*; in Marchantiales, antheridial and archegonial sessile receptacles on the thallus.

cuspidate ending abruptly in a stout, rigid point (cf. **apiculate**, **mucronate**).

cuticle an extracellular cutinized layer on leaves, stems, setae, and capsules of mosses, and capsules of hornworts.

cygneous curved like a swan's neck, e.g., setae in *Grimmia*.

cylindrical elongate and circular in transverse section.

deciduous falling off, e.g., shedding of operculum (cf. **caducous**, **fugacious**).

decumbent with stem prostrate but with ascending tips.

decurrent with basal leaf margins extending down the stem past the leaf insertion as ridges or narrow wings.

dehiscent the capsule opening regularly by means of an annulus and operculum or valves (opposed to indehiscent).

dendroid tree-like; branched above a distinct trunk-like stipe.

dentate with sharp teeth directed outward (cf. **denticulate**).

denticulate finely toothed (cf. **dentate**).

denuded stripped of leaves, made bare, or left naked, e.g., bare costa left by erosion of lamina or stem by erosion or loss of leaves.

deoperculate referring to capsule after the operculum has fallen.

depauperate poorly developed.

deuter cells see **guide cells**.

diaspore an agent of dispersal; any structure that becomes detached from the parent plant and gives rise to a new individual (syn. propagulum; cf. **brood body**, **gemma**).

dichotomous division of a structure to yield two, more or less equal, parts, e.g., forking of the axis into two branches.

dioicous (dioecious*) with archegonia and antheridia borne on separate plants. (opposed to monoicous; cf. **autoicous**, **paroicous**, **synoicous**).

diploid a cell, individual, or generation with two sets of chromosomes (2n); the typical chromosome level of the sporophyte generation.

diplolepidous (diplolepideous) form of arthrodontous peristome (originally): having main teeth with two columns of cells up dorsal face (strictly): having two concentric circles of teeth, with the outer circle (exostome) derived from thickening of the contiguous walls of the outer and primary peristomial layers and the inner circle (endostome) derived from thickening of the contiguous walls of the primary and inner peristomial layers (opposed to haplolepidous).

discoid flattened and disc-like, or plate-like.

disciform flat and circular, orbicular.

disjunct occurring in widely separated geographic areas.

distal away from the base or point of attachment; towards the apex of a leaf or stem; the outer, convex face of a spore (opposed to basal, proximal).

distichous leaves alternating in two opposite rows, e.g., *Fissidens*.

dorsal (of leaves) the abaxial, back or lower surface; (of peristome teeth) the outer face; (of stems or thalli) the outer surface, away from the substrate (opposed to ventral).

dorsal lamina part of the leaf blade opposite the sheathing base, at the back of the costa and below the apical lamina, e.g., *Fissidens*.

double peristome having both an endostome and an exostome (cf. **diplolepidous**).

dwarf male tiny male gametophyte usually borne on the female plant.

---

* See note after autoicous.

echinate roughened by blunt spiny projections.

echinulate roughened by very small spiny projections.

elater (pl. elaters, elateres; eleters) a differentiated elongate cell, dead at maturity and normally with one to three helicoidal wall thickenings, found interspersed among the spore mass in most liverwort capsules; function: to break up and subsequently help disperse the spores (cf. **pseudoelater**).

elaterophore a tuft or brush-like cluster of elater-like cells attached to either the base or apex of a capsule in a few liverworts, e.g., *Pellia*.

ellipsoidal an oval solid.

elliptical oblong with convex sides or ends.

elongate stretched out, e.g., linear.

emarginate broad, shallowly notched apex, deeper than retuse, e.g., *Tortula*.

emergent partially exposed, referring to capsules or perianths only partly projecting beyond the tips of perichaetial leaves (cf. **exserted**, **immersed**).

endostome the inner circle of a diplolepidous peristome, formed from contiguous periclinal wall-pairs of the primary and inner peristomial layers; typically a weak membranous structure consisting of a basal membrane bearing segments and cilia. The endostome is homologous with the haplolepidous peristome.

endothecium (pl. endothecia) in bryophytes the inner part of the embryonic capsule that gives rise to all tissue interior to the outer spore sac.

entire without teeth, more or less smooth on the margin, e.g., leaves, thalli.

epidermis (pl. epidermides, epiderms) the outer cell layer of a stem or a thallus.

epidermal pore opening in the upper epidermis of most complex thalloid liverworts.

equitant straddling; referring to conduplicate and strongly sheathing leaf bases, e.g., *Fissidens*.

erect with leaves directed toward stem apex; with leaf margins curved upward (adaxially); with capsules straight, not curved.

erectopatent spreading at an angle of 45° or less (cf. **patent**, **spreading**, **squarrose**, **widespreading**).

excurrent extending beyond the apical margin, e.g., an awn formed by a protruding costa.

exine the outermost wall layer of a spore that is produced by the spore.

exostome the outer circle of the diplolepidous peristome, formed from contiguous periclinal wall-pairs of the outer and primary peristomial layers; missing or rudimentary in haplolepidious peristomes.

exothecium the outermost layer of the capsule wall, consisting of exothecial cells; the capsule epidermis.

exserted projecting and exposed, e.g., capsules or perianths held clear of the tips of perichaetial leaves (cf. **emergent**, **immersed**).

falcate curved like the blade of a sickle.

falcate-secund strongly curved and turned to one side.

fastigiate with branches erect and more or less equal in length.

filament a thread-like structure, e.g., peristome teeth.

filamentous thread-like.

filiform slender and elongate, filamentous, thread-like.

fimbriate fringed, generally with radiating cell walls of partly eroded marginal cells.

flagelliform whip-like; branches with a gradual attenuation from ordinary leaves at the branch base to vestigial-leaved branch tips (cf. **stoloniferous**).

flexuose slightly and irregularly bent, twisted, or wavy (cf. **undulate**).

foot the basal portion of most bryophyte sporophytes, embedded in the gametophyte and serving as both an organ of absorption and attachment.

forked see **bifurcate**.

foveolate pitted (cf. **alveolate**).

frondose leaf-like; resembling a fern frond; in mosses, closely and regularly branched in one plane (see **pinnate**); in liverworts, a thallus that is crisped or lobed.

fugacious vanishing or readily falling away (cf. **caducous, deciduous**).

fusiform spindle-shaped; narrow (more than three times as long as wide) and tapered at both ends.

galeate hollow and vaulted; like a helmet, e.g., the lobule in certain species of *Frullania*.

gametangium (pl. gametangia) vessel bearing gametes, e.g., archegonium, antheridium (see **androecium**, **gametoecium**, **gynoecium**).

gametangiophore a specialised gametangia-bearing branch, producing either archegonia (archegoniophore) or antheridia (antheridiophore).

gamete reproductive cell; i.e., spermatozoid, ovum.

gametoecium (pl. gametoecia) gametangia and surrounding bracts (see **androecium**, **gametangium**, **gynoecium**).

gametophore a gametangium-bearing stalk (see **gametangiophore**); loosely used for the mature gametophyte plant(s) developed from the protonema.

gametophyte the haploid, sexual generation; in bryophytes, the dominant generation, consisting normally of green, leafy or thalloid plants, bearing antheridia and/or archegonia.

gemma (pl. gemmae) uni- or multicellular, filamentous, globose, ellipsoidal, cylindrical, stellate, or discoid brood bodies, relatively undifferentiated, serving in vegetative reproduction (cf. **brood body**).

gemma cup cup-shaped, gemmae-containing structure of thalline or foliar origin (cf. **splash-cup**).

gemmate budlike; loosely used to mean bearing gemmae (gemmiferous) and in describing spore sculpture of mosses, i.e., spore surface with processes (gemmae) that are constricted at the base, always higher than 0.5(−1) μm, the diameter is equal to or larger than the height.

gemmiferous bearing gemmae.

geniculate sharply bent as a knee.

gibbous swollen or bulging on one side.

glabrous smooth; not papillose, rough or hairy.

glaucous with a whitish, greyish, or bluish overcast, hue, or colour.

globose spherical.

gonioautoicous see **autoicous**.

granulate roughened with ± round bodies always shorter than 0.5 μm; referring to the surface of spores or projections on it.

granulose roughened with minute, blunt projections; also used in describing spore sculpture.

gregarious growing together in loose tufts or mixed mats.

guard cells specialised cells bordering an opening in the capsule wall (see **stoma**).

guide cells large, highly vacuolated, thin-walled and longitudinally arranged cells found in a median layer across the costa of many mosses, part of conducting parenchyma (sometimes referred to as interstereid cells, deuters, central cells, and socii).

gymnostomous lacking a peristome.

gynoecium (pl. gynoecia) the female gametoecium, consisting of archegonia and the surrounding bracts (perichaetial leaf or involucre).

hair-point a hair-like leaf tip formed by an attenuated apex or longly excurrent nerve (Smith 1978) (see **awn**).

haploid a cell, structure or organism having a single set of chromosomes (n); e.g., the normal chromosome level of the gametophyte generation.

haplolepidous (haplolepideous) form of arthrodontous peristome; originally, having main teeth formed from a single column of cells up dorsal face; strictly, having only one circle of teeth derived from thickening of the contiguous walls of the primary and inner peristomial layers [homologous to the inner diplolepidous circle, endostome]; also a member of the Haplolepidae (opposed to diplolepidous).

hoary whitish or greyish.

hyaline colourless, transparent.

hydroid tracheid-like conductive cell in the central strand of some bryophytes, especially mosses, sometimes also in the costa.

hydrophilous growing in water; loving water.

hydrophyte aquatic plant, partly or completely immersed.

hygroscopic responding to humidity changes, e.g., peristome teeth.

hymenium membrane.

hypnoid having a complete peristome; occasionally used to refer to a moss with a pleurocarpous habit.

hypodermal one or more layers of differentiated cells beneath the epidermis of the stem.

idioblast a uniquely differentiated cell, distinct from other cells of the same tissue in size, form and/or contents; e.g., ocellus, oil-cell.

imbricate closely oppressed and overlapping, e.g., with the leaf margins overlapping like shingles on a roof.

immersed submerged or below the surface; referring to a capsule or perianth exceeded by the blades or awns of the perichaetial leaves (cf. **emergent**, **exserted**), or to sunken stomata.

incised cut into sharp divisions separated by narrow sinuses.

inclined bent down; capsules that are between the erect and horizontal positions; i.e., drooping (cf. **pendent**, **pendulous**).

incrassate with thickened cell walls.

incubous lying upon; an oblique leaf insertion in which the antical (distal) leaf margins are oriented toward the dorsal stem surface; when viewed from above the antical leaf margins will overlap the postical (proximal) leaf margins of the leaves directly above.

incurved curved upward (adaxially) and inward, subjectively stronger than inflexed and weaker than involute; applied to leaf tips and margins (opposed to recurved).

indehiscent lacking distinct opening mechanism; spores shed by irregular rupture or breakdown of capsule wall, e.g., *Archidium* (opposed to dehiscent).

inflexed bent upward (adaxially) and weakly inward, applied to leaf margins or leaves on a stem (opposed to reflexed; cf. **incurved**, **inrolled**, **involute**).

inner peristome see **endostome**.

innovation a new shoot; in acrocarpic mosses, a branch formed after the formation of sex organs; in mosses subfloral innovations are produced at the base of a gynoecium; in liverworts, between bracts and perianth.

inrolled rolled upwards (adaxially) and tightly inwards; applied to leaf margins (syn. involute) (cf. **incurved**, **inflexed**).

intercalary inserted between, e.g., meristems with growth that is not apical or basal, but occurs at some distance between.

intine the innermost wall layer of a spore.

intramarginal submarginal; structures

close to or associated with the margin but not strictly on the margin.

involucre a protective sheath of tissue of thalline origin surrounding a single antheridium (e.g., some Sphaerocarpales), archegonium or sporophyte (e.g., Marchantiaceae, Metzgeriales, and Anthocerotopsida); often used loosely as a general term for any sheath-like structure surrounding sporophytes or gametangia.

involute rolled upward (adaxially) and tightly inward, applied to leaf margins (syn. inrolled; opposed to revolute; cf. **incurved**, **inflexed**).

julaceous smoothly cylindric, like a catkin, referring to stems or branches with strongly imbricate leaves.

keeled sharply folded along the middle, like the keel of a boat; V-shaped in cross-section (cf. **channelled**).

laciniate fringed with lacinia; subjectively stronger than fimbriate, but not necessarily formed by cell erosion.

lacinia (pl. laciniae) segment; appendages coarser than cilia and more than one cell wide.

lacunose perforated with holes, e.g., the surface of various thalloid liverworts.

lamella (pl. lamellae) parallel photosynthetic ridges or plates, along a leaf blade, costa, or thallus, e.g., *Pterygoneurum*.

lamina (pl. laminae) the flattened, generally unistratose and green part of the leaf blade excluding the costa and border; the expanded part of a thallus (see **apical**, **dorsal**, **vaginant lamina**).

lanceolate lance-shaped, in bryology narrow and tapered from near the base; narrowly ovate-acuminate.

lax loose; referring to large thin-walled cells, as well as to nature and spacing of leaves on stem, or of stems in a tuft.

lenticular doubly convex, lens-shaped; apical cell with two segmenting surfaces as in *Metzgeria*, *Fissidens*.

ligulate strap-shaped; narrow, moderately long, sides parallel (cf. **lingulate**).

limbate bordered by distinct, elongate, and incrassate marginal cells, e.g., *Fissidens*.

limbidium (pl. lambidia) border; differentiated leaf margin, e.g., *Fissidens bryoides*.

linear very narrow, elongate with nearly parallel sides; narrower than ligulate.

lingulate tongue-shaped; oblong with a slightly broadened apex (cf. **ligulate**).

lobule a small lobe; e.g., the smaller segment of an unequally divided leaf in leafy liverworts.

lumen (pl. lumina) the cell cavity.

lunate crescent shaped.

mammilla (pl. mammillae) strongly bulging surface of a cell; also used for various hollow papilla-like protuberances without associated local wall thickening; i.e., with the cell lumen extending into the protuberances (cf. **papillose**).

mammillose with mammillae.

marginal at the margin, especially as applied to a leaf.

median central, middle, e.g., median leaf cells are from the upper middle of a leaf, midway between costa and margin (see **mid-leaf**).

median groove furrow along thallus midline.

meristem a permanent (open system) or temporary (closed system) zone of actively dividing, undifferentiated cells; mature tissues develop from the differentiation of cells generated by mitotic divisions in this zone.

mesic moist, neither very wet nor very dry; referring to habitat.

mesophytic adapted to a moderately humid habitat.

mid-leaf middle third of a leaf; between the upper leaf and lower leaf; referring to median leaf cells between the margin and costa.

mitrate (mitriform) conic and undivided (similar to a bishop's mitre) or equally lobed at base, referring to calyptrae (opposed to cucullate; see **campanulate**).

monoicous (monoecious*) bisexual; with antheridia and archegonia on the same plant, including autoicous, paroicous, polyoicous, and synoicous (opposed to dioicous).

mucro (pl. mucrones) a short, abrupt point (see **apiculus**).

mucronate ending abruptly in a short point, usually caused by a shortly excurrent costa (apiculate is somewhat longer; cuspidate is even longer and stouter).

muticous without a point or awn.

neck the sterile basal portion of a capsule, sometimes considerably differentiated, e.g., *Bryum*); also the upper narrow part of archegonium.

nodose with short knob-like thickenings, e.g., endostomial cilia in many *Bryum* species (cf. **appendiculate**).

nodulose with nodular thickenings; minutely knobbed; sometimes referred to intracellular wall thickening, e.g., *Racomitrium* (cf. **appendiculate**).

oblate wider than long.

oblique slanted; e.g., an oblique leaf insertion is one that is between transverse and longitudinal.

obloid three-dimensional equivalent of oblong, but with reference to capsules, with rounded edges and corners.

oblong rectangular with rounded corners or ends.

obovate egg-shaped with apex broader than base.

obovoid an inversely ovoid solid.

obscure indistinct; often applied to cells with dense cytoplasmic inclusions or dense papillae making cell walls difficult to see.

obtuse broadly pointed, more than 90°; used sometimes as blunt or rounded.

ocellus (pl. ocelli) an idioblastic leaf cell having one large oil body and lacking chloroplasts, also found in underleaves, bracts, and perianths of certain leafy liverworts (*Frullania*, certain Lejeuneaceae).

oil body a membrane-bound, terpene-containing organelle unique to the cells of liverworts.

oil cell an idioblastic cell characterised by a very large oil body, common in thalloid liverworts, e.g., *Marchantia*.

opaque not transparent or translucent.

operculum (pl. opercula) the lid covering the mouth of most moss capsules; usually separated from the mouth by an annulus to open the capsule (see **stegocarpous**).

orbicular nearly circular.

ostiole a pore-like opening, e.g., the aperture through which spermatozoids are released from the antheridium in some thalloid liverworts.

outer peristome see **exostome**.

ovate outline of an egg with base broader than apex.

ovoid an egg-shaped solid.

palmate with finger-like lobes radiating from centre.

papilla (pl. papillae) cell ornamentation, a solid microscopic protuberance (see **papillose**).

papillose bearing papillae; monopapillose bearing one simple unbranched papilla on the cell surface; pluripapillose bearing several papillae, or one compound or branched papilla on the cell surface (see **C-shaped papilla**). Loosely applied to any minutely rough surface, that may be strictly mammillose.

---

* See note after autoicous.

paraphyllium (pl. paraphyllia) small green outgrowth of various shapes; produced randomly on the stems or branches of many pleurocarpous mosses (cf. **pseudoparaphyllium**).

paraphysis (pl. paraphyses) hyaline or yellowish, usually uniseriate, hair often associated with antheridia and archegonia of mosses.

parenchyma (pl. parenchymata) a tissue of relatively undifferentiated, usually thin walled and isodiametric cell, with non-overlapping end walls (opposed to prosenchyma).

paroicous (paroecious*) with antheridia and archegonia in a single gametoecium but not mixed, the antheridia in the axils of bracts just below the bracts surrounding the archegonia (cf. **autoicous**, **dioicous**, **monoicous**, **synoicous**).

patent of leaves spreading from stem at an angle of 45° or more (cf. **erecto-patent**, **spreading**, **squarrose**, **wide-spreading**).

pellucid clear, translucent or transparent.

peltate shield-like structure fixed on a central stalk.

pendent (pendant) hanging downward (cf. **inclined**, **pendulous**).

pendulous hanging, pendent, e.g., capsules drooping and inclined beyond horizontal; stems and branches that hang.

percurrent extending to the apex.

perennating tissue tissue that is capable of surviving adverse conditions and later resuming meristematic activity, e.g., tubers of *Riccia*.

perforate pierced through.

perianth organ of foliar origin enclosing the archegonia in most leafy liverworts.

perichaetium (pl. perichaetia) the gynoecium; strictly the ensheathing cluster of modified leaves or underleaves (bracts; bracteoles) and perianth, if present, enclosing the archegonia.

perichaetial leaf modified leaf or underleaf (bract; bracteole) associated with the gynoecium; collectively forming the perichaetium.

periclinal a plane of division parallel to the surface; applied to a cell division that is parallel to a structure; e.g., the free surface of an apical cell, or the long axis of a stem (opposed to anticlinal).

perigonial leaf modified leaf or underleaf (bract; bracteole) associated with the androecium; collectively forming the perigonium.

perigonium (pl. perigonia) the androecium; strictly the cluster of modified leaves or underleaves (bracts; bracteoles) enclosing the antheridia.

perigynium (pl. perigynia) a somewhat fleshy, tubular structure around the archegonial cluster and subsequent sporophyte, derived from axial cells peripheral to the archegonial cluster.

peristome a circular structure, generally divided into 2^n (i.e., 4, 8, 16, 32, or 64) teeth, arranged in a single or double (rarely multiple) row around the mouth of a capsule (see **endostome**, **exostome**, **prostome**).

persistent not falling or non-deciduous, long lasting, e.g., some calyptrae; protonema of *Ephemerum*.

piliferous with hair point (see **awn**).

pinnate with numerous, spreading branches on opposite sides of the axis and thus resembling a feather.

plagiotropic having the direction of growth oblique or horizontal.

pleurocarpous producing sporophytes laterally from a perichaetial bud or a short specialised branch (see **cladocar-**

* See note after autoicous.

pous) rather than at the stem tip; with stems usually prostrate, creeping, and freely branched mosses growing in mats rather than tufts (opposed to acrocarpous).

plicate with longitudinal furrows or pleats (plica).

pluristratose in multiple layers, e.g., denoting thickness of leaves.

polyoicous with several forms of gametoecia on the same plant (or various plants of the same species); also called heteroicous and polygamous.

pore a small aperture, the opening in the wall of some cells.

porose having pores.

postical the ventral surface of a stem; that leaf margin oriented towards the base of a longitudinal or obliquely inserted leaf (opposed to antical).

preperistome see **prostome**.

procumbent spreading, prostrate.

prolate longer than wide (opposed to oblate).

propagulum (pl. propagula) a reduced bud, branch, or leaf serving in vegetative reproduction (syn. diaspore; cf. **brood body**, **gemma**).

prorate having papillae or mammillae borne at the tips of cells, or formed by projecting cell ends, e.g., *Philonotis*.

prosenchyma (pl. prosenchymata) a tissue made up of narrow elongate cells with tapered overlapping end walls (opposed to parenchyma).

prostome (also called preperistome) a rudimentary structure outside, and usually adhering to, the main peristome teeth.

prostrate laying flat on the ground; creeping.

protonema (pl. protonemata) a filamentous, globose or thalloid structure resulting from spore germination and including all stages of development up to the production of one or more gametophores.

proximal near the base or point of attachment; the internal face of a spore (opposed to distal; cf. **basal**).

pseudautoicous see **autoicous**.

pseudoelater (pl. pseudoelateres, pseudoelaters) false elater; the unicellular or multicellular sterile cells of the hornworts developed after a few mitotic divisions and subsequent differentiation of diploid elaterocytes; protoplasmic at maturity (cf. **elater**).

pseudoparaphyllium (pl. pseudoparaphyllia) small, unistratose, filiform, or foliose structure resembling paraphyllium, but restricted to the areas of the stem around branch primordia; often found in pleurocarpous mosses (cf. **paraphyllium**).

pseudoperianth tissue of thalline origin surrounding one or several archegonia, calyptra, and subsequent sporophyte, e.g., *Marchantia.*

pyriform pear-shaped, e.g., capsules of *Bryum.*

radiculose covered with rhizoids.

receptacle a disc or wart-like mass of tissue bearing antheridia or archegonia and found directly on the thallus (e.g., *Conocephalum*, *Corsinia*), inside the thallus (e.g., *Pellia*), or elevated and terminating a gametangiophore (e.g., *Marchantia*).

recurved curved downward (abaxially) and inward; in leaves, referring to margins, apices, or marginal teeth; in peristome teeth curved outward and ± downward (opposed to incurved).

reflexed bent down (abaxially) and inward, generally referring to leaf margins or leaves on a stem (opposed to inflexed).

reticulate netted, as the network pattern produced by cell wall thickenings; often used in describing spore sculpturing.

retuse a slight indentation or notch in a broad, rounded apex (cf. **emarginate**).

revoluble rolling away, referring to an annulus that falls in a broken ring.

revolute rolled downward (abaxially) and backward, referring to a leaf margin (opposed to involute).

rhizautoicous see **autoicous**.

rhizoid hair-like structure externally and internally smooth, or tuberculate, with intracellular projections that functions in absorption and anchorage; in mosses, usually brown to reddish, simple or branched, multicellular filaments, generally with oblique end-walls; in liverworts and hornworts, unicellular and usually hyaline (cf. **tomentose**).

rhizoid furrow a longitudinal, rhizoid bearing channel or groove on the (ontogenetically) postical surface of a gametangiophore, e.g., *Marchantia*.

rhizome a slender, horizontal, subterranean stem giving rise to erect secondary stems (also called subterranean stolon).

rostellate with short beak.

rostrate beaked, narrowed into a slender tip or point.

rugulate spore surface with an irregular pattern of lumina and walls (cf. **reticulate**).

rugulose with weak, transverse wrinkles.

rupestral growing on rock or rock walls (see **saxicolous**).

saxicolous growing on rocks.

scale a thin, membranous structure.

sciophilous shade-loving.

secund turned to one side, e.g., leaves on a stem.

seriate in rows; e.g., uni-, bi-, tri-, multiseriate; applied either to adjacent rows of leaf cells, or to ranks of leaves on stem.

serrate saw-toothed; with marginal teeth pointing forward (towards apex).

serrulate minutely serrate.

sessile without a stalk or seta.

seta (pl. setae) elongated portion of the sporophyte between the capsule and foot.

sheathing surrounding and clasping the stem, base of seta, capsule, or leaves.

sheathing lamina see **vaginant lamina**.

shoulder an area of abrupt narrowing, e.g., the area on a leaf where the leaf base is abruptly narrowed to the upper lamina or blade, or a similar constriction on an exostome tooth.

sigmoid doubly curved in opposite directions, S-shaped.

sinuate (sinuose) wavy, as in leaf margin, or as in intracellular wall thickening of *Racomitrium* (see **nodulose**).

sinus the notch or indentation between two segments, as in a bifid leaf.

spathulate tapering proximally from a broad, rounded apex.

spermatozoid male gamete; in bryophytes, each gamete has two flagella.

spicate arranged in spikes.

spinose with sharp, pointed teeth; also very high, sharp leaf cell papillae or mammillae.

spinulose minutely spiny.

splash-cup any cup-shaped structure that functions as an aid in the distribution of spermatozoids or gemmae by water splash; e.g., *Marchantia, Lunularia*.

spore a reproductive unit produced in the capsule as a result of meiosis; usually minute, mostly spherical, and generally unicellular bodies that give rise upon germination to protonemata.

sporeling all structures developed between the germination of the spore and the formation of the adult gametophore; includes the protonema (variously differentiated) and the juvenile gametophore.

sporophyte the spore-bearing generation; initiated by the fertilisation of an egg; remaining attached to the gametophyte and partially dependent on it; typically consisting of a foot, seta, and capsule.

spreading forming an angle of 45° or more, e.g., the adaxial angle between a leaf and stem (cf. **erectopatent**, **patent**, **squarrose**, **widespreading**).

squarrose spreading at right angles.

stegocarpous referring to capsules with a dehiscent operculum (opposed to cleistocarpous).

stellate star-shaped.

stereids slender, elongate, thick-walled, fibre-like cells found in groups (stereid bands) in the costa or stems of some mosses.

sterile without reproductive structures or sporophytes; generally referring to absence of sexual structures but can also mean absence of asexual structures.

sterome the entire system of stereids in a moss plant.

stipe the unbranched basal part of an erect stem in a dendroid or frondose moss and some anacrogynous liverworts.

stoloniferous with slender, elongate branches with reduced or vestigial leaves throughout (cf. **flagelliform**).

stoloniform referring to stoloniferous branches or stems.

stoma (pl. stomata) minute opening in the capsule wall of hornworts, and usually in the capsule neck of mosses; usually surrounded or bordered by two guard cells.

stratose in layers, e.g., denoting thickness of leaves; i.e., uni-, bi-, multistratose (cf. **seriate**).

stria (pl. striae) fine ridge or line.

striate marked with fine ridges or lines (striae); referring to a spore surface with a regular pattern of approximately parallel lumina and walls.

striolae very fine ridges or lines.

striolate very finely ridged.

stylus (pl. styli) a column; a one-celled, uniseriate or multiseriate, subulate to triangular structure found between the lobule and the stem in certain leafy liverworts.

subfloral innovation a branch that arises in association with a perichaetial leaf (see **also subperichaetial innovation**).

subperichaetial innovation a branch that arises just below a perichaetium.

subula (pl. subulae) a long, slender point.

subulate slenderly long-acuminate (cf. **acuminate**).

succubous lying under; an oblique leaf insertion in which the antical (distal) leaf margins are oriented toward the ventral stem surface; when viewed from above the antical leaf margins will lie under or be overlapped by the postical (proximal) leaf margins of the leaves directly above; found in various leafy liverworts (cf. **incubous**).

sulcate strongly plicate, with deep longitudinal furrows or grooves.

synoicous (synoecious*) with antheridia and archegonia mixed in the same gametoecium (cf. **autoicous**, **dioicous**, **monoicous**, **paroicous**).

systyllus (systylous) with operculum remaining attached to the tip of the columella after dehiscence, e.g., *Hymenostylium*.

teeth divisions of a diplolepidous exostome or haplolepidous peristome; also sharp projections on the margins, costa, or surface of a leaf.

terete rounded in cross-section.

terminal at the apex, tip or distal end (opposed to basal).

terricolous growing on soil (terrestrial).

tessellated checkered; i.e., in a pattern of squares; applied to the basal membrane of the peristome of *Tortula*.

---

* See note after autoicous.

tetrad a group of four; i.e., spore tetrad.

tetragonal four angled.

thallose of or pertaining to a thallus; often used in conjunction with a prothallial sporeling.

thallus (pl. thalli) a more or less flattened gametophyte, not differentiated into a stem and leaves.

theca (pl. thecae) the spore-bearing portion of a moss capsule (syn. urn).

tomentose woolly, densely radiculose.

tortuose irregularly bent or twisted, e.g., leaves of *Tortula* when dry (cf. **contorted**).

trigones generally triangular or circular intracellular wall thickenings found at the point where three (or more) cells meet; especially common in leaf cells of liverworts.

trilete applied to polar spores with a triradiate ridge on the proximal face and a more or less convex distal face.

triradiate Y-shaped, often used in reference to the prominent ridge on the proximal face of a trilete spore.

truncate abruptly cut off or squared off at the apex.

tuber in mosses, gemmae born on rhizoids, found in many acrocarpous mosses, e.g., *Bryum*; in liverworts, a geotropic outgrowth from the shoot, composed of perennating tissue, allowing for aestivation and subsequent continued growth or vegetative reproduction.

tuberculate with peg-like projections.

tufa a porose limestone formed by deposition from calcareous waters.

tuft growth form with stems erect but radiating at the edges; small cushions; caespitose habit, e.g., *Orthotrichum*.

turf growth form with stems erect, parallel and close together; often covering extensive areas, e.g., *Bryum argenteum*.

underleaf ventral, variously modified leaf in most leafy liverworts (syn. amphigastria).

undulate wavy.

uniseriate in one series; applied to a hair-like structure comprised of a single row of cells; leaf cell papillae in a single row.

unistratose one-layered; comprised of a single cell layer, e.g., most bryophyte leaves.

urceolate urn-shaped; applied to capsules constricted below a wide mouth and abruptly narrowed to the seta.

urn the spore-bearing portion of a capsule (syn. theca; opposed to neck).

vaginant lamina in *Fissidens*, one of the two clasping laminae below the apical lamina.

vaginula (pl. vaginulae) a ring or sheath enveloping the base of the seta, derived from the base of the archegonium and surrounding stem tissue and remaining after the separation of the calyptra.

valve one of the parts or partially detached flap of tissue into which the capsule of most liverworts and hornworts separates upon dehiscence; rare in mosses (cf. **annulus**).

venter (pl. ventri, venters) the swollen basal portion of an archegonium, containing the ovum.

ventral (of leaves) the adaxial, top, or upper surface; (of peristome teeth) the inner face; (of stems and thalli) the lower surface, next to the substrate (opposed to dorsal).

ventral lamina see **apical lamina**.

vermicular worm-shaped; long, narrow, and somewhat wavy, commonly with rounded ends; usually applied to cells.

verrucate spore surface covered with wart-like elevations (verrucae, sing. verruca).

verrucose covered with small wart-like elevations; often used in describing spore sculpture.

verruculose irregularly roughened.

wart a small elevation or protuberance (also on spores) (cf. **papilla**).

weft a loosely interwoven, often ascending growth form; e.g., *Eurhynchium riparioides*.

whorled arranged in a ring or circle.

widespreading spreading at a wide angle, but less than 90° (cf. **erectopatent**, **patent**, **spreading**, **squarrose**).

xeric very dry; referring to habitat.

xerophilous growing in arid places (adj. xerophytic).

REFERENCES

Abramova, A. L. & I. I. Abramov (1964). *Musci e Faucibus Kondara* (*Tadzhikistania*). *Novitates Systematicae Plantarum non Vascularium* 1 : 325–341 [in Russian].

Agnew, S. & C. C. Townsend (1970). *Trichostomopsis* Card., a moss genus new to Asia. *Israel Journal of Botany* 19 : 254–259.

—& M. Vondráček (1975). A moss flora of Iraq. *Feddes Repertorium* 86 : 341–489.

Aiello, P. & M. G. Dia (2000) *Tortula israelis* (Pottiaceae, Musci) found in Sicily. *Flora Mediterranea* 10 : 377–380.

Akiyama, H. (1994). Suggestions for the delimitation of the Leucodontaceae and the infrageneric classification of the genus *Leucodon*. *Journal of the Hattori Botanical Laboratory* 76 : 1–12.

Allen, B. (1994). *Moss Flora of Central America*, Part 1: *Sphagnaceae–Calymperaceae*. *Monographs in Systematic Botany from the Missouri Botanical Garden*, Vol. 49. St. Louis, 242 pp.

Anderson, L. E., H. A. Crum & W. R. Buck (1990). List of mosses of North America north of Mexico. *The Bryologist* 93 : 448–499.

—& B. E. Lemmon (1972). Cytological studies of natural intergeneric hybrids and their parent species in the moss genera *Astomum* and *Weissia*. *Annals of the Missouri Botanical Garden* 59 : 382–416.

Arnell, S. (1957). A new species of *Riccia* from Israel. *Bulletin of the Research Council of Israel*, Sect. D, Botany 6 : 56–57.

Barbey, C. & W. Barbey (1882). *Herborisation au Levant, Egypte, Syrie et Méditerranée*. Lausanne, 183 pp.

Baudoin R., H. Bischler, S. Jovet-Ast & J. P. Hébrard (1984). Une banque de données phytoécologiques des hépatiques de la région méditerranéenne (BRYOMED). *Webbia* 38 : 385–396.

Baum, B. & S. Jovet-Ast (1962). *Riccia* récoltés en Israël. *Revue Bryologique et Lichénologique* 31 : 103.

Bilewsky, F. (1959). A further contribution to the bryophytic flora of Palestine. *Bulletin of the Research Council of Israel*, Sect. D, Botany 7 : 55–64.

—(1963). *Introduction to Bryophytes of Israel*. Tel Aviv, 78 pp. [in Hebrew]

—(1965). Moss-Flora of Israel. *Nova Hedwigia* 9 : 335–434, 18 Plates.

—(1970). Some recent bryological records for Israel. *Revue Bryologique et Lichénologique* 37 : 963–965.

—(1974). Some notes on the distribution of mosses in Israel and Palestine. *Revue Bryologique et Lichénologique* 40 : 245–261.

—(1977). New records of mosses in Israel. *Israel Journal of Botany* 26 : 93–97.

—& S. Nachmony (1955). A contribution to the Bryophytic flora of Palestine. *Bulletin of the Research Council of Israel*, Sect. D, Botany 5 : 47–58.

Bischler, H. (1978). *Plagiochasma* Lehm. & Lindenb. II. Les taxa européens et africains. *Revue Bryologique et Lichénologique* 44 : 223–300.

—(1993). *Marchantia* L. – *The European and African taxa*. *Bryophytorum Bibliotheca*, Vol. 45. J. Cramer, Borntraeger, Berlin–Stuttgart, 129 pp.

—& S. Jovet-Ast (1975). Récolte d'Hépatiques de Jérusalem à Nablus et Ein Gedi. *Revue Bryologique et Lichénologique* 41 : 17–26.

— — (1986). The hepatic flora of South-West Asia: A survey. *Proceedings of the Royal Society of Edinburgh* 89B: 229–241.

Bizot, M. (1942). Contribution à la flore bryologique du Liban. *Revue Bryologique et Lichénologique* 13 : 49–53.

—(1945). Quelques mousses de Palestine. *Revue Bryologique et Lichénologique* 15 : 68–69.

—(1954). Remarques sur *Tortula papillosissima* (Copp.) Broth. *Revue Bryologique et Lichénologique* 23 : 268–270.

—(1955). Contribution à la flore bryologique d'Asie mineure et de l'Ile de Chypre. *Revue Bryologique et Lichénologique* 24 : 69–72.

—(1956). Nouvelles remarques sur *Tortula papillosissima* (Copp.) Broth. *Revue Bryologique et Lichénologique* 25 : 268–271.

—, R. Gaume & R. Potier de la Varde (1952). Une poignée de mousses libanaises. *Revue Bryologique et Lichénologique* 21 : 11–13.

Boisselier-Dubayle, M. C. & H. Bischler (1994). A combination of molecular and morphological characters for delimitation of taxa in European *Porella. Journal of Bryology* 18 : 1–11.

— — (1998). Allopolyploidy in the thalloid liverwort *Corsinia* (Marchantiales). *Botanica Acta* 111 : 490–496.

— — (1999). Genetic relationships between haploid and triploid *Targionia* (Targioniaceae, Hepaticae). *International Journal of Plant Science* 160 : 1163–1169.

—, M. De Chaldée, L. Guérin, J. Lambourdière & H. Bischler (1995a). Genetic variability in western European *Lunularia. Fragmenta Floristica et Geobotanica* 40 : 379–391.

—, M. F. Jubier, B. Lejeune & H. Bischler (1995b). Genetic variability in three subspecies of *Marchantia polymorpha*: Isozymes, RFLP and RAPD markers. *Taxon* 44 : 363–376.

—, J. Lambourdière & H. Bischler (1998). Taxa delimitation in *Reboulia* investigated with morphological, cytological and isozyme markers. *The Bryologist* 101 : 61–69.

Bornmüller, J. (1914). Zur Flora des Libanon und Antilibanon. *Beihefte zum Botanischen Centralblatt* 31 : 177–280.

—(1931). Zur Bryophyten-Flora Kleinasiens. *Magyar Botanikai Lapok* 30 : 1–21.

Boros, A. (1925). *Funaria hungarica*, nov. spec. *Magyar Botanikai Lapok* 23 : 73–75.

Bremer, B. (1980). A taxonomic revision of *Schistidium* (Grimmiaceae, Bryophyta). *Lindbergia* 6 : 1–16, 89–117.

de Bridel, S. E. (1812). *Muscologia recentiorum*, Suppl. 2. Gotha (C.G. Ettinger), Paris, 257 pp.

Brotherus, V. F. (1893). Pottiaceae, pp. 380–439 *in*: Engler, A. & K. Prantl (eds.), *Die Natürlichen Pflanzenfamilien*, Teil I, Abteilung 3. W. Engelmann, Leipzig.

—(1904). Brachytheciaceae, pp. 1128–1166 *in*: Engler, A. & K. Prantl (eds.), *Die Natürlichen Pflanzenfamilien*, Teil I, Abteilung 3. W. Engelmann, Leipzig.

—(1924). Fissidentaceae, pp. 143–155; Pottiaceae, pp. 243–302 *in*: Engler, A. & K. Prantl (eds.), *Die Natürlichen Pflanzenfamilien*, Second Edition, Band 10. W. Engelmann, Leipzig.

—(1925). Musci (Laubmoose). *In*: Engler, A. & K. Prantl (eds.), *Die Natürlichen Pflanzenfamilien*, Second Edition, Band 11. W. Engelmann, Leipzig, 542 pp.

Bruggeman-Nannenga, M. A. (1978). Notes on *Fissidens*. I–II. *Proceedings of the Koninklijke Nederlandse Akademie van Wetenschappen*, ser. C, 81 : 387–402.

—(1982). The section *Pachylomidium* (genus *Fissidens*). *Proceedings of the Koninklijke Nederlandse Akademie van Wetenschappen*, ser. C, 85 : 59–104.

—(1987). An annotated list of *Fissidens* species from the Yemen Arab Republic and Sultanate of Oman, with *F. laxitexturatus* nov. spec. – Studies in Arab Bryophytes 7. *Nova Hedwigia* 45 : 113–117.

—& E. Nyholm (1986). *Fissidens* Hedw., pp. 8–14 *in*: Nyholm, E., *Illustrated flora of Nordic Mosses*, Fasc. 1: Fissidentaceae-Seligeriaceae. Nordic Bryological Society, Lund.

Brugués, M. (1998). The identity of *Entosthodon durieui* and *E. pallescens*. *The Bryologist* 101 : 133–136.

Brullo, S., M. Privitera & M. Puglisi (1991). Note sulla flora e vegetazione briofitica di alcune aree desertiche di Israele. *Candollea* 46 : 145–153.

Brummitt, R. K. & C. E. Powell (eds.) (1992). *Authors of plant names*. Royal Botanic Gardens, Kew, 732 pp.

Buck, W. R. (1980). Animadversions on *Pterigynandrum* with special commentary on *Forsstroemia* and *Leptopterigynandrum*. *The Bryologist* 83 : 451–465.

—(1981). A review of *Cheilothela* (Ditrichaceae). *Brittonia* 33 : 453–456.

—& H. Crum (1978). A re-interpretation of the Fabroniaceae with notes on selected genera. *Journal of the Hattori Botanical Laboratory* 44 : 347–369.

—& D. H. Vitt (1986). Suggestions for a new familial classification of pleurocarpous mosses. *Taxon* 35 : 21–60.

—& B. Goffinet (2000). Morphology and classification of mosses, pp. 71–123 *in*: Shaw, A. J. & B. Goffinet (eds.), *Bryophyte Biology*. Cambridge University Press, Cambridge.

Cano, M. J., J. Guerra & R. M. Ros (1993). A revision of the moss genus *Crossidium* (Pottiaceae) with the description of the new genus *Microcrossidium*. *Plant Systematics and Evolution* 188 : 213–235.

— — — (1994). *Pterygoneurum compactum* sp. nov. (Musci: Pottiaceae) from Spain. *The Bryologist* 97 : 412–415.

— — — (1996). Identity of *Tortula baetica* (Casas & Oliva) J. Guerra & Ros with *T. israelis* Bizot & Bilewsky. *Journal of Bryology* 19 : 183–185.

Cao, T. & C. Gao (1990). A new species of *Encalyptra* (Musci) from China. *Acta Bryolichenologica Asiatica* 2 : 1–4.

Carrión, J. S., J. Guerra & R. M. Ros (1990). Spore morphology of the European species of *Phascum* Hedw. (Pottiaceae, Musci). *Nova Hedwigia* 51 : 411–433.

—, R. M. Ros & J. Guerra (1993). Spore morphology in *Pottia starckeana* (Hedw.) C. Müll. (Pottiaceae, Musci) and its closest species. *Nova Hedwigia* 56 : 89–112.

Casares-Gil, A. & F. Beltrán (1912). *Entosthodon physcomitrioides* nov. esp. *Boletín de la Sociedad Española de Historia Natural* 12 : 375–377.

Casas, C. (1991). New checklist of spanish mosses. *Orsis* 6 : 3–26.

—& M. Brugués (1980). Nova aportacio al coneixement de la Brioflora dels Monegros. *Anales del Instituto Botánico Antonio José Cavanilles* 35 : 103–114.

—, C. Sérgio, R. M. Cros & M. Brugués (1990). Datos sobre el género *Acaulon* en la Península Ibérica. *Cryptogamie, Bryologie–Lichénologie* 11 : 63–70.

Casas de Puig, C. (1979). *Funaria pallescens* (Jur.) Broth. var. *mitratus* (Cas.-Gil) Wijk & Marg. en Menorca. *Revue Bryologique et Lichénologique* 45 : 467–470.

Castaldo-Cobianchi, R., S. Giardano & G. Cafiero (1982). Studies on *Timmiella barbuloides* (Brid.) Moenk. IV. SEM and TEM characterization of spore wall and first germination stages. *Journal of Bryology* 12 : 273–278.

Chamberlain, D. F. (1969). New combinations in *Pottia starkeana. Notes from the Royal Botanic Garden, Edinburgh* 29 : 403–404.

—(1978). *Pottia* Fürnr., pp. 234–242 *in*: Smith, A. J. E. *The Moss Flora of Britain and Ireland.* Cambridge University Press, Cambridge.

Chen, P.-C. (1941). Studien über die ostasiatischen Arten der Pottiaceae. I, II. *Hedwigia* 80 : 1–76; 141–322.

Coppey, A. (1909). *Deuxième Contribution à l'Étude des Muscinées de la Grèce. Matèriaux pour Servir à l'ètude de la Flore et de la Gèographie Botanique de L'Orient*, Fasc. 5. Imprimerie Berger-Levrault et Cie, Nancy, 50 pp.

Corley, M. F. V. (1980). The *Fissidens viridulus* complex in the British Isles and Europe. *Journal of Bryology* 11 : 191–208.

—, A. C. Crundwell, R. Düll, M. O. Hill & A. J. E. Smith (1981). Mosses of Europe and the Azores: An annotated list of species, with synonyms from the recent literature. *Journal of Bryology* 11 : 609–689.

Craig, E. J. (1939). *Aloina* (C. Muell.) Kindb., pp. 211–215 *in*: Grout, A. J. (ed.), *Moss Flora of North America North of Mexico*, Vol. I, Part 4. Newfane, Vermont [1972 facsimile edition, Hafner Publishing Company, New York].

Crosby, M. R. (1980). The diversity and relationships of mosses, pp. 115–129 *in*: Taylor, R. J. & A. E. Leviton (eds.), *The mosses of North America.* Pacific Division, American Association for the Advancement of Science, California Academy of Sciences, San Francisco.

—, R. E. Magill & C. R. Bauer (1992). *Index of Mosses. Monographs in Systematic Botany from the Missouri Botanical Garden*, Vol. 42, St. Louis, 646 pp.

Crum, H. & L. E. Anderson (1981). *Mosses of Eastern North America*, 2 Vols. Columbia University Press, New York, 1328 pp.

Crundwell, A. C. & E. Nyholm (1962). Notes on the genus *Tortella.* I. *T. inclinata, T. densa, T. flavovirens* and *T. glareicola. Transactions of the British Bryological Society* 4 : 187–193.

— — (1964). The European species of the *Bryum erythrocarpum* complex. *Transactions of the British Bryological Society* 4 : 597–637.

— — (1972). A revision of *Weissia*, subgenus *Astomum.* I. The European species. *Journal of Bryology* 7 : 7–19.

— — (1974). *Funaria muhlenbergii* and related European species. *Lindbergia* 2 : 222–229.

— — (1977). *Dicranella howei* Ren. & Card. and its relationship to *D. varia* (Hedw.) Schimp. *Lindbergia* 4 : 35–38.

Danin, A. (1996). *Plants of Desert Dunes.* Springer-Verlag, Berlin, 177 pp.

—& U. Plitmann (1987). Revision of the plant geographical territories of Israel and Sinai. *Plant Systematics and Evolution* 156 : 43–53.

—& E. Ganor (1991). Trapping of airborne dust by mosses in the Negev Desert. *Earth Surface Processes and Landforms* 16 : 153–162.

Delgadillo M. C. (1973a). A new species, nomenclatural changes and generic limits in *Aloina, Aloinella* and *Crossidium* (Musci). *The Bryologist* 76 : 271–277.

—(1973b). A quantitative study of *Aloina, Aloinella* and *Crossidium* (Musci). *The Bryologist* 76 : 301–305.

—(1975). Taxonomic revision of *Aloina*, *Aloinella* and *Crossidium* (Musci). *The Bryologist* 78 : 245–303.

Demaret, F. (1986a). Étude du matériel-type de trois taxons de *Bryum*. *Bulletin du Jardin Botanique National de Belgique* 56 : 305–314.

—(1986b). Clef des espèces européennes du groupe *Bryum capillare* Hedw. susceptibles de se rencontrer en Belgique. *Dumortiera* 34–35, 42–45.

Dickson, J. (1801). *Plantarum Cryptogamicarum Britanniae*, Fasc. 4. London, 28 pp.

Dirkse, G. M. & A. C. Bouman (1995). *Crossidium* (Musci, Pottiaceae) in the Canary Islands (Spain). *Lindbergia* 20 : 12–25.

Dixon, H. N. (1924; 1954 reprint). *The Student's Handbook of British Mosses*, Third edition. Sumfield & Day Ltd., Eastbourne, 582 pp.

Ducker, B. F. T. & E. F. Warburg (1961). *Physcomitrium eurystomum* Sendtn. in Britain. *Transactions of the British Bryological Society* 4 : 95–97.

Duell, R. (1984). *Distribution of the European and Macaronesian Mosses* (*Bryophytina*), Band I. *Bryologische Beitraege*, Vol. 4. I. Duell-Hermanns, Rheurdt, 113 pp.

—(1985). *Distribution of the European and Macaronesian Mosses* (*Bryophytina*), Band II. *Bryologische Beitraege*, Vol. 5. I. Duell-Hermanns, Rheurdt, 232 pp.

—(1992). *Distribution of the European and Macaronesian Mosses* (*Bryophytina*). *Annotations and Progress*. *Bryologische Beitraege*. Vol. 8/9. IDH-Verlag, Bad Münstereifel, 223 pp.

von der Dunk, K. & K. von der Dunk (1973). *Eucladium verticillatum* var. nov. *recurvatum*. *Herzogia* 2 : 419–422.

Edwards, S. R. (1979). Taxonomic implications of cell patterns in haplolepidous moss peristomes, pp. 317–346 *in*: Clarke, G. C. S. & J. G. Duckett (eds.), *Bryophyte Systematics*. Systematics Association, Special, Vol. 14. Academic Press, London.

—(1984). Homologies and inter-relationships of moss peristomes, pp. 658–695 *in*: Schuster, R. M. (ed.), *New Manual of Bryology*, Vol. 2. The Hattori Botanical Laboratory, Nichinan.

El-Oqlah, A. A., W. Frey & H. Kürschner (1988a). The bryophyte flora of Trans-Jordan. A catalogue of species and floristic elements. *Willdenowia* 18 : 253–279.

— — — (1988b). *Tortula rigescens* Broth & Geh. (Pottiaceae), a remarkable species new to moss flora of Jordan. *Lindbergia* 14 : 27–29.

Enroth, J. (1994). On the evolution and circumscription of the Neckeraceae (Musci). *Journal of the Hattori Botanical Laboratory* 76 : 13–20.

Feinbrun-Dothan, N. (1978). *Flora Palaestina*, Part Three : *Ericaceae to Compositae* (Text). The Israel Academy of Sciences and Humanities, Jerusalem, 481 pp.

—(1986). *Flora Palaestina*, Part Four : *Alismataceae to Orchidaceae* (Text). The Israel Academy of Sciences and Humanities, Jerusalem, 462 pp.

Field, J. H. (1963). Notes on the taxonomy of the genus *Philonotis* by means of vegetative characters. *Transactions of the Bryological Society* 4 : 429–433.

Fife, A. J. (1980). The affinities of *Costesia* and *Neosharpiella* and notes on the Gigaspermaceae (Musci). *The Bryologist* 83 : 466–476.

—(1985). A generic revision of the Funariaceae (Bryophyta: Musci). Part I. *Journal of the Hattori Botanical Laboratory* 58 : 149–196.

— (1987). Taxonomic and nomenclatural observations on the Funariaceae. 5. A revision of the Andean species of *Entosthodon*. *Memoirs of the New York Botanical Garden* 45 : 301–325.

Frahm, J.-P. (1995). Klasse Musci (Bryopsida), Laubmoose, pp. 121–318 *in* Frahm, J.-P., E. Fischer & W. Lobin, *Die Moos- und Farnpflanzen Europas*, (Gams, H. (ed.), *Kleine Kryptogamenflora*, Band IV). Gustav Fischer Verlag, Stuttgart–Jena–New York.

—& W. Frey (1983). *Moosflora*. Eugen Ulmer, Stuttgart, 522 pp.

Frey, W. (1986). Bryophyte flora and vegetation of South-West Asia. *Proceedings of the Royal Society of Edinburgh* 89B : 217–227.

—(1990). Genoelemente prä-angiospermen Ursprungs bei Bryophyten. *Botanische Jahrbücher für Systematik, Pflanzengeschichte und Pflanzengeographie* 111 : 433–456.

—, I. Herrnstadt & H. Kürschner (1990). *Pterygoneurum crossidioides* (Pottiaceae, Musci), a new species to the desert flora of the Dead Sea area. *Nova Hedwigia* 50 : 239–244.

—& H. Kürschner (1981). The bryological literature of Southwest Asia. *Journal of the Hattori Botanical Laboratory* 50 : 217–229.

— — (1983). New records of bryophytes from Transjordan with remarks on phytogeography and endemism in SW Asiatic mosses. *Lindbergia* 9 : 121–132.

— — (1984). Studies in Arabian bryophytes 3: *Crossidium asirense* (Pottiaceae), a new species from Asir Mountains (Saudi Arabia). *Journal of Bryology* 13 : 25–31.

— — (1987). A desert bryophyte synusia from the Jabal Tuwayq mountain systems (Central Saudi Arabia) with the description of two new *Crossidium* species (Pottiaceae). Studies in Arabian bryophytes 8. *Nova Hedwigia* 45 : 119–136.

— — (1991a). *Crossidium laevipilum* Thér. & Trab. (Pottiaceae, Musci), ein eigenständiges, morphologisch und standortökologisch deutlich unterscheidbares Taxon der Saharo-arabischen Florenregion. *Cryptogamie, Bryologie–Lichénologie* 12 : 441–450.

— — (1991b). Lebensstrategien von terrestrischen Bryophyten in der Judäischen Wüste. *Botanica Acta* 104 : 172–182.

— — (1991c). *Conspectus Bryophytorum Orientalum et Arabicorum*: *An Annotated Catalogue of the Bryophytes of Southwest Asia*. *Bryophytorum Bibliotheca*, Band 39. J. Cramer, Berlin–Stuttgart, 181 pp.

— — (1993). *Trichostomopsis trivialis* (C. Müll.) Robins. (Pottiaceae, Musci) eine südafrikanisch-mediterran disjunkte Sippe, neu für die Bryoflora Jordaniens. *Cryptogamic Botany* 3 : 152–156.

Froehlich, J. (1959). Bryophyten aus Vorderasien. *Annalen des Naturhistorischen Museums in Wien* 63 : 31–32.

Geheeb, A. (1903–1904). Bryophyta. *In*: Kneucker, A., Botanische Ausbeute einer Reise durch die Sinaihalbinsel vom 27. März bis 13. April 1902. *Allgemeine Botanische Zeitschrift für Systematik, Floristik, Pflanzengeographie* 9 : 185–189, 203–204; 10 : 4–5.

Geissler, P. (1985). Notulae Bryofloristicae Helveticae. II. *Candollea* 40 : 193–200.

Greene, S.W. & A.J. Harrington (1989). *The Conspectus of Bryological Taxonomic Literature*, Band 2: Guide to National and Regional Literature. *Bryophytorum Bibliotheca*, Vol. 37. J. Cramer, Berlin–Stuttgart, 321 pp.

Greven, H.C. (1995). *Grimmia Hedw. (Grimmiaceae, Musci) in Europe*. Backhuys Publishing, Leiden, 160 pp.

Griffin, D. III & W.R. Buck (1989). Taxonomic and phylogenetic studies on the Bartramiaceae. *The Bryologist* 92 : 368–380.

Grolle, R. (1983a). Nomina generica hepaticarum: References, types and synonymies. *Acta Botanica Fennica* 121 : 1–62.

—(1983b). Hepatics of Europe including the Azores: An annotated list of species, with synonyms from the recent literature. *Journal of Bryology* 12 : 403–459.

—& D. E. Long (2000). An annotated check-list of the Hepaticae and Anthocerotae of Europe and Macaronesia. *Journal of Bryology* 22 : 103–140.

Grout, A. J. (1928). *Scleropodium* Br. & Sch., pp. 51–55 *in*: *Moss Flora of North America North of Mexico*, Vol. III, Part 1. A. J. Grout, Newfane, Vermont [1972 facsimile edition, Hafner Publishing Company, New York].

—(1934). *Fabronia* Raddi., pp. 227–230 *in*: *Moss Flora of North America North of Mexico*, Vol. III, Part 4. A. J. Grout, Newfane, Vermont [*loc. cit.*].

—(1935). *Pyramidula* Brid., p. 73 *in*: *Moss Flora of North America North of Mexico*, Vol. II, Part 2. A. J. Grout, Newfane, Vermont [*loc. cit.*].

Guerra, J., J. J. Martínez, R. M. Ros & J. S. Carrión (1990). *Phascum longipes* sp. nov. on gypsum soils from Almería, Spain. *Journal of Bryology* 16 : 55–60.

—, M. N. Jimenez, R. M. Ros & J. S. Carrión (1991). El género *Phascum* (Pottiaceae) en la Península Ibérica. *Cryptogamie, Bryologie–Lichénologie* 12 : 379–423.

— —, R. M. Ros & J. S. Carrión (1992). The taxonomic status of *Tortula muralis* var. *baetica* (Musci, Pottiaceae): A comparative study. *Journal of Bryology* 17 : 275–283.

— —, J. J. Martínez-Sánchez & W. Frey (1993). *Grimmia mesopotamica* (Grimmiaceae, Musci) new to Europe. *The Bryologist* 96 : 245–247.

Haessel de Menéndez, G. (1963). Estudio de las Anthocerotales y Marchantiales de la Argentina. *Opera Lilloana* 7 : 1–297.

—(1988). A proposal for a new classification of the genera within the Anthocerotophyta. *Journal of the Hattori Botanical Laboratory* 64 : 71–86.

Hart, H. (1891). *Some Account of the Fauna and Flora of Sinai, Petra, and Wâdi Arabah*. Alexander P. Watt, London, 255 pp.

Hasegawa, J. (1988). A proposal for a new system of the Anthocerotae, with revision of the genera. *Journal of the Hattori Botanical Laboratory* 64 : 87–95.

Hedenäs, L. (1989). On the taxonomic position of *Conardia* Robins. *Journal of Bryology* 15 : 779–783.

Hedwig, J. (1791–1792). *Stirpes Cryptogamicae Novae*, Vol. 3. Leipzig, 100 pp.

Herrnstadt, I. (1992). Checklist of the bryophytes collected during Iter Mediterraneum. II. *Bocconea* 3 : 217–222.

—& C. C. Heyn (1987). Bryology in Israel. *Taxon* 36 : 772–773.

— — (1989). Variation in spore texture in the genus *Pottia* (Pottiaceae) as opposed to uniformity in some terrestrial cleistocarpous mosses (Pottiaceae and Dicranaceae), pp. 281–283 *in*: Herban, T. & C. B. McQueen (eds.), *Proceedings of the Sixth Meeting of the Central and East European Bryological Working Group (CEBWG), Liblice, Czechoslovakia, 12–16 September, 1988*. Průhonice, 283 pp.

— — (1993). New species linking *Phascum* and *Pottia* (Pottiaceae). *Nova Hedwigia* 57 : 135–139.

— — (1999). Three new taxa of Pottiaceae (Musci) from Israel: *Acaulon longifolium*, *Pottia gemmifera* and *Barbula ehrenbergii* var. *gemmipara*. *Nova Hedwigia* 69 : 229–235.

— — & M. R. Crosby (1980). New data on the moss genus *Gigaspermum*. *The Bryologist* 83 : 536–541.

— —, R. Ben-Sasson & M.R. Crosby (1982). New records of mosses from Israel. *The Bryologist* 85 : 214–217.

— — — (1991). A checklist of the mosses of Israel. *The Bryologist* 94 : 168–178.

Hill, M. O. (1981). New combinations in European mosses. *Journal of Bryology* 11 : 599–602.

—(1982). A reassessment of *Acaulon minus* (Hook. & Tayl.) Jaeg. in Britain, with remarks on the status of *A. mediterraneum*. *Journal of Bryology* 12 : 11–14.

Hilpert, F. (1933). Studien zur Systematik der Trichostomaceen. *Beihefte zum Botanischen Centralblatt* 50 : 585–706.

Holmgren, P. K., N. H. Holmgren, & L. C. Barnett (eds.) (1990). *Index Herbariorum*, Part I: The Herbaria of the World, Eighth Edition. The New York Botanical Garden, Bronx, New York, 693 pp.

Hooker, W. J. & T. Taylor (1818). *Muscologia britannica*. Longman, Hurst, Rees, Orme & Brown, London, 152 pp.

Horton, D.G. (1983). A revision of the Encalyptaceae (Musci), with particular reference to the North American taxa. II. *Journal of the Hattori Botanical Laboratory* 54 : 353–532.

—(1988). Microhabitats of the New World Encalyptaceae (Bryopsida). Distribution along edaphic gradients. *Nova Hedwigia* 90 : 261–282.

Husnot, T. (1884–1894). *Muscologia Gallica*. [1967 facsimile edition, A. Asher & Co., Amsterdam, 458 pp.]

Iwatsuki, Z. (1977). Notes on *Philonotis hastata* (Duby) Wijk et Marg. in Japan. *Proceedings of the Bryological Society of Japan* 2 : 13–14.

Jovet-Ast, S. (1961). *Targionia lorbeeriana* K. M., trois localités nouvelles. *Revue Bryologique et Lichénologique* 30 : 278.

—(1986). Les *Riccia* de la région méditerranéenne. *Cryptogamie, Bryologie–Lichénologie* 7 (Suppl.) : 287–431.

—, H. Bischler & B. Baum (1965). Hépatiques récoltées en Israël (15 mars–13 avril 1964). *Israel Journal of Botany* 14 : 36–48.

— — (1966). Les Hépatiques d'Israël: Énumération, notes écologiques et biogéographiques. *Revue Bryologique et Lichénologique* 34 : 91–126.

— — & R. Baudoin (1976). Essai sur le peuplement hépaticologique de la région méditerranéenne. *Journal of the Hattori Botanical Laboratory* 41 : 87–94.

Kanda, H. (1975). A revision of the family Amblystegiaceae of Japan. I. *Journal of Science of the Hiroshima University*, ser. B, Div. 2, 15 : 201–276.

Koponen, T. (1968). Generic revision of Mniaceae Mitt. (Bryophyta). *Annales Botanici Fennici* 5 : 117–151.

—(1971). A monograph of *Plagiomnium* sect. *Rosulata* (Mniaceae). *Annales Botanici Fennici* 8 : 305–367.

—, P. Isoviita & T. Lammes (1977). The bryophytes of Finland: An annotated checklist. *Flora Fennica* 6 : 1–77.

Kramer, W. (1980). *Tortula* Hedw. sect. *Rurales* De Not. (Pottiaceae, Musci) in der östlichen Holarktis. *Bryophytorum Bibliotheca* 21 : 1–165.

Kürschner, H. (1995). *Weissia ovatifolia* (Pottiaceae, Musci), eine neue Art aus dem Edomitischen Bergland Jordaniens. *Nova Hedwigia* 60 : 499–504.

—(1997). An annotated, corrected, and updated list of the bryological literature of southwest Asia. *Cryptogamie, Bryologie–Lichénologie* 18 : 1–46.

—(2000). *Bryophyte Flora of the Arabian Peninsula and Socotra. Bryophytorum Bibliotheca*, Vol. 55. J. Cramer, Berlin–Stuttgart, 131 pp.

Kuwahara, Y. (1978). Synopsis of the family Metzgeriaceae. *Revue Bryologique et Lichénologique* 44 : 351–410.

Lawrence, G. H. M., A. F. G. Buchheim, G. S. Daniels & H. Dolezal (eds.) (1968). *B-P-H*: *Botanico-Periodicum-Huntianum*. Hunt Botanical Library, Pittsburgh, 1063 pp.

Lewinsky, J. (1977). The family Orthotrichaceae in Greenland. A taxonomic revision. *Lindbergia* 4 : 57–103.

—(1983). Lectotypification of *Orthotrichum rupestre* Schleich. ex Schwaegrichen. *Lindbergia* 9 : 53–56.

—& J. Bartlett (1982). *Pseudoscleropodium purum* (Hedw.) Fleisch. in New Zealand. *Lindbergia* 8 : 177–180.

Lipkin, Y. & V. W. Proctor (1975). Notes on the subgenus *Trabutiella* of the Aquatic Liverwort *Riella* (Riellaceae, Sphaerocarpales). *The Bryologist* 78 : 25–31.

Litav, M. & M. Agami (1976). Relationship between water pollution and the flora of two coastal rivers of Israel. *Aquatic Botany* 2 : 23–41.

Loeske, L. (1929). *Die Laubmoose Europas*. Part 2: *Funariaceae. Fedde Sonderbeiheft* B. Max Lande, Berlin–Schöneberg, 120 pp.

Longton, R. E. (1988). Life-history strategies among bryophytes of arid regions. *Journal of the Hattori Botanical Laboratory* 64 : 15–28.

Lorentz, P. G. (1868). Über die Moose, die Hr. Ehrenberg in den Jahren 1820–1826 in Aegypten, der Sinaihalbinsel und Syrien gesammelt. *Physikalische Abhandlungen der Königlichen Akademie der Wissenschaftern zu Berlin* 1868 : 1–57.

Magill, R. E. (1981). *Bryophyta*. Part I: *Mosses*. Fasc. 1, *Sphagnaceae–Grimmiaceae*, Leistner, O. A. (ed.), *Flora of Southern Africa*. Botanical Research Institute, Department of Agriculture and Fisheries, Pretoria, 291 pp.

—(1987). Bryopyta. Part I: Mosses. Fasc. 2, *Gigaspermaceae–Bartramiaceae*, pp. 293–443 *in*: Leistner, O. A. (ed.), *Flora of Southern Africa*. Botanical Research Institute, Department of Agriculture and Water Supply, Pretoria.

—(ed.) (1990). *Glossarium Polyglottum Bryologiae*: *A multilingual glossary for bryology*. *Monographs in Systematic Botany from the Missouri Botanical Garden*, Vol. 33. St. Louis, 297 pp.

Maier, E. (1998). Zur systematischen Stellung von *Grimmia pitardii* Corb. (Musci). *Candollea* 53 : 301–308.

Malta, N. (1924). Studien über die Laubmoosgattung *Zygodon* Hook. et Tayl. 10. (Übersicht der europäischen *Zygodon*-Arten). *Acta Universitatis Latviensis (Riga)* 9 : 111–153.

—(1926). Die Gattung *Zygodon* Hook. et Tayl. *Latvijas Universitates Botaniskā Dārza Darbi* 1 : 1–185.

Manuel, M. G. (1974). A revised classification of the Leucodontaceae and a revision of the subfamily Alsioideae. *The Bryologist* 77 : 531–550.

Martínez-Sánchez, J. J., R. M. Ros & J. Guerra (1991). Briófitos interesantes de zonas yesíferas del sudeste árido de España. *The Bryologist* 94 : 16–21.

Matsui, T. & Z. Iwatsuki (1990). A taxonomic revision of the family Ditrichaceae (Musci) of Japan, Korea and Taiwan. *Journal of the Hattori Botanical Laboratory* 68 : 317–366.

Mönkemeyer, W. (1927). *Dr. L. Rabenhorsts Kryptogamen-Flora von Deutschland,*

Österreich und der Schweiz. Band IV: Die Laubmoose Europas, Andreaeales–Bryales. Akademische Verlagsgesellschaft, Leipzig, 960 pp.

Müller, K. (1951–1958). Die Lebermoose Europas. *In*: Rabenhorst, L. (ed.), *Kryptogamen-flora von Deutschland, Öesterreich und der Schweiz*, Third Edition. Akademische Verlagsgesellschaft Geest & Portig K.-G., Leipzig, 756 pp.

Nachmony, S. (1961). New additions to the moss flora of Palestine and the neighbouring countries. *Bulletin of the Research Council of Israel*, Sect. D, Botany 10 : 352–355.

Nordhorn-Richter, G. (1982). Die Gattung *Pohlia* Hedw. (Bryales, Bryaceae) in Deutschland und den angrenzenden Gebieten. 1. *Lindbergia* 8 : 139–147.

Nyholm, E. (1954–1969). *Illustrated Moss Flora of Fennoscandia*, II: *Musci*, Fasc. 1–6, The Botanical Society of Lund, Lund, 799 pp.

—(1986). *Illustrated Flora of Nordic Mosses*, Fasc. 1: *Fissidentaceae, Seligeraceae*. Nordic Bryological Society, Lund, 72 pp.

—(1989). *Illustrated Flora of Nordic Mosses*, Fasc. 2: *Pottiaceae, Splachnaceae, Schistostegaceae*. Nordic Bryological Society, Lund, 141 pp.

—(1993). *Illustrated Flora of Nordic Mosses*, Fasc. 3: *Bryaceae, Rhodobryaceae, Mniaceae, Cinclidiaceae, Plagiomniaceae*. Nordic Bryological Society, Lund, 244 pp.

—(1995). A new species of *Encalypta*. *Lindbergia* 20 : 83–84.

Ochi, H. (1980). A revision of the neotropical Bryoideae, Musci (First Part). *Journal of the Faculty of Education, Tottori University, Natural Science* 29 : 49–154.

—(1981). A revision of the neotropical Bryoideae, Musci (Second Part). *Journal of the Faculty of Education, Tottori University, Natural Science* 30 : 21–55.

Ochyra, R. (1989). Animadversions on the moss genus *Cratoneuron* (Sull.) Spruce. *Journal of the Hattori Botanical Laboratory* 67 : 203–242.

—(1992). New combinations in *Syntrichia* and *Warnstorfia* (Musci). *Fragmenta Floristica et Geobotanica* 37 : 211–214.

Paton, J. A. (1999). *The Liverwort Flora of the British Isles*. Harley Books, Colchester, 626 pp.

Perold, S. M. (1993a). Studies in Marchantiales (Hepaticae) from southern Africa. 2. The genus *Athalamia* and *A. spathysii*; the genus *Oxymitra* and *O. cristata*. *Bothalia* 23 : 207–214.

—(1993b). Studies in Marchantiales (Hepaticae) from southern Africa. 3. The genus *Targionia* and *T. hypophylla* with notes on *T. lorbeeriana* and *Cyathodium foetidissimum* (Targioniaceae). *Bothalia* 23 : 215–221.

Petit, E. (1976). Les propagules dans le genre *Philonotis* (Musci). *Bulletin du Jardin Botanique National de Belgique* 46 : 221–226.

Pierrot, R. B. (1976). *Gymnostomum calcareum* B.G. et *Gyroweisia tenuis* (Schr.) Schp. dans le Centre-Ouest. *Bulletin de la Société Botanique du Centre-Ouest*, Nouvelle sér. 7 : 135–136.

—(1978). Contribution à l'étude des espèces francaises du genre *Orthotrichum* Hedw. *Bulletin de la Société Botanique du Centre-Ouest*, Nouvelle sér. 9 : 167–183.

Potier de la Varde, R. (1956). Contribution à la flore bryologique d'Israël. *Revue Bryologique et Lychénologique* 25 : 120–123.

Proskauer, J. (1953). On a collection of liverworts from Israel. *Palestine Journal of Botany* (J ser.) 6 : 123–124.

—(1954). On *Sphaerocarpos stipitatus* and the genus *Sphaerocarpos*. *Journal of the Linnean Society*, Botany 55: 143–157.

Pursell, R. A. (1976). On the typification of certain taxa and structural variation within the *Fissidens bryoides* complex in eastern North America. *The Bryologist* 79: 35–41.

—(1987). A taxonomic revision of *Fissidens octodiceras* (Fissidentaceae). *Memoirs of the New-York Botanical Garden* 45: 639–660.

Rabinovitz-Sereni, D. (1931). Contributo alla briologia della Palestina. *Annali di Botanica*, Vol. 19, Fasc. II: 1–7.

Reimers, H. (1927). Die von Prof. Dr. K. Krause in Kleinasien besonders im Pontus, 1926 gesammelten Leber- und Laubmoose. *Notizblatt des Botanischen Gartens und Museums zu Berlin-Dahlem* 10: 27–42.

—(1957). XXXVIII. Einige bemerkenswerte Moose des östlichen Mediterrangebiets. *Willdenowia* 1: 689–703.

Richardson, D. H. S. (1981). *The Biology of Mosses*. Blackwell Science, Oxford, 220 pp.

Risse, S. (1996). *Ephemerum minutissimum* Lindb. and *E. serratum* (Hedw.) Hampe. *The Bryological Times* 90: 6 [errata emend. *loc. cit.* 92: 7. 1997].

Robinson, H. (1962). Generic revisions of North American Brachytheciaceae. *The Bryologist* 65: 73–144.

—(1970). A revision of the moss genus *Trichostomopsis*. *Phytologia* 20: 184–191.

Ros Espin, M. R. (1987). *Riella cossoniana* Trab., nueva hepática para la flora europea. *Cryptogamie, Bryologie–Lichénolologie* 8: 227–233.

Rungby, S. (1959). A contribution to the bryophyte flora of the Near East and the Middle East. *Botaniska Notiser* 112: 97–99.

Saito, K. (1975). A monograph of Japanese Pottiaceae (Musci). *Journal of the Hattori Botanical Laboratory* 39: 373–537.

—& T. Hirohama (1974). A comparative study of the spores of taxa in the Pottiaceae by use of the scanning electron microscope. *Journal of the Hattori Botanical Laboratory* 38: 475–488.

Sayre, G., C. E. B. Bonner & W. L. Culberson (1964). The authorities for the epithets of mosses, hepatics, and lichens. *The Bryologist* 67: 113–135.

Schiffner, V. (1908). Beiträge zur Kenntnis der Bryophyten von Persien und Lydien. *Österreichische Botanische Zeitschrift* 58: 225–231, 304–318, 341–349.

—(1909). Hepaticae. *In*: von Handel-Mazzetti, H., Ergebnisse einer botanischen Reise in dem Pontische Randgebirge im Sandschak Trapezunt. *Annalen des K. K. naturhistorischen Hofmuseums* 23: 133–141.

—(1913). Bryophyta aus Mesopotamien und Kurdistan, Syrien, Rhodos, Mytilini und Prinkipo. *Annalen des K. K. Naturhistorischen Hofmuseums* 27: 472–504.

Schimper, W. P. (1856). *Corollarium Bryologiae Europaeae*. E. Schweizerbart, Stuttgart, 140 pp.

von Schreber, J. C. D. (1770). *De Phasco*. Leipzig, 22 pp.

Schuster, R. M. (1969). *The Hepaticae and Anthocerotae of North America East of the Hundredth Meridian*, Vol. II. Columbia University Press, New York–London, 1062 pp.

—(1980). *The Hepaticae and Anthocerotae of North America East of the Hundredth Meridian*, Vol. IV. Columbia University Press, New York–London, 1334 pp.

—(1992a). *The Hepaticae and Anthocerotae of North America East of the Hundredth Meridian*, Vol. V. Field Museum of Natural History, Chicago, 854 pp.

—(1992b). *The Hepaticae and Anthocerotae of North America East of the Hundredth Meridian*, Vol. VI. Field Museum of Natural History, Chicago, 937pp.

Scott, G. A. M. & I. G. Stone (1976). *The Mosses of Southern Australia*. Academic Press, London, 495 pp.

Sérgio, C. (1972). Os géneros *Aschisma*, *Acaulon* e *Phascum* (Musci-Pottiaceae) em Portugal. *Boletim da Sociedade Broteriana* 46 : 457–463.

—(1984). Estudo taxonómico, ecológico e corológico de *Gymnostomum luisieri* (Sérgio) Sérgio ex Crundw. na península ibérica. *Anales de Biología, Universidad de Murcia* 2 (sección especial 2) : 357–366.

—(1988). Morphological, karyological and phytogeographic observations on *Entosthodon curvisetus* (Schwaegr.) C. Müll. as a basis for a new genus, *Funariella* Sérgio (Funariaceae): Musci. *Orsis* 3 : 5–13.

—& A. Queiroz Lopes (1972). O género *Targionia* Mich. em Portugal. *Boletim da Sociedade Portuguesa de Ciencias Naturais* 14 : 87–105.

Sharp, A. J., H. Crum & P. M. Eckel (eds.) (1994). *The Moss Flora of Mexico*. Part One: *Sphagnales to Bryales*; Part Two: *Orthotrichales to Polytrichales*. *Memoirs of The New York Botanical Garden*, Vol. 69. New York. 1113 pp.

Shaw, A. J. (1981). Taxonomic studies on *Pohlia* subgen. *Mniobryum* including *P. brevinervis* Lindb. & Arn. new for North America. *The Bryologist* 84 : 505–514.

—, L. E. Anderson & B. D. Mischler (1987). Peristome development in mosses in relation to systematics and evolution. I. *Diphyscium foliosum* (Buxbaumiaceae). *Memoirs of The New York Botanical Garden* 45 : 55–70.

— — — (1989). Peristome development in mosses in relation to systematics and evolution. III. *Funaria hygrometrica*, *Bryum pseudocapillare*, and *B. bicolor*. *Systematic Botany* 14 : 24–36.

—, B. D. Mischler & L. E. Anderson (1989). Peristome development in mosses in relation to systematics and evolution. IV. Haplolepideae: Ditrichaceae and Dicranaceae. *The Bryologist* 92 : 314–325.

Shmida, A. (1980). Vegetation and flora, pp. 97–158 *in*: Shmida, A. & M. Livne (eds.), *Mt. Hermon — Nature and Landscape*. Hakibbutz Hameuchad Publishing House, Tel-Aviv [in Hebrew].

Smith, A. J. E. (1970). *Fissidens viridulus* Wahlenb. and *F. minutulus* Sull. *Transactions of the British Bryological Society* 6 : 56–68.

—(1977). Further new combinations in British and Irish Mosses. *Journal of Bryology* 9 : 393–394.

—(1978). *The Moss Flora of Britain and Ireland*. Cambridge University Press, Cambridge, 706 pp.

—& M. E. Newton (1968). Chromosome studies on some British and Irish mosses. III. *Transactions of the British Bryological Society* 5 : 463–522.

—& H. L. K. Whitehouse (1978). An account of the British species of the *Bryum bicolor* complex including *B. dunense* sp. nov. *Journal of Bryology* 10 : 29–47.

Snider, J. A. & W. D. Margadant (1973). Nomina Conservanda Proposita (368). Proposal for the conservation of the generic name *Pleuridium* Rabenh. (1848) against *Pleuridium* Brid. (1819). (Musci). *Taxon* 22 : 691–694.

Söderström, L., K. Karttunen & L. Hedenäs. (1992). Nomenclatural notes on Fennoscandian bryophytes. *Annales Botanici Fennici* 29 : 119–122.

Stafleu, F. A. & R. S. Cowan (1976–1988). *Taxonomic Literature* (TL-2), Second

Edition, Vols. I–VII. *Regnum Vegetabile* 94, 98, 105, 110, 112, 115, 116. Bohn, Scheltema & Holkema, Utrecht–Antwerpen and dr. W. Junk b.v., Publishers, The Hague–Boston.

Stark, L. R. (1987). A taxonomic monograph of *Forsstroemia* Lindb. (Bryopsida: Leptodontaceae). *Journal of the Hattori Botanical Laboratory* 63 : 133–218.

Stone, I. G. (1989). Revision of *Acaulon* and *Phascum* in Australia. *Journal of Bryology* 15 : 745–777.

Stoneburner, A. (1985). Variation and taxonomy of *Weissia* in the southwestern United States. *The Bryologist* 88 : 293–314.

Størmer, P. (1963). Iranian plants collected by Per Wendelbo in 1959. VI. Mosses (Musci). *Årbok for Universitetet I Bergen, Mat.-Naturvitenskapelig*, ser. 11 : 3–34.

Syed, H. (1973). A taxonomic study of *Bryum capillare* Hedw. and related species. *Journal of Bryology* 7 : 265–326.

Touw, A. & H. J. Knol (1978). A note on Hedwig's plants of *Hypnum praelongum* and *H. hians*. *Lindbergia* 4 : 197–198.

Townsend, C. C. (1964–1965). Bryophytes from Cyprus. *Revue Bryologique et Lichénologique* 33 : 484–493.

—(1966). Bryophytes from Azraq National Park, Jordan. *Transactions of the British Bryological Society* 5 : 136–141.

—(1977). Bryophytes from some Greek islands. *Revue Bryologique et Lichénologique* 43 : 389–396.

Trabut, L. (1941). *Flore des hépatiques de l'Afrique du Nord*. Laboratoire de Cryptogamie, Muséum National d'Histoire Naturelle, Paris.

Vitt, D. H. (1973). *A Revision of the Genus Orthotrichum in North America, North of Mexico. Bryophytorum Bibliotheca*, Band 1. J. Cramer, Lehre, 208 pp.

—(1981). Adaptive modes of the moss sporophyte. *The Bryologist* 84 : 166–186.

—(1984). Classification of the Bryopsida, pp. 696–759 *in*: Schuster, R. M. (ed.), *New Manual of Bryology*, Vol. 2. The Hattori Botanical Laboratory, Nichinan, Miyazaki, Japan.

—& C. Hamilton (1974). A scanning electron microscope study of the spores and selected peristomes of the North American Encalyptaceae (Musci). *Canadian Journal of Botany* 52 : 1973–1981.

—, B. Goffinet & T. Hedderson. (1998). The ordinal classification of the mosses: Questions and answers for the 1990s, pp. 113–123 *in*: Bates, J. W., N. W. Ashton & J. G. Duckett (eds.), *Bryology for the Twenty-first Century*. British Bryological Society, Leeds.

Vondráček, M. (1965). Some new mosses from Iraq, collected by W. Hadač. *Bulletin de la société des amis des sciences et des lettres de Poznań*, sér. D, Sciences Biologiques 6 : 117–122.

Walther, K. (1983). Bryophytina. Laubmoose. *In*: Gerloff, J. & J. Poelt, *Engler's Syllabus der Pflanzenfamilien*, Vol. 5. Gebrüder Borntraeger. Berlin, 107 pp.

Warnstorf, C. (1916). *Pottia* Studien. *Hedwigia* 58 : 35–152.

Washbourn, R. & R. F. Jones (1937). Percy Sladen Expedition to Lake Huleh, Palestina. *Proceedings of the Linnean Society of London* 149 : 97–99.

Weitz W. & C. C. Heyn (1981). Intra-specific differentiation within the cosmopolitan moss species *Funaria hygrometrica* Hedw. *The Bryologist* 84 : 315–334.

Whitehouse, H. L. K. & H. J. During (1987). *Leptobarbula berica* (De Not.) Schimp. in Belgium and the Netherlands. *Lindbergia* 12 : 135–138.

Whitehouse, H. L. K. & A. C. Crundwell (1991). *Gymnostomum calcareum* Nees & Hornsch. and allied plants in Europe, North Africa and the Middle East. *Journal of Bryology* 16 : 561–579.

Wigglesworth, G. (1937). South African species of *Riella*, including an acount of the developmental stages of three of the species. *Journal of the Linnean Society of Botany* 60 : 309–332.

Wigh, K. (1974). The European genera of the family Brachytheciaceae (Bryophyta) and chromosome numbers published in the genus *Brachythecium*. *Botaniska Notiser* 127 : 89–103.

van der Wijk, R. (chief ed.), W. D. Margadant & P. A. Floreschütz (eds.) (1959–1969). *Index Muscorum*, Vols. I–V. *Regnum Vegetabile* 17, 26, 33, 48, 65. Utrecht.

Wilczek, R. & F. Demaret (1974). Les espèces belges du "complexe *Bryum erythrocarpum*". *Bulletin du Jardin Botanique National de Belgique* 44 : 425–438.

— — (1976a). Les espèces belges du "complexe *Bryum bicolor*" (Musci). *Bulletin du Jardin Botanique National de Belgique* 46 : 511–541.

— — (1976b). *Bryum versicolor* A. Braun ex B. S. G., espèce méconnue de la flore Belge. *Dumortiera* 4 : 14–16.

Zander, R. H. (1977). The tribe Pleuroweisiae (Pottiaceae, Musci) in Middle America. *The Bryologist* 80 : 233–269.

—(1978). New combinations in *Didymodon* (Musci) and a key to the taxa in North America north of Mexico. *Phytologia* 41 : 11–32.

—(1979). Notes on *Barbula* and *Pseudocrossidium* (Bryopsida) in North America and an annotated key to the taxa. *Phytologia* 44 : 177–214.

—(1989). Seven new genera in Pottiaceae (Musci) and a lectotype for *Syntrichia*. *Phytologia* 65 : 424–436.

—(1993). *Genera of the Pottiaceae: Mosses of Harsh Environments. Bulletin of the Buffalo Society of Natural Sciences*, Vol. 32. Buffalo, 378 pp.

Zielinski, R. (1987). *Genetic variation of the liverwort genus Pellia with special reference to Central European territory*. University of Szczecin, Szczecin.

Zohary, M. (1962). *Plant life of Palestine: Israel and Jordan*. Ronald Press, New York, 262 pp.

—(1963). On the geobotanical structure of Iran. *Bulletin of the Research Council of Israel* Sect. D, Botany, Suppl. 11 : 1–113.

—(1966). *Flora Palaestina*. Part One: *Equisetaceae to Moringaceae* (Text). The Israel Academy of Sciences and Humanities, Jerusalem, 364 pp.

—(1972). *Flora Palaestina*. Part Two: *Platanaceae to Umbelliferae* (Text). The Israel Academy of Sciences and Humanities, Jerusalem, 489 pp.

—(1973). *Geobotanical foundations of the Middle East*, Vols. 1–2. G. Fischer, Stuttgart. 739 pp.

—(1982). *Vegetation of Israel and Adjacent Areas*. Reichert, Wiesbaden, 166 pp.

INDEX OF SCIENTIFIC NAMES

Accepted names are printed in Roman type, and synonyms are printed in italics. Names of taxa not occurring in the local flora are marked by an asterisk. Excluded taxa are marked with two asterisks.

The main page reference for each valid taxon is printed in boldface type. Family names are printed in capital letters. Names above the rank of family are printed in initial capital letter and small capital letters.

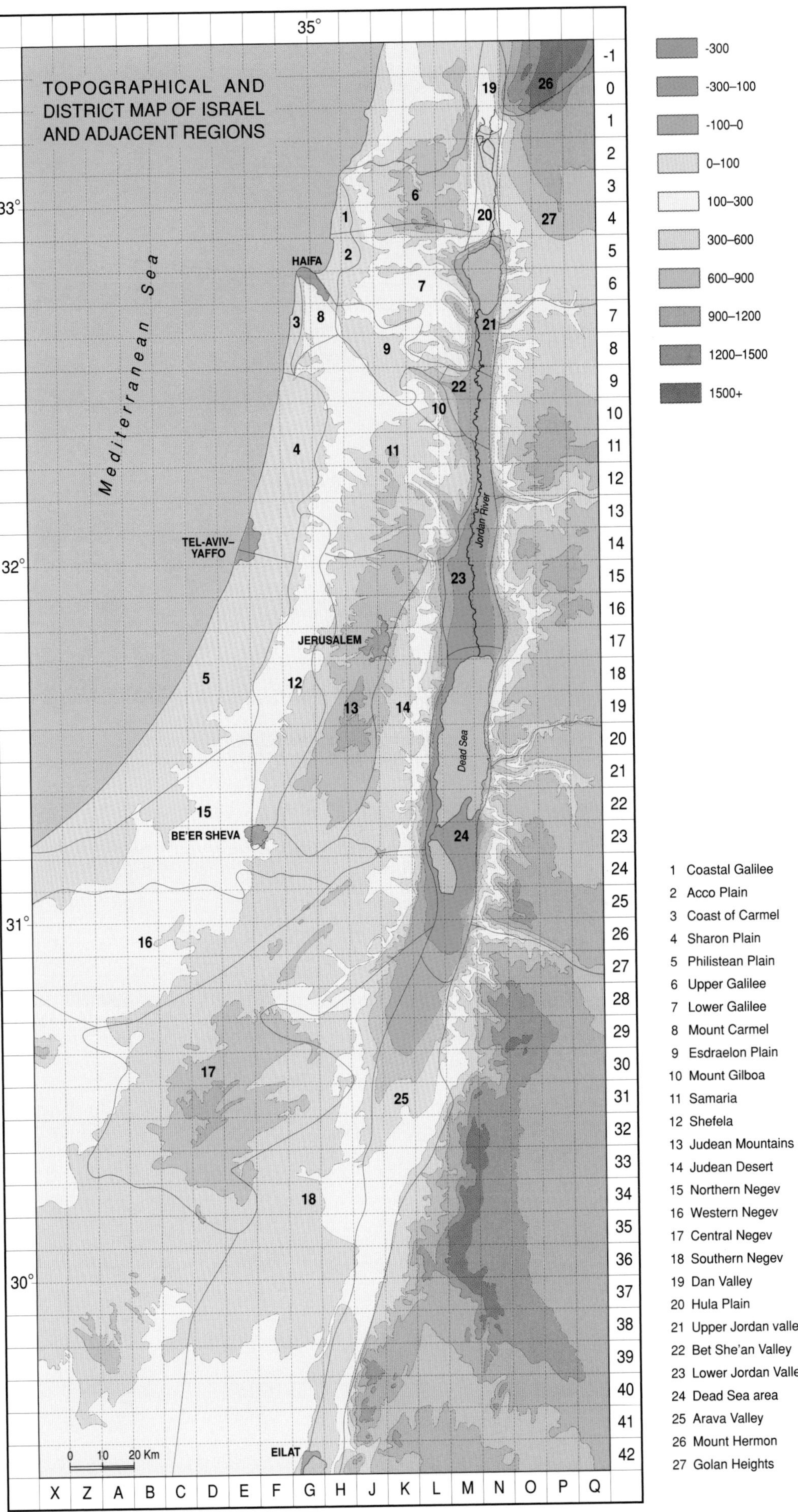

TOPOGRAPHICAL AND DISTRICT MAP OF ISRAEL AND ADJACENT REGIONS
35°
33°
32°
31°
30°
Mediterranean Sea
HAIFA
TEL-AVIV–YAFFO
JERUSALEM
BE'ER SHEVA
EILAT
Jordan River
Dead Sea
0 10 20 Km
X Z A B C D E F G H J K L M N O P Q
-1 0 1 2 3 4 5 6 7 8 9 10 11 12 13 14 15 16 17 18 19 20 21 22 23 24 25 26 27 28 29 30 31 32 33 34 35 36 37 38 39 40 41 42
-300
-300–100
-100–0
0–100
100–300
300–600
600–900
900–1200
1200–1500
1500+
1 Coastal Galilee
2 Acco Plain
3 Coast of Carmel
4 Sharon Plain
5 Philistean Plain
6 Upper Galilee
7 Lower Galilee
8 Mount Carmel
9 Esdraelon Plain
10 Mount Gilboa
11 Samaria
12 Shefela
13 Judean Mountains
14 Judean Desert
15 Northern Negev
16 Western Negev
17 Central Negev
18 Southern Negev
19 Dan Valley
20 Hula Plain
21 Upper Jordan valley
22 Bet She'an Valley
23 Lower Jordan Valley
24 Dead Sea area
25 Arava Valley
26 Mount Hermon
27 Golan Heights

נדפס בישראל במפעלי דפוס כתר, ירושלים

כתבי האקדמיה הלאומית הישראלית למדעים
החטיבה למדעי־הטבע

הצומח של ארץ־ישראל

הטחבים של ישראל

בעריכת

קלרה חן ואילנה הרנשטט

ירושלים תשס״ד